A. C. F. Colchester D. J. Hawkes (Eds.)

Information Processing in Medical Imaging

12th International Conference, IPMI '91
Wye, UK, July 7-12, 1991
Proceedings

Springer-Verlag
Berlin Heidelberg New York
London Paris Tokyo
Hong Kong Barcelona
Budapest

Series Editors

Gerhard Goos
GMD Forschungsstelle
Universität Karlsruhe
Vincenz-Priessnitz-Straße 1
W-7500 Karlsruhe, FRG

Juris Hartmanis
Department of Computer Science
Cornell University
Upson Hall
Ithaca, NY 14853, USA

Volume Editors

Alan C. F. Colchester
Department of Neurology, United Medical and Dental Schools
Guy's Hospital, St. Thomas Street, London SE1 9RT, UK

David J. Hawkes
Division of Radiological Sciences, United Medical and Dental Schools
Guy's Hospital, St. Thomas Street, London SE1 9RT, UK

CR Subject Classification (1991): I.4, I.2.5-6, J.3

ISBN 3-540-54246-9 Springer-Verlag Berlin Heidelberg New York
ISBN 0-387-54246-9 Springer-Verlag New York Berlin Heidelberg

Printing and binding: Druckhaus Beltz, Hemsbach/Bergstr.
2145/3140-543210 - Printed on acid-free paper

PREFACE

The 1991 International Conference on Information Processing in Medical Imaging (IPMI '91) was the twelfth in the series and followed the successful meeting in Berkeley, California, in 1989. The 1991 meeting was held in a glorious setting in the Kent countryside at Wye College, the Agricultural College of the University of London. The purpose of IPMI is to provide a forum for the detailed examination of methodological issues in computing which are at the heart of advances in medical image formation, manipulation and interpretation. Full-length scientific papers describing the latest techniques and results in this subject area are presented. Many of the ideas will form the basis of important journal articles in the future. This year, for the first time, the proceedings are published at the time of the conference. Papers were submitted at the end of 1990 and were reviewed in detail by the Scientific Committee. Authors of accepted papers were allowed to include their latest results before submitting the final version of their paper in March 1991. Thus, these proceedings provide a very up-to-date overview of the field. The long-paper format of IPMI means that readers and delegates can obtain a real insight into the methods in a way that can normally only be achieved from journal articles. We are very grateful for the cooperation of the authors, who were asked to submit full papers further in advance than usual, and for their adherence to the deadlines. We are also grateful to the unsuccessful authors; many papers which could not be included were of very high quality and were only excluded on space considerations.

We have decided to group papers according to information processing methodology rather than, for example, modality or clinical application. The first section contains several papers on image formation and reconstruction. Four papers on the use of prior information (especially anatomical) are grouped together in Section 2; this has emerged as an important theme this year. Registration of images from different modalities is discussed in four papers in Section 3. The excellent spatial resolution and discriminability of different tissues in MR images is exploited in several papers on segmentation in Section 4. However, there continues to be a need for improved general purpose methods of segmentation to cope with images of lesser quality and images of pathological tissues. Recent results using multi-scale approaches and methods based on analysis of surface topology are described in Section 5. In Section 6 there are papers using techniques derived from engineering, including deformable models and thin-plate splines, to represent anatomical structures and their variability. Two papers on factor analysis form Section 7. There are four papers on rule based systems and learning in Section 8; the use of uncertain classification rules increases the power of these approaches. The final section is composed of papers on image quality, display and interaction.

There is no doubt that developments in computer assisted quantitation and interpretation of medical images are being held up by fundamental problems in computer vision which are common to industrial and other application areas. Researchers in the medical area need to keep abreast of developments in the non-medical domain, and of course advances in medical image analysis can have important implications for other application areas. IPMI is a unique forum in which basic scientists and application-orientated researchers come together. We hope to see increasing input from clinical scientists with technical expertise at future conferences.

Part of the success of IPMI has been the workshop style of the meeting, which allows informal, in-depth discussions. The corollary of this is that the numbers attending the meeting need to be limited; this year we were significantly over-subscribed and unfortunately had to disappoint many people. The IPMI meetings are complementary to larger scale meetings where numerous examples of clinical applications are presented. The organisers of IPMI welcome the close cooperation with Computer Assisted Radiology (CAR) which has become established as the premier European meeting in the latter category. IPMI '91 followed CAR '91 which helped delegates travelling long distances to attend both meetings.

In recent years there have been rapid advances in the application of computing techniques to information processing in medical imaging. This volume is a timely collection of results of research by groups throughout the world including most of the major centres of activity in this area. The publication of the IPMI proceedings in the Springer Lecture Notes in Computer Science series increases its availability, and its publication at the time of the conference ensures that it is completely up-to-date. This volume will be an essential part of the library of all groups active or about to start work in this expanding area.

ACKNOWLEDGEMENTS

The meeting would not have been possible without the generous support of our colleagues in industry. We are grateful for financial sponsorship from:

Siemens AG Medical Group
Nuclear Diagnostics
IBM UK Trust
SUN Microsystems
IGE Medical Systems
Glaxo Laboratories

We are extremely grateful to the Scientific Committee for their prompt and careful refereeing of manuscripts. We had over one hundred full papers submitted and the Scientific Committee members had to referee up to ten papers each in less than one month. The quality of these proceedings and of the conference itself reflects the considerable time and effort that they all contributed.

A small meeting such as IPMI has to survive on a very tight budget and this in turn means that we have relied heavily on our long suffering Research Assistants (Glynn Robinson, Derek Hill, Lewis Griffin and Clifford Ruff) and Secretarial Staff (Margaret Micklewright, Ann Quarterman and Linda Skelton), who all worked long and hard, way beyond the call of duty, to ensure that the conference organisation ran as smoothly as possible. We also thank Nicky Colchester and Liz Hawkes for support in numerous ways over the past two years, and Liz in particular for preparing the IPMI mailing list.

Finally we thank all the participants and contributors without whom IPMI would not have its unique character.

London — A. C. F. Colchester

May 1991 — D. J. Hawkes

SCIENTIFIC COMMITTEE

Professor W. Schlegel
Institut fur Radiologie und Pathophysiologie
Deutsches Krebsforschungszentrum Heidelberg
Heidelberg
Germany

Professor C. Taylor
Department of Medical Biophysics
University of Manchester
Manchester
United Kingdom

Andrew Todd-Pokropek
Department of Medical Physics
University College London
London
United Kingdom

Dr. D. Townsend
Hôpital Cantonal Universitaire de Genève
Genève
Switzerland

Professor G. Vernazza
Department of Biophysical and
Electronic Engineering
University of Genoa
Genoa
Italy

Professor M. A. Viergever
Department of Radiology and Nuclear Medicine
University of Utrecht Faculty of Medicine
Utrecht
The Netherlands

Professor M. Yamamoto
Division of Physics
National Institute of Radiological Sciences
Chibe-shi
Japan

TABLE OF CONTENTS

4. Segmentation: Specific Applications

5. Segmentation: Multi-scale, Surfaces and Topology

6. Anatomical Models and Variability

7. Factor Analysis

8. Rule Based Systems and Learning

9. Image Quality, Display and Interaction

COLOUR PLATES

MARKER GUIDED REGISTRATION OF ELECTROMAGNETIC DIPOLE DATA WITH TOMOGRAPHIC IMAGES

P A van den Elsen and M A Viergever

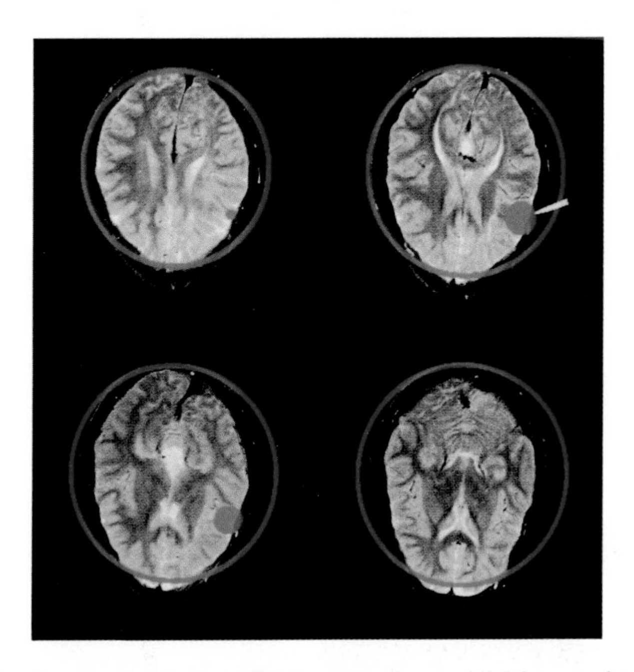

Figure 8: Dipole location visualized in four MRI slices taken from an MRI dataset, obtained from a Philips Gyroscan T5. A transverse T2 weighted Spin Echo sequence was acquired with a single acquisition (TR/TE 2000/50,100 msec). The second echo is shown here. Slice thickness is 8 mm, interslice gap is 1.6 mm. Pixel size is approximately 0.9 mm.

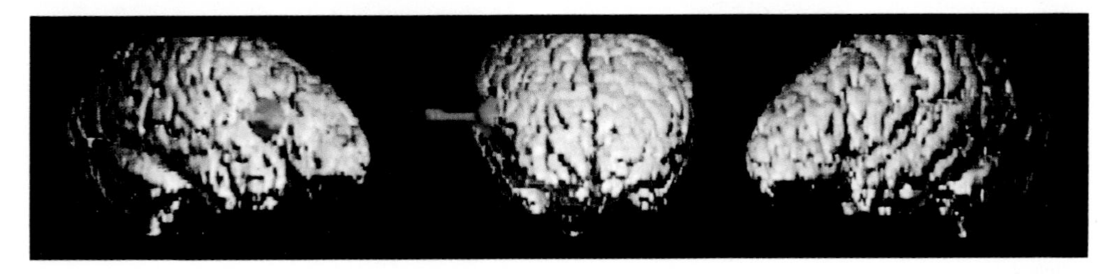

Figure 9: Right, frontal, and left rendered views of the cortex and dipole of the same patient as in Figure 8. MRI was performed on a Philips Gyroscan T5. Two interleaved T1 weighted transverse acquisitions (TR/TE 650/25 msec) with slice thickness and interslice gap of 2.5 mm were performed, thus providing a contiguous data set. Pixel size was approximately 0.9 mm.

AN ANATOMICAL-BASED 3D REGISTRATION SYSTEM OF MULTIMODALITY AND ATLAS DATA IN NEUROSURGERY

D Lemoine, C Barillot, B Gibaud and E Pasqualini

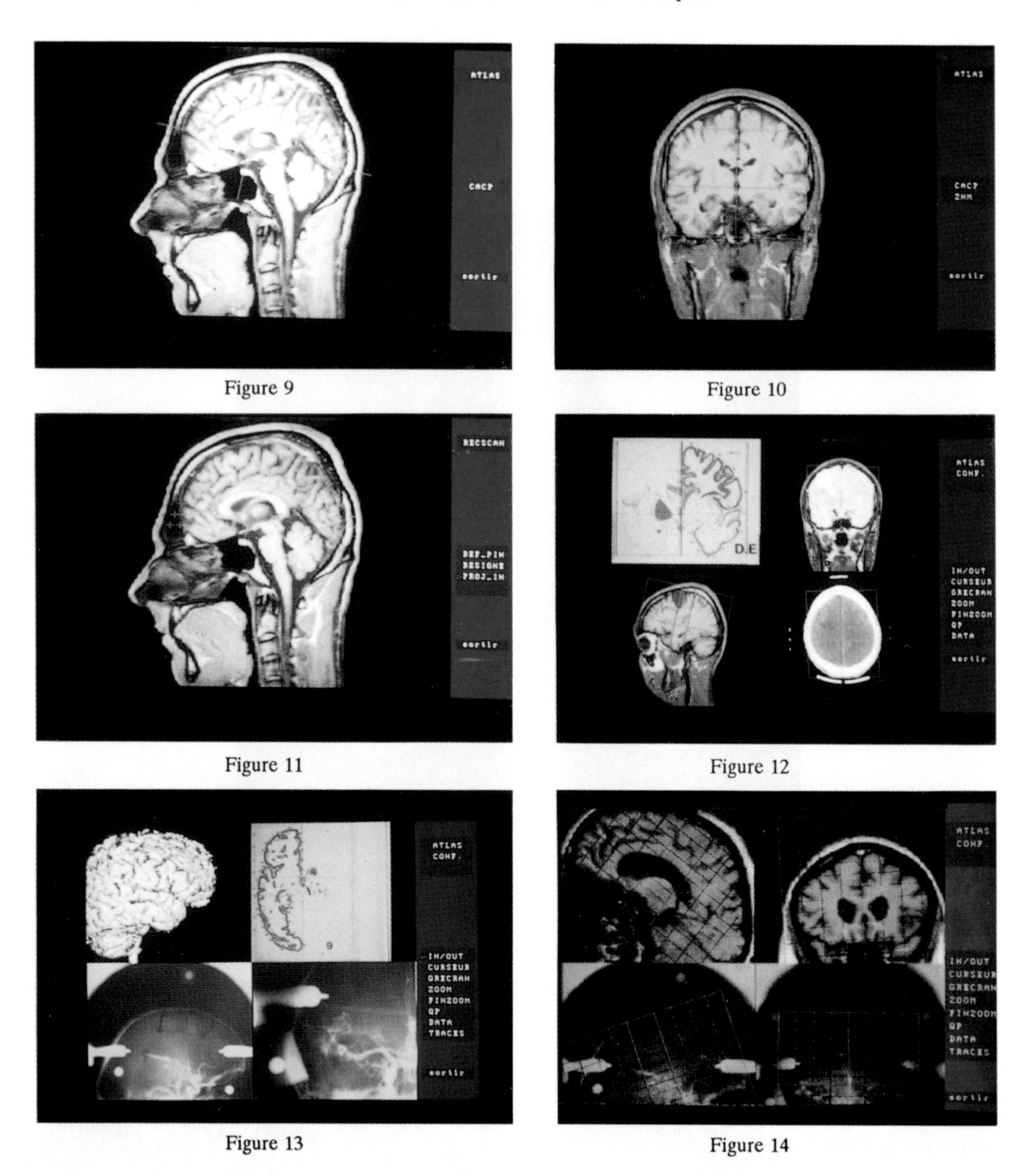

Figure 9 Figure 10 Figure 11 Figure 12 Figure 13 Figure 14

Figure 9: Illustration of the interactive determination of the AC-PC, VAC, and VPC lines upon the MRI sagittal midplane.
Figure 10: Set up of the brain encasing box upon a MRI coronal slice.
Figure 11: Assignment of graphical CT data over the MRI mid-plane during the CT-MRI registration stage.
Figure 12: Example of assignment of one point picked up upon an atlas plate (coronal view on the top left) and its corresponding positions upon three different acquisitions (two MRI and one CT).
Figure 13: Example of assignment between a 3D MR image, atlas (transaxial plate) and angiography.
Figure 14: Example of assignment between MRI (on the top) and angiography (on the bottom).

REGISTRATION OF BRAIN IMAGES BY A MULTI-RESOLUTION SEQUENTIAL METHOD

M A Oghabian and A Todd-Pokropek

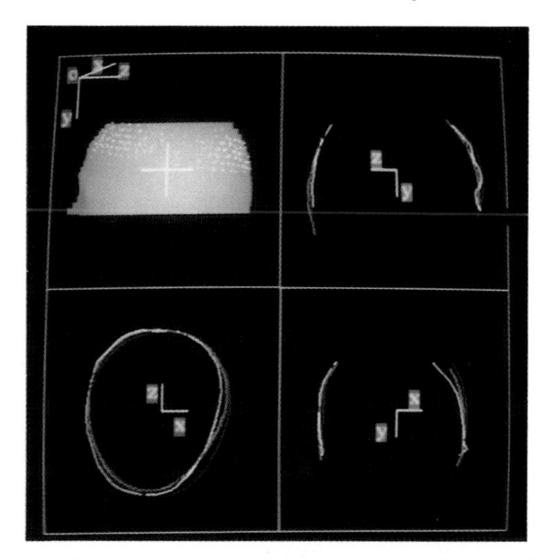

Figure 1. Sample outputs from the system when registering PET and MRI; top left showing the MRI surface rendered image where the points represent surface locations identified on the PET image. The other three images show MRI and PET contours, during the registration process, with a vector to indicate direction of shift being suggested.

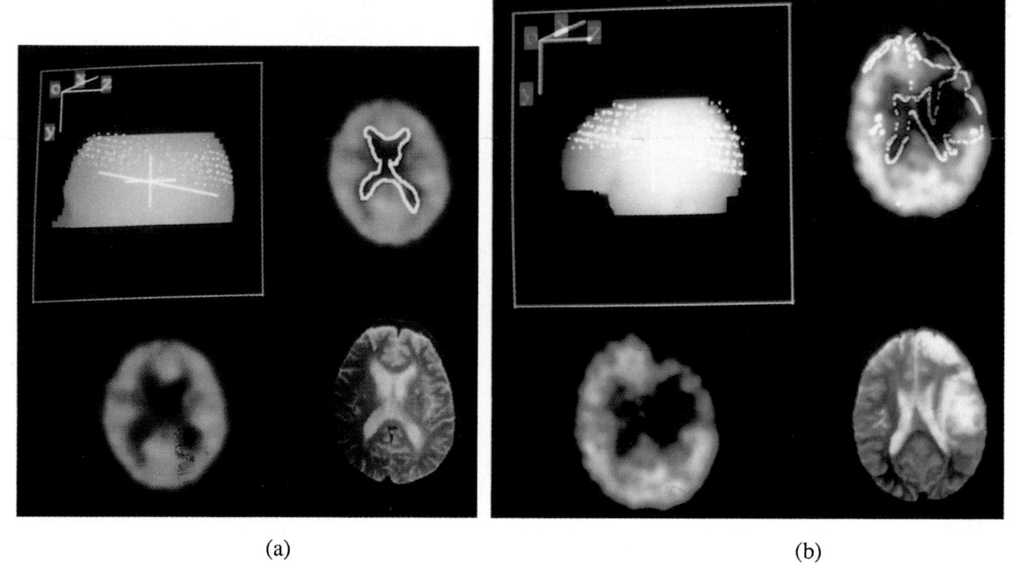

(a) (b)

Figure 6. Registered images of two different modalities. The top left image shows the MR surface model superimposed with the sampling points from the second surface image, at the end of registration process. the bottom left shows the original slice of the one of imaging modalities, bottom right is the re-section slice through 3D data set of the others. The superimposed image (contours from MRI and grey scale data from the other image) is shown in the top right corner. Figure 6a shows registration of MRI and PET. Figure 6b shows registration of MRI and SPECT.

TOWARDS AUTOMATED ANALYSIS IN 3D CARDIAC MR IMAGING

M Bister, J Cornelis and Y Taeymans

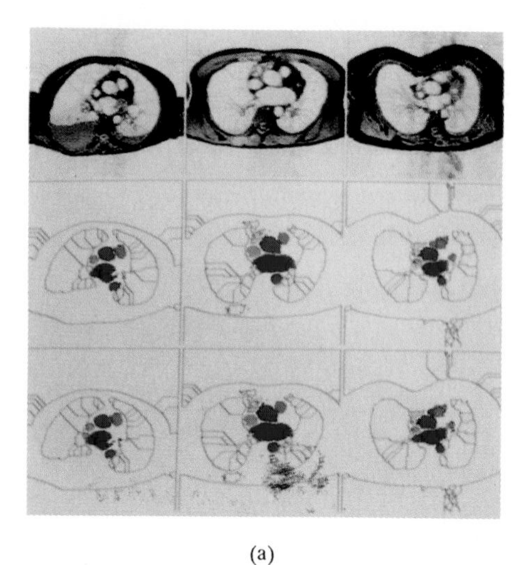

(a)

(b)

Figure 7. Labeling performances of the PDM based on a limited learning group (a) and for an independent test group (b). For both (a) and (b), the top row shows the input image with the segment borders, the mid row shows the manual labeling, and the bottom row the automatic labeling (red = Left Atrium, blue = Aorta, orange = Left Ventricle, pink = Vena Cava, green = Arteria Pulmonalis, yellow = Right Ventricle)

SEGMENTATION OF MAGNETIC RESONANCE IMAGES USING MEAN FIELD ANNEALING

W Snyder, A Logenthiran, P Santago,
K Link, G Bilbro, S Rajala

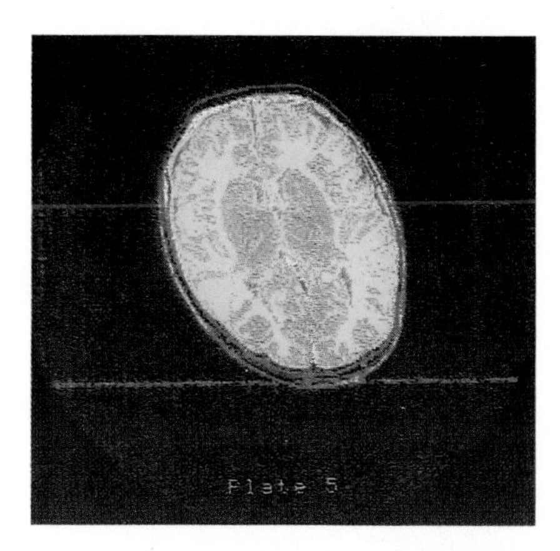

Figure 5: Original MR brain image. The pseudo-colors map brightness values are as follows: brightest are white and red, darkest are blue and black. The colors were chosen arbitrarily but distinguish areas of different density surprisingly well.

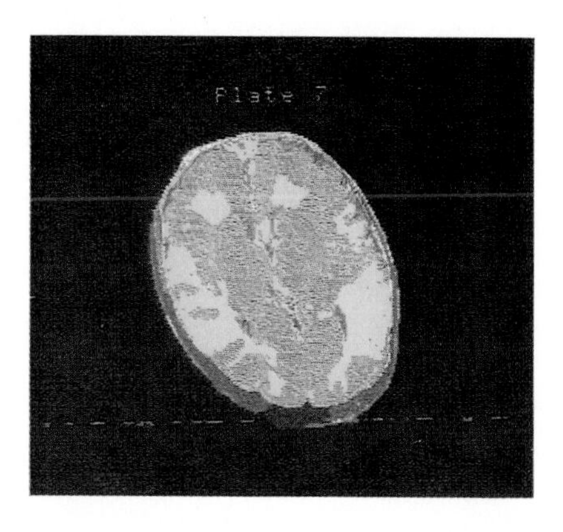

Figure 7: Brain image segmented using 3 modes and one temperature. Segmentation was chosen to emphasize CSF, white matter, and gray matter.

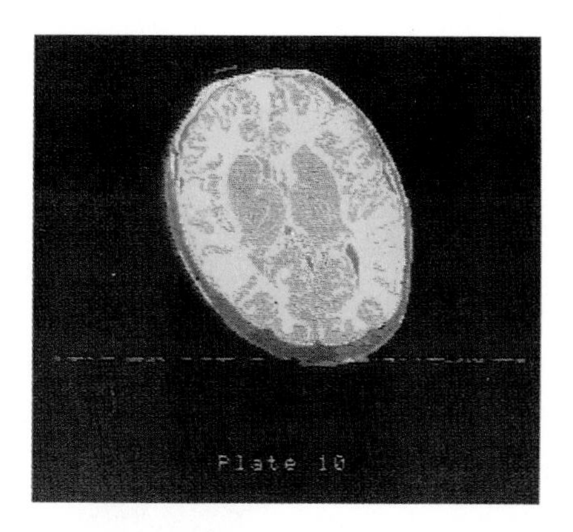

Figure 9: Segmentation of brain using 4 modes and two temperatures.

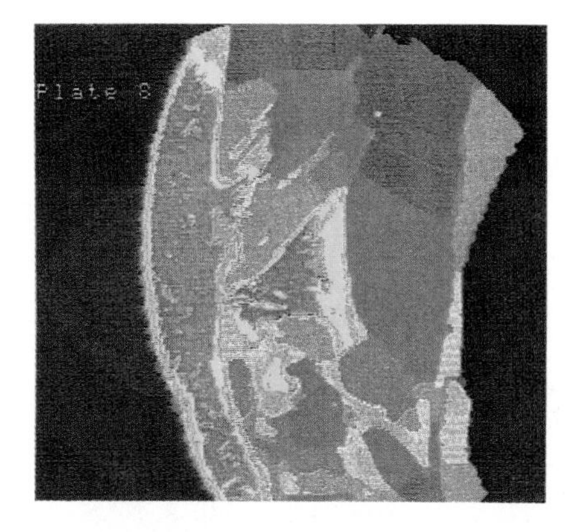

Figure 12: Segmentation of knee image emphasizing the homogeneity objective. Note the large blocks of uniform brightness.

SCALE AND SEGMENTATION OF GREY-LEVEL IMAGES USING MAXIMUM GRADIENT PATHS

L D Griffin, A C F Colchester and G P Robinson

Figure 8 - The singular points of the intensity function of the image in Figure 4. Saddles are marked by blue hexagons, maxima by pink triangles and minima by green triangles.

Figure 9 - The MGPs of the intensity function. Downhill MGPs are shown in blue and uphill MGPs in red. The arrows mark the district centroids, with the directions being the mean downhill direction within the district (this is not currently used).

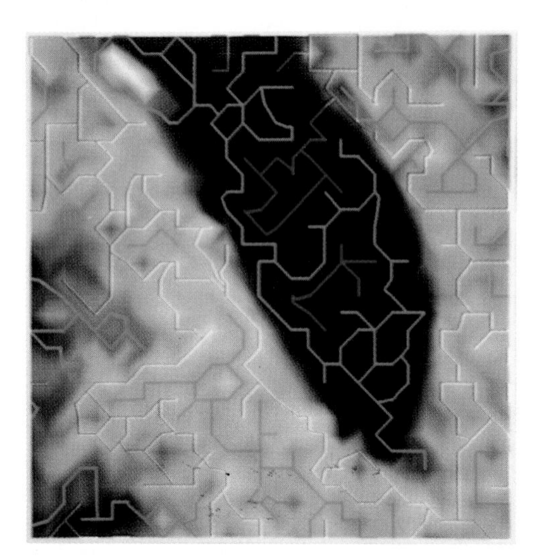

Figure 18 - The remaining MGPs after selection. The remaining downhill MGPs (blue) are the troughs and the remaining uphill MGPs (red) are the ridges.

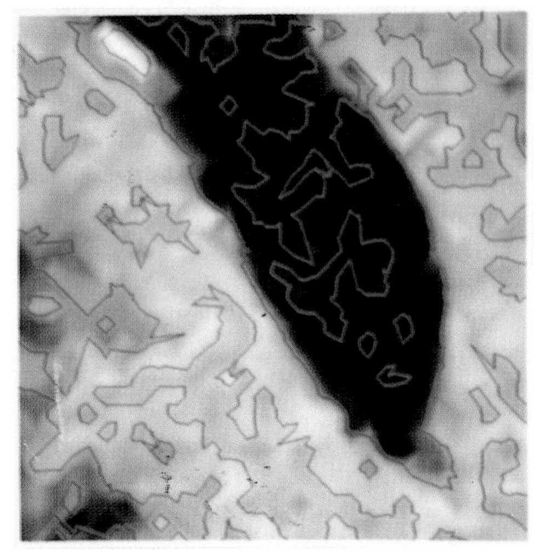

Figure 20 - The dynamic mid-grey threshold edges (section 4.2) constructed from the ridges and troughs in Figure 18.

SCALE AND SEGMENTATION OF GREY-LEVEL IMAGES USING MAXIMUM GRADIENT PATHS

L D Griffin, A C F Colchester and G P Robinson

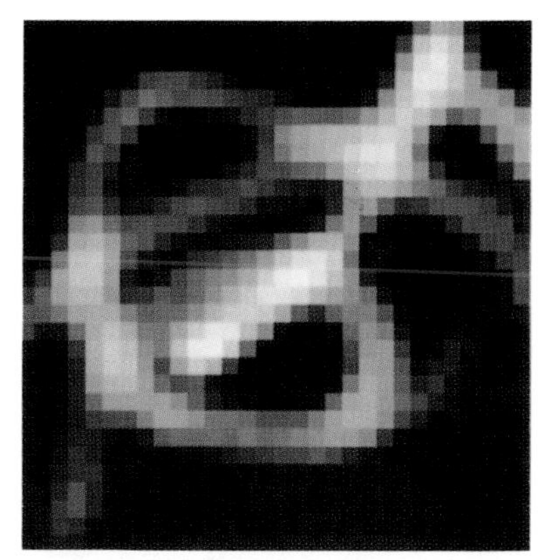

Figure 27 - A 30x30 ROI from a cerebral digitally subtracted angiogram.

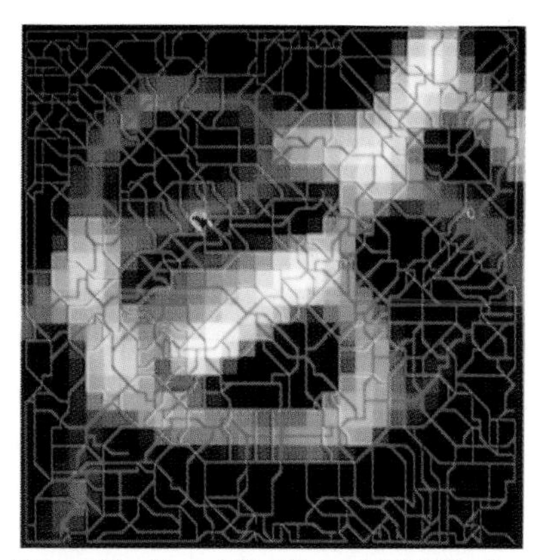

Figure 28 - Downhill differential MGPs (low gradient) are shown in blue and uphill differential MGPs (high gradient) in red.

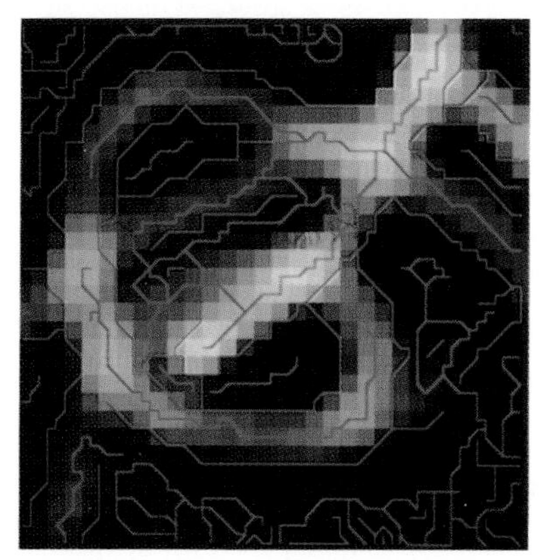

Figure 29 - The unoptimized ridges and troughs selected from the MGPs in Figure 28.

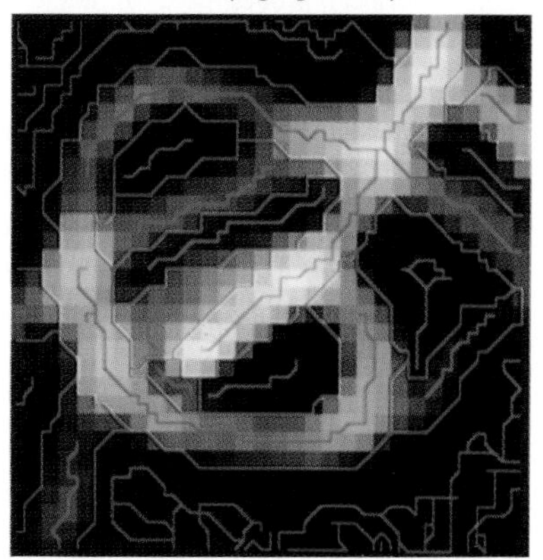

Figure 30 - The optimized ridges (edges) and troughs (grey-level skeletons) selected from the MGPs in Figure 28.

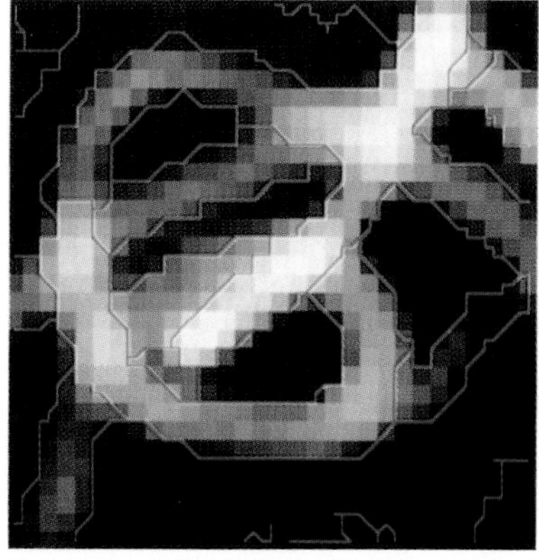

Figure 31 - The optimized ridges (edges) shown without the troughs.

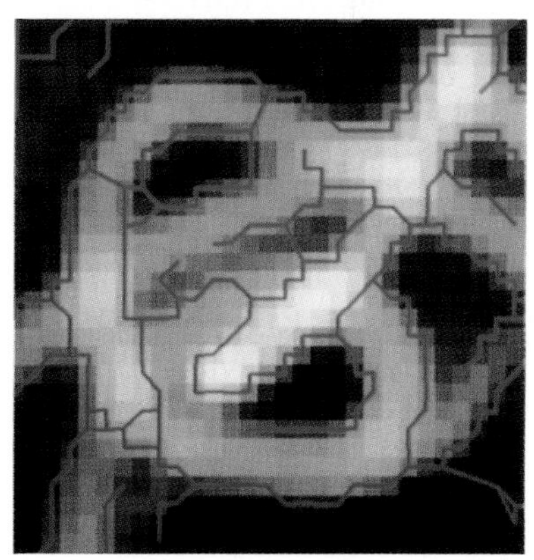

Figure 32 - Canny edges (σ=0.5).

SURGICAL PROBE DESIGN FOR A COINCIDENCE IMAGING SYSTEM WITHOUT A COLLIMATOR

J R Saffer[1], H H Barrett[1,2], H B Barber[1,2] and J M Woolfenden[2]

[1] Optical Sciences Center, University of Arizona, Tucson, AZ 85721

[2] Department of Radiology, University of Arizona College of Medicine, Tucson, AZ 85724

Abstract

Surgical probes used in nuclear medicine can be maneuvered close to a suspected tumor site, thereby achieving higher resolution and sensitivity than external gamma cameras. Tomography with handheld probes is not possible, however, so it is difficult to determine whether an increase in the photon count rate is caused by features near the probe or by variations in the distant background radiation. Our group has experimented with several probe designs that address this problem. This paper describes a probe that images nearby objects without a collimator and is insensitive to inhomogeneities in the distant background level. The probe, which is used with a radionuclide that emits multiple photons per decay, such as ^{111}In, consists of a collimatorless array of gamma-ray detectors connected by coincidence circuitry. We used a Monte Carlo routine to simulate data collection from such a system and reconstructed the images using the pseudoinverse obtained by singular value decomposition (SVD). The images show a significant suppression of distant sources when compared to a probe equipped with a conventional parallel-hole collimator.

Keywords

surgical probes; nuclear medicine; gamma rays; singular value decomposition (SVD); Monte Carlo

1. INTRODUCTION

One goal of imaging in nuclear medicine is to detect the gamma rays emitted by tumor-seeking radionuclides and identify regions of higher activity that may represent tumors. A common method is to use a gamma camera to record two-dimensional projection images of the radioisotope distribution. Tomography can be achieved by combining projections taken from many angles around the patient into a three-dimensional reconstruction. However, the projection images are often photon-starved due to insufficient uptake of the radiotracer by the tumor and the poor efficiency of the camera's collimator. In addition, scattering and attenuation of the emitted gamma rays by tissue further reduces resolution and sensitivity.

Handheld surgical probes are an alternative to external gamma cameras. Because of their small size, probes can be maneuvered close to a suspected tumor site, reducing intervening tissue and, therefore, attenuation and scatter (Woolfenden and Barber,

1989). Tomography is no longer feasible, however, so it is difficult to determine whether an increased count rate is caused by a small tumor near the probe or a change in the distant background radiation level. Our group has experimented with several probe designs to overcome this difficulty. One is a dual probe consisting of a central detector that measures the suspected tumor and a second, annular detector that measures the background, allowing background subtraction (Hickernell et al, 1988; Hickernell, 1988; Hartsough et al, 1989). A second approach is a 21-element imaging probe, which enables the surgeon to see whether there are sources in the field of view (such as blood vessels) that account for an increased count rate (Hartsough et al, 1989). These probes have been built and used in surgery. This report describes a new design for a nearsighted imaging probe which images nearby objects *without a collimator* and is insensitive to inhomogeneities in the distant background radiation because it cannot see beyond its immediate vicinity.

2. METHODS

2.1. Probe Design

The probe design is a collimatorless array of gamma-ray detectors connected by coincidence circuitry. It must be used with a radionuclide that emits two or more gamma rays per decay nearly simultaneously. An example of such an isotope is ^{111}In, which emits a 171 keV gamma ray followed 85 ns later by a 245 keV gamma ray. Unlike photon pairs produced by positron annihilation, these photons have no strong angular correlation and can emerge at any angle with respect to each other. Therefore a variety of detector geometries can be used. For our data simulations we chose a 10x10 planar array of detectors.

The coincidence circuitry triggers data collection only when two or more photons strike the detectors within a short time interval. Ideally this procedure records only photons produced by the same decay, thereby restricting the site of the decay to the common fields-of-view of the affected detectors. The low doses of radionuclide used in medicine are now an advantage, reducing the possibility of accidental coincidences from unrelated photons.

Investigation of γ-γ-coincidence imaging dates back many years (Schmitz-Feuerhake, 1970). From the start a major challenge has been overcoming the reduction in sensitivity caused by requiring coincidence. Researchers have attempted to design collimators for the optimum trade-off between resolution and sensitivity (Monahan and Powell, 1973; Chung et al, 1980; Boetticher et al, 1982). Recently, Powell (1989) suggested using time-of-flight information between arrivals at one large detector with a multihole collimator and several small, uncollimated detectors. A comprehensive comparison of multiple-photon coincidence-imaging techniques is presented by Liang and Jaszczak (1990).

Simulations of our probe system indicate that coincidence detection alone, without recourse to collimators or time-of-flight information, leads to satisfactory imaging of objects near the detector array. The reason is that a rapid falloff in sensitivity with depth is inherent in coincidence imaging, as explained below.

Figure 1 shows the geometry of our probe design. A collimatorless planar array of 100 detectors views a three-dimensional volume in which tumors may be present. If we neglect attenuation and scatter, then a photon isotropically emitted from source point **R** has a probability of reaching detector i that is related to the solid-angle $\Delta\Omega$ subtended by detector i:

$$P(1\ \text{photon}) = \frac{1}{4\pi}\,\Delta\Omega = \frac{\epsilon^2}{4\pi}\,\frac{\cos^3\theta_i(\mathbf{R})}{z^2}\,, \qquad (1)$$

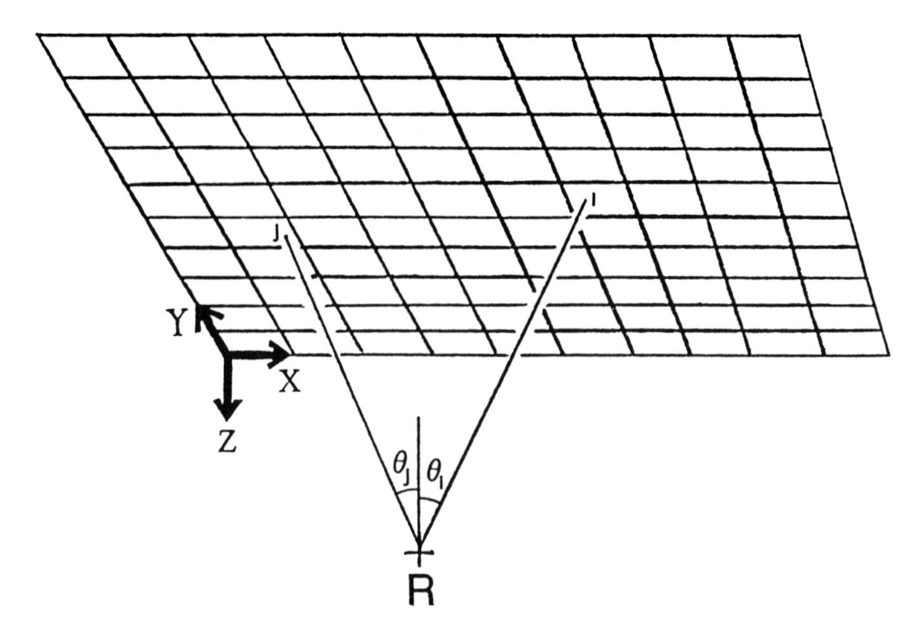

Figure 1. Geometry of the detection system.

where ϵ^2 is the detector area, $\theta_i(\mathbf{R})$ is the angle between the normal to the ith detector and location **R**, and z is the vertical distance from **R** to the detector plane. The detector efficiency is assumed to be unity.

Coincidence detection requires that two photons from the same source point reach the detector array. Because the photons are emitted independently, the probability of coincidence is proportional to the product of the solid angles subtended by detectors i and j:

$$P(2\text{ photons}) = \frac{\epsilon^4}{16\pi^2}\,\frac{\cos^3\theta_i(\mathbf{R})}{z^2}\,\frac{\cos^3\theta_j(\mathbf{R})}{z^2}\,. \tag{2}$$

The $1/z^4$ term in this equation is the source of the rapid decrease in sensitivity with depth in this design. Distant sources have a very low probability of contributing to the data, making them all but invisible to the probe.

2.2. Data Simulation

To test the utility of this probe design, we simulated data collection using a Monte Carlo approach. To avoid artifacts from discrete object representations, we modelled the object space as a continuous volume. The first object distribution tested was a point source embedded in a background having no activity. Therefore all photons originated from the same point. Random directions were selected for each emitted photon pair. If both photons from a single decay struck detectors in the array we assumed coincidence detection occurred, and we recorded the pair of detectors where the photons struck. If either photon missed striking the detectors we discarded the event.

For more complicated object distributions, points within the object volume were randomly selected. The probability of emission at that location was assigned a random number which we compared to the activity level at the corresponding point in the object model to determine whether a decay occurred. If it did, random directions for the emitted photons were selected and data collection proceeded as described above. We repeated the process until the number of disintegrations matched the activity in the field of view for an exposure time of 30 seconds. The procedure can be modified to include attenuation, scatter and accidental coincidences, although that has not been done in the data sets discussed here. The Monte Carlo technique does, however, correctly incorporate Poisson noise into the data.

Because there are 100 detectors in the array, the data vector **g** has $(100)^2$ components, one for each possible pair of detectors the photons can strike. Figure 2 shows one way of directly displaying such a data vector. The 100 larger squares represent the detector where the first photon of a photon pair struck. Each larger square is divided into 100 smaller squares. The smaller squares show which detector the second photon in the photon pair struck. Therefore each small square represents the conditional probability that a photon strikes detector j given that the other photon of that pair has already struck detector i. The data set displayed was collected for a point source at $(x,y,z) = (3.5,-1.5,2.5)$, where the origin of the coordinate system is the center of the detector array and the edges of the array are at $x = \pm 5$ and $y = \pm 5$. As might be expected, the data component with the highest value corresponds to both photons going straight up and hitting the detector directly overhead. In Figure 2 that detector combination occurs in the large square defined by the intersection of the ninth column and fourth row. Within that large square, the small square defined by the intersection of the ninth column and fourth row is the detector combination described.

The digital imaging process can be represented by the continuous-discrete model described in Barrett et al, (1991):

$$g_k = \int_\infty d\mathbf{R}\ f(\mathbf{R})\ h_k(\mathbf{R}) + n_k\ , \tag{3}$$

where g_k is the kth measurement, $f(\mathbf{R})$ is the continuous object being imaged, $h_k(\mathbf{R})$ is the average contribution of the object point at location $\mathbf{R}$ to the kth measurement, and n_k is random noise in the data. If we neglect attenuation and scatter and assume a detector efficiency of unity, h_k for our collimatorless coincidence probe is given by the right side of Eq. (2), where k lexicographically indexes all possible combinations of detectors i and j.

Although we used a continuous object volume to generate the data, a discrete object representation is necessary before image reconstruction can occur. We chose a 10x10x10 grid of object points where the spacing between points in the x, y and z planes equalled the width of one detector. In this representation h_k in Eq. (3) becomes a matrix **H**, an element of which can be written:

$$H_{kl} = H_{(ij)l} = \frac{1}{16\pi^2} \frac{\cos^3\theta_i(x_l,y_l,z_l)\ \cos^3\theta_j(x_l,y_l,z_l)}{z_l^4}\ , \tag{4}$$

where (x_l,y_l,z_l) is the lth location in object space (R_l) and the other quantities are defined in the text for Eq. (1). The detector area has been normalized to one. The dimension of **H** is 10,000x1,000.

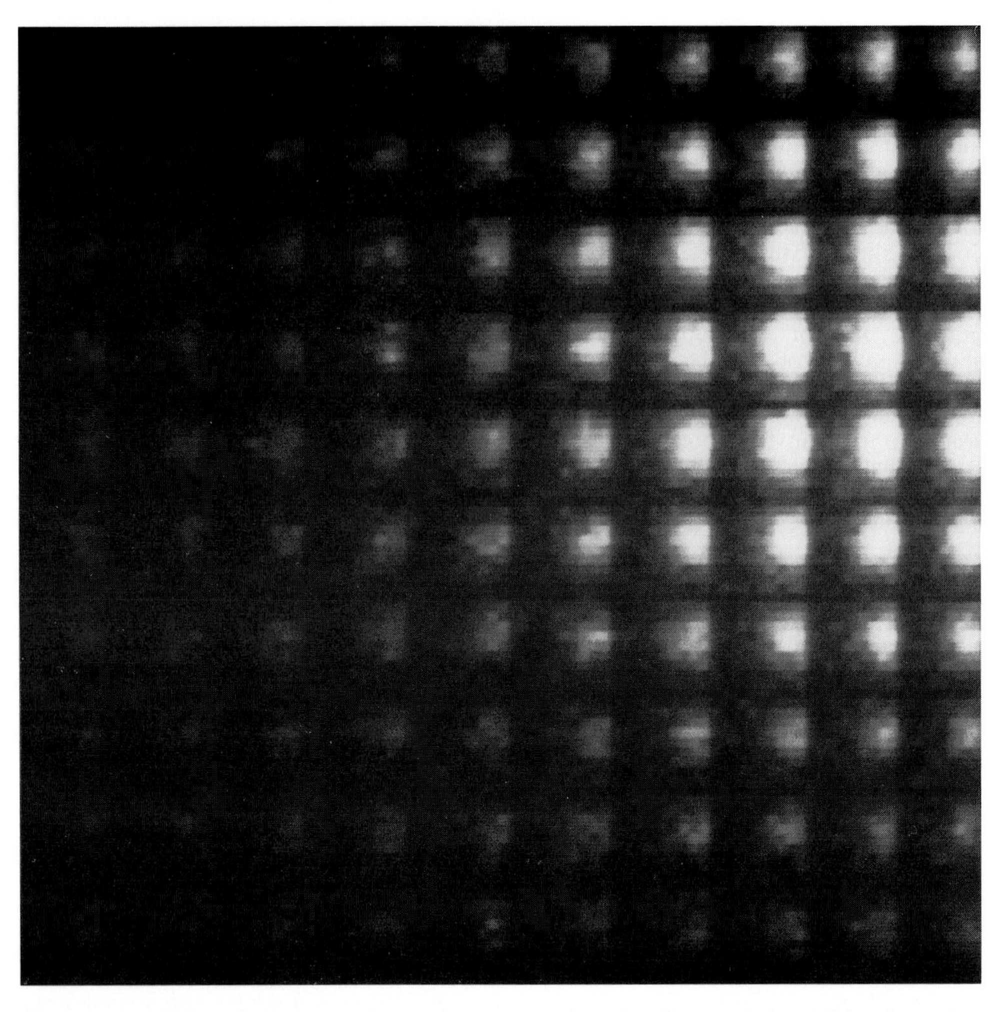

Figure 2. 4D display of data vector from point source at $(x, y, z) = (3.5, -1.5, 2.5)$. Origin of coordinate system is at the center of the detector array.

2.3. Eigenanalysis

Equipped with this analytic expression for matrix **H** we used eigenanalysis to determine the eigenvectors and eigenvalues of the square, Hermitian matrix H^tH:

$$H^tH\ \mathbf{u}_n = \lambda_n\ \mathbf{u}_n\ , \qquad n = 1,2,...,N \tag{5}$$

where $\mathbf{u}_n$ indicates the nth eigenvector, λ_n is the nth eigenvalue, and N = 1,000. The eigenvalues of H^tH are graphed in Figure 3 in order of decreasing magnitude. The abrupt falloff after the first 100 values reflects the fact that we chose an array with 100 detectors, so we cannot expect more than 100 independent contributions to the data. The change in slope around component 600 indicates the limit of machine precision.

Figure 4 shows the 10 eigenvectors with the largest eigenvalues, $\mathbf{u}_1$ through $\mathbf{u}_{10}$, reordered as images. Each row corresponds to an eigenvector. The ten subimages in each row correspond to the ten x-y planes where we evaluated **H**, at distances of z = 0.5 pixels (left column) through z = 9.5 pixels from the detector array (a pixel is the size of one detector). To compensate for the $1/z^4$ factor in **H**, each subimage has been renormalized so that its peak value is assigned the highest gray level and its minimum value the lowest gray level. Figure 5 shows eigenvectors 11 through 20. Figure 6 shows eigenvectors 91 through 100, the last eigenvectors before the falloff in the eigenvalue spectrum. All the eigenvectors display four-fold symmetry consistent with a square detector array. Note that close to the detector array (leftmost columns), there is a great deal of high frequency structure, suggesting we may expect high resolution there. However, the high frequencies disappear as distance from the detector array increases. The checkerboard pattern evident in the first subimage of eigenvector 100 (lower left corner of Figure 6) represents the Nyquist condition; the maximum recoverable frequency of the system is the inverse of twice the detector width.

2.4. Reconstruction

We used the pseudoinverse obtained by singular value decomposition (SVD) to reconstruct the object distribution. We again adopted a 10x10x10 object space discretization with the x and y grid points spaced one detector width apart, but the z grid points were spaced 1/4 pixel apart to more fully explore the region near the detector array. The SVD of the **H** matrix yields three matrices:

$$\mathbf{H} = \mathbf{U}\ \mathbf{W}\ \mathbf{V}^t\ . \tag{6}$$

The columns of matrices **U** and **V** form orthonormal basis sets for the object and image space, respectively. Matrix **U** is the same dimension as **H** (10,000x1,000), and **W** and **V** are 1,000x1,000. Matrix **W** is diagonal. Had we used the same grid spacing as we used in the eigenanalysis, the columns of **U** would equal the $\mathbf{u}_n$ calculated in Eq. (5), and the diagonal elements of **W** would equal the λ_n.

The equation for the SVD pseudoinverse reconstruction can be written [see Barrett et al, (1991)]:

$$\hat{f} = \sum_{j=1}^{R} \frac{\beta_j\ \mathbf{v}_j}{\lambda_j}\ , \tag{7}$$

where $\hat{f}$ is the estimate of the original object distribution, $\mathbf{v}_j$ is the jth column of **V**, the coefficients $\{\beta_j\}$ are defined by the dot product of columns of **U** with the data vector g, and the $\{\lambda_j\}$ are the diagonal elements of matrix **W**. The number of terms in $\hat{f}$ is R=100, the last value before the drop-off in the spectrum in Figure 3. We

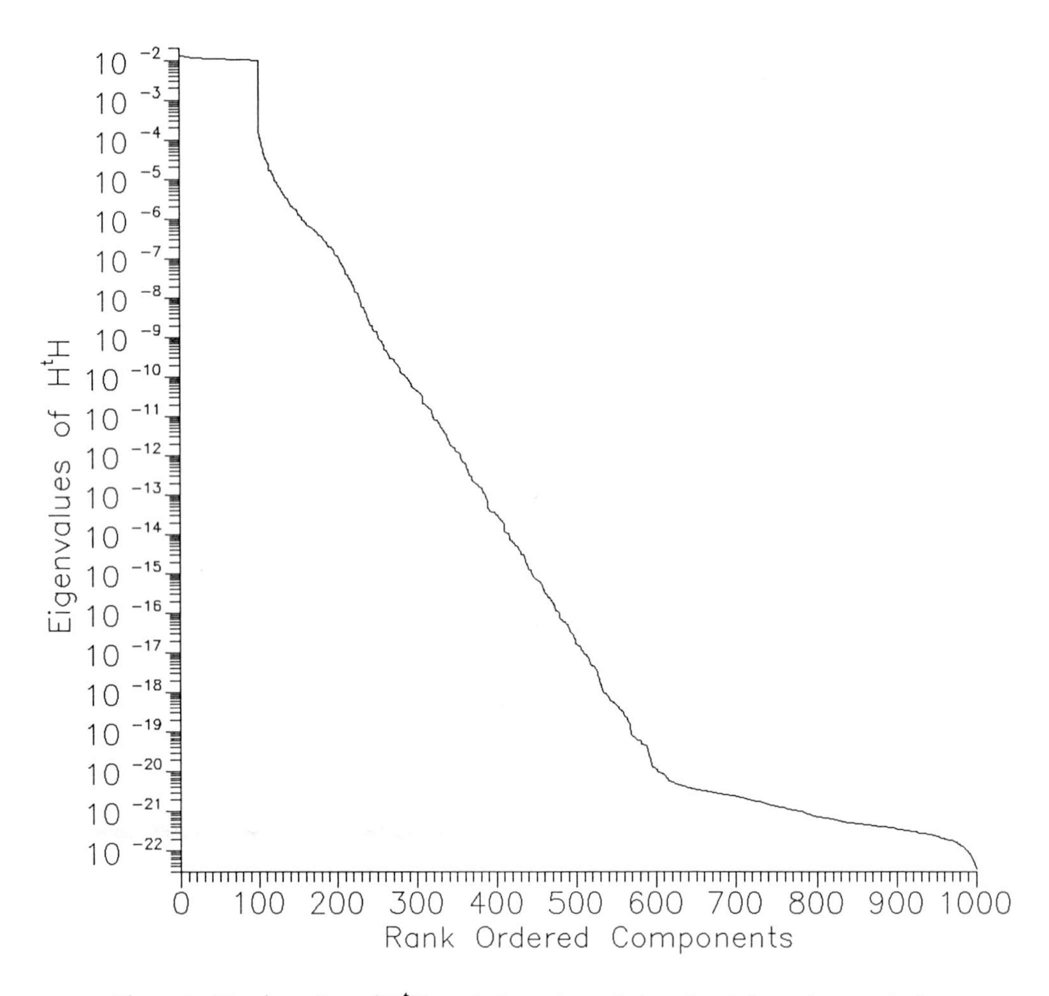

Figure 3. The eigenvalues of H^tH graphed on a log scale in order of decreasing magnitude.

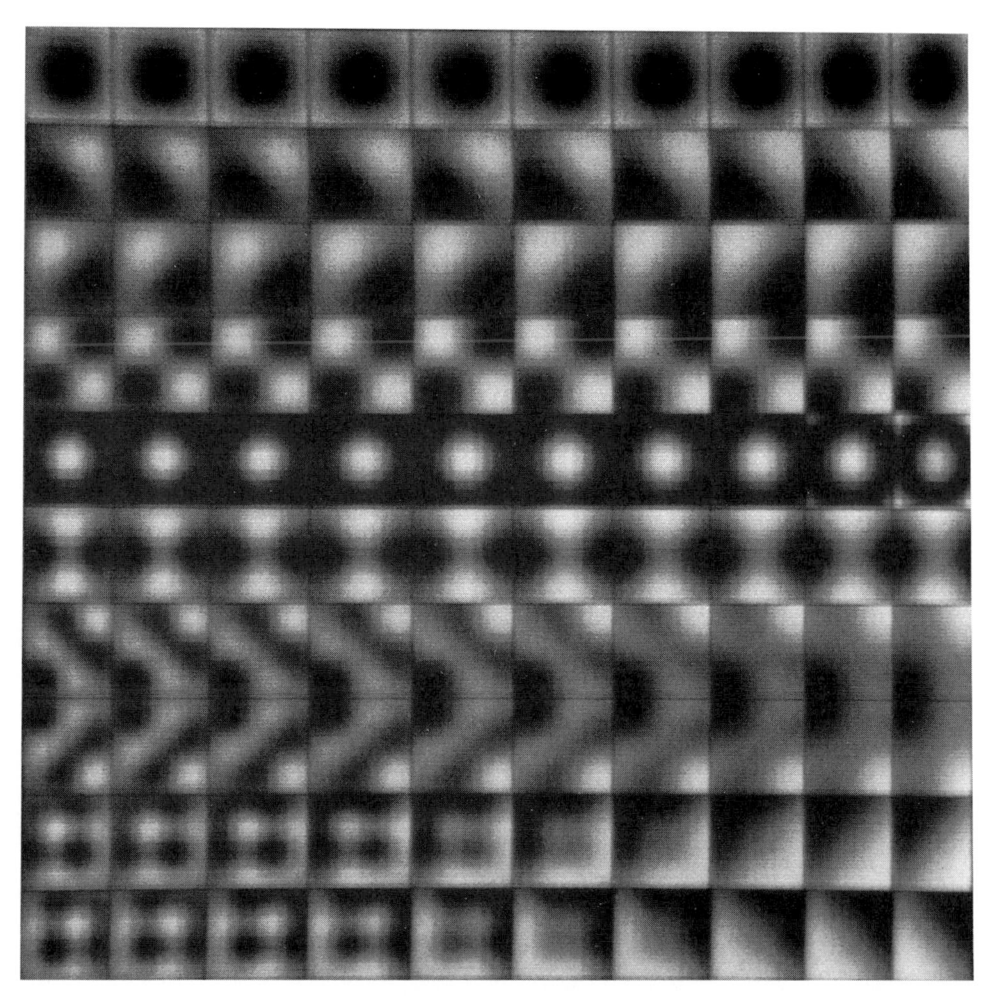

Figure 4. Eigenvectors 1 (top row) through 10.

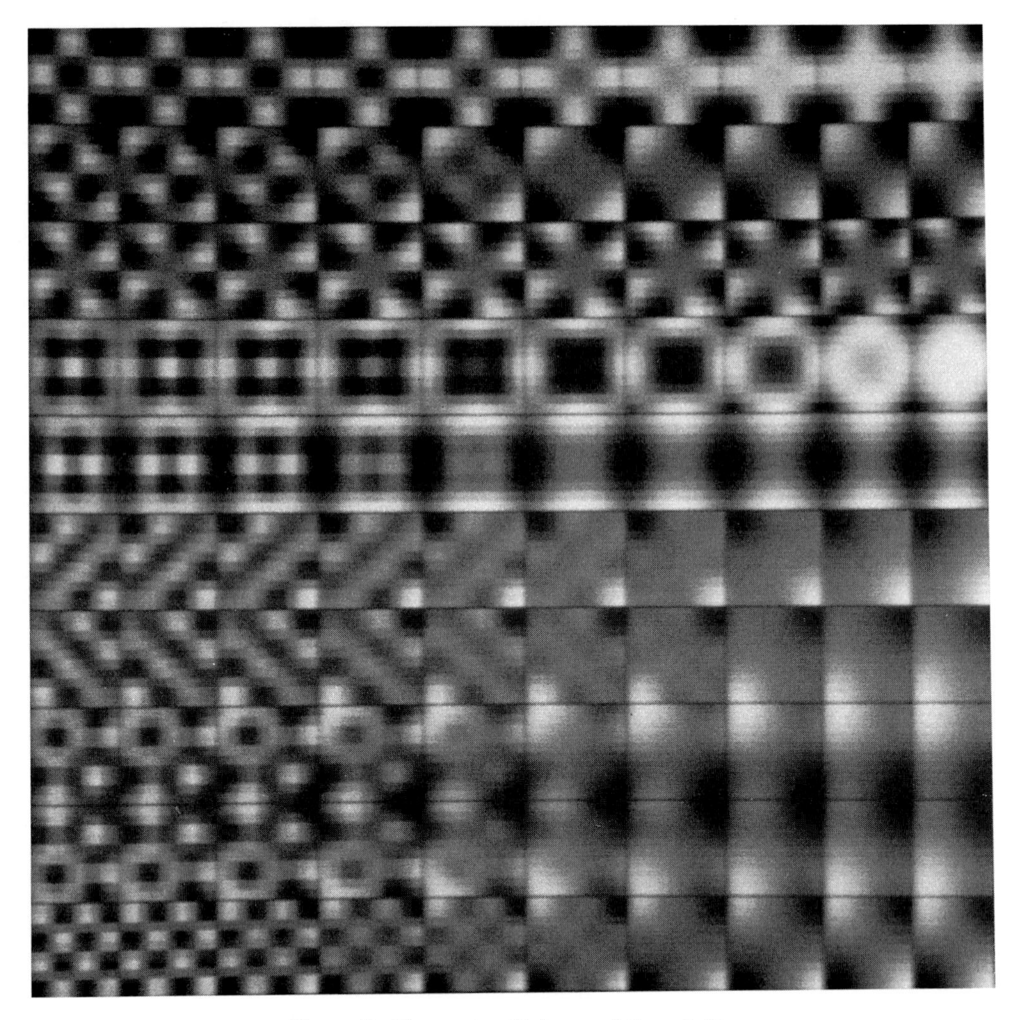

Figure 5. Eigenvectors 11 (top row) through 20.

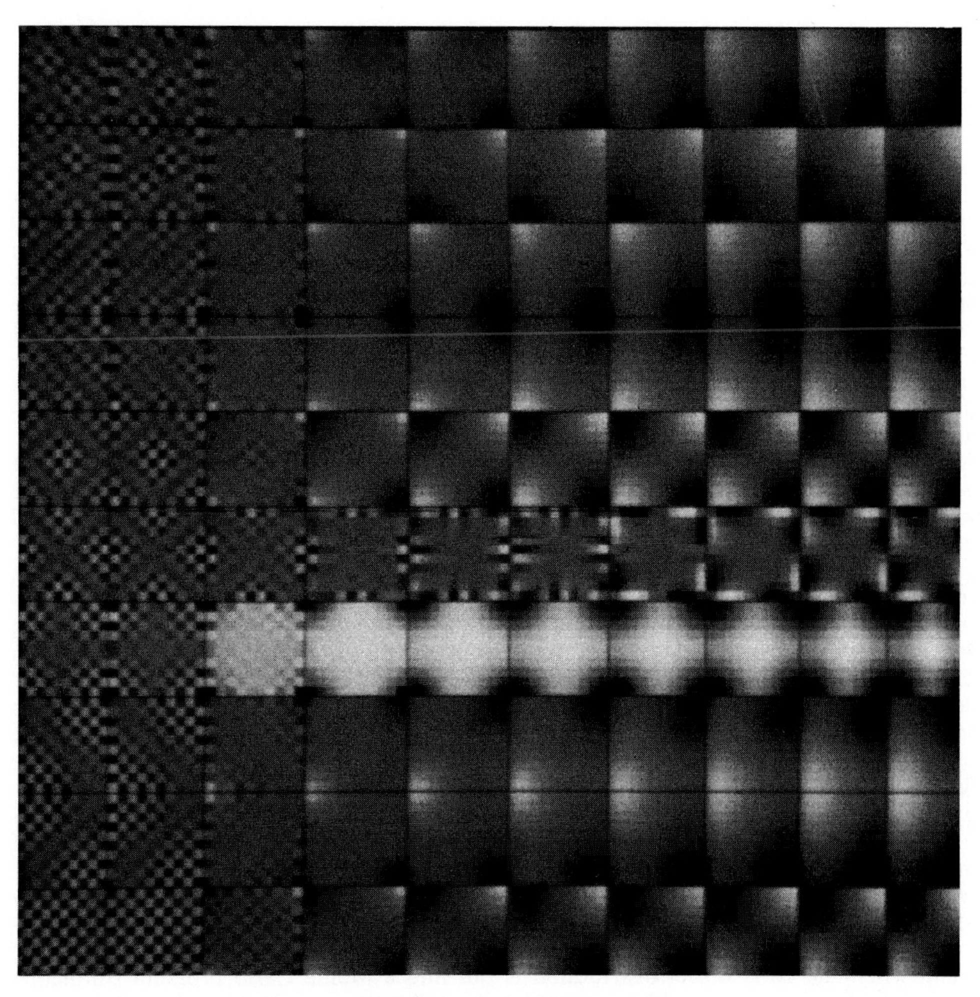

Figure 6. Eigenvectors 91 (top row) through 100.

make no attempt to restore the weak components $j > 100$.

3. RESULTS AND DISCUSSION

Figure 7 shows the reconstruction of a point source located at $(x,y,z) = (0.5,0.5,0.25)$. The ten subimages are the ten x-y planes of the reconstruction, starting with $z = 0.25$ at the upper left and proceeding to $z = 2.5$ at the lower right. Figure 8 shows the reconstruction when the point source is lowered to $(x,y,z) = (0.5,0.5,2.5)$. If this were a tomographic imaging system we would expect the point source to be best resolved in the plane of the reconstruction corresponding to its actual location. Instead we see that the best resolution occurs in the plane nearest the detector array. In Figure 9 the pixel values within each of the ten subimages of Figure 7 (top line) and Figure 8 (bottom line) are summed and plotted on a logarithmic scale along with the results from reconstructions of point sources located in the eight intervening planes. Note that there is a falloff of nearly three orders of magnitude within 2.5 pixels of the detector array. This is indeed a nearsighted probe. Also, because the plane nearest the detector array dominates the reconstruction, there is little reason to display 10 separate x-y planes. In the remaining reconstructions we sum corresponding pixels in the 10 x-y planes to yield a single two-dimensional 10x10 image.

To test the probe design on a more complicated detection task, we used the configuration shown in Figure 10. The two spheres touching the detector array represent the tumors we want to detect. A distant inhomogeneous background interfering with the detection task is simulated by four larger spheres centered 12.5 pixels from the detector array. The rest of the object volume is uniform. The pixel (and detector) size is 0.2x0.2 cm, appropriate for surgical probes. The activity in the spheres is 3700 Bq/cm^3, which is typical for tumor-seeking radiotracers, and the uniform background activity is 370 Bq/cm^3. As mentioned previously, the exposure time is 30 seconds.

The top gray-level image in Figure 11 represents a perfect, noise-free reconstruction of the two tumors near the detector array. To the right is a surface map of the gray levels in the image. The middle gray-level image is the SVD pseudoinverse reconstruction of data obtained from the object distribution in Figure 10 using the Monte Carlo program described previously. A surface map of this simulated collimatorless coincidence image appears on the right. The bottom gray-level image is the result when the same object distribution is put into a Monte Carlo program simulating non-coincidence imaging with a parallel-hole collimator attached to the 10x10 array (one collimator bore per detector). Its surface map is on the right. In the coincidence image the $1/z^4$ falloff in the number of coincident photons greatly suppresses contributions from the four distant spheres, leaving the two nearby tumors clearly visible. In the parallel-hole collimator image, however, neither nearby tumor is discernable. Even if the number of photons in the image is increased by doubling the activity or the exposure time, the variations in the distant background continue to swamp the tumor signal.

4. CONCLUSIONS

In summary, collimatorless coincidence imaging appears feasible for objects near the detector array. The region of greatest resolution for this imaging system occurs within $z = 2.5$ pixels. The advantages of collimatorless coincidence imaging are: 1) high resolution near the face of the detector array, and 2) suppression of background radiation. Evaluation of the effects of accidental coincidences is needed along with further investigation of the dependence of system performance on background radiation levels and the number of counts collected.

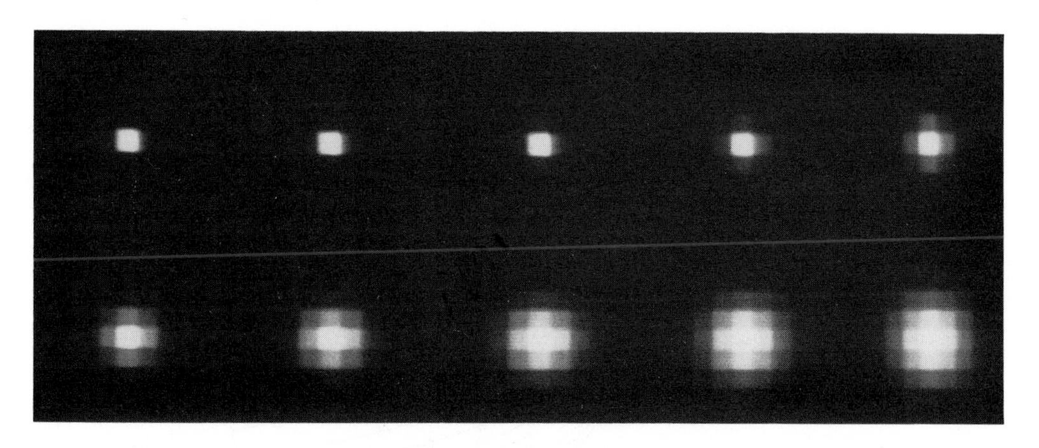

Figure 7. SVD pseudoinverse reconstruction of point source at $(x, y, z) = (0.5, 0.5, 0.25)$.

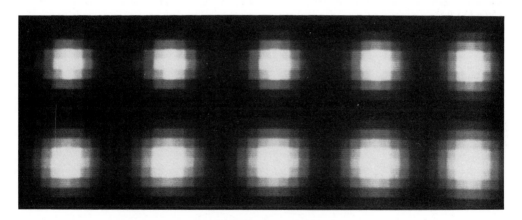

Figure 8. SVD pseudoinverse reconstruction of point source at $(x, y, z) = (0.5, 0.5, 2.5)$.

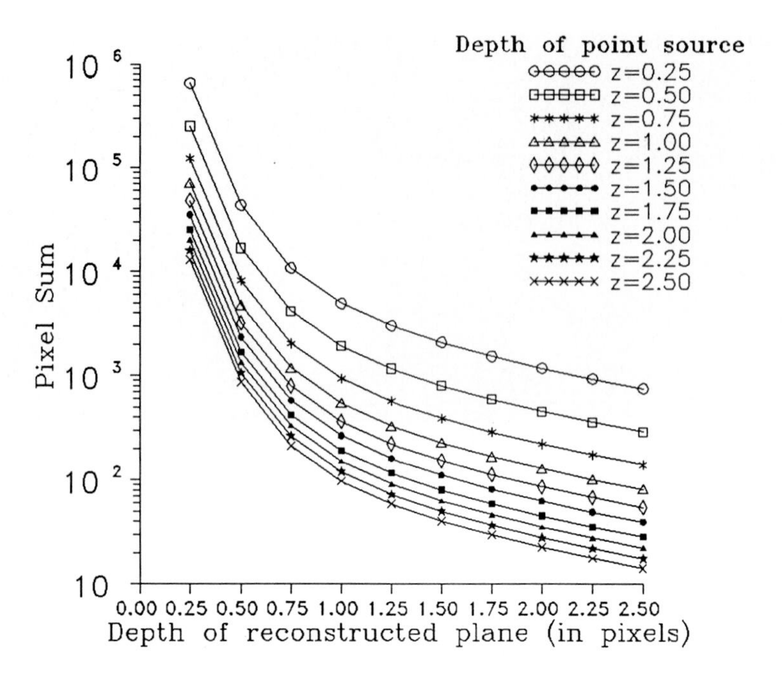

Figure 9. Sum of the pixels within each x-y plane in the reconstructions of point sources at (x, y) = (0.5, 0.5).

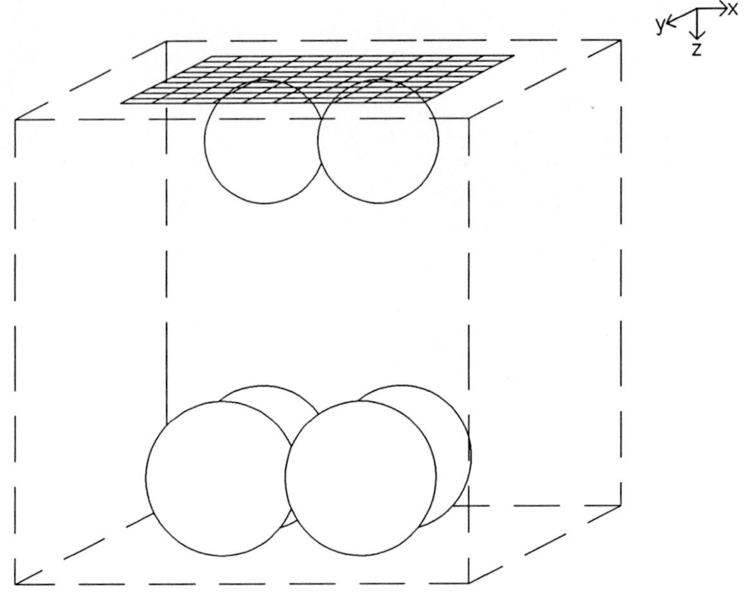

Figure 10. Object distribution used in generating data. The object volume is a cube 15 pixels on a side (one pixel is the size of one detector), centered beneath the 10x10 detector array. The spheres touching the detector array are centered at (x, y, z) = (-1.5, 0.5, 2.0) and (2.5, 0.5, 2.0) and are four pixels in diameter. The four distant spheres are centered at (x, y, z) = (±2.5, ±2.5, 12.5) and are each five pixels in diameter. The rest of the object volume is uniform and has 1/10th the activity level of the six spheres.

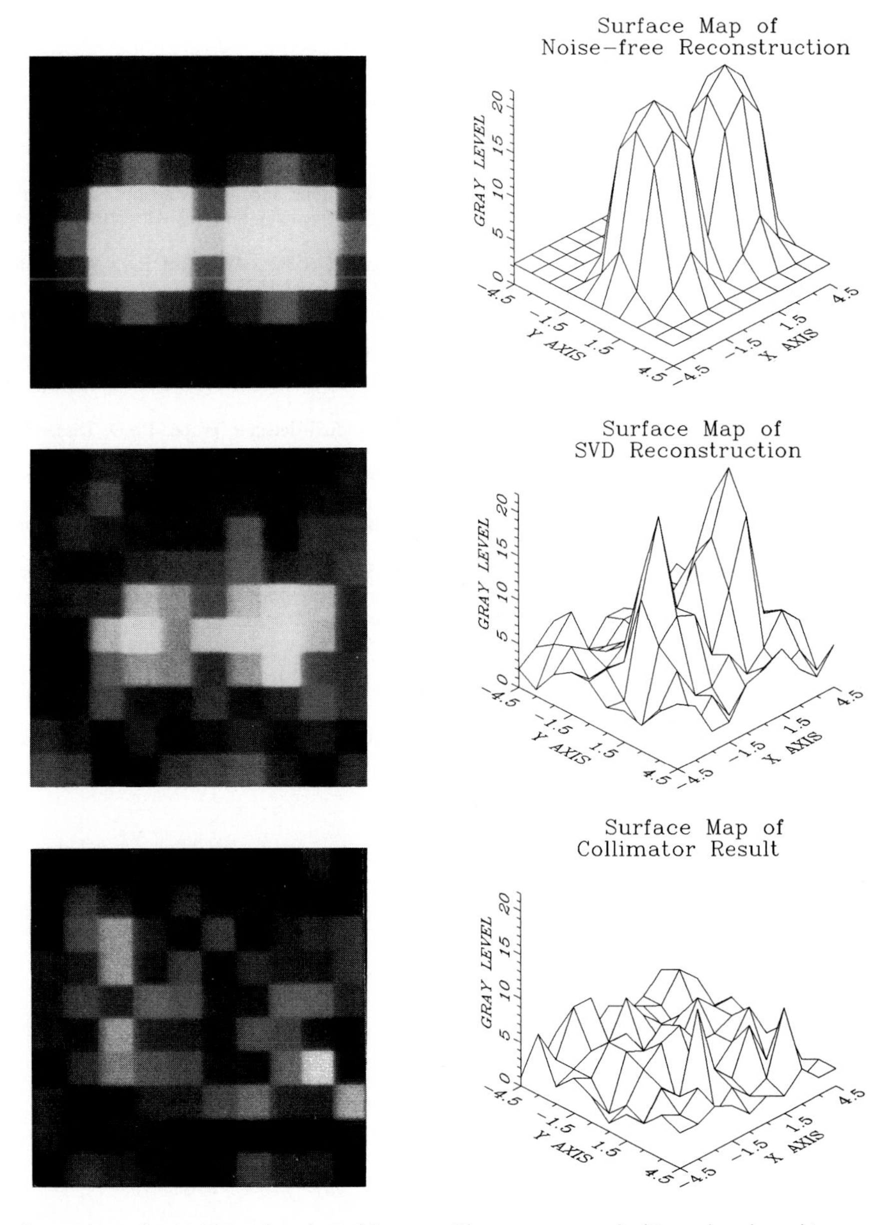

Figure 11. Comparison of coincidence imaging without a collimator to non-coincidence imaging with a parallel-hole collimator. The top image is a perfect, noise-free reconstruction of the two nearby tumors in Figure 10. The middle gray-level image is the SVD reconstruction. The gray-level images are normalized so that the peak gray level in the noise-free image equals the peak gray level in the SVD reconstruction. The bottom image is the result obtained when a parallel-hole collimator is used instead of collimatorless coincidence imaging. The collimator bores are 10 pixels long and 0.9 pixels in diameter, yielding an efficiency of 3.22×10^{-4}.

Acknowledgements

We thank Ted Bowen for discussions and helpful advice. This work was supported by National Cancer Institute grants P01 CA23417 and R01 CA52643.

References

Barrett HH, Aarsvold JN and Roney TJ (1991). Null functions and eigenfunctions: tools for the analysis of imaging systems. In: Information Processing in Medical Imaging. Ortendahl DA and Llacer J (eds), Wiley-Liss, New York, pp. 211-226.

Boetticher H, Helmers H, Schreiber P, Schmitz-Feuerhake I (1982). Advances in γ-γ-coincidence scintigraphy with the scintillation camera. Phys. Med. Biol. **27**:1495-1506.

Chung V, Chak KC, Zacuto P and Hart HE (1980). Multiple photon coincidence tomography. Semin. Nucl. Med. **10**:345-354.

Hartsough NE, Barber HB, Woolfenden JM, Barrett HH, Hickernell HS and Kwo DP (1989). Probes containing gamma radiation detectors for *in vivo* tumor detection and imaging. Proc. SPIE **1068**:92-99.

Hickernell TS (1988). Statistical decision making with a dual-detector probe. Ph.D. Dissertation, University of Arizona.

Hickernell TS, Barber HB, Barrett HH and Woolfenden JM (1988). A dual-detector probe for surgical tumor staging. J. Nucl. Med. **29**:1101-1106.

Liang Z and Jaszczak R (1990). Comparisons of multiple photon coincidence imaging techniques. IEEE Trans. Nucl. Sci. **37**:1282-1292.

Monahan WG and Powell MD (1973). Three-dimensional imaging of radionuclide distribution by gamma-gamma coincidence detection. In: Tomographic Imaging in Nuclear Medicine. Freedman GS (ed), Society of Nuclear Medicine, New York, pp. 165-175.

Powell MD (1989). Multiphoton, time-of-flight three-dimensional radionuclide imaging. Med. Phys. **16**:809-812.

Schmitz-Feuerhake I (1970). Studies on three-dimensional scintigraphy with γ-γ-coincidences. Phys. Med. Biol. **15**:649-656.

Woolfenden JM and Barber HB (1989). Radiation detector probes for tumor localization using tumor-seeking radioactive tracers. Am. J. Roentgenol. **153**:35-39 (invited paper).

VOXEL BASED MONTE CARLO CALCULATIONS OF NUCLEAR MEDICINE IMAGES AND APPLIED VARIANCE REDUCTION TECHNIQUES

George Zubal and Chuck Harrell

Division of Imaging Science
Department of Diagnostic Radiology
Yale University, New Haven, CT 06510

Abstract

Due to the availability of digitally stored human anatomy images, 3-dimensional surfaces of internal structures of the body can be stored in computer volume arrays. Such a volume based software phantom delineates internal human organs with millimeter resolution and lends itself to fully 3-dimensional Monte Carlo simulations. Our simulation models 45 internal human organs (each with an associated radioisotope concentration and attenuation coefficient), calculates gamma radiation histories through these structures, and accepts gamma events onto a collimated planar camera. Variance reduction techniques are applied to decrease the time required to compute a given number of events at the detector. Stratification and two implementations of forced detection variance reduction techniques are compared to "brute force" calculations for their efficiency speed-ups in this heterogeneous geometry. Simulated clinical images of the liver are shown.

Keywords

Simulation; software phantom; stratification; forced detection

1. INTRODUCTION

Monte Carlo modeling of the transmission and attenuation of internal radiation sources has led to a better understanding of the image formation process in diagnostic radiology. Historically, computer models of the physical distributions of radio-isotopes and attenuating materials have been highly simplified in order to achieve reasonable turn around times of the Monte Carlo simulations. Due to the current availability of inexpensive memory and disk storage as well as ever increasing execution speeds, these software models may be made more complex in order to model the human anatomy more realistically and permit statistical Monte Carlo computations to be completed within acceptable time limits.

The largest volume of literature describing Monte Carlo calculations for photon interactions in tissue equivalent materials concerns health physics and therapeutic dose calculations for internal or external radiation sources (Snyder et al, 1967, Bice et al, 1985, Williamson et al, 1983, Dale, 1983, Williamson 1983, Dale 1982, Williams et al, 1986, Boyer and Mok, 1986, Cunningham et al, 1986, Kijewski et al, 1986, Mohan et al, 1986, Mackie et al, 1985, Persliden and Carlsson, 1984). More recently Monte Carlo techniques have been applied to better understand the image formation process in radiology (Barnea and Dick, 1986, Chan and Doi, 1985, Kanamori et al, 1985, Dance and Day, 1984,) and particularly for analyzing scatter and attenuation problems in nuclear medicine. Most recently, variance reduction techniques have been applied to fairly homogeneous phantoms for single slice tomography calculations (Haynor et al, 1990) The software phantoms modeled in these simulations have been limited almost exclusively to simple point, rod, and slab shapes of homogeneous sources and attenuating media. Such simple geometries are useful in studying more fundamental issues of scatter and attenuation, but have the shortcoming that clinically realistic distributions cannot be fully appreciated by such simple geometries. They can only be inferred or experimentally measured but not directly studied by this mathematical simulation. We have constructed an "anatomically correct" heterogeneous human geometry for use in these types of radiologic calculations (Zubal et al, 1990).

Monte Carlo simulations with such a detailed human software phantom allows us to investigate several clinically significant applications of detection and image reconstruction techniques. Previously evaluated attenuations and scatter correction techniques applied to tomographic reconstruction algorithms for nuclear medicine (Logan and Bernstein, 1983, Floyd et al, 1984, Floyd et al, 1985, Floyd et al, 1986, Floyd et al, 1987, Beck et al, 1982, Acchiappati et al, 1989) and by extension, new imaging concepts and novel instrumentation geometries (Hickernell et al, 1990, Woolfenden and Barber, 1990, Rowe et al, 1990) can be aided in their development through a close scrutiny of the image formation process as modeled in a true-to-human geometry. We present here a general simulation tool for understanding the physics involved in detecting radiation from internal sources in a patient anatomy and ways to increase the speed of calculation.

2. METHODS, PHANTOM AND SIMULATION

2.1 Anatomic Data

CT supplies the required high resolution 3-dimensional human anatomy necessary to construct an anthropomorphic phantom suitable for Monte Carlo calculations, since the CT Hounsfield numbers are closely related to attenuation coefficients. CT is used to study diffuse diseases in the whole body and a considerable number of patients are routinely imaged from head to midthigh. One such patient served as our anatomical 3-dimensional data set. A total of 78 slice images were acquired from neck to midthigh with a 1 cm slice thickness using a 48 cms field of view. During a second imaging session, 72 slices of the same patient were acquired of the head and neck region in 5 mm slices thickness with a field of view of 25 cms. The body and head slices were transferred to our image processing lab by reading the reconstructed transverse slices from the CT archive magtape, decompressing the images, and storing them on disk. The transverse images were displayed on a microVAX 3500 with color workstation for viewing and interactive manipulation. A serial line high resolution Summagraphics bitpad provided high resolution cursor control.

2.2 Organ Delineation

In order to make the 3-dimensional anatomical data suitable for use in radiologic calculations, the internal structural surfaces must be known in order to define source distributions and global attenuation coefficients. Members of the medical staff outlined 45 separate internal organs contained in the transverse slice data and used a region of interest (ROI) coloring routine to fill the inside of each organ outline with a unique index value. More than 1200 ROIs were drawn with 1 millimeter resolution to define this fully 3-dimensional voxel phantom of the human torso.

An in-house program was developed to read the transverse slices from disk, display them on the color workstation monitor, and permit outlining of organs under bitpad cursor control. The x and y integer positions of all of the organ outlines are stored on disk with a resolution of 512 by 512 pixels. A separate variable size file exists for each transverse slice and contains coordinates all of the contours drawn on that slice. These contours serve as the input to the filling routine, which creates a fixed size organ index image. The organ index image is a 512 by 512 byte matrix filled with integer values which delineate the internal structures (organs) of the body. The organ index image is therefore in effect the original CT transverse slice in which the Hounsfield numbers are replaced by integers corresponding to the organ.

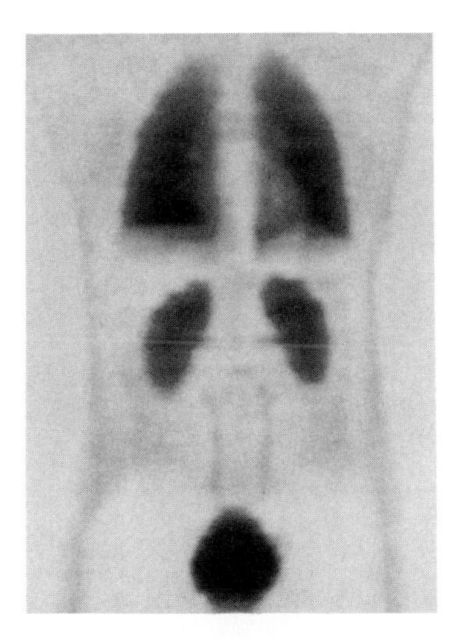

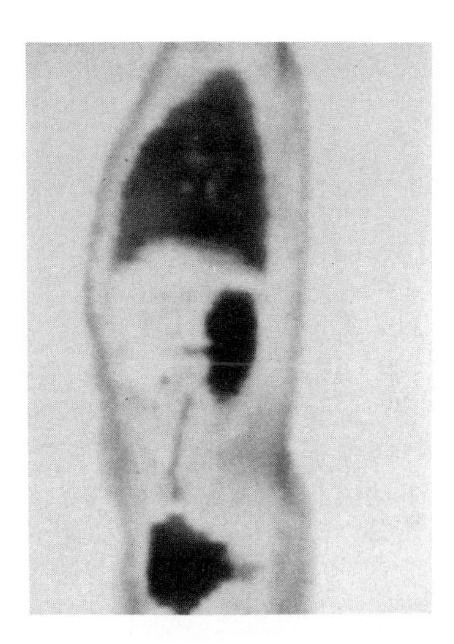

Figure 1: Anterior and lateral projection images of the anthropomorphic phantom showing the urine, bladder, kidneys, lungs and skin.

The 2-dimensional organ index images can be read into a 512 by 512 by 78 voxel 3-dimensional matrix, in which the x,y resolution is 1 millimeter per pixel and the z-resolution is 10 millimeters. In order to make the data more manageable and to have consistent resolution along all three axes, we routinely transform the original data into a 128 x 128 x 384 matrix where the cubic voxel resolution is 2 millimeters on each side. This resolution is achieved by averaging pixels in the x,y plane and by duplicating slices along the z-direction.

In order to appreciate the detail of the anthropomorphic phantom, we filled various structures in the body with varying relative concentrations of activity and projected the primary radiation onto anterior and lateral camera positions; this in effect eliminates all scatter events and shows the intrinsic detail and resolution of the phantom. The images for relative concentrations of 7.0, 9.0, 7.0, 3.0, and 1.0 in the urine, bladder, left and right kidney, lungs, and skin surface are shown in Figure 1. , and concentrations of 3.0, 3.0, 3.0, 3.0, 3.0, and 1.0 for long bones, pelvis, ribcage and sternum, spine, spinal cord and skin in Figure 2. This digitized human torso serves as an input to the Monte Carlo program for simulating planar nuclear medical images.

2.3 The Monte Carlo Simulation Procedure

An existing Monte Carlo program (Snyder et al, 1967, Eckerman and Ryman), originally developed for calculating internal radiation doses, was modified to permit a more detailed analysis of the radiation escaping from the anthropomorphic phantom and to include the model of a parallel hole collimated nuclear medicine camera. The modified code models the following physical properties (see Figure 3):

- A typical male human torso digitized into 128 x 128 x 384 voxel elements. A total of 45 internal organs corresponding to the major organs of the human body are delineated within this voxel phantom. Each voxel is indexed to an organ name and linear attenuation coefficient.
- Variable concentrations of diagnostically relevant radioactive sources distributed homogeneously throughout each organ volume, and emission of source gamma rays from these source sites.

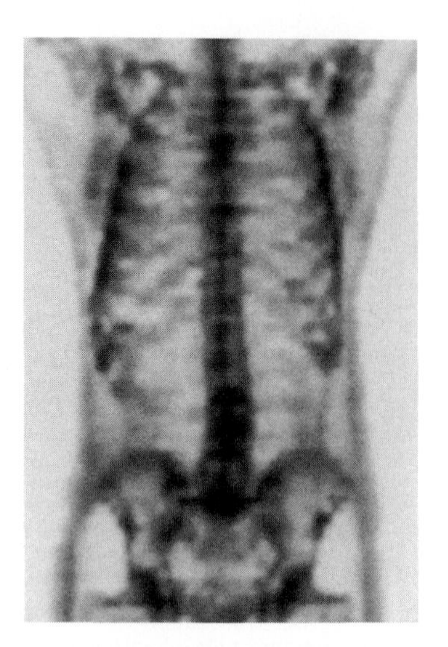

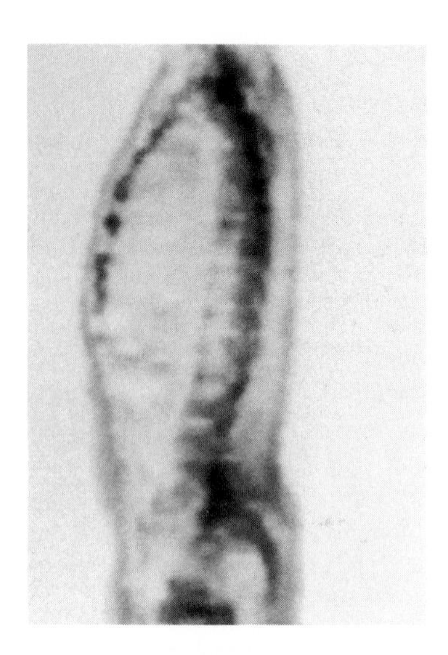

Figure 2: Anterior and lateral projection images of the anthropomorphic phantom showing the long bones, pelvis, ribcage, sternum,spine, spinal cord, and skin.

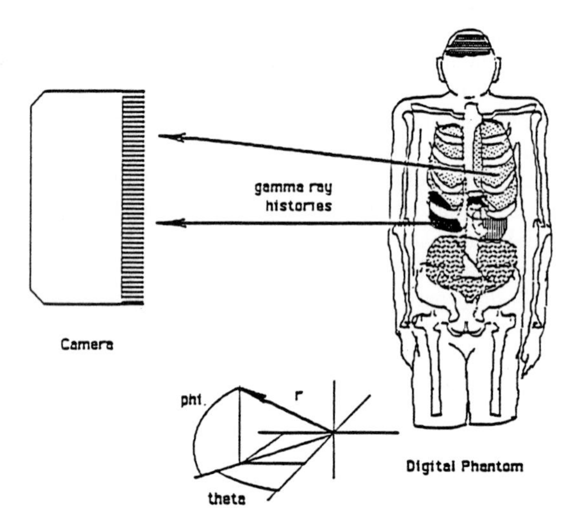

Figure 3: Schematic representation of the anthropomorphic phantom, gamma ray histories and nuclear camera.

- Compton scatter events are modeled according to the Klein Nishina equation.
- Attenuation properties of the internal organs are calculated from the organ's composition and density, and is a function of the gamma radiation's current energy.
- Passage of the gamma rays out of the anthropomorphic phantom, through a parallel hole collimator for detection in the position sensitive camera.

2.4 Generation of Source Distribution

In order to correctly model the statistical generation of source gamma rays within the organs of the phantom, random numbers (distributed between 0 and 1) are generated. The first random number selects the organ in which the gamma ray originates. Subregions between the numerical values of 0 and 1 are attributed to the various organs in the body. The size of each numerical region corresponds to the fraction of the total radioactivity found in that organ. Each time a random number falls within an organ's region, the program selects that organ for the generation of the next source gamma ray. The voxel within the organ which is taken as the site of the gamma emission is also selected randomly. The 3-dimensional initial direction of the gamma ray is calculated through 3 random numbers which are applied to the unit direction vectors.

2.5 Generation of Interaction Sites

In order to calculate the position at which the source gamma ray interacts, a potential interaction site is calculated by using an attenuation coefficient μ_0 which is greater than or equal to any other attenuation coefficient in the phantom and calculating $d = -\ln(n)/\mu_0$ where n is a random number between 0 and 1. The x,y,z position at a distance, d, along the initial gamma ray direction is taken as a site of interaction only if a newly generated random number falls within the range 0 to μ/μ_0 where μ is the attenuation coefficient of the voxel at position x,y,z. If this potential interact site is not accepted, the gamma ray is allowed to continue its flight. If the interaction site is accepted, a new direction and energy of the scattered gamma ray is calculated from tabulated atomic cross sections and according to a statistical treatment of the Klein Nishina equation (McMaster et al, Berger and Seltzer, Kahn, 1954).

2.6 Emerging Radiation and Detection

The 3-dimensional flight history of each gamma ray is traced until the gamma ray emerges from the surface of the phantom or its energy falls below the cut-off energy. The parameters associated with each emerging gamma ray are its 3-dimensional direction, energy, and the number of times it has scattered. The 3-dimensional direction vector is used to determine the probability with which the gamma ray successfully passes through the simulated parallel hole collimator and is counted by the imaging camera (Sorenson and Phelps, 1987). Compton scatter is not currently modeled in the detector. A lower cut-off energy is set to model the energy acceptance window used in cameras to reject scattered radiation. When a gamma ray falls below this user selectable energy value, its history is terminated.

2.7 Simple Phantom Experiment

In order to assess the speed up in calculation time for variance reduction techniques, a simplified phantom consisting of a single voxel containing a 140 keV source emitter centered in a 30 cm diameter water sphere was input into the Monte Carlo simulation. A source energy of 140 keV was selected because it is the emission energy of 99mTc -the most widely used isotope for nuclear medicine imaging studies. The CPU time required to gather 100,000 gamma events by the camera from this point source was used as the calculation timing measure; these simulations were carried out for lower cut-off energy values between 135 and 40 keV.

3. METHODS, VARIANCE REDUCTION TECHNIQUES

3.1 Stratification

The energy that a gamma has after a scatter event can be easily calculated by

$$\frac{E'}{E_0} = \frac{1}{1+\alpha(1-cos\theta)}$$

where E' is the energy of the gamma ray after the scatter,

E_0 is the energy of the gamma ray before the scatter,
θ is the scatter angle, and
α is the ratio of E_0/(rest energy of the electron).

Knowing the position and size of the imaging camera for a given simulation, and the lower energy cut-off, we can define a 3-dimensional angular direction within which gamma rays can finally reach the camera without falling below the lower energy cut-off. Source gammas whose initial direction lies outside of this 3-dimensional solid angle can only reach the camera by scattering below the energy window.

In our implementation of stratification, we sample source gammas within an angle of $cos^{-1}[1-(E_0/E'-1)/\alpha]$ of the normal to the camera; source gamma rays outside of this 3-dimensional region are not allowed. The speed up attributed to this uneven source sampling techniques is dependent on the lower cut-off energy (see Figure 4). For 140 keV source radiation, and a lower cut-off of 110 keV no speed up due to source stratification is expected since for θ=90 degrees, the allowable solid angle fills the complete sphere.

3.2 Forced Detection

Forced detection relies on the concept that after a gamma scatters in the phantom, and whichever direction it scatters into, there is a finite probability that it could have been detected in the camera. We increment the camera image at that possible detection site by the probability that the scattered gamma would have been detected.

We calculate forced detection events by two approaches. In our first approach, when a gamma ray scatters and its new direction generally points toward a camera, it will either leave the phantom and be counted in the camera (image is incremented by the probability of passing through the collimator) or it will scatter again. If the gamma ray leaves the phantom, we allow it to be counted in the camera. If it is determined that it will scatter again, we increment the camera image by the probability that it would not have this future interaction. This probability is calculated by integrating the attenuation coefficients along the current ray direction from the current scatter site out to the edge of the phantom and applying this integrated length to the law of exponential attenuation:

$$exp(-\int_{c.p.}^{e.o.p.} \mu(r)rdr)$$

where *exp* is the exponential function,
$\mu(r)$ is the attenuation coefficient of the heterogeneous phantom along the path of r,
r is the 3-dimensional ray length through the phantom.
The integral is carried out along the ray from the current position ($c.p.$) in the phantom out to where the ray intersects the outside surface of the phantom ($e.o.p.$).
We refer to this variance reduction technique as simple "forced detection" or FD1.

In our second approach, we assume that regardless of the direction that a gamma ray scatters into, there is a finite probability that it could have scattered directly into the camera. Although scattering probabilities for gamma rays at and below 140 keV are moderately forward peaked, we currently assume the scatter probabilities to be isotropic for ease of computation. The probability of scatter into the camera is determined by the ratio of the collimator acceptance angle to the full 4π spherical geometry. The product of this forced angle scatter is multiplied by the previously described exponential nonattenuation probability to increment the camera image. We tersely refer to this variance reduction technique as "forced angle" or FD2.

4. RESULTS AND DISCUSSION

4.1 Performance Improvement

As is typical for most variance reduction techniques, additional computation is required to compute a photon's history. This must be overcompensated by the fact that the variance in the final image is strongly reduced, otherwise the application of the techniques is not merited. That is to say, that variance reduction techniques effectively permit many more gamma events are counted at the camera per unit time (Jenkins et al, 1988, Kalos and Whitlock, 1986, Morin, 1988).

A final efficiency can be quantified by looking at the computation time (CPU time) required to generate a given total number of gamma events detected at the camera. Figure 4 is a comparison of the computation efficiency improvements for three variance reduction techniques discussed above. Times are normalized such that 1.0 corresponds to the brute force technique generating 40,000 camera events in 19 minutes. We notice that none of the techniques show much improvement when gamma histories are calculated down to energy cut-off values below approximately 100 keV. We believe this to be due to the fact that as more and more scatter events per gamma history are computed, the scatter physics dominate the computation expense. As long as reasonably high cut-off energies are selected, speed up factors of up to 2 are achievable.

It is interesting to note that forced angle technique results in a slightly decreased efficiency in calculating camera events. We believe this to be due to the high expense of calculating an independent distance to the edge of the phantom only to increment the camera with a low probability. The low probability is due to the fact that the ratio of the collimator acceptance angle to 4π geometry is approximately 0.0005.

In order to demonstrate that the variance reduction techniques are not influencing the characteristics of our image, we generated profiles through the simulated point source (Figure 5). If we assume the brute force technique (no variance reduction applied) to be best representation

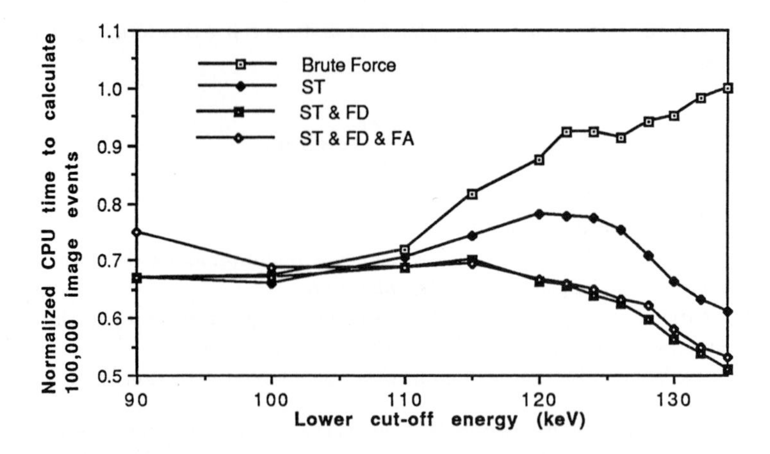

Figure 4: Graph of computation efficiencies showing time to calculate 100k image gamma events vs. lower energy cut-off shown for 4 cases: brute force, stratification (ST), ST+forced detection (FD1), and ST+FD+forced angle (FD2).

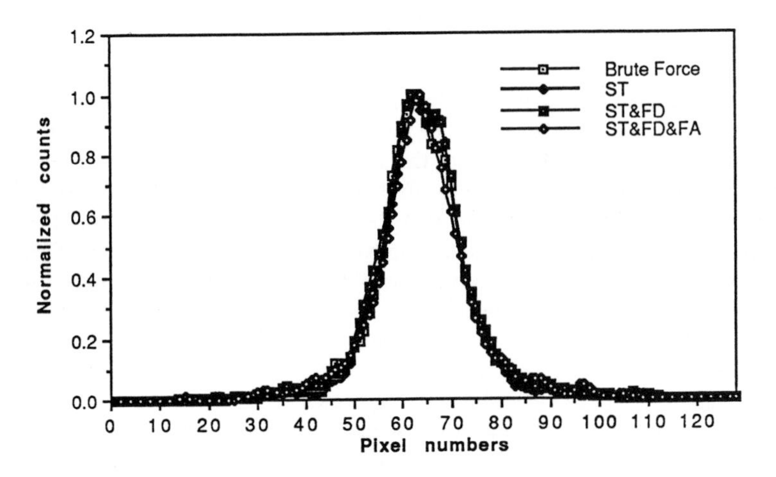

Figure 5: Profiles through point source calculation using methods of variance reduction techniques.

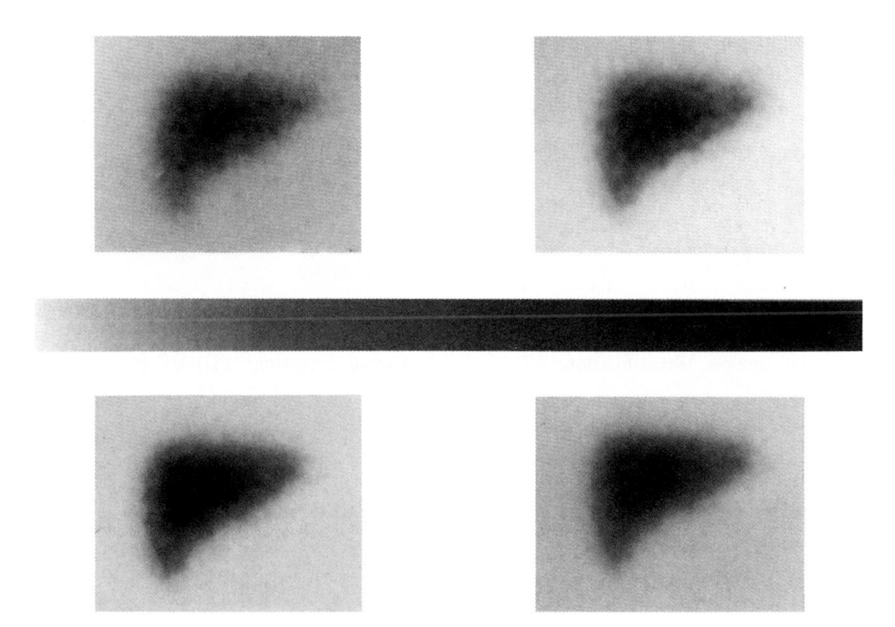

Figure 6: Simulated liver images computed by the BF (top left), ST (top right), FD1 (bottom left), and FD2 (bottom right).

of the "true" distribution of events; then by applying efficiency increasing techniques, we should not see any appreciable change in the point source distribution of the simulated image. We see that the profiles are indeed virtually identical to each other.

4.2 Clinical Example

Once we were comfortable with the benchmark simulations which we ran for the simplified phantom, we replaced the simple phantom with the high resolution anthropomorphic phantom. The liver of the phantom was filled with 100 per cent of the 140 keV source energy radiopharmaceutical and a lower energy cut-off of 115 keV was set. Forty million source events were generated and an anterior view simulated image was computed. The images were computed using the brute force technique (BF), source stratification (ST), forced detection (FD1) and forced angle (FD2) (Figure 6). A total of 23655, 23765, 48256, and 57456 image counts respectively were calculated in each image; CPU time was 1025, 995, 1651, and 2002 minutes respectively. To be certain that the variance reduction techniques do not inappropriately influence the final images of distributed sources, we generated difference images by subtracting the brute force generated image from the ST, FD1, and FD2 images (Figure 7).

5. CONCLUSIONS

Variance reduction techniques of stratification, forced detection, and forced scatter angle can be applied to voxel-based Monte Carlo simulations in order to achieve significantly improved image statistics for planar images of fully 3-dimensional source distributions in complicated heterogeneous phantoms. These methods of speeding up the computation of planar images do not significantly degrade or alter the structure or noise characteristics of simulated nuclear medicine images.

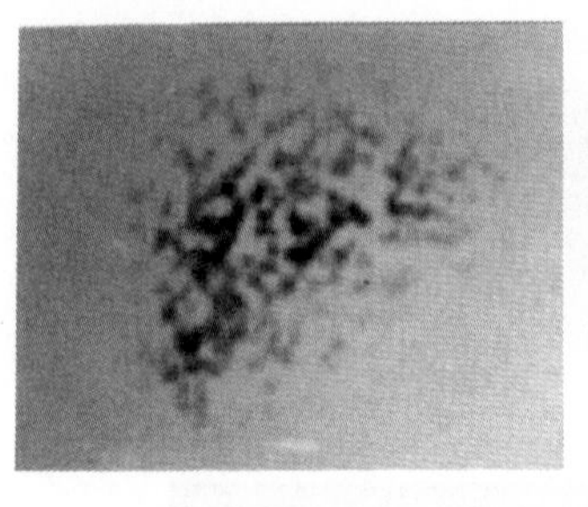

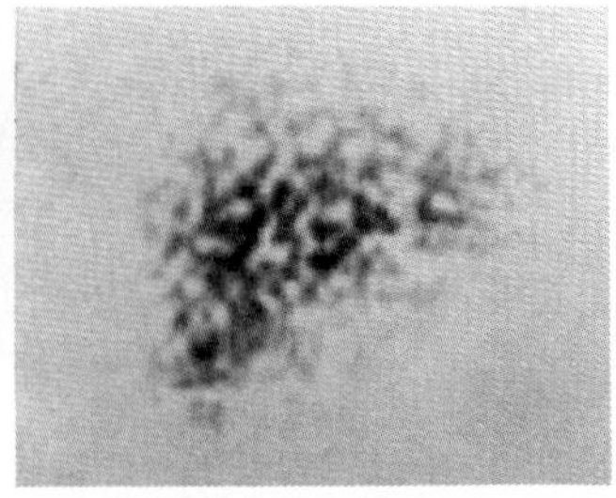

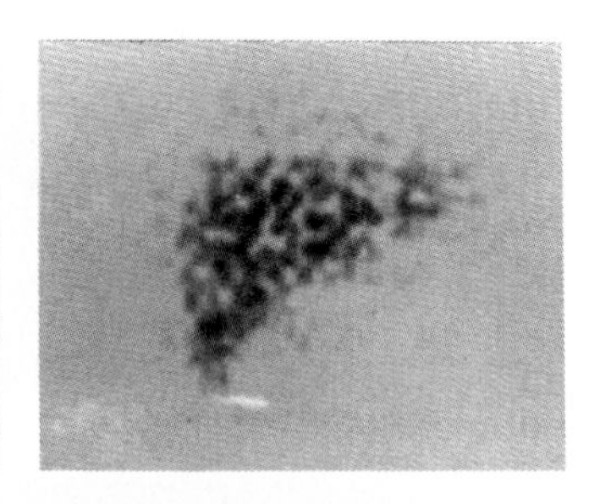

Figure 7: Simulated liver differences images computed by subtracting the brute force image (BF) from the variance reduction images. Stratification: ST-BF (left); Forced detection: FD1-BF (center), and Forced angle: FD2-BF (right)

Acknowledgements

Work performed under Contract No. DE FG02-88ER60724 with the U.S. Department of Energy.

References

Acchiappati D, Cerullo N, Guzzardi R (1989). Assessment of the Scatter Fraction Evaluation Methodology using Monte-Carlo Simulation Techniques. European Journal of Nuclear Medicine N11:683-686.

Barnea G, Dick CE (1986). Monte Carlo studies of x-ray scatterings in transmission diagnostic radiology. Med Phys 13(4):490-5.

Beck JW, Jaszczak RJ, Coleman RE, Starmer CF, Nolte LW (1982). Analysis of SPECT Including Scatter and Attenuation Using Sophisticated Monte Carlo Modeling Methods. IEEE Transactions on Nuclear Science, NS-29(1):506-511.

Berger MJ, Seltzer SM. National Academy of Science - National Research Council Publ. 1133 (Nuclear Science Series Report no. 39) pp. 205-268.

Bice AN, Links JM, Wong DF, Wagner HN (1985). Absorbed fractions for dose calculations of neuroreceptor PET studies. Eur J Nucl Med 11(4):127-131.

Boyer AL, Mok EC (1986). Calculation of photon dose distributions in an inhomogeneous medium using convolutions. Med Phys 13(4):503-9.

Chan HP, Doi K (1985). Physical characteristics of scattered radiation in diagnostic radiology: Monte Carlo simulation studies. Med Phys 12(2):152-65.

Cunningham JR, Woo M, Bielojew AF (1986). The dependence of mass energy absorption coefficient ratios on beam size and depth in phantom. Med Phys 13(4):496-502.

Dale RG (1982). A Monte Carlo derivation of parameters for use in the tissue dosimetry of medium and low energy nuclides. Br J Radiol 55(658):748-57.

Dale RG (1983). Some theoretical derivations relating to the tissue dosimetry of brachytherapy nuclides with particular reference to I-125. Med Phys 10(2):176-183.

Dance DR, Day GJ (1984). The computation of scatter in mammography by Monte Carlo methods. Phys Med Bio 29(3):237-47.

Eckerman K, Ryman J: ALGAMP, Health and Safety Research Division, Oak Ridge National Laboratory, Oak Ridge, Tenn.

Floyd CE, Jaszczak RJ, Greer KL, Coleman RE (1985). Deconvolution of Compton scatter in SPECT. J Nucl Med 26(4):403-8.

Floyd CE, Jaszczak RJ, Greer KL, Coleman RE (1986). Inverse Monte Carlo as a Unified Reconstruction Algorithm for ECT. J nucl Med 27(10):1577-85.

Floyd CE, Jaszczak RS, Harris CC, Coleman RE (1984). Energy and spatial distribution of multiple order Compton scatter in SPECT. A Monte Carlo investigation. Phys Med Bio 29(10):1217-30.

Floyd CE, Jaszczak RJ, Greer KL, Coleman RE (1987). Brain Phantom: high resolution imaging with SPECT and I-123." Radiology 164(1):279-81.

Haynor DR, Harrison PL, Lewellen TK, Bice AN, Anson CP, Gillespie SB, Miyaoka RS, Pollard KR, Zhu JB (1990). Improving the Efficiency of Emission Tomography Simulations Using Variance Reduction Techniques. IEEE Transactions on Nuclear Science. 37(2):749-753.

Hickernell TS, Barrett HH, Barber HB, Woolfenden JM, Hall JN (1990). Probability modeling of a surgical probe for tumor detection. Phys Med Bio, 35(4):539-559

Jenkins TM, Nelson WR, Rindi A (1988). Variance Reduction Techniques in Monte Carlo Transport of Electrons and Photons. Plenum Press.

Kahn H (1954). Report AECU - 3259 pp. 64-65.

Kalos MH, Whitlock PA 1986. Monte Carlo Methods Volume I. John Wiley and Sons.

Kanamori H, Nakamori N, Inone K (1985). Effects of Scattered x-rays on CT images. Phys Med Bio 30(3):239-49.

Kijewski PK, Bjarngard BE, Petti PL (1986). Monte Carlo calculations of scatter dose for small field sized in a Co-60 beam. Med Phys 13(1):74-7.

Logan J, Bernstein HJ (1983). A Monte Carlo simulation of Compton scattering in positron emission tomography. J Comput Assist Tomogr 7(2):316-20.

McMaster WH, Del Grande NK, Mallett JH, Hubble JH: UCRL 50174, II Rev. I.

Mackie TR, Scrimger JW, Battista JJ (1985). A convolution method of calculating dose for 15 MV x-rays. Med Phys 12(2):188-96.

Mohan R, Chui C, Lidofsky L (1986). Differential pencil beam dose computation model for photons. Med Phys 13(1):64-73.

Morin RL 1988. Monte Carlo Simulation in the Radiological Sciences. CRC Press.

Persliden J, Carlsson GA (1984). Energy imparted to water slabs by photons in the energy range 5-300 keV. Calculations using a Monte Carlo photon transport model. Phys Med Bio 29(9):1075-88.

Rowe RK, Barrett HH, Patton DD (1990). Design Study For a Stationary 3-D Spect Brain Iamging System. J Nucl Med, 31(5):769

Snyder W, Ford MR, Warner G (1978). Estimated of Specific Absorbed Fractions for Photon Sources Uniformly Distributed in Various Organs of a Heterogeneous phantom. NM/MIRD Pamphlet No. 5 Society of Nuclear medicine Publication.

Sorenson JA, Phelps ME (1987). Physics in Nuclear Medicine, Second Edition. Grune and Stratton.

Williams G, Zankl M, Abmayr W, Yeit R (1986). The calculation of dose from external photon exposures using reference and realistic human phantoms and Monte Carlo methods. Phys Med Bio 31(4):449-52.

Williamson JF, Morin RL, Khan FM (1983). Monte Carlo evaluation of the Sievert integral for brachytherapy dosimetry Phys Med Bio 28(9):1021-1032

Williamson JF, Morin RL, Khan FM (1983). Dose calibrator response to brachytherapy sources: A Monte Carlo and analytic evaluation. Med Phys 10(2):135-140.

Woolfenden JM, Barber HB (1990). Design and Use of Radiation Detector Probes for Intraoperative Tumor Detection Using Tumor-Seeking Radiotracers. Nuclear Medicine Annual. Raven Press

Zubal G, Gindi G, Lee M, Harrell C, Smith E (1990). High Resolution Anthropomorphic Phantom for Monte Carlo Analysis of Internal Sources. Proceedings of the Third Annual IEEE Symposium on Computer-Based Medical Systems, Chapel Hill, NC. June 3-6:540-547.

SPECT SCATTER CORRECTION IN NON-HOMOGENEOUS MEDIA

S R Meikle[1], B F Hutton[1], D L Bailey[1], R R Fulton[1] and K Schindhelm[2]

[1]Department of Nuclear Medicine, Royal Prince Alfred Hospital, Sydney 2050 and

[2]Centre for Biomedical Engineering, University of New South Wales, Sydney 2033, Australia.

Abstract

Single photon emission computed tomography (SPECT) has the potential for quantitation of absolute activity concentration in vivo. The accuracy of activity estimates depends to a large extent on the accuracy of the attenuation and scatter corrections performed. This is particularly so in the thorax, where assumptions about regular object shape and constant density are inappropriate. We have developed a method of scatter correction based on convolution subtraction (CS) which takes account of variable tissue density. Rather than assuming the scatter fraction to be constant, the scatter fraction is determined at each point in the image based on measured photon transmission through the object. For comparison, a modified lower window subtraction (LWS) technique has also been developed, involving convolution of lower window data with a theoretically derived kernel, which more accurately relates lower window scatter to photopeak scatter compared with conventional LWS.

Both methods have been assessed in a phantom study using a non-uniform medium and distributed activity. The methods described were compared with the two conventional methods on which they are based. The accuracy for determining activity concentration was calculated for a 5cm diameter "hot" cylinder, the "warm" background and a 10cm diameter "cold" air cylinder. Quantitative accuracy of >95% was achieved in the "hot" cylinder and in the "warm" background for both the TDCS and MLWS methods, whilst all 4 methods yielded no significant reconstructed counts in the region of the "cold" air cylinder, indicating very good cold contrast.

Keywords

Quantitation; attenuation; transmission; tomography; convolution subtraction; lower window subtraction.

1. INTRODUCTION

The most commonly employed method of image reconstruction in single photon emission computed tomography (SPECT) is filtered backprojection. This method assumes that the measured projections represent line integrals of the radioactivity distribution in the object being imaged. However, due to photon attenuation and scattering within the object, the data do not conform exactly to line integrals. Without compensation for these effects, the technique is limited to qualitative rather than quantitative analysis of the reconstructed images.

The problem of correcting for attenuation and scattering is particularly difficult in the thorax where tissue density varies markedly. To date, most of the work on quantitative SPECT in the thorax has concentrated on the attenuation problem. There are a number of ways in which the accuracy of attenuation correction in non-homogeneous media may be improved. One example is to use measured attenuation coefficients (μ) directly rather than using assumed values. This may be achieved by performing a separate CT scan (Moore, 1982) or by acquiring a transmission study using a flood source which can be acquired simultaneously with the emission scan (Bailey et al, 1987). These methods are normally used in combination with iterative or so-called 2nd order techniques (Murase et al, 1987) which generally involve refining the correction based on errors calculated from the initial, or 1st order, estimate. Using these techniques, simulation has demonstrated that quantitative accuracy of >95% can be achieved in the absence of scatter and other degrading factors (Hutton et al, 1988). However, a rigorous quantitative regime must also account for scattered photons in the correction procedure.

There are two commonly employed methods of scatter correction in SPECT. One approach is to estimate the scatter distribution by convolution of the photopeak image with a previously measured scatter function, which usually has a mono-exponential nature. The resulting image is scaled and then subtracted from the photopeak image (Axelsson et al, 1984). The scatter function is obtained by measuring the slope of the exponential "tails" of the gamma camera response from a point source of activity placed in a scattering medium. It can be demonstrated that, in general, this method over-estimates scatter and may be improved by using the first estimate of scatter as the input to a second convolution-subtraction step to yield an improved estimate of the scatter distribution (Bailey et al, 1989). This procedure can be repeated iteratively and results in increased quantitative accuracy as well as improved contrast. These approaches assume that the scatter function (ie the exponential convolution kernel) is spatially invariant and that the scatter fraction, taken to be the ratio of scattered to total recorded counts, is constant over the entire object volume. Whilst this gives an acceptable estimate of the scatter distribution in a uniform object, recent studies using Monte Carlo simulation have shown that scatter is highly dependent on the homogeneity of the object (Ljunberg and Strand, 1990). This suggests that density values derived from transmission measurements may provide additional information which can be used to improve the estimate of scatter.

Alternatively, a second image may be recorded in a Compton scatter window (eg, 92-125keV for ^{99m}Tc) and a fraction of this image (0.4-0.5) subtracted from the photopeak image (Jaszczak et al, 1984). This method assumes that the spatial distribution of scattered photons recorded in the lower energy window is the same as the scatter distribution in the photopeak. This is clearly not the case, since photons which are recorded in a lower energy window will have lost more energy through scattering collisions than those recorded in the photopeak. Therefore, they are likely to have undergone multiple scattering interactions or may have been scattered through greater angles giving rise to a broader scatter distribution. It has been previously demonstrated (Todd-Pokropek et al, 1984), however, that there is a relationship between the spatial distribution of scattered photons recorded in different energy windows and that the information from these data may be used to obtain an improved estimate of the photopeak scatter distribution. Lower window methods have the advantage that they inherently account for object inhomogeneity since scattered photons are recorded directly. Therefore, such techniques may be useful for comparison with other scatter correction methods, particularly in non-uniform media.

The aim of this study was to develop a scatter correction technique, based on the convolution subtraction method, which makes use of transmission measurements to improve the accuracy of the scatter estimate in non-uniform media. A modified lower window subtraction method has also been developed for comparison with the new technique.

1.2. Theory

1.2.1. Transmission Dependent Convolution Subtraction (TDCS)

The scatter correction method proposed by Axelsson et al (1984) involves estimating the photopeak scatter image, g_s, by convolution of the measured image, g_o, with an experimentally determined mono-exponential scatter function, s:

$$g_s = k g_o * s \tag{1}$$

where k is the scatter fraction, assumed constant throughout the image. It is also assumed that the scatter point response, s, is spatially invariant, in order to satisfy stationarity. Recent simulation studies using Monte Carlo methods have shown that the space invariance of s is a reasonable assumption, even in media with variable attenuation (Ivanovich and Weber, 1990). Intuitively, however, one might expect the scatter fraction to vary spatially according to regional tissue density.

The scatter fraction, k, at each point in the image is defined as the ratio of scattered to total recorded counts:

$$k = \frac{C_{broad} - C_{narrow}}{C_{broad}} \tag{2}$$

where C_{broad} are broad beam counts (including scatter) and C_{narrow} are narrow beam counts (excluding scatter). The difference between broad beam and narrow beam counts represents the scatter component.

We can also write this in terms of the buildup factor, B, which is defined as the ratio of broad beam to narrow beam counts at depth d (Hubbell, 1963):

$$k = 1 - \frac{1}{B(d)} \tag{3}$$

Now, using Wu and Siegal's formulation of the depth dependent buildup factor (Wu and Siegal, 1984), we can write:

$$k = 1 - \frac{1}{A\text{-}Be^{-\kappa d}} \tag{4}$$

where A, B and κ are constants. In their formulation, Wu and Siegal (1984) assumed that buildup is dependent only on depth, whereas, in the more general case, buildup will be dependent on the effective attenuation path length, μd, in tissue. Therefore, we can express Eq. (4) in terms of narrow-beam transmission values, $e^{-\mu d}$, and, hence, estimate scatter fraction from the relationship:

$$k = 1 - \frac{1}{A\text{-}Be^{-\xi\mu d}} \tag{5}$$

where ξ is a constant. Therefore, the proposed method utilizes the measured transmission scan to estimate the scatter fraction spatially for all projections.

1.2.2. Modified Lower Window Subtraction (MLWS)

The lower window subtraction (LWS) technique inherently accounts for object inhomogeneity by directly recording scattered photons. Therefore, it provides a useful comparison with the transmission dependent method. However, LWS has limitations which need to be recognized. The scatter image represents photons which have lost significantly more energy than those recorded in the photopeak. Therefore, a different scatter distribution is expected. This was confirmed by imaging a ^{99m}Tc point source in 8cm of water in various energy windows (Figure 1). As expected, the scatter distribution was found to broaden as energy decreased. This is suggestive of a convolution relationship between lower window data and photopeak scatter.

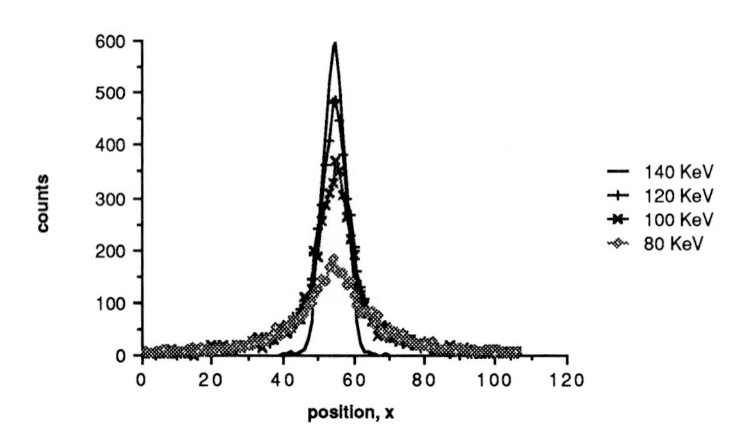

Figure 1: Scatter profiles for a ^{99m}Tc point source in 8cm of water. Images were acquired in 5% energy windows at 80KeV, 100KeV, 120KeV and 140KeV. The profiles demonstrate broadening of the scatter distribution as energy decreases with a concommittant decrease in peak amplitude, suggesting a convolution relationship between data recorded in different energy windows.

As a result of this finding, a modified lower window subtraction (MLWS) method was derived as follows. If we model scatter recorded in the photopeak as a convolution of the primary photon distribution with a mono-exponential function, and scatter recorded in a lower energy window as a convolution with a different mono-exponential function then we can obtain the following relationship (see Appendix) between the image recorded in a lower energy window and the photopeak scatter image:

$$g_s = g_{ul} * \{\alpha(\delta + \beta e^{-br})\} \tag{6}$$

where g_{ul} are the lower window data, δ is the Dirac delta function in 2D, α, β and b are constants and $*$ denotes convolution.

Therefore, the distribution of photopeak scatter can be expressed in terms of the lower energy window data, g_{ul}, and some constants which may be experimentally determined. In order for this formulation to be useful, it must be shown that the convolution kernel is independent of source depth and geometry.

2. METHODS

2.1. Transmission Dependent Convolution Subtraction (TDCS)

In order to determine the constants in Eq. (5), a ^{99m}Tc point source was imaged at various depths in water (0-16cm) and a function of the form $A\text{-}Be^{-\xi\mu d}$ was fitted to the measured scatter buildup using non-linear least squares. A large field of view gamma camera* was used with a low energy general purpose (LEGP) collimator and a 140KeV±10% energy window.

The proposed method for scatter compensation using measured transmission values makes an important assumption, namely that the scatter fraction is dependent on object density and thickness, as measured by photon transmission, whilst the shape of the scatter function remains reasonably constant for a given photon energy and specified imaging conditions (gamma camera, collimator etc). This hypothesis was tested by imaging a ^{99m}Tc point source centrally located between 2 media whose density could be varied while keeping the geometry constant. The material used consisted of polystyrene foam blocks with dimensions 25cmx20cmx13cm (LxWxD) placed in plastic containers of the same dimensions. Varying volumes of water were added to the foam blocks and uniformly distributed to achieve any desired density. For each measurement, the effective attenuation path length (μd) was determined by recording the transmitted counts from a collimated ^{99m}Tc line source. Images were obtained for μ values ranging from 0.006cm^{-1} to 0.15cm^{-1}.

The shape of the scatter function was characterized by determining the slope of the exponential tails of the log-linear count profiles. Scatter fraction was calculated by subtracting the counts in air, after applying the appropriate narrow beam attenuation factor, from the counts in scatter and taking the ratio of the result (ie scatter counts/total recorded counts).

2.2. Modified Lower Window Subtraction (MLWS)

To determine the values of the constants in Eq. (6) and their variation with source depth, geometric mean images of a ^{99m}Tc point source in 16cm of water, with the point source positioned at a range of depths (3-13cm), were acquired in two energy windows (140KeV±10% and 100KeV±10%). The "true" photopeak scatter distribution, g_s, was obtained by subtracting the point spread function obtained in air, after applying the appropriate attenuation factor, from that obtained in water. The values of the unknown constants were determined by fitting the estimated scatter profile to the function of Eq. (6) using non-linear least squares.

An elliptical phantom with major and minor axes 28.5cm and 20cm respectively was used to test the stability of the kernel under different imaging geometries. The phantom was filled with water and a line source of ^{99m}Tc placed centrally in the phantom. The phantom was imaged (i) with the detector parallel to the major axis and (ii) with the detector parallel to the minor axis. Finally, a cylinder of 5cm diameter containing air was placed off-centre in the phantom adjacent to the centrally located line source and imaged with the detector parallel to the major axis. The values of the convolution constants were determined for these 3 imaging situations using the same method as outlined above. All imaging was performed on a large field of view gamma camera+ with a LEGP collimator.

2.3. Phantom Study

The quantitative accuracy of these methods was assessed by reconstructing absolute activity concentrations in an elliptical phantom with major and minor axes 28.5cm and 20cm respectively. A 5cm diameter cylinder was filled with water with a ^{99m}Tc concentration of 0.191MBq.ml^{-1} and placed off-centre in the phantom. A 2nd cylinder of 10cm diameter containing air was placed in the contralateral position within the phantom. The remainder of the phantom was filled with water with lower activity concentration (0.043MBq.ml^{-1}). Tomographic images of the phantom, acquired in 2 energy windows (140KeV±10% and 100KeV±10%), were obtained using 64 projections over 360° at 20 seconds per projection angle. A separate transmission scan was performed using a collimated ^{99m}Tc line source (Tan et al, 1989).

Four methods of scatter correction were applied to the emission data, including convolution subtraction (CS) using a scatter fraction of k=0.45, transmission dependent convolution subtraction (TDCS), lower window subtraction (LWS) using a scatter fraction of k=0.45 and modified lower window subtraction (MLWS). The scatter-corrected images were then reconstructed using standard convolution backprojection. The transmission data, acquired with the collimated line source, were transformed to projections of μ values and reconstructed (Bailey et al, 1987). These data were used to affect 2nd order attenuation correction using a hybrid of the Chang (1978) and Morozumi (1988) algorithms. Briefly, the reconstructed emission image is multiplied point by point by a correction map calculated from the reconstructed μ values to yield an initial (1st order) estimate. This is then forward projected once through air and once through the reconstructed μ values. The latter results in projections, p', which may be directly compared with the original measured projections, p. A 2nd order correction is then applied by modifying the projections according to the equation:

$$p'(t,\theta) = \frac{p_{\mu=0}(t,\theta)\ p(t,\theta)}{p_{\mu}(t,\theta)} \qquad (7)$$

where $p_{\mu=0}$ and p_{μ} are the data forward projected through air (μ=0) and through the measured attenuation respectively, and t is the position along the projection at angle θ.

Using the measured gamma camera efficiency and voxel scaling factors, the scatter and attenuation corrected reconstructions were converted to absolute activity concentrations. Values for activity concentration in the 5cm cylinder, the "warm" background and the "cold" air cylinder were compared with the known concentrations.

3. RESULTS

3.1. Transmission Dependent Convolution Subtraction (TDCS)

The scatter buildup measured using a ^{99m}Tc point source in various depths of water is shown in Figure 2. The measured buildup factors were very nearly linear with depth over the range of depths measured. The values obtained for the constants in Eq. (5) were A=2.86±0.03, B=1.87±0.03 and ξ=0.25±0.01. These were used to calculate the theoretical curve for scatter fraction determination shown in Figure 3.

The results from the foam block experiment to test the assumptions of the transmission method are presented in Figure 3, along with the theoretical transmission derived scatter fractions. The measured scatter fractions varied with effective attenuation path length (hence, transmission factor) as predicted by the theory, with values ranging from 6.4% to 42.3% for transmission factors of 0.92 and 0.13 respectively. (A transmission factor of 0.13 corresponds to a 13.5cm depth of water (μ=0.15cm^{-1})). The slope of the mono-exponential scatter tails, however, was constant over the range of transmission factors measured with a mean slope of 0.15±0.01.

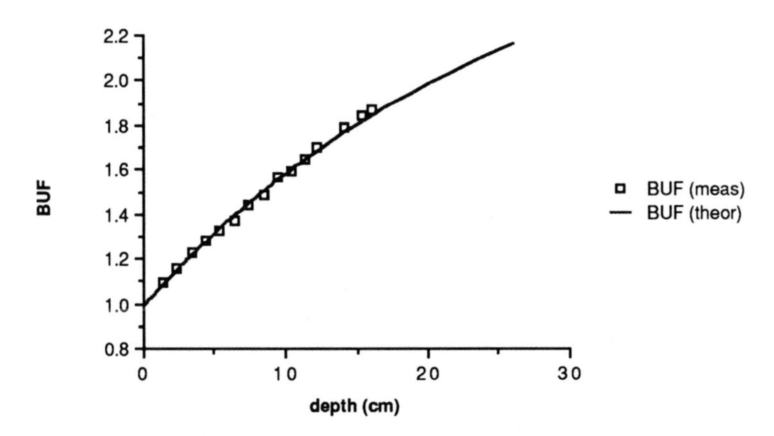

Figure 2: Scatter buildup as a function of depth in water using a ^{99m}Tc point source. The total recorded counts at each depth represent broad beam measurements. Corresponding narrow beam measurements were made by applying an attenuation factor to an air measurement, taking the known attenuation into account. Buildup factors (BUF) were then determined by calculating the ratio of broad beam to narrow beam counts at each depth.

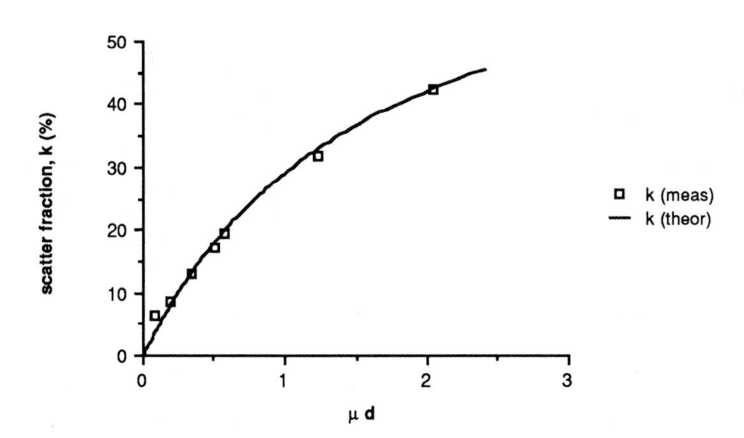

Figure 3: Measured scatter fractions as a function of effective attenuation path length (μd) for a ^{99m}Tc point source at a depth of 13cm in a variable μ material. The solid curve represents the theoretical determination of scatter fraction derived from the depth data (Figure 2).

3.2. Modified Lower Window Subtraction (MLWS)

The estimated photopeak scatter profile using MLWS is shown in Figure 4 along with the measured scatter and that estimated by conventional LWS. The values of the constants in Eq. (6), obtained by non-linear least squares, were α=5.5±0.31, β=0.14±0.013 and b=0.40±0.03. These were found to be independent of source depth and relatively independent of geometry. The data are presented in Table 1.

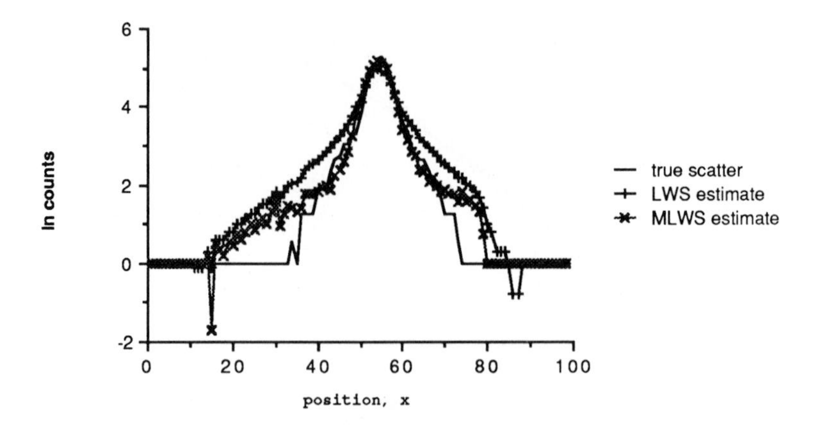

Figure 4: Photopeak scatter profile of a ^{99m}Tc point source in 8cm of water along with the estimate obtained by modified lower window subtraction (MLWS) and conventional lower window subtraction (LWS). MLWS yields an improved estimate of the photopeak scatter, particularly near the peak of the distribution. There is a tendency by both methods to overestimate the scatter towards the extremities, although the differences are exaggerated by the log transformation.

Table 1: Fitted values of the constants in the MLWS convolution kernel for a range of depths (after taking the geometric mean of opposing images) in a water tank (I) and for different geometries using an elliptical tank (II, III and IV).

	depth (cm,cm)	$\alpha(\sigma)$	$\beta(\sigma)$	b(σ)
I	GM(3,13)	5.1 (0.32)	0.14 (0.014)	0.39 (0.033)
	GM(4,12)	5.4 (0.29)	0.14 (0.012)	0.40 (0.028)
	GM(5,11)	5.4 (0.33)	0.14 (0.014)	0.41 (0.034)
	GM(6,10)	5.2 (0.37)	0.14 (0.017)	0.42 (0.042)
	GM(7,9)	5.9 (0.26)	0.14 (0.010)	0.39 (0.024)
	GM(8,8)	5.7 (0.28)	0.14 (0.011)	0.40 (0.026)
	mean:	5.5 (0.31)	0.14 (0.013)	0.40 (0.03)
II		5.5 (0.34)	0.16 (0.03)	0.35 (0.06)
III		5.9 (0.34)	0.15 (0.04)	0.34 (0.09)
IV		5.9 (0.33)	0.15 (0.04)	0.32 (0.08)

3.3. Phantom Study

The counting efficiency of the gamma camera was measured to be 156 counts.sec^{-1}.MBq^{-1}. Using this and the voxel scaling factors, reconstructed activity concentrations for the "hot" cylinder, the "warm" background and the "cold" air cylinder were calculated and are presented in Table 2 for four different scatter correction schemes; (i) convolution subtraction with a constant k (CS); (ii) convolution subtraction with k dependent on transmission (TDCS); (iii) lower window subtraction (LWS) and (iv) modified lower window subtraction (MLWS). Each was followed by 2nd order attenuation correction using the method described.

4. DISCUSSION

The basic assumptions inherent in the transmission dependent scatter correction method are validated by the results shown in Figures 2 and 3. Firstly, the shape of the scatter function measured in various attenuating media was found to be constant with a mean slope of 0.15±0.01. This supports the hypothesis that the shape of the scatter function is independent of object density. Furthermore, a spatially invariant

Table 2: Activity concentrations calculated from the reconstructed phantom study using 4 different scatter correction methods; convolution subtraction (CS), transmission dependent convolution subtraction (TDCS), lower window subtraction (LWS) and modified lower window subtraction (MLWS). Following scatter correction, a hybrid 2nd order attenuation correction algorithm was applied using measured narrow beam transmission measurements (see text). The known concentrations were 0.191MBq.ml^{-1} in the "hot" cylinder, 0.043MBq.ml^{-1} in the "warm" background and 0MBq.ml^{-1} in the "cold" air cylinder.

Method	"hot" cyl (MBq.ml^{-1}) [% accuracy]	"warm" bkg (MBq.ml^{-1}) [% accuracy]	"cold" cyl (MBq.ml^{-1})
CS	0.185 [97%]	0.040 [93%]	<0.0001
TDCS	0.189 [99%]	0.044 [103%]	0.003
LWS	0.199 [104%]	0.049 [114%]	0.001
MLWS	0.184 [96%]	0.044 [103%]	0.002

scatter function is a necessary criterion for the implementation of a stationary convolution. Secondly, the measured scatter fractions varied for materials with different μ values, in agreement with the theory, from 6.4% at low transmission factors up to 42.3% at an equivalent depth in water of 13.5cm. This validates the assumption that scatter fraction varies with effective attenuation path length and fits the proposed model of a μd-dependent buildup factor. Both of these findings are in agreement with those of Ivanovich and Weber (1990), obtained by Monte Carlo simulation.

Data from the second experiment (using water only as the scattering medium) were recorded using similar geometry to the variable μ experiment, except that in this case the depth (d) was made to vary rather than the density (μ) of the material. These data were used to determine the values of the constants in Eq. (5). If the transmission method for determination of scatter fraction is valid then these two sets of data should yield similar results since the theory predicts that scatter fraction depends only on the effective attenuation path length (μd). Indeed, the scatter fractions measured from the variable μ data were in excellent agreement with the theoretical predictions derived from the variable depth data (indicated by the solid curve in Figure 3).

The limitations of LWS, outlined in the text, are demonstrated by Figure 4. The scatter estimate obtained by LWS is significantly broader than the "true" scatter profile and of lower amplitude. On the other hand, the MLWS method demonstrates a good fit to the "true" scatter profile, particularly in the region about the peak of the distribution. There is still a slight tendency to overestimate scatter towards the extremes of the tails, although the differences are exaggerated by the log transformation used in Figure 4. This method, then, provides an improved model of photopeak scatter compared with conventional LWS. In addition, the convolution kernel was found to be constant with depth after applying the geometric mean to opposing images as well as being relatively independent of source geometry. Whilst this is not a rigorous validation of the technique, it demonstrates that the method is reasonably robust under different scattering conditions and provides a useful comparison for the TDCS method.

The use of transmission measurements to determine the scatter fraction, k, resulted in an improved estimate of the activity concentration in the phantom compared with conventional CS, both in the "hot" cylinder (99% cf 97%) and in the "warm" background (103% cf 93%). The reduced count recovery in the region of the distributed background suggests that the scatter estimate obtained by CS is broader than the true photopeak scatter distribution, resulting in over-subtraction. The improved recovery in the case of TDCS indicates that this is corrected by the use of measured photon transmission to determine regional scatter fraction. The LWS method under-estimates the photopeak scatter contribution, particularly in the "warm" distributed background where the count recovery was 114%. This finding is consistent with the results shown in Figures 1 and 4 where lower window scatter was found to be lower in amplitude and slightly broader than the photopeak scatter distribution, probably due to multiple scattering and scattering through large angles. The MLWS method, on the other hand, yields an accurate estimate of activity concentration both in the "hot" cylinder (96% accuracy) and in the "warm" background (103%). The reconstructed counts in the region of the "cold" air cylinder were very close to zero, as expected, in all the scatter correction methods applied, indicating very good cold contrast.

To ensure that the low reconstructed counts in the region of the "cold" cylinder are not due to oversubtraction, the processing was modified to allow the reconstruction of negative values. One might be suspicious that this is the case, particularly for CS and LWS, where a constant scatter fraction is applied to the images, independent of object density. However, no oversubtraction was demonstrated in any of the resulting reconstructions following application of the four scatter correction techniques described. Therefore, the errors in the scatter predicted by CS and LWS appear to manifest mainly in the region of a "hot" object in a high density surround and not in regions of low activity concentration and low density. This is analogous to the situation encountered in SPECT myocardial perfusion studies. The methods developed in this paper, on the other hand, particularly TDCS, predict scatter distributions which are more structured and more highly object dependent. This is reflected in the improved estimate of activity concentration in the region of the distributed background.

As a practical solution to the problem of correcting for scatter in clinical SPECT studies, the transmission dependent method is attractive since it makes use of transmission measurements which, in our current quantitative SPECT regime, are also made for the purpose of attenuation correction. Therefore, additional information about the object being imaged is utilized to affect more accurate scatter correction in non-homogeneous media. In addition, this method may have an advantage over lower window methods in the application to emission radionuclides of low photon energy such as ^{201}Tl, where the acquisition of data in a window of lower energy than the main emission window (72 KeV) may prove difficult in practice. Also, in our current implementation of the transmission method, a collimated line source is used (Tan et al, 1989) which uses two acquisition windows (which may have the same energy settings) to collect emission and transmission data simultaneously, making acquisition into a third window, for the purpose of recording lower window scatter, difficult. Therefore, narrow beam transmission measurements may be used to affect accurate attenuation **and** scatter correction for quantitative SPECT in a routine clinical setting.

5. CONCLUSIONS

The determination of absolute activity concentrations in vivo using SPECT may have important clinical benefits, especially in studies of absolute myocardial perfusion and dosimetry calculations. These applications require accurate attenuation and scatter correction which is particularly difficult in the thorax because of the wide variation in tissue densities. This paper describes a method of scatter compensation based on the convolution subtraction technique, however scatter fraction is varied spatially according to regional tissue density as determined by transmission measurements. For the purpose of comparison, a method involving convolution of lower window data with a theoretically derived kernel has also been developed, which provides an alternative, accurate model of photopeak scatter in non-homogeneous media.

The two methods have been assessed for their quantitative accuracy in a non-homogeneous phantom. Activity concentration was more accurately estimated following scatter correction using the transmission dependent (TDCS) and modified lower window subtraction (MLWS) methods compared with the conventional methods on which they are based. The TDCS method has particular appeal because it makes use of transmission measurements which are also used in our attenuation correction procedure and may have advantages over lower window techniques in the application to low photon energy radionuclides.

Notes

* GE-400AT, General Electric Corporation Medical Systems, Milwaukee, WI, USA.
\+ Philips Diagnostomo, Philips Medical, Hamburg, FRG.

APPENDIX

Image data recorded in a photopeak window, g_0, is usually modelled as a summation of primary, g_p, and scattered, g_s, photons;

$$g_0 = g_p + g_s$$

If the scatter distribution is approximated as the convolution of primary photons with a mono-exponential scatter function, then we can write

$$g_0 = g_p + g_p * ae^{-br}$$

where a and b are constants.

Similarly, data recorded in a lower energy window, g_{ul}, may be expressed as

$$g_{ul} = g_p * ce^{-dr}$$

where c and d are constants.

We seek some function which relates the lower window image to the photopeak scatter. Analysis of the scatter profiles obtained from a point source in various energy windows indicates that the scatter distribution becomes broader as energy decreases (see text). This is suggestive of a convolution relationship between g_{ul} and g_s:

$$g_{ul} * h = g_s$$

$$\Rightarrow \quad (g_p * ce^{-dr}) * h = g_p * ae^{-br} \qquad (1)$$

where h is the function whose form we wish to determine. Now, taking Laplace transforms of both sides of the equation:

$$P(s)\left(\frac{c}{s+d}\right)H(s) = P(s)\left(\frac{a}{s+b}\right)$$

$$H(s) = \left(\frac{a}{s+b}\right)\left(\frac{s+d}{c}\right)$$

$$= \frac{a}{c}\left(1+\frac{d-b}{s+b}\right)$$

$$L^{-1}: \qquad h(r) = \frac{a}{c}\{\delta + (d-b)e^{-br}\}$$

where δ is the Dirac delta function. Substituting for h in Eq. (1):

$$g_s = g_{ul} * \{\alpha(\delta + \beta e^{-br})\}$$

where $\alpha = \frac{a}{c}$ and $\beta = d-b$. Therefore, photopeak scatter may be expressed in terms of g_{ul} and the constants α, β and b.

References

Axelsson B, Msaki P and Israelsson A (1984): Subtraction of compton-scattered photons in single-photon emission computerized tomography. J Nucl Med 25:490-494.
Bailey DL, Hutton BF, Walker PJ (1987). Improved SPECT using simultaneous emission and transmission tomography. J Nucl Med 28:844-851.
Bailey DL, Hutton BF, Meikle SR, Fulton RR and Jackson CB (1989): Iterative scatter correction incorporating attenuation data (abstract). Eur J Nucl Med 15:452.
Chang LT (1978). A method for attenuation correction in radionuclide computed tomography. IEEE Trans Nucl Sci NS-25:638-643.
Hubbell JH (1963). A power series buildup factor formulation. Application to rectangular and off-axis disk source problems. J Res NBS 67C:291-306.
Hutton BF, Bailey DL, Fulton RR and Meikle SR (1988). Use of a patient derived computer simulation for assessment of SPECT attenuation correction algorithms (abstract). Aust NZ J Med 18:508.
Ivanovich M and Weber DA (1990). Monte Carlo study of compton scattering in SPECT (abstract). Eur J Nucl Med 16:404.
Jaszczak RJ, Greer KL, Floyd CE Jr, Harris CC and Coleman RE (1984). Improved SPECT quantitation using compensation for scattered photons. J Nucl Med 25:893-900.
Ljungberg M and Strand S (1990). Accurate scatter and attenuation correction of SPECT projections for extended sources in a non-homogeneous object: a Monte Carlo study. In: Development and evaluation of attenuation and scatter correction techniques for SPECT using the Monte Carlo method. Ljunberg M (thesis). University of Lund press, Lund, pp. 151-167.
Moore SC (1982). Attenuation compensation. In: Computed emission tomography. Ell PJ and Holman BL (eds), Oxford University Press, London, pp. 339-360.
Morozumi T, Nakajima M, Ogawa K and Yuta S (1988). Attenuation correction methods using the information of attenuation distribution for single photon emission CT. Med Imag Tech 2:20-28.
Murase K, Itoh H, Mogami H, Ishine M, Kawamura M, Lio A, Hamamoto K (1987). A comparative study of attenuation correction algorithms in single photon emission computed tomography (SPECT). Eur J Nucl Med 13:55-62.
Tan P, Bailey DL, Hutton BF, Fulton RR, Meikle SR, Keyser R and Barbagallo S (1989). A moving line source for simultaneous transmission/emission tomography (abstract). J Nucl Med 30:964.
Todd-Pokropek A, Clarke G and Marsh R (1984). Preprocessing of SPECT data as a precursor for attenuation correction. In: Information processing in medical imaging. Deconninck F (ed), Martinus Nijhoff, Boston, pp130-150.
Wu RK and Siegal JA (1984). Absolute quantitation of radioactivity using the buildup factor. Med Phys 11:189-192.

CONE BEAM SINGLE PHOTON EMISSION COMPUTED TOMOGRAPHY USING TWO ORBITS

R Clack[†], G L Zeng, Y Weng, P E Christian, and G T Gullberg

Medical Imaging Research Laboratory, Department of Radiology, University of Utah, Salt Lake City, UT 84132, USA,
[†]Present address: Mathematisches Institut, Universitat Freiburg, Hebelstrasse 29, D 7800 Freiburg im Breisgau, Germany

Abstract

It is known that cone-beam projection measurements from a single planar orbit of the focal point do not satisfy Tuy's sufficiency condition for exact reconstruction. It is also known that two such orbits, oriented orthogonally, do satisfy the condition. In this paper, we present a fast convolution-and-backprojection algorithm to perform reconstructions from two orbits of cone-beam data. The algorithm has been applied to simulated data, and phantom data taken on a clinical SPECT system.

Keywords

Image reconstruction; three-dimensional reconstruction; cone beam; x rays; single photon emission computed tomography.

1. INTRODUCTION

Cone-beam collimators for single-photon emission imaging have been under investigation for several years now (Jaszczak, 1986, Floyd, 1986, Jaszczak, 1988). The principal advantage of a cone-beam collimator is that a large fraction of the detector crystal face can be utilized even when imaging smaller organs. This efficient utilization results in a combination of resolution and sensitivity gain over images formed using standard parallel-hole or fan-beam collimators. Comparisons of cone-beam, parallel-hole, and fan-beam collimators for heart imaging have been investigated (Hahn, 1988, Gullberg, 1990, Gullberg, 1991).

For single photon emission computed tomography (SPECT), multiple views are collected as the collimator is rotated in a circular or elliptical orbit around the long axis of the patient, and these data are processed by reconstruction algorithms to form a three-dimensional image of the tracer distribution in the patient. The situation for parallel-hole or fan-beam collimators differs considerably from the cone-beam case. For parallel-hole and fan-beam collimators the reconstruction procedure is usually the respective "filtered-backprojection" technique (Herman, 1980), and these two algorithms are mathematically equivalent. No such equivalent algorithm can exist for the cone-beam case however, as the cone-beam data set differs from the fan-beam or parallel-hole data sets. Specifically, for the parallel-hole or fan-beam cases, the data consists of all line-integrals of activity lying in separate transaxial planes, and the reconstruction methods take advantage of the independence of transaxial planes to process the data as a set of two-dimensional problems. For the cone-beam case the only transaxial slice for which there are measured line-integrals is the slice containing the focal point of the cone. All the other data consists of oblique line-integrals that cross several transaxial planes. A suitable reconstruction algorithm must handle this data set in a "truly" three-dimensional fashion.

Currently, the most popular algorithm used to process cone-beam projections is that of Feldkamp et al (1984). This algorithm was originally derived using approximations to the fan-beam formulas, and has since been extensively analyzed mathematically as it appears to be very effective and robust (Singh, 1989, Leahy, 1990). The algorithm is a convolution-and-backprojection method and admits a fast efficient implementation. Iterative Expectation Maximization (EM) methods (Shepp, 1982, Lang, 1984) are alternative approaches currently under investigation (Gullberg, 1989). Although these methods are

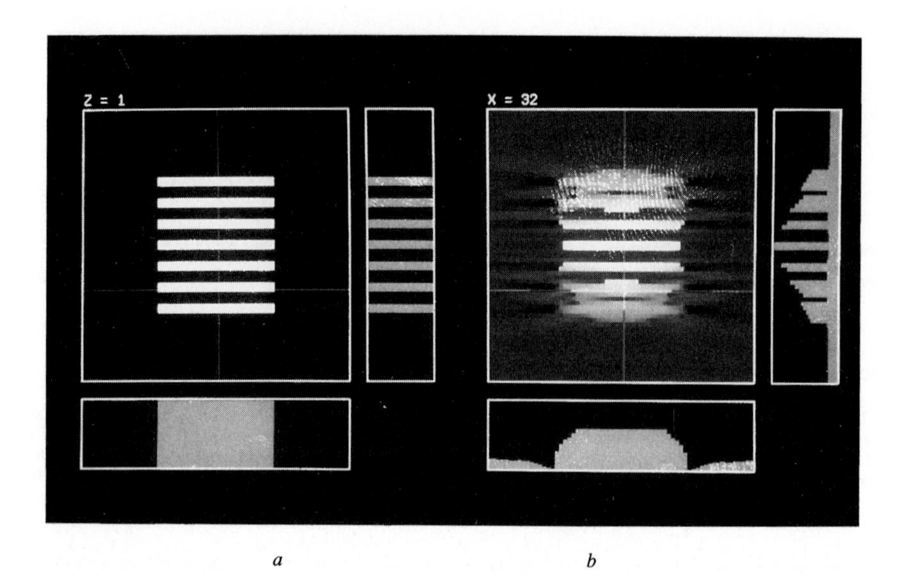

a *b*

Figure 1. Single orbit reconstruction of the "seven disks" phantom. Each disk lies in a transverse plane and has diameter 32 *cm* and height 2 *cm*. (*a*) A sagittal cut through the center of the phantom. The position of the profiles is indicated by the cross-hairs on the image. (*b*) The result of the Feldkamp reconstruction applied to ideal cone-beam projection data with an orbit radius of 50 *cm*. The scale of the profiles is from -20 to 100, where the activity level of the phantom is 100.

considerably slower they have certain advantages, particularly the ability to include physical effects such as attenuation into the model. A drawback of the EM methods (and other iterative approaches) is the lack of a definitive stopping criterion. The procedure is often halted after a predetermined number of iterations, with different researchers reporting values ranging from tens to thousands of iterations.

A common difficulty that any algorithm for single-orbit cone-beam SPECT must address is data insufficiency. It is known that if the focal point of the collimator follows a single planar orbit, then the data obtained does not satisfy Tuy's sufficiency condition (Tuy, 1983) for exact three-dimensional reconstruction. Different algorithms will cope with this problem in different ways, but for any algorithm there will be some source distributions which will be difficult to recover exactly. As an illustration of the effects of the data sufficiency problem, we have simulated exact projections of a phantom consisting of seven hot disks evenly spaced in transaxial planes, and used the Feldkamp algorithm to reconstruct the phantom. Figure 1 shows the familiar distortions: (a) elongation of structure in the axial direction, (b) the corresponding artificial reduction in activity in these regions, and (c) negative undershoots at the sharp edges in the transaxial planes. These distortions become progressively worse for planes further from the central plane (containing the orbit), and only the central plane is reconstructed free of artifacts. In Figure 2 we present some cone-beam patient data and compare it to data taken from the same patient using a conventional fan-beam collimator and reconstruction algorithm. In both cases, 128 views were taken with approximately 30 seconds per view. The apparent elongation of the brain in the vertical direction for the cone-beam study may well be the manifestation of distortion (a) in a practical imaging situation.

In this work we propose to use a second orbit of measurements in order to fulfill Tuy's condition. Tuy's data sufficiency condition requires that every plane which passes through the imaging field of view must also cut through the orbit at least once. As mentioned above, the conventional single planar orbit does not satisfy this condition whereas two orbits arranged as illustrated in Figure 3, does satisfy the condition.

This dual orbit geometry has been proposed by other researchers, notably Tuy (1983), Smith (1985), and Grangeat (1987), who all point out the data sufficiency condition and provide inversion formulas suitable for more general orbits. Tuy and Smith based their work on an equation relating processed cone-beam projections to a filtered form of the three-dimensional Radon transform. Both Tuy and Smith give inversion formulas which would be awkward to implement as an algorithm, although Smith did derive a convolution-and-backprojection procedure for orbits satisfying a more stringent

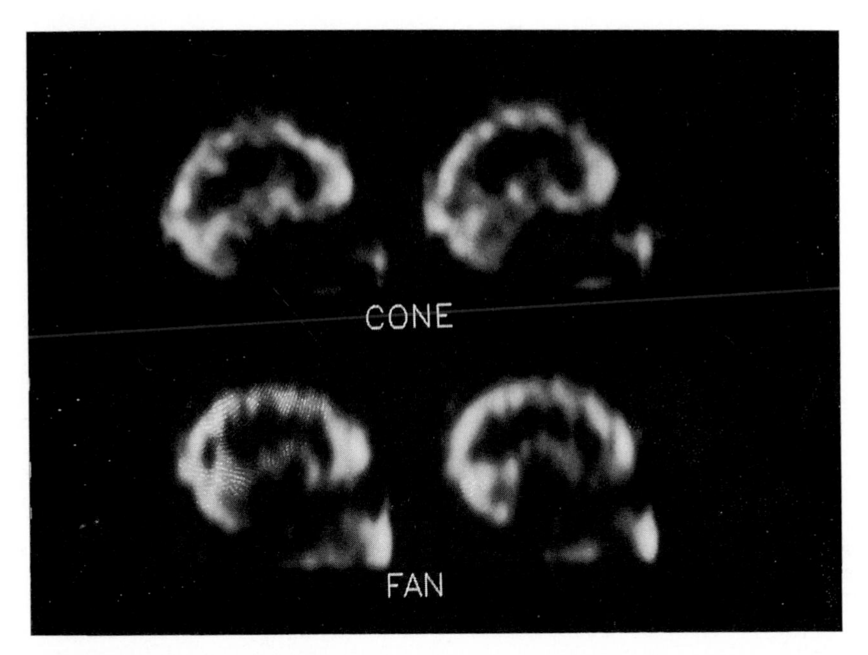

Figure 2. HMPAO brain study of a patient imaged under identical conditions using a Fan-Beam collimator (bottom) and a Cone-Beam collimator (top). The same sagittal sections are displayed from each scan.

requirement than Tuy's condition. The proposed two orbit geometry does not satisfy Smith's requirement. Grangeat showed that certain values of the three-dimensional Radon transform can be accessed directly from a single cone-beam projection. He suggests a general algorithm which applies a form of the Radon inversion formula to the sorted Radon values. One drawback of this approach is that it involves the sorting (or rebinning) of the cone-beam data.

In the following section we describe a fast algorithm tailored to the two-orbit geometry of Figure 3. The algorithm does not represent an exact inversion formula, but our goal here is only to demonstrate that there exist practical, effective methods to reconstruct two-orbit cone-beam data. In section 3, we present results of tests using simulation studies and phantom data measurements made on our SPECT system.

2. ALGORITHM DESCRIPTION

The algorithm we propose is based on the three-dimensional (3D) Radon inversion formula. The 3D Radon inversion procedure recovers the original function from its integrals over planes. Our approach is to convert the cone-beam projection data into plane integrals, backproject these values, and use a form of the inversion formula for reconstruction.

The Radon transform of a function f is the set of all plane integrals $p_\theta(t)$ where θ is the unit normal vector to the plane, and t is its distance from the origin:

$$p_\theta(t) = \int_{(x \cdot \theta = t)} f(x)\, dx \quad . \tag{1}$$

For a fixed θ the one dimensional set of values p_θ is called a Radon projection. A slightly generalized form of the inversion formula can be expressed as

$$\hat{f} = \hat{b}\hat{h} \, , \tag{2}$$

where $\hat{f}$ denotes the 3D Fourier transform of f and similarly for b and h. The function b represents the backprojection of the Radon projections over all directions S^2, and we allow the possibility of some projections being backprojected with a positive multiplicity $c(\theta)$. Note that the multiplicity varies only in θ, and not in t:

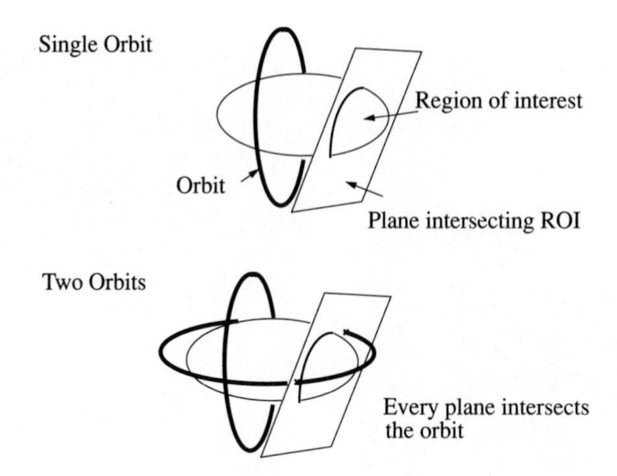

Figure 3. Tuy's sufficiency condition is not satisfied for the case of a single planar orbit of the collimator focal point (upper). It is possible to find a plane passing through the field of view which does not intersect the orbit. For two orthogonal orbits (below), Tuy's condition is satisfied.

$$b(x) = \int_{S^2} c(\theta)\, p_\theta(x \cdot \theta)\, d\theta \,. \tag{3}$$

The function $\hat{h}$ is the 3D filter function and, in polar coordinates with radius ρ and direction θ, is given by

$$\hat{h}(\rho\theta) = \frac{\rho^2}{c(\theta)} \,. \tag{4}$$

The multiplicities, $c(\theta)$, will account for an overlap of measurements of $p_\theta(t)$ arising from the two orbits.

We use g_a to represent the cone-beam projection with the focal point at position a. Without loss of generality we consider a cone-beam projection to be the two-dimensional set of measured values on the plane containing the origin and orthogonal to the vector a:

$$g_a(x) = \int_{-\infty}^{\infty} f(a + s(x-a))\, ds \qquad x \in a^{\perp}. \tag{5}$$

Since the two-orbit geometry satisfies Tuy's condition, any plane $P_{\theta t} = \{x\colon x \cdot \theta = t,\ \|\theta\| = 1\}$ which passes through the support of f must intersect the dual orbit at some point a. We obtain an approximation of $p_\theta(t)$ by summing values measured along the line of intersection of $P_{\theta t}$ and the cone-beam projection (See Figure 4.):

$$p_\theta(t) \approx \int_{(P_{\theta t} \cap a^{\perp})} g_a(x)\, dx \,. \tag{6}$$

This well-known approximation was published by Grangeat (1987) who also studied carefully the error bounds on it.

For a single cone-beam measurement, many plane integrals $p_\theta(t)$ can be obtained using the summation described by Eq. (6). All plane integrals are available by combining those from each cone-

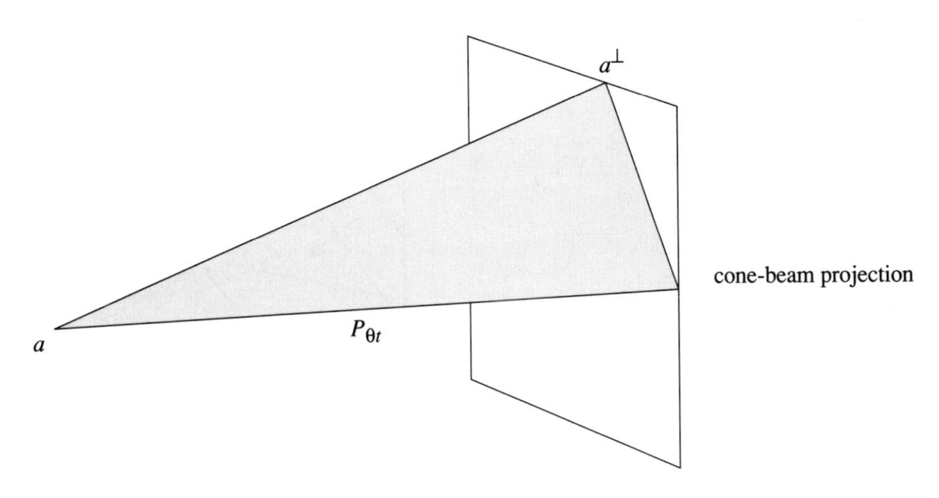

Figure 4. A plane integral is approximated by a line integral on the cone-beam detector plane. The focal point of the cone-beam projection is at a. The integrated activity in the plane $P_{\theta t}$ passing through a is approximately the sum of the cone-beam measurements along the line of intersection of $P_{\theta t}$ and the cone-beam projection.

beam projection of the two orbits. Note that some $p_\theta(t)$ values are available from both orbits and some from only one orbit, but in order to use Eq. (2) for reconstruction these multiplicities must be controlled to have only angular dependence.

A "complete" Radon projection p_θ is a projection which is measured for all t. The set of complete Radon projections available from a single orbit will depend on the size of the object (the support of f) and on the orbit radius r. Assuming $d < \sqrt{2}\, r$ where d is the object diameter, complete Radon projections can be obtained for all θ in C_i, where C_i is the angular region outside the cone of aperture 45° whose axis coincides with the normal direction of the orbit plane:

$$C_i = \{\theta \colon |\theta \cdot O_i| \le \frac{1}{\sqrt{2}},\ \|\theta\| = 1\} \qquad \text{for } i = 1, 2\,, \tag{7}$$

where O_i is the unit normal of the ith orbit plane.

We obtain the backprojection b (Eq. (3)) as the sum of backprojected Radon projections p_θ from the two orbits. For the ith orbit,

$$b_i(x) = \int_{C_i} p_\theta(x \cdot \theta)\, d\theta\,, \tag{8}$$

and, because the region $C_1 \cup C_2$ covers the unit ball for orthogonal orbits,

$$b(x) = b_1(x) + b_2(x) = \int_{S^2} c(\theta)\, p_\theta(x \cdot \theta)\, d\theta\,, \tag{9}$$

where

$$c(\theta) = \begin{cases} 2 & (\theta \in C_1 \cap C_2) \\ 1 & (\theta \notin C_1 \cap C_2) \end{cases} \tag{10}$$

It only remains now to describe how $b_i(x)$ (Eq. (8)) is obtained from the cone-beam projections. For a fixed projection g_a, define a direction u on the projection plane ($\|u\| = 1$, $u \in a^{\perp}$) and consider the two-dimensional convolution

$$g_a^u = g_a * h_u\,, \tag{11}$$

where

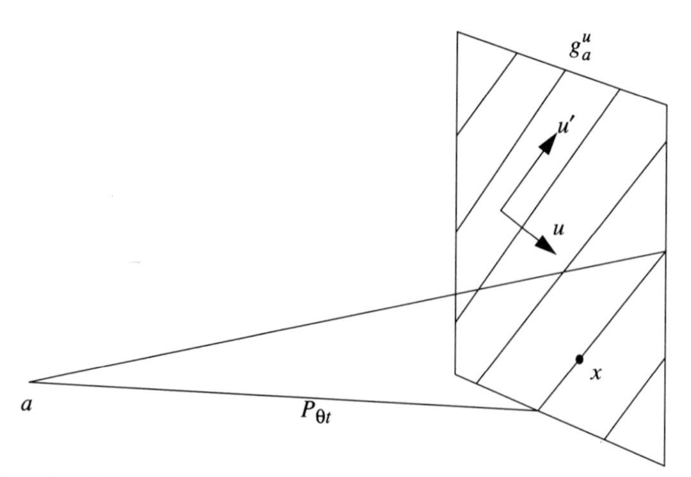

Figure 5. A Radon backprojection is formed from a cone-beam backprojection of a modified cone-beam projection. The modified projection g_a^u is constant in the u' direction, and its value is the sum of the original measurements along the line. When this line of values is backprojected through a, the equivalent Radon backprojection of the value $p_\theta(t)$ is achieved in the plane $P_{\theta t}$.

$$h_u(x) = \delta(x \cdot u) \qquad (x \in a^\perp) \ . \tag{12}$$

This convolution has no effect in the u direction, but in the orthogonal direction u' ($u' \in u^\perp \cap a^\perp$, $\|u'\| = 1$) each value has been replaced with the sum of the values along that line. Thus g_a^u is constant in the u' direction and $g_a^u(x)$ is the approximation of $p_\theta(t)$ (Eq. (6)) where $P_{\theta t}$ is the plane passing through x, and spanned by u' and $x - a$ (see Figure 5). Consider now the effect of backprojecting each value of g_a^u along the line through x to the focal point a. This cone-beam backprojection gives the equivalent Radon backprojection for the "wedge-shaped" set of planes $P_{\theta t}$ defined by u' and $x \in a^\perp$.

In order to extract the required set of $p_\theta(t)$ values from the cone-beam projection g_a, this procedure should be performed for a range of values of u. The exact range has some dependency on $x \in a^\perp$ (the position in the projection plane) which we ignore, and we use all vectors u that lie within 45° of the orbit plane:

$$U = \{u: |u \cdot O_i| \leq 1/(\sqrt{2}),\ u \in a^\perp,\ \|u\| = 1\} = C_i \cap a^\perp \ . \tag{13}$$

We now combine the multiple convolutions and cone-beam backprojections into a single step, by summing over all $u \in U$.

$$g_a^* = \int_U g_a^u du = g_a * h^* \tag{14}$$

$$h^*(x) = \int_U h_u(x)\, du \quad (x \in a^\perp) \quad = \begin{cases} \dfrac{1}{\|x\|} & (x/\|x\| \notin U) \\ 0 & (x/\|x\| \in U) \end{cases} \ . \tag{15}$$

We perform a single cone-beam backprojection of g_a^* to complete the processing of this cone-beam projection. When each projection in the orbit has been processed we obtain $b_i(x)$ of Eq. (8).

In summary, the algorithm consists of a convolution-and-backprojection step followed by three-dimensional filtering.

1. For each cone-beam projection g_a, form $g_a^* = g_a * h^*$ using h^* given in Eq. (15).

2. Obtain b (Eq. (3)), by performing a cone-beam backprojection of each of the modified projections g_a^* from both orbits.

3. Filter b (in three-dimensions) to obtain f (see Eqs. (2), (4), (7), (10)).

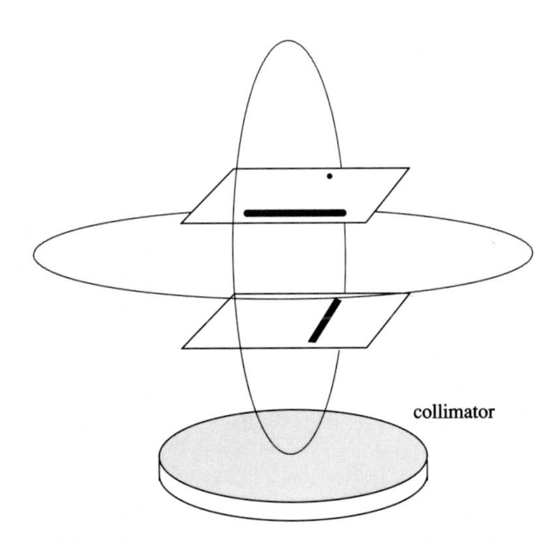

Figure 6. Illustration of the real data set up. A point source and a line source are on the upper plane. There is a line source on the lower plane lying perpendicular to the other line source.

We point out that the formation of b using steps (1) and (2) is not strictly accurate. In particular, the exact form of the filter h^* is not quite shift-invariant, and also the correct procedure to backproject g_a^u has a weak u dependence. We do not analyze these small approximations here as the errors appear to be undetectable in practical imaging situations and in our simulations.

3. IMPLEMENTATION AND RESULTS

The algorithm described in the previous section was implemented in FORTRAN on a SUN workstation in the Medical Imaging Research Laboratory at the University of Utah. The cone-beam projection data was assumed to be digitized into a 64×64 grid with n projections per orbit. The two-dimensional convolution with the function h^* was performed using a 2D FFT routine, and applied after the projection data was zero-padded to 128×128 to avoid the classical wrap-around phenomenon. These modified data were backprojected into a $64 \times 64 \times 64$ array using a standard voxel-driven backprojector with bilinear interpolation. When all the cone-beam projections had been processed, the final backprojection image was filtered using a 3D FFT routine and the filter $\hat{h}$ described in Eqs. (4) and (10). To avoid amplification of high-frequency noise, the 3D image was smoothed using a Hanning filter, rolled off to zero at the Nyquist frequency.

On a SUN-4 computer, the reconstruction times were approximately 3-4 minutes for the 3D filtering, and 15-20 seconds for the processing (2D filtering and backprojection) of a single cone-beam projection. For two orbits of $n = 64$ projections each, the total processing time was 30-45 minutes.

We have tested our algorithm with phantom data taken from our Picker SX-300 SPECT system. The phantom consisted of two 5 *cm* line sources and a point source lying in parallel x-y planes and oriented as shown in Figure 6. For each orbit, $n = 128$ projections were measured, with approximately 50,000 events per projection. The collimator orbit radius was 35 cm and the collimator focal length was 51 cm. We illustrate the results in Figure 7. In Figure 7*a* (upper left) the first projection set is shown with the collimator positioned parallel and below the planes of the phantom. The magnification of the line and point source data is caused by that plane being closer to the focal point. Figure 7*b* (lower left) shows the same data after the two-dimensional filter has been applied. In Figures 7*c* and 7*d* (right) the two x-y planes of the reconstruction which contain the sources are shown. The reconstruction shows good separation and uniformity of the activity, with no apparent artifacts.

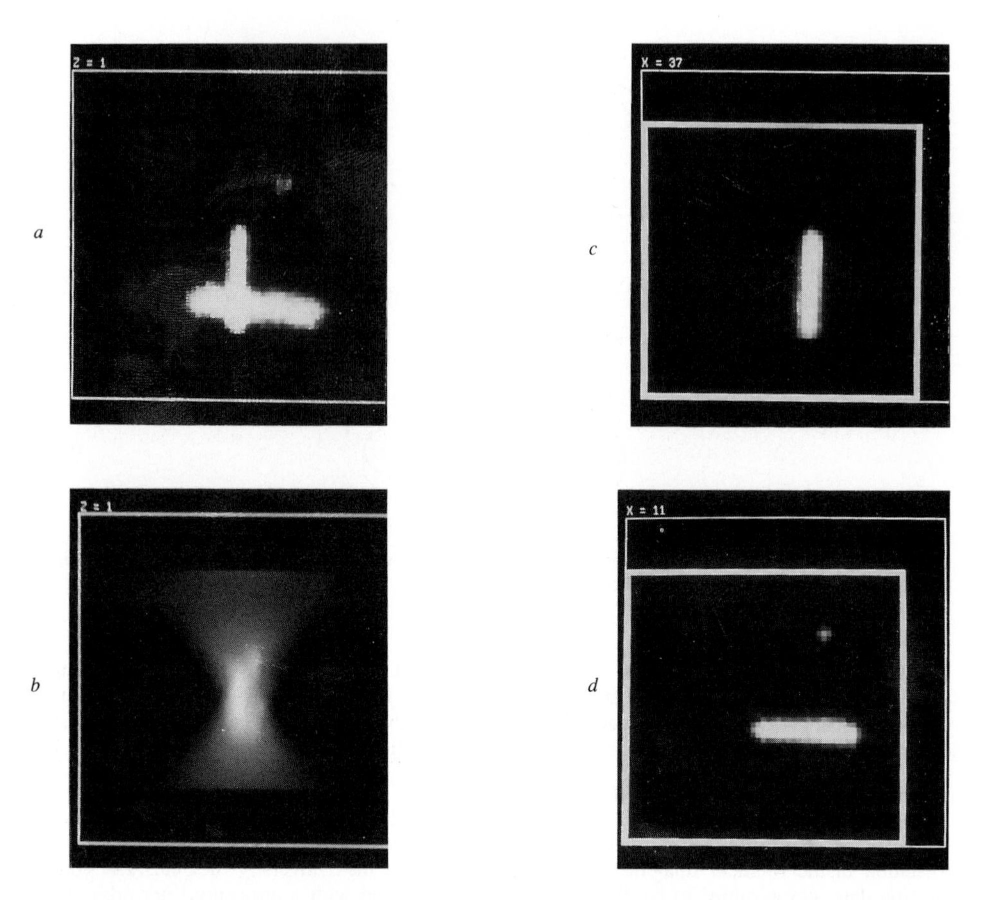

Figure 7. Point-line phantom results: (*a*) the first projection set with the collimator positioned parallel and below the planes of the phantom, (*b*) the same data after the two-dimensional filtering, and (*c*), (*d*) the two x-y planes of the reconstruction which contain the sources.

In order to examine the effects of the algorithm more closely and to verify that the single-orbit distortions have been eliminated, we applied the algorithm to ideal (simulated) data from the "seven disks" phantom. For this phantom we took $n = 64$ projections per orbit with the same size of phantom and orbit radius (distance between focal point and center of rotation) as described in Figure 1. Figure 8 shows the result of our reconstruction for the same slice as was shown in Figure 1. The improvement over the single orbit construction of Figure 1(*b*) is dramatic. The displayed profiles indicate that none of the three classical distortions (elongation of structure, reduction in activity, undershoots) appear in the two-orbit reconstruction. Each of the disks seem to be reconstructed with comparable accuracy.

The algorithm derived in section 2 was based on several approximations. To evaluate the effect of these approximations we performed the following test. We generated ideal Radon projections (Eq. (2)) of the "seven disks" phantom, and reconstructed these data according to Eqs. (2) and (3) using $c(\theta)$ as given in Eq. (10). As this reconstruction does not involve the approximations of Eqs. (6), (14), and (15) we expect the result to indicate the best image quality that our technique could achieve. In Figure 9 we show the usual slice through this ideal reconstruction of the "seven disks" phantom. Comparison with Figure 8 suggests that there is little loss of image quality in our cone-beam reconstruction compared to the optimal reconstruction. This indicates that the approximation errors in converting cone-beam projections to Radon projections are small.

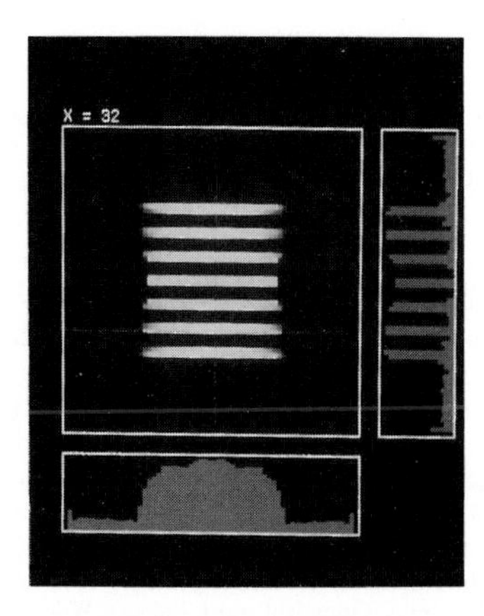

Figure 8. The "seven disks" phantom reconstruction using the two-orbit algorithm. The geometry and displayed slice are exactly as given in Figure 1. The horizontal and vertical profiles are positioned as indicated by the cross hairs.

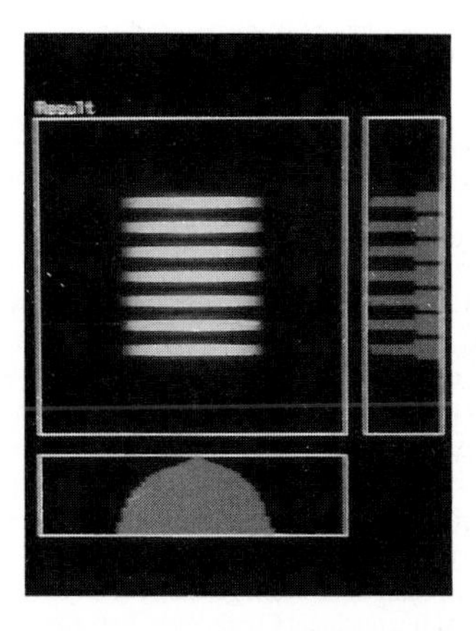

Figure 9. The "seven disks" phantom reconstruction using three-dimensional Radon projections and Radon inversion. The displayed slice and profiles are the same as given in Figure 8.

4. DISCUSSION AND CONCLUSIONS

We have developed a fast convolution-and-backprojection algorithm to process cone-beam data taken from two orthogonal orbits. The algorithm does not represent an exact inversion formula because obtaining $b_i(x)$ in Eq. (8) efficiently involves some approximations in the processing of the cone-beam projection data.

We have implemented the algorithm and made preliminary tests using simulated data and using phantom data taken from a clinical SPECT system. The images obtained using two orbits are superior to those of a single orbit geometry. The approximations made in the theory do not result in discernible artifacts in the images.

We now make two remarks of theoretical interest. First, in our algorithm, the tomographic filtering occurs after backprojection. In classical tomography this approach is unfavorable as the backprojection image has infinite extent and large arrays are required to avoid errors during the subsequent filtering stage. In the case of the three-dimensional Radon inversion formula (Eq. (2) for example), a large backprojection array is not required as the filtering is a local operation. Usually, the three-dimensional convolution kernel corresponding to filtering is unbounded, and distant spatial values contribute to a point in the reconstruction. For local filtering this is not the case, and it is actually possible to reconstruct small subregions of activity without artifacts. Thus the usual objection to post-backprojection filtering does not apply to our approach.

A second point of theoretical interest involves increasing the data usage in the current algorithm. It can easily be shown (using the central-slice theorem) that only half of the two-dimensional frequency information is being accessed from each cone-beam projection measurement. Although this is sufficient for reconstruction, improved images might be obtained if the filter h^* were adjusted to have a smaller non-zero region. This can be accomplished if the regions C_i (see section 2) are enlarged, which is possible with more stringent requirements on the orbit geometry and object size. For example, if $d < (2\sqrt{3}/3)r$ then $C_i = \{\theta\colon |\theta \cdot O_i| \le 1/2, \|\theta\| = 1\}$ for $i = 1, 2$ with the corresponding increase in U (Eq. (13)) and consequently the fraction of frequencies used in the projection data increases from 1/2 to 2/3. We are studying the effects of this improved utilization.

We are also investigating other extensions of this work motivated by practical considerations. For various reasons related to the geometry of the collimator motion and the patient, we are studying an implementation whereby the two orbits are not orthogonal and yet still satisfy Tuy's condition. We are also studying a variation where one and one-half orbits are measured rather than the two full orbits. This orbit arrangement may be particularly suitable for brain imaging. Our preliminary investigations have been encouraging.

Acknowledgments

We thank P. Harvey of Queen's University, Canada for the use of the SUN-based display program, G. Besson of the University of Utah for his useful discussions, and M. Defrise of the Vrije University of Brussels for suggesting the use of the seven disks phantom and for correcting several small mistakes in the equations.

References

Feldkamp LA, Davis LC, and Kress JW (1984). Practical cone-beam algorithm. J. Opt. Soc. Am. A 1:612-619.

Floyd CD Jr, Jaszczak RJ, Greer KL, and Coleman RE (1986). Cone beam collimation for SPECT: simulation and reconstruction. IEEE Trans. Nucl. Sci. NS-33:511-514.

Grangeat P (1987). Analysis d'un Systeme D'Imagerie 3D par reconstruciton a partir de X en geometrie conique. Ph.D. Thesis, l'Ecole Nationale Superieure Des Telecommunications.

Gullberg GT, Zeng GL, Tsui BMW, Hagius JT (1989). An iterative reconstruction algorithm for single photon emission computed tomography with cone beam geometry. Int J of Imag Sys & Techn 1:169-186.

Gullberg GT, Zeng GL, Christian PE, Tsui BMW, Morgan HT (1990). Single photon emission computed tomography of the heart using cone beam geometry and noncircular detector rotation. In: Information Processing in Medical Imaging, XIth IPMI International Conference, Berkeley, California, June 19-23. Eds: D.A. Ortendahl and J. Lalacer, Wiley-Liss, New York, 1990, pp. 123-130.

Gullberg GT, Zeng GL, Christian PE, Datz FL, Morgan HT (1991). Cone beam tomography of the heart using SPECT. Invest. Radiol. (in press).

Hahn LJ, Jaszczak RJ, Gullberg GT, Floyd CE, Manglos SH, Greer KL, and Coleman RE (1988). Noise characteristics for cone beam collimators: a comparison with parallel hole collimator. Phys. Med. Biol. 33:541-555.

Herman GT (1980). Image Reconstructions From Projections: The Fundamentals of Computerized Tomography, Academic Press, New York.

Jaszczak RJ, Floyd CE, Manglos SH, Greer KL, and Coleman RE (1986). Cone beam collimation for single photon emission computed tomography: analysis, simulation, and image reconstruction using filtered backprojection. Med. Phys. 13:484-489.

Jaszczak RJ, Greer KL, Coleman RE (1988). SPECT using a specially designed cone beam collimator. J. Nucl. Med. 29:1398-1405.

Lang K, Carson R (1984). EM reconstruction algorithms for emission and transmission tomography. J. Comp. Assist. Tomogr. 8:306-316.

Leahy RM and Yan XH (1990). Derivation and analysis of a filtered backprojection algorithm for cone beam projection data. Submitted to IEEE Trans. Med. Imag.

Shepp LA and Vardi Y (1982). Maximum likelihood reconstruction for emission tomography. IEEE Trans. Med. Imag. MI-1:113-122.

Singh M, Leahy R, Brechner R, and Yan XH (1989). Design and imaging studies of a position sensitive photomultiplier based dynamic SPECT system. IEEE Trans. Nucl. Sci., 36(1):1132-1137.

Smith BD (1985). Image reconstruction from cone-beam projections: necessary and sufficient conditions and reconstruction methods," IEEE Trans. Med. Imag. MI-4:14-25.

Tuy HK (1983). An inversion formula for cone-beam reconstruction. SIAM J. Appl. Math. 43:546-552.

SPLINE-BASED REGULARISATION FOR DISCRETE FBP RECONSTRUCTION

J-P V Guedon[1,2] and Y J Bizais[1]

[1] Imagerie Medicale Multimodalite, Faculte de Medecine de Nantes, France.
[2] Center for Devices and Radiological Health, FDA, Rockville MD, USA.

Abstract

In this paper, we show that tomographic images are degraded by the unsuitable discretisation of continuous schemes, and the non-trivial null space in the case of angular sampling. Usually these two types of degradations are not studied separately. However, discretisation can be performed properly, while the null space is irreducible. For this reason, we study the relationships between continuous and discrete versions of a direct reconstruction method (FBP). They are characterized by an interpolation / sampling kernel, called the Pixel Intensity Distribution Model (PIDM). By defining the latter as B-spline functions, the existence and the uniqueness of the solution is guaranteed. It follows that projections must be oversampled. We test the robustness of this exact solution (for an infinite number of projections) by decreasing the number of angles. PIDM results are much better then FBP ones, showing that FBP reconstructed images are degraded not only by the null space, but also by unsuitable discretisation.

We also analyze the influence of degradations induced by an imaging device (mechanical instability and blur) and by projection noise in SPECT. Discretisation-related degradations depend on projection sampling. For this reason, proper oversampling is achieved when the corresponding degradations are negligible in comparison to the ones induced by the imaging device.

Our algorithm is constrained by the amount of information input to the system and controlled by the number of projection angles and the PIDM order. Optimal values of these parameters could be found for a predefined task using the ROC curve methodology.

Keywords

B-splines - Filtered Back-projection - Pixel Intensity Distribution Model

1. INTRODUCTION

Tomographic reconstruction plays a major role in medical imaging. It can be considered as a subset of medical image processing. Consequently it can be described as such :
- Projection data are acquired by an image source, processed by a computer algorithm, and tomographic slices are analyzed according to a specific detection task.
- Processing and analysis are performed on digital data. In particular, involved operands (projections, slices) and operators (projection, reconstruction) are discrete.

Moreover the reconstruction algorithm has specific features :
- It is derived from a continuous mathematical theory (Radon, 1917). The inverse problem is known to be unstable (ill-posed), even in the continuous case.
- In the case of angular sampling, no unique solution can be found (because of the null-space of the inverse operator). (Katz, 1978. Louis, 1981).

It follows that special attention should be paid to the discretisation of the reconstruction scheme (relationships between continuous and discrete versions). In this paper we show that careful discretisation leads to better quality reconstruction. Nevertheless, optimal quality can be achieved only according to a predefined detection task. In other words, parameters controlling the reconstruction algorithm must be tuned according to acquisition constraints and detection task characteristics (Hanson, 1998. Myers and Hanson, 1990) as shown in Figure 1 :

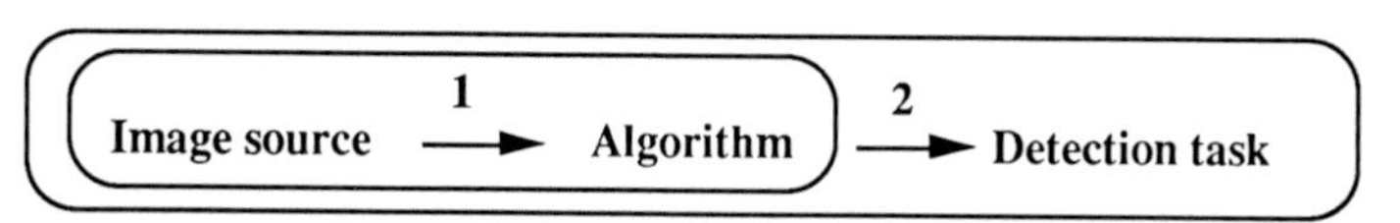

Figure 1 : The entire process of tomographic reconstruction

The paper is organised as follows :

1- Reconstruction algorithm. Relationships between continuous and discrete schemes are analysed. Proper discretisation of the reconstruction algorithm is derived for the Filtered Back-Projection (FBP) algorithm. The influence of angular sampling is then studied in order to understand the role of the irreducible null-space. Finally control parameters of the reconstruction are derived.
2- Image source. Every image source has specific characteristics degrading projection data. We shall analyse one detector (rotating γ-camera for SPECT) in order to understand the interaction between projection characteristics and the reconstruction algorithm. Moreover, for a given image source, a global constraint can be defined (amount of information input into the system).
3- Detection task. An optimally reconstructed image can be produced by tuning control parameters according to the global constraint and the predefined task.

2. CONTINUOUS AND DISCRETE RECONSTRUCTION ALGORITHMS

2.1. Introduction

For all (direct and iterative) reconstruction methods, an explicit or implicit relationship between the continuous version (the only theoretically exact one) and the discrete version (the only one which can be implemented) must be defined. This is obvious for iterative methods in which operators are discretised (Gordon, 1974). It is not so clear for direct methods in which brute force discretisation is usually applied.

All direct methods are equivalent (Barrett et al, 1976), which can be demonstrated using the reconstruction of a point source (linear, shift-invariant system is assumed). Consequently our work is based on the FBP algorithm.

It must be remembered that :
- The degree of ill-possedness of the inverse problem depends on the class of functions to be reconstructed. The inverse problem is very ill-posed in L^1 ($\mathbb{R}^2$), mildly ill-posed in L^2 ($\mathbb{R}^2$), and a unique solution can be found in Sobolev spaces (Louis, 1981).
- In the case of angular sampling, the null space of the inverse operator is non-trivial. Consequently a unique solution may not be found even by imposing regularity conditions on the function to be reconstructed.

Following these considerations, we propose the following strategy to discretise the inverse problem :
- For an infinite number of projection angles, we choose a class of functions for which the inverse problem is stable. We properly discretise corresponding operands and operators, in order to obtain a complete equivalence between continuous and discrete versions.
- We study the influence of angular sampling to understand how the non-trivial null space degrades the reconstruction.

In this section we apply the above strategy to first order spline functions (PIDM-0), and we extend it to higher order spline functions (PIDM-n).

2.2. Exact reconstruction of PIDM-0

2.2.1. Relationships between continuous and discrete schemes

Our approach is fully described in Figure 2. A class of discrete functions to be reconstructed ($f(k,l)$) is defined. An equivalent continuous function ($f(x,y)$) is generated from it, using an interpolating kernel (Pixel Intensity Distribution Model, noted PIDM). $f(x,y)$ is projected using the continuous Radon operator ($\mathcal{R}$). The resulting projections $p(t,q)$ can be discretised using a sampler consistent with the PIDM. These three steps define the discrete Radon operator. In the same way, a discrete inverse Radon operator can be specified as the combination of an interpolator generating $p(ti,\theta)$ from $p(t,\theta)$, a continuous inverse Radon operator ($\mathcal{R}^{-1}$), and a sampler consistent with the PIDM. A discrete reconstructed function is thus obtained. As in the continuous case, the discrete inverse operator $\boldsymbol{R}^{-1}$ can be decomposed into a discrete backprojection step and a discrete filter.

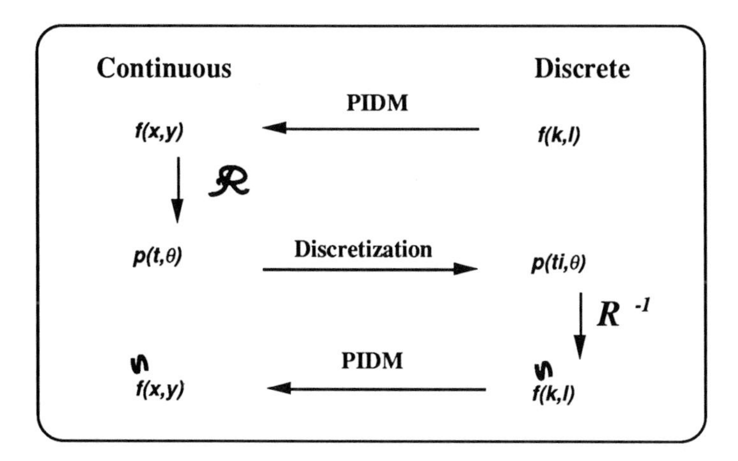

Figure 2 : Relationships between continuous and discrete reconstruction

2.2.2. Pixel intensity distribution model (PIDM)

In a first step, we choose the simplest PIDM (PIDM-0), a uniform function within a pixel. The PIDM can be seen as a tensor product of two 0-degree B-spline functions (Haar system) as defined in Eq. (1). The pixel size length is Δ and to ensure continuity, a Dirichlet condition is defined at the boundaries of the one dimensional spline function of degree 0 $s_{0,\Delta}$ as in Eq. (2).

$$f_0(x,y) = s_{0,\Delta}(x) \otimes s_{0,\Delta}(y) \quad (1)$$

with

$$s_{0,\Delta}(x) = \begin{cases} 1 \;\; if \; |x| < \Delta/2 \\ 1/2 \; if \; |x| = \Delta/2 \\ 0 \;\; if \; |x| < \Delta/2 \end{cases} \quad (2)$$

2.2.3. Projections of PIDM-0

As we showed in an earlier publication (Guédon and Bizais, 1990), the projections of PIDM-0 can be computed analytically and have an important feature : they vary with the projection angle (contrary to a delta function model) and are of trapezoidal shape. A fruitful expression for these projections is derived by considering the trapezoidal form as a convolution of two rectangular shapes of size length $\Delta cos\theta$ and $\Delta sin\theta$:

$$p_0(v,\theta) = s_{0,\Delta cos\theta}(t) * s_{0,\Delta sin\theta}(t) \quad (3)$$

2.2.4. Discretisation of projections

We have also shown previously (Guédon and Bizais, 1990) that projections must be sampled at a higher rate that the reconstruction plane, in order to preserve the spectral (angularly varying) content of the continuous projections. This will be referred to as oversampling. More precisely, if the sampling rate of the reconstruction plane is Δ and the projection sampling rate is τ, the oversampling ratio is defined as $\rho = \Delta/\tau$.

2.2.5. Discrete Backprojection

To be consistent with the above definition, the discrete backprojected value in a pixel must be defined as the integral of continuous backprojection values inside the pixel (Joseph, 1982). As shown in Figure 3, the 2D integral can be computed by integrating 1D integrals along backprojection lines inside the pixel. This is equivalent to convolving the projections by a kernel w_0, which happens to be equal to the PIDM-0. This result provides a very efficient way to compute the discrete backprojection (Guédon, 1990).

$$w_0(v,\theta) = s_{0,\Delta cos\theta}(t) * s_{0,\Delta sin\theta}(t) \quad (4)$$

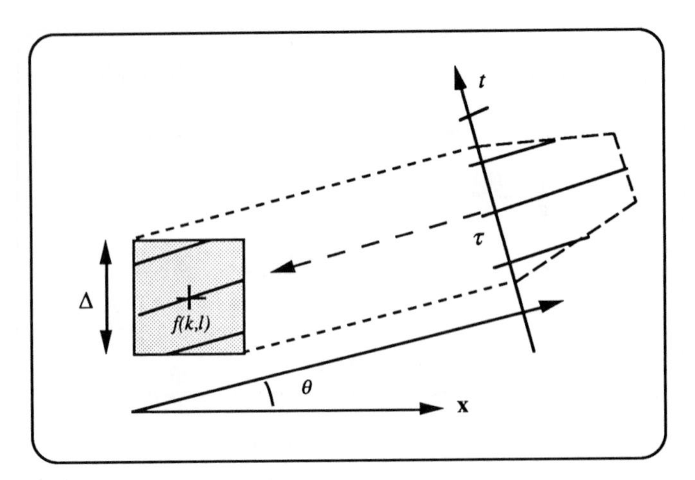

Figure 3 : Discrete Backprojection

2.2.6. Discrete filtering

The continuous filter is the well-known Ramp filter which is usually regularised by using a low-pass filter. In our case the convolution filter defining the discrete backprojection can be used to stabilise the reconstruction filter as written in Eq. (5). It should be noticed that the obtained low-pass filter varies with angle and depends on oversampling.

$$k\,(t,\theta) = w_0(t,\theta) * \mathcal{F}_1^{-1}\left[\pi\left|\nu\right| A_\rho(\nu)\right] \tag{5}$$

with the oversampling-dependant apodisation window :

$$A_\rho(\nu) = \begin{cases} 1 & \text{if } \left|\nu\right| < \dfrac{1}{2\tau} = \dfrac{\rho}{2\Delta} \\ 0 & \text{elsewhere} \end{cases} \tag{6}$$

2.2.7. Discrete reconstruction

The above described algorithm leads to a discrete reconstructed function (Eq. (7)) consistent with the used PIDM. The corresponding reconstruction scheme is exact, provided that the oversampling ratio is large enough (see below). It must be stressed that continuous and discrete schemes are different, as already shown by Beylkin (Beylkin, 1987).

$$\tilde{f}_0\,(x,y) = \frac{1}{\pi}\int_0^{\pi} p_0(t,\theta) * \left\{ w_0(t,\theta) * \mathcal{F}_1^{-1}\left[\pi\left|\nu\right| A_\rho(\nu)\right]\right\} d\theta \tag{7}$$

2.3. Extension to PIDM-n

In this section the above approach is extended to consider the class of B-spline functions as Pixel Intensity Distribution Models. Depending on the selected order of the model, different reconstructions may be obtained. The existence and the uniqueness of the solution are recursively demonstrated for each model order.

PIDM-1 is defined as the tensor product of two B-spline functions of degree 1 :

$$f_1(x,y) = s_{1,\Delta}(x) \otimes s_{1,\Delta}(y) \quad . \tag{8}$$

Through the recursive relationship between spline functions

$$s_{n,\Delta}(x) = s_{n-1,\Delta}(x) * s_{0,\Delta}(x) \quad , \tag{9}$$

we see that the intensity inside the pixel has a pyramidal form which vanish at the center of the neighboring pixels. Projections of this model can be easily found using the Fourier transform of Eq. (8) :

$$F_1(\lambda,\mu) = \mathcal{F}_2[f_1(x,y)] = S_{1,\Delta}(\lambda) \,.\, S_{1,\Delta}(\mu) \quad . \tag{10}$$

Applying Eq. (9) yields an equivalent form for $F_1(\lambda,\mu)$:

$$F_1(\lambda,\mu) = S_{0,\Delta}(\lambda) \,.\, S_{0,\Delta}(\mu) \,.\, S_{0,\Delta}(\lambda) \,.\, S_{0,\Delta}(\mu) \quad , \tag{11}$$

and we can invoke the Central Slice Theorem to write the Fourier transform of a projection in terms of $F_1(\lambda,\mu)$:

$$P_1(\nu,\theta) = \mathcal{F}_1^{-1}[p_1(t,\theta)] = F_1(\nu\cos\phi, \nu\sin\phi)\Big|_{\phi=\theta} \quad . \tag{12}$$

Using the inverse 1D Fourier transform of Eq. (12) with Eq. (11) leads to:

$$p_1(t,\theta) = p_0(t,\theta) * w_0(t,\theta) \quad . \tag{13}$$

2.3.1. PIDM-n

When the original function is assumed to be a tensor product of two B-splines of degree n :

$$f_n(x,y) = s_{n,\Delta}(x) \otimes s_{n,\Delta}(y) \quad , \tag{14}$$

its 2D Fourier transform can be written, using the 1D Fourier transform of Eq. (9) as :

$$F_n(\lambda,\mu) = S_{n-1,\Delta}(\lambda) \,.\, S_{n-1,\Delta}(\mu) \,.\, S_{0,\Delta}(\lambda) \,.\, S_{0,\Delta}(\mu) \quad . \tag{15}$$

Now, using the recurrent property proved for n=1 in Eq. (11), we obtain :

$$F_n(\lambda,\mu) = F_{n-1}(\lambda,\mu) \,.\, F_0(\lambda,\mu) \quad . \tag{16}$$

Using the Central Slice Theorem in the same manner as for the PIDM of degree 1, the projections can be written :

$$p_n(t,\theta) = p_0(t,\theta) * w_0^n(t,\theta) \quad , \tag{17}$$

where the superscript n refers to a product of n convolutions of the kernel w_0.

The filtered projections are given by convolution with the stabilized Ramp :

$$\tilde{p}_n(t,\theta) = p_0(t,\theta) * w_0^n(t,\theta) * \left\{ w_0(t,\theta) * \mathcal{F}_1^{-1}\left[\pi|\nu| A_\rho(\nu)\right] \right\} \quad , \qquad (18)$$

and the reconstructed function is obtained after the backprojection step :

$$\tilde{f}_n(x,y) = \frac{1}{\pi}\int_0^{\pi} \tilde{p}_n(t,\theta)\, d\theta \quad . \qquad (19)$$

2.3.3. Properties of PIDM-n solutions

In summary, the projection / reconstruction process can be described for the continuous and discrete versions as shown in figure 4.

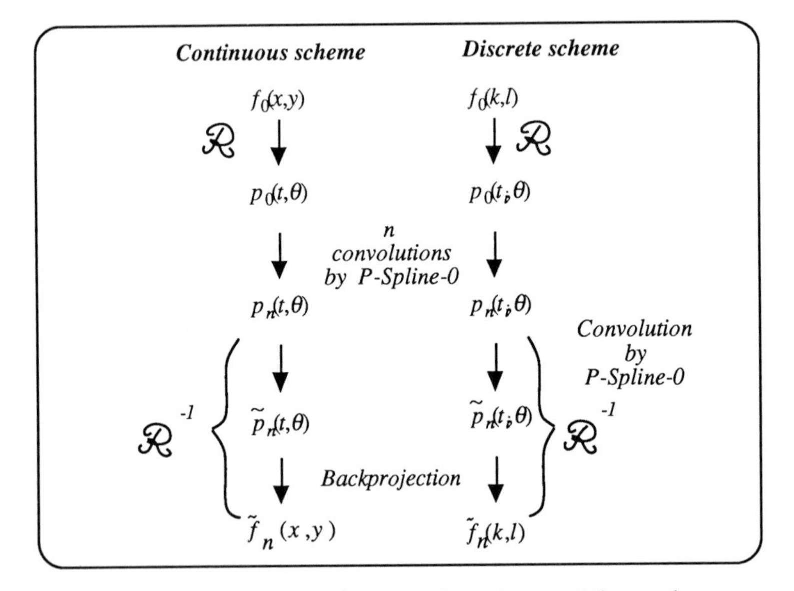

Figure 4 : Projection / reconstruction process for continuous and discrete schemes

The PIDM-n solution has the following properties :

- It is smoothed according to the selected PIDM order. The order can be seen as a control parameter to be tuned according to projection data features and the predefined detection task.
- The Ramp filter is naturally regularised through the angle dependent low-pass filter defined by the PIDM.
- The additional convolution by PIDM-0 in the discrete case implies the proper implementation of discrete backprojection. This additional operator for the discrete case corresponds to the results obtained by Unser et al (Unser and Alroubi, 1990, Unser et al, 1991) for discretisation of 1D B-spline functions.
- Convolution by the PIDM-n can be considered as a matched filtering, constraining the solution to belong to the appropriate space.

2.3.4. Discrete source reconstruction

To illustrate the above theory, a discrete point source equivalent to the delta function, was defined as a 17^2 square ("big pixel") in a 512^2 reconstruction plane. Figure 5 shows the reconstruction using PIDM-0, PIDM-3, and the standard FBP algorithm. It can be seen that drastic improvements are obtained by the PIDM algorithm over the FBP method. In particular the Dirichlet condition is perfectly satisfied for PIDM (note the 1/2 amplitude value reconstructed at the object side).

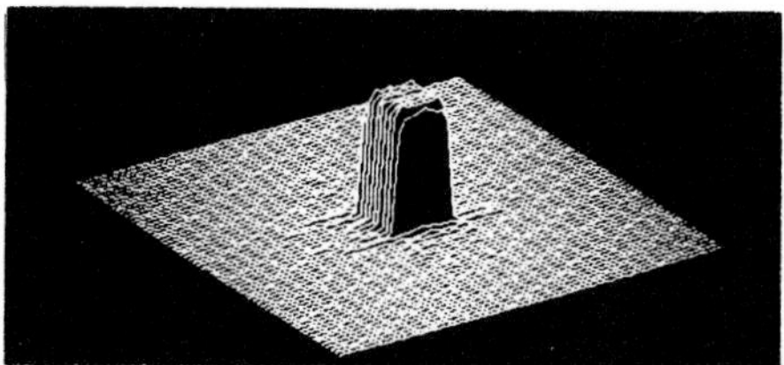

Standard FBP algorithm

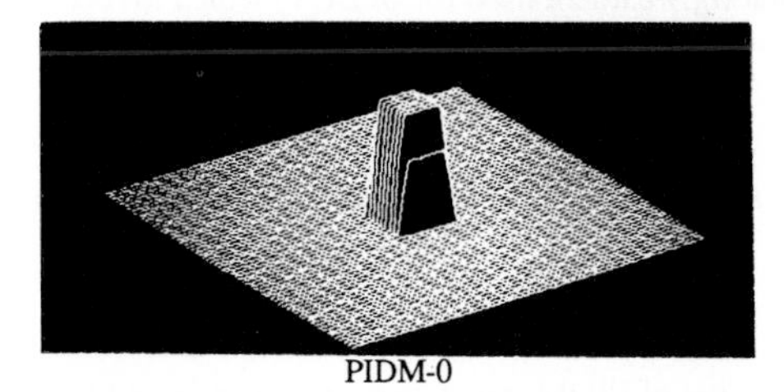

PIDM-0

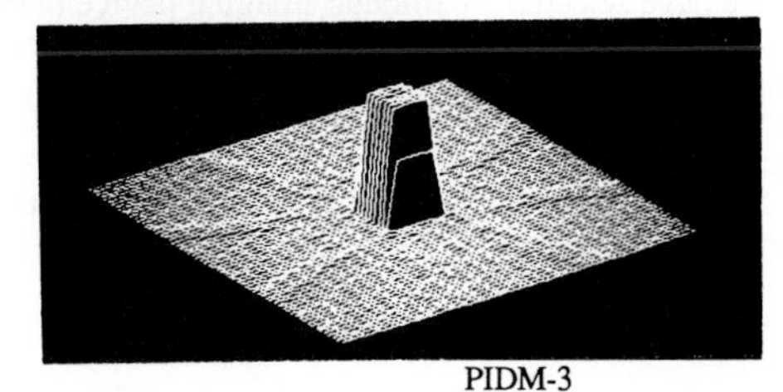

PIDM-3

Figure 5 : Discrete point source reconstructions

2.4 Angular sampling

For an infinite number of projections, the angular information is preserved and the solution is exact. For a finite number of projections, the solution cannot be exact because the null space is non trivial. However, for the PIDM algorithm, degradation of the results are only induced by the null space and not by discretization errors (as for FBP).

We analytically computed some functions of the null space and found that they are too complex to be understood. For this reason we numerically computed them, showing that :

- The convolution kernel is not perfectly stable for small angles (0° to 20°) : some poles of the Z-transform are outside the unit circle.
- Line artifacts appear at the edges of the discrete reconstructed square object for FBP but not for PIDM. It is predicted by our theory, but not by the delta function basis approach (Brooks et al, 1978).

The number of projection angles can be seen as a control parameter for acquisition and reconstruction processes.

Figure 6 illustrates the influence of the number of projections showing reconstructions of the Shepp-Logan phantom for 512 angles for PIDM-0, 64 angles for PIDM-0, and 64 angles for FBP. By comparing the 64 angle reconstruction, it can be seen that FBP artifacts come from both true null space ghosts and discretization errors.

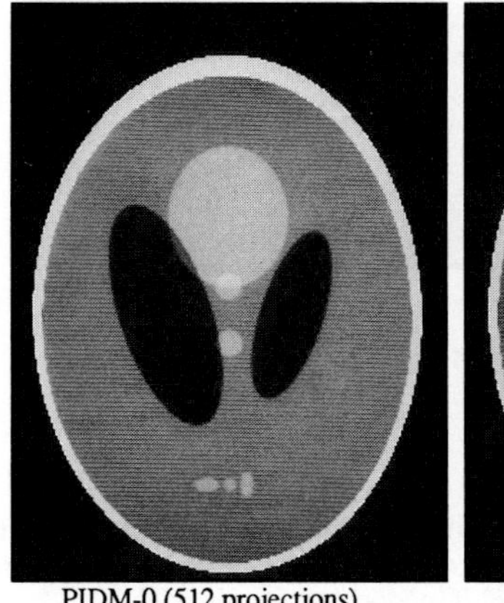

PIDM-0 (512 projections)

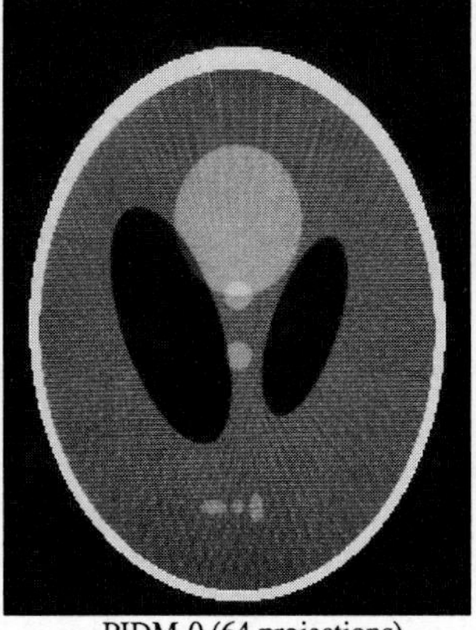

PIDM-0 (64 projections)

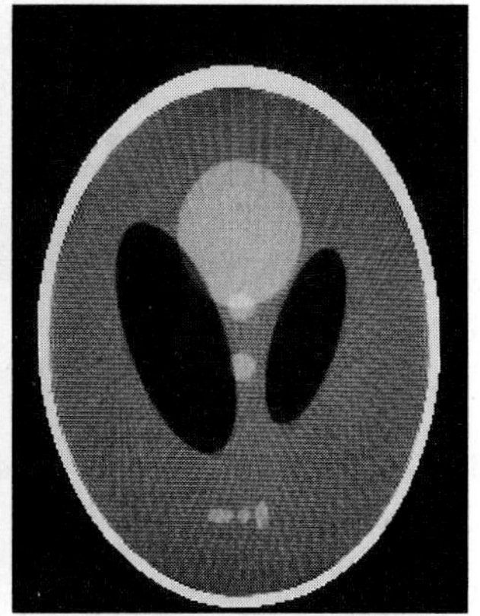

Standard FBP (64 projections)

Figure 6 : Shepp-Logan phantom reconstructions

3. ROLE OF THE IMAGE SOURCE

3.1. Introduction

In this section, we introduce the characteristics of projection data related to the imaging system. Through this we can analyze :
- The role of the imaging system in the degradation of reconstructed images, as compared to that induced by the algorithm,
- The interaction between these two types of degradation.

We have selected a particular imaging device (a rotating γ-camera used for SPECT) which involves many kinds of degradations for our investigations. This system gives images degraded by :
- quantum noise,
- mechanical stability and resolution,
- attenuation and scatter.

Moreover, the total amount of information input to such a device (the acquired number of photons) can be easily constrained.

3.2. Noise in the projections

Thanks to the smoothing variable provided by the degree of PIDM, projection noise can be reduced according to its estimated level.

We generated a Gaussian white noise, the variance of which was typical of SPECT data (128 photons/pixel). The noise was added to the data from the square object for an infinite number of projections. Images reconstructed with PIDM-0, PIDM-3 and FBP show the same ranking as in the noiseless case (PIDM better then FBP), and the smoothing effect of PIDM order.

A similar experiment was carried out for a finite number of projections. Results for 64 projections are shown in Figure 7. The comparison between FBP and PIDM-0 images shows the effects of discretization (Dirichlet condition is not reconstructed for the FBP case). Comparison between PIDM-0 and PIDM-3 images shows the effects of PIDM order on smoothing.

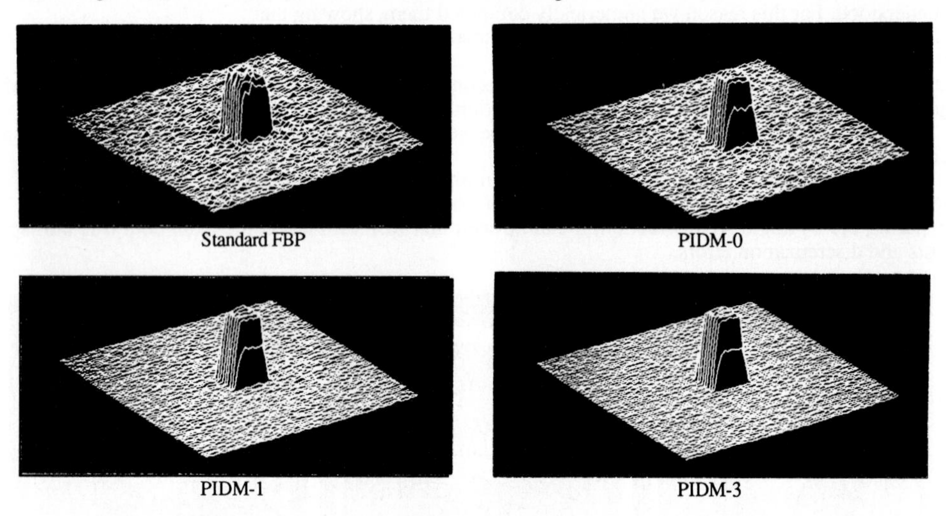

Figure 7 : Discrete point source reconstructions using 64 projections

3.3. Mechanical instability and resolution

Image degradations induced by mechanical instability and limited resolution of γ-cameras are well-understood (Todd-Pokropek, 1982). We analyzed the effects of the projection of the center of rotation and simulated blurring by convolving projections with a Gaussian kernel (Guédon and Bizais, 1991).

The results can be summarized as follows :
- The projection of the center of rotation can be estimated with an accuracy of about 1.5 mm. When projections are not oversampled (pixel size in projection equal to 6 mm), the limiting factor is the projection sampling and large errors are induced by it. When projections are oversampled ($\rho = 4$), the limiting factor is the uncertainty about the projection of the center of rotation. For this reason, there is no need to oversample beyond this value.
- Blurring induces smoothing of projection data. Our results obtained with PIDM appear to be better than FBP ones . Moreover, because of the oversampling, partial deconvolution can be carried out.

It must be stressed that both mechanical instability and blurring give an upper value of oversampling. There is no need to sample projections with a sampling distance smaller than the uncertainty about the projection of the center of rotation, or much smaller than the detector resolution. For SPECT, $\rho = 4$ is a typical upper boundary.

3.4. Other degradations

Other important degradations such as attenuation and scattering do occur. These are more difficult to simulate. For this reason, we tested our algorithm on actual data of a Jaszczack phantom, in which all degradations are integrated. Despite the poor quality of the camera we used, we found that PIDM results are significantly better than FBP ones.

4. TASK DEPENDENT OPTIMIZATION

To more objectively assess the quality of the reconstruction, an optimization step is needed. This optimization problem can be summarized as follows :
1) A total amount of information (Q) is input into the system (Q is the total number of detected photons in SPECT).
2) Q is distributed into P projections. Consequently the signal to noise ratio (SNR) in the projections increases with Q and decreases with P.
3) A further parameter to optimize is the PIDM degree (n), which determines the smoothness of the reconstruction.

As it is well-known (Myers and Hanson, 1990. Myers et al, 1990. Wagner, 1983), there exists no set (Q,P,n) for which the solution is always optimal. On the contrary for a predefined detection task to be performed on reconstructed images, an optimal set for this task can be found experimentally using the ROC curve methodology. This work has not been carried out yet but is planned for future investigation.

5. CONCLUSION

In this paper we showed that tomographic images are degraded by :
1) the unsuitable discretization of continuous schemes, and
2) the non-trivial null space in the case of angular sampling.

Usually these two types of degradations are not studied separately. However, discretization can be performed properly while the null space is irreducible. For this reason, we studied the relationships between continuous and discrete versions of a direct reconstruction method (FBP). This work can be easily extended to other direct methods. The relationships are characterized by an interpolation / sampling kernel, called PIDM. By defining the latter in terms of B-spline functions, the existence and the uniqueness of the solution is guaranteed. It follows that projections must be oversampled. We tested the robustness of the exact solution (for an infinite number of projections) by decreasing the number of angles, in order to study the influence of the null space. PIDM results are much better then FBP ones, showing that FBP reconstructed images are degraded not only by the null space but also by unsuitable discretization.

We also analyzed the influence of degradations induced by an imaging device (SPECT) : projection noise, mechanical instability, and blur. Discretization-related degradations depend on the projection sampling rate. For this reason, proper oversampling is achieved when the corresponding degradations are negligible in comparison to the ones induced by the imaging device. Finally overall effects were studied experimentally, using a physical phantom.

Our algorithm is constrained by the amount of information input into the system, and controlled by the number of projection angles and the PIDM order. The next step is to determine optimal values of these parameters for a predefined task using the ROC curve methodology.

Acknowledgements

The authors would acknowledge the great help of Kyle Myers in rewieving the manuscript (with her native English language) and thanks the rewievers for their comments. Special thanks to Mikael Unser and Akram Aldroubi for their helpful comments about splines theory. Jean-Pierre Guédon is now with the Center for Devices and Radiological Health / Food and Drug Administration, Rockville MD, under a NRC postdoc position.

References

Radon J. (1917). Uber die Bestimmung von Functionen durch ihre Integralwerte langs gewisser Mannigfaltigkeiten. Berichte Sachsische Academie der Wissenchaften, Leipzig, Math. - Phys. Kl, Vol-69:262-267.

Beylkin G. (1987). Discrete Radon Transform. IEEE Trans. Acoust., Speech Signal Proces. ASSP-35(2):162-172.

Brooks R.A., Weiss G.H. and Talbert A.J. (1978). A new approach to interpolation in Computed Tomography. J. Comp. Ass. Tomography Vol 2:577-585.

Gordon R. (1974). A tutorial on ART. IEEE Trans. Nucl. Sci., NS-21:78-93.

Guédon J-P. and Bizais Y. (1990). Projection and Backprojection models and projection sampling in Tomography. SPIE Medical Imaging IV, Vol 1231:206-217.

Guédon J-P (1990). Sampling problems in Tomography, Ph.D. dissertation (in French), ENSM, Nantes.

Guédon J-P. and Bizais Y. (1991). A Spline based tomographic reconstruction method. SPIE Medical Imaging V, Vol 1443.

Hanson K. (1989). Optimisation for object localization of the constrained algebraic reconstruction technique. SPIE., Medical Imaging III , Vol 1090.

Hanson K. (1989). Optimization of the constrained algebraic reconstruction technique for a variety of visual tasks. Proc. of the XIth IPMI Intern. Conference , Progress in clinical and biological research Vol. 363, Orthendhal & Llacer Eds., Wiley-Liss.

Joseph P.M. (1982). An improved algorithm for reprojection rays through pixel images. IEEE Trans. on Med. Imaging, TMI-1:192-196.

Katz M.B. (1978). Questions of uniqueness and resolution in reconstruction from projections. Lect. Notes Biomath. 26, Ed S. Levin, Springer-Verlag, Berlin.

Louis A.K. (1981). Orthogonal function series expansions and the null-space of the Radon transform. medical image processing group, Technical report MIPG52, Buffalo, Univ. of New-York.

Ludwig D. (1966). The Radon transform on Euclidean space. Comm. on pure and applied mathematics, Vol XIX:49-81.

Myers K.J. and Hanson K.M. (1990). Comparison of the algebraic reconstruction technique with the maximum entropy reconstruction technique for a variety of detection tasks. SPIE Medical Imaging IV, Vol 1231:176-187.

Myers K.J., Rolland J.P., Barrett H.H. and Wagner R.F. (1990). Aperture optimization for emission imaging : effect of a spatially varying background. JOSA, Vol. 7:1279-1293.

Schmidlin P. and Doll J. (1989). Implementation of iterative reconstruction in PET. Proc. of the XIth IPMI Conference, Berkeley, CA, Progress in clinical and biological research. Vol. 363, Orthendhal & Llacer Eds., Wiley-Liss.

Todd-Pokropek A. (1982). The mathematics and physics of emission computed tomography. Radionuclide Imaging 3-31.

Unser M., Aldroubi A. and Eden M. (1990). A sampling Theory for Polynomial Splines. Intern. Symp. on Infor. Theory and its Appl. ISITA90:279-282.

Unser M., Aldroubi A. and Eden M. (1991). Recursive Regularization Filters:Design, Properties, and Applications. IEEE trans. on pattern anal. mach. intel. PAMI-13-3.

Wagner R.F. (1983). Low contrast sensitivity of radiologic, CT, nuclear medicine, and ultrasound medical imaging systems. IEEE Trans. on Medical Imaging, TMI-2:105-121.

RECONSTRUCTION OF 3-D BRANCHING STRUCTURES

C J Henri[2], D L Collins[2] and T M Peters[1,2]

McConnell Brain Imaging Centre, Montreal Neurological Institute

Depts of Medical Physics[1] and Biomedical Engineering[2], McGill University

3801 University Street

Montreal, Quebec, CANADA

H3A 2B4

Abstract

We describe a new approach to the problem of reconstructing 3-D branching objects from a small number of projections. Given the object-to-image transformations and an assumption of structural connectivity, we describe how to elucidate the imaged structure from the multitude of artifacts resulting from all possible triangulations. The method involves generating two or more intermediate reconstructions from different pairs of projections, then comparing structural similarities/differences to identify the artifacts. Since the intermediate reconstructions are assured to be connected (i.e. not fragmented), it becomes possible to refine portions of the structure that previously were considered ambiguous. Simulations were performed using a mathematically defined test object having 10 branches. In every case, the true structure was able to be distinguished from the artifacts using as few as three projections.

Keywords

Few view reconstruction; projection angiography; structural connectivity; consistency.

1. INTRODUCTION

We describe the theoretical basis for a new approach to the problem of reconstructing branching three-dimensional (3-D) objects from a small number of projections. In so doing, we establish a clear understanding of how certain constraints are best employed and how the notion of consistency is achieved.

The motivation for this work is derived from a desire to reconstruct cerebral vasculature from digital subtraction angiograms. In particular, our interests lie in surgical planning applications and multi-modality image analysis and integration. We have addressed these issues previously and demonstrated some degree of success using stereoscopic techniques (Henri et al, 1990 and Peters et al, 1989), but the lack of freedom in generating arbitrary or optimal views often precludes all but the most basic applications.

This paper does not summarize or directly extend previous methods used to reconstruct vascular networks, but instead re-examines fundamentals that are common to all techniques and addresses the problem of deriving a solution that is consistent with the given image data. The approach described here is conceptually simple and is built only on an assumption of connectivity.

The method involves generating two or more 3-D reconstructions from different pairs of projections and then uses geometrical constraints to eliminate improbable or impossible structures in each one. The so-called correspondence problem in stereo-matching paradigms is circumvented by generating 'intermediate reconstructions' that comprise all possible structures that could have yielded the given projections. We demonstrate that certain realizations (in particular, unlikely ones), often provide mechanisms for resolving ambiguities that would otherwise have been absent had they not been considered to begin with. We emphasize the relevance of generating intermediate reconstructions from pairs of projections rather than simultaneously from three or more. Here, structural continuity is important since connectivity constraints cannot be applied optimally when the structure is fragmented. Subsequent processing is an exercise in distinguishing between true and artifactual structures by seeking a solution that is consistent with the set of acquired images and the knowledge that the structure is connected.

We begin by addressing the reconstruction methodology. This is followed by a description of the simulations that were performed to investigate key aspects of the methods we are about to present. We conclude with a discussion concerning refinements to the procedure and its potential implementation in practice.

2. METHODS

In order to facilitate the following presentation, we neglect all sources of measurement error, including image digitization and resolution considerations. Although this leads to a discussion that may at times appear tenuous, we prefer to delay addressing the more realistic case until later (Section 4). For now, we consider only an idealized branching 3-D structure; one that has a skeletal representation and is connected. We assume further that the entire structure appears in each acquired image and that the imaging geometry is known; i.e. the position and orientation of the camera relative to the imaged object is known.

Let us consider an example in which two images of a 'Y'-shaped structure are acquired, as illustrated in Figure 1. For a plane that intersects the 'Y' just above its vertex, lines have been drawn from the centres of projection, L and R, through each branch to determine where they appear in the image. The resulting positions are labeled $L1$, $L2$, $R1$ and $R2$.

Figure 2 illustrates the plane from Figure 1, where the cross-sections through the branches have been numbered 1 and 2. Given only the points in each image and the centres of projection, the problem is to determine that only points 1 and 2 are present and that they are located exactly as depicted in Figure 2. For the given geometry, however, the points in each image could have been generated by a number of alternative configurations. In fact, as many as four points could have been present, each located at the line intersections between the image plane and the centres of projection (see also points 3 and 4 in Figure 2). Consequently, it is impossible to obtain a unique solution to this problem without additional information. Such information, for example, might include a third projection from a new viewpoint, or the knowledge that there are only two points, and that number 1 is one of them.

For a plane that intersects the 'Y' close to its vertex, points 1 and 2 move closer together as do points 3 and 4 where the lines from the centres of projection intersect. For now, consider the structure that is reconstructed by stacking successive planes through the 'Y' and including in each plane all such points arising from line intersections. In 3-D we no longer have a 'Y'-shaped structure, but instead one that has two additional branches above the vertex: one anterior and one posterior.

At this point, we note that the reconstructed object remains connected (i.e. not fragmented) and that the artifactual branches join the true structure at its vertex. In fact, any reconstruction generated using only two projections, as above, will always be connected. This assumes that the imaged object itself is connected, and neglects the effects of image digitization, etc.. This observation has important ramifications that will become apparent below.

Consider now a plane through the 'Y' structure higher up (see Figure 3). Because the branches are farther apart at this level, points 1 and 2 are also farther apart, as are the other points, 3 and 4. In fact, point 3 has ventured behind the image plane, giving a configuration that

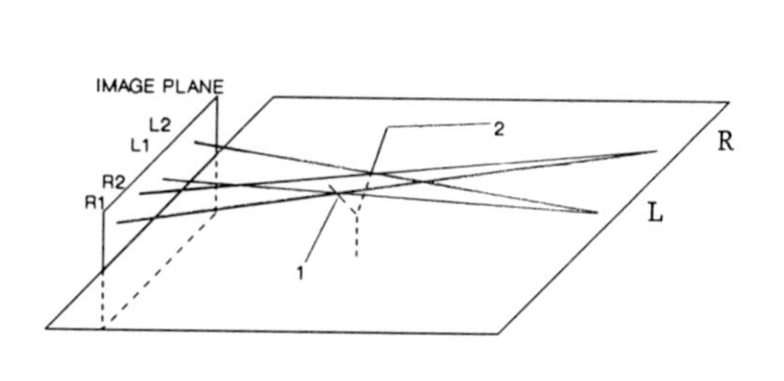

Figure 1: Two projections are acquired of a 'Y'-shaped structure. Lines have been drawn from the centres of projection, **L** and **R**, through branches 1 and 2 of the 'Y' to determine where they appear in the image plane (giving points $L1$, $L2$, $R1$ and $R2$).

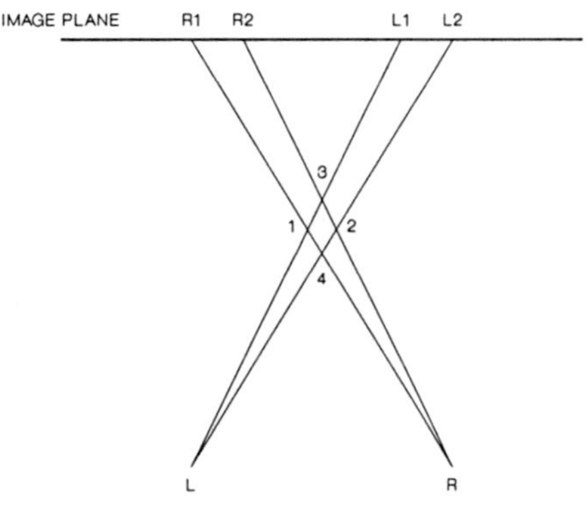

Figure 2: A plane through the 'Y' structure illustrated in Figure 1 yields the given cross-sectional view. The branches of the 'Y' have been labeled 1 and 2. Points 3 and 4 identify additional intersections of the lines between the centres of projection and the image plane.

is clearly not one of the possible configurations. This information, together with the knowledge that the structure is connected, enables us to not only discard point 3 when it is reconstructed behind the image plane, but also to associate with every point connected back to the vertex a high likelihood of being artifactual. In other words, we have accrued evidence that the branch arising from point number 3 at each level through the 'Y' is artifactual simply because a portion of it ventured outside the allowable bounds of reconstruction. Later we will strengthen this notion of 'evidence' by considering the circumstances under which such a branch might, in fact, actually be part of the imaged structure. At this point, we note that by discarding the branch in question in whole (i.e. back to the vertex of the 'Y'), the resulting structure remains consistent with its projections. Although this is reassuring, consistency alone is not sufficient to prevent us from doing so in error and, therefore, should be further substantiated.

In the discussion surrounding branch number 3, we assumed that the allowable boundaries for reconstruction were defined by the space between the image plane and the centres of projection. In practice it is often possible to impose more specific limits, increasing the likelihood of branch number 4 venturing outside the allowable boundaries (see Figure 4). Of course it is possible that the artifactual branches never venture outside these boundaries, in which case they are impossible to distinguish from true ones without additional assumptions or information. The likelihood of this occurring will depend upon several factors, including the relative separation between the centres of projection; the separation of the branches in the imaged structure, and the degree to which magnification is employed in the projection.

Before extending the principles outlined above to include three or more projections, we wish to acknowledge some parallels between the concepts we have described and those that have been used by others to resolve the inherent ambiguities. Consider, for instance, the correspondence problem in stereo-matching where the aim is to identify corresponding points on the object in each acquired image, then triangulate to obtain 3-D coordinates (Barnard et al, 1980, Duda et al, 1973, Horaud et al, 1989 and Prazdny, 1983). The prospect of searching for corresponding

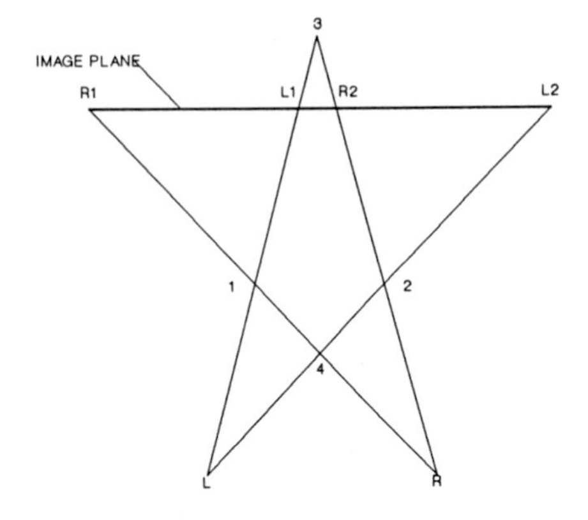

Figure 3: A plane through the top of the 'Y' structure: One of the four line intersections (number 3) now occurs behind the image plane. Since this configuration is not possible, we are able to discard this point and make inferences about the associated branch in planes at lower levels.

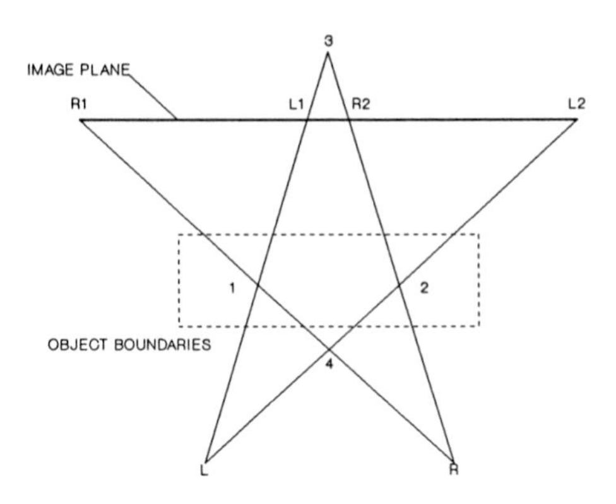

Figure 4: By specifying allowable reconstruction boundaries, we are able to distinguish between true and artifactual branches. At the level illustrated here, it is apparent that branches 3 and 4 are artifactual over their extent outside these limits.

points randomly in two images is unattractive and is usually avoided by invoking the 'epipolar line' constraint (Prazdny, 1983, Baker et al, 1981, Barnard et al, 1987 and Mohan et al, 1989). The search space is consequently reduced to a single line through an image. In the discussion above, we have implicitly used such a constraint as well. By considering separate planes through different levels of the imaged structure, we are in fact isolating the search space for corresponding points to a single line through each image. The triangulation process amounts to calculating the coordinates of the points at the intersections of the lines between the image plane and centres of projection.

Although the epipolar line constraint simplifies the correspondence problem substantially, there remain ambiguities: Unless there is a single point on the epipolar line, there will be more than one possible configuration that could have yielded the given pair of projections. It is common, therefore, to invoke a constraint on the allowable range of disparities in an image pair (Barnard et al, 1980 and Levine et al, 1973). Such a constraint effectively reduces the length along the epipolar line that is searched for correspondences. If it is narrow enough to isolate a single point, then the resulting match is unique and the overall ambiguity of the problem is reduced. This constraint is equivalent to our notion of recognizing limits on the boundaries of the reconstructed object. However, there is a difference in philosophy between our formulation and the typical stereo-matching techniques. It lies in how the disparity constraint is applied: Instead of enforcing disparity limits during reconstruction, and thus considering only object configurations that lie within the allowable boundaries, we reconstruct all possible configurations, even those violating allowable disparities, and delay imposing the constraint until later. We elaborate on the significance of this difference below.

2.1. Additional Projections

In the discussion above, we considered the reconstruction problem using stereoscopic projections. We note that, in principle, any two views may be employed and that it is often advantageous

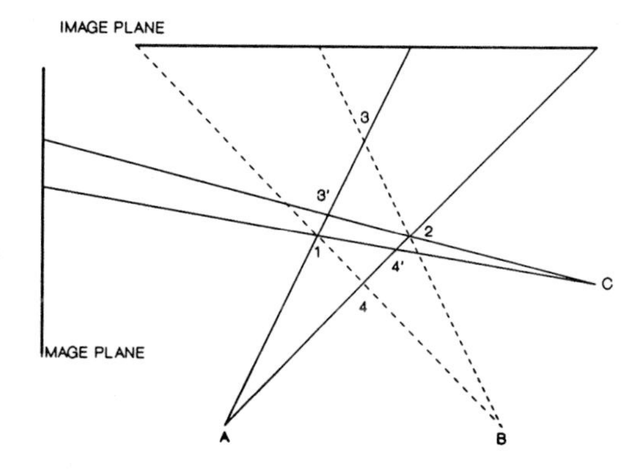

Figure 5: A new pair of projections yields a different 4-point configuration than those illustrated in Figures 1-4. We note, however, that points 1 and 2 have remained at the same positions, while the artifactual points have shifted (from 3 and 4 to $3'$ and $4'$, respectively).

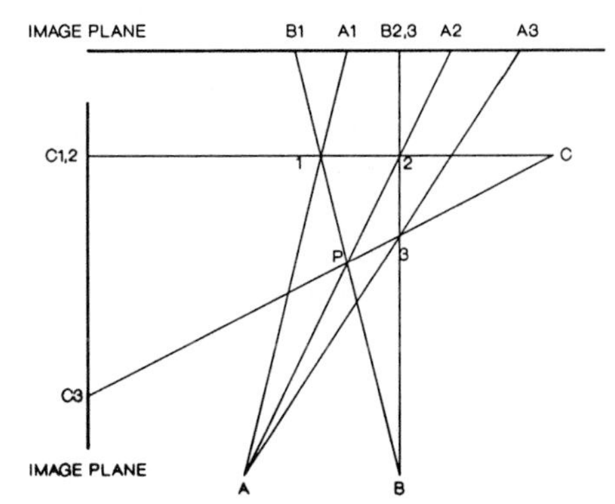

Figure 6: Three projections are acquired (from centres of projection A, B and C) of a structure whose cross-section appears as the given 3-point distribution (points 1, 2 and 3). Triple line intersections occur at each point as well as at one artifactual point, 'P'.

to increase the separation between the centres of projection in order to improve the spatial resolution of the triangulation calculations when there are associated measurement uncertainties (Fencil et al, 1990 and Mawko, 1989). We now consider how additional projections are incorporated into the reconstruction problem.

For simplicity, we assume we are given only three projections. Using any two to generate a 3-D reconstruction as described above, we may obtain three different structures. Let us examine any two of these and refer to this approach as 'method 1'. (Below we have purposely neglected to include constraints on the reconstruction boundaries.)

For the example of the 'Y' given above, one reconstruction might be the same four-branch structure that was derived from the imaging geometry used earlier. When we perform a second reconstruction using a different pair of projections, a new object is obtained. Figure 5 illustrates a hypothetical imaging geometry for a plane through the 'Y', along with its artifactual points: $3'$ and $4'$. We note immediately that the true points, 1 and 2, have remained in the same positions while the artifactual points have shifted. We conclude, therefore, that points 3 and 4, or alternatively, $3'$ and $4'$ are false and may be discarded. This step is delayed, however, until the entire reconstruction (containing true and false points at each level) has been generated. The reason for this will become apparent below.

In this particular example, the same conclusion could have been reached by employing the three projections simultaneously and keeping only points that occurred at line intersections that are common to the three views. We call this 'method 2'. With this approach, however, it is possible to incorrectly identify artifactual points as true ones when dealing with more complex structures. For example, the three point distribution illustrated in Figure 6 is assumed to represent a cross-section through an arbitrary 3-D branching structure. Lines have been drawn from the centres of projection, A, B and C, through each point to their respective positions in each image plane. We notice that exactly three lines pass through each of the true points, and that lines $\overline{AA2}$, $\overline{BB1}$ and $\overline{CC3}$ also intersect, but at a false point, 'P'. Clearly we do not want to include this point in our reconstructed object, but we are unable to discard it using the criterion of method 2. Let us examine what happens to P in the cross-sections just above or below the one in Figure 6. The point will appear in adjacent cross-sections only if lines $\overline{AA2}$, $\overline{BB1}$ and $\overline{CC3}$ continue to intersect, albeit slightly above or below the plane considered above. For this to happen, the positions of the true points can change in only very specific directions and must do so cooperatively. Since this will not always be the case, an object reconstructed using method 2 is not guaranteed to be connected; i.e. the above point 'P' might be isolated from the remaining structure. Pursuing this issue farther, one might suggest that a fragmented structure arrived at using method 2 could be cleaned-up using connectivity relationships. Although this would provide a mechanism for discarding isolated branch segments, it would overlook those that might remain connected to the true structure.

We now return to method 1, where we have two reconstructions generated from different pairs of projections. From Figure 6 we note that 'P' is common to each reconstruction and, in fact, occurs where an artifactual branch in one coincides with an artifactual branch in an other. Thus, the point can only be recognized as artifactual by establishing the association with its branch in either reconstruction. Clearly, this is only possible when P is not isolated. For this reason, method 1 is better suited for handling the problem of distinguishing between true and artifactual branches and forms the basis of our approach to solving the reconstruction problem.

2.2. Intermediate Reconstructions

We refer to the reconstructions generated from different pairs of projections as 'intermediate reconstructions' (IRs) and recall that each is generated, as in the above examples, by collecting all points in each plane resulting from possible triangulations. Although the true object is known to be present and common to each IR, we are faced with the problem of distinguishing between it and the multitude of artifactual branches. From the discussion surrounding Figure 5, we noted that certain differences exist between each pair of reconstructions, specifically in the positions of the artifactual branches. Although we are reminded that it is possible for portions of the artifactual branches to coincide and thus be indistinguishable at these points, the connectivity assumption acts powerfully in our favour since it is highly unlikely that artifactual branches coincide at more than a few adjacent points (recall the discussion surrounding Figure 6). Consequently, support is effectively guaranteed for the 'evidence' we referred to earlier in the discussion of branch number 3 in Figures 1 - 3; particularly when there are more than two different IRs (i.e. when three or more projections are available).

It is enlightening to consider the difference between this approach (i.e. method 1) and method 2 from the last section. The distinction lies in how connectivity is employed and is made clear by considering the effect of performing a logical AND on any pair of IRs. In so doing, we obtain the object that would have been reconstructed using method 2; i.e. all points that are simultaneously common to each projection used in generating the IRs. We recall, also, that this object may be fragmented.

In performing the logical AND, we are assured that the resulting structure is consistent with each pair of projections. In other words, this structure, when re-projected into each image, will align perfectly with the original data. Although this is a necessary characteristic of the desired solution, it is not sufficient to ensure that we obtain the correct one. Upon imposing connectivity, we are able to refine the solution without losing its consistency. As mentioned earlier, however, we are no longer able to identify potentially artifactual branches that may still be attached to the true structure. This is because we have neglected the evidence that they are connected to segments that are inconsistent with the given projections (a consequence of the logical AND). It is for this reason, with our approach, that we delay the use of connectivity and avoid directly applying a logical AND to the IRs.

2.3. Summary

The approach we have described above is intended for reconstructing branching 3-D structures based on three or more distinct projections. Although we have described the application of knowledge concerning allowable reconstruction boundaries, such knowledge is not necessary. The primary assumption is one of connectivity. We have further assumed that the transformations between object-oriented coordinates and the image planes are known, permitting epipolar lines to be determined and triangulation to be performed.

The first step is to generate two or more IRs using different pairs of projections. This is accomplished by generating a point for every possible triangulation in the image pair, subject to epipolar line constraints. We note that, in each case, the true structure is guaranteed to be part of the IR and that this object is known to be connected.

The second step involves discarding from each IR the portions of any branch venturing outside the boundaries within which the true object is known to be confined. Since only artifactual branches are capable of existing outside this region, we are able to make inferences concerning their validity inside the boundaries as well. Depending on how precisely we are able to specify the allowable reconstruction volume, this step could substantially simplify further processing.

The key step in the procedure involves comparing the two (or more) intermediate reconstructions. Here we examine whether a given branch in one IR also exists in the other(s) at the same location. If a given branch in one IR occupies an area that is void in an other, then it is discarded. If only a portion of the branch deviates between IRs, then connectivity assumptions allow us to accrue evidence concerning the validity of the remaining portions; specifically that they may be artifactual. This procedure ensures that only artifactual branches are discarded since trues will always be present at the same position in each IR. Because different imaging geometries are used to generate the IRs, only the positions of artifactual branches change. It follows that if every artifactual branch assumes a different position in at least one IR, then we are guaranteed to isolate the true structure by pursuing this line of reasoning. The validity of this statement is central to our discussion and will be considered in greater detail in the following sections.

2.4. Experiments

Simulations were performed to study several aspects of the procedure outlined above. Although the results presented here are preliminary, they serve to corroborate a number of the observations noted above. Again we found it useful to consider only an idealized 3-D branching structure; one whose form was known in terms of the coordinates of several points along each branch. The aim of the simulation was to assess the ability to distinguish between true and artifactual branches in the intermediate reconstructions described above. In so doing it was expected that we might also gain some insight into how the procedure might be best applied in practice; specifically with regard to the number of projections and preferred imaging geometry. Although we recognize the importance of addressing the issue of reconstruction accuracy, we postpone this topic for a proper treatment at a later date.

2.4.1. Simulation Methodology.

A test object consisting of ten connected branches of varying lengths was created using a stereoscopic display system (Henri et al, 1990 and Peters et al, 1989). Points defining a branch were selected using a stereoscopic cursor that was freely maneuverable in 3-D. Four projections of the structure were generated mathematically by simulating the positions of each centre of projection and image plane for two stereo-pairs, as illustrated in Figure 7. This particular configuration corresponds to the geometry with which we routinely acquire digital subtraction angiograms at our institute. We note, however, that the stereo-pairs are actually acquired with stereo-shifts that are perpendicular to each other. This was also true of our simulated projections. The test

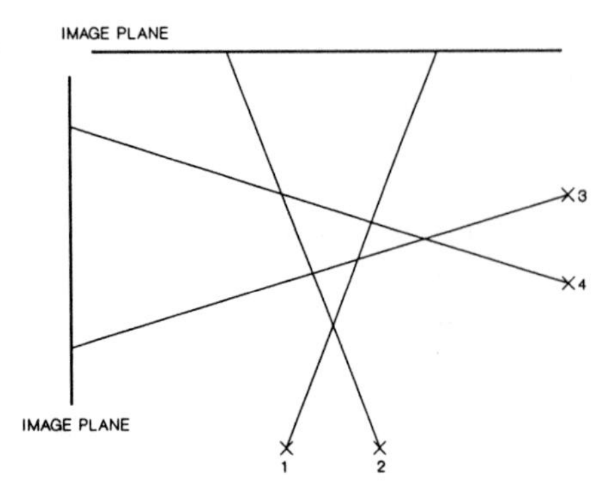

Figure 7: Biplane stereoscopic imaging geometry. Four projections are acquired: 2 stereo-pairs at 90° to each other.

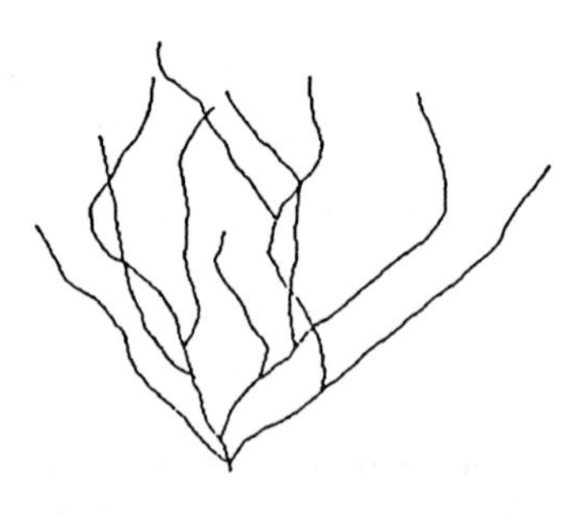

Figure 8: A test object with 10 branches was employed in the simulations. Points along each branch were selected using the mouse-controlled cursor of a stereoscopic display system.

object, as it appeared in one projection, is shown in Figure 8. The angulation between projections comprising a stereo-pair was 7°, while it was 90° between the two pairs themselves. In each case, a 100*cm* focal distance was employed with the object's centre located approximately 35*cm* from the image plane. These parameters are representative of those used in practice.

Four IRs were generated using the following pairs of projections (see Figure 7): (1,3), (1,4), (2,3) and (2,4). Each pair comprised projections acquired at approximately 90° to each other. This assured that measurement uncertainties due to image digitization were minimal. The following pseudo-code illustrates the procedure for generating an IR from the image pair (1,3):

```
/* Calculate 3-D coordinates of each point on each branch (true or artifactual)
   by considering all possible triangulations. */

main()
{
   FOR i = 1: 10
      project branch[i] into image #1;
      FOR j = 1: 10
         project branch[j] into image #3;
         do_triangulations( i, j );
         store intermediate_reconstruction_branch[i,j];
      END
   END
}

SUBROUTINE do_triangulations( i, j )
{
   FOR ( each point along branch[i] in image #1 )
      calculate epipolar line in image #3;
      IF ( epipolar line intersect branch[j] in image #3 ) XYZ = triangulate;
   END
}
```

The algorithm listed above generates a separate list of 3-D coordinates for each branch in the IR. For reference purposes, each list was indexed according to the branch-pair from which it was generated. For example, list[i,j] corresponds to the branch (in the IR) that was generated from the projection of the test object's branch[i] in one image and the projection of the test object's branch[j] in the other image. Clearly it is not so straightforward to obtain separate lists for each branch in practice. (This is discussed later.)

Upon being faced with the exercise of comparing each IR, it became apparent that there was no obvious criterion for doing so. Here we examined how the positions of each branch varied among the IRs. Specifically, for each point in list[i,j] comprising a branch in one IR, we determined the distance in 3-D to the closest point in a second IR. Subsequently, the mean of these distances for each branch was computed and used as a measure of its difference in position between the two IRs. (This procedure was repeated for six different pairs of IRs.) The rationale was that the points along a true branch should have consistently small (ideally zero) closest-point distances between IRs, while it was expected that the converse would not be true for artifactual branches. In other words, the artifactual branches should be distinguishable from trues since only the former are expected to vary in position among the IRs. Indeed, this was confirmed in our results.

3. RESULTS

For presentation here, we have selected the results obtained by comparing the IR generated from projections 1 and 3 with the one generated from projections 2 and 4. These two reconstructions are referred to below as IR_A and IR_B, respectively.

In Table 1 we list the mean closest-point distance for points along each branch in IR_A when compared to IR_B. Row and column indices correspond respectively to the indices i, j in the branch-lists described above. (The 10 branches in the test object yield a maximum of 100 branches in each IR.) Thus, each entry corresponds to a branch in IR_A with the main diagonal representing trues. In Figures 9 and 10 we plot the point-by-point closest-point distances for two branches; one true (list[1,1]) and one artifactual (list[0,3]). The corresponding entries in Table 1 represent the means of these plots.

From Table 1, it is readily observed that true branches are easily distinguished from the others. We note that these results were obtained without imposing any constraints on the reconstruction boundaries.

Tables 2-7 summarize the results obtained for the six different combinations of intermediate reconstructions, A and B. We list the mean and maximum mean closest-point distances for true branches, and the three smallest mean closest-point distances for artifactual branches. In every case, the ten lowest values correspond to the true branches of the test object.

3.1. Observations

The results presented above indicate that it is possible to distinguish between true and artifactual branches on the sole basis of mean closest-point distance between branches in IRs. In every case, it was possible to set a threshold to isolate the true structure from a given IR. While this is encouraging, we emphasize that the results are likely to be affected by the imaging geometry; particularly if the angle between projections is small. This would result in triangulation measurements that are likely to be corrupted severely by noise due to image digitization, hence making the distinction between true and artifactual branches less clear.

A related issue is the effect of using IRs that are generated from significantly different pairs of projections. In our simulations we employed four different images comprising two stereo-pairs. We do not consider the 7° angulation between each stereo-pair to constitute a significant difference and, consequently, we did not generate IRs from any stereo-pair. Using three projections, only three distinct IRs are available, with any pair possessing a common projection. As can be seen in Tables 2-7, this degeneracy appears as a decrease in contrast between the statistics for true and artifactual branches. Here the greatest difference between the maximum mean closest-point

Table 1: The 100 entries in this table represent the mean closest-point distance (in mm) of all points comprising each true and artifactual branch between 2 intermediate reconstructions (generated using projections (1,3) and (2,4), respectively). Null entries occur when triangulations were not possible; i.e. when there were no epipolar line intersections for the given branches. Note that the entries along the diagonal (corresponding to true branches) are one order of magnitude smaller than the others. This table is summarized in Table 4.

i\j	0	1	2	3	4	5	6	7	8	9
0	0.150	1.090	1.948	1.891	3.259	2.609	2.244	0.772	1.455	1.590
1	0.652	0.148	3.041	2.084	3.106	2.942	3.283	1.532	1.420	1.819
2	1.910	2.420	0.137	1.946	2.603	1.251	1.279	2.068	1.498	1.125
3	1.529	1.650	2.231	0.080	3.214	1.396	1.043	1.128	0.732	2.511
4	2.804	2.914	2.430	2.670	0.207	1.218	2.037	3.394	2.956	1.871
5	2.542	2.659	1.620	1.406	1.795	0.115	1.636	2.123	1.537	1.683
6	2.095	2.883	1.974	1.062	4.718	1.451	0.112	1.546	-	-
7	0.610	1.421	2.621	1.138	4.606	2.630	1.658	0.108	1.347	3.117
8	1.533	1.364	1.440	0.585	2.635	1.400	-	1.353	0.068	2.301
9	1.962	2.248	1.309	2.323	1.667	0.893	-	2.929	2.304	0.136

distance for a true branch and the minimum mean closest-point distance for an artifactual branch occurred when IR_A and IR_B were generated using distinct projections. This can be seen, for instance, by comparing this measurement in Tables 4 and 5 ($0.378mm$ and $0.366mm$, respectively) with Tables 2, 3, 6 and 7 ($0.040mm$, $0.034mm$, $0.063mm$ and $0.045mm$, respectively). We suspect that there are several mechanisms responsible for this behaviour; in particular, the fact that the projection axes were close to being within the same plane. This problem is less severe when the three projections are out of plane, but the extent to which the imaging geometry can be optimized is not presently clear.

Finally, we reiterate that the mean closest-point measurement may not be optimal for quantifying the differences between the IRs. By considering the mean, each point on a given branch influences the measurement equally. Since we are more concerned with points that deviate the most, it seems that these should receive a greater weighting and have more influence in quantifying the differences.

Since the relative spacing and number of points along a branch varies, a fitted curve or spline might be more robust in terms of describing its position, orientation, etc.. In this case, for example, separate branches in IRs could be compared in terms of the coefficients of their polynomials and significance tests performed to assess the variations. Ultimately we seek a criterion that will indicate whether a given branch exists in each reconstruction at the same position, with the same shape. Here the ability to recognize small deviations is important since it would allow us to handle more complex structures with closely packed branches. It is clear that a proper treatment of the issue will have to characterize the measurement uncertainties in each step of the procedure. This way we will be better able to select a criterion that is optimal in terms of distinguishing between true and artifactual branches.

4. DISCUSSION

Although we have demonstrated some encouraging results through the simulations described above, we feel it prudent to consider some issues concerning the possible implementation of this method in practice. A good perspective of these issues may be obtained by considering the difficulties in directly extending the methods used in our simulations to reconstruct more realistic data, like x-ray angiograms.

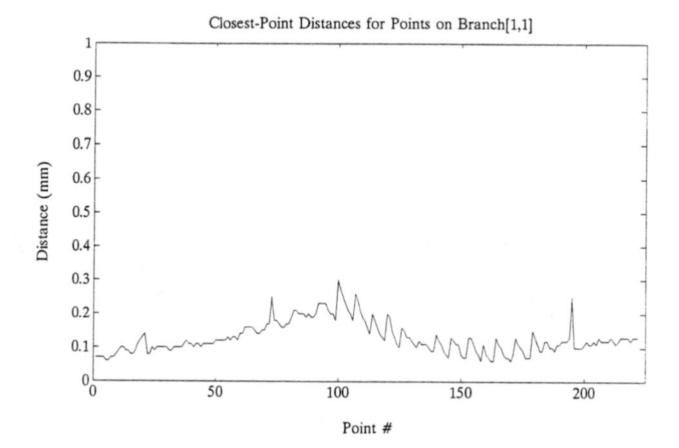

Figure 9: Plot of the closest-point distances for successive points along the (true) branch [1,1] in IR_A when compared to IR_B. The mean of this plot corresponds to the entry [1,1] in Table 1.

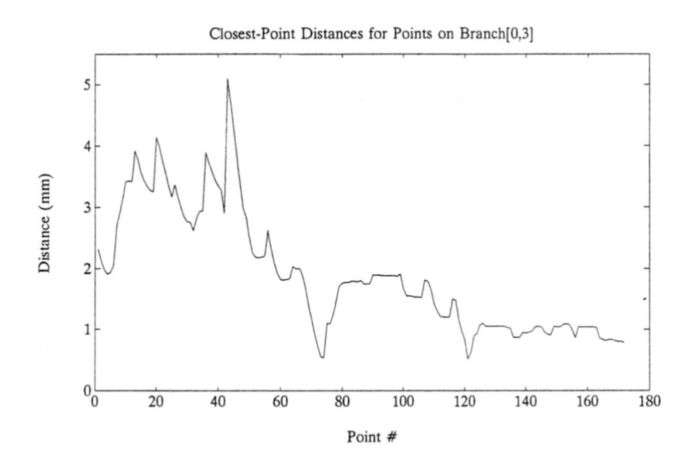

Figure 10: Plot of the closest-point distances for successive points along the (artifactual) branch [0,3] in IR_A when compared to IR_B. The mean of this plot corresponds to the entry [0,3] in Table 1.

Tables 2 - 7: Data for mean closest-point (C.P.) distance, between intermediate reconstructions (IR_A and IR_B), of points comprising true and artifactual branches. IR_A and IR_B were generated using the 2 projections indicated. In every case, the 10 true branches (numbered 0-9) had mean C.P. distances less than any artifactual branch. The 3 lowest mean C.P. distances for artifactual branches are given, along with the mean and maximum value observed for true branches.

2	$IR_A = (1,3)$ $IR_B = (1,4)$	
	Distance [mm]	Branch Indices
mean C.P. (trues)	0.004	N/A
maximum C.P. (trues)	0.008	5,5
1st min C.P. (artifacts)	0.048	2,0
2nd min C.P. (artifacts)	0.096	0,2
3rd min C.P. (artifacts)	0.106	3,5

3	$IR_A = (1,3)$ $IR_B = (2,3)$	
	Distance [mm]	Branch Indices
mean C.P. (trues)	0.119	N/A
maximum C.P. (trues)	0.191	4,4
1st min C.P. (artifacts)	0.225	1,0
2nd min C.P. (artifacts)	0.273	0,1
3rd min C.P. (artifacts)	0.428	8,3

4	$IR_A = (1,3)$ $IR_B = (2,4)$	
	Distance [mm]	Branch Indices
mean C.P. (trues)	0.126	N/A
maximum C.P. (trues)	0.207	4,4
1st min C.P. (artifacts)	0.585	8,3
2nd min C.P. (artifacts)	0.610	7,0
3rd min C.P. (artifacts)	0.652	1,0

5	$IR_A = (1,4)$ $IR_B = (2,3)$	
	Distance [mm]	Branch Indices
mean C.P. (trues)	0.121	N/A
maximum C.P. (trues)	0.190	4,4
1st min C.P. (artifacts)	0.556	1,0
2nd min C.P. (artifacts)	0.578	8,3
3rd min C.P. (artifacts)	0.610	7,0

6	$IR_A = (1,4)$ $IR_B = (2,4)$	
	Distance [mm]	Branch Indices
mean C.P. (trues)	0.129	N/A
maximum C.P. (trues)	0.198	4,4
1st min C.P. (artifacts)	0.261	1,0
2nd min C.P. (artifacts)	0.324	0,1
3rd min C.P. (artifacts)	0.456	3,8

7	$IR_A = (2,3)$ $IR_B = (2,4)$	
	Distance [mm]	Branch Indices
mean C.P. (trues)	0.013	N/A
maximum C.P. (trues)	0.028	4,4
1st min C.P. (artifacts)	0.073	2,0
2nd min C.P. (artifacts)	0.104	3,5
3rd min C.P. (artifacts)	0.115	7,0

Clearly the performance of the algorithm will be affected by the relative sparsity of the imaged structure. It is reasonable to expect, therefore, that the number of projections required to ensure accurate results should take into account the complexity of the imaged structure. Our particular interests lie in reconstructing the internal carotid arterial system of the brain as imaged using digital subtraction angiography. The ten branch model employed in the simulations above is, therefore, less complex than we expect to encounter in practice, but might be representative of a sub-system or less complex vascular structure.

The approach we have described here bears some resemblance to methods used in deriving the wire frames of objects in computer aided design and computer aided manufacturing (CAD/CAM) applications. Specifically, our notion of generating multiple solutions and discarding those that are inconsistent among two or more IRs is similar in concept to the methods described by Wesley et al (1981) for deriving wire frames of topologically legitimate objects that are consistent with the given projections. However, unlike our more general problem, there are a number of constraints in CAD/CAM applications that help to resolve ambiguities. These are typically related to the fact that the objects being reconstructed are composed of planar or curved surfaces and inherently possess some degree of symmetry in their arrangement. Markowsky et al (1980) and Wesley et al (1981) have formulated the reconstruction problem in terms of topological relationships between vertices and edges. The grouping of these entities to form an object is guided by the knowledge that certain configurations yield meaningless geometrical forms that cannot exist in the real world. Although this has allowed the reconstruction problem to be posed in a more tractable framework than has been possible in our case, the introduction of specific knowledge concerning characteristics of the structure to be reconstructed might facilitate the application of CAD/CAM methods to solve our problem. This could be accomplished, for instance, by characterizing branching patterns in some manner (as bifurcations only, for example), or by introducing model-based information to reduce the number of allowable configurations.

Throughout our discussion we have assumed that the object-to-image transformations are known for each projection, allowing each IR to be generated in absolute coordinate space. In practice, these transformations may be obtained by calibrating the imaging apparatus before acquiring the projections (Sobel, 1974), or by deriving the geometry based on the positions of six or more landmarks in each image (Sutherland, 1974). (We employ the latter method in stereotactic surgery planning.) Alternatively, a 3-D object-oriented coordinate system may be derived from two images in which at least eight corresponding points are visible (Fencil et al, 1990 and Metz et al, 1989). Since these transformations have a direct bearing on the accuracy of triangulation calculations, it will be necessary to establish the relationship between the transformations and the subsequent criterion for comparing IRs.

The statement that there is a low likelihood of artifactual branches coinciding perfectly in two different reconstructions presumes a certain degree of sparsity and lack of symmetry in the imaged structure. Clearly, branch length will also influence the ability to recognize artifactual structures: The greater the number of points whose positions deviate between two IRs, the greater the likelihood that the branch will be recognized as artifactual. When the complexity of the imaged structure is such that there are artifactual branches that cannot be distinguished from trues, it becomes necessary to introduce additional knowledge or constraints beyond those already considered. In reconstructing coronary vessels from projections, the use of model-based information has been applied successfully, although using an approach different from the one described here (Smets et al, 1989 and Smets, 1990). In other applications it may be possible to employ less specific knowledge such as allowable branching configurations or some sort of hierarchical relationship.

The importance of acquiring projections that are 'significantly different' is best appreciated by considering the result of generating an IR when a branch of the imaged object is in the same plane as the axes of the two projections. In this case, the triangulation procedure yields a group of points comprising a plane instead of a branch. Since the true branch (or a portion of it) can occur anywhere in this plane, it is necessary for at least one additional projection to be acquired out-of-plane. This way, even if the true branch lies within a plane in a second IR, the common intersection of the two planes allows it to be isolated. Here, as elsewhere, *a priori* knowledge concerning the imaged structure might allow certain 'optimal' projections to be acquired. In this manner, the occurrence of such planes might be minimized in order to facilitate comparisons between IRs.

A third issue concerns the accurate delineation of an imaged structure that may actually have varying thickness. This presumes that we wish to reconstruct only a 3-D skeletal representation of the structure. Since several methods for delineating vessels have been described in the literature (Collorec et al, 1988, Eichel et al, 1988, Hoffmann et al, 1987, Nguyen et al, 1986, Smets et al, 1988, Suetens et al, 1987 and Sun, 1989), more relevant questions concern the performance of the algorithm in the presence of artifacts (objects that are not part of the imaged structure), or given imperfectly delineated branches. Since artifacts are expected to be distinct from the main delineated structure, they may be discarded before generating the IRs. When this is not possible, the artifact will yield a branch in one IR that is highly unlikely to exist in the other(s), thus allowing it to be discarded in the usual fashion.

When the IR is fragmented, it becomes necessary to consider each segment individually in terms of its relation to each IR. Thus, the connectivity constraint can no longer be applied as effectively, resulting in branch segments that are difficult to distinguish as true or artifactual. An inaccurately delineated vessel, on the other hand, will introduce a bias in the positions of the true and artifactual branches it spawns. While this may not affect the recognition of artifactual branches, it may lead to the interpretation of the corresponding true branch as artifactual. However, it is not expected that only one branch would be inaccurately delineated in this manner; meaning that all branches in each IR would be affected to the same degree; i.e. raising only the background noise in the measurements between IRs.

Until now we have assumed that it is possible to isolate any list of points comprising a given branch from the entire set of points describing an IR. In reality reconstruction via triangulation yields a single list of 3-D coordinates, without order, generated without knowledge of point-to-point connectivity or branch identity. We recall that this problem was circumvented in our simulations by generating separate lists for each branch. In practice we require some method for identifying points as comprising branches; i.e. to establish connectivity. To this end, we are currently investigating the potential for using a minimum spanning tree (MST) (Shamos, 1978), or some adaptation thereof. Conceptually, the MST is well suited to this problem since the point-to-point spacing along the structure is considerably smaller than the spacing between branches (assuming a relatively sparse branching structure). Alternatively, performing the reconstruction within a finely sampled volume would circumvent the problem of directly parameterizing the imaged structure and allow a high-level 3-D object detection scheme to elucidate the separate branches. This approach seems better suited to handle the problem of reconstructing grey-scale images containing branches that have varying thicknesses. In any case, it is clear that this problem must be addressed in greater detail.

Finally, although the simulation results indicate that it is possible to distinguish between true and artifactual branches, it is not clear how to select the threshold between the two. If the number of true branches is known *a priori*, then there is no problem. However, it is more realistic to assume that the number is not known and that the threshold will vary depending on such factors as the measurement uncertainties, imaging geometry and the number of projections employed. Nevertheless, depending on the application, it may not be crucial to discard every artifactual branch, in which case the selection of a threshold could be quite flexible, or even interactive.

5. CONCLUSIONS

The problem of reconstructing 3-D data from a small number of projections is inherently underdetermined and must be solved by introducing *a priori* knowledge or assumptions concerning structural properties of the imaged object. Here we have established a conceptual basis for reconstructing in 3-D a branching object from three or more of its distinct projections. Fundamental to this technique is the assumption of connectivity. Although a constraint on the boundaries of reconstruction is easily accommodated, it is not essential. In fact, the technique readily admits artifactual structures and exploits their properties to resolve several ambiguities in the reconstruction process.

Unlike techniques that isolate portions of the data during reconstruction or attempt to obtain a direct solution from the images, we generate two or more intermediate results (complete reconstructions) and impose connectivity constraints in 3-D. Here we emphasize that connectivity cannot be applied optimally when dealing with fragmented structures. For this reason, we generate two or more reconstructions from different pairs of projections, then discard branches that are inconsistent in the sense that they do not occupy the same position in each reconstruction. Since only artifactual branches will be inconsistent, we are assured that the true structure will remain intact. Furthermore, we have suggested that true and artifactual branches may be distinguished using as few as three distinct projections. We base this assertion on the apparent low likelihood of an artifactual branch coinciding, for its whole length, with an other branch in each intermediate reconstruction.

The simulation results presented support our line of reasoning. It was demonstrated that every artifactual branch exhibited a larger deviation in position between two reconstructions than the worst corresponding measurement for a true branch. We note, however, that these results are expected to vary depending upon the projection geometry and complexity of the imaged object (both not studied here). Further refinements in the technique are expected to be obtained by formally characterizing several sources of measurement error.

In anticipating the practical application of these ideas, we foresee the need to further address the issue of connectivity. In particular, it is not presently clear how to isolate individual branches within intermediate reconstructions. The extension to include branch thickness and cross-section is also unclear, but might be accomplished using adjunct techniques that first require estimates of branch centre-lines. Despite these issues, the approach we have described is straightforward and flexible, requiring a minimum of constraints. The route to obtaining a solution is clear and avoids forming premature or subjective decisions based on an incomplete analysis of the data.

At present, we are investigating the practical application of the concepts described in this paper to reconstruct wire frame outlines of vascular structures from their projections. Because several aspects of this problem are common to other areas as well, notably CAD/CAM applications, we expect the approach described here to be one of several used in concert to obtain a solution.

Acknowledgements

We wish to thank Drs. A. Evans and A. Olivier for their enthusiastic participation and helpful insights into several aspects of the work described here. We gratefully acknowledge the financial support provided by Siemens Medical Systems, the Medical Research Council of Canada, Fonds pour la Formation de Chercheurs et L'Aide à la Recherche du Québec, and Fonds de la Recherche Santé du Québec.

References

Henri CJ, Peters TM, Lemieux L and Olivier A (1990). Experience with a computerized stereoscopic workstation for neurosurgical planning. In: Proc. First Conf. on Visualization in Biomedical Computing. IEEE Computer Society Press, Los Alamitos CA, pp. 450-457.

Peters TM, Clark JA, Pike GB, Henri C, Collins L, Leksell D and Jeppsson O (1989). Stereotactic neurosurgery planning on a personal-computer-based work station. Journal of Digital Imaging 2:75-81.

Baker HH and Binford TO (1981). Depth from edge and intensity based stereo. In: Proc. Seventh Intl. Joint Conf. on Artificial Intelligence. William Kaufmann Inc., Los Altos CA, pp. 631-636.

Barnard ST and Thompson WB (1980). Disparity analysis of images. IEEE Trans. Pattern Anal. Machine Intell. PAMI-2:333-340.

Barnard ST and Fischler MA (1987). Stereo vision. In: Encyclopedia of Artificial Intelligence. Shapiro SC (ed), Wiley, New York, pp. 1083-1090.

Collorec R and Coatrieux JL (1988). Vectorial tracking and directed contour finder for vascular network in digital subtraction angiography. Pattern Recognition Letters 8:353-358

Duda RO and Hart PE (1973). Pattern Recognition And Scene Analysis. Wiley, New York.

Eichel PH, Delp EJ, Koral K and Buda AJ (1988). A method for a fully automatic definition of coronary arterial edges from cineangiograms. IEEE Transactions on Medical Imaging 7:313-320.

Fencil LE and Metz CE (1990). Propagation and reduction of error in three dimensional structure determined from biplane views of unknown orientation. Medical Physics 17:951-961.

Hoffmann KR, Doi K, Chan HP and Chua KG (1987). Computer reproduction of the vasculature using an automated tracking method. Proceedings of SPIE Medical Imaging 767:449-453.

Horaud R and Skordas T (1989). Stereo correspondence through feature grouping and maximal cliques. IEEE Trans. Pattern Anal. Machine Intell. PAMI-11:1168-1180.

Levine MD, O'Handley DA and Yagi GA (1973). Computer determination of depth maps. Computer Graphics and Image Processing 2:131-150.

Markowsky G and Wesley MA (1980). Fleshing out wire frames. IBM J. Res. Develop. 24:582-597.

Mawko GM (1989). Three-dimensional analysis of digital subtraction angiograms for stereotactic neurosurgery planning. PhD Thesis, McGill University.

Metz CE and Fencil LE (1989). Determination of three-dimensional structure in biplane radiography without prior knowledge of the relationship between the two views:theory. Medical Physics 16:45-51.

Mohan R, Medioni G and Nevatia R (1989). Stereo error detection, correction, and evaluation. IEEE Trans. Pattern Anal. Machine Intell. PAMI-11:113-120.

Nguyen TV and Sklansky J (1986). Computing the skeleton of coronary arteries in cineangiograms. Computers and Biomedical Research 19:428-444.

Prazdny K (1983). Stereoscopic matching, eye position, and absolute depth. Perception 12:151-160.

Shamos MI (1978). Computational Geometry. PhD Thesis, Yale University.

Smets C, Verbeeck G, Suetens P and Oosterlinck AJ (1988). A knowledge-based system for the delineation of blood vessels on subtraction angiograms. Pattern Recognition Letters 8:113-121.

Smets C, Suetens P, Oosterlinck A and Van der Werf F (1989). A knowledge-based system for the labeling of the coronary arteries. In: Computer Assisted Radiology, Proc. 3rd Intl. Symp. CAR. Lemke HU (ed), Springer-Verlag, Berlin, pp. 322-326.

Smets C (1990). A knowledge-based system for the automatic interpretation of blood vessels on angiograms. PhD Thesis, Leuven University.

Sobel I (1974). On calibrating computer controlled cameras for perceiving 3-d scenes. Artificial Intelligence 5:185-198.

Suetens P, Van Cleynenbreugel J, Fierens F, Smets C and Oosterlinck A (1987). An expert system for blood vessel segmentation on subtraction angiograms. Proceedings of SPIE Medical Imaging 767:454-459.

Sun Y (1989). Automated identification of vessel contours in coronary arteriograms by an adaptive tracking algorithm. IEEE Transactions on Medical Imaging 8:78-88.

Sutherland IE (1974). Three-dimensional data input by tablet. Proceedings of the IEEE 62:453-471.

Wesley MA and Markowsky G (1981). Fleshing out projections. IBM J. Res. Develop. 25:934-954.

PRELIMINARY EXAMINATION OF THE USE OF CASE SPECIFIC MEDICAL INFORMATION AS "PRIOR" IN BAYESIAN RECONSTRUCTION

J Llacer,[1] E Veklerov[1] and J Nuñez[2]

[1]Engineering Division

Lawrence Berkeley Laboratory, University of California

Berkeley, CA 94720

and

[2]Departament d'Astronomia

Facultat de Fisica, Universitat de Barcelona

Barcelona, Spain

Abstract

In this paper we attempt to answer one specific question regarding the role of "prior" distributions in Bayesian reconstructions in Emission Tomography. The question is: Can prior information on some areas of an imaging field improve the results of a reconstruction in other areas of the same image? We answer the question for the specific case of a simple image with a large square of activity surrounding a smaller square, with different ratios of activity in the two regions. For the case of ratios of 10:1 we find that feasible MLE reconstructions exhibit some effect in the internal region due to the presence of the external region and that those effects can be reduced by a factor of ~ 0.5 by using as prior information the known mean of activity in the large outer region. Two methods of obtaining reconstructions with pixel dependent constraints are developed in order to obtain the above results. The conclusion of the study is, principally, that "priors" should have a local action, i.e., should apply directly to the region of interest if a substantial improvement in reconstruction quality is to be achieved.

Keywords

Bayesian reconstruction; Prior information; Medical prior information; Maximum a posteriori (MAP); Successive Substitutions.

1. INTRODUCTION

In this paper we examine some preliminary questions that arise when one attempts to use medical evidence that is specific to the patient case under consideration as "prior" in Bayesian reconstruction. The use of prior information in medical image reconstruction by statistical parameter estimation methods has been principally restricted to the context of regularization: an entropy prior can be used to obtain images which are feasible at convergence (Nuñez and Llacer, 1989 and Nuñez and Llacer, 1990), the use of a smoothed Filtered Backprojection (FBP) image from a given set of projection data can be used as prior to require that a Maximum Likelihood Estimator (MLE) image from the same data set does not deviate more than a certain amount from the prior (Levitan and Herman, 1987), and a Gibbs energy function can be used as prior to select solutions that are smooth, except at boundaries, a direction of research that was initiated by Geman and Geman (1984) and has been followed by numerous workers.

More recently, however, attempts are being made to utilize anatomical information of a specific patient obtained by X-ray Computerized Tomography (XCT) or Magnetic Resonance Imaging (MRI) techniques in order to obtain prior information that can result in improved Emission Tomography (ET) reconstructions of the same patient (Chen et al., 1990). Work in that direction can go beyond regularization and, although it can meet with strong difficulties, we feel it promises to be able to tap medical knowledge that, at present, is not utilized at the time of image reconstruction.

One question that comes up early in trying to utilize medical knowledge of a specific patient as prior for a Bayesian reconstruction is whether that knowledge should be about the specific Region-of-Interest (ROI) that a physician may want to concentrate on, or it can be knowledge about other regions in the image and still benefit the results in the ROI. We can rephrase the question as: Can prior information on some areas of an imaging field benefit the reconstruction of other areas? Since the tomographic image reconstruction problem is totally connected (all data points influence the reconstructed values of any given pixel), one would expect the answer to the above question to be affirmative.

In this paper we attempt to answer the above question and determine the extent to which prior knowledge about a region can influence the quality of reconstruction in another region. We begin by showing that the existence of large, high activity areas in an image does influence pixel values in low activity regions of an image, although the effect can be relatively small with MLE reconstructions. We proceed by modifying the Bayesian algorithm of Levitan and Herman (1987) so that it can use prior information on a particular region that does not include the ROI, allowing the ROI to be reconstructed freely. We then examine the results of such reconstructions. A similar investigation is carried out with a modified form of the Bayesian algorithm of Nuñez and Llacer (1990) in which a cross-entropy prior is allowed to act in the region where prior information is well known, leaving a ROI to be reconstructed by an essentially unconstrained MLE. We conclude by discussing the implications of the findings.

2. INTER-REGION EFFECTS

In this section we show our findings regarding the effect of the activity in one region on the standard deviation of another region, which we will take to be smaller than the first one. Figure 1 shows the phantom that we have simulated by computer, with the parameters of a CTI-831 PET tomograph (single slice, no attenuation). The larger outer square will be labeled region "a" and the smaller inner square will be region "b". The areas sampled, in order to determine the statistical parameters resulting from the reconstructions, were large regions away from boundaries by at least five pixels. Eight different activity ratios have been simulated, keeping a fixed activity per pixel in region "b". The first phantom (0:1 ratio) had a mean activity of 9.92 counts per pixel in "b", with a total of 17,506 counts, and 0 counts in "a". The specific count distribution of region "b" in the 0:1 phantom and the resulting projection data were kept intact when generating the other seven phantoms. The latter ones were obtained by adding counts to the data set only in region "a". In a reconstruction with no influence of one region upon the other, we would expect the statistical parameters and the two-dimensional "noise shapes" recovered in region "b" to be invariant with changes in the number of counts in region "a". Table I shows the statistical parameters measured in the two regions for the eight phantoms as a function of the activity ratios between regions "a" and "b". The columns for "Source distribution" correspond to the results of the specific Poisson realizations of the phantom of Figure 1, with σ values very close to the square root of the region means. The FBP reconstructions were carried out with a Butterworth filter with parameters such that the edge response, measured by averaging 100 rows of the reconstructed images, showed practically identical slopes to the MLE images in the 5% to 95% region of the activity step. The MLE images were feasible images obtained by the Post-filtering method, i.e., iterating past the onset of feasibility and filtering back to feasiblity with a small two-dimensional Gaussian kernel. We have defined feasible images as those images that, if they were true radioisotope distributions in a patient, could have given the initial scan data by Poisson processes with means defined by the image values (Veklerov and Llacer, 1987 and Llacer and Veklerov, 1989). For activity ratios 0:1 to 2:1 we iterated to 150% iterations with respect to the onset of feasibility,

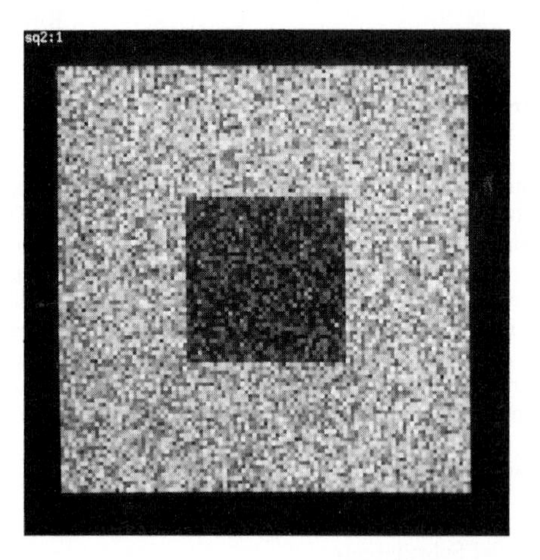

Fig. 1: Phantom used in the simulations, showing outer region "a" and inner region "b". The picture corresponds to the case with a ratio of 2:1 in the number of counts in the two regions.

filtering with σ = 0.65 pixels. At higher activity ratios, we had to iterate to 250% in order to keep the bias in the sampled area of region "b" below that of the FBP reconstruction. Filtering was with a kernel of σ = 1.0 pixel. Figure 2 compares the average edge response of the FBP - Butterworth method of reconstruction with the MLE at 250%, σ = 1 pixel, for the 10:1 activity ratio. The edge response for the MLE at the onset of feasibility is also shown.

With regard to the measurement of the onset of feasibility in the MLE method, we would like to point out that one gets an unambiguous indication of that onset during the iterative process by using the test of Veklerov and Llacer (1987) with computer simulated data, as the ones shown here. The actual range of iterations signaled by the test depends, however, on whether one choses to test all the tubes (detector pairs) in a data set or only those that have counts in the projection data. The results described in this section have used the latter criterion, which results in earlier stopping but requires iterating beyond it for reasonably unbiased images.

Table I

Statistical parameters measured in the different reconstructions of the phantom of Fig. 1
(mean m and standard deviation σ are in counts)

Activity ratio	Source distribution				FBP (Butterworth)				MLE Post-filtered			
	Region "a"		Region "b"		Region "a"		Region "b"		Region "a"		Region "b"	
a:b	m	σ	m	σ	m	σ	m	σ	m	σ	m	σ
0:1	0.0	0.0	9.958	3.177	-0.006	1.91	10.10	2.66	.041	0.092	10.32	1.22
0.1:1	.978	.991	9.958	3.177	1.04	2.36	9.95	3.09	1.01	0.32	10.06	1.19
0.2:1	2.066	1.48	9.958	3.177	1.99	2.75	10.10	3.11	1.96	0.53	10.08	1.07
0.5:1	4.90	2.166	9.958	3.177	4.86	3.40	10.36	3.80	4.72	0.71	9.98	0.82
1:1	10.04	3.14	9.958	3.177	9.73	4.36	10.37	4.36	9.74	1.26	10.13	0.97
2:1	19.77	4.44	9.958	3.177	19.07	6.37	10.60	4.73	19.66	2.66	10.20	1.15
5:1	48.78	7.12	9.958	3.177	47.91	9.38	10.46	8.38	49.76	5.60	10.41	1.75
10:1	98.99	9.65	9.958	3.177	95.37	14.01	11.64	12.78	98.82	10.78	11.48	2.35

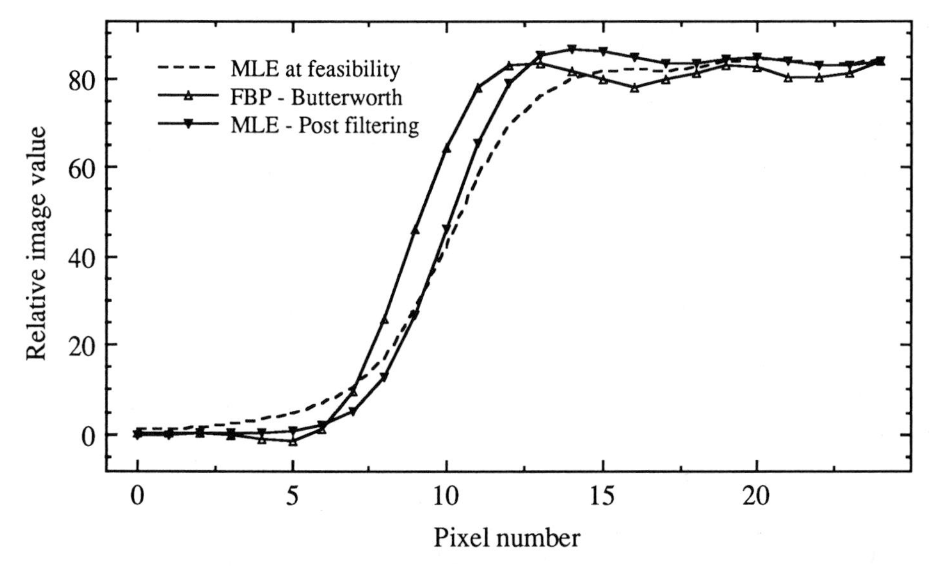

Fig. 2: Edge response of MLE at feasibility, FBP with Butterworth filter and MLE at 250% number of iterations with respect to the onset of feasibility, after post-filtering with a two dimensional Gaussian kernel of σ=1 pixel. The edges for MLE reconstructions are displaced to the right with respect to the FBP results because the former does not assume space invariance of the point response function and the reconstructed objects are somewhat different in size when comparing the two methods of reconstruction.

Figure 3 shows the mean values obtained in region "b" with error bars indicating ± 1 standard deviation, for the different activity ratios. The sensitivity of σ in region "b" to counts in "a" is very strong in the FBP case. It is possible that using sophisticated methods of interpolation in the backprojection step may result in more favorable reconstructions. The backprojection algorithm used (subroutine BIN) (Huesman et al., 1977) calculates the contributions to the backprojection at the center of pixels by linear interpolation between the projection bin values. The feasible MLE results show a considerably smaller value of σ, but it does increase when the ratio of activities between regions increases above 2:1. Using a larger filter kernel in the Post-filtering step would result in decreased values of σ but the step response of the system would be slower than in the FBP case.

From the above results we can draw the conclusion that by selecting the parameters of a feasible MLE reconstruction with care, the influence of a high activity, large region in an image on the statistical characteristics of a low activity region can be minimized, but not removed. We will now proceed to investigate whether prior knowledge of the true mean of region "a" can result in a further reduction in the standard deviation in region "b".

3. THE LEVITAN-HERMAN ALGORITHM

3.1. EM implementation

In this section we abstract the work of Levitan and Herman (1987) and indicate a possible limitation when we attempt to use the resulting algorithm for the purpose of Bayesian reconstruction in which prior information is only known for part of the image. The Maximum *a posteriori* probability (MAP) $P_B(x|y)$ and the maximum likelihood (ML) $P_L(y|x)$ are related by Bayes' equation:

$$P_B(x|y) = P_L(y|x) \cdot P_A(x) / P(y), \tag{1}$$

with $P_A(x)$ and $P(y)$ being the *a priori* probability distributions of the image and measurements vectors x and y, respectively. The logarithm of the *a posteriori* probability is, excluding terms independent of x,

$$\ln P_B(x|y) = \sum_{i=1}^{I}\left[y_i \ln\left(\sum_{j=1}^{J} a_{ij}x_j\right) - \sum_{j=1}^{J} a_{ij}x_j\right] - \tfrac{\gamma}{2}\sum_{u,v=1}^{J}(x_u - m_u)H_{uv}(x_v - m_v) \tag{2}$$

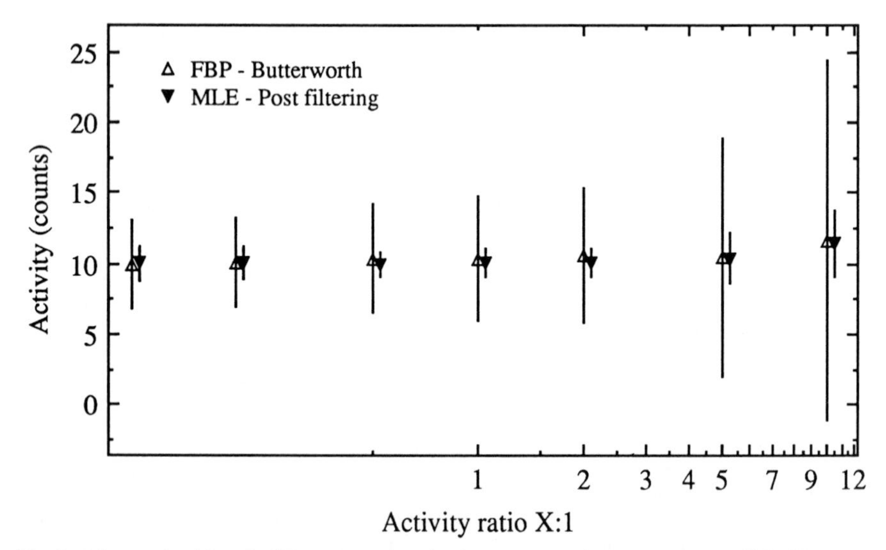

Fig. 3: Mean and ± 1 Standard Deviation values for the estimates of activity in Region "b" of Fig. 1 as a function of activity ratios.

where a_{ij} are the elements of the transition matrix of the reconstruction problem, I is the number of data points and J is the number of pixels to be estimated. The first summation in (2) corresponds to the log likelihood for the estimation problem and the second term corresponds to a penalty function of quadratic form (Gaussian before taking the logarithm) which consists of a nonsmoothness measure of the image x which has to be minimzied simultaneously with the maximization of the likelihood function. The weight factor γ provides an adjustable parameter determining the relative strength of the likelihood vs. the penalty function and the values of vector m correspond to the means of the known prior distribution from which we do not want to deviate more than a certain amount. The amount of deviation allowed is controlled by the elements of the matrix H, which, in fact, specify the covariance matrix of a Gaussian distribution. Like Levitan and Herman, we consider the case in which matrix H is diagonal, so that the penalty function:

$$\sum_{u,v=1}^{J}(x_u - m_u)H_{uv}(x_v - m_v) = \sum_{j=1}^{J} h_j\left(x_j - m_j\right)^2 . \tag{3}$$

The implementation of the EM algorithm from the resulting log *a posteriori* probability requires solving a set of uncoupled quadratic equations for x_j

$$-1 + \frac{x_j^{(k)}}{x_j}\sum_{i=1}^{I}\frac{a_{ij}y_i}{\sum_{n=1}^{J} a_{in}x_n^{(k)}} - \gamma h_j(x_j - m_j) = 0, \tag{4}$$

where the superscript (k) indicates the value of an estimate in the kth iteration. The iterative formula resulting is the following:

$$x_j^{(k+1)} = \left[\gamma h_j m_j - 1 + \sqrt{\left(\gamma h_j m_j - 1\right)^2 + 4\gamma h_j \hat{x}_j^{(k+1)}}\right] / 2\gamma h_j \tag{5}$$

where

$$\hat{x}_j^{(k+1)} = x_j^{(k)}\sum_{i=1}^{I}\frac{a_{ij}y_i}{\sum_{n=1}^{J} a_{in}x_n^{(k)}} . \tag{6}$$

The intermediate estimate $\hat{x}$ of (6) results in a step identical to the Maximum Likelihood Estimator (MLE) method of Shepp and Vardi (1984). The final value x is a modification of the intermediate estimate depending on the values of γ, h_j and m_j.

The algorithm of (5) and (6) does not conserve counts. As Levitan and Herman indicate, a Lagrange multiplier can be introduced in the target function to be maximized to impose the condition

$$\sum_{j=1}^{J} x_j^{(k)} = \sum_{i=1}^{I} y_i . \tag{7}$$

The solution of the resulting imaging problem by the EM algorithm requires the solution of J cubic equations, which is not attempted by those authors. Nevertheless, using the algorithm of (5) and (6) for reconstructions in which the covariance parameter h_j is a constant independent of j, i.e., allowing the same deviation from the means m_j for all the pixels, results in the desired regularized results, as published in Levitan and Herman (1987). The image can be normalized after the procedure has converged, if desired.

Our intention has been to use the Bayesian formulation of Levitan and Herman to test the case in which the mean values m_j are well known for some region or regions of the image, but not known in a particular ROI. In that case we can attempt a reconstruction in which the values h_j are large for the regions that are well known (allowing for small differences between the estimates and the means) and small for the pixels in the ROI. Under conditions of non-conservation of counts, the meaning of the specified means m_j and parameters h_j may be lost if there is a general drift in the reconstructed pixel values. For that reason we formulated a solution to the maximization problem by the method of Successive Substitutions (SS) which can easily incorporate the requirement of conservation of counts.

3.2. Successive Substitution Method

The SS method (Hildebrand, 1974) was used successfully by Nuñez and Llacer (1990) to obtain an iterative formula for a Bayesian reconstruction algorithm with entropy prior. In the present context, it has allowed the maximization of the log *a posteriori* expression (2) after incorporating the condition of conservation of counts. The new function to maximize is:

$$\ln P_B(x|y) = \sum_{i=1}^{I}\left[y_i \ln\left(\sum_{j=1}^{J} a'_{ij}x_j\right) - \sum_{j=1}^{J} a'_{ij}x_j\right] - \frac{\gamma}{2}\sum_{u,v=1}^{J}(x_u - m_u)H_{uv}(x_v - m_v) - \mu\left(\sum_{j=1}^{j} x_j - \sum_{i=1}^{I} y_i \Delta y_i\right) \tag{8}$$

in which the matrix elements a'_{ij} have been corrected for γ-ray absorption and detector non-uniformity by the introduction of the parameters Δy_i, as follows:

$$a'_{ij} = a_{ij} / \Delta y_i \tag{9}$$

and the term from the conservation of counts has been likewise corrected.

Setting the partial derivatives with respect to x_j equal to zero, we obtain the general equation

$$\sum_{i=1}^{I}\left[-a'_{ij} + \frac{y_i a'_{ij}}{\sum_{n=1}^{J} a'_{in} x_n}\right] - \gamma\sum_{u=1}^{J} H_{iu}(x_u - m_u) - \mu = 0 \tag{10}$$

Separating μ to the right hand side, and adding a constant C, raising to a power P and multiplying by x_j both sides, we obtain

$$x_j\left[\sum_{i=1}^{I}\frac{a'_{ij}y_i}{\sum_{n=1}^{J} a'_{in}x_n} - \sum_{i=1}^{I} a'_{ij} - \gamma\sum_{u=1}^{J} H_{iu}(x_u - m_u) + C\right]^P = x_j(\mu + C)^P \tag{11}$$

The iterative algorithm is then

$$x_j^{(k+1)} = Kx_j^{(k)}\left[\sum_{i=1}^{I}\frac{a'_{ij}y_i}{\sum_{n=1}^{J} a'_{in}x_n^{(k)}} - \sum_{i=1}^{I} a'_{ij} - \gamma\sum_{u=1}^{J} H_{iu}(x_u - m_u) + C\right]^P. \tag{12}$$

The constants C and P are arbitrary, in principle, but they have a well defined role, as discussed in detail in Nuñez and Llacer (1990). C is a constant that insures that the expression between brackets remains positive for all pixels and P is an acceleration parameter. The final image is independent of C and P, but both affect convergence rates. Convergence is fastest when C is as small as possible consistent with not having negative values for the expression between brackets in (12). We have obtained excellent convergence for the range $1 \le P \le 3$, with a speed up which is approximately proportional to that value. The constant K is given by:

$$K = \frac{1}{(\mu + C)^P}. \tag{13}$$

Evaluation of K at each iteration to preserve the total number of counts is equivalent to evaluating the Lagrange multiplier μ as one more parameter to be estimated during the reconstruction. The value of K changes slightly during the initial iterations and it becomes stable further on. We have implemented the algorithm (12) for the case of diagonal H, using a value of the acceleration parameter P equal to unity.

3.3. Results of Simulations

The phantom of Figure 1 with a 10:1 activity ratio between regions "a" and "b" has been used for the experiments to be described in this section. Table I and Figure 3 show that at that activity ratio, the standard deviation in region "b", σ_b, is larger than at lower activity ratios. The value of σ_b can always be made smaller by filtering, but that would result in a worse edge response than that shown in Figure 2, which would be undesirable. We want to show here that knowledge of the correct mean values in region "a", m_a, can result in a decreased value of σ_b.

We use the iterative formula (12) for the case in which matrix H is diagonal, but with different values for regions "a" and "b", namely h_a and h_b. Since γ and the elements of H are always multiplied together, we set $\gamma = 1$, without loss of generality, and modify only the values of H. In all cases we have set $h_b = 0$, i.e., we have not imposed any prior on the reconstructed values for pixels in region "b". The measurements of statistical parameters in the two regions have been carried out in large areas away at least 5 pixels from the region boundaries. The results of the reconstructions have not been post-filtered, so as not to introduce one more variable which would obscure the results we are trying to observe. The hypothesis testing parameter H for feasibility has been calculated for complete data sets, rather than only for tubes with counts in the projection data.

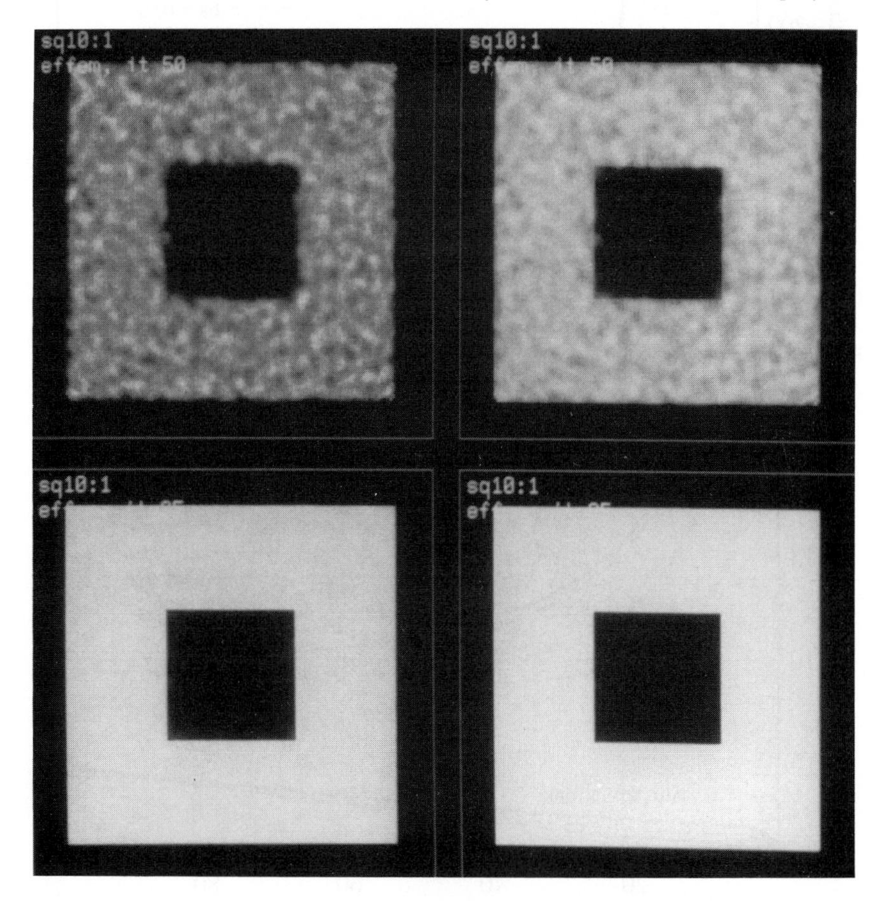

Fig. 4: Reconstructions of the phantom of Fig. 1 with four different values of the covariance parameter h_a. a) top left is for $h_a = 0.0$, which is the standard MLE method, b) top right is for $h_a = 0.001$, c) bottom left is for $h_a = 0.01$ and d) bottom right is for $h_a = 0.03$.

3.3.1. Effect of h_a on feasibility, m_b and σ_b

Figures 4a, b, c and d show the reconstructions obtained with h_a = 0.0 (standard MLE), 0.001, 0.01 and 0.03, respectively. They correspond to an iteration number near the onset of feasibility (it. 50 for h_a = 0.0 and 0.001 and it. 25 for h_a = 0.01 and 0.03). The mean value m_a that has been given as prior information is 99.24 counts, which is the mean value used in the generation of the phantom. Figure 4 shows the tightening of the estimate of activity in region "a" about its mean as h_a is made larger. The effect of h_a on the statistical parameters of region "b" will be shown below.

Since the original phantom image is feasible (it generated the projection data by a Poisson process), feasibility is obtained at a lower iteration number when h_a is made larger. Figure 5 shows the hypothesis testing parameter H described in Veklerov and Llacer (1987) as a function of iteration number for the four reconstructions of Figure 4. When the value of H is below ~ 36, the image can be considered feasible with 99% confidence level.

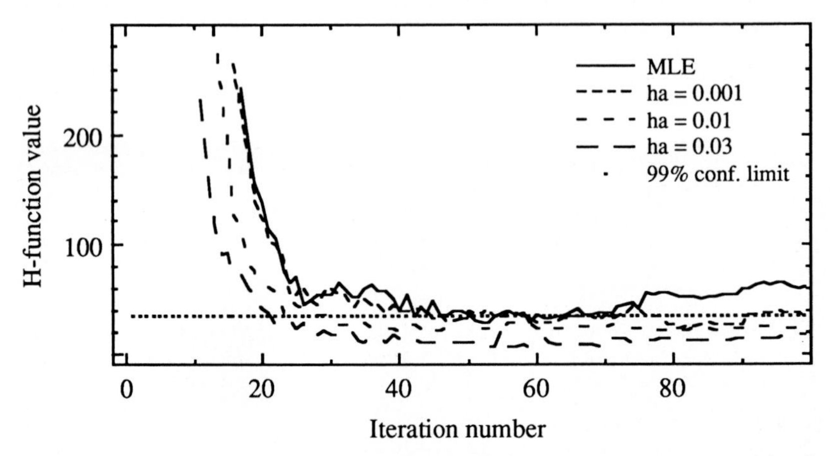

Fig. 5: Value of the feasibility hypothesis testing function H as a function of iteration number for the four reconstructions of Fig. 4. Iteration numbers in which H is below the 99% confidence level can be considered to result in feasible images.

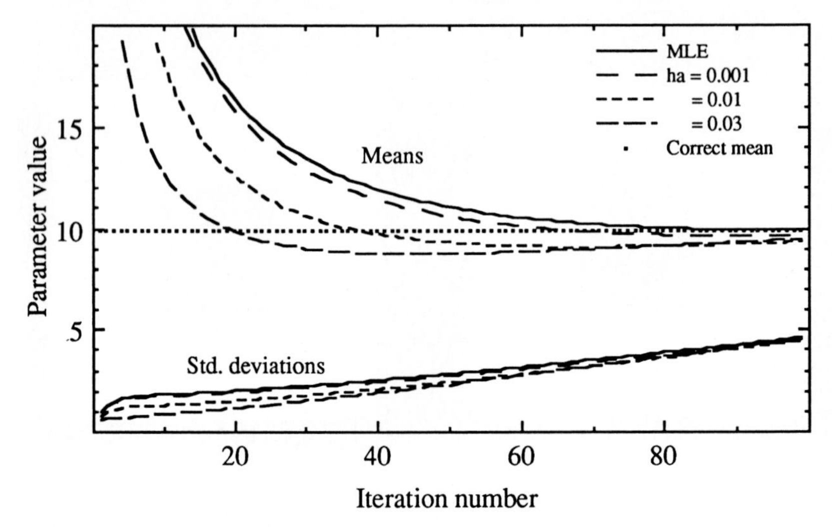

Fig. 6: Means and std. deviations measured in region "b" for the reconstructions of Fig. 4 as a function of iteration number for the different values of the covariance parameter for region "a".

Figure 6 shows the means and standard deviations obtained in region "b" for the different reconstructions. Errors in the estimated means are found in all cases. The standard MLE reaches the correct value for m_b in the vicinity of it. 100, but it does not stay at that value. There is an upwards creeping effect that continues up to it. 500, at least. When using prior information in region "a", it appears that there is a region of acceptably low bias at iteration numbers near the onset of feasibility (calculated with all tubes). At higher iteration numbers it also follows an upward trend, at least up to it. 500. It must be indicated that these results are for one single realization of the phantom of Figure 1. A proper study of bias would require reconstructions of a number of independent realizations of the phantom.

When the prior information is used with substantial strength, the standard deviations σ_b are noticeably smaller in the regions near feasibility, although at high iteration values the differences become small. With $h_a = 0.03$, corresponding to constraining the counts of region "a" to a std. deviation of 4.08 counts around a mean of 99.24, σ_b improves by a factor of two compared with the standard MLE, both at feasibility.

3.3.2. Sensitivy of results on m_a

It is reasonable to expect that specifying an erroneous value for m_a will result in bias in the reconstructed value of m_b in an algorithm that conserves counts. This effect has been observed experimentally: we have specified a value of $m_a = 98.0$, which is 1.24 counts lower than the correct average. For a covariance parameter $h_a = 0.01$, at feasibility, the reconstructed mean m_a was approximately one count lower and the reconstructed mean m_b was approximately 1 count higher than the results with the correct m_a. The standard deviation σ_b was found to have increased slightly, but the signal-to-noise ratio m_b/σ_b was practically unchanged. Possible implications of this findings will be discussed in the conclusions.

3.3.3. Using the EM formulation without conservation of counts

As indicated above, the iterative formulas (5) and (6) obtained by the EM algorithms do not conserve counts and we were not confident of obtaining meaningful results in the case of pixel dependent h_j's. Nevertheless, we have tested the EM algorithm for the two region case of Figure 1 as originally described by Levitan and Herman. For $h_a = 0.01$, we found that the means and standard deviations obtained in region "b" are not significantly different from the results obtained with the SS approach which conserves counts. We do not know at this time why that is the case, or whether it is a general result.

4. A BAYESIAN CROSS-ENTROPY ALGORITHM

4.1. The FMAPE Algorithm With Cross-Entropy

In Nuñez and Llacer (1990) a Maximum a Posteriori algorithm with entropy prior (FMAPE) that could be made to converge to feasible images by adjusting one parameter was presented. The FMAPE algorithm can be easily cast into a cross-entropy form (the nonuniform prior of Liang et al, 1989) by introducing the cross-entropy prior values Q_j which can be the expected values of the image pixels, if available. Using the notation of Levitan and Herman, the new log *a priori* function to be maximized is

$$\ln P_B(x|y) = \sum_{i=1}^{I}\left[y_i \ln\left(\sum_{j=1}^{J} a_{ij}x_j\right) - \sum_{j=1}^{J} a_{ij}x_j\right] - \sum_{j=1}^{J}\frac{x_j}{\Delta x}\ln\left(\frac{x_j}{\Delta x Q_j}\right) - \mu\left(\sum_{j=1}^{j} x_j - \sum_{i=1}^{I} y_i \Delta y_i\right) \tag{14}$$

where Δx is the adjustable parameter whose selection controls the point of convergence of the resulting algorithm. The iterative formula is

$$x_j^{(k+1)} = K x_j^{(k)}\left[\Delta x \sum_{i=1}^{I} a'_{ij}\left(\frac{y_i}{\sum_{j=n}^{J} a'_{in} x_n^{(k)}} - 1\right) - \ln\left(\frac{x_j^{(k)}}{Q_j}\right) + C\right]^P . \tag{15}$$

A somewhat different manipulation of the terms in the SS method leads to a different form of the iterative formula which separates clearly the likelihood and the regularization parts of the iterative steps:

$$x_j^{(k+1)} = Kx_j^{(k)}\left[\sum_{i=1}^{I}\frac{a'_{ij}y_i}{\sum_{j=n}^{J}a'_{in}x_n^{(k)}} - \sum_{i=1}^{I}a'_{ij} - \frac{1}{\Delta x}\ln\left(\frac{x_j^{(k)}}{Q_j}\right) + C\right]^P . \quad (16)$$

If prior information in the form of mean values Q_j is available and parameter Δx is small enough, the algorithm (14) will converge to the values Q_j, i.e., the cross-entropy term in (13) dominates. In the opposite case, a ML estimation results. By adjusting Δx it is then possible to control the degree to which the solution converges towards the prior information Q_j.

We have thought of using a pixel dependent parameter Δx_j in the Bayesian formulation (14) in order to force those pixels whose prior information Q_j is well known to converge to that prior, while letting the other pixels converge to a much less constrained relationship with their Q_j's.

4.2. The FMAPE With Cross-Entropy and Variable Convergence Parameter

In order to derive a version of the FMAPE that contains a variable paramer Δx_j, we will require an approach that is consistent with the concept of entropy for an ensemble of N photons that have to be counted a number Δx_j at a time. We start considering the case with prior information and totally independent photons, for which Liang et al,[10] for example, give a multiplicity expression

$$W(x) = \frac{N!}{\prod_{j=1}^{J} x_j!}\prod_{j=1}^{J}(P_j)^{x_j} \quad (17)$$

where P_j is the prior probability of a photon belonging to pixel j. If Q_j is the prior photon distribution, a natural choice for P_j is

$$P_j = \frac{Q_j}{N} \quad (18)$$

in which we have implicitly assumed that $\sum_{j=1}^{J} Q_j = N$.

We consider next the case in which photons can only be detected in groups Δx_j. The number of groups to be distributed into J pixels is now

$$N' = \sum_{j=1}^{J}\frac{x_j}{\Delta x_j}. \quad (19)$$

N' depends on the values Δx_j and it not strictly a constant. The right hand side of (19) will be used in the following to reflect that fact. The multiplicity expression corresponding to (17) will now be

$$W(x) = \frac{\left(\sum_{j=1}^{J}\frac{x_j}{\Delta x_j}\right)!}{\prod_{j=1}^{J}\left(\frac{x_j}{\Delta x_j}\right)!}\prod_{j=1}^{J}(P_j)^{\frac{x_j}{\Delta x_j}} \quad (20)$$

With the prior probability $P_A(x)$ proportional to the multiplicity and using Sterling's approximation,

$$\ln P_A(x) = \left(\sum_{j=1}^{J}\frac{x_j}{\Delta x_j}\right)\ln\left(\sum_{j=1}^{J}\frac{x_j}{\Delta x_j}\right) - \sum_{j=1}^{J}\frac{x_j}{\Delta x_j}\ln\left(\frac{x_j}{\Delta x_j P_j}\right) + \text{constants} \quad (21)$$

A reasonable choice for the prior probability P_j is now

$$P_j = \frac{Q_j / \Delta x_j}{\sum_{n=1}^{J} (Q_n / \Delta x_n)} \tag{22}$$

Maximization of (21) with the values of (22) results in estimates x_j which are proportional to Q_j. This would correspond to the imaging problem in the absence of projection data.

With the above expressions for the cross-entropy terms, the log *a posteriori* expression to be maximized is

$$\ln P_B(x|y) = \sum_{i=1}^{I}\left[y_i \ln\left(\sum_{j=1}^{J} a'_{ij} x_j\right) - \sum_{j=1}^{J} a'_{ij} x_j \right] + \left(\sum_{j=1}^{J} \frac{x_j}{\Delta x_j}\right) \ln\left(\sum_{j=1}^{J} \frac{x_j}{\Delta x_j}\right)$$

$$- \sum_{j=1}^{J} \frac{x_j}{\Delta x_j} \ln\left(\frac{x_j \sum_{n=1}^{J} \frac{Q_n}{\Delta x_n}}{Q_j}\right) - \mu\left(\sum_{j=1}^{J} x_j - \sum_{i=1}^{I} y_i \Delta y_i\right) \tag{23}$$

By the SS method, we arrive at the following iterative procedure:

$$x_j^{(k+1)} = K x_j^{(k)} \left[\sum_{i=1}^{I} \frac{a'_{ij} y_i}{\sum_{j=n}^{J} a'_{in} x_n^{(k)}} - \sum_{i=1}^{I} a'_{ij} - \frac{1}{\Delta x_j} \ln\left(\frac{x_j^{(k)} \sum_{n=1}^{J} \frac{Q_n}{\Delta x_n}}{Q_j \sum_{n=1}^{J} \frac{x_n^{(k)}}{\Delta x_n}} \right) + C \right]^P . \tag{24}$$

As in the case of the SS form of the Levitan and Herman algorithm and in the simple FMAPE, the constant K can be evaluated at each iteration to insure conservation of counts. As in the previous cases, K should remain stable after the first few iterations.

4.3. Results of Simulations

We have investigated the results of reconstructing the phantom of Figure 1 with a 10:1 count ratio by the iterative formula of (24). We have used the correct known means for regions "a" and "b" as prior pixel values Q_a and Q_b, respectively. Values Δx_a = 20, 5, 2 and 1 have been used, corresponding to increasingly tighter fitting to the known mean of region "a". The value Δx_b=100 was used in region "b", which corresponds to a virtually unconstrained MLE reconstruction. The constant C of (24) has been adjusted to values ranging from 5.0 to 11.5, using the smallest possible value for each reconstruction to maintain positivity of all the pixel values during the iterative process. The acceleration parameter P has been set to 2.0. All those values have been determined empirically by observing the reconstructions obtained with uniform $\Delta x_a = \Delta x_b$.

Convergence rates for these calculations are different from those resulting from the Levitan-Herman criterion. Therefore, in order to compare the performance of that method with the current one we plot the standard deviation in region "b" as a function of the error from the known mean in the same region. Figure 7 shows the results for the standard MLE, the Levitan-Herman method with h_a = 0.01 and the cross-entropy method with Δx_a= 1.0. The latter two corresponding to a very similar constraint on region "a": the obtained values of σ_a were 1.41 to 1.42 for both cases.

In comparing the two methods of using pixel dependent constraints, we find that the method of Levitan and Herman appears more effective in decreasing the standard deviation of the low activity region than the cross entropy method, at least for the example that we have analyzed. The first method also uses parameters which have a clear physical meaning: h_j's are the reciprocals of $2\pi\sigma^2$ of Gaussian distributions determining the degree of tightness about the prior means that one wants to obtain in the reconstructions. The parameters Δx of the cross-entropy formulation have a less well-defined physical meaning. Since the two methods are actually different, we can expect that the resulting reconstructions will have different properties. We have not attempted, at this time, to find those differences.

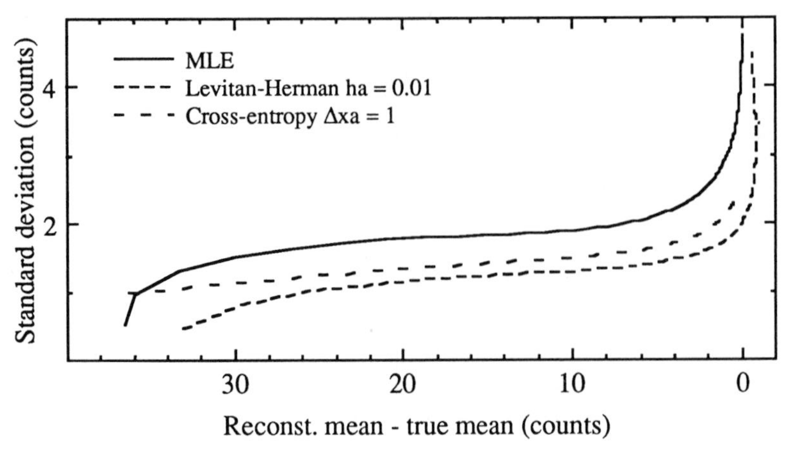

Fig. 7: Comparison of the standard deviation in region "b" vs. error in the mean of the same region for comparable levels of constraint in the reconstruction of region "a".

5. CONCLUSIONS

In this paper we have attempted to find the answer to a specific question: Can prior information on some areas of an image benefit the reconstruction of other areas? We have focused our attention on a simple type of image in which the area "a" whose mean is known is substantially larger than the area "b" whose characteristics we want to estimate. In Sect. 2 we have seen that as the ratio of activities in region "a" vs. "b" grows larger than 1:1, the Filtered Backprojection reconstructions suffer from a strong influence of one region upon the other. Using feasible MLE images, with appropriate post-filtering, this effect is diminished considerably without loss of image sharpness, but it is not eliminated totally. We have then shown that placing a strong constraint on region "a" in the worst of the cases studied (ratio 10:1), can bring the standard deviation in region "b" to a value approximately 1/2 of the MLE results, before filtering.

Then we can answer the above question in the affirmative, but a strong qualification has to be added. Using MLE feasible images with post-filtering (images with low bias) (Llacer and Bajamonde, 1990), the effect is already small for the phantom that we have studied. One would expect to find few real medical images that are as extreme as the phantom of Figure 1 in activity and area ratios and we would conclude, therefore, that the technique may only apply to a few special cases. One such case would be when a sequence of similar images is obtained from a patient and details of a particular time frame is desired for a particular ROI. Then, one can use the sum total of all time frames as an estimate of the means for the rest of the image and use a Bayesian method as the ones discussed in this paper for the reconstruction of single time frames. If the mean is a good representation of the image in the selected time frame for the image, except for the ROI, an improvement in lesion detectability and/or quantification of isotope uptake could be expected in that ROI.

A more important conclusion, however, is the realization that prior information has to act locally, i.e., the prior information has to be about the ROI selected.

Acknowledgments

This work has been supported by the Director, Office of Energy Research, Office of Health and Environment Research, Physical and Technological Division, of the U.S. Department of Energy under Contract No. DE-AC03-76SF00098. Jorge Nuñez would like to acknowledge the support of a grant by the Gaspar de Portola Catalonian Studies Program at the University of California and the Generalitat de Catalunya, Spain.

References

Chen C-T, Ouyang X, Ordonez C, Wong WH, Hu X and Metz CE (1990). Sensor Fusion in Image Reconstruction. pres. 1990 IEEE Nucl. Sci. Symp., Arlington, VA.

Geman S and Geman D (1984). Stochastic relaxation, Gibbs distributions, and the Bayesian restoration of images. IEEE Trans. Pattern Anal. Machine Intel., PAMI-6, 721-741.

Hildebrand FB (1974). Introduction to Numerical Analysis, 2nd ed., McGraw-Hill, New York.

Huesman RH, Gullberg GT, Greenberg WL and Budinger TF (1977). Donner Algorithms for Reconstruction Tomography. Publ. 214, Lawrence Berkeley Laboratory.

Levitan E and Herman GT (1987). A Maximum *A posteriori* Probability Expectation Maximization Algorithm for Image Reconstruction in Emission Tomography. IEEE Trans. Med. Imaging, MI-6, No. 3, 185-192.

Liang Z, Jaszczak R and Greer K (1989). On Bayesian Image Reconstruction from Projections: Uniform and nonuniform *a priori* source information. IEEE Trans. Med. Imaging, MI-8, 227-235.

Llacer J and Veklerov E (1989). Feasible Images and Practical Stopping Rules for Iterative Algorithms in Emission Tomography. IEEE Trans. Med. Imaging, MI-8, 186-193.

Llacer J and Bajamonde AC (1990). Characteristics of Feasible Images Obtained from Real PET Data by MLE, Bayesian and Sieve Methods. To be publ. Proc. of SPIE 1990 Int. Symp. on Optical and Optoelectronic Applied Science and Eng, San Diego, CA.

Nuñez J and Llacer J (1989). Maximum Entropy and the Concept of Feasibility in Tomographic Image Reconstruction. Proc. SPIE Medical Imaging III Conf., 1090, 359-372.

Nuñez J and Llacer J (1990). A Fast Bayesian Reconstruction Algorithm for Emission Tomography with Entropy Prior Converging to Feasible Images. IEEE Trans. Med. Imaging, MI-9, No. 2, 159-176.

Shepp LA and Vardi Y (1984). Maximum Likelihood Reconstruction for Emission Tomography. IEEE Trans. Med. Imaging, MI-1, 113-122.

Veklerov E and Llacer J (1987). Stopping rule for the MLE algorithm based on statistical hypothesis testing. IEEE Trans. Med. Imaging, MI-6, 313-319.

ON RECONSTRUCTION AND SEGMENTATION OF PIECEWISE CONTINUOUS IMAGES

Z Liang, R Jaszczak and R Coleman

Department of Radiology, Duke University Medical Center
Durham, North Carolina 27710

Abstract

We evaluate an image model for simultaneous reconstruction and segmentation of piecewise continuous images. The model assumes that the intensities of the piecewise continuous image are relatively constant within contiguous regions and that the intensity levels of these regions can be determined either empirically or theoretically before reconstruction. The assumptions might be valid, for example, in cardiac blood-pool imaging or in transmission tomography of the thorax for non-uniform attenuation correction of emission tomography. In the former imaging situation, the intensities or radionuclide activities within the regions of myocardium, blood-pool and background may be relatively constant and the three activity levels can be distinct. For the latter case, the attenuation coefficients of bone, lungs and soft tissues can be determined prior to reconstructing the attenuation map. The contiguous image regions are expected to be simultaneously segmented during image reconstruction. We tested the image model with experimental phantom studies. The phantom consisted of a plastic cylinder having an elliptical cross section and containing five contiguous regions. There were three distinct activity levels within the phantom. Projection data were acquired using a SPECT system. Reconstructions were performed using an iterative maximum *a posteriori* probability procedure. As expected, the reconstructed image consisted of contiguous regions and the acitivities within the regions were relatively constant. Compared with maximum likelihood and a Bayesian approach using a Gibbs prior, the results obtained using the image model demonstrated the improvement in identifying the contiguous regions and the associated activities.

Keywords

Maximum *a posteriori* probability; *A priori* intensity-level information; Region-of-interests; Information criteria.

1. INTRODUCTION

The importance of identifying isolated regions-of-interests (*ROIs*) (e.g., organs and/or tumors) within the body in medical image processing is well known (Goris and Briandet, 1983; Margulis and Gooding, 1989; Pohost et al, 1989). For example, in cardiac blood-pool imaging, if the boundaries of the blood-pool within the ventricle can be identified at different times, then the blood-volume curve within the ventricle can be determined.

Conventionally, *ROIs* are identified by two steps. First, an image is reconstructed from acquired data using a reconstruction method (Budinger and Gullberg, 1974; Herman, 1980). Then a segmentation or enhancement technique is applied to the reconstructed image (Derin and Elliott, 1987; Hall, 1979). After the first step, the reconstructed regions of different objects (or tissues within the body) are usually interlaced and the boundaries are blurred due to low-count density of acquired data and artifacts relating to object motion, detection errors, etc. In the second step, although segmentation techniques can be used to minimize the problem associated with the interlacing and blur, the segmented image might not be consistent with the acquired data. Hence the artifacts in the reconstructed image may not be eliminated.

In our previous work (Liang et al, 1991a), a multinomial image model for simultaneous reconstruction and segmentation of piecewise continuous images was proposed. In this paper, we extend the model and evaluate an exponential image model for piecewise continuous images with experimental phantom studies.

2. THEORY

The model evaluated in this paper assumes that the intensities of the piecewise continuous image are relatively constant within contiguous regions and that the intensity levels of these regions are distinct and can be determined either empirically or theoretically prior to reconstruction. These assumptions might be valid, for example, in cardiac blood-pool imaging or in transmission tomography of the thorax for non-uniform attenuation correction of emission tomography. In the former situation, the intensities or radionuclide activities within the regions of myocardium, blood-pool and background (or surrounding tissues outside the heart) may be relatively constant. The three activity levels of these regions may be different. The activity level of the blood can be measured empirically by taking a blood sample from the patient. The activity of the heart muscle may be assumed to be negligible. The activity of the background may be determined from previous studies of typical blood-pool/background ratios and from the acquired data. A theoretical discussion for determining the activity levels is included in the Appendix. For the latter case, the attenuation coefficients of spine bone, lungs, and soft tissues are distinct and can be determined prior to reconstructing the attenuation map for non-uniform attenuation correction.

In order to facilitate the discussion of the image model, the source space is partitioned into J small cubic volumes or voxels (or pixels in a two dimentional case). Each voxel is assigned a value ϕ_j which represents the average of source intensity over the cubic volume j. In emission tomography, ϕ_j reflects the average photon emission in voxel j per unit time at time=0 (at time=0 means that the decay of radioisotope should be considered and corrected, if necessary). Then $\Phi = \{ \phi_j \}_{j=1}^{J}$ represents the average emission map over the source-voxel array at time=0, which is the true image to be reconstructed. For transmission tomography, ϕ_j represents the attenuation coefficient of voxel j.

Mathematically the assumptions of the image model are characterized by a set of assumed intensity levels $\{\phi_s^e\}_{s=1}^{K}$ and/or some relating parameters (e.g., variance σ_s^2 about the level ϕ_s^e), where K is the number of the levels (Liang et al, 1991a). These levels and relating parameters may be determined either empirically or theoretically prior to reconstruction. The reconstructed image $\{\phi_j\}$ which represents the true image is expected to be consisting of contiguous regions. The intensity of each region is expected to be relatively constant and be close to one of the levels $\{\phi_s^e\}$.

Under the assumptions, voxel j would have its value ϕ_j distributed following the probability density function

$$p_j(\phi_j) = p_j(\phi_j \,|\, \phi_s^e) = \frac{1}{K} \sum_{s=1}^{K} \delta(\phi_j - \phi_s^e) . \tag{1}$$

In practice the assumed intensity levels $\{\phi_s^e\}$ are not accurate, and an exponential function rather than the delta function is used to reflect the variation of ϕ_j around the levels $\{\phi_s^e\}$

$$p_j(\phi_j) = \sum_s c_s \, exp[- (\phi_j - \phi_s^e)^2 / 2 \sigma_s^2] \tag{2}$$

where c_s is the normalization constant and σ_s reflects the variation of ϕ_j around ϕ_s^e. In general, each photon would have different chances of being emitted from the K different intensity levels $\{\phi_s^e\}$, function (2) then becomes

$$p_j(\phi_j) = \sum_s w_s \, c_s \, exp[- (\phi_j - \phi_s^e)^2 / 2 \sigma_s^2] \tag{3}$$

where w_s reflects the ratio of the number of voxels having the similar intensity ϕ_s^e to the total number J of voxels in the source array (Liang and Hart, 1987; Liang et al, 1991a).

The theoretical determination of these level parameters $\{w_s, \phi_s^e, \sigma_s^2\}$ is discussed in the Appendix. In the following, we assume that these parameters are obtained empirically.

Since the image is expected to consist of contiguous regions, each voxel would be classified to a region having an intensity level of $\{\phi_s^e\}$. The classification of voxel ϕ_j to a region is usually determined by its neighbors $\{\phi_{l \neq j}\}$ (Young and Fu, 1986). Let $w_s = w_s(\phi_j \,|\, \phi_{l \neq j})$ represent the probability of classifying voxel ϕ_j to a region with a level ϕ_s^e. Although the probability $w_s(\phi_j \,|\, \phi_{l \neq j})$ could be specified on a Markov random field, here we use a very simple technique to capture the neighborhood-interaction information

$$w_s(\phi_j \,|\, \phi_{l \neq j}) = \text{median} \{ exp[- (\phi_r - \phi_s^e)^2 / 2 \sigma_s^2] \}_{r=(j,l)} \tag{4}$$

where the index r covers the voxel j and its neighbors ($l \neq j$). In the phantom studies as discussed later, l is chosen to cover the 8 nearest neighbors of voxel j.

If we neglect the non-negativity condition for each voxel value, the image model can be expressed as

$$P(\Phi) = \prod_{j=1}^{J} \sum_s w_s \, c_s \, exp[-(\phi_j - \phi_s^e)^2 / 2\sigma_s^2]\,. \tag{5}$$

This expression is the generalization of the model reported in (Liang and Hart, 1988a) from one dimension to multidimensions. If the non-negativity condition is considered, the image model would be expressed as

$$P(\Phi) = \frac{N}{\prod_j \phi_j!} \prod_{j=1}^{J} [p_j]^{\phi_j}\,, \text{ and } \sum_j p_j = 1 \tag{6}$$

where p_j is the function (3) and $N = \sum_j \phi_j$ is a constant. It is noted that function (6) requires the values $\{\phi_j\}$ to be integers. When the actual source values are real, $\{\phi_j\}$ would be scaled to corresponding integers by a multiplier during image reconstruction. After reconstruction, the obtained image values will then be scaled back to real values. The scaling is usually done in the computer program (Liang and Hart, 1988b).

Up to now, we have discussed the image model expressed by function (5) or (6). This model is treated, in this paper, as *a priori* source information to supplement the measurement information or data likelihood. In the following, we will discuss only the measurement information of emission tomography.

In emission tomography, the Poisson likelihood of a data distribution is well established as (Shepp and Vardi, 1982)

$$P(\mathbf{Y}|\Phi) = \prod_{i=1}^{I} exp(-\sum_j R_{ij}\,\phi_j)(\sum_j R_{ij}\,\phi_j)^{Y_i} / Y_i! \tag{7}$$

where R_{ij} is the probability of detecting photons from voxel j for projection ray i, Y_i is the number of photons detected for projection ray i, and $\mathbf{Y} = \{Y_i\}_{i=1}^{I}$ (I is the number of contributing rays). The assumption made in the likelihood (7) is that each data element Y_i is a Poisson variable with a mean $\sum_j R_{ij}\,\phi_j$ and all elements $\{Y_i\}$ are independent. The means $\{\sum_j R_{ij}\,\phi_j\}$ of the Poisson variables $\{Y_i\}$ reflect the relation between the measured data and the source $\{\phi_j\}$ via the probability matrix $\{R_{ij}\}$.

Since $\{R_{ij}\}$ are real values continuously distributed over the interval [0,1], the means $\{\sum_j R_{ij}\,\phi_j\}$ are continuous real values over $[0,\infty)$ regardless of ϕ_j being real or integer. Since the emissions are relatively uniform over a $4\,\pi$ solid angle, the values $\{R_{ij}\}$ are very small for a conventional ring-detector system, and the range of $\{\phi_j\}$ values is relatively large on $[0,\infty)$. In this case, the truncation error between real and integer is negligible when function (6) is used.

For parallel ray geometry, the value R_{ij} can be assumed to be proportional to the intersection length of voxel j and projection ray i, if the attenuation, scattering, and collimation variation are neglected (Budinger and Gullberg, 1974; Herman, 1980). This assumption is used in this paper. If $\{R_{ij}\}$ are chosen as the intersection lengths of the voxels and the projection rays, the voxel values $\{\phi_j\}$ will be divided by the proportionality factor, if the measured data are used directly. In practice, the proportionality factor is usually neglected for image display because $\{\phi_j\}$ have to be scaled anyway. However, for quantitative evaluation of images, as in single photon or positron emission tomography (SPECT or PET), the proportionality factor must be considered. In this paper, we will consider the attenuation correction by including the attenuation factors in $\{R_{ij}\}$ (Budinger and Gullberg, 1974).

3. METHODS

The method used in this paper for simultaneous reconstruction and segmentation of piecewise continuous images is a maximum *a posteriori* probability approach. It is expressed, via Bayes' law of $P(\Phi|\mathbf{Y}) = P(\mathbf{Y}|\Phi)\,P(\Phi)\,/\,P(\mathbf{Y})$, as: $P(\Phi^*|\mathbf{Y}) \approx P(\mathbf{Y}|\Phi^*)\,P(\Phi^*) = \text{maximum}$, or

$$\ln P(\mathbf{Y}|\Phi^*) + \ln P(\Phi^*) = \text{maximum}\,. \tag{8}$$

Directly solving Eq.(8) for Φ^* is very difficult due to the large size of the matrix $\{R_{ij}\}$. For example, if 120 projections each with 128 rays are acquired from a source array of 128×128, the size of the matrix is 15360×16384. An iterative approach may be therefore more practical. The expectation-maximization (EM) technique (Dempster et al, 1977) is then employed to compute the solution Φ^* iteratively (Liang and Hart, 1988a)

$$\phi_k^{(n+1)} = \phi_k^{(n)} \frac{\sum_i R_{ik} Y_i / \sum_j R_{ij} \phi_j^{(n)}}{\sum_i R_{ik} + \xi_k^{(n)} Z_k} \tag{9}$$

where

$$Z_k(\Phi) = -\frac{\partial}{\partial \phi_k} \ln P(\Phi) |_{\Phi = \Phi^{(n)} + \lambda d^{(n)}}, \text{ and } \xi_k^{(n)} = \frac{a \sqrt{n}}{b+n} \sum_i R_{ik} > 0. \tag{10}$$

In Eq.(10), $\lambda \approx 1$ and $d_k^{(n)} = \phi_k^{(n)} - \phi_k^{(n-1)}$ can be chosen to facilitate computation; when $\phi_k^{(n)} + \lambda d_k^{(n)} = \phi_k^{(n)} [\sum_i R_{ik} Y_i / \sum_j R_{ij} \phi_j^{(n)}]/[\sum_i R_{ik}]$ is assumed, the convergence of the Bayesian image processing (BIP) algorithm (9), at $(n+1)$-th iteration, is along the maximum likelihood (ML) gradient direction at n-th iteration (Dempster et al, 1977), and the *posteriori* probability increases monotonically. The adjustable parameter $\xi_k^{(n)}$ increases smoothly as the iterative index n increases, and $\xi_k^{(n>m)}$ is set to be the maximal value $\xi_{\max}^{(m)}$ after m iterations. The values $a = 0.05$ and $b = 50 = m$ were used in the phantom studies. It is noted that the a, b, and m are related to each other. There is only one parameter $\xi_{\max}^{(m)}$. The role of a, b, and m is to increase $\xi_k^{(n)}$ from very small to $\xi_{\max}^{(m)}$ smoothly. A detailed discussion of the empirically determined parameter $\xi_k^{(n)}$ is available in references (Liang and Hart, 1988a,b). The expression of Z_k is, for function (5)

$$Z_k(\Phi) = [\sum_s (\phi_k - \phi_s^e) H_s / \sigma_s^2] / \sum_s H_s \tag{11}$$

and for function (6), Z_k becomes

$$Z_k(\Phi) = [(\ln\phi_k + \phi_k/2) - \ln\sum_s H_s] + [(\phi_k / \sum_s H_s) \sum_s (\phi_k - \phi_s^e) H_s / \sigma_s^2] \tag{12}$$

where

$$H_s = w_s \, c_s \, exp[-(\phi_k - \phi_s^e)^2 / (2\sigma_s^2)]. \tag{13}$$

The BIP algorithm (9) has been tested with experimental phantom studies, and compared to the ML-EM algorithm (Shepp and Vardi, 1982) [i.e., dropping the term $\xi_k^{(n)} Z_k$ in Eq.(9)] and the Bayesian algorithm with a Gibbs prior (Bayesian-Gibbs) (Johnson, 1989).

The phantom design was particularly relevant to the model of piecewise continuous images. The phantom consisted of a plastic cylinder having an elliptical cross section containing five contiguous regions (see Figure 1). The two "lung" regions and the "bone" region have no activity. The activity ratio of the hot sphere and the background regions is measured as 6.7/1.3 ($\mu Ci/ml$) (Gilland et al, 1991).

Projection data were acquired using a three-headed SPECT system[1] with parallel-hole collimators. There were 120 projection angles equally spaced over 360 degrees. Each projection had 128×128 samples.

The non-uniform attenuating properties of the phantom was considered during image reconstruction (Liang et al, 1991b). The effects of scattering and collimation variation were neglected. The image size was 128×128.

The projection data were first backprojected into image space with a filtered-backprojection method (Budinger and Gullberg, 1974; Herman, 1980). From the histogram of the backprojected image, four intensity levels $\{\phi_s^e\}_{s=1}^4$ were obtained: 0.25/voxel for the "lung" regions, 2.5/voxel for the background, 1.25/voxel for the "bone" region, and 12.5/voxel for the hot sphere region. The parameter $\sigma_s = 0.1$ was assumed for all the 4 levels based on the observation of the histogram. The estimated activity levels of the "lung" and "bone" regions were not zero because of scattering. The level of the "bone" was not the same as the level of the "lungs" due to the variation of collimation. These four "assumed" levels were then used by the BIP algorithm (9) to simultaneously reconstruct and segment the image.

1) Trionix Triad SPECT system

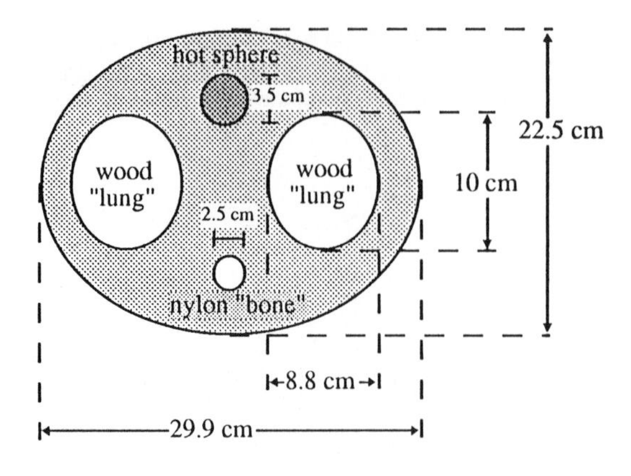

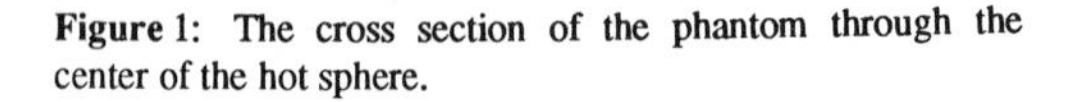

Figure 1: The cross section of the phantom through the center of the hot sphere.

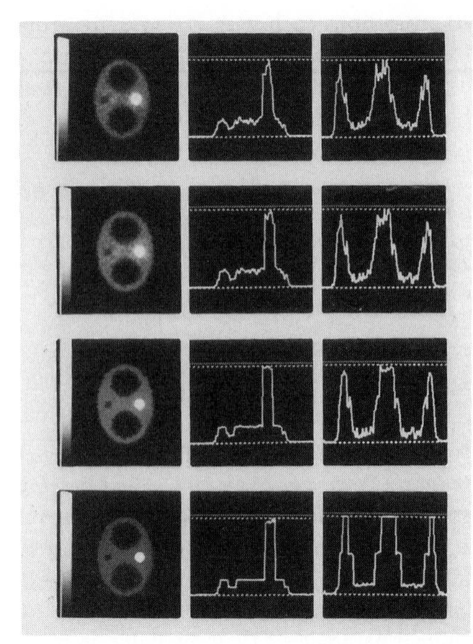

Figure 2: The images reconstructed from the central-slice projections using the BIP algorithm after 20 (top), 30 (second row), 50 (third row), and 100 (bottom) iterations, respectively. The curves shown in the central column represent the activity profiles drawn through the center of the images horizontally. The curves on the right column are the vertical profiles through the center of the images. These profiles is one-pixel wide.

4. RESULTS

In this section, we present the images reconstructed by using the BIP algorithm (9). These images are compared with those obtained using the ML-EM algorithm (Shepp and vardi, 1982) and the Bayesian-Gibbs approach (Johnson, 1989).

Figure 2 shows the images reconstructed from the central-slice projection data using the BIP algorithm (9) and (11). The counts of the central-slice projection data were approximately half million. The top image was obtained after 20 iterations, and the subsequent images were obtained after 30, 50, and 100 iterations, respectively. The curves shown in the central column are single-pixel-wide profiles drawn through the center of the images horizontally. The curves on the right column are single-pixel-wide profiles drawn through the image centers vertically. The images reconstructed using the Eqs.(9) and (12) [which considers the non-negativity condition see function (6)] were similar to the images of Figure 2; hence, these images are not shown here.

In performing the reconstruction with the BIP algorithm (9), we tried a few values for the free adjustable parameter $\xi_{\max}^{(m)}$ in Eq.(10) by changing the values a and b. Keeping b fixed, when $a = 0.01$, the resulting images were noiser, as compared with the images of Figure 2. When $a > 0.1$, distortion and artifacts in the reconstructed images were observed. Since $\xi_{\max}^{(m)}$ is dependent on the true image to be reconstructed, it is very important to try a few phantom studies before applying the BIP algorithm (9) to real data which are acquired from an object having similar properties as the phantom does.

For comparison the reconstructed images by the Bayesian-Gibbs approach after 20, 50, 100, and 200 iterations, respectively, are shown in Figure 3. The curves are the horizontal and vertical profiles through the image centers, respectively. There are two free parameters in the Bayesian-Gibbs algorithm. One controls the smoothness of the contiguous regions, and the other adjusts the line process for edge enhancement. These two parameters are also object dependent. If they are not appropriately chosen, presence of artifacts in reconstructed image is likely. Figure 4 shows the reconstructed images using the ML-EM algorithm after 10, 20, 30 and 50 iterations, respectively.

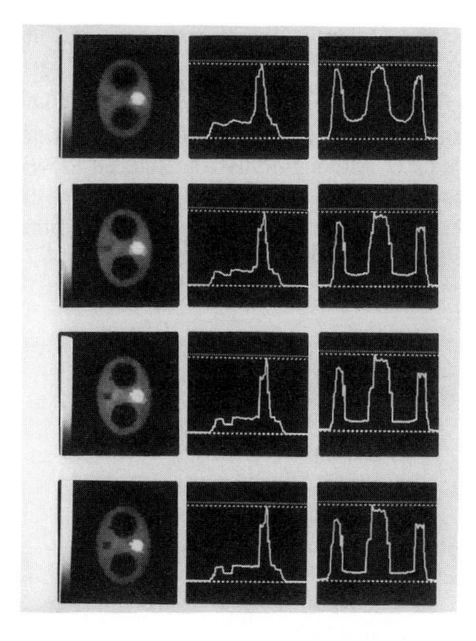

Figure 3: The images of the Bayesian-Gibbs algorithm for the projection data after 20, 50, 100, and 200 iterations, respectively. The curves are the activity profiles as described in Figure 2.

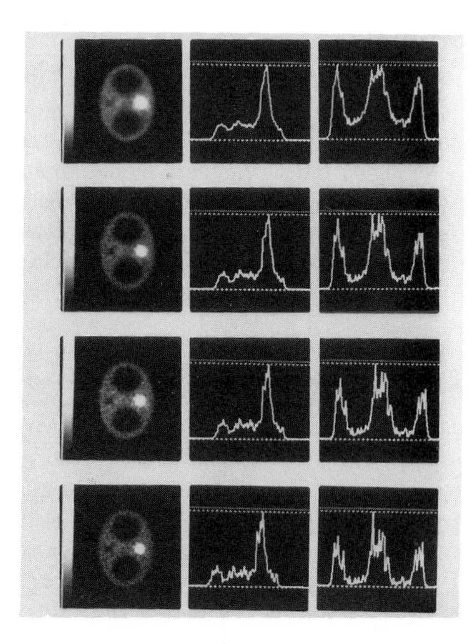

Figure 4: The images of the ML algorithm for the projection data after 10, 20, 30, and 50 iterations, respectively. The curves are the activity profiles as described in Figure 2.

Quantitative comparisons in terms of mean and standard deviation for the *ROIs* are given in the table. The images evaluated were the ML-EM image after 50 iterations, the Bayesian-Gibbs image after 200 iterations, and the BIP image after 100 iterations (see the bottom images of Figures 2, 3, and 4). For the hot sphere region, a square size covering 30 pixels was chosen to calculate the mean and the standard deviation. A rectangular size of 180 pixels was used to compute the mean and the standard deviation for the center region of the background. Two square sizes of 15 and 140 pixels were used to compute the means and the standard deviations of the "bone" and "lung" regions, respectively. The maximum and minimum pixel values within the chosen sizes are also shown in the table.

The extra computation time for the *a priori* information Z_k in the BIP algorithm (9) is about 2-3% of that of the ML-EM algorithm [i.e., dropping the term $\xi_k^{(n)} Z_k$ in Eq.(9)]. The extra memory needed to store Z_k is negligible.

Both the BIP and Bayesian-Gibbs approaches showed significant improvement in image quality, as compared with the ML method. Compared to the Bayesian-Gibbs approach, the BIP algorithm (9) demonstrated an improvement in identifying the *ROIs* and their activities, except near the edge of the "lung" regions. The extra activity level near the edge of the "lung" regions is the same as that of the "bone". The activity levels for the "lung" and "bone" regions were not zero since scattering was not included in the reconstruction method. Also, the effect of finite spatial resolution may have contributed to the activity observed in the "bone" region.

It is noted that the reconstructed activity levels of the contiguous regions using the BIP algorithm are different from the assumed levels $\{\phi_s^e\}$. This is because of the requirement that the segmented contiguous regions must satisfy the data constraints. The activity ratio of the reconstructed hot sphere and background regions (i.e., 5.135) is closer to the actually measured ratio (i.e., 5.153), as compared to the assumed ratio of 12.5/2.5 = 5.0.

5. DISCUSSIONS

We have described an image model for piecewise continuous images. In the experimental phantom studies, the regions of interests (i.e., the "lungs", "bone" and hot sphere) were simultaneously reconstructed and segmented. Noise within the regions was suppressed. A quantitative improvement in identifying the *ROIs* and their activities was observed (see the Table).

In formulating the image model as expressed by function (5) or (6), we used a very simple model for local correlations of $\{w_s\}$ as given by expression (4). More sophisticated modeling of local correlations with higher order neighbors should improve the detection of edges. Although only four intensity levels were considered in this paper, it is anticipated that it would be straightforward to include more intensity levels. If the difference in intensity between two adjoining regions is very low, as compared with noise level, the reconstructed BIP image would have only one intensity distributed over the two regions.

Table

Definitions	ML(image)	Bayesian-Gibbs	BIP(image)
# iteration	50	200	100
hot sphere:	size(30)	size(30)	size(30)
	max(14.63)	max(14.69)	max(13.90)
	min(5.86)	min(6.56)	min(12.25)
	mean(10.345)	mean(10.447)	mean(13.100)
	s.d.(2.302)	s.d.(1.878)	s.d.(0.359)
background:	size(180)	size(180)	size(180)
	max(8.17)	max(3.00)	max(2.56)
	min(0.99)	min(2.46)	min(2.54)
	mean(2.885)	mean(2.737)	mean(2.551)
	s.d.(1.097)	s.d.(0.106)	s.d.(0.004)
"bone":	size(15)	size(15)	size(15)
	max(2.13)	max(1.51)	max(1.38)
	min(0.78)	min(1.39)	min(1.32)
	mean(1.453)	mean(1.441)	mean(1.348)
	s.d.(0.394)	s.d.(0.032)	s.d.(0.018)
"lungs":	size(140)	size(140)	size(140)
	max(0.89)	max(0.38)	max(0.39)
	min(0.05)	min(0.29)	min(0.20)
	mean(0.294)	mean(0.322)	mean(0.306)
	s.d.(0.155)	s.d.(0.017)	s.d.(0.033)

The computed values for the ML-EM image after 50 iterations, Bayesian-Gibbs image after 200 iterations, and the BIP image after 100 iterations. The term "size" refers to the number of pixels within the *ROI* selected to calculate the mean, standard deviation (s.d.), maximum pixel value (max), and minimum pixel value (min).

Acknowledgments

The authors appreciate Dr. T. Turkington providing the display program and Dr. D. Gilland and Mr. K. Greer for their assistance in data acquisition. This investigation was supported by Grant #HL44194, awarded by the National Heart, Lung, and Blood Institute, and in part by Grant #CA33541, awarded by the National Cancer Institute, and Grant #DE-FG05-89ER60894, awarded by the Department of Energy.

Appendix

In this section, a theoretical discussion for determining intensity levels and/or relating parameters by applying information criteria to acquired projection data is given. Since the intensity levels and relating parameters are determined in image space rathe than in projection space, the acquired projection data are backprojected first to the image space. This step can be accomplished with a conventional backprojection method (Budinger and Gullberg, 1974; Herman 1980). Let $\mathbf{X} = \{x_r\}_{r=1}^{J_0}$ be the backprojected data vector with J_0 elements in image space. The information criteria discussed here are the Akaike information criterion (*AIC*) (Akeike, 1974)

$$AIC(\Theta) = -2\ln[\,\Psi(\mathbf{X}\,|\,\Theta)\,] + 2K_0 \tag{A.1}$$

and the minimum description length (*MDL*) information criterion (Rissanen, 1978; Schwarz, 1978)

$$MDL(\Theta) = -\ln[\,\Psi(\mathbf{X}\,|\,\Theta)\,] + (1/2)\,K_0 \ln J_0 \tag{A.2}$$

where $\Psi(\mathbf{X}\,|\,\Theta)$ is the maximum likelihood of backprojected data vector $\mathbf{X}$ given the parameter vector Θ, and K_0 is the number of freely adjustable parameters in Θ.

The information criteria are usually deployed, post hoc, to select one from a number of models having different dimensions of Θ, given $\mathbf{X}$. The models are defined by the likelihood $\Psi(\mathbf{X}\,|\,\Theta)$ with different forms characterized by the parameter vector Θ, or with one and the same form but with different restrictions on the parameter vector Θ. In the former case, there are a number of models defined by the different likelihood forms. For each different form, the corresponding parameter vector Θ is determined by maximizing the corresponding likelihood. The information criterion *AIC* or *MDL* is then used to select one of these models. The one selected minimizes the criterion (A.1) or (A.2). In the latter case, there are a number of models defined by the same likelihood form with different restrictions on the parameter vector Θ or with different parameter vectors. These parameter vectors are determined by maximizing the likelihood with the restrictions on Θ. One vector (or one model in this case) is then selected from these vectors by minimizing the criterion (A.1) or (A.2). The latter approach may be appropriate for selecting the intensity levels from the backprojected data. There may be a number of models of the same likelihood form with different numbers of intensity levels. The model which minimizes *AIC* or *MDL* is selected, and this selected model will have a specific number of intensity levels.

Using the notations of this paper, the likelihood is formulated as follows. The backprojected data vector $\mathbf{X}$ has J elements, or $J_0 = J$. Let $f(x_r\,|\,\theta_s)$ represent the exponential function (2), or

$$f(x_r\,|\,\theta_s) = (2\pi\sigma_s^2)^{-1/2}\, exp\,[\,-(x_r - \phi_s^e)^2 / (2\sigma_s^2)\,]\,,\quad r=1,2,...,J\,,\quad s=1,2,...,K\,, \tag{A.3}$$

where $\Theta = \{\theta_s = (\phi_s^e, \sigma_s^2)\}_{s=1}^K$ is a row vector of $2K$ elements, and K is the number of intensity levels. Assume that $W = \{w_s\}_{s=1}^K$, $(0 < w_s < 1\,,\ \sum_s w_s = 1)$, are the weights of $f(x_r\,|\,\theta_s)$. The likelihood function for $\mathbf{X}$ is then expressed as (Lei and Sewchand, 1988)

$$L(\mathbf{X}\,|\,W,\Theta) = \prod_{r=1}^{J} g(x_r\,|\,W,\Theta) = \prod_{r=1}^{J}\sum_{s=1}^{K} w_s\, f(x_r\,|\,\theta_s) \tag{A.4}$$

and

$$\Psi(\mathbf{X}\,|\,\Theta) = \max\,\{\,L(\mathbf{X}\,|\,W,\Theta)\,\}\,.$$

Now we have a number of models of an exponential likelihood form $\{\theta_s = (\phi_s^e, \sigma_s^2)\}$ with different dimensions (or levels) K. The problem of determining the intensity levels is to select the K value and therefore the values of $\{\theta_s = (\phi_s^e, \sigma_s^2)\}$. This problem is the same as that of the second case mentioned above. For different K, there are different models of the exponential likelihood form with different parameter vectors $\Theta(K)$. That model with K_c levels, which minimizes the information criterion *AIC* or *MDL*, will be selected.

The numerical calculations for K_c can be performed as follows: (1) determine the relation between K and K_0; (2) assume a K value, which is in a prespecified range $[K_{min}, K_{max}]$, where $K_{min} \le K_c \le K_{max}$; (3) maximize $\Psi(\mathbf{X}\,|\,\Theta)$ with respect to Θ for the assumed K value; (4) compute the criterion value of *AIC* or *MDL*; (5) repeat steps (2)-(4) for $K \pm 1$ values; (6) if the calculated criterion value for K using *AIC* or *MDL* is less than those for $K \pm 1$, then $K_c = K$ is selected, otherwise, $K+2$ or $K-2$ are assumed and the steps above are repeated. In (A.4), there are $K_0 = 3K - 1$ freely adjustable parameters: $2K$ free parameters of $\{\theta_s = (\phi_s^e, \sigma_s^2)\}$ and $K-1$ free parameters of $\{w_s\}$; hence step (1) is accomplished. The starting value K in the step (2) is chosen arbitrarily, depending on one's prior knowledge about the backprojected data $\{x_r\}$. If the starting value K is chosen closer to K_c, then less computation time is needed.

Many numerical techniques can be used to maximize the likelihood with respect to the parameters $\{w_s, \theta_s\}$ (Luenberger, 1973). We will use the EM technique (Dempster et al, 1977) to derive an iterative algorithm for the parameters of the image model (5) or (6).

In terms of EM terminology, (A.4) represents the family of sampling densities for the incomplete data $\{x_r\}_{r=1}^J$. Each observed datum x_r is associated with an unobserved level θ_s. There exists an unobserved vector $\mathbf{T} = \{T_r\}_{r=1}^J$, whose element T_r is an indicator vector of length K. The K components of the indicator vector T_r are all zero except for one equal to unity indicating the unobserved level θ_s associated with the datum x_r. Thus, the complete data are $\{x_r, T_r\}_{r=1}^J = \{\chi_r\}_{r=1}^J$. The family of sampling densities for the complete data $\chi = \{\chi_r\}$ is then (Dempster et al, 1977)

$$\psi(\chi\,|\,W,\Theta) = \prod_{r=1}^{J}\prod_{s=1}^{K} w_s^{T_{rs}}\, f(x_r\,|\,\theta_s)^{T_{rs}}\,. \tag{A.5}$$

The E-step is expressed, at the n-th iteration, as (Dempster et al, 1977)

$$Q(W,\Theta \mid W^{(n)},\Theta^{(n)}) = E[\ \ln\psi(\chi \mid W,\Theta) \mid \mathbf{X}, W^{(n)},\Theta^{(n)}\ ] \tag{A.6}$$

$$= \sum_r \{ \sum_s [\ T_{rs}^{(n)} \ln(w_s) + T_{rs}^{(n)} \ln f(x_r \mid \theta_s)\ ]\ \}$$

where (Redner and walker, 1984; Titterington et al, 1985)

$$T_{rs}^{(n)} = E[\ T_{rs} \mid x_r, W^{(n)},\Theta^{(n)}\ ] = \frac{w_s^{(n)} f(x_r \mid \theta_s^{(n)})}{g(x_r \mid W^{(n)},\Theta^{(n)})}\ . \tag{A.7}$$

The M-step generates the $(n+1)$-th iterated result which maximizes $Q(W,\Theta \mid W^{(n)},\Theta^{(n)})$ respect to W and Θ respectively. For parameter W, function Q is maximized, subject to the constraint $\sum_s w_s = 1$. By introducing a Lagrange multiplier κ, the solution $w_s^{(n+1)}$ is given by

$$\left.\frac{\partial \{ Q - \kappa(\sum_s w_s - 1)\}}{\partial w_s}\right|_{w_s^{(n+1)}} = 0$$

or

$$\kappa = J \quad \text{and} \quad w_s^{(n+1)} = \frac{1}{J} \sum_r T_{rs}^{(n)}\ . \tag{A.8}$$

When $f(x_r \mid \theta_s)$ is the exponential (or Gaussian) function of (A.3), the solutions $(\phi_s^e)^{(n+1)}$ and $(\sigma_s^2)^{(n+1)}$ are (Lei and Sewchant, 1989; Redner and Walker, 1984; Titterington et al, 1985)

$$(\phi_s^e)^{(n+1)} = \frac{\sum_r T_{rs}^{(n)} x_r}{\sum_r T_{rs}^{(n)}} \tag{A.9}$$

and

$$(\sigma_s^2)^{(n+1)} = \frac{\sum_r T_{rs}^{(n)} [\ x_r - (\phi_s^e)^{(n+1)}\ ]^2}{\sum_r T_{rs}^{(n)}}\ . \tag{A.10}$$

The mathematical derivations above can be applied to other distributions of $f(x_r \mid \theta_s)$ as shown below:

(1): If a Poisson distribution is considered for $f(x_r \mid \theta_s)$, the level parameter $\theta_s^{(n+1)} = (\phi_s^e)^{(n+1)}$ is given by (A.9). The weight parameter $w_s^{(n+1)}$ is expressed by (A.8).

(2): Since in most imaging situations, the backprojected data $\{x_r\}$ by non-linear backprojection methods are non-negative (where the Gaussian distribution is not suitable) and the variance σ_s^2 is not equal to the mean ϕ_s^e (as the Poisson distribution does), a Gamma distribution might be more appropriate. The $(n+1)$-th iterated results w_s and ϕ_s^e for the Gamma distribution are given by (A.8) and (A.9), and the variance σ_s^2 is determined by

$$(\sigma_s^2)^{(n+1)} = \alpha_s^{(n+1)} [\ \beta_s^{(n+1)}\ ]^2 \tag{A.11}$$

where

$$\alpha_s^{(n+1)} = (1/6) + (1/2) \left\{ ln[\ (\phi_s^e)^{(n+1)}\ ] - [\ \sum_r T_{rs}^{(n)}\ ln(x_r)\ ] / [\ \sum_r T_{rs}^{(n)}\ ] \right\}^{-1}$$

and

$$\beta_s^{(n+1)} = (\phi_s^e)^{(n+1)} / \alpha_s^{(n+1)}\ .$$

(3): For multivariate normal distributions of

$$f(x_r \mid \theta_s) = (2\pi)^{-\eta/2} \mid \Omega_s \mid^{-1/2} exp[\ -(1/2)(\mathbf{x_r} - \boldsymbol{\phi}_s^e)^\tau\ \Omega_s^{-1} (\mathbf{x_r} - \boldsymbol{\phi}_s^e)\ ]$$

where $\mathbf{x_r}$ and $\boldsymbol{\phi}_s^e$ are vectors of length η, and Ω_s is a positive-definite symmetric $\eta \times \eta$ matrix, the $(n+1)$-th iterated weight w_s is expressed as (A.8) and the mean and variance matrix are given by

$$(\boldsymbol{\phi}_s^e)_i^{(n+1)} = \frac{\sum_r T_{rs}^{(n)} (\mathbf{x_r})_i}{\sum_r T_{rs}^{(n)}} \tag{A.12}$$

and

$$(\Omega_s)_{ij}^{(n+1)} = \frac{\sum_r T_{rs}^{(n)} [\ (\mathbf{x_r})_i - (\boldsymbol{\phi}_s^e)_i^{(n+1)}\ ]\ [\ (\mathbf{x_r})_j - (\boldsymbol{\phi}_s^e)_j^{(n+1)}\ ]}{\sum_r T_{rs}^{(n)}}\ . \tag{A.13}$$

(4): If the multivariate distributions have a common covariance matrix, $\Omega_s = \Omega$, then (A.13) becomes

$$\Omega_{ij}^{(n+1)} = \frac{1}{J} \sum_r (\mathbf{x}_r)_i (\mathbf{x}_r)_j - \sum_s w_s^{(n)} (\phi_s^e)_i^{(n+1)} (\phi_s^e)_j^{(n+1)} . \quad \text{(A.14)}$$

(5): If the backprojected data on the image-voxel array are distributed following a Gauss-Markov random process, the function $f(.)$ can be expressed as (Besag, 1974)

$$f(x_r \mid \theta_s, \zeta_s) = (2\pi\sigma_s^2)^{-1/2} \exp\{ - [(x_r - \phi_s^e) - \sum_{l \varepsilon M_r} \zeta_{sl} (x_l - \phi_s^e)]^2 / (2\sigma_s^2) \}$$

where ζ_{sl} reflects the correlations between voxel r and its M_r neighbors. If only first-order neighbors in a two dimensional case are considered,

$$\sum_l \zeta_{sl} (x_l - \phi_s^e) = \zeta_{sH} \sum_H (x_H - \phi_s^e) + \zeta_{sV} \sum_V (x_V - \phi_s^e)$$

where the indices H and V cover the pixels within horizontal and verticle cliqes respectively (Besag, 1974). The weight parameter w_s, at $(n+1)$-th iteration, is given by (A.8). The mean and variance parameters are determined by

$$(\phi_s^e)^{(n+1)} = \frac{\sum_r T_{rs}^{(n)} [1 - \sum_l \zeta_{sl}^{(n+1)}] [x_r - \sum_l \zeta_{sl}^{(n+1)} x_l]}{\sum_r T_{rs}^{(n)} [1 - \sum_l \zeta_{sl}^{(n+1)}]^2} \quad \text{(A.15)}$$

and

$$(\sigma_s^2)^{(n+1)} = \frac{\sum_r T_{rs}^{(n)} \{ [x_r - (\phi_s^e)^{(n+1)}] - \sum_l \zeta_{sl}^{(n+1)} [x_l - (\phi_s^e)^{(n+1)}] \}^2}{\sum_r T_{rs}^{(n)}} . \quad \text{(A.16)}$$

The correlation parameters $\zeta_{sl} = (\zeta_{sH}, \zeta_{sV})$ are expressed as

$$\zeta_{sH}^{(n+1)} = \frac{(a_V b_{HV}) - (a_H b_{VV})}{(b_{HV})^2 - (b_{HH} b_{VV})} \quad \text{and} \quad \zeta_{sV}^{(n+1)} = \frac{(a_H b_{HV}) - (a_V b_{HH})}{(b_{HV})^2 - (b_{HH} b_{VV})} \quad \text{(A.17)}$$

where

$$a_H = \sum_r \{ T_{rs}^{(n)} [x_r - (\phi_s^e)^{(n+1)}] \sum_H [x_H - (\phi_s^e)^{(n+1)}] \} , \quad a_V = \sum_r \{ T_{rs}^{(n)} [x_r - (\phi_s^e)^{(n+1)}] \sum_V [x_V - (\phi_s^e)^{(n+1)}] \} , \text{(A.18)}$$

$$b_{HH} = \sum_r T_{rs}^{(n)} \{ \sum_H [x_H - (\phi_s^e)^{(n+1)}] \}^2 , \quad b_{VV} = \sum_r T_{rs}^{(n)} \{ \sum_V [x_V - (\phi_s^e)^{(n+1)}] \}^2 ,$$

$$\text{and} \quad b_{HV} = \sum_r \{ T_{rs}^{(n)} \sum_H [x_H - (\phi_s^e)^{(n+1)}] \sum_V [x_V - (\phi_s^e)^{(n+1)}] \} .$$

The $(\phi_s^e)^{(n+1)}$ in (A.18) can be replaced by $(\phi_s^e)^{(n)} + \rho [(\phi_s^e)^{(n)} - (\phi_s^e)^{(n-1)}]$ as Eq.(10) did, where $0 \le \rho \le 1$.

Note that the information criteria provide a possible means to theoretically determine the intensity levels. The level parameter values can be computed using the iterative algorithms above. The convergence of the EM technique was discussed by Dempster et al (1977). The iterative process can be terminated by

$$\sum_s [w_s^{(n+1)} - w_s^{(n)}]^2 / \sqrt{K-1} \le \varepsilon \quad \text{(A.19)}$$

where ε is a prespecified small value. Both the information criteria measure the Kullback-Leibler mean distance [or the directed Kullback divergence (Kullback, 1959)] between the estimated parameter model and the parameter model which generates the measured data, and selected that model which minimizes the criterion (A.1) or (A.2).

It has been shown that when *AIC* is used in model selection, that model with minimum *AIC* gives the minimum mean squared error of prediction (Efron, 1986). If the maximum likelihood $\Psi(.)$ is identical for two models, *AIC* selects that one with fewer freely adjustable parameters $\{\theta_s\}$.

The difference between *AIC* and *MDL* is due to the penalty terms of $(2K_0)$ for *AIC* and $(K_0 \ln J_0)$ for *MDL*. The former assumes a uniform prior to all model candidates and proposes to select the model which yields the maximum likelihood, so the penalty term of *AIC* is independent of the number of measured data. The latter one, however, assigns each competing model an *a priori* probability and selects the model which yields the maximum *a posteriori* probability. Detailed comments on the two criteria can be found in references (Akeike, 1978; Baxter, 1990; Stone, 1979).

References

Akaike H (1974). A New Look at the Statistical Model Identification. IEEE Trans. Auto. Control, vol.19, pp.716-723.

Akaike H (1978). A Bayesian Analysis of the Minimum AIC Procedure. Ann. Inst. Statist. Math., vol.30, pp.9-14.

Baxte r L (1990). Futures of Statistics. The Amer. Statistician, vol.44, pp.128-129.

Besag J (1974). Spatial Interation and the Statistical Analysis of Lattice Systems. J. R. Stat. Soc., series B, vol.26, pp.192-236.

Budinger T and Gullberg G (1974). Three-Dimensional Reconstruction in Nuclear Medicine Emission Imaging. IEEE Trans. Nucl. Sci., vol.21, pp.2-20.

Dempster A, Laird N and Rubin D (1977). Maximum Likelihood from Incomplete Data via the EM Algorithm. J. R. Stat. Soc., vol.39, series B, pp.1-38.

Derin H and Elliott H (1987). Modeling and Segmentation of Noisy and Textured Images Using Gibbs Random Fields. IEEE Trans. Pattern Anal. Machine Intell., vol.9, pp.39-55.

Efron B (1986). How Biased is the Apparent Error Rate of a Prediction Rule. J. Am. Stat. Assoc., vol.81, pp.461-470.

Gilland D, Jaszczak R, Greer K. Coleman R (1991). Quantitative SPECT Reconstruction of Iodine-123 Data. J. Nucl. Medicine, to appear.

Goris M and Briandet P (1983). *A Clinical and Mathematical Introduction to Computer Processing of Scintigraphic Images.* Raven Press, New York.

Hall E (1979). *Computer Image Processing and Recognition.* Academic Press, New York.

Herman G (1980). *Image Reconstruction from Projections: the fundamentals of computerized tomography.* Academic Press, New York.

Johnson V (1989). Bayesian Restoration of PET Images Using Gibbs Priors. In: *Information Processing in Medical Imaging*, vol.11, Berkerley, CA.

Kullback S (1959). *Information Theory and Statistics.* John Wiley & Sons, Inc., New York.

Lei T and Sewchand W (1988). A New Stochastic Model-Based Image Segmentation Technique for X-ray CT Images. In: SPIE, *Visual Communication and Image Processing* 3, Boston, MA.

Lei T and Sewchand W (1989). Image Segmentation by Using Finite Normal Mixture Model and EM Algorithm. In: *IEEE Intern. Conf. on Image Processing*, Singapore.

Liang Z and Hart H (1987). Bayesian Image Processing of Data from Constrained Source Distributions: fuzzy pattern constraints. Phys. Med. Biol., vol.32, pp.1481-1494.

Liang Z and Hart H (1988a). Source Continuity and Boundary Discontinuity Considerations in Bayesian Image Processing. Med. Physics, vol.15, pp.754-756.

Liang Z and Hart H (1988b). Bayesian Reconstruction in Emission Computerized Tomography. IEEE Trans. Nucl. Sci., vol.35, pp.877-885.

Liang Z, Jaszczak R, Coleman R and Johnson V (1991a). Simultaneous Reconstruction, Segmentation, and Edge Enhancement of Relatively Piecewise Continuous Images with Intensity-Level Information. Med. Physics, vol.18, in press.

Liang Z, Gilland, Jaszczak R and Coleman R (1991b). Implementation of Non-Linear Filters for Iterative Penalized Maximum Likelihood Image Reconstruction. In: *IEEE Conf. Med. Imaging*, Arlington, VA.

Luenberger D (1973). *Introduction to Linear and Non-Linear Programming.* Addison-Wesley, Reading, MA.

Margulis A and Gooding C (1989). *Diagnostic Radiology.* J. B. Lippincott Comp., Philadelphia.

Pohost G, Higgins C, Morganroth J, Ritchie J and Schelbert H (1989). *New Concepts in Cardiac Imaging.* Year Book Med. Publish., Inc., Chicago.

Redner R and Walker H (1984). Mixture Densities, Maximum Likelihood and the EM Algorithm. SIAM Review, vol.26, pp.195-239.

Rissanen J (1978). Modeling by Shortest Data Description. Automatica, vol.14, pp.465-471.

Schwarz G (1978). Estimating the Dimension of a Model. Annal. Stat., vol.6, pp.461-464.

Shepp L and Vardi Y (1982). Maximum Likelihood Reconstruction for Emission Tomography. IEEE Trans. Med. Imaging, vol.1, pp.113-122.

Stone M (1979). Comments on Model Selection Criteria of Akaike and Schwarz. J. Royal Statist. Soc., vol.41, pp.276-278.

Titterington T, Smith A and Makov V (1985). *Statistical Analysis of Finite Mixture Distributions.* John Wiley & Sons, Inc., New York.

Young T and Fu K (1986). *Handbook of Pattern Recognition and Image Processing.* Academic Press, New York.

INCORPORATION OF ANATOMICAL MR DATA FOR IMPROVED FUNCTIONAL IMAGING WITH PET*

R Leahy and X Yan

Signal and Image Processing Institute,
Department of Electrical Engineering-Systems,
University of Southern California,
Los Angeles, California 90089-0272

Abstract

A statistical approach to PET image reconstruction offers several potential advantages over the filtered backprojection method currently employed in most clinical PET systems: (1) the true data formation process may be modeled accurately to include the Poissón nature of the observation process and factors such as attenuation, scatter, detector efficiency and randoms; and (2) an *a priori* statistical model for the image may be employed to model the generally smooth nature of the desired spatial distribution and to include information such as the presence of anatomical boundaries, and hence potential discontinuities, in the image. In this paper we develop a Bayesian algorithm for PET image reconstruction in which a magnetic resonance image is used to provide information about the location of potential discontinuities in the PET image. This is achieved through the use of a Markov random field model for the image which incorporates a "line process" to model the presence of discontinuities. In the case where no *a priori* edge information is available, this line process may be estimated directly from the data. When edges are available from MR images, this information is introduced as a set of known *a priori* line sites in the image. It is demonstrated through computer simulation, that the use of a line process in the reconstruction process has the potential for significant improvements in reconstructed image quality, particularly when prior MR edge information is available.

Keywords

Positron Emission Tomography; Markov Random Field; MAP Reconstruction.

1 INTRODUCTION

Magnetic Resonance (MR) imaging is capable of producing high resolution images of the human anatomy. An appropriate choice of pulse sequence can produce T1 or T2 weighted images (Mansfield and Morris, 1982) with excellent soft tissue contrast. Positron emission tomography (PET) (Muehllehner and Colsher, 1982) is capable of producing somewhat lower resolution images of functional activity through the use of radiolabeled pharmaceuticals. While PET and MR produce images of different parameters, the underlying anatomical structures and the boundaries between them are common to both images; in other words, the spatial distribution of functional activity is dependent on the underlying anatomical structure in which the activity occurs. Thus if one could extract high resolution structural information from the MR image, this could be used as important *a priori* information about potential boundaries between structures in the PET image. In this paper we describe a method by which this information, extracted from MR images, may be incorporated into the PET image reconstruction process.

*This work was supported by the Z. A. Kaprielian Innovative Research Fund at the University of Southern California, the National Science Foundation (grant MIP-8708708) and the Whitaker Foundation.

The PET imaging problem may be formulated as one of tomographic image reconstruction. Reconstruction of PET images in most commercial PET systems is based on the well known filtered backprojection (FBP) algorithm (Shepp and Logan, 1979) initially developed for transmission x-ray computed tomography. Although this reconstruction algorithm is computationally straight forward, the quality of the images is severely limited by the failure of the algorithm to take into account the random nature of the data. The publication (Shepp and Vardi, 1982) of a new method for PET image reconstruction based on a maximum likelihood (ML) formulation sparked considerable interest in statistical approaches to PET reconstruction. In their seminal paper, Shepp and Vardi develop a statistical model for the data and apply the EM (expectation maximization) method (Dempster et al, 1977) to maximize the resulting likelihood function. The images obtained in this manner are generally considered qualitatively better than those from FBP. However, it has been widely noted that after several iterations of the EM algorithm the image quality begins to deteriorate due to the ill-conditioned nature of the ML problem in PET. Methods to overcome this problem include early termination of the EM iterations using a statistical hypothesis test (Hebert et al, 1988), (Veklerov and Lacer, 1987) and regularization using sieves (Miller et al, 1986), smoothing (Silverman et al, 1990) or Bayesian methods (Geman and McClure, 1985), (Hebert and Leahy, 1989) and (Johnson et al, 1990). All these techniques have been shown to alleviate the divergence problem to some extent. The stopping rule (Hebert et al, 1988), (Veklerov and Lacer, 1987) is theoretically unappealing since one begins by searching for an optimal (ML) solution, but then terminates the search before reaching this point. Similarly, the method of smoothing (Silverman et al, 1990) results in a suboptimal solution. The method of sieves effectively solves the problem by restricting the solution space using a set of regularizing functions. While this method is statistically sound, one often suffers from edge artifacts or over-smoothing due to the choice of sieve kernel. In the following we will restrict our attention to Bayesian approaches to the reconstruction problem.

Through the use of Bayes theorem, one can introduce a prior distribution to describe the statistical properties of the unknown image and thus produce a posterior probability distribution for the image conditioned upon the data. Maximization of some function of this distribution over the set of allowed images results in a Bayesian estimate of the PET image. Although one can select from a number of alternative cost criteria within this framework, by far the most popular, and that used in the following, is the maximum *a posteriori* (MAP) estimator in which the solution is chosen as that image which maximizes the posterior probability. Liang and Hart have developed MAP estimators based on the EM algorithm for Gaussian and Poisson priors (Liang and Hart, 1987). Levitan and Herman have also developed an EM algorithm for an uncorrelated Gaussian prior in (Levitan and Herman, 1987), using a smoothed FBP reconstruction as an estimated mean.

An alternative class of models, the Gibbs distribution or Markov random field (MRF) (Besag, 1986), are based on the observation that a property common to almost all images is one of local correlation, i.e. there is a high degree of spatial correlation in image intensity between pixels and their neighbors. This property is captured by the MRF by specifying the conditional probability for each pixel as a function only of those pixels in the neighborhood of the pixel. It has been shown that the joint distribution for an MRF has a special form termed the Gibbs distribution (Besag, 1986) which facilitates both the formulation of problems involving these models and efficient numerical computation of the resulting estimator . MRFs have been applied to PET image reconstruction by (Geman and McClure, 1985), (Hebert and Leahy, 1989), (Green, 1990), (Johnson et al, 1990) and (Yan and Leahy, 1991).

A problem which arises with the use of MRF models is that in choosing a prior which favors strong local correlations, large variations in image intensity are penalized. The resulting effect is that MAP estimators employing MRFs may produce over smoothed reconstructions particularly across true boundaries in the image. This problem may be alleviated to some extent by choosing the smoothing term to penalize intensity differences only up to a certain threshold (Hebert and Leahy, 1989), (Geman and McClure, 1985). A more satisfying approach to this problem, and one which allows us to incorporate the *a priori* MR information as described in section 4, is the use of "line sites" in the model (Geman and Geman, 1984). In this model, in addition to the pixel sites representing the image intensity, one introduces an additional variable between each pair of neighboring pixels - this variable should take values of either "1" to denote the presence of a boundary between the two pixels or "0" to represent the absence of a boundary. In (Yan and Leahy, 1991) we constructed a prior which is a function of both the pixel intensities and the line sites. Then when computing the interaction terms between pixels, one only considers those pixel pairs for which the associated line site is "0" (i.e. no boundary); in this way one avoids smoothing across boundaries in the reconstructed image. In addition we added a penalty function to avoid forming too many boundaries. Finally a smoothing term was added to the line sites themselves, to attempt to obtain closed and connected boundaries while discouraging redundant boundaries and very small regions. A similar approach was employed by (Johnson et al, 1990) for PET imaging. However, in that case no penalty term was employed to limit the formation of too many boundaries (i.e. line sites = 1). It can be shown that in the absence of such a penalty term, the algorithm (Johnson et al, 1990) will converge to a ML solution.

In the case where MR images are available, edge information may be extracted from these images and used as known *a priori* line sites in the MRF formulation. In the following we describe an algorithm for combining PET data and MR edge information in a Bayesian framework using an MRF prior with both intensity and line processes. An alternative algorithm for performing this task was developed independently by (Chen et al, 1990) based on a direct extension of the method described in (Johnson et al, 1990).

2 PET SYSTEM MODELING

An important requirement for quantitative PET imaging is that the system be accurately modeled. For algorithms such as ML and MAP that are based on probabilistic approaches, this means that one should have an accurate statistical description for the data as a function of the unknown parameters $\lambda(i,j)$ *i.e.* the image. Let y_k, the k-th element of observed data Y, denotes the number of coincidences detected between the kth detector pair, and λ denote an $N \times N$ array of mean emission rates from the source, thus $\lambda(i,j)$ represents the mean emission rate from pixel (i,j) of the source image.

Consider the number of photons y_k detected between detector pair k. This number is the sum of three components: t_k – the number of true coincidence events detected; s_k – the number of events detected after one or both of the primary photons has undergone scattering; and r_k – the number of random coincidences, *i.e.* two independent photons arriving at the detector pair within the coincidence window and counted as a true coincidence. Then:

$$y_k = t_k + s_k + r_k. \tag{1}$$

The true, unscattered coincidences, t_k, form the desired data, *i.e.* the tomographic projection data or 'sinogram', the latter two components introduce errors that must be corrected for. In the majority of commercial PET systems randoms and scatter are corrected for by a subtraction, *i.e.* the randoms rate and scatter fraction are estimated (Muehllehner and Colsher, 1982) for each detector pair, this estimate is then subtracted from the observed data to yield an estimate of the true coincidences (Bergstrom et al, 1980). More sophisticated methods such as correction for scatter by deconvolution of the data (King et al, 1981) have also been proposed. However, all such forms of data correction pose a major problem for statistically based iterative methods such as those to be developed here. This problem is that, although the original data are well modeled as a set of independent Poisson random variables (Shepp and Vardi, 1982), any corrections applied to this data change its statistical distribution and hence approaches based on the Poisson model become inappropriate. This problem has been largely overlooked in the majority of papers dealing with ML and MAP estimation, *i.e.* the assumption is often made that the data y_k are all true, unscattered coincidences. Let us consider how one can incorporate these other factors into a statistical approach.

True coincidences: the probability, $P_k^T(i,j)$ of detecting a true coincidence (photon pair) due to a positron emission from pixel (i,j) at detector pair k is dependent on the following factors (Muehllehner and Colsher, 1982): positron range and variations in angular separation from 180°, the probability of attenuation, the spatial and energy resolution of the detectors, detector efficiency and dead time and the position of pixel (i,j) relative to the detector pair k. These factors are, with the exception of detector dead time, independent of the source activity, thus if $\lambda(i,j)$ denotes the mean emission rate from pixel (i,j), one can model t_k as a Poisson random variable with mean:

$$\mathbf{E}\{t_k\} = \sum_{i,j} P_k^T(i,j)\lambda(i,j). \tag{2}$$

This is the basis for the model developed by (Shepp and Vardi, 1982) and is used by most other researchers in the development of statistical PET reconstruction methods. Techniques for calibrating PET systems in which the above factors are measured have been developed extensively in the nuclear medicine instrumentation community and are available on most commercial PET systems, *e.g.* calibration for detector efficiency is performed on a routine basis and corrections for attenuation are found from measurements obtained using a rotating external transmission source (Huesman et al, 1988). These factors can be included in the probability $P_k^T(i,j)$ with little theoretical problem. An important issue is how to incorporate them so that the matrix, $P_k^T(i,j)$, may be handled efficiently (note that this matrix typically has of the order of 10^7 elements). This may be achieved by a factorization of the elements $P_k^T(i,j)$ into a product of probabilities, each due to one of the factors (efficiency, attenuation, etc.) described above. This approach was taken in a recent paper on electronically collimated SPECT (Hebert et al, 1990) in which we were able to reduce the associated computer storage requirements for this matrix by several orders of magnitude.

Scattered coincidences: these are coincidence detections which occur when one or both of the primary photons produced by a positron/electron annihilation is scattered within the body. If one does not consider the possibility of scatter in developing the data model, then the annihilation is considered to have occurred along the strip joining the two detectors - as would be the case for unscattered coincidences. Since scatter is a significant fraction of the total number of detections (e.g. $16\% - 18\%$ in the Siemens ECAT953) these events cannot be ignored. As with true unscattered coincidences, the mean scattered radiation scales linearly with source activity and the number of detected scattered events is well modeled as a Poisson variable with mean:

$$\mathbf{E}\{s_k\} = \sum_{i,j} P_k^S(i,j)\lambda(i,j). \tag{3}$$

Again, one must calculate the matrix, whose elements are $P_k^S(i,j)$. This is difficult since scatter is dependent on the density of the material along the photon path. Since the scatter fraction is a small, though not insignificant, component of the data one need probably not be as accurate in modeling the scatter as in modeling the true coincidences. In fact, reasonable qualitative, but not quantitative, results have been obtained by simply ignoring scatter. A reasonable compromise between accuracy and computational complexity is to model scatter using a spatially dependent convolutional blurring of the data at the detectors. A similar method has been applied for data correction before reconstruction (King et al, 1981), however, since the scatter varies as a function of depth, this is not strictly correct and the correction is more appropriately included in $P_k^S(i,j)$.

Random Coincidences: these coincidence detections arise from the simultaneous detection of two photons produced by two different positron annihilations. These differ from true and scattered coincidences in that the randoms rate scales as approximately the square of the source activity as opposed to linearly (Muehllehner and Colsher, 1982). In most PET scanners randoms are corrected for by measuring a randoms rate using a delayed timing window (Holmes et al, 1984) and subtracting this from the measured data. This correction is inappropriate for our statistical model since although both the data and the measured randoms are Poisson processes, their difference is not. One way in which randoms may be included in a statistical formulation is to maintain the off-timing randoms counts r_k as an additional set of Poisson data with unknown mean, q_k.

We can combine the above models to arrive at a complete probabilistic description of the data. Each of the three components in equation (1) is well modeled as a Poisson random variable. Since the sum of Poissons is itself Poisson, we obtain as data a set of Poisson random variables with mean:

$$\mathbf{E}\{y_k\} = \sum_{i,j} P_k^T(i,j)\lambda(i,j) + \sum_{i,j} P_k^S(i,j)\lambda(i,j) + q_k \tag{4}$$

To develop a statistical algorithm incorporating all of the above factors, the observed data, y_k, may be combined with the randoms data, r_k, to form a joint probability distribution. In the following section we give an EM algorithm for PET image reconstruction using this approach. An alternative is to proceed in the normal fashion, subtract the randoms counts from the observed coincidence data and assume the resulting measurements are Poisson, although this clearly introduces some degree of modeling error into the reconstruction process.

3 STATISTICAL APPROACHES TO PET

In this section we describe several statistical approaches to PET reconstruction based on the generalized EM algorithm. We begin in section 3.1 with a definition of the generalized EM algorithm for parameter estimation. In the following subsections we then apply this method to maximum likelihood and MAP estimation of PET images.

3.1 The GEM Algorithm for PET

The MAP estimate of a set of parameters A given a set of data B is the set which maximizes the posterior probability $P(A|B)$, i.e.

$$\max_A P(A|B) = \max_A \left\{ \frac{P(B|A)P(A)}{P(B)} \right\}.$$

where $P(B|A)$ denotes the conditional probability for the data given A and $P(A)$ denotes a prior probability distribution on A. The EM algorithm is a general method for solving this problem by choosing an intermediate set of unobserved "complete" data C such that there is a known many-to-one mapping from C to B. The problem may then be solved by choosing some feasible initial estimate A^0

of A and applying the iteration:

$$\begin{array}{ll} \text{E-step:} & \mathbf{E}_C \{\log P(C|A)|A^n, B\} \\ \text{M-step:} & A^{n+1} = \arg\{\max_A Q(A|A^n)\}, \quad n \Leftarrow n+1, \end{array} \tag{5}$$

where the function $Q(A|A^n)$ is defined as

$$Q(A|A^n) = \mathbf{E}_C \{\log P(C|A)|A^n, B\} + \log P(A).$$

A more general form of this algorithm, the generalized EM (GEM) algorithm, replaces the M-step with an updating procedure for A which satisfies the following inequality:

$$Q(A^{n+1}|A^n \geq Q(A^n|A^n).$$

The convergence properties of this method are discussed in (Dempster et al, 1977) and (Wu, 1982).

3.2 Maximum Likelihood via EM

For PET image reconstruction, the parameters to be estimated are the emission intensities, i.e. $A = \lambda$ where λ denotes the set of pixel intensities $\lambda(i,j)$, $i,j = 1,\cdots,N$. The observed data are the set of coincident detections denoted by $B = Y$, with elements y_k $k = 1,\cdots,M$. Let $P_k(i,j)$ denote the known probability that an emission from source point (i,j) is detected at the detector pair corresponding to index k. The unobserved complete data C, as defined in (Shepp and Vardi, 1982), are denoted $x_k(i,j)$ and are defined to be the number of emissions from pixel (i,j) which are collected in detector pair k. The variables $x_k(i,j)$ are well modeled as conditionally independent Poisson random variables with mean $P_k(i,j)\lambda(i,j)$. The incomplete data y_k are related to the complete data $x_k(i,j)$ as $y_k = \sum_{i,j} x_k(i,j)$ and hence are also conditionally independent Poisson random variables with mean $\sum_{i,j} P_k(i,j)\lambda(i,j)$ (Vardi et al, 1985).

Maximum likelihood estimation is equivalent to MAP estimation with a uniform prior, i.e. $P(\lambda)$ is a constant. Applying the EM algorithm to this problem we obtain the result

$$\mathbf{E}_X \{logP(X|\lambda)|\lambda^n, Y\} = \sum_{i,j} \{-a(i,j)\lambda(i,j) + b^n(i,j)\log\lambda(i,j)\} + consant, \tag{6}$$

where

$$\begin{array}{rcl} a(i,j) & = & \sum_k P_k(i,j) \\ b^n(i,j) & = & \lambda^n(i,j) \sum_k \frac{y_k P_k(i,j)}{\sum_{i',j'} P_k(i',j')\lambda^n(i',j')}. \end{array} \tag{7}$$

Substituting this result into the M-step and optimizing over λ, we obtain the well known EM algorithm for MLE:

$$\lambda^{n+1}(i,j) = \lambda^n(i,j) \frac{\sum_k \frac{y_k P_k(i,j)}{\sum_{i',j'} P_k(i',j')\lambda^n(i',j')}}{\sum_k P_k(i,j)}. \tag{8}$$

In the case where the random measurements are treated as an additional data set one can modify the above EM algorithm by choosing the data as $B = \{Y, R\}$, where R denotes the set of random measurements, and the unknown parameters to be estimated as $A = \{\lambda, Q\}$, where Q denotes the mean of the randoms process. In this case, one chooses initial feasible estimates, λ^0 and Q^0, of λ and Q

respectively and performs the following iteration:

$$q_k^{n+1} = \frac{1}{2}\left[r_k + \frac{y_k q_k^n}{\sum_{i,j} P_k(i,j)\lambda^n(i,j) + q_k^n}\right], \tag{9}$$

$$\lambda^{n+1}(i,j) = \frac{\lambda^n(i,j)}{\sum_k P_k(i,j)} \sum_k \frac{y_k P_k(i,j)}{\sum_{i',j'} P_k(i',j')\lambda^n(i',j') + q_k^n}. \tag{10}$$

We expect the above iteration to suffer from the same ill-conditioning which usually occurs in conventional ML PET reconstruction. This problem arises due to the high dimensionality of the estimated parameters relative to the data. To overcome this problem we could employ MAP estimators similar to those described below, but with an additional prior distribution to perform spatial smoothing of the randoms process, Q.

3.3 MAP using MRFs with Intensity Process Only

In (Hebert and Leahy, 1989) we described a GEM algorithm for MAP estimation where the prior distribution for λ is a Markov random field:

$$P(\lambda) = \frac{1}{Z} e^{-\gamma U(\lambda)}, \tag{11}$$

where Z a normalizing constant, γ is a prior parameter and $U(\lambda)$ denotes a Gibbs energy function defined on the set

$$\Omega = \{\lambda(i,j) \in \mathcal{R} \mid \lambda(i,j) \geq 0; i,j = 1,\cdots,N\}. \tag{12}$$

Applying the GEM algorithm to this problem one obtains the following procedure:

$$\begin{array}{ll} \text{E-step:} & \mathbf{E}_X\{logP(X|\lambda)|\lambda^n, Y\} = \sum_{i,j}\{-a(i,j)\lambda(i,j) + b^n(i,j)\log\lambda(i,j)\} + constant \\ \text{M-step:} & \max_\lambda Q(\lambda|\lambda^n) = \mathbf{E}_X\{\log P(X|\lambda)|\lambda^n, Y\} - \gamma U(\lambda). \end{array} \tag{13}$$

The E-step remains the same as for the ML-EM algorithm. A suitable M-step is described in (Hebert and Leahy, 1989). In this case the definitions of the parameters, A, and data, B, remain the same as for the original form of the ML-EM algorithm. Although with a suitable choice of $U(\lambda)$, this method avoids the ill-posedness of the ML approach, the algorithm tends to over-smooth the reconstructed image particularly across true boundaries in the image. To overcome this problem, we now consider the use of MRFs with an additional line process.

3.4 MAP using MRFs with Intensity and Line Process

Two algorithms are described in (Yan and Leahy, 1991) for solving the MAP reconstruction problem, in which the unknown image was modeled using a joint Gibbs distribution of emission intensities and line processes. The line process is used to model discontinuity-like edges in images (Geman and Geman, 1984), and can be regarded as an indicator variable which is set to zero wherever neighboring pixels lie in the same region (*i.e.* have similar values) and is otherwise set to one. Let $\{l(i,j)\}$ represent line sites lying between each pair of neighboring intensities, as illustrated in Figure 1. For the purposes of computation, each $l(i,j)$ may be either binary valued $\{0,1\}$ or take on any real value on the interval $[0,1]$. Then the parameters to be estimated in the GEM algorithm are $A = \{\lambda, l\}$. The joint Gibbs distribution of λ and l may be written in the form (Geman and Geman, 1984):

$$P(\lambda, l) = \frac{1}{Z} e^{-\gamma U(\lambda, l)}, \tag{14}$$

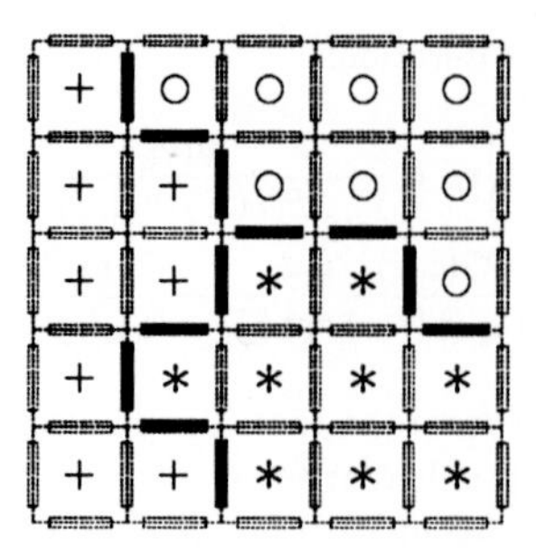

Figure 1: Line sites are introduced between each pair of neighboring pixels, and take the value "1" to indicate the presence of a boundary between adjacent pixels and otherwise take the value "0". Three different regions are represented using '+', '*' and 'o'. The line sites are equal to "1" (solid bars) on the boundaries of these regions.

where γ is a prior parameter. The Gibbs energy function $U(\lambda, l)$ is defined on the set $\Omega \times \mathcal{I}^L$, where Ω is defined in (12). The notation $\mathcal{I}$, for the case where the line processes $l(i,j)$ are binary random variables, represents the set $\{0,1\}$; for the case where $l(i,j)$ are treated as continuous valued random variables defined on the interval $[0,1]$, then $\mathcal{I} = [0,1]$. L denotes the total number of line process variables.

The GEM algorithm for the MAP problem in this case becomes

$$\begin{array}{ll} \text{E-step:} & \mathbf{E}_X\left\{logP(X|\lambda)|\lambda^n, Y\right\} = \sum_{i,j}\left\{-a(i,j)\lambda(i,j) + b^n(i,j)\log\lambda(i,j)\right\} + constant \\ \text{M-step:} & \max_{\lambda,l} Q(\lambda, l|\lambda^n, l^n) = \mathbf{E}_X\left\{\log P(X|\lambda)|\lambda^n, Y\right\} - \gamma U(\lambda, l). \end{array} \tag{15}$$

Note that in this case, the E-step remains identical to that in (6), only the M-step is changed. The parameters to be estimated in the M-step are the intensities $\lambda(i,j)$ and the lines $l(i,j)$. In (Yan and Leahy, 1991) the M-step in the GEM algorithm is selected as an updating procedure for both λ and l that satisfies the following condition at each iteration

$$Q\left(\lambda^{n+1}, l^{n+1}|\lambda^n, l^n\right) \geq Q\left(\lambda^n, l^n|\lambda^n, l^n\right).$$

This is achieved in two steps:

- M1-step: find estimates of intensities λ^{n+1} satisfying

$$Q\left(\lambda^{n+1}, l^n|\lambda^n, l^n\right) \geq Q\left(\lambda^n, l^n|\lambda^n, l^n\right); \tag{16}$$

- M2-step: find estimates of lines l^{n+1} satisfying

$$Q\left(\lambda^{n+1}, l^{n+1}|\lambda^n, l^n\right) \geq Q\left(\lambda^{n+1}, l^n|\lambda^n, l^n\right). \tag{17}$$

  Note that for this M2-step, $\mathbf{E}_X\left\{\log P(X|\lambda)|\lambda^n, Y\right\}$ is a constant with respect to l^{n+1} and thus (17) is equivalent to the following:

$$-U(\lambda^{n+1}, l^{n+1}) \geq -U(\lambda^{n+1}, l^n). \tag{18}$$

In (Yan and Leahy, 1991) we constructed two energy function and described two algorithms: GEM-1 and GEM-2, for solving this problem by following the general recipe (15). The GEM-1 algorithm is a simple gradient-type method that uses a continuous valued line process. The GEM-2 algorithm, in which line processes are treated as binary random variables, replaces the use of the gradient in the M2-step with an updating rule based on the conditional posterior probability.

3.5 MAP Incorporating Prior Edge Information

Although the above approach has produced encouraging reconstructions (Yan and Leahy, 1991), we have observed through computer simulation that the estimated line sites are not always consistent with the original image and are sometimes inaccurately located. This is due to both the finite resolution of the simulated PET system, and the random nature of the data. A MAP reconstruction approach, including line-site estimation, has the potential for further enhancing the quality of PET reconstructions if accurate line locations can be extracted from images obtained from an anatomical imaging modality (MR or CT) and incorporated in the PET algorithm.

Let us now assume that the line sites, l, consist of two parts: 1) l_a: the set of known edges obtained through some edge detection process from an MR image; and 2) $l_F = \{l - l_a\}$. This second set may contain additional edges due to the presence of discontinuities in the PET image which do not correspond to anatomical boundaries. It is important to include and estimate the set l_F since it is possible to obtain additional edges in the PET image due to sharp changes in function within a single anatomical region. For this problem, the parameters to be estimated are $A = \{\lambda, l_F\}$; the observed data are $B = \{Y\}$ and l_a are treated as *a priori* information. The GEM algorithm (15) in this case can then be written as:

$$\begin{array}{ll} \text{E-step:} & \mathbf{E}_X\left\{logP(X|\lambda)|\lambda^n, Y\right\} = \sum_{i,j}\left\{-a(i,j)\lambda(i,j) + b^n(i,j)\log\lambda(i,j)\right\} + constant \\ \text{M-step:} & \max_{\lambda, l_F} Q(\lambda, l_F|\lambda^n, l_F^n) = \mathbf{E}_X\left\{\log P(X|\lambda)|\lambda^n, Y\right\} - \gamma U(\lambda, l_F, l_a). \end{array} \quad (19)$$

In section 4, a GEM algorithm for solving this problem is described.

4 MAP FOR PET WITH INCORPORATION OF MR DATA

In this section, we describe a MAP reconstruction algorithm for incorporation of anatomical MR boundaries. This algorithm is a modified version of the GEM-2 algorithm for the case where *a priori* edge information is available. In order to simplify the notation, we consider only first-order neighbors for the intensity and horizontal and vertical line fields in the following derivation. The model can be easily extended to higher-order neighbors and vector line fields. Due to the relatively small and complex boundaries appearing in our phantom (Figure 2(d)), we have used a second-order neighborhood and four-direction line fields in our simulations.

We denote the line process by l_s, where $s = \{F, a\}$. Let l_a denote the known locations of edges extracted from the segmented MR image and l_F the unknown functional boundaries to be estimated. Where appropriate we add a second index to l_s to denote the orientation of the line site: $l_{p,s}$ where $p = x$ denotes a horizontal line site and $p = y$ denotes a vertical line site. A suitable energy function for first order interactions is then[1]:

$$\begin{aligned} U(\lambda, l_s) &= \sum_{i,j}\left\{\beta\lambda_x^2(i,j)(1 - l_{y,s}(i,j)) + \alpha l_{y,s}(i,j) + \mathcal{H}_y(l_s)\right\} + \\ &\quad \sum_{i,j}\left\{\beta\lambda_y^2(i,j)(1 - l_{x,s}(i,j)) + \alpha l_{x,s}(i,j) + \mathcal{H}_x(l_s)\right\}, \end{aligned} \quad (20)$$

The line sites $l_{p,s}(i,j)$ are binary random variables. In each pair of braces, the first term reflects our

[1]For a finite lattice, care must be taken in defining potential functions on cliques near the image boundaries. For notational simplicity, in the following we assume that appropriate modifications are made to the potential functions at the image boundary.

belief that the image values should not change abruptly except at discontinuities. The second term penalizes each non-zero line site in the image; $\mathcal{H}_p(l_s)$ denotes a measure of interaction between lines which encourages the estimated edges to be connected while suppressing the formation of very small regions and redundant lines. $\lambda_x(i,j)$ and $\lambda_y(i,j)$ are the differences between adjacent horizontal and vertical intensities respectively, defined as

$$\begin{aligned} \lambda_x(i,j) &= \lambda(i-1,j) - \lambda(i,j) \\ \lambda_y(i,j) &= \lambda(i,j-1) - \lambda(i,j). \end{aligned} \tag{21}$$

We employ the same line interaction terms used in (Yan and Leahy, 1991):

$$\begin{aligned} \mathcal{H}_x(l_s) &= -\epsilon\alpha\left[l_{x,s}(i-1,j) - l_{x,s}(i,j-1)\right] l_{x,s}(i,j) \\ \mathcal{H}_y(l_s) &= -\epsilon\alpha\left[l_{y,s}(i,j-1) - l_{y,s}(i-1,j)\right] l_{y,s}(i,j). \end{aligned} \tag{22}$$

This function serves the purpose of increasing the penalty on a horizontal (vertical) line if either of its neighbors on adjacent rows (columns) is "on" and decreasing the penality on a horizontal (vertical) line if either of its horizontal (vertical) neighbors on the same row (column) is "on". This can be seen clearly later in the line updating scheme. The parameter ϵ controls the relative penality associated with this term. We note that as the neighborhood size is increased, the function above can be modified to allow better control of the formation of boundaries.

We now describe an updating procedure for both λ and l_F by following a similar derivation to that described for the GEM-2 algorithm in (Yan and Leahy, 1991). Here two types of edge processes are included: the known prior edges $l_{p,a}$ obtained from the MR image, and the unknown functional edges, $l_{p,F}$, to be estimated. The known prior anatomical boundaries $l_{p,a}$ are unchanged throughout the updating process.

In our simulation we found that the overall computational cost was reduced by updating each intensity (line) site more than once at each M1-step (M2-step). We also found that the line processes converge faster than the intensity processes and are therefore updated less often. We thus restate the M1-step and M2-step in the following way:

- M1-step: Condition (16) for this case can be rewritten as:

$$Q\left(\lambda^{n+1}, l_F^n|\lambda^n, l_F^n\right) \geq Q\left(\lambda^n, l_F^n|\lambda^n, l_F^n\right). \tag{23}$$

  Here λ^{n+1} denotes the estimates from this step, which are obtained by updating each intensity site an arbitrary number of times, $r \geq 1$.

- M2-step: Based on the result λ^{n+1} from the M1-step, the estimates l_F^{n+1} should satisfy condition (18), rewritten as:

$$Q\left(\lambda^{n+1}, l_F^{n+1}|\lambda^n, l_F^n\right) \geq Q\left(\lambda^{n+1}, l_F^n|\lambda^n, l_F^n\right),$$

  or equivalently

$$-U(\lambda^{n+1}, l_F^{n+1}, l_a) \geq -U(\lambda^n, l_F^n, l_a). \tag{24}$$

  where l_F^{n+1} are the line estimates obtained by updating each line site an arbitrary number of times, $q \geq 0$.

The Gibbs energy function $U(\lambda, l_s)$ as defined in (20), results in a function $Q(\lambda, l_F^n|\lambda^n, l_F^n)$ that is concave with respect to each $\lambda(i,j) \geq 0$. Hence in the M1-step, we use the concave nature of this function to perform a sequence of one dimensional line searches, updating each of the intensity processes in turn. This simple approach ensures that the function $Q(\lambda^{n+1}, l_F^n|\lambda^n, l_F^n)$ is monotonically nondecreasing at each M1-step.

M1-step: Setting the partial derivative of $Q(\lambda, l_F^n|\lambda^n, l_F^n)$, defined in (19), with respect to each $\lambda(i,j)$ equal to zero and simplifying, yields the quadratic equation:

$$C_1\lambda^2(i,j) - [C_2 - a(i,j)]\,\lambda(i,j) - b^n(i,j) = 0. \tag{25}$$

where C_1 and C_2 are defined as follows:

$$\begin{aligned} C_1 &= 2\beta\gamma\left[(1 - l_{x,s}^n(i,j)) + (1 - l_{y,s}^n(i,j)) + (1 - l_{y,s}^n(i+1,j)) + (1 - l_{x,s}^n(i,j+1))\right] \\ C_2 &= 2\beta\gamma\left[\lambda(i-1,j)(1 - l_{y,s}^n(i,j)) + \lambda(i,j-1)(1 - l_{x,s}^n(i,j)) + \right. \\ &\quad \left. \lambda(i,j+1)(1 - l_{x,s}^n(i,j+1)) + \lambda(i+1,j)(1 - l_{y,s}^n(i+1,j))\right]. \end{aligned} \tag{26}$$

When computing C_1 and C_2, the most recent estimates of $\lambda(i \pm 1, j \pm 1)$ should be used since the pixel intensities are updated sequentially. Since $\lambda(i,j) \geq 0$, we must choose the non-negative root of (25). Assuming C_1 is non-zero, the updating scheme for $\lambda(i,j)$ becomes:

$$\begin{aligned} \lambda^{n_r}(i,j) &= \frac{[C_2 - a(i,j)] + \sqrt{[C_2 - a(i,j)]^2 + 4C_1 b^n(i,j)}}{2C_1}, \quad n_r \Leftarrow n_r + 1; \\ \lambda^{n+1} &\Leftarrow \lambda^{n_r}, \quad \text{after } r \text{ iterations.} \end{aligned} \tag{27}$$

In this scheme we update each pixel r times. The estimates λ^{n_r} are then assigned to λ^{n+1}.

The above procedure, when applied to all intensity sites in turn, results in an increase in the function $Q(\lambda, l_F^n|\lambda^n, l_F^n)$, *i.e.* condition (23) is satisfied. The case where $C_1 = 0$ implies that all lines involved in C_1 are "on". In this case C_2 is also equal to zero and the M1-step at this site reduces to the ML solution. In other words, one would expect no intensity smoothing between neighbors to occur if all the lines around the pixel are "on".

We now turn to the problem of updating the unknown functional boundaries $l_{p,F}$ while keeping the known prior anatomical boundaries $l_{p,a}$ fixed. Our goal is to obtain a result in which the estimated functional boundaries $l_{p,F}(i,j)$ converge to the values 1 or 0 that, jointly with λ, maximize the posterior density function. We rule out a stochastic search as too computationally expensive and restrict our attention to deterministic algorithms. During the M2-step we wish to update the unknown line sites so as to satisfy (24). The method of iterated conditional modes (ICM) (Besag, 1986) can be shown to produce a sequence with an associated energy function that is nonincreasing and would therefore satisfy (24). Combining this ICM approach with the M1-step defined above, we would expect convergence of this algorithm to a local maximum of the posterior distribution. However, when the ICM method is applied to the estimation of unknown line sites, it has been found that the line estimates converge faster than the image intensities (Rangarajan and Chellappa, 1990), (Johnson et al, 1990) and (Yan and Leahy, 1991). It is intuitively more appealing to have an updating procedure in which the rate of convergence of the unknown line process can be controlled by allowing the variables $l_{p,F}(i,j)$ to assume values on the continuous interval $[0,1]$ but eventually converge to one of the end points. In this case, a zero-one decision about the presence of a boundary is delayed until later in the iteration process. By replacing the conditional mode in the ICM algorithm with a conditional average (ICA), (Johnson et al, 1990) developed an iterative algorithm which does allow the estimated line site to take values in the interval $[0,1]$. However, there is no guarantee in the case of ICA that the estimated line process converges to the values 0 or 1. Furthermore, is it not proven that ICA necessarily converges to a (local) maximum.

In developing the M2-step below, we adopt a third alternative proposed in (Yan and Leahy, 1991) to ICM and ICA. Recall that the goal of the M2-step is to update the unknown line sites such that $U(\lambda^{n+1}, l_F^{n+1}, l_a) \leq U(\lambda^{n+1}, l_F^n, l_a)$. Here we use the iterated conditional average of a modified energy function with fixed $\lambda = \lambda^{n+1}$: $U_T(\lambda^{n+1}, l_F, l_a) = U(\lambda^{n+1}, l_F, l_a)/T$, where $U(\lambda^{n+1}, l_F, l_a)$ is defined in (20). Fixing $T = 1$ would produce an iteration equivalent to the ICA algorithm of (Johnson et al, 1990). However, as $T \to 0$, it can be shown that the variables $l_{p,F}(i,j)$ converge to either 0 or 1 and this algorithm would converge to a local minimum of the energy function $U(\lambda^{n+1}, l_F, l_a)$ with respect to l_F on the set $\{0,1\}^L$. The goal of this procedure is to use the parameter T to control the rate of convergence of the unknown line process relative to the intensity process and hopefully to then converge on a more desirable local maximum of the posterior distribution.

M2-step: Let $N_{p,F}(i,j)$ denote the set of intensities and lines which interact with $l_{p,F}(i,j)$. The local conditional probability of $l_{x,F}(i,j)$ and $l_{y,F}(i,j)$ for the energy function $U_T(\lambda^{n+1}, l_F, l_a)$ is given by

$$\begin{aligned}
P_T(l_{x,F}(i,j)|N_{x,F}(i,j)) &= \frac{e^{-\frac{1}{T}\{\beta\lambda_y^{n+1}(i,j)^2(1-l_{x,F})+\alpha l_{x,F}-\epsilon\alpha[l_{x,s}(i-1,j)+l_{x,s}(i+1,j)-l_{x,s}(i,j-1)-l_{x,s}(i,j+1)]l_{x,F}\}}}{e^{-\frac{1}{T}\beta\lambda_y^{n+1}(i,j)^2}+e^{-\frac{1}{T}\{\alpha-\epsilon\alpha[l_{x,s}(i-1,j)+l_{x,s}(i+1,j)-l_{x,s}(i,j-1)-l_{x,s}(i,j+1)]\}}} \\
P_T(l_{y,F}(i,j)|N_{y,F}(i,j)) &= \frac{e^{-\frac{1}{T}\{\beta\lambda_x^{n+1}(i,j)^2(1-l_{y,F})+\alpha l_{y,F}-\epsilon\alpha[l_{y,s}(i,j-1)+l_{y,s}(i,j+1)-l_{y,s}(i-1,j)-l_{y,s}(i+1,j)]l_{y,F}\}}}{e^{-\frac{1}{T}\beta\lambda_x^{n+1}(i,j)^2}+e^{-\frac{1}{T}\{\alpha-\epsilon\alpha[l_{y,s}(i,j-1)+l_{y,s}(i,j+1)-l_{y,s}(i-1,j)-l_{y,s}(i+1,j)]\}}}.
\end{aligned} \tag{28}$$

The local conditional mean of the zero-one variable $l_{p,F}(i,j)$ is then equal to the probability that $l_{p,F}(i,j) = 1$. Using this posterior conditional mean, the updating procedure for $l_{p,F}(i,j)$ is defined as

$$\begin{aligned}
l_{x,F}^{n_q}(i,j) &= \frac{1}{1+e^{\frac{1}{T}[\alpha-\beta\lambda_y^{n+1}(i,j)^2-\epsilon\alpha(l_{x,s}(i-1,j)+l_{x,s}(i+1,j)-l_{x,s}(i,j-1)-l_{x,s}(i,j+1))]}}, \quad n_q \Leftarrow n_q+1 \text{ and reduce } T, \\
l_{x,F}^{n+1} &\Leftarrow l_{x,F}^{n_q}, \quad \text{after } q \text{ iterations;} \\
l_{y,F}^{n_q}(i,j) &= \frac{1}{1+e^{\frac{1}{T}[\alpha-\beta\lambda_x^{n+1}(i,j)^2-\epsilon\alpha(l_{y,s}(i,j-1)+l_{y,s}(i,j+1)-l_{y,s}(i-1,j)-l_{y,s}(i+1,j))]}}, \quad n_q \Leftarrow n_q+1 \text{ and reduce } T, \\
l_{y,F}^{n+1} &\Leftarrow l_{y,F}^{n_q}, \quad \text{after } q \text{ iterations.}
\end{aligned} \tag{29}$$

When computing (29) the known prior edges and the most recent estimates are used for $l_{p,s}(i \pm 1, j \pm 1)$.

In general, the updating procedure (29) defined above does not guarantee that $U_T(\lambda^{n+1}, l_F, l_a)$ decreases at each iteration of the M2-step for a fixed T. In addition, the estimates from (29) vary with the parameter T. However in the limit as $T \to 0$, it can be shown that (29) results in a decrease in $U(\lambda^{n+1}, l_F, l_a)$ and therefore satisfies condition (24). Also (29) converges to a (local) minimum of the function $U(\lambda^{n+1}, l_F, l_a)$ as $T \to 0$ at the M2-step. The advantage of (29) is that we are able to control this optimization process through T and we may find a better (local) maxima as T is gradually reduced to zero.

Lemma 1 *(i). The line estimate $l_{p,F}^n(i,j)$ obtained using (29) satisfies the condition*

$$0 \leq l_{p,F}^n(i,j) \leq 1 \quad \textit{for any } T,\ n = 0, 1, 2, \cdots. \tag{30}$$

(ii). For fixed λ, the iteration (29) converges to either 0 or 1 as $T \to 0$. (iii). For fixed λ, as $T \to 0$ the iteration (29) becomes the iterated conditional modes (ICM) algorithm with the associated energy function $U(\lambda, l_F, l_a)$, and therefore converges to a local minimum of $U(\lambda, l_F, l_a)$ at the M2-step on the set $\{0,1\}^L$ where L denotes the number of line process variables.

The proof for Lemma 1 is straightforward and omitted here.

5 RESULTS

To illustrate the potential performance of the methods described above for PET imaging we have conducted some preliminary studies using a computer generated phantom. We began with a real MR image of a patient with a large white matter lesion, Figure 2(a). Using the segmentation method described in (Wu and Leahy, 1991) we segmented this image into four distinct tissue types: white matter, grey matter, lesion and ventricles. The resulting segmentation is shown in Figure 2(b) using the mean intensity for each class; the boundaries between different regions are shown in Figure 2(c). The result of this segmentation was used as a template for generating a phantom PET image, i.e. we assume that a radiopharmaceutical is used in which there will be different amounts of activity in each of the four anatomical regions. The PET phantom is shown in Figure 2(d). To generate a more realistic image, the intensity within each region varies according to a first order Gauss Markov random field model (Chellappa and Kashypa, 1985). In addition to the variations in activity in different anatomical regions, we added a hot spot in the left side of the white matter region to demonstrate the ability of the new algorithm to detect functional boundaries not present in the MR (anatomical) image. We will refer to this as a functional lesion below.

The computer generated PET data were based on a simplified model of a single ring of the Siemens ECAT system. The source was assumed to be of maximum dimension 22cm diameter, the detector ring of diameter 76cm with 384 detectors. Data were generated as a set of fan beam projections obtained by pairing each of the 384 detectors with sufficient of the remaining detectors to cover the required 22cm diameter field of view. Factors such as randoms, scatter, attenuation and detector efficiency were not considered here - however, since the comparisons are all based on the same data, the comparison should indicate the relative potential performance of the various reconstruction methods. After generating the means of the projection data, a pseudo-random Poisson generator was used to generate Poisson data for each data sample with a mean of 200 counts per sample.

The results are shown in Figures 2(e) though 2(i). Figure 2(e) is a filtered backprojection (FBP) reconstruction and shows the typical noisy image obtained with low count rates. Note that there is virtually no distinction between the mean activity in the grey and white matter and the additional functional lesion is barely visible. Figures 2(f) and 2(g) show ML and MAP reconstructions from the same data - the MAP reconstruction does not include line sites. There is in both cases a qualitative improvement over the FBP method with the MAP result smoother than the ML image, but neither of the results shows good white/grey matter contrast.

Figure 2(h) shows a reconstruction using the GEM-2 algorithm (Yan and Leahy, 1991) in which all line sites are estimated. Figure 2(i) shows the MAP reconstruction using line sites in which the anatomical line sites are included from the MR image - the other line sites are then estimated. These results indicate that there is a clear benefit to the use of line sites in MAP PET reconstruction in comparison to the other methods described above. Even when a priori edge information is not available, this methods performs surprisingly well - there is now noticeable grey/white matter contrast and the functional lesion is clearly visible. When the MR information is included the situation becomes even better. Note also, that although the functional lesion, in the left-middle portion of the PET phantom, does not correspond to any anatomical boundaries, the MAP estimator, with a line process, is able to clearly detect the presence of this lesion. In this case the shape of the lesion is somewhat distorted, however, this is to be expected as it arises from the finite resolution of the PET system. The importance of this observation is that the use of prior anatomical information does not preclude the presence of additional structure in the reconstructed PET image. Furthermore, the presence of a boundary in the

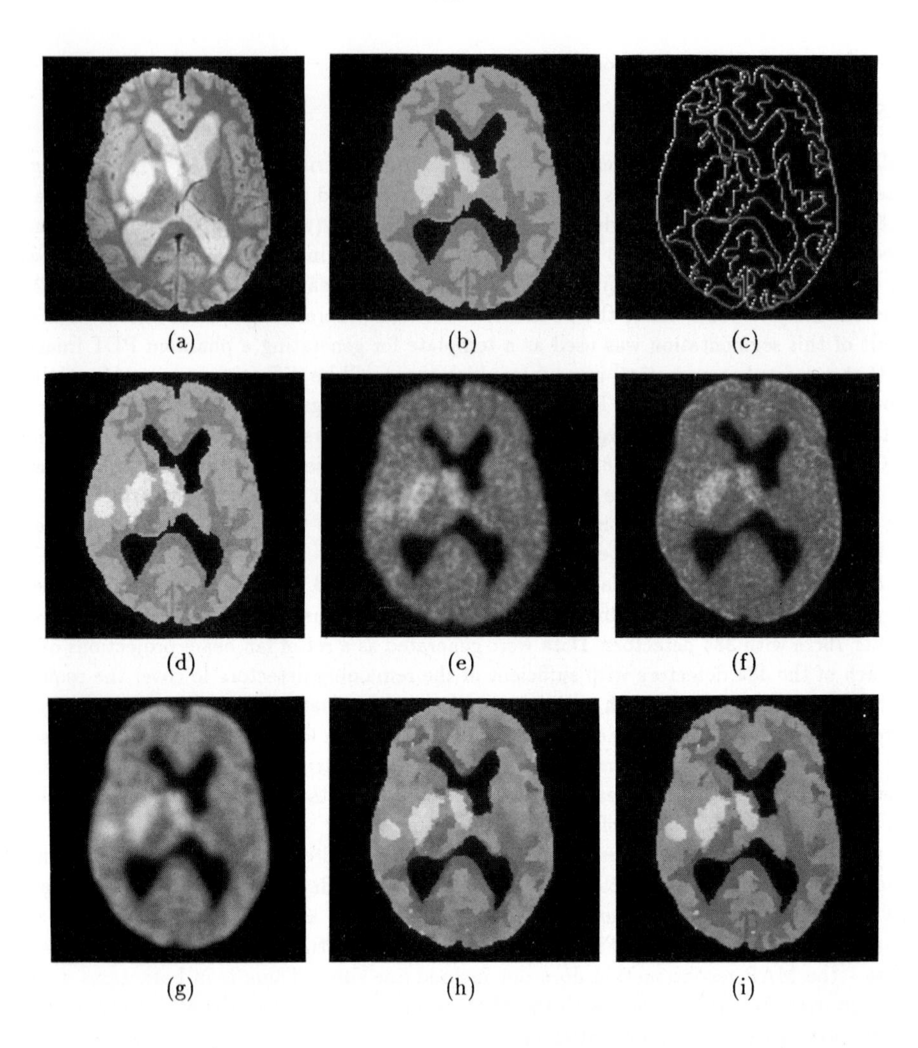

Figure 2: (a): Original MR brain image. (b): Segmentation of (a) into four tissue types. (c): Anatomical boundaries extracted from (b). (d): Computed PET phantom generated from MR template (b). (e): PET reconstruction using filtered backprojection. (f): PET reconstruction using maximum likelihood estimation. (g): PET reconstruction using MAP estimation - no lines. (h): PET reconstruction using the GEM-2 algorithm - estimated lines. (i): PET reconstruction using this proposed algorithm - some line sites from MR image.

anatomical image does not force a corresponding boundary in the reconstructed PET image - if the data shows no evidence for such a boundary then it will not be present in the reconstructed PET image. This is because the function of the anatomical boundary information is to avoid smoothing across anatomical boundaries rather than to force the formation of corresponding boundaries in the PET image.

The next stage of this work is to apply the methods described above to real PET and MR images. Additional problems arise when using real data, namely i) one cannot ignore the presence of randoms, attenuation and scatter and ii) the imaging systems must be registered to a common coordinate system. The first of these may be overcome to some extent using the modeling techniques described in section 2. The problem of image registration has been addressed by a number of groups using both anatomical and external landmarks and we will adapt these methods to the MR/PET image registration problem.

References

Bergstrom M, Bohm C, Ericson K, Eriksson L, and Litton J, (1980). Correlations for attenuation, scattered radiation, and random coincidences in a ring detector positron emission transaxial tomography. *IEEE Transactions on Nuclear Science*, 7(1):549–554.

Besag JE, (1986). On the statistical analysis of dirty pictures. *J. Royal Statist. Soc., B*, 48:259–302.

Chellappa R and Kashypa R, (1985). Texture synthesis using 2-D noncausal autoregressive models. *IEEE Transactions on Acoustic, Speech and Signal processing*, ASSP-33(1):194–203.

Chen CT, Ouyang X, et al., (1990). Sensor fusion in image reconstruction. *IEEE Nuc. Sci. Symp.*

Dempster AP, Laird NM, and Rubin DB, (1977). Maximum likelihood from incomplete data via the EM algorithm. *Journal of Royal Statistical Society, Series B*, 39(1):1–38.

Geman S and Geman D, (1984). Stochastic relaxation, Gibbs distributions, and the Bayesian restoration of images. *IEEE Transactions on Pattern Analysis and Machine Intelligence*, PAMI-6(6):721–741.

Geman S and McClure DE, (1985). Bayesian image analysis: An application to single photon emission tomography. *Proceedings of the American Statistical Association*, pages 12–18.

Green PJ, (1990). Bayesian reconstructions from emission tomography data using a modified EM algorithm. *IEEE Transactions on Medical Imaging*, 9(1):84–93.

Hebert T and Leahy R, (1989). A generalized EM algorithm for 3D bayesian reconstruction from poisson data using Gibbs priors. *IEEE Transactions on Medical Imaging*, 8(2):194–202.

Hebert T, Leahy R, and Singh M, (1988). Fast MLE for SPECT using an intermediate polar representation and a stopping criterion. *IEEE Transactions on Nuclear Science*, 34:615–619.

Hebert T, Leahy R, and Singh M, (1990). 3D MLE reconstruction for a prototype SPECT system. *J. Opt. Soc. Amer. (A)*, 7(7):1305–1313.

Holmes T, Ficke D, and Snyder D, (1984). Modeling of accidental coincidences in both conventional and TOF PET. *IEEE Transactions on Nuclear Science*, 31:627–631.

Huesman R, Derenzo S, et al., (1988). Orbiting transmission source for positron tomography. *IEEE Transactions on Nuclear Science*, 35:735–739.

Johnson VE, Wong WH, Hu X, and Chen CT, (1990). Bayesian restoration of PET images using Gibbs priors. *XIth International Conference on Information Processing in Medical Imaging.*

King PH, Hubner K, and Holloway E, (1981). Noise identification and removal in positron imaging systems. *IEEE Transactions on Nuclear Science*, 28(1):148–151.

Levitan E and Herman GT, (1987). A maximum a posteriori probability expectation maximization algorithm for image reconstruction in emission tomography. *IEEE Transactions on Medical Imaging*, MI-6(3):185–192.

Liang E and Hart H, (1987). Bayesian image processing of data from constrained source distributions-non-valued, uncorrelated and correlated constraints. *Bull. Math. Biol.*, 49:51–74.

Mansfield P and Morris PG, (1982). *NMR imaging in Biomedicine*. New York: Academic.

Miller M, Snyder D, and Moore S, (1986). An evaluation of the use of sieves for producing estimates of radioactivity distribution with the EM algorithm for PET. *IEEE Transactions on Nuclear Science*, 33.

Muehllehner G and Colsher JG, (1982). *Instrumentation in Computed Emission Tomography*. New York.

Rangarajan A and Chellappa R, (1990). Generalized graduated non-convexity algorithm for maximum a posteriori image estimation. *Proceedings of The Tenth International Conference on Pattern Recognition.*

Shepp LA and Logan BF, (1979). The fourier transform of a head section. *IEEE Transactions on Nuclear Science*, NS-21:21–43.

Leahy, 1989, Johnson et al, 1989, Leahy et al, 1989, Liang et al, 1989) The model for the object priors were motivated by assumptions of object smoothness with occasional discontinuities.

These smoothness assumptions are generic in that they apply to the general class of objects under consideration. One might well suppose that prior information concerning the smoothness and discontinuities of the specific object being imaged should provide some additional advantage relative to generic information. We note here that some work by (Chen et al, April 1990, Chen et al, June 1990) along these lines has been previously reported, but no details of their formulation were presented.

An opportunity to obtain such prior information is provided by recent results in the 2-D registration of anatomical (computerized tomography (CT) or magnetic resonance imaging (MRI)) images with functional (SPECT or PET) images from the same patient (Chen et al, 1987, Gerlot and Bizais, 1987, Kessler et al, 1986). To the extent that boundaries of some anatomical objects correspond to edges in the spatial distribution of radiopharmaceuticals, we indeed get just the sort of nongeneric smoothness information mentioned above.

In this paper, we first comment in more detail on the use of anatomical images as priors, then offer a Bayesian reconstruction model. Some results on simple mathematical phantoms are presented.

2. THE ANATOMICAL IMAGE AS A PRIOR

Figure 1 shows autoradiographs of monkey brains after injection of an neuroreceptor seeking pharmaceutical. These autoradiographs would serve as "ground truth" images for the corresponding SPECT reconstructions. The correlation to anatomy is clear. The boundary between white and grey matter is distinct, the LGN (round thing in the middle) is clearly demonstrated as is the striatum (structure on top of the LGN). Furthermore, the distribution within certain regions, such as white and grey matter, is reasonably uniform. Since these and other organs can be delineated on anatomical images, there is some hope that selected boundaries obtained from CT and MRI can indeed serve as priors in a reconstruction.

In the work here, we model the prior information obtained from the anatomical image by an edge map $e(i,j), 0 \leq e \leq 1$ whose value reflects our confidence that a significant anatomical edge exists at location (i,j). (A value of unity corresponds to the highest confidence.) For the present work, we have two edge maps, e^h for horizontal edges and e^v for vertical edges. A diagonal line can be approximated as a staircase of vertical and horizontal edge segments. The values of the edge map cannot be derived simply by running an edge operator over the anatomical image; instead, the value of e may depend on which anatomical boundary is being considered. We need both a strong anatomical edge and a reasonable expectation that the edge is likely to correspond to a functional edge in order to set e to a high value. On the other hand, we demand that the edge map not overrule the data – an edge in the functional image should be reconstructed even if it not promoted by the edge map, and a strong vote by the edge map should not create a false edge in the functional image. We address this question further in the next section. At present, though, there is no principled way to choose $e(i,j)$.

A spatial blurring of e can also be used to model registration errors once these have been calibrated. For example, if a registration system has an uncertainty of 1 mm, we can use that value to determine the variance of a Gaussian kernel that convolves e^v in the x-direction.

In practice, the edge map can be derived and directly applied to the reconstruction matrix of the functional image if some external registration scheme is used, such as fiducial markers. If landmark registration is used, then a crude reconstruction, based on available projection data only, can be used to obtain registration information. The edge map is imposed and the crude reconstruction improved by the prior information.

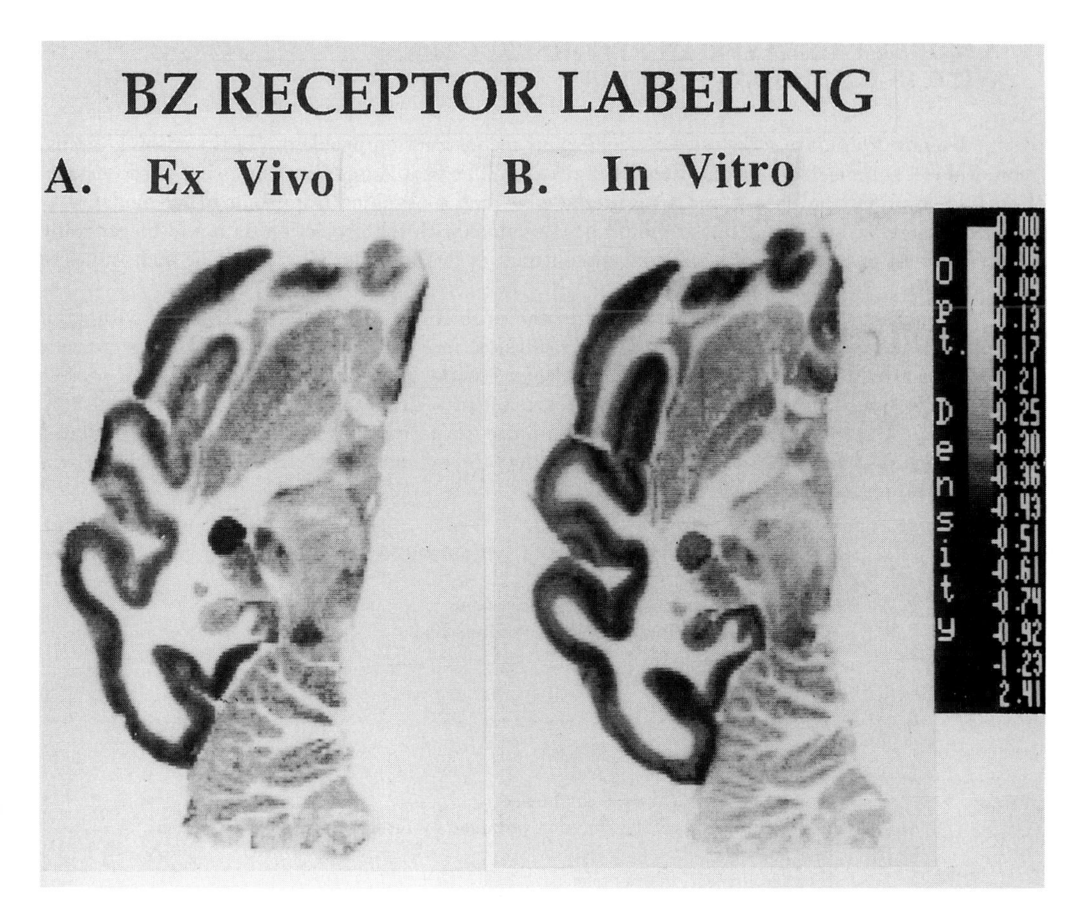

Figure 1: Autoradiograph of monkey brain shows radioisotope uptake corresponding to concentration of neuroreceptors in the brain.

3. A MODEL FOR BAYESIAN RECONSTRUCTION WITH INTERMODALITY PRIORS

We now formulate a Bayesian model for our reconstruction problem. (In the ensuing discussion, Table 1 summarizes our notational definitions.) The problem is to estimate a 2-D reconstruction $\hat{\mathbf{f}}$ given projection data $\mathbf{g}$ and a registered edge map $\mathbf{e}$ obtained from some other modality as described previously. The estimate should be consistent with the projection data and be generally smooth except at locations of significant discontinuities in the underlying object or high values of $\mathbf{e}$.

Though our model nominally addresses the problem of SPECT reconstruction with prior information obtained from registered CT, our physical modeling of the SPECT process is crude and ignores attenuation and scatter models. Also, we model noise as an additive Gaussian process on the projection data instead of the correct version in which each object pixel is governed by a Poisson process of unknown mean that is sampled to derive the number of photons radiated towards each detector. The supposition here is that the advantages of prior edge knowledge will carry over into a more accurate model.

Table 1: Table of symbols used

Symbol	Meaning
$\mathbf{g}, \mathbf{G}$	Projection data, associated random field
$\mathbf{f}, \mathbf{F}$	2–D source function, associated random field
$\hat{\mathbf{f}}$	optimal estimate for $\mathbf{f}$
$\mathbf{f}_x$	$\frac{\partial \mathbf{f}}{\partial x}$
$\mathbf{f}_y$	$\frac{\partial \mathbf{f}}{\partial y}$
$\mathbf{l}^h, \mathbf{l}^v$	horizontal, vertical line process
$\mathbf{L}$	Random field for line process
$\mathbf{e}^h, \mathbf{e}^v$	Horizontal, vertical edge confidence
$\phi_{ij}(f_x)$	nonquadratic potential function obtained by integrating out line process
$V_C(\mathbf{l})$	line potential
$\mathcal{H}$	forward projection operator
κ, κ', A	constants
$\alpha(\mathbf{e})$	coupling term

3.1. Formulation in terms of line processes

We first formulate our problem in a transparent manner with the aid of line processes as proposed by Geman and Geman (1984). The resulting energy function is then converted to one that is more easily minimized by integrating out the line processes.

The problem is formulated as a search over the space of 2-D intensity images $\mathbf{f}$ and (unobservable) line processes $\mathbf{l}$. In our formulation, the line process $\mathbf{l}$ consists of vertical and horizontal binary (0 or 1) variables $l^v(i,j)$ and $l^h(i,j)$, respectively, located at each lattice site (i,j). A value $l^v(i,j) = 1$ corresponds to the presence of a vertical edge located between pixels at (i,j) and $(i+1,j)$ (similar argument for $l^h(i,j) = 1$). If $\mathbf{F}$ and $\mathbf{L}$ are the 2-D random fields for intensity and lines, then a probabilistic version of the reconstruction problem is to search for that $\hat{\mathbf{f}}, \hat{\mathbf{l}}$ that maximizes the joint probability $P(\mathbf{F}=\mathbf{f}, \mathbf{L}=\mathbf{l}|\mathbf{G}=\mathbf{g})$. By Bayes Theorem:

$$P(\mathbf{F}=\mathbf{f}, \mathbf{L}=\mathbf{l}|\mathbf{G}=\mathbf{g}) = P(\mathbf{G}=\mathbf{g}|\mathbf{F}=\mathbf{f}, \mathbf{L}=\mathbf{l})P(\mathbf{F}=\mathbf{f}, \mathbf{L}=\mathbf{l}) \tag{1}$$

We may convert the maximization problem to the minimization of an energy function by invoking the fact that a Gibbs distribution may be completely specified by an energy function E:

$$P(\mathbf{X}=\mathbf{x}) = \frac{1}{Z}\exp(-\beta E(\mathbf{x})) \tag{2}$$

where Z, the partition function, is a suitable normalization, and β is a positive constant identified as the inverse of the temperature T in a subsequent annealing process. Here, we model both the likelihood and priors of Eq. (1) as Gibbs distributions. If both the likelihood and priors are Gibbs, then we may combine Eq. (1) and Eq. (2)

$$P(\mathbf{F}=\mathbf{f},\mathbf{L}=\mathbf{l}|\mathbf{G}=\mathbf{g}) = \frac{1}{Z}\exp(-\beta(E_D+E_P)) \tag{3}$$

where the net energy $E = E_D + E_P$ is composed of terms associated with the Gibbs likelihood and prior, respectively.

In particular, we model the likelihood energy as

$$E_D(\mathbf{f}) = A||\mathcal{H}\mathbf{f} - \mathbf{g}||^2 \tag{4}$$

The symbol $\mathcal{H}$ is the forward projection operator (Radon transform) that integrates $\mathbf{f}$ along parallel rays to form projections; A is a constant. The vector norm definition implies summation over all elements in the projection space. The definition of E_D in Eq. (4) is tantamount to the assumption that the projection data $\mathbf{g}$ differs from the line integrals $\mathcal{H}\mathbf{f}$ by the addition of independent Gaussian noise of variance $\frac{1}{A}$ at each detector location. Also, the likelihood is obviously independent of the line process.

Of more interest is the prior, also Gibbs. We may write the associated energy as:

$$\begin{aligned} E_P(\mathbf{f},\mathbf{l}) &= \sum_{(i,j)} \left(f_x^2(i,j)(1-l^h(i,j)) + \alpha^h(i,j)l^h(i,j)\right) \\ &+ \sum_{(i,j)} \left(f_y^2(i,j)(1-l^v(i,j)) + \alpha^v(i,j)l^v(i,j)\right) \\ &+ V_C(\mathbf{l}) \end{aligned} \tag{5}$$

The two terms involving f_x and f_y encourage smoothness except where discontinuities (l^h or $l^v = 1$) occur. (Note: we take for our definitions of partial derivatives $f_x(i,j) = f(i+1,j) - f(i,j)$ and $f_y(i,j) = f(i,j+1) - f(i,j)$.) The term $V_C(\mathbf{l})$ encourages certain local configurations of the line process. This term involves interactions among neighboring elements of the line process. Since we shall eventually dispense with this term, we do not discuss it in great detail, but see, for example (Geman and Geman, 84, Johnson et al, 1989) for details.

The term of interest, $\alpha(i,j)$, expresses the coupling between modalities:

$$\begin{aligned} \alpha^h(i,j) &= \kappa e^h(i,j) + \kappa'(1-e^h(i,j)) \\ \alpha^v(i,j) &= \kappa e^v(i,j) + \kappa'(1-e^v(i,j)) \end{aligned} \tag{6}$$

Here κ and κ' are constants with $\kappa < \kappa'$. Recall that elements of $\mathbf{e}$ $(0 \leq e \leq 1)$ measure the confidence of an edge location in the registered anatomical image.

The effect of the coupling term can be seen with reference to Eq. (5). (In the following argument, the term $V_C(l)$ is suppressed (since we later throw it out) and for brevity we drop the h superscript from e^h and l^h.) If $e(i,j)$ is close to unity, then the coupling term becomes κ . In that case, energy is reduced by turning on a line process $l(i,j) = 1$ only if $f_x^2(i,j) > \kappa$. Conversely, if $e(i,j)$ is close to zero, it is worth imposing an edge only if $f_x^2(i,j) > \kappa'$. Since $\kappa < \kappa'$, the effect is to switch on $l(i,j)$ with high expectation of an edge and moderate edge in the data, or low expectation of an edge and strong edge in the data. In this manner, the data is allowed to overrule the expectation if it is strong enough.

A coupling term of this sort was proposed first by Poggio et al. (1988) and Gamble et al. (1989) in the context of data fusion for computer vision. In their work, an attempt was made to fuse edges from registered intensity, stereo, texture, color,and motion images of the same scene.

4. DERIVATION OF THE EFFECTIVE ENERGY FUNCTION

In the absence of the interaction term $V_C(\mathbf{l})$, the line processes are independent of each other in the conditional distribution $P(\mathbf{L} = \mathbf{l}|\mathbf{F} = \mathbf{f})$ Interaction terms are usually added in order to enforce additional constraints such as hysteresis and non-maximum suppression (Blake and Zisserman, 1987, Geiger and Girosi, 1989). Hysteresis aids in the formation of unbroken contours whereas non-maximum suppression helps in reducing both the number of spurious edges and multiple responses to a single edge. The important point is that these constraints are still implicitly enforced even in the absence of $V_C(\mathbf{l})$ (Blake and Zisserman, 1987). The additional information present in $e^v(i,j)$ and $e^h(i,j)$ compensates for the lack of explicit hysteresis and non-maximum suppression. When interaction terms are present, the overall effect is the variation of the threshold for creating an edge. In hysteresis, the threshold for creating an edge is lowered in proportion to the confidence that the edge is part of a contour. In non-maximum suppression, the threshold for creating an edge is increased in proportion to the extent to which the multiple response criterion is being violated. In our approach, these space varying thresholds are represented by $e^v(i,j)$ and $e^h(i,j)$ which are externally provided as opposed to being internally generated by additional constraints.

Maximum a posteriori (MAP) estimation still needs to be performed over the intensities and the line processes. However, the independence of the line processes in the conditional distribution $P(\mathbf{L} = \mathbf{l}|\mathbf{F} = \mathbf{f})$ can be used to find an analytical solution for the MAP estimate of each line process while keeping the intensities fixed. Blake and Zisserman (1987) show that this reduces to a simple solution for each line process

$$\hat{l}^v(i,j) = \begin{cases} 0 & f_x^2(i,j) \leq \alpha^v(i,j) \\ 1 & f_x^2(i,j) > \alpha^v(i,j) \end{cases} \tag{7}$$

with a similar expression for the horizontal line process $l^h(i,j)$. This provides a basis for the earlier argument used to motivate the introduction of edge confidence parameters. When this MAP estimate is substituted back into the energy function, we obtain an effective energy function,

$$E(\mathbf{f}) = E_D + \sum_{(i,j)} (\phi^*(f_x(i,j)) + \phi^*(f_y(i,j))) \tag{8}$$

where

$$\phi^*(z) = \begin{cases} z^2 & z^2 \leq \alpha \\ \alpha & z^2 > \alpha \end{cases} \tag{9}$$

We have suppressed the space indices for the sake of convenience.

When the conditional-mean estimate (CM) is desired, we need to evaluate the expected value of the intensities and the line processes (given the data) and not their most probable values. These expected values can be evaluated if the partition function Z is known, however, the partition function calculation is intractable. Instead, we can use the independence of the line processes in the conditional distribution and integrate them out while keeping the intensities fixed. Geiger and Girosi (1989) show that the marginal distribution $P(\mathbf{F} = \mathbf{f}|\mathbf{G} = \mathbf{g})$ can be evaluated as follows.

$$\begin{aligned}
P(\mathbf{F} = \mathbf{f}|\mathbf{G} = \mathbf{g}) &= \frac{1}{Z_f}\exp(-\beta E_D) \\
&\quad \Pi_{(i,j)} \left(\Sigma_{l^h(i,j)\in\{0,1\}} \exp(-\beta(f_x^2(i,j)(1 - l^v(i,j)) + \alpha^v(i,j)l^v(i,j)))\right) \\
&\quad \Pi_{(i,j)} \left(\Sigma_{l^h(i,j)\in\{0,1\}} \exp(-\beta(f_y^2(i,j)(1 - l^h(i,j)) + \alpha^h(i,j)l^h(i,j)))\right) \\
&= \tfrac{1}{Z_f}\exp\Big[-\beta\Big\{E_D + \Sigma_{(i,j)}(-\tfrac{1}{\beta})\log\left[\exp(-\beta f_x^2(i,j)) + \exp(-\beta\alpha^v(i,j))\right] \\
&\quad + \Sigma_{(i,j)}(-\tfrac{1}{\beta})\log\left[\exp(-\beta f_y^2(i,j)) + \exp(-\beta\alpha^h(i,j))\right]\Big\}\Big] \\
&= \tfrac{1}{Z_f}\exp\Big[-\beta\Big\{E_D + \Sigma_{(i,j)}(\phi(f_x(i,j)) + \phi(f_y(i,j)))\Big\}\Big]
\end{aligned} \tag{10}$$

where

$$\phi(z) \stackrel{def}{=} -\frac{1}{\beta}\log(\exp[-\beta z^2] + \exp[-\beta\alpha]) \tag{11}$$

and Z_f is the partition function corresponding to the marginal distribution of $\mathbf{F}$. The conditional distribution of the line process is:

$$P(L^v(i,j) = l^v(i,j)|\mathbf{F} = \mathbf{f}) = \frac{1}{Z_l}\exp\left[-\beta(f_x^2(i,j)(1 - l^v(i,j)) + \alpha^v(i,j)l^v(i,j))\right] \tag{12}$$

where $Z_l = \exp(-\beta f_x^2(i,j)) + \exp(-\beta\alpha^v(i,j))$ is the partition function of each vertical line process with the intensities held fixed. The line process can be recovered at each temperature by evaluating the expected value of the line process from its conditional distribution.

$$\mathcal{E}\{l^v(i,j)\} = \frac{\exp(-\beta\alpha^v(i,j))}{\exp(-\beta\alpha^v(i,j)) + \exp(-\beta f_x^2(i,j))} \tag{13}$$

where $\mathcal{E}$ is the expectation operator. The expected value of the line process can also be obtained from the expression

$$\mathcal{E}\{l^v(i,j)\} = -\frac{1}{\beta}\frac{\partial \log Z_l}{\partial\alpha^v(i,j)} = \frac{\partial\phi}{\partial\alpha^v(i,j)} \tag{14}$$

An equivalent result holds for $l^h(i,j)$.

Only as the temperature approaches zero, does the expected value of the line process converge to zero or one. This follows intuitively from the peaking of the distribution around its most probable values at $T = 0$. We have found an expression for the expected value of the line process given the intensities. We still need to find the expected value of the intensities. Once again, Z_f cannot be evaluated. Instead, we can approximate the CM estimate of the intensities by their MAP estimate and continue to use Eq. (13) for the CM estimate of the line processes. The difference between the MAP and CM estimates can be clearly seen by comparing Eq. (7) and Eq. (13). The MAP estimate does not involve the temperature whereas the CM estimate is temperature dependent. As the temperature $T = \frac{1}{\beta}$ is reduced, $\phi(z)$ in Eq. (11) approaches the Blake and Zisserman function $\phi^*(z)$ Eq. (9). There is also a closer qualitative relationship between the sequence of $\phi(z)$ generated by varying the temperature and the sequence of $\phi(z)$ functions generated through the graduated non-convexity approach of Blake and Zisserman (1987).

Simulated annealing is the global optimization strategy adopted in this approach. As mentioned previously, we are first investigating the no interactions case. We have found both $\phi(z)$ functions (Eq. (9) and Eq. (11)) to be unsuitable for our approach for different reasons. The Blake and Zisserman function results in a highly non-convex function with very narrow valleys (LeClerc, 1989). The sequence of $\phi(z)$ functions in Eq. (11) implies that the energy function itself changes as annealing progresses. This is harder to use than a single $\phi(z)$ function. Both functions vary as z^2 for small z causing the gradient of the ϕ function to vanish. When the gradient vanishes, the MAP estimate of $\mathbf{F}$ is governed only by the data term causing instabilities. Our approach is to use a single $\phi(z)$ function which implicitly incorporates the line process and whose gradient does not vanish as z tends to zero. The $\phi(z)$ function used is from the family of $\phi^{(k)}(z)$ suggested by Geman and McClure (1987).

$$\phi^{(k)}(z) = \frac{\alpha|z|^k}{\alpha + |z|^k} \tag{15}$$

The particular ϕ function chosen corresponds to $k = 1$. This function has been used by Geman and Reynolds (1990) in association with simulated annealing. An advantage of this function is its non-interpolative nature. $\phi^1(z)$ is concave and therefore the gradient with respect to z does not vanish as z tends to zero. Line processes are addressed implicitly. However, an explicit expression for the line process can be obtained by considering the following $\phi_2(z,l)$ function (Ranarajan and Chellappa, 1990)

$$\phi_2(z,l) = |z|(1-l)^2 + \alpha l^2 \tag{16}$$

The line process can be eliminated from Eq. (16) by minimizing it with respect to l and then substituting the expression for l back into Eq. (16). Obviously, this is possible only if $\phi(z,l)$ is differentiable with respect to l. l in Eq. (16) is the analog line process.

$$\hat{l} = l(z) = \frac{|z|}{\alpha + |z|} \tag{17}$$

With $\hat{l}$ as in Eq. (17), we get $\phi_2(z, l(z)) = \phi(z)$.

Therefore, the effective energy function used is:

$$E(\mathbf{f}) = E_D + \sum_{(i,j)} \left(\phi_{ij}^v(f_x(i,j)) + \phi_{ij}^h(f_y(i,j)) \right) \tag{18}$$

where we have used ϕ_{ij}^v and ϕ_{ij}^h to stress the space variant form of the ϕ function. At each location, the ϕ function is

$$\phi_{ij}^v(f_x(i,j)) = \frac{(\kappa e^v(i,j) + \kappa'(1 - e^v(i,j)))|f_x(i,j)|}{\kappa e^v(i,j) + \kappa'(1 - e^v(i,j)) + |f_x(i,j)|} \tag{19}$$

with a similar expression for ϕ_{ij}^h.

Since stochastic relaxation is being employed, we can approximate the expected values of the intensities by running the algorithm at constant temperature and taking the average over the samples. In this paper, we have restricted ourselves to the MAP estimate of the intensities. Experimental details involved in the implementation of simulated annealing are discussed next.

5. SIMULATION RESULTS

The optimization of the nonconvex energy function was carried out with via simulated annealing procedure that used the Metropolis algorithm (Metropolis et al, 1953). In our implementation, the intensity of a pixel in the current image estimate is altered and the change in energy Δ due to the alteration is calculated. A move that reduces energy is accepted always, and a move that increases energy is accepted with probability $exp(-\Delta E/T)$ where T is the current temperature. Below we describe the actual annealing schedule and method of altering intensities.

The results of our experiment were obtained with an annealing schedule given by Simchony et al (1990)

$$T_k = \frac{T_0}{\log(1+k)} \tag{20}$$

where k is the iteration number, and $T_0 = 2.55$. A total of 1807 temperatures were used. At each temperature, each pixel is visited once using a raster scan. At each visit, a new intensity value is sampled from a Gaussian density of standard deviation (Simchony et al, 1990)

$$\sigma_k = \frac{\sigma_0}{1 + 0.025k} \tag{21}$$

with $\sigma_0 = 25$. Then the change in the forward projections and the energy due to the new pixel value are calculated.

In our first set of results Figure 2, the original 32x32 noise free image (upper left corner of Figure 2) has one large square of pixel value 100 that contains three smaller rectangles. The outside of the square has pixel values of zero. The upper left rectangle has pixel values of 130, the upper right rectangle has pixel values of 150 and the lower rectangle has the values of 75. Poisson noise was added to the projections of the noise free image. In all cases, only 8 projections of 32 elements each were used, so we in effect starved reconstruction of projection data. The far right reconstruction in the middle row of Figure 2 is obtained from optimizing only the consistency term in the energy function. The middle picture in the middle row is the reconstruction obtained by

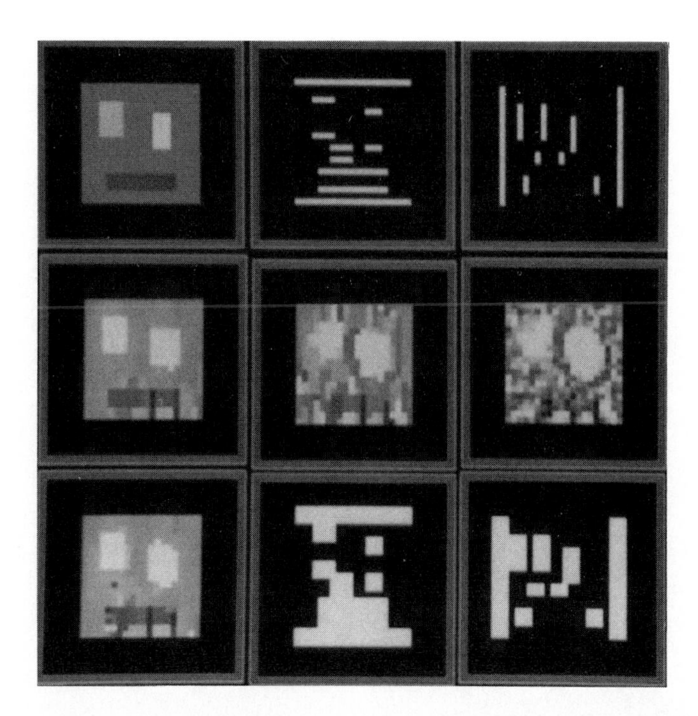

Figure 2: Simulations on a mathematical phantom : The first row shows the original figure, and the horizontal and vertical edge maps (complete with defects). The second row shows (from left to right) a reconstruction obtained using the edge maps in the first row, a reconstruction with all $e = 1$, and a reconstruction using data consistency only (no priors). The bottom row shows from left to right, a reconstruction obtained using the blurred edge maps, the horizontal blurred edge map and the vertical edge map.

optimizing the energy function with e^h and e^v of value unity everywhere. Here the probability of forming an edge is high everywhere if the difference of the neighboring pixel values exceeds the threshold. This reconstruction thus corresponds to the use of a generic piecewise smoothness constraint (though no V_C term is present.) The far left reconstrunction in the middle row is obtained from using the edge confidence maps of the top two pictures on the right in Figure 2. These maps are modified versions of the real edge maps so that they include a fake region (which is the smallest region in the middle) and also leave out an edge (the right vertical edge of the upper right rectangle). The map on the left is the horizontal edge confidence map and the right is the vertical edge confidence map. The results show that the data overruled the case of a fake region and of a missing edge. The two maps of the bottom row are the blurred version of two maps of the top row. The reconstruction using these blurred edge confidence maps is shown in the lower left corner of Figure 2. The parameters used in the energy function Eq. (18) are $A = 0.01$, $\kappa = 15$ and $\kappa' = 50$.

Note that the brightest rectangle is reconstructed in all cases with a width that is too large. This is an artifact apparently caused by having too little projection data. The excess width should cancel with the double dark line artifact in the lower bottom of each image. Also, the edge map seems to be doing some good in that the reconstructions that use the edge maps compare favorably with the ones that do not utilize the edge priors.

Our second set of results shows a 64x64 noise free brain phantom in the upper left corner of Figure 3. This was obtained from tracing edges of an actual MRI image of a human brain. The phantom has three regions that represent grey matter, white matter and generalized internal

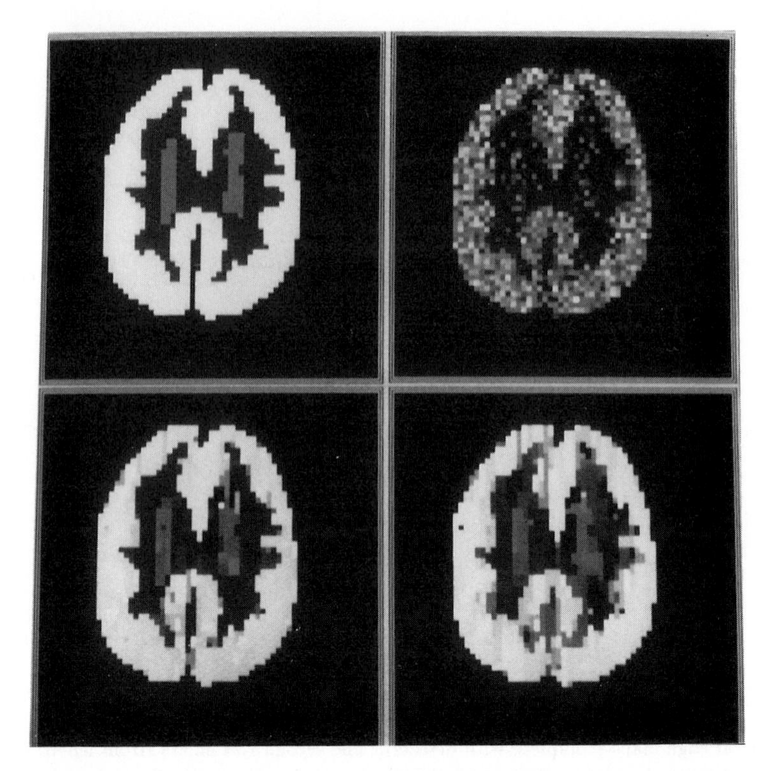

Figure 3: Simulations on a brain phantom : Upper left: 64x64 original image. Upper right: Reconstruction using data consistency only. Lower left: Reconstruction obtained using the blurred edge maps. Lower right: Reconstruction with all $e = 0.5$.

structure (thalamus or caudate). The outer region (grey matter) has values of 100, the intermediate region (white matter) has values of 25 and the two inner regions have values of 50. Again we used only 16 projections of 64 elements and added Poisson noise to the projections for all reconstructions. Parameter values of $A = 0.01$, $\kappa = 10$ and $\kappa' = 100$ were used for Figure 3 results. The upper right corner of Figure 3 is a reconstruction with only the consistency term in the energy function. The reconstruction in the lower left corner of Figure 3 is obtained by using blurred edge confidence maps. These maps are derived in such a way that they have values of 1 at true edges and of 0.5 at neighbors of the true edges. The lower right of Figure 3 is the reconstruction obtained by optimizing the energy function with e^h and e^v of value 0.5 everywhere.

6. CONCLUSIONS

We have demonstrated a model for incorporating prior knowledge of discontinuity location into a reconstruction algorithm. The results appear favorable so far as shown in the results. Validation studies showing the correlation of anatomical and functional edges are necessary, of course, but the case shown in the autoradiograph seems encouraging.

If this type of intermodality information proves useful, one might envision an integrated modality system such as that reported in (Bailey et al, 1987) for simultaneous CT and SPECT acquisition. Here, an external source mechanically linked to a rotating gamma camera provides attenuation (CT) measurements in register with the transmission measurements. Not only is registration automatically ensured, but the transmission image provides an attenuation map for more accurate SPECT reconstruction. With no extra effort, the edge map can be derived to aid in the reconstruction.

Acknowledgements

We gratefully acknowledge Dr. Robert Innis for providing the monkey brain autoradiographs. We thank Joachim Utans for discussions and assistance with the paper. This work was supported by DOE grant DE FG02-88ER60724.

References

Bailey DL, Hutton BF and Walker PJ (1987). Improved SPECT using simultaneous emission and transmission tomography. The Journal of Nuclear Medicine 28(5):845–851.

Blake A and Zisserman A (1987). Visual Reconstruction. MIT Press, Cambridge, MA.

Chen C, Pelizzari CA, Chen GTY, Cooper MD and Levin DN (1989). Image analysis of PET data with the aid of CT and MR images. In CN de Graaf and MA Viergever, editors, Information Processing in Medical Imaging pp. 601, Plenum Press.

Chen CT, Johnson VE, Wong WH, Hu X, and Metz CE (April 1990). Bayesian image reconstruction in positron emission tomography. IEEE Transactions on Nuclear Science 37(2):636–641.

Chen CT, Johnson VE, Hu X, Wong WH, and Metz CE (June 1990). PET image reconstruction with the use of correlated structural images (abstract). In JNM, Proceedings of the 37th Annual Meeting pp. 748.

Gamble EB, Geiger D, Poggio T, and Weinshall D (1989). Integration of vision modules and labeling of surface discontinuities. IEEE Transactions on Systems, Man and Cybernetics. 19(6):1576–1581

Geiger D and Girosi F (1989). Parallel and deterministic algorithms for MRFs: surface reconstruction and integration. Technical Report A. I. Memo, No. 1114, Artificial Intelligence Lab, M.I.T.

Geman S and Geman D (1984). Stochastic relaxation, Gibbs distributions and the Bayesian restoration of images. IEEE PAMI 6:721–741.

Geman S and McClure DE (1985). Bayesian image analysis: an application to single photon emisson tomography. Proceedings of the American Statistical Association.

Geman S and McClure DE (1987). Statistical methods for tomographic image reconstruction. In Proceedings of the 46th Session of the ISI, Bulletin of ISI.

Geman D and Reynolds G (1990). Constrained restoration and the recovery of discontinuities. Technical Report, University of Massachusetts at Amherst.

Gerlot P and Bizais Y (1987). Image registration: a review and a strategy for medical applications. In CN de Graaf and MA Viergever, editors, Information Processing in Medical Imaging pp. 81–89, Plenum Press.

Hebert T and Leahy R (1989). A generalized EM algorithm for 3-D Bayesian reconstruction for Poisson data using Gibbs priors. IEEE Transactions on Medical Imaging 8(2):194–202.

Johnson VE, Wong WH, Hu X, and Chen C (1989). Bayesian reconstruction of PET images using Gibbs priors. In DA Ortendahl and Jorge Llacer, editors, Information Processing in Medical Imaging pp. 15–28, Wiley-Liss.

Kessler ML, Pitluck S, and Chen GTY (1986). Frontiers of radiation oncology, proceedings of WCCF workshop on advances in treatment planning.

Leahy R, Hebert T, and Lee R (1989). Applications of Markov random fields in medical imaging. In DA Ortendahl and Jorge Llacer, editors, Information Processing in Medical Imaging pp. 1–14, Wiley-Liss.

LeClerc YG (1989). Constructing simple stable descriptions for image partitioning. International Journal of Computer Vision vol.3:73–102.

Liang Z, Jaszczak R, Floyd C, and Greer K (1989). A spatial interaction model for statistical image processing. In DA Ortendahl and Jorge Llacer, editors, Information Processing in Medical Imaging pp. 29–44, Wiley-Liss.

Metropolis N, Rosenbluth A, Rosenbluth M, Teller A and Teller E (1953). Equation of state calculations by fast computing machines. Journal of Physical Chemistry 21:1087–1091.

Poggio T, Gamble EB, and Little JJ (1988). Parallel integration of vision modules. Science 242:436–439.

Rangarajan A and Chellappa R (1990). Adiabatic approximation as a tool in image estimation. In Parallel Architectures for Image Processing SPIE 1246, Santa Clara, CA.

Simchony T, Chellappa R, and Lichtenstein Z (1990). Relaxation algorithms for map estimation of grey-level images with multiplicative noise. IEEE Transactions on Information Theory 36(3):608–613.

INTERACTIVE 3D PATIENT - IMAGE REGISTRATION

C A Pelizzari[1], K K Tan[2] *, D N Levin [2],
G T Y Chen[1], J Balter[1]

[1]Department of Radiation Oncology and [2] Department of Radiology
The University of Chicago, Chicago, Illinois 60637, USA
* Present address: Philips Research Laboratory, Briarcliff Manor, NY, USA.

Abstract

A method has been developed which allows accurate registration of 3D image data sets of the head, such as CT or MRI, with with the anatomy of the actual patient. Once registration is accomplished, the patient and image spaces may be interactively explored, and any point or volume of interest in either space instantly transformed to the other. This paper demonstrates the use of this technology in accurately transferring radiation therapy treatment plans from the 3D image space in which they are simulated, to the physical patient. This provides a heretofore missing objective link between 3D image-based simulations and actual treatment delivery.

Keywords

Radiation therapy; treatment planning; localization; multimodality imaging; coordinate transformations.

1. INTRODUCTION

1.1. Overview

Numerous applications take advantage of the three-dimensional nature of modern tomographic medical image data sets. X-ray computed tomography (CT), magnetic resonance imaging (MRI), and (in principle) ultrasound may be used to produce 3D models of a subject's anatomy. MRI and ultrasound may also be used to image flow or to isolate regions of significant flow, as in MR angiography. Single photon (SPECT) or positron (PET) emission tomography provide 3D information concerning physiological functions such as blood flow, metabolism, blood-brain barrier breakdown, and receptor binding. Under certain circumstances, two or more 3D image volumes may be registered with each other to form a hybrid multimodality data set which may provide enhanced diagnostic or therapy planning utility compared to the individual scans (Levin et al 1989; Hu et al 1990).

Highly developed 3D application areas include planning of radiation therapy (Goitein et al 1983; McShan et al 1979), neurosurgery (Kelly 1986; Watanabe et al 1987; Zinreich et al 1990) and craniofacial surgery (Vannier et al 1984); design of bone grafts and prostheses; and others. Each application involves some type of simulation carried out on a 3D patient model derived from image data. It is frequently necessary to transfer the results of the simulation to the actual patient. For example, in the case of neurosurgery planning the positions of burrholes for a simulated craniotomy may need to be transferred to the patient's head. Unfortunately, it is by no means simple to arrange the imaging procedure so that coordinates defined with respect to the image-derived patient model may be transferred accurately onto the real patient at a later time.

Applications requiring guaranteed congruence of computer-simulated and actual patient coordinate systems frequently utilize stereotactic frames rigidly fixed to the patient's skull (Lutz et al 1988, Kelly 1986). We have developed a technique (Tan et al 1990) for retrospective patient-image registration by which a coordinate transformation directly relating the computer patient model and the real patient can be established without the use of special measures during the imaging procedure. This technique has been applied to the planning of neurosurgery and localization of surface electroencephalography electrodes with respect to gyral anatomy (Tan et al 1990), to radiographic localization of subdural electrodes with respect to gyral anatomy (Grzeszczuk et al 1990), and transfer of radiation treatment plans from computer model to treatment machine. The radiation treatment planning application is described in the present paper.

1.2. Radiation Therapy Planning

Of the various application areas utilizing 3D image-based simulation, radiation treatment planning is perhaps the most highly developed in terms of widespread incorporation of the technique into clinical practice at a large number of institutions. In this case, a 3D patient model including both the tumor volume to be irradiated and normal anatomical structures which must be avoided is constructed. The geometry of simulated external radiation beams is then optimized to fully irradiate the modeled target while minimizing irradiation of critical normal tissues. Elaborate software systems for performing these simulations have been developed (Fraass et al 1987; Purdy et al 1987; Sontag et al 1987) and a number are commercially available.

Computer-simulated beam targeting can readily be refined to the level of one pixel (i.e., about 1mm) translations, and changes in beam direction of one degree or less. This apparent high precision, unfortunately, only exists in the virtual world of the image-based model. The problem of actually delivering to the real patient the exact treatment that has been simulated on the computer model remains unsolved except in a few special situations. Errors in patient setup for head and neck irradiation over a course of radiation therapy, which may take several weeks, have been observed to average 8mm using generally accepted methods for patient positioning (Rosenthal et al 1990).

There are essentially two methods for ensuring geometric consistency of simulated and delivered treatments. First, the patient may be immobilized identically during scanning and treatment. If a coordinate system is defined with respect to the immobilization device, the planes of a CT or MRI scan may be known relative to that system. Then image-space coordinates may be directly related to the immobilization device, and thus to the patient. The more severe the immobilization, the greater the accuracy with which the computer-simulated treatment can be delivered (Sherouse et al 1990) . The ultimate development of this approach is the use of a rigidly attached stereotactic frame, as is done for high-dose radiosurgery when accuracy at the 1mm level is required (Lutz et al 1988) . Such a solution is not practical for radiotherapy in general, however. Stereotactic localization frames attached to fitted masks have also been used (Schad et al 1987) which are more suitable for general use but still somewhat cumbersome.

A second approach to ensuring consistency of computer-simulated and delivered treatment is the comparison of projection images (digital reconstructed radiographs) from the image data with actual radiographs made with the megavoltage therapy beam (verification images) or with a diagnostic X-ray beam on a therapy simulator. The patient position and beam orientation are adjusted until the predicted and observed radiographs match. In practice, a combination of these methods is usually employed, with modest immobilization for scan and treatment and retrospective comparison of predicted with observed projections.

We have developed a technique for image-patient registration whereby the direct coordinate transformation between the real and simulated patient may be established retrospectively, and applied it to several problems including that of patient localization for radiation therapy. Setup accuracy approaching that of rigid stereotactic devices is obtainable; however, no immobilization is required during scanning. The system of patient marking used facilitates repeated accurate setup so that computer-simulated treatment plans may be correctly reproduced throughout the course of treatment.

2. METHOD

As pointed out by Siddon (1981) , it is useful to cast the beam-aiming problem in terms of transformations between several coordinate systems. In the present context, three coordinate systems are of interest: one attached to the treatment machine ("beam" coordinate system), one attached to the patient ("patient" coordinate system) and one defined with respect to the image data ("image" coordinate system). To allow accurate reproduction of computer-simulated beam geometry at treatment, we need only arrange that the coordinate transformation between the actual radiation beam and the actual patient is identical to that between the computer-simulated beam and the image-based patient model. This beam-to-patient transformation is not directly accessible, since the simulated beam geometry is defined with respect to the image-based patient model and not the actual patient. However, we can decompose the beam-to-patient transformation into the product of two other transformations, each of which can be objectively defined. The image-to-beam transformation is known from the computer simulation; it represents the geometry of the simulated beams relative to the image-based patient model. An image-to-patient transformation is also required; it is to define this transformation that stereotactic frames are used during scanning. In the present technique, we measure the image-to-patient transformation *a posteriori* using image-patient registration. Both transformations being known, it is straightforward to concatenate them to recover the beam-to-patient transformation, and to apply marks which demonstrate the beam coordinate system on the patient's surface. Figure 1 schematically illustrates the required transformations and the overall structure of the method. Each phase of the procedure is discussed in detail below.

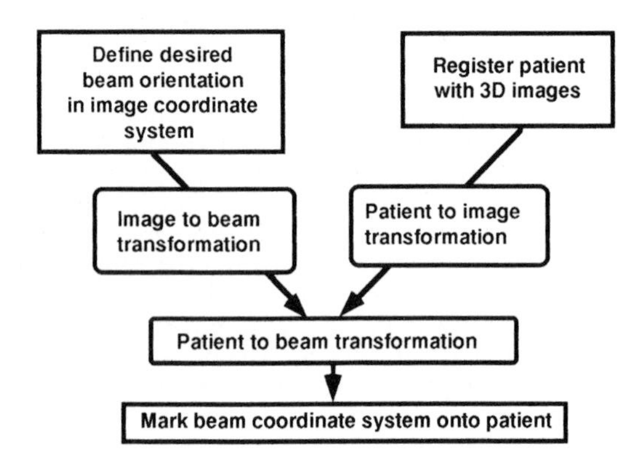

Figure 1. Schematic diagram of the image-patient registration procedure as applied to patient localization for radiation therapy.

2.1. Image-to-Beam Transformation

The image-to-beam transformation is defined by the treatment planner during the image-based simulation. By choosing beam entry direction and target point so as to achieve optimal tumor irradiation and normal tissue sparing, the user defines translational and rotational components of a rigid-body transformation between a coordinate system fixed relative to a simulated treatment machine, and a system fixed with respect to the image-based model patient. The user views a 3D perspective-corrected model of relevant anatomy as if from the source of radiation; hence the common terminology "beam's eye view" for this method of simulation (Goitein et al 1983). Such a

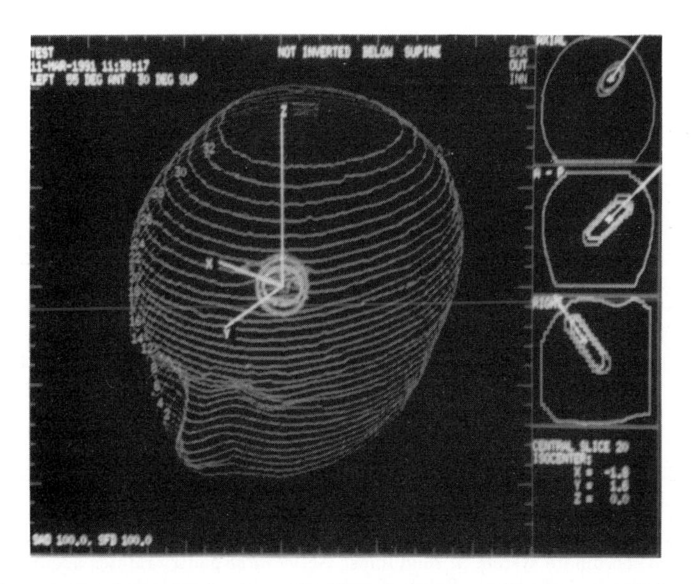

Figure 2. Beam's eye view targeting display: the phantom's external contours, and the inner and outer walls of the cylindrical target are shown in perspective as if from the source of radiation. Viewing direction is exactly along the cylinder axis. Perfect setup of the computer-modeled treatment plan on the therapy machine would place the radiation source at precisely this viewpoint, thus the term "beam's eye view."

beam's eye view for the phantom described below is shown in Figure 2. The viewing transformation from the image coordinate system to this viewpoint is exactly the transformation one would need to apply to the radiation source to achieve the simulated irradiation geometry, if the patient and image coordinate systems were identical and the patient coordinate system were initially aligned with the treatment room coordinate system. As discussed by Siddon, the rotational part of this transformation may be decomposed into a set of rotations which correspond to the motions of which the therapy machine is capable. These include rotation of the patient on the treatment couch about a vertical axis in the treatment room (illustrated schematically in figure 3), rotation of the accelerator gantry about a horizontal axis, and finally rotation of the collimator about the beam direction. The translational part of the image-to-beam transformation is simply the vector from the center of rotation of the simulated gantry (the machine's isocenter, at the origin of the unprimed coordinate system in Figure 3) and the origin of the 3D image coordinate system. The rotational part of the beam's eye view transformation is

$$R_{BEV} = R_{Z'}(\chi) \bullet R_Y(\gamma) \bullet R_Z(-\tau)$$

where R_Y is a rotation of the coordinate system about the treatment room y (horizontal) axis, γ is the gantry rotation away from vertical, R_Z is a rotation of the patient about the treatment room z (vertical) axis, τ is the turntable (patient couch) rotation angle, $R_{Z'}$ is a rotation of coordinates about the final viewing direction and χ is the collimator rotation angle. Given any 3D rotation matrix, it is straightforward to solve for the angles γ and τ and so to calculate the required machine settings to achieve the specified direction of incidence. (We ignore the collimator angle here since it does not affect the beam direction.) If the actual patient could be positioned with the image coordinate axes exactly aligned with the treatment room axes, then by applying the translation, gantry and couch angles the simulated beam geometry should be exactly reproduced. Thus we next define the transformation between the actual patient and the images.

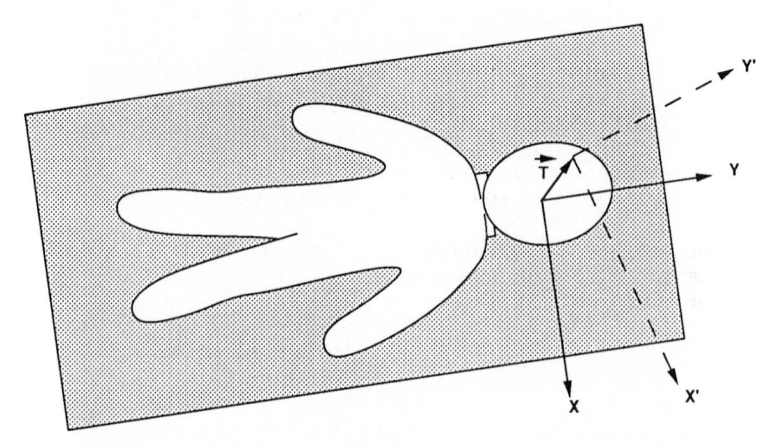

Figure 3. Schematic diagram of patient on movable couch. Room-fixed coordinate system is shown in dashed, patient-fixed coordinate system in solid lines. Accelerator gantry rotates about the Y' axis. Couch rotates about the Z' axis (out of plane). Setup of patient involves translation, gantry rotation, and couch rotation to align the dashed axes with patient as specified by the computer-modeled treatment plan.

2.2. Image-to-Patient Transformation

The image-to-patient coordinate transformation is found by applying the surface fitting method developed for multimodality image registration (Pelizzari and Chen 1987; Pelizzari et al 1989). This method determines the transformation between two coordinate systems by finding the transformation which best matches models of the same surface as defined in each of the coordinate systems. Since anatomical objects are not perfectly spherical and differ in shape from one subject to another, this simple technique has proven useful for registration of CT, MRI, PET and SPECT image volumes where either the outer surface of the head or the surface of the brain can be defined in each of two scans. The present method uses the same surface-matching software to find the transformation between a patient and an image volume, using a 3D model of the external surface measured with a three-dimensional digitizer. The present results were obtained using a magnetic digitizer, the 3Space Tracker supplied by Polhemus Navigation Science,

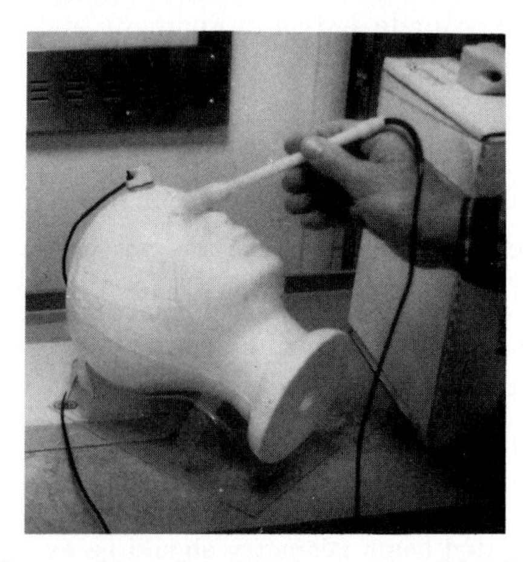

Figure 4. Magnetic digitizer in use. Magnetic source is at upper right, "tracker" on phantom forehead, "localizer" in operator's hand.

Colchester, VT, USA. The magnetic digitizer is shown in use in Figure 4. As configured here, it consists of a control unit and three magnetic elements: a source which produces a magnetic field with well-known spatial and temporal variations; a "tracker" which can report its position and orientation in 3D relative to the source; and a "localizer" which can report its position and direction with respect to the source. Software of our design running on a Sun workstation is used to interrogate the position and orientation of the active elements as required. In Figure 4 the tracker is shown attached to the forehead of the phantom, while the localizer wand is shown digitizing the phantom surface. The tracker defines a set of three orthogonal axes fixed to the patient, whose position and orientation are known relative to the source. As the localizer is swept over the surface, acquired coordinates relative to the source are projected onto the tracker axes, which are simultaneously acquired. Thus coordinates are generated in a patient-fixed coordinate system, independent of patient motion relative to the source. One to two hundred points are acquired over the surface to produce a 3D model for fitting to the image-defined surface. The rotation and translation of the digitized model required to match it with the image-defined model is precisely the image-patient transformation we require.

The surface matching software (Pelizzari et al 1989) utilizes a nonlinear least-squares search to find the rigid-body transformation which when applied to the digitized patient surface model, minimizes the average distance between the digitized points and the image-based 3D model of the patient surface. Since this is the same image-based model used for beam's eye view targeting, this step directly links the patient in the laboratory to the modeled radiation treatment.

2.3. Transfer of Beam Coordinate System to Patient

Once the image-patient and image-beam transformations are known, the localizer is again used, this time to mark the patient for setup. The simplest marking strategy is to trace out the intersections of the principal planes of the image coordinate system with the patient surface. This results in a set of lines in planes containing the unprimed axes of Figure 3, extending across the entire patient surface. The image coordinate system may then be aligned with the treatment machine coordinate system, which is defined in the treatment room by three orthogonal planes of laser light. Then the translation, gantry and table rotations determined from beam's eye view targeting (the image-beam transformation) may be applied to achieve the correct beam orientation relative to the patient. We have found that more accurate positioning is obtained when the translation and table rotation components of the image-beam transformation are applied to the image coordinate system before tracing the intersections of its principal planes with the patient surface. This results in a set of lines on the patient surface in planes containing the primed axes in Figure 3, which are to be aligned with the treatment machine coordinate system, i.e. the laser planes. After this alignment, only the gantry and collimator rotations need to be applied to arrive at the final correct beam geometry.

Principal planes of the beam coordinate system are marked on the patient surface as follows: the localizer is moved along the surface of the patient. As each point is acquired, it is transformed first into the image coordinate system by applying the image-patient coordinate transformation as found by the surface fitting procedure. The image-space coordinate is then rotated about the beam's eye view origin, around an axis parallel to the vertical axis of the CT slice planes (the antero-posterior axis, for transverse slices), by the negative of the turntable angle. This corresponds to rotating the patient couch around the beam's eye view isocenter, by the turntable angle - the difference between the unprimed and primed axes in Figure 3. Whenever any component of the resulting coordinate is zero, the point is marked on the surface. By this means points on the surface lying on the three principal planes of the simulated beam coordinate system are traced out. These planes correspond to the primed coordinate system in Figure 3.

3. RESULTS

We demonstrate the technique using an anthropomorphic phantom. This consists of a low-density plastic head containing a target constructed from a dense plastic cylinder 25mm in diameter and 100mm long. A crosshair of 1.5mm diameter solder wire is placed 6mm from each end of the cylinder, with the center of the crosses on the axis. This is intended not as a realistic tumor shape, but as a sensitive indicator of beam aiming errors. The phantom with the target in place was imaged in a GE9800 CT scanner with 5mm slice thickness and 0.9mm pixel size. Contours of the external surface of the phantom and inner and outer walls of the target cylinder were extracted from each slice.

Beam's eye view targeting was used to align a simulated radiation beam direction with the axis of the cylindrical target as shown in Figure 2. The origin of the simulated beam coordinate system (the isocenter of the simulated therapy machine) was placed at the centroid of the target contour on one of the CT slices. This corresponds to a point on the cylinder axis 42mm in from the entrance surface. Variation in beam orientation of 0.5 degree, or translation of 0.5 mm normal to the axis noticeably degraded the symmetry of the projected target contours on the beam's eye view; thus we estimate the view in Figure 2 to be within 0.5 degree and 0.5 mm of optimum irradiation geometry. The gantry and turntable angles corresponding to the chosen view are 45.2 and 44.7 degrees respectively.

A surface model consisting of 150 points measured with the Polhemus digitizer was used in determining the image-patient transformation. Figure 5 shows the image-defined model and the digitized model being fitted together. The mean squared misfit of the digitized points onto the image-defined surface was 0.7 mm^2. The image-patient transformation parameters are shown on the figure.

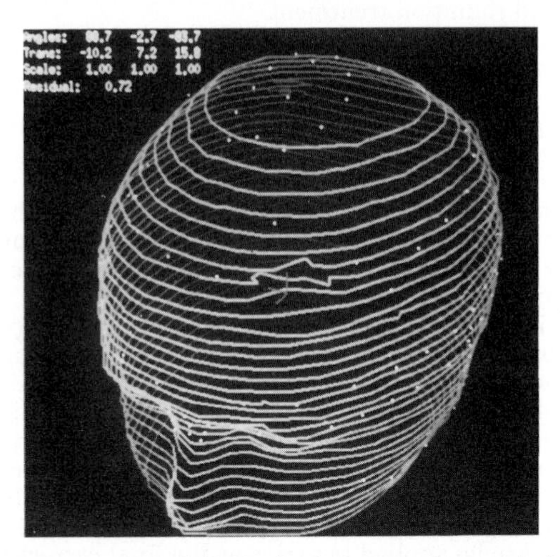

Figure 5. CT-defined surface (contours) and digitized points from phantom (points) at conclusion of least-squares surface fitting program. Coordinate transformation which places points optimally onto surface is the image-patient transformation we require.

Principal planes were marked on the phantom surface as described above, with both translation and couch rotation included from the image-to-beam transformation. Figure 6 shows the phantom with principal planes marked.

The phantom was then placed on the couch of a Varian Ximatron radiation therapy simulator, and the lines on the surface carefully aligned with the laser planes in the simulator room. The gantry was rotated away from vertical by the calculated angle of 45.2 degrees, and a radiogaph taken. Figure 7(a) is a photograph of the resulting image, taken from the screen of the simulator's fluoroscope. Note the asymmetry in the image of the cylindrical target, indicating imperfect alignment of the beam direction and the cylinder axis. Rotating the gantry one degree results in an image with the same amount of asymmetry in the opposite direction; we thus conclude that the beam alignment is correct to within one degree. Rotating the gantry by 90 degrees from the specified orientation yields the radiograph in Figure 7(b). The crosshair in the simulator aperture

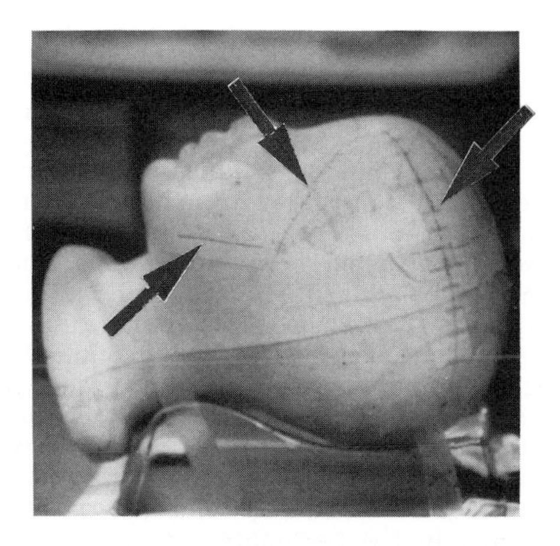

Figure 6. Phantom with localizer marks (arrows) in place, aligned with treatment room laser planes. Crosshair shadow with hatch marks is projected from collimating aperture of radiation therapy simulator.

is seen to be misaligned with the cylinder axis by approximately 1.2 degrees (n.b.: the orientation of the crosshair in the aperture is itself uncertain by at least 0.5 degree due to imprecision in the digital readouts on the simulator). The position of the isocenter is 42.5 mm from the beam entrance surface of the cylinder, 0.5 mm from the "correct" value of 42 mm. The radial distance of isocenter from the cylinder axis is less than 0.5 mm.

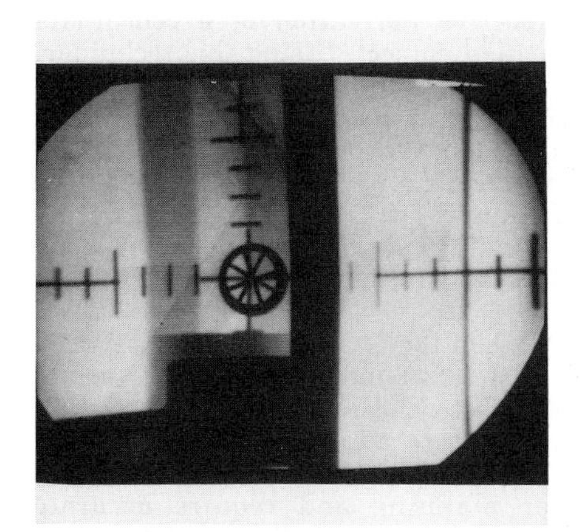

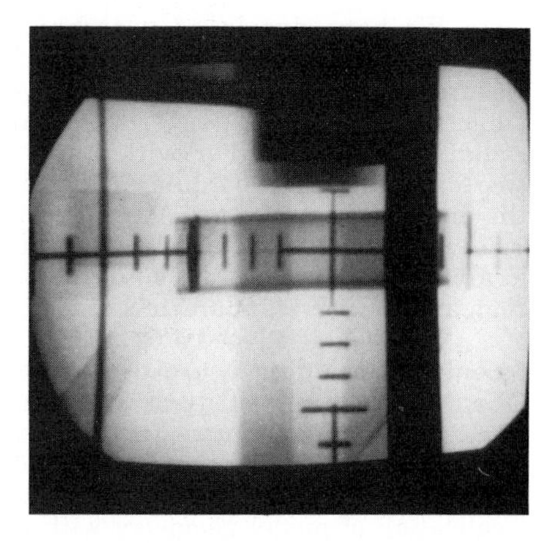

Figure 7. Left: radiograph of phantom after alignment of localizer lines with treatment room laser planes, and setting of gantry angle to calculated value. Note asymmetry of target wall image indicating imperfect alignment. Right: radiograph with gantry rotated 90 degrees from nominal position, thus viewing target normal to axis.

4. DISCUSSION

The technique presented here offers several advantages. Accuracy at the level usually associated with stereotactic frame-based methods has been demonstrated for phantom localization. The present technique however requires no special-purpose imaging study and no difficulty for the patient. The patient is required to wear the tracker sensor without removing it for the period while the surface is digitized, the image-patient transformation is determined using the surface fitting program, and the beam coordinate system is marked on the patient surface. This represents a total time of approximately 30 minutes during which the sensor is taped to the forehead. The patient may move more or less freely during this time since the technique is self correcting for patient motion. Potential drawbacks include the possible sensitivity of the magnetic localizer to metal objects near the apparatus, which unfortunately may preclude its use as a monitor of patient motion on the therapy couch, and its relatively high cost (circa $12,000). We plan to investigate the use of alternative 3D localizers such as mechanical arms. The present hardware has the considerable advantage of being self-correcting for patient motion when the tracker element is used to define a patient-fixed coordinate reference. There is also the need for a moderately powerful computer workstation to perform the surface fitting. The technique as implemented here can only deliver full accuracy when rigid external anatomy such as the cranial surface is available to register the patient and images. Thus applicability in other parts of the body is questionable. In our experience the nonlinear least-squares surface fitting procedure works well in this and other applications of image-patient registration, when care is taken in digitizing points to assure reasonably complete coverage of the cranial surface. Convergence to local minima in the parameter space is avoided in our software by providing visual feedback as shown in Figure 5, and allowing the operator to steer the fitting procedure as necessary.

The most accurate localization is obtained when a marking strategy is adopted which leaves as few motions as possible yet to be applied, after aligning the marks with the treatment room lasers. This is due in large part to limited precision and accuracy in the readouts of therapy machine motions - the fewer adjustments left to the therapy machine the better.

5. CONCLUSIONS

We have developed a technique for retrospective derivation of a coordinate transformation between CT or MR images and the actual subject. Using this technique, we are able to transfer coordinate information from a computer simulation to the subject's surface. This permits marking a patient with the position of simulated radiation beams, and/or with lines which should lie on the principal planes of the treatment machine coordinate system if the patient is set up perfectly according to the the computer simulation. Accuracy on the order of 1 mm in 3D translation and 1 degree in beam orientation have been demonstrated in phantom studies. Tests using bony landmarks as targets in volunteers, and actual setup of radiotherapy patients for comparison with standard techniques are planned.

We have developed a number of other applications of the image-patient registration technique, including neurosurgery planning (Tan et al 1990) and localization of surface (Tan et al 1990) and subdural (Grzeszczuk et al 1990) electroencephalography electrodes. For transfer of computer-simulated craniotomy locations to patients, accuracy on the order of 3-5 mm has been demonstrated. We anticipate that other applications which utilize 3D image-based models for diagnosis or planning and require accurate localization of model geometry with respect to the actual patient will benefit from the application of this type of technique.

References

Fraass BA and McShan DL (1987). 3-D treatment planning. In: The use of computers in radiation therapy, Bruinvis IAD (ed), Elsevier, Amsterdam, pp. 273-276.

Goitein M (1983). Multidimensional treatment planning. Int. J. Rad. Onc. Biol. Phys. 9: 777-797.

Grzeszczuk R, Tan KK, Levin DN, Pelizzari CA, Chen GTY, Spire J-P, Towle VL (1990). Three-dimensional localization of cortical electrodes with respect to gyral anatomy. Radiology 177(P):290.

Hu X, Tan KK, Levin DN, Pelizzari CA, Chen GTY (1990). A volume-rendering technique for integrated three-dimensional display of MR and PET data. In: 3D imaging in medicine (Proc. NATO Advanced Research Workshop), Hohne KH (ed), Springer, Berlin, pp. 379-397.

Kelly PJ (1986). Computer-assisted stereotaxis: new approaches for the management of intracranial and intra-axial tumors. Neurology 36: 427-439.

Levin DN, Hu X, Tan KK, Galhotra S, Pelizzari CA, Chen GTY, Beck RN, Chen C-T, Cooper MD, Mullan JF, Hekmatpanah J, Spire J-P (1989). The brain: integrated three-dimensional display of MR and PET images. Radiology 172: 783-789.

Lutz W, Winston KR, Malecki N (1988). A system for stereotactic radiosurgery with a linear accelerator. Int. J. Rad. Onc. Biol. Phys. 14: 373-381.

McShan DL, Silverman A, Lanza DN, Reinstein LE, Glicksman AS (1979). A computerized three-dimensional treatment planning system utilizing interactive color graphics. Brit. J. Radiology 52: 478-481.

Pelizzari CA and Chen GTY (1987). Registration of multiple diagnostic imaging scans using surface fitting. In: The use of computers in radiation therapy, Bruinvis IAD (ed), Elsevier, Amsterdam, pp. 437-440.

Pelizzari CA, Chen GTY, Spelbring DR, Weichselbaum RR, Chen C-T (1989). Accurate three-dimensional registration of CT, MR and PET images of the brain. J. Comput. Assist. Tomogr. 13: 20-26.

Purdy JA, Wong JW, Harms WB, Dryzmala RE, Emami B, Matthews JW, Krippner D, Ramchandar PK (1987). Three dimensional treatment planning system. In: The use of computers in radiation therapy, Bruinvis IAD (ed), Elsevier, Amsterdam, pp. 277-279.

Rosenthal SA, Galvin JM, Goldwein JW, Smith AR, Blitzer PH (1990). Measurement of variability in patient positioning for radiation therapy with use of serial portal film measurements. Radiology 177(P): 297.

Schad LR, Boesecke R, Schlegel W, Hartmann GH, Sturm V, Strauss LG, Lorenz WJ (1987). Three dimensional image correlation of CT, MR and PET studies in radiotherapy treatment planning of brain tumors. J. Comput. Assist. Tomogr. 11: 948-954.

Sherouse GW, Bourland JD, Reynolds K, McMurry HL, Mitchell TP, Chaney EL (1990). Virtual simulation in the clinical setting: some practical considerations. Int. J. Rad. Onc. Biol. Phys. 19: 1059-1066.

Siddon RL (1981). Solution to treatment planning problems using coordinate transformations. Med. Phys. 8: 766-774.

Sontag MR, Altschuler MD, Bloch P, Reynolds RA, Wallace RE, Waxler GK (1987). Design and implementation of a second generation three-dimensional treatment planning system. In: The use of computers in radiation therapy, Bruinvis IAD (ed), Elsevier, Amsterdam, pp.285-288.

Tan KK, Levin DN, Pelizzari CA, Chen GTY (1990). Interactive stereotaxic localization of brain anatomy.
Radiology 177(P): 217.

Vannier MW, Marsh JL, Warren JO (1984). Three-dimensional CT reconstruction images for craniofacial surgical planning and evaluation. Radiology 150: 179-184.

Watanabe E, Watanabe T, Manaka S, Mayanagi TY, Takakura K (1987). Three-dimensional digitizer (neuronavigator): new equipment for computed tomography-guided stereotaxic surgery. Surg. Neurol. 27: 543-547.

Zinreich SJ, Dekel D, Leggett B, Greenberg M, Kennedy DW, Long DM, Bryan RM (1990). Three-dimensional CT interactive "surgical localizer" for endoscopic sinus surgery and neurosurgery. Radiology 177(P): 217.

MARKER GUIDED REGISTRATION OF ELECTROMAGNETIC DIPOLE DATA WITH TOMOGRAPHIC IMAGES

P A van den Elsen and M A Viergever

Computer Vision Research Group, University Hospital Utrecht, The Netherlands

Abstract

Anatomical interpretation of EEG and MEG derived source estimations is difficult, and their correspondence with pathology as revealed by medical images is hard to assess. In this study a method is presented to register electromagnetic source data with tomographic image data of the same patient, thus facilitating the interpretation of the dipole characteristics with respect to the patient's anatomy (MRI, CT) or metabolism (SPECT). The method utilizes external triangular markers that are easy to apply to the skin and indicate reference points with subslice accuracy, even if these are located slightly outside the scanned volume. In this way accurate matching is ensured not only in high resolution images but also in standard CT and MR imaging protocols employing thick slices and/or large interslice gaps. While a similar triangular marker can be used for SPECT imaging, point-like radioactive markers have been considered as well because of their simplicity. At present no final conclusions can be drawn about the optimal design of the SPECT marker. The clinical potential of dipole source modelling in epilepsy and other neurological applications has not yet been established, mainly because the accuracy of the source estimations is still uncertain. The registration method proposed in this paper is much more accurate than the present-day source estimations, and hence will keep its value when improved dipole models are developed.

Keywords

CT; EEG; Epilepsy; MEG; MRI; SPECT; Volume Visualization.

1. INTRODUCTION

In neurology, as well as in many other medical fields, a variety of signal modalities are used to establish a basis for diagnosis and treatment. Different modalities usually provide complementary information. For example, anatomical information is provided by CT (computer tomography) and MRI (magnetic resonance imaging), whereas SPECT (single photon emission computed tomography), PET (positron emission tomography), EEG (electroencephalography) and MEG (magnetoencephalography) provide functional information. In CT images bones and calcifications are optimally delineated, whereas MRI best depicts soft tissue. SPECT and PET provide information on local metabolism while EEG and MEG measure electric and magnetic activity of the brain.

For epilepsy EEG is an important modality, providing key information on the localization of the focus. MEG may prove to be of equivalent value. It is difficult to relate these recordings, or even brainmaps derived from these recordings, to the patient's anatomy. A first step towards improvement being conducted in research centres throughout the world, is to estimate the position,

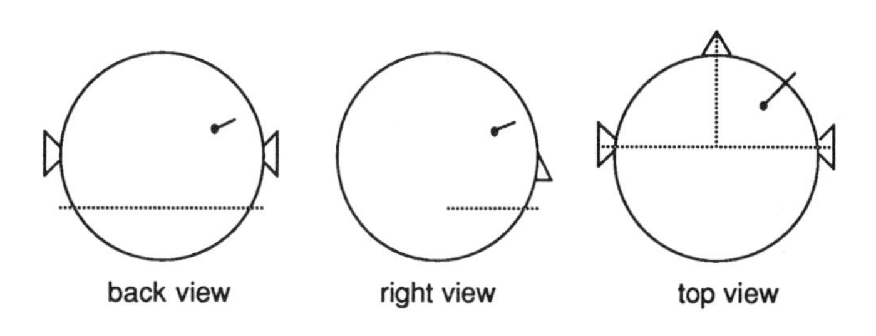

Figure 1: Visualization of a calculated dipole in three projections (back, right, and top view) of the sphere that is used to model the head. The dipole location is indicated by a little dot and its orientation by a line segment attached to it. The coordinate axes through the preauricular points and the nasion are indicated by dotted lines.

orientation, and strength of the sources of the electromagnetic brain activity, which are assumed to have a dipole-like nature (Barth et al, 1989; Salustri et al, 1989; De Munck, 1989, 1990; Gevins et al, 1990).

A commonly used simplification in source estimation models is the representation of the head by a layered spherical volume conductor with homogeneous and isotropic properties (Frank, 1952; Witwer et al, 1972; De Munck, 1989). Projections of this sphere, as depicted in Figure 1, are frequently used for visualization of the dipoles (Wong, 1989; Ueno and Iramina, 1990; Weinberg et al, 1990). The most important limitations of this type of visualization are the lack of correlation with anatomical structures, and the fact that there is no feedback on how the sphere fits the surface of the head. The nose and ears which are often added to the circles to facilitate the differentiation between the three views, are even deceiving in this respect. This is shown by the dotted lines in Figure 1, representing a coordinate system drawn through three reference points on the patient's head, being the nasion (the connection between nasal bones and the frontal bone of the skull) and points just in front of the earholes. The lines show that the depicted ears and nose are not placed well, except in the top view.

Registration of dipole data with CT and MRI solves these visualization problems. There are various ways to register dipoles with tomographic images. Gevins (1990) brings the spatial positions of the electrodes measured by a 3-D digitizer, into agreement with skin surface data obtained from MRI, by minimizing the mean distance to the skin surface for all electrodes. Another approach is to define anatomical landmarks on the patient, e.g. the nasion and preauricular points, which determine a coordinate system that is used in the dipole calculations. Locating these points visually in the tomographic image slices is a subjective method with low accuracy. The reference points can also be located by using external markers placed on the skin at the reference points (Yamamoto et al, 1988; George et al, 1989; Pantev et al, 1990). Another marker guided registration method utilizes external markers placed at the estimated surface location of the calculated source (Barth et al, 1984; Orrison et al, 1990). This method has the prerequisite that dipole calculations must preceed imaging. In marker guided methods special positioning requirements in the imaging protocol are usually necessary, and generally accuracy is highly diminished by thick slices and/or large interslice gaps as frequently used in standard MRI and CT protocols.

Although the registration method described in this paper makes use of external markers, it does not have these drawbacks. A marker design is presented that indicates the reference points on the skin with subslice accuracy, even when located in an interslice gap. An accurate match is thus provided in standard protocols with low resolution in the direction perpendicular to the slices as well as in high resolution images.

Results of the registration are shown in 2D and 3D. Both methods have their specific advantages. For example, showing the dipole in the original 2D tomographic slices (CT, MRI) facilitates correlation with pathological areas as seen on those slices. On the other hand, volume visualization of a cortex (MRI) containing a dipole facilitates the interpretation of the dipole with respect to the patient's gyral anatomy.

2. METHODS

2.1. Registration

In the interpretation of the EEG data an orthonormal coordinate system is used for the source estimation. The coordinate system is based on the aforementioned reference points: the line through the preauricular points is taken as the y-axis, the line through the nasion which intersects the y-axis perpendicularly is taken as the x-axis, and the line through the origin perpendicular to the x- and y-axes is taken as the z-axis. When for each electrode the linear distances to all three reference points are measured, the coordinates of the electrodes with respect to the described coordinate system can be determined. Then the origin and the radius of the sphere that fits the electrodes best are calculated. This sphere is used as a model of the head in the source estimation (De Munck 1989, 1990).

In order to register the dipole data with tomographic images, one needs to find the anatomical reference points in the images. Since the size of the voxels is known from the acquisition protocol, the dipole coordinate system can then be reconstructed in the images, and the dipole data can be transferred. Automatic detection of the reference points in the images is not feasible, and interactive definition by an operator is subjective, inaccurate, and time–consuming. Therefore external markers were chosen for this task.

2.2. Markers

Figure 2 shows that a small 'point-like' marker cannot achieve subslice accuracy, and that this type of marker might be missed completely when located in an interslice gap, which is present in many standard imaging protocols. An additional problem is that in CT imaging the nasion is often beneath the lowest slice, because scanning through the eyes is avoided. To improve accuracy and to solve these problems, we designed a marker that meets the following criteria:

- able to designate points with subslice accuracy
- capable of finding reference points slightly outside the scanned volume
- easily applicable to the patient's skin
- accurately replaceable
- MRI and CT compatible
- (semi-)automatic detection feasible

The marker design is presented in Figure 3. The vertex indicated as 'pointer' is used to designate the reference point, thus providing accurate replaceability. The marker is attached to the skin of the patient with sticky tape. Figure 4 depicts how the marker accurately locates points of interest, even outside the scanned volume. Our first model, made of PVC with tubes filled with a Gd-DTPA (1 : 200) solution, is 4 cm high, and has a pointing angle of 30 degrees. It can handle slice spacings of approximately 1 cm. Figure 5 shows how the marker appears in CT and MR images.

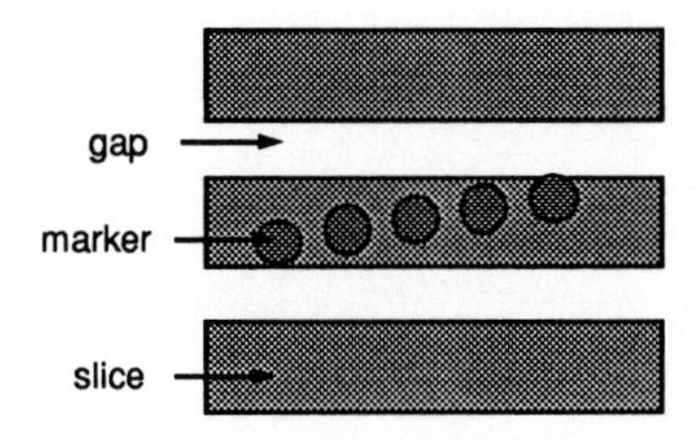

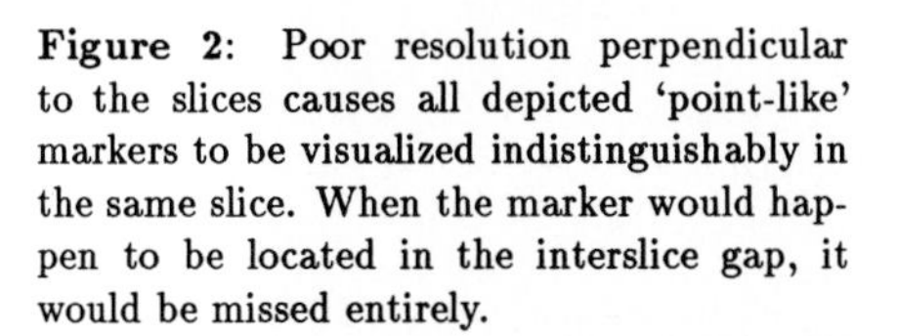
Figure 2: Poor resolution perpendicular to the slices causes all depicted 'point-like' markers to be visualized indistinguishably in the same slice. When the marker would happen to be located in the interslice gap, it would be missed entirely.

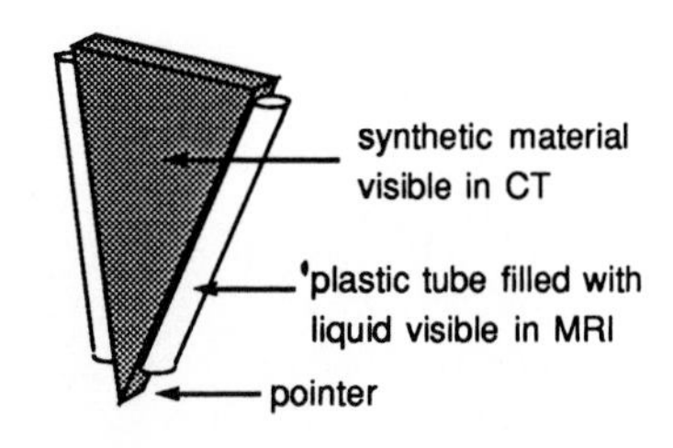

Figure 3: Design of an external marker for MRI and CT imaging

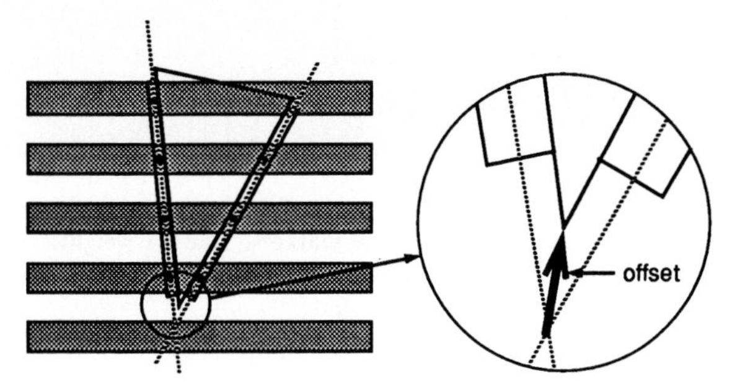

Figure 4: Schematic outlining how the reference points are located using the triangular marker. In MRI the centres of the two liquid columns are determined in each pertinent slice. Two intersecting lines are constructed through these centres. The reference point can be found from the intersection by taking into account an offset. In CT the centres of the liquid columns or the sides of the synthetic triangle can be localized, depending on the materials used.

A similar marker design with tubes filled with a radioactive liquid can be used for SPECT imaging. The low resolution of SPECT however, requires a large scale marker to prevent the tubes from merging. The disadvantages of 'point-like' markers are not as strong for SPECT as they are for CT and MRI. The reference points are always within the scanned volume (because there are no interslice gaps in the 3D reconstruction), and subslice accuracy can be achieved by computing the centre of gravity of the response caused by a point-like radioactive marker (which because of the width of the point spread function is visible in more than one slice). We have therefore started to investigate small, 'point-like' radioactive markers for SPECT. The marker consists of a circular piece of filtering-paper with a diameter of approximately 5 mm, fastened to a piece of plaster. The filtering-paper is soaked with a droplet of radioactive liquid (see section 2.3 for technical details). The markers are simple to prepare while the patient is positioned, and are easy to use. Figure 6 shows two SPECT slices in which the markers are visible.

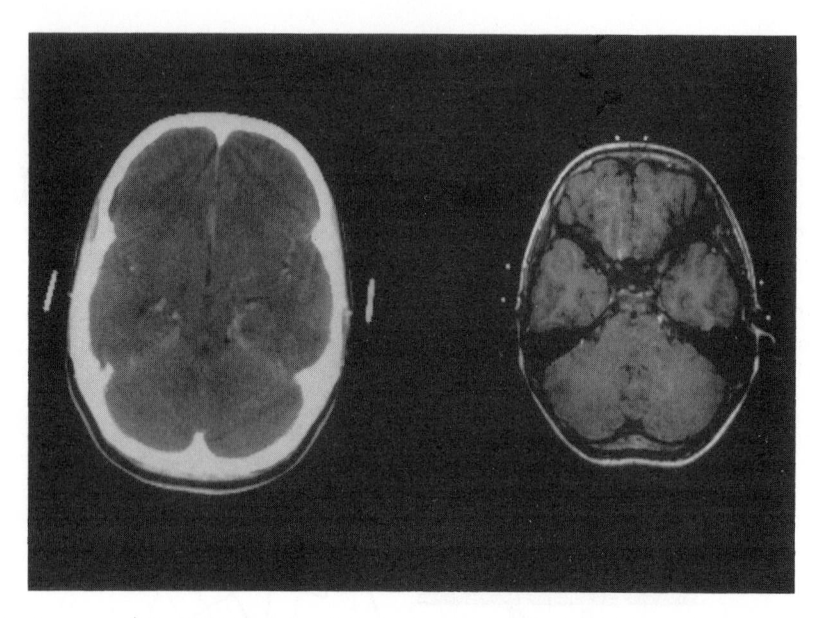

Figure 5: Transverse CT and MRI slices of the brain, showing the appearance of the markers. The CT slice on the left shows cross sections through the lateral markers. The T1 weighted MRI slice on the right shows cross sections through all three markers. In MRI only the liquid in the tubes is detectable.

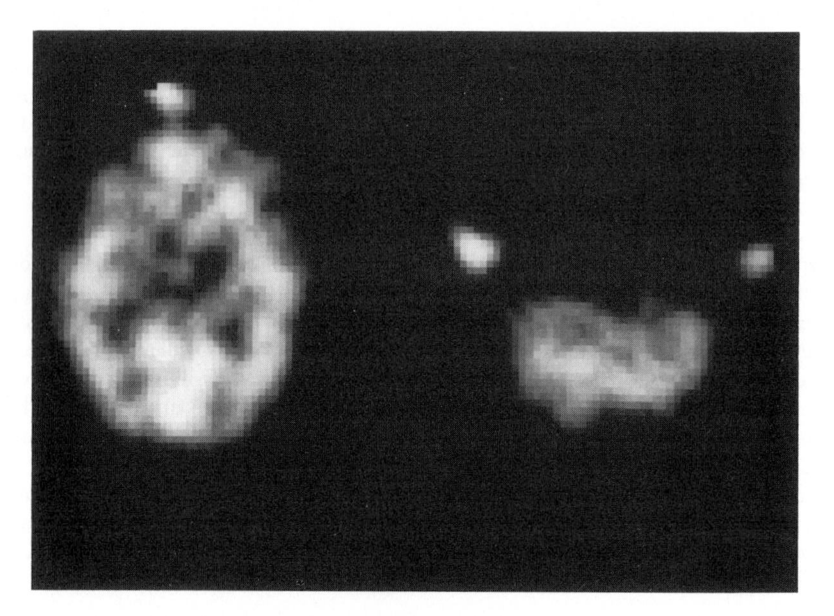

Figure 6: Two transaxial SPECT slices of the brain, showing the appearance of 'point-like' markers, the left slice showing the nasion marker, the right slice showing the preauricular markers.

2.3. Data acquisition

The MR images presented in this paper were obtained from a Philips Gyroscan S15 which is a 1.5 Tesla machine, and a Gyroscan T5 which is a 0.5 Tesla machine. The CT scan was recorded on a Philips Tomoscan LX. In all acquisitions the slices are oriented approximately parallel to the skull base, although the angulation in CT is somewhat less than in MRI. The parameter settings of the MRI and CT machines are provided in the figure legends.

The SPECT images were acquired from an Elscint round field-of-view 409 rotating gamma camera with a longbore high resolution collimator. The adult injection dose for epilepsy studies is 185 MBq of ^{123}I-Iomazenil, a specific benzodiazepine antagonist (Samson et al, 1985; Van Huffelen et al, 1990). Acquisition starts 80 minutes post injection (360 degrees, 60 steps, 40 sec/view). Markers are soaked with a droplet of a technetium-99m solution (specific activity 20 MBq/ml). Slice reconstruction is done with normalization and correction to biological half-life, using a Butterworth filter. The in-plane resolution is 12 mm FWHM.

The source estimation software used at present models the head as a sphere with three layers for scalp, skull, and brain. A time window is chosen, and within that window the location, orientation, and strength of a fixed set of dipoles is estimated (De Munck, 1989, 1990). The input data are obtained with 32 scalp electrodes placed according to the international 10–20 system, completed with electrodes at interposed positions according to the method of Buchsbaum (Buchsbaum et al, 1986).

2.4. Visualization

Both methods for visualization model the dipole as a small sphere at the location of the estimated source, with a small rod protruding in the direction of the dipole orientation, the length of the rod representing its strength. Multiple dipoles, as well as time sequential dipole data, can be visualized. In addition, the sphere used to model the head, important electrode positions, and the axes of the dipole coordinate system can be pictured. The first visualization method shows the dipole in the original slices, which results in cross sections through the sphere and rod. The cross sections can be shown as opaque discs, as transparent discs, or as circles, the latter two types leaving brain structures visible. The sequential slices can also be displayed successively in one window (cine-mode), for more 3D insight. The second method is a volume visualization method (Levoy, 1988), resulting in 3D views of the cortex (MRI) or skull (CT) combined with the sphere and rod.

3. RESULTS

At the University Hospital in Utrecht two groups of epilepsy patients are candidates for EEG derived source estimations: children with benign rolandic epilepsy, and patients with pharmacologically refractory focal epilepsy. In rolandic epilepsy patients, but also in some children without epilepsy, centrotemporal spikes occur in the EEG. Discrimination between those children through spike analysis is investigated (Van der Meij et al, 1990). In the preoperative evaluation of the second group, patients with medically intractable focal epilepsy, various modalities are used to decide whether surgery is advisable, and to determine the exact location and size of the area of the brain to operate upon. The modalities which are currently employed are MRI, CT, PET, SPECT, scalp EEG, and intracranial EEG, while shortly MEG will be available. In this group of patients, especially in temporal focal epilepsy, a considerable number of cases do not qualify for dipole calculations from interictal scalp EEG, for lack of recorded spikes. For these patients EEG recordings with sphenoidal and/or nasopharyngeal electrodes or long term EEG recordings with subdural and depth electrodes may be needed to gather sufficient EEG information. At this time only the cases that do show obvious spikes in scalp EEG recordings are registered with image data. Results of the registration method

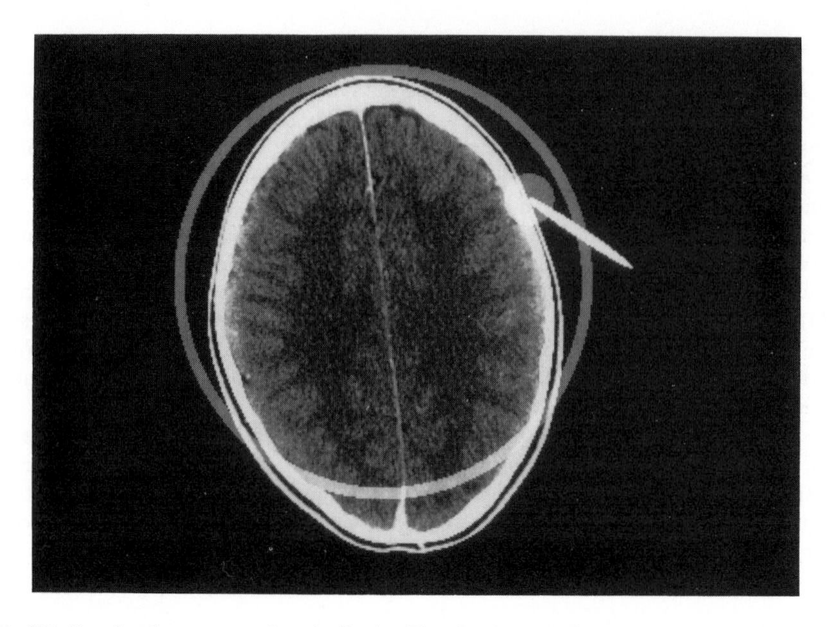

Figure 7: CT slice showing a neuro-electric dipole. The slice is part of a contrast material enhanced study obtained on a Philips Tomoscan LX at tube settings of 120 kV and 200 mA. Slice spacing is 6 mm, slice thickness and interslice gap both being 3 mm. Pixel size is approximately 0.7 mm.

will be shown for both groups of epilepsy patients. We will show not only promising examples, but also examples that clearly show the limitations of the present-day source estimations.

Figure 7 shows the CT slice nearest to the source calculated from EEG spikes in the left centrotemporal region in a 9-year-old boy suffering from benign rolandic epilepsy. The large circle is the intersection of the slice and the sphere that models the head. The bad fit of the sphere with respect to both radius and position clearly shows the limitations of sphere models. The sphere can be too wide in parietal and temporal regions, which can account for registered dipoles to be located in the skull or skin instead of in the cortex. It is expected that when the head model is improved, these dipoles will move inwards. In this example the position of the sphere is badly chosen, which could for example be caused by a bad sphere fitting algorithm, or by inaccuracies in measuring the linear distances between electrodes and reference points.

Figure 8 (Colour Plate ,p. 1) shows MRI slices obtained from another patient, suffering from symptomatic rolandic epilepsy. In this study a pathological area can be observed subcortically in the frontal and temporal lobe of the right hemisphere. The calculated source is located within the pathological area. Figure 9 (Colour Plate,p. 1) shows three 3D views of the brain of the same patient, rendered from different MRI data. The brain is depicted in grey shades, the neuro-electric dipole is shown in colour. There seems to be less structure present at the brain surface of the right frontal lobe compared to the left one. This agrees with the flattened brain contours in the right frontal lobe in the corresponding slices (Figure 8: Colour Plate ,p. 1).

Adequate 3D display requires a high resolution sampling in the direction perpendicular to the slices. This is well examplified in Figure 10, which shows the results of volume visualization in a case in which the best available data consisted of slices with 6 mm slice spacing. The patient did not suffer from seizures, although frontotemporal spikes were present in the EEG.

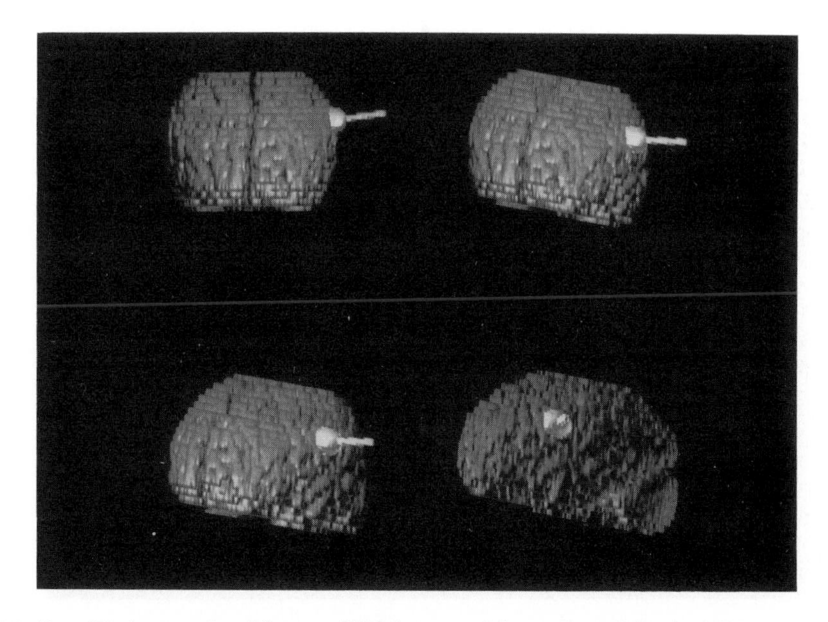

Figure 10: Four 3D views rendered from an MRI data set with a registered dipole. MRI was performed on a Philips Gyroscan S15. A transverse T2 weighted Spin Echo sequence was acquired with a single acquisition (TR/TE 2838/50,100 msec). Slice thickness is 5 mm, interslice gap is 1 mm. Pixel size is approximately 0.9 mm. The second echo was used for the rendering, because the difference in intensity between brain and cerebral fluid facilitates segmentation of the brain which is needed for visualization.

Figure 11 shows the results of EEG/SPECT registration for an 8-year-old girl, evaluated for epilepsy surgery. She suffered from seizures since the age of 14 months. In view of her youth, EEG recording with intracerebral electrodes was not performed. MRI was normal, CT images showed a hyperdense lesion in the frontal lobe of the right hemisphere. The CT scan had already been performed (without markers) before registration with EEG was part of the protocol and could therefore not be used for matching. Interictal EEG showed abundant focal spikes in the right frontal lobe. The SPECT images are not optimal, because the dose this patient received was halved because of her age. SPECT showed diminished receptor binding in the right frontotemporal region, with the area of maximal decreased receptor activity located somewhat more frontobasal than the registered dipole, which lies well within the abnormal area. This is the same patient and dipole as depicted in Figure 1.

4. DISCUSSION

The clinical value of dipole calculations in epilepsy and other applications is not yet determined, notably because the accuracy of the source estimations is still uncertain.

In the process of EEG recording, source estimation, imaging, and registration, errors are encountered at each step. For example, inaccuracies are made in manually measuring the electrode positions with respect to the reference points. These errors may be reduced by using a pair of digital callipers which were recently constructed. Alternatively a 3D-digitizer could be used (Gevins 1990). In our hospital, generally not all electrode positions are included in the procedure; in order

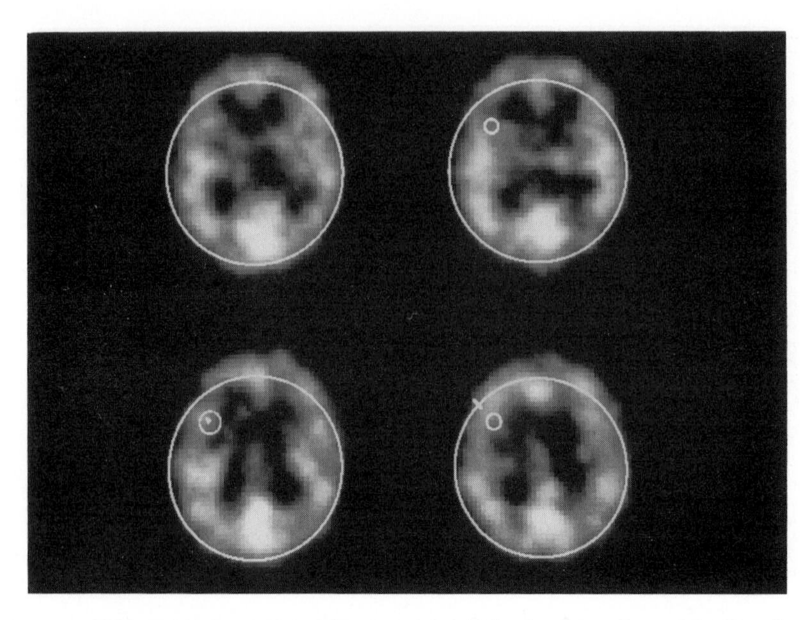

Figure 11: Four attenuated transaxial SPECT images with dipole. Part of a series of 21 slices with slice thickness of approximately 7 mm and pixel size of approximately 4 mm, obtained by the procedure described in section 'Data Acquisition'.

to gain time only the ones pertaining to the 10–20 system are taken into account. The sphere fit consequently is just based on these actually measured electrode positions. For the source estimation the position of the unmeasured electrodes is interpolated from the known positions of surrounding electrodes. Sometimes some of the (occipital and/or frontal) electrodes are deliberately left out of the sphere fitting, to arrive at a sphere that better fits the head near the dipole. The sensitivity of the model to measuring and interpolation inaccuracies as well as to the number of electrode positions measured should be investigated. The used source estimation software (De Munck 1989, 1990) implicitly assumes the origin of the coordinate system defined by the preauricular points and the nasion to be equidistant to both preauricular points. The error introduced by this restriction is currently reduced by redefinition of the reference points if the measured linear distances from the preauricular points to the nasion are not equal. The assumed nature of the model (the spherical model for the head, the localized nature of the dipole, the isotropy of the conductivity) and the signal noise all account for errors in the dipole calculations. The existing source estimation methods require further study as concerns the effects of inaccuracies in the various parameters.

The placement of the markers can introduce errors in locating the anatomical reference points on the patient's head when the imaging and EEG recording are not done successively. In this case it is advisable to record the exact locations, for example on the patient (non-washable ink) or on film. The susceptibility and chemical shift artefacts in MRI could cause distortion of the reference points designated by the marker with respect to the brain. The Appendix describes an experiment which was conducted to investigate the severity of the MR distortions. The experiment, in which readout gradient amplitudes were widely varied, demonstrates that the reference points indicated by the marker shift over a slight distance only. For carefully placed triangular markers, the matching process is much more accurate than the used dipole calculations, and will therefore still be useful when better models (like models that make use of surface information of the head, e.g. Meijs et al, 1987) will become available.

The registration procedure described in this paper does not depend on a particular type of coordinate system used for the dipole calculations. As long as the EEG coordinate system can be reconstructed from a number of reference points that can be localized and indicated on the head, the technique described here can be applied. The reference points used in our studies have the advantage that at these points the skin is relatively stable under moderate facial muscle movement. Although only EEG derived dipoles have actually been matched, the method can be used without modification for MEG derived dipoles, inasmuch as these can be estimated in a similar coordinate system.

When the three anatomical reference points that were discussed above are employed, transverse image data are preferred, because they allow all markers to be placed approximately perpendicularly to the slices so as to maximize the number of slices that intersect the marker. However, the markers that are not perpendicular to the slices in coronal or sagittal studies, can still be localized with at least the same accuracy as achieved by point–like markers.

At present the markers are pointed out in the image interactively, but we are currently working on (semi-)automatic detection for CT and MRI. The marker design has not yet been optimized as far as materials are concerned. Other marker models based on the same principle will be tested for visibility and susceptibility artifacts in MRI. Plexiglass might be a good alternative for the triangles, whereas a copper sulphate solution might be a good substitute for Gd-DTPA. Another approach would be to fill the tubes with a mixture of two liquids, one visible in CT and one in MRI, while using a light material with low visibility in both MRI and CT for the triangles. For the SPECT markers the radioactive dose is being optimized. The amount of radioactivity obviously depends directly on the specific activity and the volume of the droplet, which gave us some trouble at first. At present, the amount of radioactivity is checked by the gamma camera before the marker is attached to the patient's skin with sticky tape.

Finally, we note that any two tomographic images that were recorded with the markers in place can be registered. We are currently developing software for MRI/CT/SPECT registration based on three or more markers.

5. CONCLUSIONS

Visualization of electromagnetic sources in tomographic images greatly enhances the anatomical interpretation of the estimated locations, which increases the understanding of the used models, and may lead to clinical applications. The clinical potential of dipole calculations in epilepsy and other applications has not yet been assessed; however, the accuracy of the source estimations is still unsatisfactory. At present, the registration method described in this article is much more accurate than the source estimations; consequently, it will still be useful when better source estimation models have become available.

Acknowledgements

This research was supported in part by the Dutch ministry of Economic Affairs through a SPIN grant, and by the industrial companies Hewlett–Packard, Agfa–Gevaert, Philips Medical Systems, KEMA, and Tektronix. The authors gratefully acknowledge the cooperation with the Departments of Clinical Neurophysiology, Nuclear Medicine, and Radiology. The authors in particular want to thank A.J. van Dongen for her help with the development of the SPECT marker, and L.C. Meiners for her contribution in defining a suitable MRI protocol. The authors are indebted to E.J.D. Pol, prof. A.C. van Huffelen, J.W. van Isselt, and dr. C.J.G. Bakker for their critical comments on earlier versions of this manuscript.

AN ANATOMICAL-BASED 3D REGISTRATION SYSTEM OF MULTIMODALITY AND ATLAS DATA IN NEUROSURGERY

D. Lemoine[†], C. Barillot[†], B. Gibaud[†], E. Pasqualini[‡]

[†]INSERM U-335, Laboratoire SIM, Faculté de Médecine, 35043 Rennes Cedex, France.

[‡]Service de Neurochirurgie, Hôpital de Pontchailou , 35033 Rennes Cedex, France.

Abstract

The knowledge of patient neuro-anatomy is key information, at least in the understanding of the pathological processes and in the elaboration of precise treatment strategies in neurosurgery. In addition to classical radiology systems like angiography, CT scanner or MRI have greatly contributed to the improvement of the patient anatomy investigation . Each examination modality still carries its own information and the need to make a synthesis between them is obvious but still makes different problems hard to solve. There is no unique imaging facility which can bring out the whole set of known anatomical structures, brought together in a neuro-anatomical atlas. Nevertheless, it is very important for the physician to assign location to these structures from the images delivered by the studies. Only an accurate fusion of these data may help the physician to recognize the precise anatomical structures involved in the therapeutic process he has to set up.

The aim of this study is to provide an understanding of heterogeneous data. We propose a method to register multimodality data according to a common referential system called Proportional Squaring. Upon this geometrical basis, the deformation model is built up allowing the transfer of different patient data including the atlas within the same referential.

Keywords

Image Registration, Data Fusion, Brain Atlas, Multimodality Imaging, Stereotactic Neurosurgery, Deformation Model, MRI, CT, Angiography.

1. INTRODUCTION

Modern neurology and neurosurgery make extensive use of medical images for both diagnostic and therapeutic purposes. Imaging modalities are now quite numerous, since the new ones have not always replaced the former ones: namely, they are complementary because they are suitable to display the various anatomical structures: CT and conventional radiology are relevant for the skull and ventricular system, angiography (film-based or digital) is used to display the blood vessels, MRI is suitable to visualize cerebral tissues and nuclear medecine (PET, SPECT) for functional imaging. Nevertheless, this large number of imaging modalities (we have only mentioned the major ones) can only bring a partial answer to the problem of localizing cerebral structures. On the one hand, no single modality can actually provide information to accurately identify cerebral structures like, for instance, the brain stem or the gyri of the cortex; complex fusion of several modalities (mainly angiography and MRI data) is necessary to reach this latter goal. A common geometrical reference is needed in order to be able to mix data from the various modalities. Furthermore, the ability to correlate in-vivo data with an anatomical model, for instance a 3D atlas, brings additional information that significantly helps the understanding of brain anatomy (Bookstein 1988, Chen et al. 1985, Christopher et al. 1991, Dann et al. 1988, Evans et al. 1991, Fox et al. 1985).

Our objective is to propose solutions to these problems : our approach consists of validating and quantitatively assessing the use of anatomical references; this choice offers the advantage of remaining independent of any external landmarks leading in most cases to study protocols which may be difficult to carry out and unpleasant for the patient. The second objective is to be able to refer to anatomical models by means of a comprehensive deformation model which enables the transfer of 3D points from a patient's data set to the atlas or from the atlas to a patient's data set or between two patient data sets. The registration and deformation models are intrinsically 3D : the actual representation made by means of 2D sections, coming either from in-vivo studies or from the atlas, is only related with the capabilities of the workstation currently used to run the application (PC AT). Namely, the algorithms can be applied to 3D volumetric data sets (with an isotropic resolution in the 3 directions of space) without modification. We also made a choice for the Talairach Proportional Squaring Model (PSM). It is based on a very simple principle and has proven to be quite accurate, especially for the deep cerebral structures, as demonstrated by Talairach's statistical study on cerebral anatomy.

Our primary aim was to realize a computer platform in order to facilitate and to increase the accuracy of the registration procedures which are currently manually performed by superposing films, drawing proportional squarings and reading atlas plates. Beyond this straightforward application in surgical practice, this platform could be used for research purposes, especially for quantitatively evaluating the variability of cerebral structures with the final aim of managing this information within a computerized 3D brain atlas.

2. PRINCIPLE OF THE TALAIRACH'S PROPORTIONAL SQUARING MODEL

2.1. Use of a Ventricular-Based Line

Cerebral anatomy studies have demonstrated that bone based lines were not very accurate in modelling the proportions existing between the intra-cerebral structures. As an example, a skull with a large antero-posterior dimension does not automatically imply a wide thalamus, equally the angle between the Francfort line and the bi-commissural line is too inconstant to be used as a good reference among patients (between 11°5' and 18°5' for a sample of 5 people). Thus, bone based lines can not provide an accurate anatomical referential system (Longuet and Higgins, 1981). On the contrary, the ventricular system is a good referential since its shape and its orientation are tightly related to brain stem motion. The necessity to find out a general orientation map of the structures and to define individual ratios for the structure dimensions resulted in the choice of a ventricular and proportional base line : the bi-commissural line AC-PC (Naidich et al. 1986, Talairach et al. 1967-1988, Vanier et al. 1985).

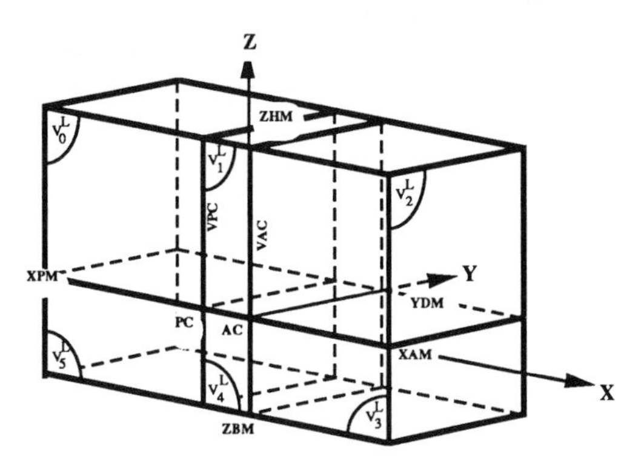

Figure 1 : Definition of the Proportional Squaring Model

2.2. Definition of the Proportional Squaring Model

The AC-PC line is defined as a straight line tangent to the upper border of the anterior commissura and tangent to the lower border of the posterior commissura within the sagittal mid-plane. From this line, two verticals are defined, *i)* VAC as the perpendicular to AC-PC at the posterior border of AC and *ii)* VPC as the perpendicular to AC-PC at the anterior border of PC as shown in Figure 2. These three lines and the mid-plane divide the cerebral volume into 12 parts (6 per hemisphere) with regards to the brain dimensions (Figure 1). For each hemisphere, we define :

- the upper point of the parietal cortex (along VAC) : *zhm*,
- the most posterior point of the occipital cortex (along AC-PC) : *xpm*,
- the lower point of the temporal cortex (along VAC) : *zbm*,
- the most anterior point of the frontal cortex (along AC-PC) : *xam*,
- the most lateral point of the parieto-temporal cortex (along the perpendicular to the mid-plane): *ydm* for the right hemisphere and *ygm* for the left hemisphere.

Within each of these sub-volumes, the deformation model defines the location of a structure as statistically proportional to the size of this sub-volume. It is a piecewise linear deformation model. For a particular patient, the calculation of the PSM from the brain extrema enables the computation of anamorphosis coefficients which make the transfer of information between a patient and a reference possible.

2.3. Computation of the Proportional Squaring Model

Let $P_{patient} = (x_{pat}, y_{pat}, z_{pat})$ be the coordinates of a point according to the referential centered at the anterior commissura AC where the axes X_{pat}, Y_{pat} and Z_{pat} are respectively defined along the AC-PC and VAC lines and the normal to the mid-plane (Figure 1); let $P_{model} = (x_{model}, y_{model}, z_{model})$ be the coordinates of the corresponding point within the referential built upon a model; let $M_{patient\text{-}model}$ be the transformation matrix between the "patient" referential and the "model" referential and considering Vj_i as the sub-volumes of the Proportional Squaring Model, where $j = r$ and $j = l$ for the right and the left hemisphere respectively, and $i = (0, ..., 5)$, the sub-volume number which includes the current point; the transfer of information between the "patient" and the "model" can be calculated as :

$$P_{model} = M_{patient\text{-}model} \bullet P_{patient}$$

with :

$$M_{patient\text{-}model} = \begin{bmatrix} Sx_i^j & 0 & 0 & Tx_i^j \\ 0 & Sy_i^j & 0 & Ty_i^j \\ 0 & 0 & Sz_i^j & Tz_i^j \\ 0 & 0 & 0 & 1 \end{bmatrix}$$

where $S_i^j = \left[Sx_i^j Sy_i^j Sz_i^j \right]$ is the scaling vector and $T_i^j = \left[Tx_i^j Ty_i^j Tz_i^j \right]$ is the translation vector.

The $M_{patient\text{-}model}$ transformation matrix is computed according to the sub-volume the point $P_{patient}$ belongs to. Thus, for the V_i's of the right hemisphere, the computation is performed as follows:

$$V_0^r \Rightarrow S_0^r = \left[\alpha_0, \alpha_3, \alpha_5 \right], \ T_0^r = \left[-(ac - pc)_{model} + (ac - pc)_{patient} \cdot \alpha_0, \ 0 \ , \ 0 \right]$$

$$V_1^r \Rightarrow S_1^r = \left[\alpha_1, \alpha_3, \alpha_5 \right], \ T_1^r = [0, \ 0 \ , \ 0]$$

$$V_2^r \Rightarrow S_2^r = \left[\alpha_2, \alpha_3, \alpha_5 \right], \ T_2^r = [0, \ 0 \ , \ 0]$$

$$V_3^r \Rightarrow S_3^r = \left[\alpha_2, \alpha_3, \alpha_6 \right], \ T_3^r = [0, \ 0 \ , \ 0]$$

$$V_4^r \Rightarrow S_4^r = \left[\alpha_1, \alpha_3, \alpha_6 \right], \ T_4^r = [0, \ 0 \ , \ 0]$$

$$V_5^r \Rightarrow S_5^r = \left[\alpha_0, \alpha_3, \alpha_6 \right], \ T_5^r = \left[-(ac - pc)_{model} + (ac - pc)_{patient} \cdot \alpha_0, \ 0 \ , \ 0 \right]$$

where:

$$\alpha_0 = \frac{xpm_{model} + (ac - pc)_{model}}{xpm_{patient} + (ac - pc)_{patient}} \qquad \alpha_1 = \frac{(ac - pc)_{model}}{(ac - pc)_{patient}} \qquad \alpha_2 = \frac{xam_{model}}{xam_{patient}}$$

$$\alpha_3 = \frac{ydm_{model}}{ydm_{patient}} \qquad \alpha_4 = \frac{ygm_{model}}{ygm_{patient}} \qquad \alpha_5 = \frac{zhm_{model}}{zhm_{patient}} \qquad \alpha_6 = \frac{zbm_{model}}{zbm_{patient}}$$

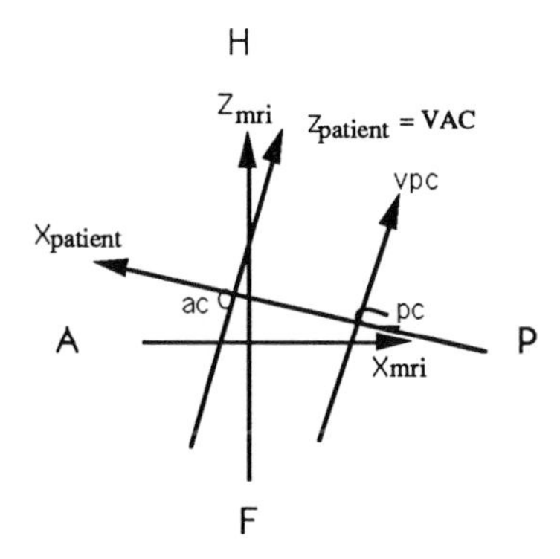

Figure 2 : Definition of the AC-PC base-line upon the mid-plane sagittal slice

3. REGISTRATION OF MULTIMODAL STUDIES

A registration procedure consists of defining the 3D geometrical relationships between in-vivo studies (CT, MRI, angiograms,...) and the proportional squaring system previously defined. MRI is actually the only modality upon which the bi-commissural line as well as the brain extrema can be detected. For this reason, the proportional squaring of a particular patient, including the definition of the AC-PC line and the computation of the scaling coefficients, are set up during the MRI registration procedure.

In the work described below and according to the hardware limitations (PC workstation under MS-DOS), we have chosen to use directly (without preprocessing) non-isotropic sets of images which may have been acquired from multiple incidences (transaxial, sagittal or coronal orientation for MRI). It should be noted that the registration stage between multi-incidence data from the same modality would no longer be necessary if 3D isotropic data were available. The following text presents the registration principles for different modalities (MRI, CT, angiography).

3.1. Registration of a MRI Study

3.1.1. Principle

As specified before, a standard MRI study may involve several analysis incidences. Several data are necessary to accurately define the patient referential. Consequently this study procedure requires specific data : a sagittal mid-plane slice, a sagittal or coronal slice including *zhm*, a sagittal or coronal slice including *zbm*, an axial or sagittal slice including *xam*, an axial or sagittal slice including *xpm*, an axial or coronal slice including *ydm* and an axial or coronal slice including *ygm*.

The registration of the different MRI incidences with respect to each others is performed by using the numerical values carried out by the MRI system along with the images. These data make computation of the geometrical transformations between an image and a basic referential attached to the machine (called "machine referential") easy. This directly issues the transformation matrix between one image of one series (*I of S*) and another image of another series (*I' of S'*).

3.1.2. Definition of the "Patient Referential"

The registration problem under consideration is to define as exactly as possible the base lines of the patient geometrical reference (AC-PC line and the two vertical lines VAC and VPC). The geometrical referential, as defined in *§2-2*, cannot be detected by using an automatic procedure involving image processing algorithms. Consequently, we made a choice for an interactive designation of the PS referential, whatever the MRI data base is (isotropic or multi-incidence). Obviously, the accuracy of this procedure will be highly dependent on the sampling quality of the cerebral volume resulting from the MRI study. The methodology proposed here involves four stages :

1. *Definition of approximated base lines (Figure 2) :* The AC-PC line and its two associated vertical lines are interactively defined upon a sagittal slice showing the anterior and posterior commissuras (as explained on *§2-2*), (Figure 9: Colour Plate , p. 2).

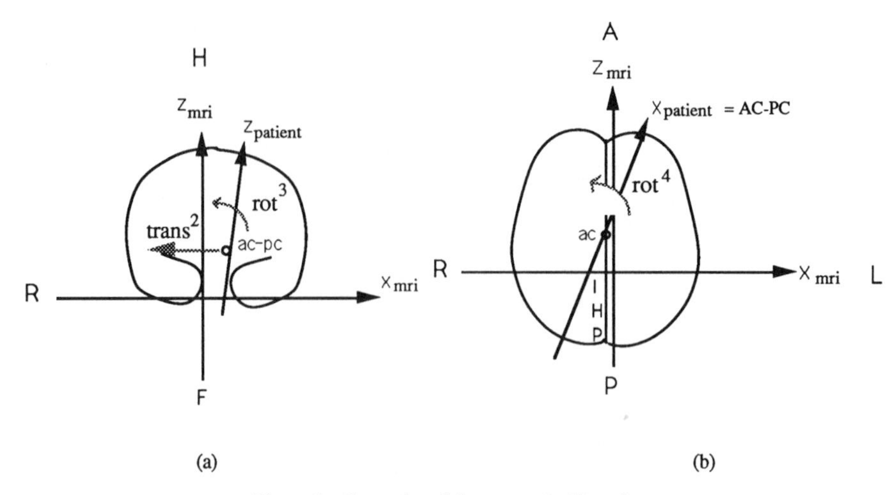

Figure 3 : Correction of the geometrical base lines
a : rotation of VAC about the AC-PC line on a coronal slice
b : rotation of AC-PC about the VAC line on a transaxial slice.

2. *Translation of the AC location (Figure 3a)* **:** A translation of the AC location can be performed (if necessary) in order to replace it into the mid-plane estimated from a coronal slice containing (or very close to) the anterior commissura (*trans*^{2} on Figure 3a).
3. *Rotation of the VAC base line about the AC-PC line (Figure 3a)* : On the same coronal slice, the VAC and mid-plane orientation are corrected by making a rotation of VAC about the AC-PC axis (*rot*^{3} on Figure 3a).
4. *Correction of the AC-PC line (Figure 3b)* : A correction of the mid-plane defined by the AC-PC and VAC lines is performed by estimating a rotation of this plane about VAC upon a transaxial slice containing (or very close to) AC (*rot*^{4} on Figure 3b).

The four respective transformation matrices involved in the processing can be summerized as follows :

$$M^1_{patient\text{-}MRI} = \begin{bmatrix} [rot^1] & [trans^1] \\ 000 & 1 \end{bmatrix}$$

$$M^2_{patient\text{-}MRI} = M^1_{patient\text{-}MRI} \bullet \begin{bmatrix} [Id] & \begin{bmatrix} 0 \\ 0 \\ 0 \end{bmatrix} \\ [trans^2] & 1 \end{bmatrix}$$

$$M^3_{patient\text{-}MRI} = M^2_{patient\text{-}MRI} \bullet \begin{bmatrix} [rot^3] & \begin{bmatrix} 0 \\ 0 \\ 0 \end{bmatrix} \\ 000 & 1 \end{bmatrix}$$

$$M_{patient\text{-}MRI} = M^4_{patient\text{-}MRI} = M^3_{patient\text{-}MRI} \bullet \begin{bmatrix} [rot^4] & \begin{bmatrix} 0 \\ 0 \\ 0 \end{bmatrix} \\ 000 & 1 \end{bmatrix}$$

The transfer of information between the "patient referential" and MRI data is calculated as :

$$P_{patient} = M_{patient\text{-}MRI} \bullet P_{MRI}$$

Once the geometrical basis of the patient referential is defined, the last stage consists of determining the 3D box encasing the brain. This is done by using the slices previously selected during the MRI study protocol (Figure 10: Colour Plate, p. 2). Finally, the transfer of information between the "model" referential and MRI data is calculated as :

$$P_{model} = M_{model\text{-}patient} \bullet M_{patient\text{-}MRI} \bullet P_{MRI}$$

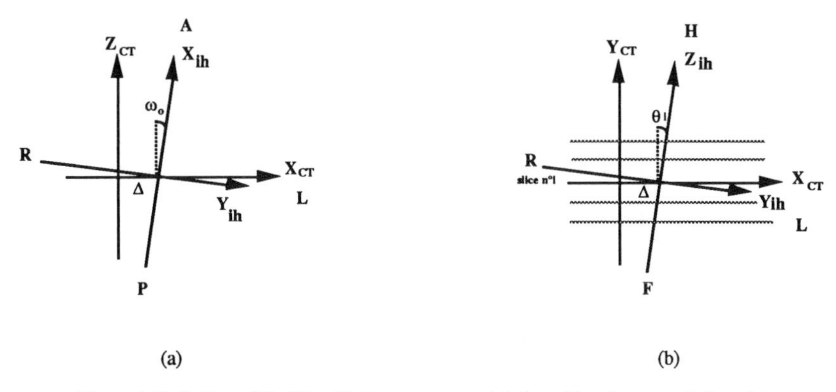

(a) (b)

Figure 4: Definition of the CT mid-plane upon an axial plane (a) and a coronal plane (b)

3.2. Registration of a CT Study

3.2.1. Principle

Cerebral structures such as the anterior and posterior commissuras are not visible on transaxial CT slices. The reconstruction of the mid-plane from CT has not proven suitable to deduce the AC-PC line, especially when using a standard 5mm slice thickness. The method proposed here is based on the localization of some specific information which can be localized upon both the CT and the MRI mid-planes. The registration of these data will effect the geometrical transformation between CT and MRI. The registration between CT data and the "patient referential" is performed by using the results of the MRI registration as linking information.

3.2.2. Estimation of the CT Sagittal Mid-Plane

The definition of the CT sagittal mid-plane is interactively performed by the designation of the estimated mid-plane position on each slice. Three steps are involved in this definition :

1. *Learning stage*: On a reference slice, the estimation of the position and the orientation of the mid-plane is performed by using two parameters Δ and ω_0 (Figure 4a). The initial value of θ_0 is set to zero.
2. *Update stage* : On the other slices, new values (ω_i, θ_i) can be assigned for correction of the mid-plane orientation (Figure 4b).
3. *Optimization stage :* The set of the estimated parameters ω and θ allows the estimation of an average position of the CT mid-plane by using the following relations:

$$\omega = \frac{1}{N}\sum_{i=1}^{N} \omega_i \quad , \quad \theta = \frac{1}{(N-1)}\sum_{i=2}^{N} \theta_i$$, where N is the number of CT slices.

3.2.3. CT-MRI Assignment

On the mid-plane defined as above, a set of points featuring structures, which can be seen on both MRI and CT, is interactively designated. These structures can be the skin, the middle of the bone, the limits of the ventricles, or others. The registration between CT and MRI is then performed by interactively carrying the graphical representations of structures of interest from CT over to their corresponding position on the MRI mid-plane. The geometrical transformation is directly derived from this procedure (Figure 11: Colour Plate , p. 2).

It seems rather difficult to use automatic pattern recognition techniques to detect reference structures. The selection of these structures may change from one CT study to another since the sampling of cerebral volume (by the CT) may also change (the CT study does not usually cover the entire head). Nevertheless, it is still possible to think about particular procedures which would help the assignment stage between CT and MRI.

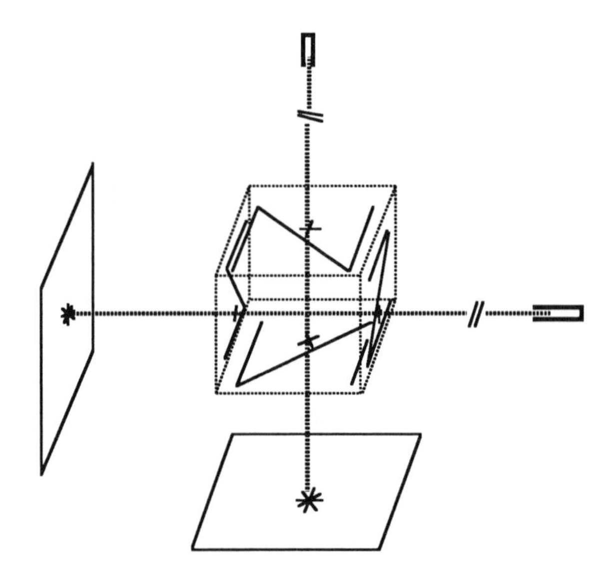

Figure 5 : Geometrical conditions of the stereotactic angiography study protocol

3.3. Registration of Angiographic Data

As opposed to the two other modalities, the angiography study protocol is strict and its completion, in stereotactic conditions, rather awkward. Pairs of radiographs at a distance of 5 metres (face and profile) are performed from two co-planar and perpendicular sources. An instrumental framework (on which referential landmarks can be attached) is fixed to the patient's head (Figure 5). The advantage of taking film at a distance of 5 metres is that the X-Rays beams are almost parallel (in a first approximation), since the magnification ratio varies between 1.02 and 1.08 from side to side of the frame.

The referential associated to the patient, called the object referential, is composed of referential elements. These elements are flat plates including radio-opaque landmarks shaped as N's and crosses allowing the neurosurgeon to correctly position the frame. Under scopy, it is possible to check along the face and profile orientations whether the crosses are correctly placed and aligned or not. This procedure makes the determination of the transformation matrices between the object referential and the two film referentials easier. These matrices are a function of the geometrical features of the stereotactic room and have been experimentally determined (Le Gall 1990, Longuet and Higgins 1981, Metz et al. 1989).

Since no particular caution is taken during the film digitization (by using a video camera connected to the PC), the first stage consists of computing the transformation matrices between the radiograph referentials and the projection referentials, which are considered as the acquisition referentials (one for each incidence). Three coins have been placed on the two film holders which can be automatically detected after digitization in order to perform this preliminary 2D registration.

As for CT, the AC and PC structures can not be recognized on the angiograms, thus the registration of the patient AC-PC referential can not be done directly. The MRI mid-plane slice is used once again as the connection between angiography and the patient referential. The structures used to carry the angiography information over to MRI are bone contours (to get a good idea of the general shape) along with blood vessels elements (arteries and veins) like the pericalleus artery or the great cerebral vein of Galeus which surround the extremities of the corpus callosum, easily identified on MRI.

3.3.1. Definition of the Angiographic Mid-Plane

Concerning the definition of a 3D point within the object referential, two corresponding points upon face and profile films have to be designated in order to find out the correct 3D position. The way to find out the 3D location of a point within the object referential is to firstly select its position on the face or profile view and then to adjust this position along the epipolar line upon the other view (profile or face) (Figure 6).

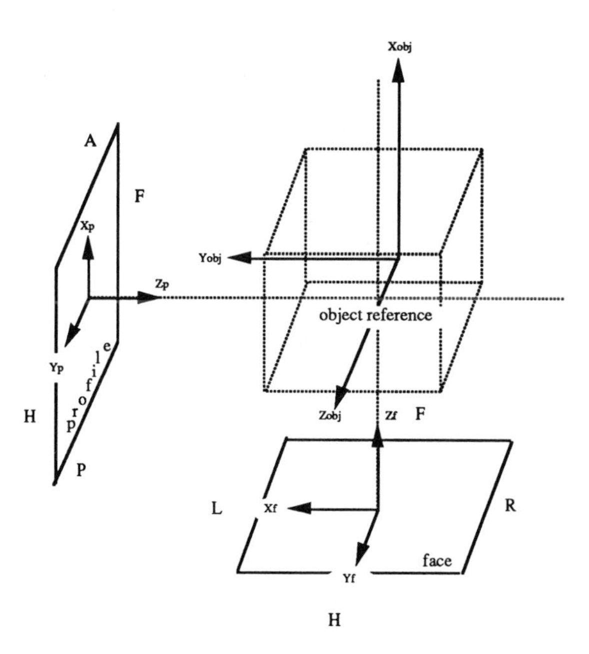

Figure 6 : Principle of the projection geometry between the object referential and the two orthogonal projections (face and profile).

The location of the mid-plane is interactively defined on a face view angiogram. This method has a drawback : the face incidence is supposed to be perpendicular to the real mid-plane, which is usually the case in most studies. A slight rotation cannot be easily detected. Further work is needed to estimate its influence with regards to the registration accuracy.

The projection matrices between the object referential and the orthogonal views are :

$$\begin{bmatrix} t\,x_p \\ t\,y_p \\ t \end{bmatrix} = \left[P_{i,j}\right] \cdot \begin{bmatrix} x_{obj} \\ y_{obj} \\ z_{obj} \\ 1 \end{bmatrix} \quad \text{and} \quad \begin{bmatrix} t\,x_f \\ t\,y_f \\ t \end{bmatrix} = \left[F_{i,j}\right] \cdot \begin{bmatrix} x_{obj} \\ y_{obj} \\ z_{obj} \\ 1 \end{bmatrix} \text{ with } i = \{1, 2, 3\} \text{ and } j = \{1, 2, 3, 4\}$$

3.3.2. Registration between Angiography and MRI

As with CT, several points viewed on both angiography and MRI are used to transfer the information between both modalities : *i)* the pericalleus artery and the great cerebral vein of Galeus which are attached to the corpus calleus extremas and *ii)* the bone contours can be viewed on angiography and estimated upon the MRI mid-plane slice. From this registration, the geometrical transformation between angiography and MRI and then between angiography and patient can be derived.

This work was intentionaly done without involving segmentation techniques in vessel detection on angiograms. This domain is widely discussed in the literature and the complexity of the anatomical structures to be dealt with does not simplify this problem (spatio-temporal filtering, segmentation of overlapping vascular structures, ...). Nevertheless, as for CT, it can be possible to work out automatic processes to carry pre-recognized vessel structures over MRI (registration of vessels surrounding cortex gyri for instance).

3.4. Registration of 3D Representations

The aim of this part is to make possible the use of 3D representation of pre-registered MRI or CT data sets. This is not exactly a registration problem in the sense of detecting reference structures and registering them. Nevertheless, it is necessary to compute the geometrical transformations making up the relation between a point on a 3D representation and its position within the registered data base. The actual solution is to compute the 3D representations of pre-registered 3D data sets by means of external software (Barillot 1988, Robb and Barillot 1989). These CT or MRI data bases have to be re-sampled in order to make isotropic volumes. The 3D display software performs the surface segmentation and brings out a "3D image" which includes a 2D projection (with surface shading) and the depth information of the displayed points (Z-buffer). In addition with the "3D image", the display software carries out a transformation matrix which makes the transfer of a "3D image" point possible (x_i, y_i, $Zbuff_i$) to the referential system of the original data base.

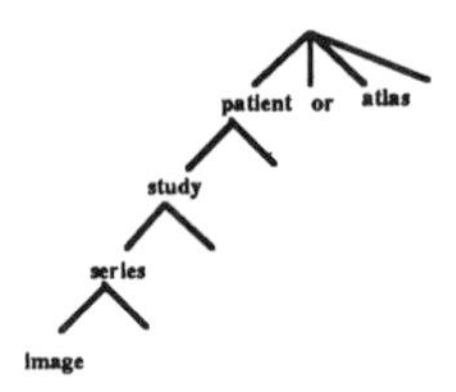

Figure 7 : Archiving Data Hierarchy.

4. APPLICATION

4.1. Multimodality Images Fusion

In order to validate and actually exploit these registration techniques we built a multi-modal point transfering application : this application allows point to point coordinate transfer between data related to a patient and a digitized atlas or between the atlas and the patient data, or between a patient data set and another patient. These assignments are performed by using the proportional squaring deformation model. This application is made of two parts : *i)* an acquisition and registration module and *ii)* a point to point transfer module. Concerning the second one, the principle consists of selecting an input and output data from the *patient/ study / series / image* hierarchy (Figure 7) (Barillot et al. 1989).

The display screen is split into four parts, the upper left quadrant is used to display the input image. A point is then selected by moving a cursor with the mouse on this image in order to define a point in the 3D space, then the system computes the 3 output images, by selecting within three series the closest planes to the selected 3D point; planes are displayed and cursors show the localization of the transfered points upon the output images. This application has been implemented on a PC AT with a 512 x 512 x 12 bits bit map (8 bits + 4 bits for graphics) and connected to a video camera.

4.2. Data Acquisition

The Talairach atlas is composed of anatomical plates in the three orthogonal planes : sagittal, axial and coronal. The application considers this atlas as a particular patient with only one study and three series. All data (except for angiographic images which have to be digitized on the workstation) are retrieved from the central database of the SIRENE PACS (CT, MR, 3D images) (Gibaud et al. 1989).

4.3. Results

All software developments have already been done, this actually makes the registration of all of the imaging modalities previously mentioned possible. The completion of all kinds of point to point assignments can also be done as patient to atlas, atlas to patient or patient to patient (Figures 12, 13, 14: Colour Plates , p. 2). A number of additional tools have been developed in order to enhance the readability of the images and to facilitate the use of the system : zoom of an area of interest, full screen display, superposition of the proportional squaring, printing of geometric data and others.

The primary use of the system has produced good results and no obvious registration errors could be detected. However, a major concern is how to assess the accuracy of the registration and to study how inaccuracies may be propagated. Any application, and especially in neurosurgery, which uses such registration techniques must be able to give quantitative measurements of the uncertainty related to any point transfer between the registered data sets.

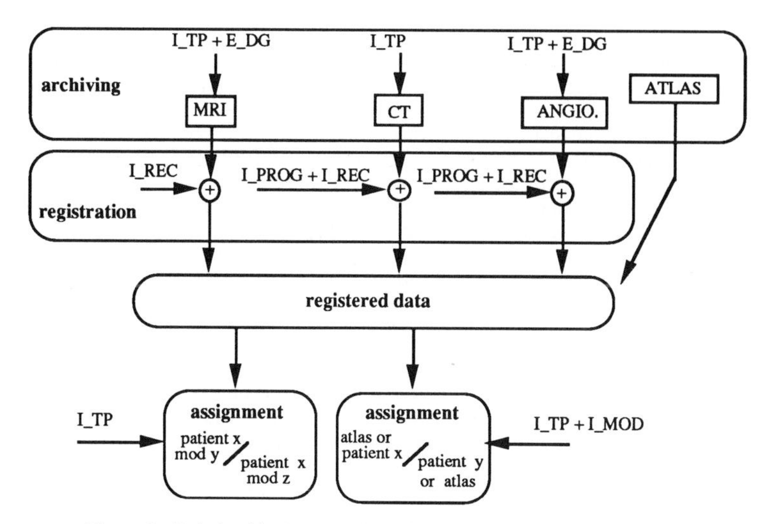

Figure 8 : Relationships between the different inaccuracy and error sources

5. EVALUATION

5.1. Position of the Problem

The assessment of registration has to be considered at two levels. With respect to the registration of data concerning one patient, the problem is essentially related to the accuracy of the registration between modalities. Namely, the anamorphosis coefficients are obviously the same for all modalities and do not introduce additional errors. Regarding the registration between different patients, the problem is also to evaluate the accuracy of the deformation model itself. It is well known that the proportional squaring model is quite suitable for the deep cerebral structures but can only provide approximate statistical information about cerebral gyri, for instance.

5.2. Errors and Inaccuracy Causes (Figure 8)

In order to measure registration uncertainties, we have designed a model showing the various error origins and the way they cumulate (Figure 8). A primary cause of errors concerns the pixel size (field of view / number of pixels) : *E_TP*, the pixel size in the third direction has also to be considered. Such an error applies to all modalities. Additional distorsion may come up with some modalities, like a geometric distorsion in MR due to the non homogeneity of the magnetic field, or like distorsions due to the digitization camera for angiographic images. The second origin is related to user interaction. The designation of anatomical structures is obviously subjective : an expert may vary in designating a single structure as well as different experts may also vary in designating the same structure (Dann et al. 1988, Seitz et al. 1990). Studies on consistency and variability will helps us to quantitatively evaluate this uncertainty (*I_REC*). Such an error may induce a bad definition of a structure or a plane, especially when the study does not perfectly fit the study protocol (for instance, designation of the inter-hemispheric plane in a coronal angiogram). Finally, it is clear that the registration accuracy of a CT or an angiographic data set is highly dependent on the MR registration quality. Consequently, a cumulative effect has to be taken into account (*I_PROG*). Another issue is to evaluate the quality of the deformation model itself, and the intrinsic quality of the digitized Talairach atlas.

6. CONCLUSION

We have described in this paper a straightforward way to register multimodality and multipatient data by means of anatomical reference structures. This geometrical referential is widely acknowledged as one of the most accurate, in the anatomical sense, in the context of stereotactic neurosurgery. We have focused our work on finding reliable procedures which help to define this referential using data gathered from a patient. Beyond the prevalence given to the MRI modality in this approach, we wanted to put stress on the establishment of systematic procedures which take into account the particularity of the different modalities (nature of the information carried out, sampling of the cerebral medium, etc...) and ensure maximum exactness along the different registration stages.

Further research is needed concerning the validation of these procedures. It is actually crucial to be able to quantify the uncertainty of the registration processes between modalities and between patients. This project is going to be the main part of our work during the coming months. Based on these validations, a more fundamental study will be to improve the deformation model in order to extend its application field, which is usually limited to the central region of the brain.

The application field of such a system is huge. It could be helpful in the neurosurgical domain (biopsy, definition of intra-cerebral entry ways, etc...), in the functional anatomy domain (concerning the localization of epilepsy crises generators), in radiotherapy for the treatment of cerebral lesions or even in the research domain like, for instance, the constitution (or the improvement) of an anatomical brain atlas.

7. REFERENCES

Barillot C. (1988). "Interactive Ray-Tracing for Volume Rendering of Medical Data", INSERM U335, Internal Report.

Barillot C., Lemoine D. , Gibaud B., Toulemont P.J. and Scarabin J.M. (1990) "A PC Software Package to Confront Multimodality Images and a Stereotactic Atlas in Neurosurgery", SPIE Vol 1232 Med. Im. IV, pp 188-199.

Bookstein F.L. (1989) "Principal Warps : Thin-Plate Splines and the Decomposition of Deformation" IEEE PAMI Vol.11(6), pp 567-585.

Chen G.T.Y., Kessler M. and Pitluck S. (1985) "Structure Transfer in Three-Dimensional Medical Imaging Studies", Nat Comput. Graph. Ass., pp 171-175.

Christopher J.H., Pike G.B., Collins D.L. and Peters T.M. (1991) "Three-Dimensional Display of Cortical Anatomy and Vasculare : Magnetic Resonance Angiography Versus Multimodality Integration", J Digital Im., Vol 4 (1), pp 21-27.

Dann R., Hoford J., Kovacic S., Reivich M. and Bajcsy R. (1988) "Three Dimensional Computerized Brain Atlas for Elastic Matching : Creation and Initial Evaluation", SPIE Med. Im. II, Vol 914, pp 600-608.

Evans A.C., Dai W., Collins L., Neelin P. and Marrett S. (1991) "Warping of a Computerized 3-D Atlas to Match Brain Image Volumes for Quantitative Neuroanatomical and Functional Analysis", SPIE MI V Vol.1445, (to appear).

Fox P.T., Perlmutter J.S. and Raichle M.E. (1985) "Stereotactic Method of Anatomical Localization for Positron Emission Tomography", J Comput. Assist. Tomogr., Vol.9, pp 141-153.

Gibaud B., Picand F. et al. (1989) "SIRENE : Design and Evaluation of a PACS Prototype" Proc. of CAR'89, Springler Verlag Ed., pp 484-488.

Le Gall G. (1990) "Informatisation des Données Stéréotaxiques et Dosimétriques pour l'Irradiation de Lésions Cérébrales : une Approche Robotisée", Ph.D. thesis, Université Paul Sabatier, Toulouse, France.

Longuet H.C. and Higgins (1981) "A Computer Algorithm for Reconstructing a Scene from Two Projections", Nature Vol.293, pp 133-135.

Metz C.E. and Fencil L.E. (1989) "Determination of Three-Dimentional Structure in Biplane Radiography without Prior Knowledge of the Relationship between the Two Views : Theory", Med Phys. Vol 16(1), pp 45-51.

Naidich T.P., Daniels D.L., Pech P. et al. (1986) "Anterior Commissure : Anatomic-MR Correlation and Use as a Landmark in Three Orthogonal Planes", Radiology, Vol 158 (2).

Robb R.A. and Barillot C. (1989) "Interactive Display and Analysis of 3D Medical Images", IEEE TMI Vol.8(3), pp.217-226.

Seitz R.J., Bohm C., Greitz T., Roland P.E. et al. (1990) "Accuracy and Precision of the Computerized Brain Atlas Programme for Localisation and Quantification in Positron Emission Tomography", J. Cereb. Blood Flow Metab., Vol. 10 (4), pp. 443-457.

Talairach J., Szikla G., Tournoux P. et al. (1967) "Atlas d'Anatomie Stéréotaxique du Télencéphale", Masson Ed., Paris.

Talairach J. and Tournoux P. (1988) "Co-Planar Stereotactic Atlas of the Human Brain", Georg Thieme Verlag Ed., Stuttgart.

Vanier M., Roch Lecours A., Ethier R. et al. (1985) "Proportional Localization System for Anatomical Interpretation of Cerebral Computed Tomograms", J. Comput. Assist. Tomogr., Vol 9(4), pp 715-724.

REGISTRATION OF BRAIN IMAGES BY A MULTI-RESOLUTION SEQUENTIAL METHOD.

M A Oghabian and A Todd-Pokropek

Department of Medical Physics, University College London
London WC1E 6BT, U.K.

Abstract

In many circumstances the clinical interpretation of an imaging from a single image modality is inadequate. A number of registration techniques have been introduced in the literature in order to correlate the clinical information obtained from two different imaging modalities. All methods based on some distance measure suffer from the presence of multiple local minima when minimization algorithms are used to reduce the distance between two edges, or surfaces. A multi-resolution technique has been developed in conjunction with a sequentially improved distance function in order to register sets of MRI, PET, and SPECT images. A global search is initially performed on coarse resolution 3D surface images of each modality where a variable threshold is used to select any likely match location for finer resolution levels. An adaptive termination of the computation of the distance function is possible due to the sequential nature of its evaluation. The superimposed images of MRI and HMPAO images, displayed as slices and in 3-D, were clinically helpful.

Keywords

Medical imaging; multimodality imaging; three dimensional graphics; surface detection; surface fitting; superimposition; minimization; least square; image display.

1. INTRODUCTION

In many circumstances the clinical interpretation of an image from a single image modality is inadequate. There are potentially many situations where a clinician would like to inspect a volume of interest by using more than one image modality. Viewing different (raw) images side by side provide little detailed understanding of similarities and differences between them. Accurate alignment of such images provides anatomical and functional information in the set of superimposed data.

Any registration technique attempts, firstly, to find some set of corresponding points (control points or pixels) in the different images of a common object. Secondly, the geometric differences such as translation, rotation and scaling are computed by minimizing the differences (e.g. a distance measure), or from a different point of view maximizing the similarities, between these points measured on the two sets of images that have been recorded under different patient positioning and imaging conditions.

The most widely used approaches to registering 3D volume data are based on the use of the (small) set of points defined by internal or external landmarks on the patient (Merickel and Carthy, 1985). In general, positioning and fixing the landmarks on a patient is not technically easy and it is inconvenient for both technician and patient. Some other techniques have been developed using control points, computed from specific common features of both sets of images (Hu, 1962, Maitra, 1979, Mitiche and Aggarwal, 1983, Abu-Mostafa and Psaltis, 1984, Kall and Kelly, 1985). Computer generation of such control points is often not suitable for complex medical images in which the structures of the images are dissimilar in the two studies, which is a requirement of those methods.

Other frequently used registration methods are based on edge information obtained in a preprocessing stage. These methods have been used by a number of workers (Andrus and Campbell, 1975, Svedlow and Gillem, 1978, Medioni and Nevatia, 1984, Bhanu and Faugeras, 1984) and overcome some of the difficulties associated with the other techniques. Surface fitting algorithms have been derived directly from edge-based methods in 3-D space. In this approach, a distance function, normally the least squares distance (LSD), is constructed to evaluate the fit between common surfaces computed from the two sets of images. The mean value of the LSD is then minimized to determine the best fit transformation. This technique works well in registering brain images using skin surfaces or brain surfaces, even when there are big differences in the internal structures of the two set of images.

The choice of minimization method is very important in computer cost (time and memory required) and the accuracy of the determination of the global minimum of the function. Most fast minimization algorithms, as used by the other workers (Pelizzari and Chen, 1987, Chen et al, 1988, Pelizzari and Chen, 1989) to minimize the LSD function, suffer from the presence of multiple local minima preventing the determination of the true global minimum. Even if perfect segmentation were possible, local minima can still occur. The minimization process can easily fail as a result to register accurately.

In the surface fitting method as developed here, an iterative method is used, since proper corresponding points are not known in advance and are re-estimated at each iteration. The transformation matrix is updated in respect of the latest promising surface match, which in turn leads to generating a new correspondence point set.

It is possible to try to find a global minimum by a global search through all possible transformation parameters. Checking all search locations by performing a grid search (changing one variable parameter at a time) is very robust, but is extremely slow. Therefore, most minimization algorithms use some strategy to search for a downhill trend to find the (often local) minimum.

Registration techniques based on a distance function such as the LSD, generally evaluate all the n extracted features (distances between corresponding points) at each stage, prior to any decision on the correctness of the fit. On the other hand, they rely on the mean value of the distance function arising from all n points. Thus, evaluating all transformations (in a global search) is not normally possible due to time and space requirement of the process. It is obvious that evaluating the function at an insufficient number of search locations (transformation parameters) would not give satisfactory correct match locations, that is, give proper transformation parameters for accurate registration. Thus a problem would occur when the cost of evaluating all points at all search locations is excessive. However, building up the cumulative error point by point is a sequential operation. Thus a test can be made as each point is added to see if a threshold has been exceeded and the corresponding transformation abandoned. This enables methods to be implemented which are less dependent on, or independent of the evaluation of the distance function, in order to determine the next direction in which to search. In addition, a multi-resolution technique can be used for setting limits to the transformation parameters and creating a smaller search space, in order to speed up the global search and to improve the sequential matching process. It has been shown by Mandava and Fitzpatrick (1989) that two global search algorithms, Simulated Annealing and the Genetic Algorithm, lead to faster convergence when implemented with search space scaling.

This paper has been organized as follows. The currently available types of registration method have been reviewed in this introduction. Section 2 describes the algorithms for registering 3-D brain data. Sequential registration methods on surface fitting is described in section 2.1. Section 2.2 includes a description of the multi-resolution method. Section 2.3 describes the material and data used. Some results, and the performances of the algorithms compared to those of the other commonly used algorithms are outlined in section 3. Finally, the conclusions are presented in section 4.

2. MATERIAL AND METHODS

A surface fitting algorithm was implemented where the fit was evaluated by a least square distance function (LSD) minimized by a sequentially terminated multi-resolution technique. The distance function Σd^i which is to be minimized can be expressed as:

$$\Sigma d^i = \Sigma_i \, (P_t^i[T] - P_o^i)^2 \text{ For points (i..n),}$$

$$\Sigma d^i = \Sigma \, \{(X_t^i, Y_t^i, Z_t^i) - (X_o^i, Y_o^i, Z_o^i)\}^2 \qquad (2.1)$$

where X_t, Y_t and Z_t are the transformed coordinates for point P_t. P_t and P_o are vectors representing the corresponding points on each surface: the surface to be transformed and the target (objective) surface respectively. Σd^i is the mismatch value (Least Square Value) which is the sum of the squares of the distances between each individual pairs of points.

Sets of 2-D and 3-D MR brain images taken under different imaging conditions and noise levels were used primarily. Simulated PET images were also generated by adding Gaussian noise to the Gaussian blurred MR images. One set of such data was geometrically modified by a pre-defined transformation to yield a known misregistered data set used for the evaluation of the procedures being tested.

The external surfaces of the objects to be registered were generated by the application of a surface detection algorithm on the serial slices of each 3-D grey level image data set. This includes extraction of 2-D contours from individual slices using grey level thresholding and/or a gradient operator, and applying an interpolation algorithm to generate a series of intermediate contours between each pair of consecutive cross-sections. An 'edge circularity check and neighbouring comparison' is also performed when necessary (Cappelletti and Rosenfeld, 1989). The binary surface (being the set of voxels lying on the surface as defined by the surface detection algorithm in 3-D space) was used as the primitive to be registered. For the surface to be fitted to, the objective surface $\mathbf{P}_o$, a 3-D binary surface (solid model) is used. For the other data set, the transforming surface $\mathbf{P}_t$ (the surface which is transformed at each stage of the fitting process), only a series of 3-D points selected from the 3-D surface are used for registration. A 3-D box or volume of interest (VOI) is placed around each target surface to be fitted, either manually or automatically, using the information obtained in the surface detection process. The centroid of such a VOI becomes the origin of the coordinate system of the objective surface. Alignment of the two centroids brings the two surfaces into a close position which makes possible the application of the surface fitting algorithm as described below.

The choice of correspondence points on the transforming surface (the position and number of selected points) included in the registration process should satisfy the assumption that the sampling points for the transforming surface are representative of all its image points and reflect all potential misregistration information about the whole data set (Nagel and Rosenfeld, 1972). This was attempted, here, by dividing the area of the surface into m regions (James, 1980), each having different distinct locations and where possible, different slopes (surface normals). One point was selected in each region. The coordinates x,y,z of each selected point were then compared with those of previously selected points in order to reject any point having at least two coordinates similar or close to other such points. Sampling was performed on those regions of the transforming surface image for which there were corresponding on the objective surface data set. For example the region of the eyes in an MR head image was not used for sampling since the SPECT data usually does not adequately represent the complexity involved in this region. The number of points and their location (where they are selected) are two important factors in the sensitivity of the Least Square Distance Function (LSD) to an applied transformation. As suggested by Nagel (1972), the points have been ordered in such a way that any sub-set of them reflect similar misregistration information as the whole set.

The distance between the sampling points on the surface to be transformed with respect to corresponding points on the surface to be fitted (objective surface) are determined by ray tracing. The corresponding pairs of points are assumed to be at the intersection points of rays originated at the centroid and passing through the sampling points of the surface to be transformed intersecting with the objective surface. The distances between the corresponding points are then computed using the L2 norm distance function (see Eq. 2.1) for each transformation. The transformation parameters can be determined either by the minimization process in an iterative manner (for example by using the direction search method of Powell, 1964) or by a global search strategy. Only the points selected on transforming surface are shifted whereas the objective surface is kept constant during the process. The transformation parameters (the three shifts in x, y, and z, and the three rotations around the x, y, and z axes) are termed the search location in the next section where a global grid search algorithm is used to find the best transformation parameters giving the minimum of the distance function.

Having found the appropriate transformation, it may be used to transform any structure in either image set (transforming or objective data) into the coordinate of the other. The images can then be fitted to generate a useful superimposed image for clinical interpretation.

2.1. Sequential Methods in Registration

While a global grid search was initially applied to test all possible transformation parameters, a widely used minimization method (e.g. Powell, 1964) was employed thereafter. Checking all the promising transformation parameters and terminating a sequential process when an predetermined cumulative error is exceeded after evaluating only a number of the points (m) is the main feature of the new technique. However, there will be a trade-off between the possible misregistration error and the number of points for which the process is allowed to continue before terminating the test location (in transformation space). This termination is actually controlled by a preset n-dependent variable threshold which defines the sufficiency and desirability of the cumulative errors. The threshold T for stopping the evaluation of the LSD, as defined by a number of workers (Barnea and Silverman, 1972, Rosenfeld and Vanderbrug, 1977, Vanderbrug and Rosenfeld, 1977, Wong and Hall, 1978, Wong, 1978 Wong and Hall, 1979) can be derived from the value of the expected cumulative error R_m. This is the unavoidable error generated during the registration process due to any factor other than the misregistration error, such as image noise, surface extraction and transformation noise etc. Since this threshold should take account of some possible variation in R_m, a criterion based on the number of deviations (g) from expected error (R_m) is suggested (Barnea and Silverman, 1972). The desired threshold sequence can then be derived from the following equation:

$$T_m = R_m + g\sigma = R_m + gr\sqrt{m}, \qquad (2.2)$$

where r is the amplitude of the average distance errors at the true match location (considered as the expected error), σ is the S.D. of the error distribution of the individual errors, and R_m is the sum of the error terms (expected cumulative error) after evaluating m points of the registered surface data set.

Since the evaluation of errors is sequential, the order in which the points participate in the evaluation is important. They should be ordered such that, firstly, evaluation in such an order causes the termination decision earlier, and secondly, a small number of selected points mimic the entire response of all surface points (e.g. they should selected from all parts of the data set and not from regions grouped together).

2.2. Multi-Resolution Registration

To improve the results, the method was modified and implemented in multi-resolution (scale) space, where the fitting started in low resolution surface images, and stepped to a higher resolution when the desired fit passed an appropriate threshold (Barnea and Silverman, 1972, Wong and Hall, 1978, Wong, 1978). The input to the multi-resolution image registration system are two sets of binary surfaces obtained by the surface extraction technique. A sequence of lower resolution surface data set of sizes 128*128, 64*64, 32*32 are generated from two original surface data of size 256*256. The criterion for selecting a surface point at a lower resolution level (having a bigger voxel size) from a higher resolution level depends on the location and number of neighbourhood surface points (for example eight adjacent voxels). One of the simplest criteria can be expressed as follows:

Let a voxel f_{L-1} in low resolution level L-1 be set as a surface point if any two of the eight adjacent voxels in the higher resolution image (level L) is on the surface.

$f_{L-1}(i,j,k)=1$, if at least any two voxels of $f_L(2i+\alpha,2j+\beta,2k+\gamma)=1$, for $\alpha,\beta,\gamma=0,1$
otherwise;
$f_{L-1}(i,j,k)=0$. (2.3)

where α, β and γ are scalar values 0 or 1 in order to define the eight neighbouring voxels for each surface point (2i,2j,2k) of higher resolution level L. (for i,j,k between 1 and n, where n is the size of 3D binary surface data set at lower resolution level L-1).

The most likely transformation parameters are applied to each surface data set of lower resolution, and the distance errors are measured and cumulated sequentially until they exceed a pre-defined threshold. Since at a lower level of resolution both the size of data set and the number of possible search locations

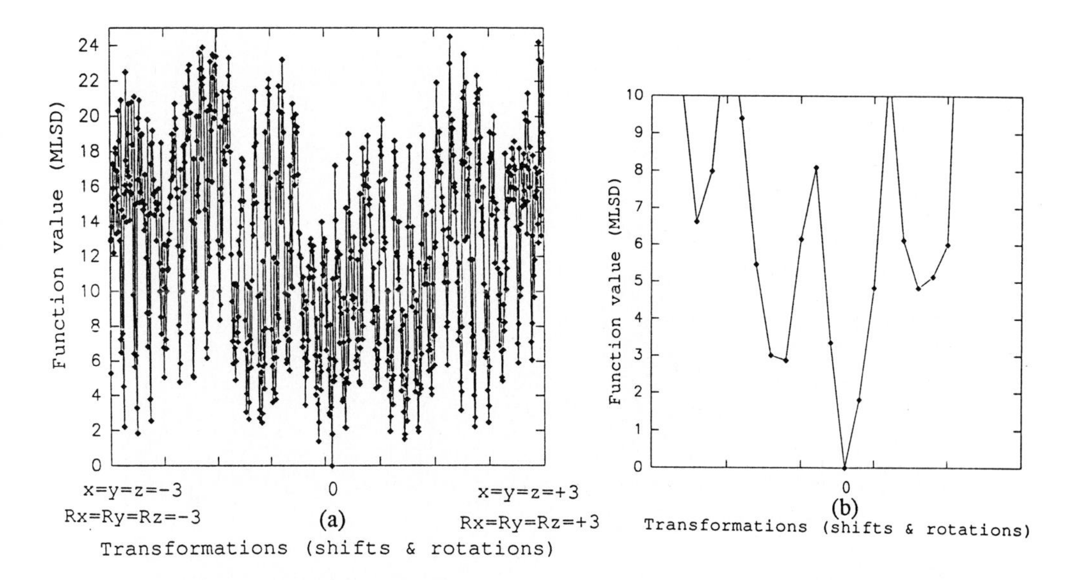

Figure 2. The function values (mean LSD) for fitting MRI with MRI versus a range transformation parameters showing the local minima occurred at these locations. a) ±3 degree shifts and rotations were tested. b) Only 24 mismatch locations near the true match location were tested.

(shifts in x, y and z direction) decreases by a factor of M^3 (M is the ratio of the size of the higher resolution image to the lower resolution image), the LSD function converges to an appropriate outcome (either rejection or acceptance as a potential minimum) for each transformation in a very short time (see table 1). At low resolution, the minimization process also converges rapidly to a overall minimum, even if the entire transformation space is searched globally, since the search space is also much smaller. The number of correspondence points were chosen here as 128, 64, 32, and 24 for levels $\underline{4}$ (256*256), $\underline{3}$ (128*128), $\underline{2}$ (64*64) and $\underline{1}$ (32*32), respectively. The number of required points decreases at coarser resolution (lower level) which implies a higher convergence speed for the algorithm. The method uses the best transformation parameters, and the most promising search location found at a given level L-1, to pass them as seed points to the next higher level L for further investigation. Therefore the performance computed on the basis of those few pixels matched in a coarse resolution image is used as representative of the next resolution level, and in turn up to the original high resolution image.

The variable threshold T can then be modified to account for the termination of the function evaluation at different resolution levels **k**.

$$T^k_m = R^k_m + g\sigma = R^k_m + gr^k\sqrt{n} \tag{2.4}$$

It is desirable that this threshold should not eliminate those locations that are likely to be important as the true match location at a higher resolution level.

2.3. Material

The technique was used to register the MR with MR images, PET with MR images, and SPECT (HMPAO) with MR brain images of a number of patients. Brain surfaces were usually used for the registration of MRI with HMPAO brain images. A STIR sequence was used to create 2-D MR images of 5mm thickness, which give a fat-suppressed image (skin surface) which makes the detection of the brain surface easier. The skin surface was used where there was a brain surface deformity in one of the

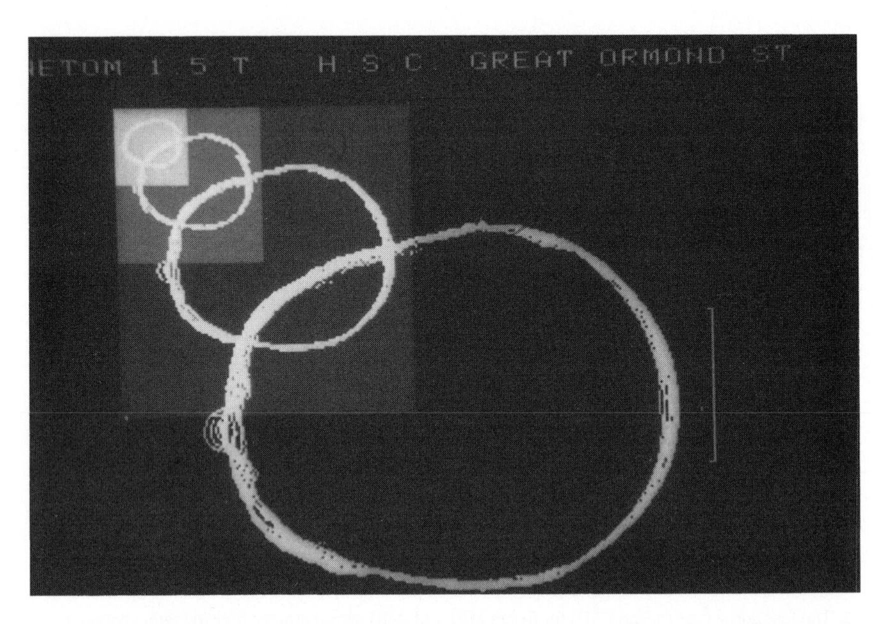

Figure 3. Contours determined for skin surface in the multi-resolution study. The size of the data set in the current work was 256*256*120 (at levels 4), 128*128*60 (level 3), 64*64*30 (level 2), and 32*32*15 (level 1), respectively.

images by setting a lower threshold to detect the SPECT surface, and by using a shorter echo time in the MR sequence to visualize the skin for a better detection. Skin surfaces were used for registration of MRI with MRI or PET images. These surfaces were obtained from the transmission data set, in the PET images, which are usually acquired during scanning for attenuation correction purpose. All the experiments were performed on a Micro VAX II computer interfaced with an image processing work station (KONTRON) on which the two surfaces can be visually inspected, and the fitting parameters modified, or the program terminated if the fit appeared to be inappropriate (or it started to diverge). The contours of the ROI extracted from the re-sectioned MRI were superimposed on the original 2-D SPECT or PET images as requested by the physicians. Each display screen shows four 256*256 images, being the SPECT slice and the MR re-sectioned image side by side in one half, and the 3-D surface model showing the position of the visible spect image (and the sampled transforming points) and the superimposed image in the other half, as shown in Figure 6 (Colour plate, p. 3).

3. RESULTS AND DISCUSSION

These algorithms were initially tested by attempting to register two arbitrary misregistered MR images generated by applying a known geometric transformation on a set of original MR brain images of a patient. The same experiments were performed on PET simulated images generated by adding a Gaussian distributed noise to a degraded MRI data set. This technique has also been tried for registration of PET with MRI, and SPECT with MRI for a number of patients, but the numerical results obtained by those experiment are out of the scope of this paper since these methods are still under investigation. However, some of the clinical results are shown at the end of this section in order to illustrate the reliability of the technique for other imaging modalities.

Figure 1 (Colour plate, p. 3) shows a sample output from the display system which runs in parallel on the microVax and linked Kontron workstation, when registering PET and MRI. The superimposed 2-D contours with respect to the three main axes visualizes the progress of registration during the process and

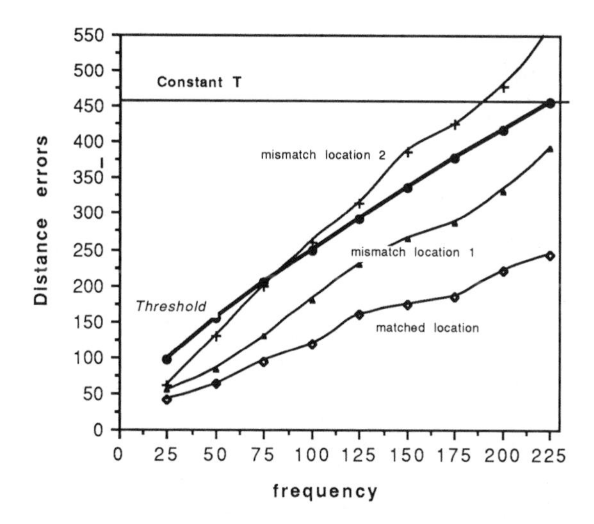

Figure 4. The error growth curve showing the cumulative error values at 3 locations (3 different transformation settings). The superiority of the variable threshold over the constant threshold level is indicated from these graphs. The thick line plotted shows the predetermined threshold value for g=3 S.D. The error is accumulated as each of the points is evaluated, and the search location is rejected if the mismatch value is bigger than the threshold.

allows modification of the parameters, or possibly manual interaction in selecting one of the best minima occurred in the global search.

As shown in figure 2, a great number of local minima could occur in near-match locations (transformation parameters which are close to the optimal transformation). Although, the values of local minima at mismatch locations, as shown in this figure, are usually bigger than the true global minimum, the relative values of the various minima (including the global value) change in the presence of noise, that is, when surface detection algorithms fail to give the exact (perfect) location of surfaces on two different types of images (also a result of noise in the images). In this situation, finding the true global minimum is very difficult and sometime not possible, even with a global search through transformation space.

However, performing a global search produce all potential minima and thus the comparison and then the selection of the best minimum is possible interactively by viewing the superimposed contours on the display console.

As expected, more detailed structures are contained in the image matrix size of 256*256 than 32*32. On the other hand, all the sharp edges including most of the noisy structures fade out in the scaled (low resolution) surfaces. As shown in figure 3 the surfaces at the lowest level tend towards a simpler shape than those at the highest level when the degrading resolution criterion (Eq. 2.3) is applied to the original high resolution images. The success of this method can be attributed to the fact that in spite of the error in the low-level image creation, the two corresponding images (e.g. MRI and PET) at each resolution are consistent, so that their alignment leads to a minimum distance error measure which is not far from the true global minimum at the highest resolution level.

Figure 4 indicates the value (superiority) of the variable threshold over the constant threshold level. The thick line plotted shows the predetermined threshold value for g=3 S.D. deviations from the mean. The error is accumulated as each of the points is compared, and the search location is rejected if the mismatch value is bigger than the threshold. It may be seen that the mismatch location 2 is rejected after testing 75 sampling points, whereas the mismatch location 1 is not rejected at any point, and thus accepted as a candidate of the true match location for further evaluation in the next higher resolution level. The number of these accepted (unrejected) mismatch locations usually increases as the resolution level decreases. The threshold itself only plays the role in setting the trade off between the number of such

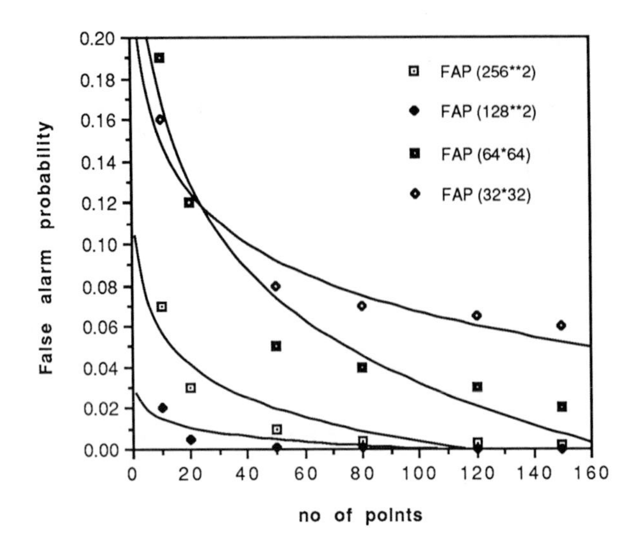

Figure 5. False alarm probability (FAP) of the matching process versus the number of points at levels 1 to 4 (size 32*32 to 256*256, respectively). The points were sorted before evaluation. g=3 S.D.s used in threshold setting process.

rejections, and the speed and accuracy of match.

As shown in Figure 5 the False Alarm Probability (FAP) decreases as the number of selected points increases. The response is highly dependent on the threshold. This threshold can achieve a reasonable match probability, even at the lowest resolution level, if set properly. The ability of a low resolution image to provide more geometric information from a smaller number of the sampling points can be deduced from this figure. As illustrated, the lower resolution level (L=1) is less sensitive to the number of selected points than the higher level (L=4) in which the FAP decreases very rapidly to zero with the number of points. The effect is more noticeable where the number of examined points n is below a certain limit (e.g. 50). Although the FAP decreases rapidly at all resolution levels in the range $0<n<50$, the FAP does not undergo a substantial change at level 1 (or 2) whether or not a small number of points is evaluated. Consequently the change in the FAP at levels 1 and 2 is not as significant as at levels 3 and 4. One can also see that the FAP increases as the resolution level decreases. In general, errors introduced in the creation of low resolution images depend primarily on the decision rule used for setting the surface points at a lower resolution level from a higher level. The speed of the registration algorithm also depends on the number of points which need to be tested before any transformation parameter is rejected as an invalid match location, or selected as a true match location. As shown in the figure 5, a smaller number of points are required at the lower resolution level (size 32*32) than the higher level (256*256). Therefore at the lowest resolution level all possible combinations of transformation parameters can be tested for the best LSD distance measure.

The results of applying the different minimization algorithms shown in Table 1, which indicates that the proposed technique guarantees the detection of the global minimum (true match location) in a reasonable time, whereas other algorithms may converge only after a long time due to several re-initializations of the minimization process, needed to avoid local minima.

4. CONCLUSIONS

The registration of data from different image modalities may be assisted by using multi-resolution sequential methods. This largely eliminates the human operator interaction, often required by the presence of local minima, and leads to a faster convergence among global search methods. In addition the speed of the process can be improved and the occurrence of any local minima during the search can be eliminated by appropriately setting threshold values for each resolution level. The technique was tried on various sets of MRI, PET and SPECT images in number of patients and showed a good convergence in registering of any of the two data set brain images. A total of 8 patients have so far had SPECT and MRI data registered as part of an ongoing clinical trial. In only two cases, manual operator interaction was required to select the appropriate global minimum given several apparent local minima, in both cases probably a result of the presence of large 'defects' in the images. A full clinical assessment of this trial is planned for the end of the current year.

Table 1. The results of applying 3 different minimization algorithms indicating a reasonable accuracy and duration of time with the new technique. In the multi-resolution experiment, 10 degrees of rotation and 32 pixels shift were tested around the true match location.
TP= transformation parameters found by registration process. MI=Manual Interaction applied RI= Re-Initialization MLSD= Mean Least Square Distance. (Powell and FRPRMN are two gradient based minimization methods modified and used in the current work).

Minimization Method	Duration of time	MLSD	Error for TPs	No of MI or RI
Powell				
NMR &NMR	130 Sec	0 90	0,0,1,-1,1,0	1
NMR & PE	180 Sec	1 52	0,-1,1,1,1,1	1
FRPRM				
NMR & NMR	200 Sec	0 80	1,1,0,1,0,-1	2
NMR & PE	222 Sec	1 95	0,0,2,1,2,0	3
Multi-Res				
NMR & NMR	405 Sec	1 00	1,0,0,2,0,0	-
NMR & PET	433 Sec	1 24	0,0,0,2,1,0	-

Acknowledgements

The authors would like to acknowledge the help of staffs in the X-ray department and NMR unit in Great Ormond Street hospital for sick children, The MRC Cyclotron of Hammersmith hospital, and department of Radiological Sciences at Guy's hospital in providing data and advice for this project.

References

Abu-Mostafa Y S, Psaltis D (1984). Recognition aspects of moment invariants. IEEE Trans. Pattern Anal. Mach. Intel. Vol.6: 698-706

Andrus J, Campbell C (1975).Digital image registration method using boundary maps: IEEE Trans. on Comp. Vol.19:935-940.

Barnea D, Silverman H (1972). A class of algorithm for fast digital image registration. IEEE Trans. on Comp. Vol. C21, No.2: 179-186

Bhanu D, Faugeras O (1984), Shape matching of two dimensional objects. IEEE Trans. Pat. Anal. Mach. Intel. Vol.6: 137-155

Cappelletti J D, Rosenfeld A (1989). Three-Dimensional boundary following: Comp. Vision Grap. and image processing, Vol.48: 80-92.

Chen C-Tu, Pelizzari C A, Chen G T Y (1988). Correlating Functional Nuclear Medicine Images with Structural CT or MR Images: Proceedings of 1988 meeting of Asian Society of Nuclear Medicine.

Hu M K (1962).Visual pattern recognition by moment invariants: IRE Trans.on Information Theory. Vol.8: 179-187

James F (1980). Monte carlo theory and practice: Rep. Prog. Phys. Vol.43: pp. 1173-1189.

Kall B A, Kelly P J (1985). Cross registration of points and lesion volumes from MR and CT: IEEE 7th annual conference of the engineering in medicine and biology society. 937-942

Maitra S (1979). Moment invariants: Proc.IEEE. Vol.67: 697-699

Mandava V R, Fitzpatrick J M (1989). Adaptive search space scaling in digital image registration: IEEE Trans. on Med. Imag. Vol.8, No.3: 251-262

Medioni S, Nevatia R (1984). Matching images using linear features: IEEE Trans. Pattern Anal. Mach. Intel. Vol.6:675-685.

Merickel M, Mc Carthy M (1985). Registration of contours for 3D reconstruction: IEEE 7th annual conference of the engineering in medicine and biology society. 616-620

Mitiche A, Aggarwal J K (1983). Contour registration by shape-specific points for shape matching. Comp. Vision Grap. and Image Processing. Vol.22: 396-408

Nagel R N, Rosenfeld A (1972). Ordered search techniques in template matching: Proc. IEEE. Vol.60: pp.242-244.

Pelizzari C A, Chen G T Y (1987). Registration of multiple diagnostic imaging scans using surface fitting: Proceeding of the 9th ICCR, pp.437-440.

Pelizzari C A , Chen G T Y (1989). Accurate Three-Dimensional registration of CT, PET and NMR Images of the brain: J. of Comp. Assis. Tomog. Jan 1989.

Powell M.J.D (1964). An efficient method of finding the minimum of a function of several variables without calculating derivatives: The Comp. J. Vol.7: 155-162

Price K, Reddy R (1979). Matching segments of image: IEEE Trans. Pattern Anal. Mach. Intel. Vol.1: 110-116.

Rosenfeld A, Vanderbrug G J (1977). Coarse-Fine template matching, IEEE Trans on Sys. Man and Cybernetics. Feb 1977: 104-107.

Svedlow M, Gillem C D (1978). Image registration: Similarity measure and preprocessing method comparisons. IEEE Trans. Aerospace. Elec. Sys. Vol.14: 141-149

Vanderbrug G J, Rosenfeld A (1977). Two-stage template matching, IEEE Trans. on Comp. Apr. Vol.C26, No.4:384-392.

Wong R Y, Hall E L (1978). Sequential Hierarchical Scene Matching, IEEE Trans. on Comp. Apr 1978, Vol C27,No.4:359-366.

Wong R Y (1978). Sequential scene matching using edge features, IEEE. Trans. on Aerospace and Elec. Sys. Jan 1978, Vol.Aes14, No.1: 128-140

Wong R Y, Hall E L (1979). Performance comparison of scene matching techniques. IEEE Trans. on Pattern. Anal. & Mach. Intel. Jul 1979. Vol.PAMI-1, No.3: 325-330

AUTOMATING SEGMENTATION OF DUAL-ECHO MR HEAD DATA

[1] Guido Gerig, [2] John Martin, [2] Ron Kikinis, [1] Olaf Kübler,
[2]Martha Shenton, [2] Ferenc A. Jolesz

[1] Institute for Communication Technology, Image Science Division,
ETH-Zentrum CH-8092 Zurich, Switzerland
[2] Department of Radiology, Brigham and Women's Hospital,
Boston MA 02115, USA

Abstract

Multiecho MR acquisition yields various information about tissue and csf characteristics. The analysis of two-dimensional scatterplots generated from dual-echo MR data turns out to be a useful tool. It allows the optimization of MR acquisition parameters for later specific recognition or segmentation tasks. Multivariate statistical classification techniques are applied to dual-echo MR data to segment volume head data into anatomical objects and tissue categories (brain white and gray matter, ventricular system, outer csf space, bone structure, tumor).

To overcome the sensitivity of voxel-based classification to noise we applied a preprocessing technique based on anisotropic diffusion. This preprocessing increases the separability of clusters. We illustrate the robustness of supervised classification with the segmentation of a series of MR head data in a research study.

For a given set of MR parameters we show that the configuration of clusters in feature space is comparable between studies. This allows us to develop an automated clustering technique that considers a priori knowledge about cluster attributes and their configuration in feature space. The automated classification technique omits subjective criteria in the training stage of supervised classification and yields reproducible segmentation results. A new study will check the reliability of the automated classification in comparison with supervised classification.

The segmentation of head MR data in routine clinical applications gives some important qualitative and quantitative information about brain atrophy, brain volume, volume of csf spaces and morphological changes of local brain areas. Combined 3D views of multiple anatomical objects are shown. They clarify the perception of 3D relationship and highlight the locations and types of structural abnormalities.

Keywords

Medical image processing, 3D segmentation, statistical classification, clustering, magnetic resonance volume data, multiecho image data, 3D display

1 Introduction

Magnetic Resonance Imaging (MRI) is capable of providing both structural and functional information of biological tissue. This noninvasive imaging technique permits examinations without disturbing the physiology of the organ system under study, a highly desired quality for a radiological method.

Multiecho acquisition allows better discrimination of different tissue types and anatomical functional units because specific characteristics are enhanced in multiple spectral channels. Recent progress in acquisition techniques has resulted in a remarkable increase of scanning speed. This results in either a decrease in acquisition time or an increase in the number of slices, both of which are to the benefit of the patient. An increased number of images holds an enormous amount of information, but poses the problem of visual analysis by radiologists. The human visual system is extremely effective at the interpretation of 2D images, but the act of 3D imagination out of sectional information is a highly specialized human vision task and requires intensive training. In complex situations, and if quantitative as well as qualitative results are desired, a purely interactive process by far exceeds practical time limits and must be supported by automated procedures.

In order to parallel the rapid progress in data acquisition, it is very important to provide efficient analysis tools which are appropriate for the complex information inherent in the data. There is growing interest in the utilization of computerized methods for the examination of digital medical data, permitting the extraction of quantitative and qualitative information [Kübler and Gerig, 1990]. The requirements for computerized medical image analysis are:

- minimal user interaction (high degree of automation)
- efficient when running on standard workstations
- robustness (insensitive to noise and to variations among patients)

Statistical classification of multivariate image data, if certain assumptions on the underlying image model are fulfilled, satisfies these criteria. The present paper describes a multistage processing scheme to segment 3D MR image data into anatomically and functionally distinct regions. We use the established statistical classification methods to segment dual-echo MR data of the human head. The supervised approach still includes human interaction for the training of the sought categories. This has the advantage of being guided by a human expert, permitting immediate correction through a visual feedback via the classification results. But it has the problems of generating results which are not fully reproducible and of being biased by subjective criteria of different users. Unsupervised classification is put in competition with supervised techniques in order to assess consistency and reliability of both methods.

2 Related work

Magnetic resonance imaging is intrinsically multispectral. This fact stimulated several attempts to apply statistical software packages to multiecho magnetic resonance image data [Vannier et.al., 1985], [Vannier et.al., 1988 a], [Vannier et.al., 1988 b], [Merickel et.al., 1988]. The major problem turned out to be the data quality. First, MR images are corrupted by a considerable noise component. Since statistical classification is based on the multispectral properties of single pixels or voxels, noisy input data lead to a noisy classification image. Second, MR image data are often radiometrically inhomogeneous due to RF field deficiencies. While this is not a problem for the visual analysis if the local contrast is large enough, the variability of tissue intensity values with respect to image location severely affects a segmentation based on absolute pixel measurement. Vannier [Vannier et.al., 1988 a] describes a correction method based on a line by line histogram evaluation and a subsequent radiometric correction. The method assumes that the histogram of each image line reflects identical tissue information (a questionable approach). Furthermore only vertical (or horizontal) radiometric distortions are taken into account, while the actual distortions due to RF field inhomogeneities are more accurately described by two-dimensional polynomial functions of second degree.

Improvements in scanner technology have overcome the homogeneity problem, and a new smoothing technique especially adapted to multivariate 3D MR data has significantly reduced the noise problem [Gerig et.al., 1989], [Gerig and Kikinis, 1990], [Gerig et.al., 1990]. The resulting improvement in data quality favors a segmentation approach based on statistical classification. In our discussion in section 3.4 and appendix A we demonstrate the low variability in detected signal strength for the same tissues within the slices and across the slices.

3 Background

3.1 Individual segmentation steps

The segmentation of MRI data into anatomically and functionally distinct regions requires the application of a multistage processing scheme:

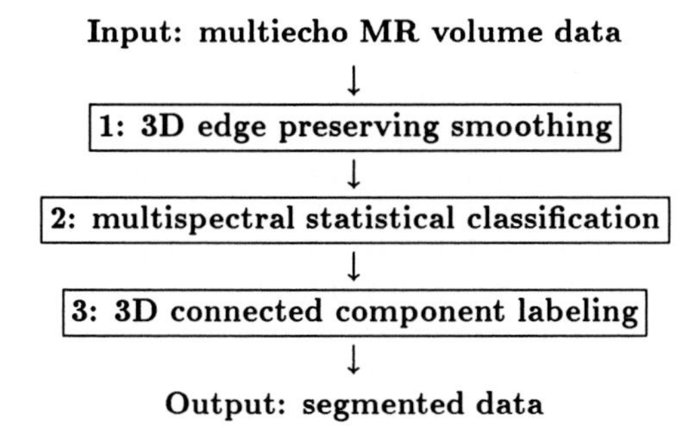

In the present paper we focus our discussion on step 2, the statistical classification. The smoothing technique mentioned above is the preprocessing (step 1). Step 3 represents a 3D connectivity method that uniquely labels connected voxel-groups to anatomical objects.

3.2 MRI methods

Spin echo images of the brain were obtained as part of routine clinical studies. Imaging was performed at 1.5 Tesla using a General Electric SIGNA MRI system. Data acquisition used a double echo sequence (TR 3000ms, TE 30/80ms) with 3mm axial slice distance and thickness. Contiguous slices were obtained by combining two interleaved sequences. Imaging time was reduced by using half-Fourier sampling (0.5 NEX). We used flow compensation (gradient moment nulling) and presaturation of a slab inferior to the head to reduce flow related artifacts and to obtain low intra-arterial signal intensity. Fiftyfour double echo slices were used to cover the brain volume. The imaging time was under 8 minutes.

Five cases using identical instrument parameters and one case with the same parameters, but acquired on a scanner representing 1985 technology, formed the database used in the present paper.

3.3 Image model requirements for statistical classification

In a simple segmentation approach, the image is broken up into regions with constant intensities degraded by noise. For images of our natural environment this model is obviously inaccurate. Images generated by a well-defined physical process may match the model assumption, as the information of each picture element expresses a physical measurement. If additionally the structure being imaged can be characterized by the model of being piecewise constant, computer vision algorithms designed for the simple model are suitable. The MR imaging process and the tissue characteristics in certain anatomical regions match the model assumption in a first approximation.

3.4 Radiometric variability of volume data

A classification based on absolute intensity values assumes that a given tissue class yields voxels of constant values. To test the radiometric variability we decided to use white matter as our reference class. White matter regions are represented as large and fairly homogeneous areas in several sections within the data volume. At different locations within the slices and in different slices we selected ROI's representing white matter areas. The multivariate statistical parameters within the single regions and of the complete population were compared. The analysis is described in appendix A. The results demonstrate the excellent homogeneity of the MR data and the correspondence of the structures in the image volume with the model assumption.

3.5 Scatterplots

A scatterplot is the multidimensional extension of a histogram, each variable (or channel) forming a coordinate axis. The multiple measurements of each voxel address one cell of the scatterplot. By accumulating votes each cell counts the number of pixels or voxels with identical multispectral characteristics. Ideal homogeneous areas with constant dual-echo measurements are transformed into one single cell, its vote defining the size of the area. If corrupted by noise, homogeneous areas result in clusters scattered around the ideal means. If additionally the absolute voxel values are degraded by radiometric distortion, clusters are spread even more due to the greater variability of values of each class. Depending on the noise level, clusters can overlap or even be merged together. Figure 1 shows two scatterplots, the horizontal and vertical axes corresponding to the first and second echo of a dual-echo MR study. The MR parameters were identical in both acquisitions. The input data for the left scatterplot were acquired by an older MR scanner, whereas the right plot demonstrates the progress in noise reduction, contrast enhancement and field homogeneity available with the latest technology.

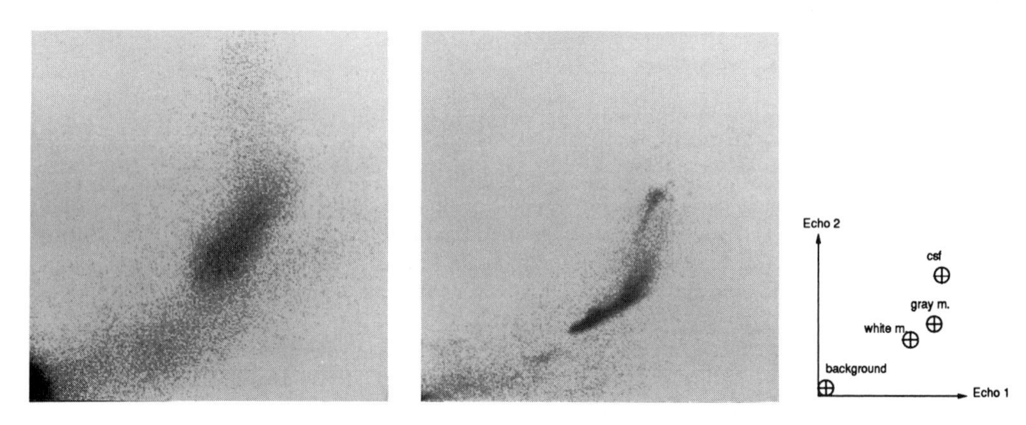

Figure 1: Scatterplots: Left: MRI technology 1985, Right: MRI technology 1990

A scatterplot can be interpreted as follows. We assume that different tissue types are represented by characteristic signatures. If several sharp clusters can be distinguished we can conclude that a large contrast and low noise exist. In such a situation there is a good chance to estimate meaningful cluster parameters. Also the error rate of a statistical classification will be significantly decreased. If, on the other hand, data are scattered and clusters overlap, caused by either a large amount of noise or radiometric distortions, then the error rate will be much greater and consequently it will be difficult, or even inappropriate, to segment the images using statistical classification methods.

class	mean		principal axes		rotation	correlation	a priori prob.
ω_i	μ_1	μ_2	σ_1	σ_2	(deg)	ρ	$P(\omega_i)$
background	12.7	8.6	3.0	2.4	3.8	0.03	0.25
white matter	489.1	221.2	26.4	15.0	47.4	0.51	0.25
gray matter	603.9	305.9	35.4	18.3	64.7	0.48	0.25
csf	618.7	527.8	25.0	12.9	-4.5	0.23	0.25

Table 1: Cluster statistics from supervised training

The scatterplot can be used to assess image quality and to optimize the MR imaging parameters. It does not, however, contain information about local contrast, which is another important cue for visual analysis.

4 Supervised statistical classification

The design of a statistical classifier requires complete knowledge about the probabilistic structure of the problem. An excellent discussion of statistical decision theory can be found in the book of [Duda and Hart, 1973]. One approach, called supervised learning, is to use samples to estimate the parameters. With supervised learning the state of nature of the samples is known. A user applies his knowledge about anatomy to draw training regions representing typical image areas for each class. The number of samples should be sufficient to obtain valid estimates for the statistical parameters. This number depends on the dimensionality of the problem (number of features) and on the type of the density functions to be estimated. We decided to keep the number of dimensions minimal, thereby decreasing the cost and complexity of both the feature extractor and the classifier.

For simplicity, we assume that the samples are drawn from a normal population with mean vector μ and covariance matrix Σ. It can be shown [Duda and Hart, 1973] that the maximum likelihood estimate of the mean vector is the sample mean. Similarly, the maximum likelihood estimation of Σ is essentially identical to the sample covariance matrix.

Figure 2 displays the polygonal regions of interest for the classes background, white matter, gray matter and csf. The resulting sample statistics are illustrated on the right. For the segmentation of volume data we selected training regions not only in one slice, but in several slices equally distributed throughout the volume. The class parameters are calculated based on the set of voxels within the training regions. Table 1 lists the parameters for two-dimensional Gaussian density functions, the principal axes such as the rotation angles are given. The a priori probabilities were calculated based on the number of training samples per class.

In addition to the parametric classification, we have also used the non-parametric Parzen-window technique [Duda and Hart, 1973]. These supervised techniques have been applied to over 60 cases. From our two years of experience we conclude that statistical classification applied to dual-echo MR volume data represents a robust and efficient segmentation method.

5 Automated segmentation using unsupervised classification

Although being an efficient segmentation method, the supervised classification technique still requires a certain amount of interaction. An experienced user needs about 15 minutes for a careful interactive selection of training regions. Another important factor favoring an unsupervised classification is the bias introduced by interactive training. A training may be done differently by different users, making it difficult to reproduce classification results. Even the same user, performing a training at

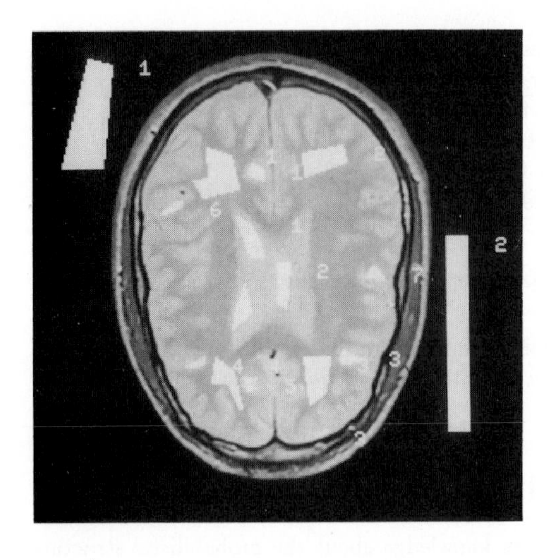

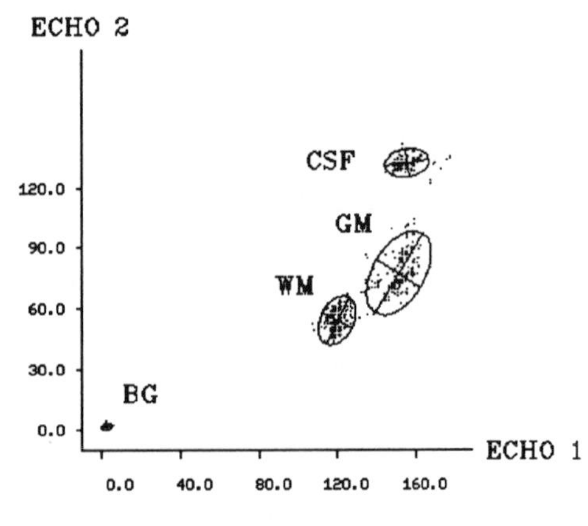

Figure 2: Training regions and statistics for the classes background (BG), white and gray matter (WM, GM) and CSF (axes scaled by 0.25)

two different times on the same data, will produce slightly different results. Finally, an unsupervised approach may reveal characteristics in the data that were unobserved by the human interpreter. This may alter the criteria used in designing the classifier.

Unsupervised parameter estimation differs from a supervised approach in that the samples are not labelled. The set of samples comes either from a typical region of the image containing an appropriate representation of all the classes, or from the complete collection of voxels.

5.1 Unsupervised clustering

The objectives for any clustering procedure are the identification of the cluster centers, the estimation of the scatter parameters and the determination of the a priori probabilities. One difficulty with unsupervised clustering is that it is sometimes difficult to choose the best criteria for splitting and merging classes, and the best criteria for determining the number of classes that can be separated [Vannier et.al., 1988 a]. There is even a discussion of "peculiar" characteristics of different clustering procedures [Coleman and Andrews, 1979]. The operator runs the algorithm with a set of parameters and hopes to obtain a clustering which results in the correct classification. Due to the complexity of the mathematical packages and their tuning parameters, it is difficult to correct classification errors. The influence of the parameters on the classification result is often not clear.

Two special conditions with respect to dual-echo MR images simplify the clustering. First, from our large experience with supervised classification on dual-echo MR head data, we know the number of clusters and their approximate parameters. Second, a two-dimensional clustering can be more easily controlled than clustering in a higher-dimensional space.

For our procedure the scatterplot of the complete data or of just a subset of the data serves as input. The scatterplot is a discrete estimation of the mixture density function

$$f(\mathbf{x}) = \sum_i f(\mathbf{x}|\omega_i)\, P(\omega_i)$$

where each ω_i is a separate class, $f(x|\omega_i)$ is the class specific density function and $P(\omega_i)$ is the a priori probability of class i.

The clustering procedure consists of the following steps:

class	mean		principal axes		rotation	correlation	a priori prob.
ω_i	μ_1	μ_2	σ_1	σ_2	(deg)	ρ	$P(\omega_i)$
background	11.5	7.6	3.2	2.7	16.4	0.11	0.51
white matter	500.2	225.4	28.5	19.2	33.7	0.35	0.12
gray matter	596.1	297.8	37.0	25.6	57.9	0.32	0.12
csf	622.9	512.3	32.2	21.6	71.5	0.24	0.02

Table 2: Cluster statistics from unsupervised training

1 Approximation of the density function $f(\mathbf{x})$ using a Parzen window approximation (estimation of the continuous density function from the discrete scatterplot counts).

2 Detection of relative maxima of $f(\mathbf{x})$, which serve as cluster center candidates.

3 Selection of desired cluster centers by comparing the candidate list with a list of cluster centers known from previous experience. Additional clusters representing altered tissues or unexpected changes are detected by selecting all cluster candidates with significant energy.

4 Estimation of cluster parameters within zones around each cluster center, the size of the zones being based on the distances between the cluster centers and on the a priori knowledge about the clusters to be detected.

5 Calculation of the a priori probabilities by counting the number of samples within the zones around each cluster center.

This simple procedure results in a list of clusters with a priori probabilities $P(\omega_i)$ and class-conditional densities $f(\mathbf{x}|\omega_i)$. Table 2 lists the results for the same case that was used for supervised classification (Table 1). Although the problem of unsupervised learning is more difficult, the results are very similar. The mean values of overlapping clusters (i.e. white and gray matter) differ slightly. This can be explained by the different estimation method. In supervised learning, parameters are estimated from test samples which are known to belong to a specific class. In the unsupervised approach, the parameters are extracted from the mixture of class-conditional densities, estimation errors can occur when clusters overlap. The parameter estimation for the csf class differs significantly, especially in the second echo mean μ_2. These differences are consistent with our experience that observers try to label voxels only partially filled with csf (mainly in the sulci) as csf. This biases the classifier to shift the discrimination function towards gray matter, resulting in a better defined brain surface.

5.2 Results of unsupervised parameter learning

The automated clustering performed well in three cases (see Table 3), and with some limitations in the other two cases. The cluster statistics are absolutely comparable to the parameters obtained by supervised learning. In two cases the gray matter centers show remarkably smaller values (marked with *) than obtained by manual training. An interesting correlation can be found between the success rate of automatic clustering and the acquisition date. The algorithm performed well on the first three cases, which are all acquired on the 13th of May. In the cases 4 and 5, a significant change of cluster center locations can be detected (see means of csf class). Furthermore in both cases the white and gray matter clusters are merged together. A comparison of the scatterplots of all the five cases supports the assumption that the MR scanner tuning was suboptimal before the 13th of May, and excellent after that date. This leads to the hypothesis that the shimming of the machine has changed between the 12th and the 13th of May.

case	background		white matter		gray matter		csf		acquisition
	μ_1	μ_2	μ_1	μ_2	μ_1	μ_2	μ_1	μ_2	date
case 1	11.5	7.6	500.2	225.4	596.1	297.8	622.9	512.3	13-May-90
case 2	11.6	7.6	487.1	217.7	594.3	298.0	625.7	518.9	13-May-90
case 3	11.5	7.7	494.5	221.1	599.9	294.6	646.5	533.3	13-May-90
case 4	11.7	7.9	460.2	214.0	537.2*	258.4*	581.4	483.9	06-May-90
case 5	11.2	7.7	459.4	204.2	547.3*	265.8*	589.3	479.7	12-May-90
case 1'	11.2	7.4	502.6	230.7	596.9	302.6	-	-	

case 1': parameter estimation based on a different slice pair

Table 3: Cluster means from unsupervised learning

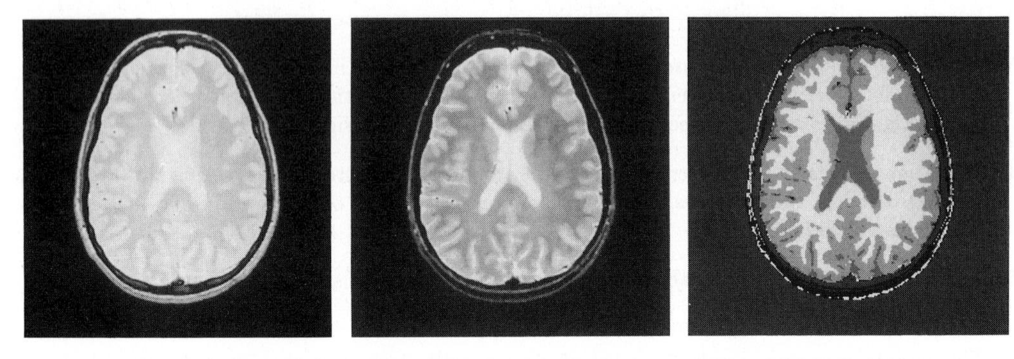

Figure 3: Original image data (left and middle) and classification result (right)

The fact that the imaging parameters (TR, TE, FOV, slice thickness) are kept fixed raises the the idea of reusing the classification map once generated. Vannier, for example, [Vannier et.al., 1988 a] asks if the discrimination functions could be applied to image data of the same subject at later times, or even be applied to other subjects. The results of Table 3 clearly show that each case has to be treated individually. For each class, the MR signals between patients are not identical. The large variations show that we are not simply dealing with sample sets of an identical population.

5.3 Classifier design

Based on the estimated parameters the classifier can be designed. The different class conditional densities favor a maximum likelihood classifier, resulting in quadratic discriminance functions. The sometimes peculiar subdivision of the feature space by the hyperquadric decision functions is overcome by creating a rejection class for the most unlikely samples. The decision is based on rejecting a fixed percentage of the samples of each cluster.

The knowledge about the a priori probabilities and probability densities is used to design an optimal classifier. Optimal means choosing either a Bayes decision rule if minimum overall classification error rate is to be achieved, or a maximum likelihood classification if small classes have to be given a chance to compete against classes with large a priori probabilities.

An important implementation aspect is the extremely efficient classification of the image data by defining the decision regions in feature space as a look up table. The classification of one voxel in a 3D data set only requires one array access, independent of the type of classifier.

Figure 3 shows two original MR slices (first and second echo) and the resulting classification.

6 Validation of segmentation results

A common approach to validate segmentation results is to compare a manually segmented image with the classification result. MR image data, even if significantly improved by the new filtering method, represent complex image structures. It is doubtful whether a manual tracing of the csf along the brain surface can give a clear segmentation to be used as a ground truth. Most of the surfaces contain finely detailed elements appearing as partial volumes. Furthermore tissue boundaries are not always clearly defined. After some experiments we found that the error rate of the manually segmented image is very high.

Most often in statistical classification, a confusion matrix based on the labelling of the training samples (or of an independent second set of training samples) is calculated. Again, subjective criteria are involved. The confusion matrix of carefully selected areas with high probability to be assigned to a specific class comes out much better than a matrix based on uncertain training regions. In the case shown in Figure 2 it was possible to classify 100% of the training samples correct, resulting in an excellent confusion matrix.

To eliminate subjective criteria, a phantom simulating the tissue characteristics of the brain has been constructed [Kikinis et.al., 1990], [Cline et.al., 1990]. Using agarose gel doped with $NiCl_2$ allows independent control of T1 and T2 relaxation times. The phantom is shaped like a typical axial brain slice. Since the size of the regions are exactly known the performance of MR acquisition processes and of segmentation methods can be controlled. A first analysis is done by Cline et.al. [Cline et.al., 1990]. They found a maximum error of 6% between measured and calculated areas of brain tissue. The results reflect not only segmentation errors, but also geometric distortions of the MR imaging process. We plan to use phantom measurements as a reference data set to estimate the classification error rate and to check the reliability of supervised and unsupervised classification.

7 3D renderings

3D renderings are used to visualize 3D segmentation results. After segmentation, each voxel is labeled with its most probable class. Since the assignment of a class label does not define anatomical objects, we run a 3D connected component labeling. Connected groups of voxels with a common tissue category are assigned a unique label. The largest groups of white matter and gray matter type regions are selected to represent white matter and gray matter of the brain. The csf type regions divide into the ventricular system and the subarachnoid space. 3D renderings of the objects segmented with the unsupervised statistical classification method are shown in Figure 4. The process starting from MR volume data and resulting in 3D surfaces was completely automatic, without any human interaction. The renderings were obtained using the MRconsole software package [Cline et.al., 1990].

8 Conclusions

Our statistical classification scheme that uses an unsupervised parameter estimation was completely successful in three of the five cases presented here. The clustering procedure located the clusters of the four most important categories (background, white matter, gray matter, and csf). The estimated parameters are very similar to those obtained by supervised parameter learning.

In the other two cases the estimated parameters of background, white matter and csf were still close to those obtained by supervised parameter learning. Gray matter, however, was affected by its proximity (in feature space) to the white matter. The scatterplots of these two cases clearly show a decreased separability. From the acquisition dates we assume that the MR scanner was not optimally tuned.

The first three cases, however, illustrate that optimal image data can be acquired. Since our automated clustering performs well on high quality image data, we believe that the method proposed

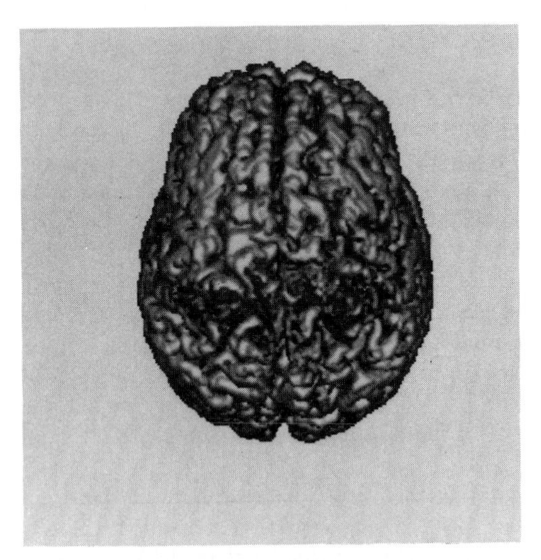

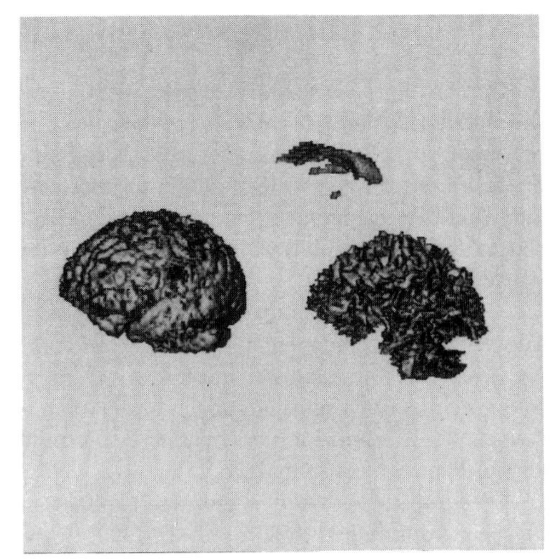

Figure 4: 3D surface rendering of the brain (left) and of the set of surfaces created by the automated procedure (gray matter, ventricular system, white matter)

here can successfully be applied if the MR scanner is carefully shimmed. A simple test of data quality is the visual analysis of the two-dimensional scatterplot.

In a second study, 14 of a total of 16 cases could be analyzed using unsupervised clustering. In the remaining two cases white and gray matter could not be separated due to overlapping clusters. A further refinement of the clustering technique is necessary to accurately find the parameters of density mixtures. Future tests will show, if ultimately unsupervised classification can minimize or even replace supervised techniques.

There are two important advantages of the unsupervised procedure. First, the process is completely automatic and can run as a batch job in the background, without any tedious manual interaction. Second, the automated parameter learning is fully reproducible because subjective variations created by selecting different types of training areas are avoided.

The segmentation procedure is simple, but powerful. It runs on inexpensive workstations, which more and more are connected to scanners as additional research consoles. The segmentation of dual-echo MR head data (54 double echo slices) into outer csf, white and gray matter and ventricular system and the generation of 3D renderings takes a total of nearly one hour (SUN Sparc station). This time divides into 20 minutes for edge preserving smoothing, 10 minutes for supervised training or automatic clustering, 2 minutes for classification and 20 minutes for connected component labeling and the definition object surfaces. The simplicity of the two-dimensional clustering procedure allows control of each step if necessary.

In a NIH study with 800 MR acquisitions, a comparison between unsupervised and supervised classification will be made. This study will, we hope, be an important step in bringing computer assisted procedures into the routine clinical analysis of MR data. The great benefit of computerized analysis is the new qualitative and quantitative information about morphology and tissue characteristics which becomes accessible. Examples are the measurement of brain volume and the volumes of csf spaces, and the use of the brain surface for surgical planning.

Acknowledgements

The authors wish to thank Harvey E. Cline and William E. Lorensen, General Electric Corporation, Schenectady NY for supplying the 3D surface rendering software.

References

[Cline et.al., 1990] Cline, H.E., Lorensen, W.E., et.al., *3-D Segmentation of MR Images of the Head using Probability and Connectivity*, J Comput Assist Tomogr 1990; 14(6):1037-1045.

[Coleman and Andrews, 1979] Coleman G.B. and Andrews, H.C., *Image Segmentation by Clustering*, Proc. of the IEEE, Vol. 67, No. 5, May 1979, pp. 773-85

[Duda and Hart, 1973] Duda R.O. and Hart, P.E., *Pattern Classification and Scene Analysis*, by John Wiley & Sons,Inc, 1973

[Gerig et.al., 1989] Gerig, G., Kuoni, W., Kikinis, R. and Kübler, O., *Medical Imaging and Computer Vision: An integrated approach for diagnosis and planning*, 11. DAGM-Symposium Mustererkennung, 2.-4. Oct. 1989, Informatik Fachberichte IFB 219, Springer Verlag Berlin, pp. 425-432

[Gerig and Kikinis, 1990] Gerig,G. and Kikinis, R., *Segmentation of 3D Magnetic Resonance Data*, in Progress in Image Analysis and Processing, Cantoni, Cordella, Levialdi, Sanniti di Baja edts., World Scientific Singapore, Proceedings of 5th Int. Conf. on Image Analysis and Processing, 1990, pp. 602-609

[Gerig et.al., 1990] Gerig, G., Kikinis, R., Kübler, O., *Significant Improvement of MR Image Data Quality using Anisotropic Diffusion Filtering*, Technical Report BIWI-TR-112, Institute for Communication Technology, Image Science Division, ETH-Zurich, Switzerland, March 1990

[Kikinis et.al., 1990] Kikinis, R., Jolesz, F., Gerig, G., Sandor, T., Cline, H., Lorensen, W., Halle, M., Benton, St., *3D Morphometric and Morphologic Information derived from Clinical Brain MR Images*, in: 3D Imaging in Medicine, Höhne K.H., Fuchs H., Pizer St.M. editors, NATO ASI Series, Serie F: Computer and Systems Science Vol. 60, Springer-Verlag, Proceedings of the NATO Advanced Research Workshop, 25-29 June 1990, Travemünde, FRG, pp. 441-454

[Kübler and Gerig, 1990] Kübler, O. and Gerig, G., *Segmentation and Analysis of Multidimensional Data-Sets in Medicine*, in: 3D Imaging in Medicine, Höhne K.H., Fuchs H., Pizer St.M. editors, NATO ASI Series, Serie F: Computer and Systems Science Vol. 60, Springer-Verlag, Proceedings of the NATO Advanced Research Workshop, 25-29 June 1990, Travemünde, FRG, pp. 63-81

[Merickel et.al., 1988] Merickel, M.B. et.al., *Multispectral Pattern Recognition of MR Imagery for the Non-invasive Analysis of Atherosclerosis*, Proc. of 9th Int. Conf. on Pattern recognition, Rome, Italy, Nov. 1988, pp. 1192-1197

[Vannier et.al., 1985] Vannier, M.W., Butterfield, R.L., Jordan, D., et.al., *Multispectral Analysis of Magnetic Resonance Images*, Radiology 154, 1985, pp. 221-224

[Vannier et.al., 1988 a] Vannier, M.W., Speidel, Ch.M., and Rickman, D.L., *Magnetic Resonance Imaging Multispectral Tissue Classification*, NIPS Volume 3, August 1988, pp. 148-154

[Vannier et.al., 1988 b] Vannier, M.W. et.al., *Validation of Magnetic Resonance Imaging (MRI) Multispectral Tissue Classification*, Proc. of 9th Conf. on Pattern Recognition, Rome, Italy, Nov. 1988, pp. 1182-1186

A Radiometric variation within data volume

The radiometric variability is tested by selecting ROI's in distinct areas and comparing their statistics. We have chosen white matter areas because they are represented as large regions. In a first test, we selected ROI's within the slices which are placed against the four corner directions, as far from the slice center as possible. The statistical parameters reflect a mixture between the variability in anatomy (white matter is not completely homogeneous), a human factor introduced by the interactive selection of the ROI's, and the radiometric distortion. A typical result is illustrated in Table 4.

location in	ROI size	mean		principal axes		rotation	correlation
slice pair 61/62	# voxels	μ_1	μ_2	σ_1	σ_2	(deg)	ρ
upper left	20	477.1	220.4	30.7	7.2	38.4	0.89
upper right	20	487.5	201.9	20.5	7.1	41.4	0.78
lower left	20	483.6	234.7	22.3	7.6	40.6	0.79
lower right	20	510.7	230.7	23.3	9.5	39.4	0.71
equally distributed	200	487.2	216.1	26.5	15.1	48.4	0.51

Table 4: Radiometric homogeneity of white matter within slice

There is an increase of first echo mean value in the lower right and a decrease of the second echo mean value in the upper right region. If compared to the statistics of the equally distributed population, the maximal changes in μ_1 are $+23.5(+4.8\%)$ and $-10.1(-2.1\%)$, and in μ_2 are $+18.6(+8.6\%)$ and $-14.2(-6.6\%)$.

In a second test we took equally distributed samples from different slice levels:

axial slice number	ROI size	mean		principal axes		rotation	correlation
	# voxels	μ_1	μ_2	σ_1	σ_2	(deg)	ρ
slices 41/42	200	479.5	213.0	30.5	20.6	30.2	0.33
slices 51/52	200	487.7	219.2	29.0	17.8	26.9	0.38
slices 61/62	200	487.2	216.1	26.5	15.1	48.4	0.51
slices 71/72	200	492.2	228.1	15.4	14.8	-9.2	-0.02
slices 81/82	200	484.0	226.2	21.0	14.7	7.9	0.10

Table 5: Radiometric homogeneity of white matter along axial direction

The two tables allow comparison of mean values of different sample groups. Since the variations are based not only on noise, but also on anatomical variability and image plane distortions, statistical tests will not give complete information about the segmentation quality.

More important are comparisons of these variations with the distances between the clusters, because they give some insight into the separability of classes. A useful measure is the criterion function proposed by Fisher (see [Duda and Hart, 1973] pp. 114):

$$J(\mathbf{w}) = \frac{|\tilde{m}_1 - \tilde{m}_2|^2}{\tilde{s}_1^2 + \tilde{s}_2^2} = \frac{\mathbf{w}^t S_B \mathbf{w}}{\mathbf{w}^t S_W \mathbf{w}}$$

$\mathbf{w}$: projection direction for sample points

$\tilde{m}_i$: sample mean for the projected points

$\tilde{s}_i$: standard deviation of the projected samples

The *Fisher linear discriminant* is defined as the direction $\mathbf{w}$ which maximizes $J(\mathbf{w})$. The matrices S_B and S_W are called between- and within-class scatter matrices. The ratio of between class scatter to within class scatter can be calculated for each cluster pair, a maximum ratio signifying a good discrimination. A critical pair in our MR-head segmentations are the classes white and gray matter. With s_1 and $s_2 \approx 20$ and a cluster distance of 145 (rough estimates from Table 1), we obtain $J(\mathbf{w}) \approx \frac{145^2}{20^2+20^2} \approx 28$. This result illustrates the minimum overlap between these two clusters (see Figure 2). The J values for the two cases in Table 3 with lower gray matter means are 10 for case 4 and 9 for case 5.

As long as the variation of the means (see tables 4 and 5) is sufficiently small compared to the between class scatter S_B and the within class scatter S_W, the separability of the clusters and therefore the quality of segmentation will not be significantly affected.

AUTOMATIC DETECTION OF BRAIN CONTOURS IN MRI DATA SETS

M E Brummer[1], R M Mersereau[2], R L Eisner[3], R R J Lewine[4]

[1]Department of Radiology, Emory University School of Medicine, Atlanta, GA, USA

[2]School of Electrical Engineering, Georgia Institute of Technology, Atlanta, GA, USA

[3]Carlyle Fraser Heart Center, Emory University Hospital at Crawford Long, Atlanta, GA, USA

[4]Department of Psychiatry, Emory University School of Medicine, Atlanta, GA, USA

Abstract

An algorithm is presented for fully automated detection of brain contours from single-echo 3-D coronal MRI data. The technique detects structures in a head data volume in a hierarchical fashion. Detections consist of histogram-based thresholding operation, followed by a morphological cleanup procedure of the binary threshold mask images. Anatomic knowledge, essential for the discrimination between desired and undesired structures, is implemented through a sequence of conventional and new morphological operations. Innovative use of 3-D distance transformations allows implicit evaluation of anatomic relationships for structure recognition. Overlap tests between neighbouring slice images are used to propagate coherent 2-D brain masks through the third dimension. A summary of results of testing the algorithm on 23 test data sets is presented, with a discussion of potential for clinical application and generalization to other problems, and of limitations of the technique.

Keywords

Image segmentation, contour detection, MRI, brain imaging.

1. INTRODUCTION.

Three-dimensional (3-D) rendering, and many image analysis problems using tomographic medical image data, require some form of tissue classification, or more generally, image segmentation. Few examples are known where this segmentation problem is a trivial task for a computer. Currently, operator interaction is encountered even in most practical implementations for such seemingly obvious segmentation problems as bone classification with CT data. One of the current challenges in medical image processing is to establish useful techniques for segmentation of Magnetic Resonance Imaging (MRI) data. With MRI, one of the principal regions of interest is the brain. The literature is abundant on both quantitative analysis (Lewine et al., 1990), and 3-D rendering (Hoehne et al., 1990, Levoy et al., 1990) of MRI brain data.

The segmentation problem for MRI data depends on a large number of factors. Image contrast, resolution, signal to noise ratio, slice thickness, complexity of the scene, data set size, and RF-coil uniformity are just a few of the important ones. Based on some or all of these factors, an approach must be selected for the segmentation task. For example, for analysis problems involving only a few slice images, completely manual image segmentation can be a feasible option. Accurate volume estimates and realistic 3-D renderings, however, often require small voxel sizes, hence large data sets. The need for higher levels of automation of the segmentation process grows with the size of the data set. With a data set as shown in Figure 1, with 40 contiguous slices, most of which intersect the brain, a solution of generating contours of the brain entirely by hand is no longer viable.

A range of approaches has been proposed for detection of various structures in the head, with varying levels of automation and practical applicability. MRI's ability to generate different contrasts has been exploited by analysis techniques for multispectral MRI data (Windham et al., 1988). Levin et al. (1990) report on an interactive edge-following technique, operating on individual slice images. Pizer et al. (1990) designed an algorithm which divides the image into a hierarchy of elementary regions using intensity extrema. Bomans et al. (1990) investigate optimal tissue border localization. Menhardt (1990) introduces fuzzy logic and fuzzy topological concepts to encode elementary anatomical knowledge into scalar or vector-valued 2-D MRI segmentation. Raya (1990) reports on the implementation of a rule-based approach for automatic segmentation of 3-D brain data.

In this paper we contribute an algorithm, operating on scalar (single-echo) 3-D MRI data of the brain, which alternates different levels of histogram-based thresholding with sequences of morphological operations, into a completely automatic procedure for detection of brain contours. The overall approach is the following: thresholding is used to generate an initial binary mask with connected components, which are considered candidates for a certain region of interest. Subsequent morphological processing discards erroneous candidates, through encoding of elementary anatomical information and correction of thresholding errors.

The algorithm was designed to operate on anisotropically sampled MRI data, as produced by multi-slice scan techniques. This anisotropy, with voxel size ratios x:y:z of approximately 1:1:5, causes poor performance of many 3-D image operations. Most operations were therefore implemented as computationally less costly 2-D neighbourhood operations, applied to the slice images. An exception is the Distance Transformation, discussed in section 2.4.1, which is performed in 3-D, in order to encode essential 3-D information. In addition, the third dimension is handled by overlap tests between mask images of neighbouring slices.

2. METHODS

2.1. MRI Data.

Our brain contour detection algorithm was designed to operate on MRI scans, designed for measuring volumes of the brain and the lateral ventricles. A MRI pulse sequence with RF-recalled echoes is used, since these must be considered superior to gradient echo techniques in terms of image contrast and signal to noise ratio.

The effect of partial volume effects on volume estimates can be reduced a priori by minimizing the relative number of voxels with partial volume effects (Brummer et al., 1989). This can achieved by minimizing voxel dimensions, which amounts to pursuing the minimum achievable slice thickness, and through optimal alignment of the voxels with structures of interest. Considering the lateral ventricles as the critical structure prone to partial volume errors, a coronal scan orientation was therefore adopted.

The most important contrast requirement is a good contrast between each structure of interest in the image and the surrounding structures. In our case, the structures of interest are the brain and the ventricles, and thus good contrast between brain and Cerebro-Spinal Fluid (CSF) is the main issue. Flow artifacts in the CSF are reduced by minimizing its signal intensity.

The resulting scan protocol, further constrained by clinical scan time limitations, has the following parameters: a repetition time T_r=650 ms, an echo time minimized to T_e=30 ms, slice thickness of 5 mm, field of view minimized to encompass the head, usually 220 mm, and a number of slices sufficient to contain the whole brain in the data volume, usually 35-40 slices. The total

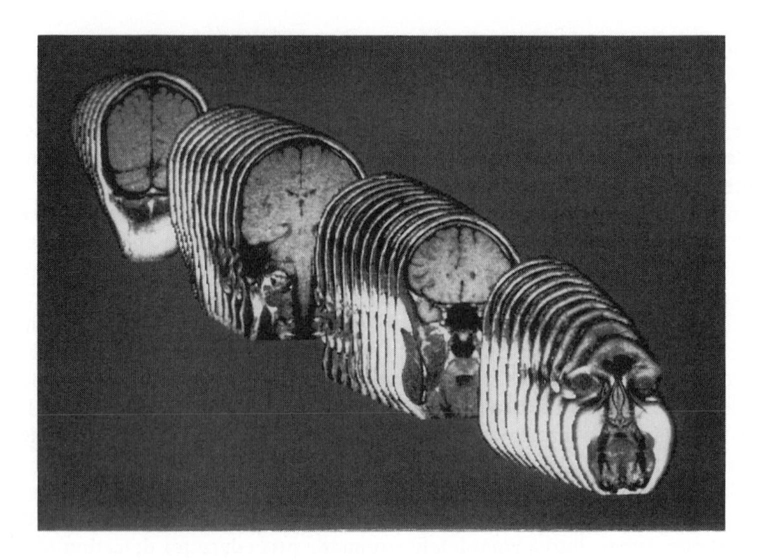

Figure 1: Example of a coronal MRI data set.

scan time was 11 minutes per excitation. Data shown in this paper were acquired on Philips Gyroscan equipment, at field strengths of 0.5T and 1.5T, which required 2 and 1 excitations respectively for clinically acceptable signal to noise ratios. An example data set is shown in Figure 1.

2.2. Algorithm Description.

Figure 2 shows a block diagram illustrating the complete algorithm for brain contour detection. The algorithm can be divided into two distinct functional steps: head contour detection, and brain contour detection. The first step, head contour detection, yields a 3-D mask, which contains essential information for the second step.

For the head contour detection, a greylevel histogram is generated of the entire data set. From this histogram, noise parameters are computed, and a background threshold is generated, using elementary knowledge about the physics of the imaging process and probability theory. This part is detailed in section 2.3.2. A thresholding operation generates an initial set of 2-D masks for the entire head. Subsequently, a sequence of morphological operations is applied to correct thresholding errors, and to condition the mask for the remainder of the algorithm. These operations are described in greater detail in section 2.4.2.

The result of the above is a binary 3-D mask for the head. This mask is used at multiple instants in the second part of the algorithm. This part also starts with a histogram analysis, with as aim determination of upper and lower thresholds of the greylevel range representing brain. For generating the greylevel histogram, a large number of voxels, which we know with good probability not to be part of the brain, is excluded, to better approximate a greylevel distribution of brain only. For some RF coils of MR imagers, an extra processing step must be included here, to correct for intensity variations in the image, which result from non-uniform RF-power distribution over the data volume (Brummer, 1991). Upper and lower greylevel thresholds are computed automatically for the greylevel range of the brain. Section 2.3.3 describes this step of the algorithm. Again, thresholding yields an initial 3-D mask.

The image is considerably more complex near brain boundaries than near head boundaries. In addition, the greylevel range of brain is shared by some other structures and border regions,

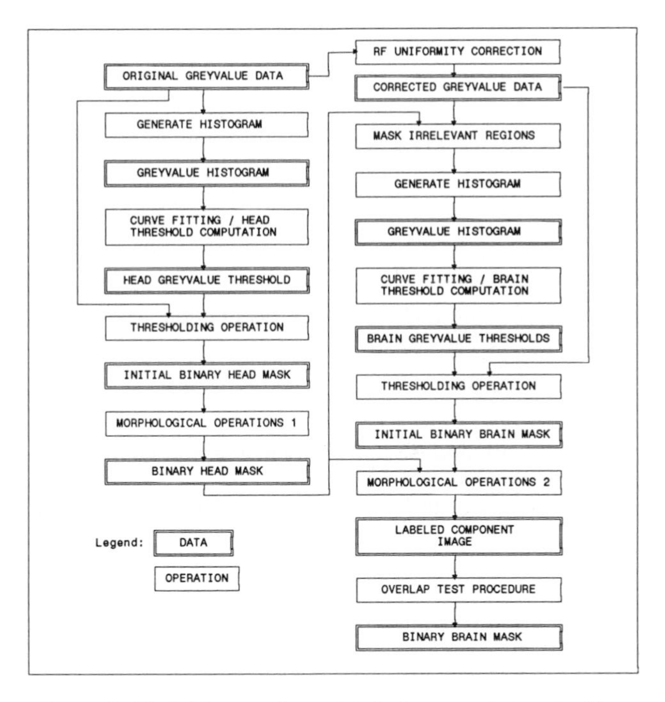

Figure 2. Block Diagram of complete brain segmentation algorithm.

which appear as erroneous regions in these initial brain masks. These problems are addressed by another sequence of morphological operations, described in section 2.4.3. This part of the algorithm produces a set of slice images with labeled connected objects as candidate brain regions.

An overlap test algorithm between these labeled regions in adjacent slices incorporates 3-D connectivity information for decisions about which regions from these label images should be marked as belonging to the brain in a final brain mask image. This procedure is described in section 2.5.

2.3. Histogram Analysis.

In this section we present the methods, used in our algorithm, for automatic threshold selection. We will start with an overview of relevant aspects of random image noise, encountered in MRI data.

2.3.1. Noise Model.

Complex quadrature MRI data are polluted by thermal noise which by good approximation behaves as additive white noise (Papoulis, 1984), with normal distribution. Hence we may consider the complex image $\hat{f}_c(\vec{x})$ after reconstruction as a superposition of a "true" complex image function $f_c(\vec{x})$ and a complex noise sequence $n_c(\vec{x})$:

$$\hat{f}_c(\vec{x}) = f_c(\vec{x}) + n_c(\vec{x}) \tag{1}$$

where $\vec{x}$ is a location vector. In order to eliminate phase artifacts, usually only the modulus $f(\vec{x}) = \|\hat{f}_c(\vec{x})\|$ is displayed.

In the background of the image $f_c(\vec{x})=0$. In the remainder of the section, we will omit the spatial indexing since it is not relevant here. A value f in the modulus image is defined as

$$f = \|n_c\| = \sqrt{(n_r^2 + n_i^2)} \tag{2}$$

with n_r = Re n_c, and n_i = Im n_c. Considering a normal distribution $p(n_r)$ of the noise in the real channel

$$p(n_r) = \frac{1}{\sqrt{2\pi\sigma^2}} e^{-\frac{n_r^2}{2\sigma^2}} \tag{3}$$

and an identical distribution of n_i, we find a Rayleigh distribution (Papoulis, 1984) of f:

$$p(f) = \frac{f}{\sigma^2} e^{-\frac{f^2}{2\sigma^2}} \tag{4}$$

This distribution is clearly recognized in the low-greyvalue range of the greyvalue histogram of the data (light shading in Figure 3).

In non-background regions in the image, if we define an additive noise term n in the modulus image

$$= \|f_c\| + n \tag{5}$$

it can be shown that, in the high image intensity limit $\|f_c\| \gg \sigma$, n has a distribution identical to that of the channel noise.

2.3.2. Background threshold determination.

A greyvalue threshold for separation of image objects and background can be generated as follows. First, a greyvalue histogram h(f) of the data set is generated. A least-squares fit of a scaled Rayleigh curve is performed on the low-greyvalue range of the histogram. The best-fit Rayleigh function r(f) is hypothesized to be the histogram of the noise pixels in the data. Subsequently, a function g(f)=h(f)-r(f) must represent a greylevel histogram of the rest of the data. Thus a bimodal greylevel distribution is generated, and a minimum-error threshold (Rosenfeld and Kak, 1982) can be determined. The result of this thresholding operation on several slices of a typical coronal data set is shown in Figure 4. Procedures for correction of misclassified regions are described in section 2.4.2.

2.3.3. Brain threshold determination.

Two observations lead us to a thresholding approach for initial brain segmentation. Firstly, the brain is a large, homogeneous object. This rather intuitive statement translates into a good probability to find a well-defined peak in the greylevel histogram for brain. In addition, regardless of whether the MRI scan protocol maximizes or minimizes contrast between white and grey brain matter, the image intensity of the CSF that surrounds the brain is generally either higher or lower than both kinds of brain tissue. Therefore, a single range of greyvalues can be determined to correspond with the brain in a particular scan, and separate it from CSF. This does not mean, however, that any such greyvalue range will represent exclusively brain tissues. Problems related to this issue will be addressed in section 2.4.3.

The light shaded histogram in Figure 3 shows a greyvalue histogram of the whole data set. Better definition of the greylevel range of brain is achieved by exclusion of all background pixels from the histogram (medium shading), through masking the data with the head contour mask. In addition, we know the brain to be located predominantly in the top portion of the head. If the pixels which are taken into account in generating the histogram are limited to that region, the fraction of brain pixels in the histogram is increased dramatically. From the head mask we can

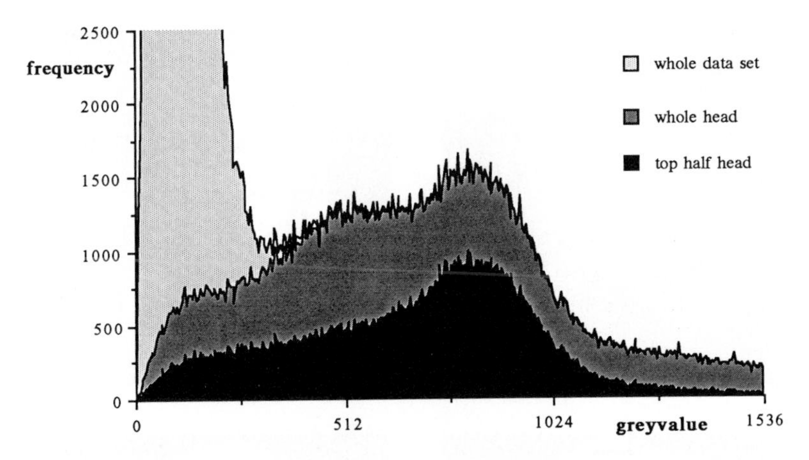

Figure 3: Greyvalue histogram of the whole data set, and modifications for determination of thresholds for the greyvalue range of brain (see text).

generate an estimate for what is e.g. the top half of the head, and mask off the remainder. This operation may seem crude, but is remarkably effective (dark shaded histogram of Figure 3). The above technique generates a histogram which shows a well-defined absolute maximum peak in the greyvalue range of the brain, which is not always encountered in the histogram of the whole data set.

A least-squares fit of a Gaussian curve is performed through a range of the histogram, limited on either side of the peak by the greyvalue distance to the nearest half-maximum. Upper and lower greyvalue thresholds are computed on both sides of the peak of the Gaussian, at two standard deviations away. Figure 4 shows some typical initial brain masks generated with these thresholds for slices of a data set. Note that indeed most brain pixels are included in the mask, but also various parts of other tissues throughout the head.

2.4. Morphological Operations.

The thresholding operations described in the preceding sections do not exploit any a priori knowledge about the anatomical features revealed by the data. They generate a raw binary mask, indicating which regions in the image are candidates to represent structures of interest, based on greylevel information alone. In order to refine this mask, we need tools which help us match information from shape and location of these regions with knowledge about anatomy and the imaging process. These tools are morphological operations.

Before proceeding with a description of morphological algorithms, implemented for various tasks in the brain segmentation, we will briefly review the most important concepts and operations.

2.4.1. Morphological Concepts and Operations.

Connectivity. Each pixel in a 2-D image domain, ignoring edge pixels, has eight direct neighbours. Considering the pixels as adjacent squares, neighbouring pixels which share a side of the square are called 4-adjacent neighbours; these, augmented with those neighbours which only share a corner, will be called 8-adjacent. Similarly, in a 3-D image domain, we can define 6-, 18-, and 26-adjacent neighbours, since neighbouring voxels can share a face, an edge, or a corner of a cube. A subset of an image is i-connected, if for any two pixels in the subset a path between these pixels exists, which consists of a sequence of i-adjacent neighbours in this subset. For a

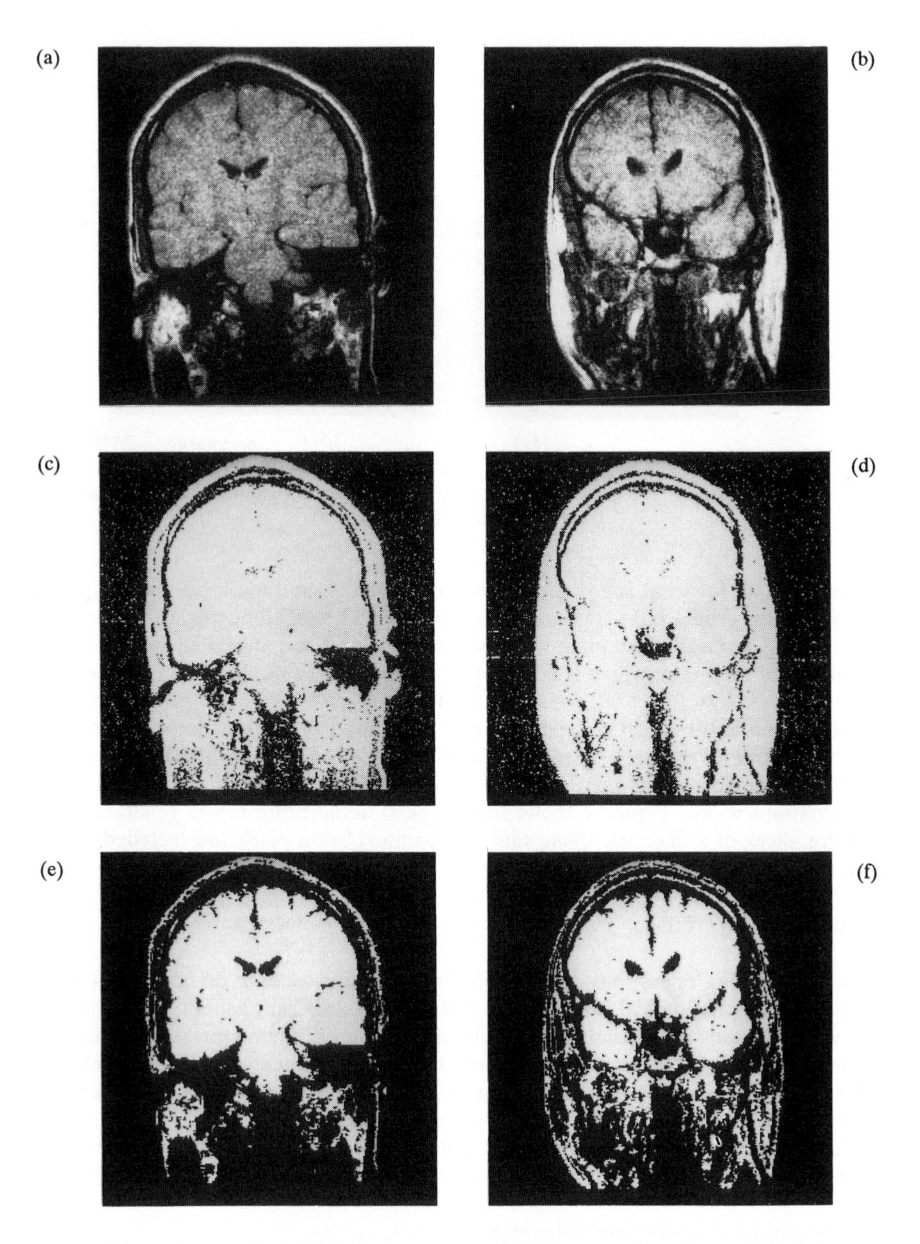

Figure 4: (a,b) Original greyvalue images.
(c,d) Threshold masks for initial head segmentation (see text).
(e,f) Threshold masks for initial brain segmentation (see text).

thorough treatment of the connectivity concept we refer to Rosenfeld (1970).

Binary Erosion. Haralick et al. (1987) define erosion, also known as thinning, as a transformation C = A $\ominus$ B, which combines sets A and B in N-space $\mathbf{R}^N$ following

$$C = \{ c \in \mathbf{R}^N : a = c + b \in A \ \forall \ b \in B \} \tag{6}$$

B is called the "structuring element" of the erosion. The operation essentially "peels away" from set A an outer layer of elements, determined by the size and shape of the structuring element.

Binary Dilation. This operation, also known as dilatation or thickening, is a dual to erosion. It is defined (Haralick et al., 1987) as a transformation C = A $\oplus$ B:

$$C = \{ c \in \mathbf{R}^N : c = a + b, \ a \in A \ , \ b \in B \} \tag{7}$$

It grows a layer, defined by structuring element B, onto the boundary of image objects A.

Opening. An opening is defined (Haralick et al., 1987) as a sequence of an erosion, followed by a dilation with the same structuring element. As a coarse description of its functionality, we could state that this operation removes components, smaller than the structuring element, and "opens" thin, elongated bridges between larger components.

Closing. Conversely, a closing (Haralick et al., 1987) follows a dilation by an erosion with the same structuring element. A closing operation closes small holes in objects, and fills narrow gaps in or between connected components.

Labeling of Connected Components. Component labeling (Groen, 1978) is a crucial step in moving from neighbourhood operations to operations on image objects. This operator detects which regions in a binary image are connected, and assigns each connected region a unique label.

Dilation of Label Image. This novel operator is a modification of the binary dilation. As in the binary dilation, background points directly adjacent to connected objects are included in the object. However, the input and output of the operation are here label images, and included pixels are now assigned the label value of the object. A tie-breaker is needed for background points, which are adjacent to more than one object. An efficient solution for this issue is assigning the first label that is encountered while scanning the neighbour points.

Pepper-and-Salt Noise Removal. Single-pixel objects in a background and single-pixel holes in an object of a binary image, referred to as pepper-and-salt noise (Groen, 1978), are often the result of propagation of uncorrelated noise in an original image through some binarization process. They can be detected as pixels with binary value different from all neighbours, and they can be removed by changing their binary value.

Distance Transformations. A distance transformation is a function which maps each point of a binary input image to its distance from the nearest background point in the image. Mappingsto true Euclidean distances are computationally extremely costly, and many approximating alternatives have been proposed for digital images. An overview of techniques is presented by Borgefors (1986). We have used 2-D and 3-D implementations of the fast chamfer-distance algorithms.

Max-feature Operation. This operation maps connected components of a label image to the maximum value found in corresponding regions in a feature image.

2.4.2. Morphological Operations for Head Contour Detection.

The thresholding procedure of section 2.3.2 generates a set of binary slice masks, some examples of which are shown in Figure 4c-d. Pepper-and-salt noise removal will eliminate most of the misclassified singular points in the background. Figure 5 shows a block diagram of the complete sequence of morphological operations.

Low-intensity structures in the head will appear as holes in these threshold masks. Since these structures are inside the head, it is desirable to fill these holes. We remove the holes by first inverting the initial binary head mask. This makes the background and the holes the connected objects in the image. These objects are labeled, and subsequently all but the largest one (the

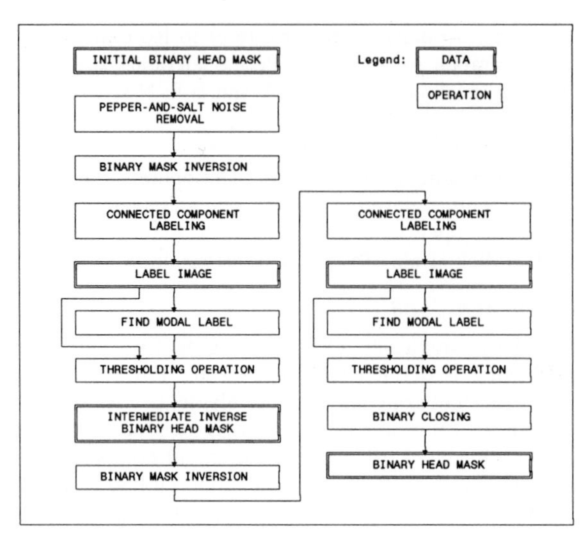

Figure 5: Block diagram of the morphological operations for head contour detection.

"modal label"), invariably the background, are discarded. Inverting the mask again yields the head as a connected object without holes.

If there is correlated noise in the slice image, e.g. motion artifacts from swallowing or eye motion, the mask may contain larger misclassified regions in the background. These artifact objects in the head mask images are generally smaller than the object representing the head, and can therefore be removed by the same procedure described above, without the mask inversions.

In coronal scans some head contours follow signal voids inside the head formed by the ear cavities and adjacent bony structure (Figure 4c). An important purpose of the head mask is to provide a distance measure between candidate brain regions and the outside border of the head. It is therefore desirable to fill these cavities in the mask. This task is performed by a closing operation. We found empirically that a circular structuring element of diameter 0.2 times the maximal left-right cross-section of the head, computed from the head mask, is adequate.

2.4.3. Morphological Operations for Brain Contour Detection.

After the initial threshold segmentation step described in section 2.3.3, we have a set of binary slice images, which contain fairly "thick", homogeneous objects which represent the brain, as well as multiple object regions which represent structures other than brain tissue. Examples for two different sections of the brain are shown in Figure 4e-f.

The non-brain object regions in these mask images can be attributed to two phenomena: structures in the image with similar greyvalue range, and boundary regions between two structures with higher and lower image intensities (partial volume voxels), weighted such that the resulting intensity equals that of brain. An example of the first phenomenon are the tissues related to the nasal mucosa and muscles in the mask of Figure 4f below the temporal lobes of the brain. The second phenomenon can be observed on either side of the skin and adjoining fatty tissues. Some of these transition regions form false connections between brain and other tissues in the masks. This is illustrated by Figure 6a, which shows the same mask of Figure 4f, and Figure 6b, which shows a single 8-connected component in this image, in which brain appears connected to the nasal mucosa.

Our decision algorithm for brain object detection relies on overlap tests between connected components in neighbouring slices. Therefore, brain objects may not be connected to non-brain

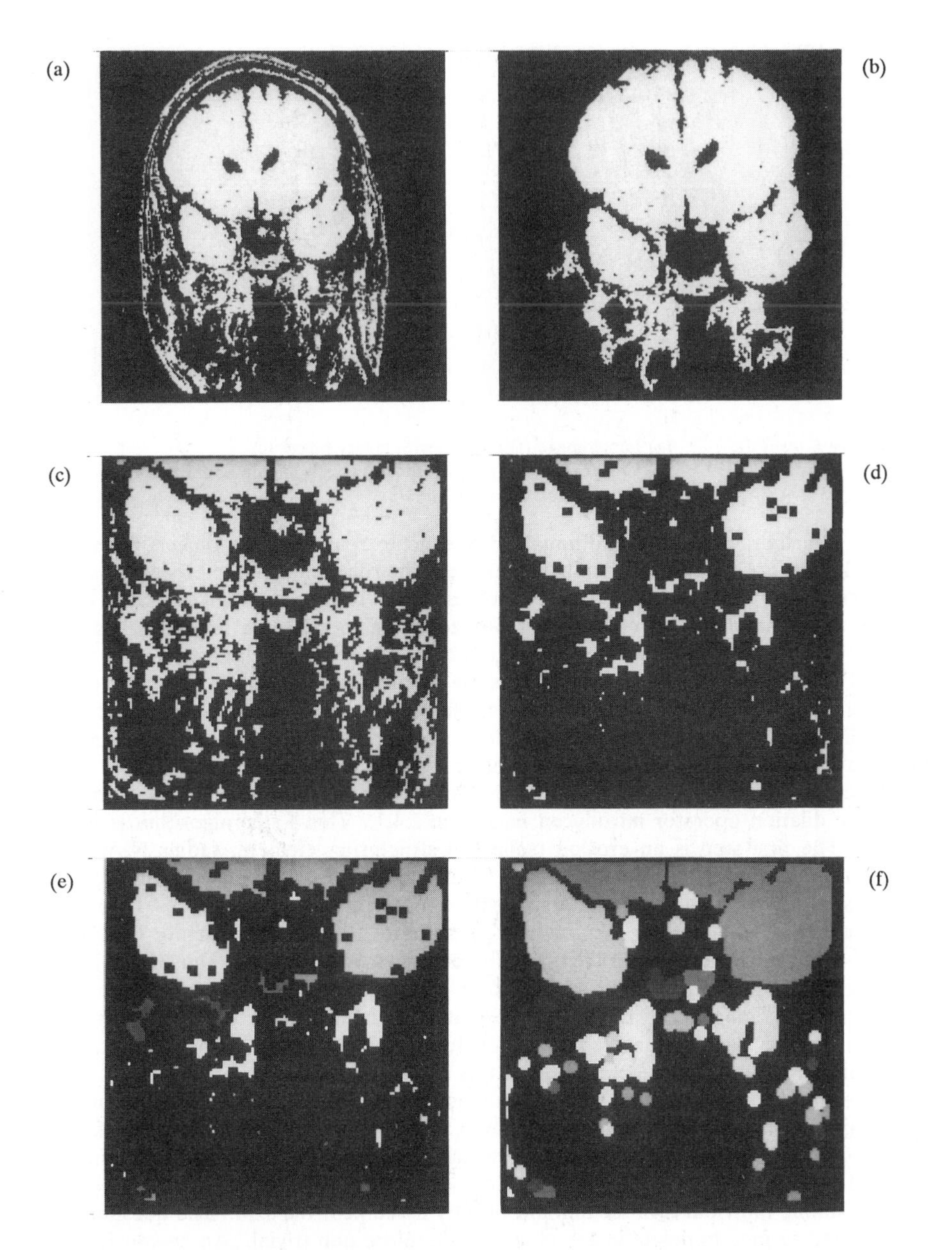

Figure 6: (a) Initial brain mask.
(b) A single connected component in image (a).
(c) Detail of (a) magnified by a factor 2.
(d) Result of erosion on image (c).
(e) Result of component labeling in (d).
(f) Result of label image dilation on (e).

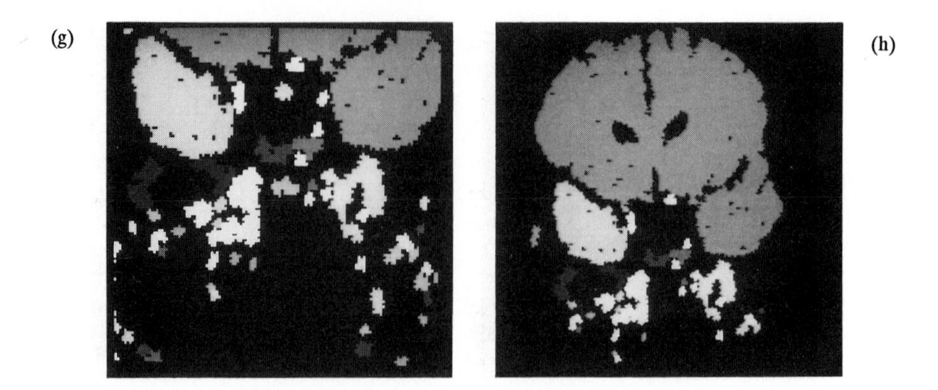

Figure 6 (cont'd): (g) Label image (f) masked with binary image (c)
(h) Components in (g), corresponding with initially connected component (b).

structures. Further, exploiting additional knowledge to remove those objects from the masks, which are known not to represent brain, will improve the robustness of the procedure significantly.

A conventional approach for removal of undesired bridges between objects in binary mask images is the opening. This operation is not adequate in our case for the following reason. Openings with a small structuring element remove elongated thin bridges between connected regions. The undesired bridges in our images are usually thin, but not necessarily elongated. As a result, a bridge may be removed by the erosion, but is often re-established by the subsequent dilation. Opening with a larger structuring element will remove more undesired connections, but introduces severe distortion in the brain objects.

These problems are avoided by a morphological region-splitting algorithm, which uses the label image dilation operator introduced in section 2.4.1. This 4-step algorithm is illustrated in Figure 6. The first step is an erosion (spherical structuring element, radius N, with N=2 for Figure 6d). Subsequently, the components in the eroded image are labeled (Figure 6e). Next, this label image is dilated with a slightly larger structuring element (e.g. radius N+1), so the dilated objects are slightly larger than the original ones before the erosions (Figure 6f). Finally this dilated label image is masked with the original binary image (Figure 6g). Thus the object contours of the original image are restored. At this point, object regions which were weakly connected in the original mask images, are represented by different labels in the label image, regardless of whether or not they are disconnected in the result. Figure 6h shows regions corresponding to the component of Figure 6b after application of this algorithm.

The transition region with partial volumes of high-intensity skin and fat tissues and low-intensity air or bone invariably contains voxels in the greyvalue range of brain. We know that all these object regions in the initial brain mask, which are close to the head surface, do not represent brain, and should be removed as candidate objects for such. However, there are considerable variations among individuals in the amount of fatty tissue near the skin. The question of how far from the head surface to delete image objects is therefore non-trivial. An answer is provided by the following algorithm. First, a 3-D distance transformation (Figure 7c) is computed for all pixels, which evaluates the 3-D distance to the head surface (Figure 7b). A histogram is evaluated of these distance values, restricted to those pixels in the initial brain mask (Figure 7d). In this distance histogram (Figure 8), distance zero represents voxels right at the surface of the head. A first local maximum represents all partial volume skin regions discussed above, and a subsequent local minimum shows how far from the head surface the skull is located. This minimum in the distance histogram is used to define a distance threshold for removal of image objects near the boundary of the head.

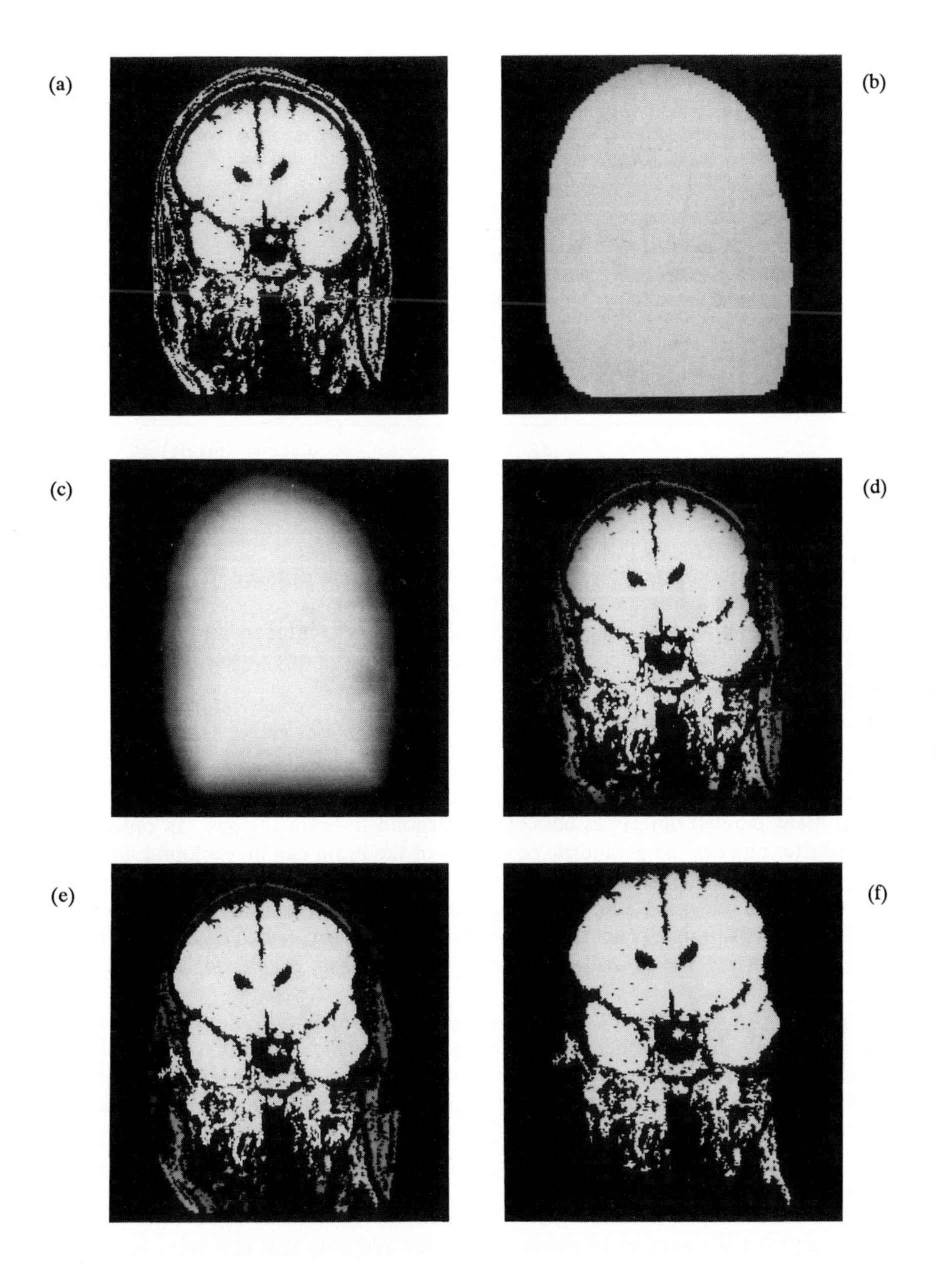

Figure 7: (a) Initial brain mask of slice image of Figure 4b.
(b) Head mask of the same slice image.
(c) 3-D Distance transform of the head mask for the same slice.
(d) Distance transform (c) masked with the initial brain mask (a).
(e) Max-feature operator result for connected components of (a) with (c) as feature image.
(f) Result of distance thresholding on (e); see text for distance threshold.

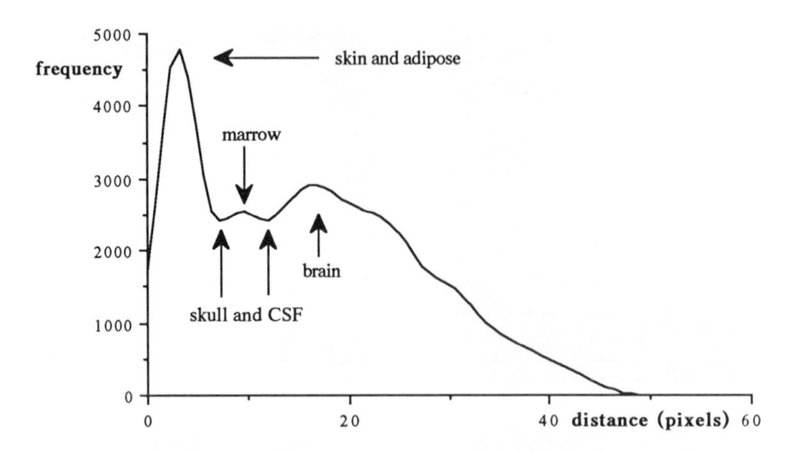

Figure 8: Histogram of distance values in the masked distance map of Figure 7d.

Our implementation removes all objects with maximum distance to the skin (Figure 7e) of less than 1.5 times this histogram minimum (Figure 7f).

Figure 9 shows a block diagram of the above steps combined into a single sequence of operations. Note that the distance thresholding operation is integrated into the disconnection algorithm.

2.5. Mask Image Overlap Test Algorithm.

The morphological algorithms for brain contour detection generate a set of label images, in which as many labeled objects as possible correspond to brain regions. In order to eliminate remaining clutter objects, the geometric continuity of the brain can be exploited: in neighbouring slices, the brain will be represented by relatively similar subsets of pixels. If the brain has been detected correctly in one slice, an overlap test can be used to determine which objects in the next slice should represent brain. A heuristic procedure can be designed to determine the brain mask in one or more starting slices, and subsequently overlap tests propagate the mask through the slices.

Overlap Index. This slice-to-slice overlap test must provide an answer to the question, given an image with labeled objects for a certain slice, which combination of labeled objects corresponds best with a reference mask (of a previous slice) according to an overlap criterion. Our algorithm maximizes a formal overlap index $\nu_{R,T}$, which compares two object masks R and T, and is defined as

$$\nu_{R,T} = \frac{N(R \cap T)}{N(R \cup T)} \tag{8}$$

where N(A) denotes the number of pixels in a set A. We note that this index is symmetric in R and T. The index assumes values in the interval [0,1]. It is easily verified that value $\nu=0$ corresponds with completely disjoint masks, and that only identical masks result in an index value $\nu=1$. Using this overlap index, we can find a subset of image objects in a mask image, which maximally overlaps a reference mask. If the reference mask is a set R, and test mask T consists of n objects t_k, (k=1,..,n), we first order the objects according to their individual relative overlap ratio $\omega_{k,R}$ with R, defined by

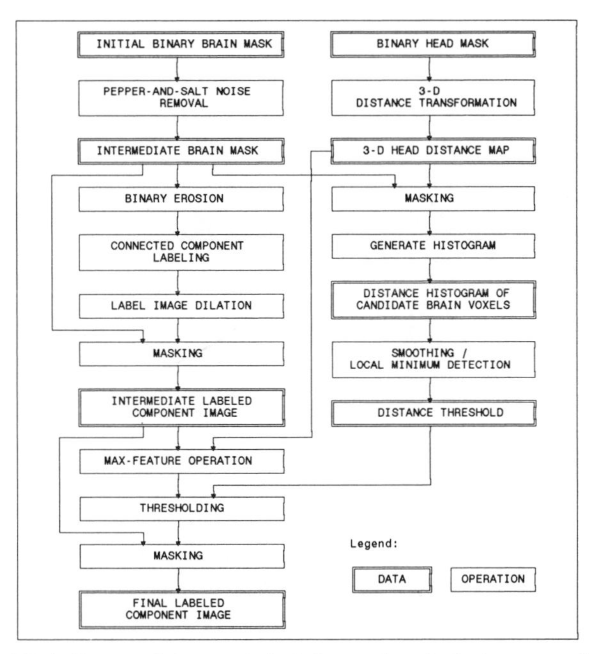

Figure 9: Block diagram of the morphological operations for brain contour detection.

$$\omega_{k,R} = \frac{N(t_k \cap R)}{N(t_k)} \tag{9}$$

Hence we obtain an ordering $\{k_1,..,k_i,..,k_n\}$, with $k_i \neq k_j$ for $i \neq j$ and $\omega_{ki,R} \geq \omega_{kj,R}$ if $i<j$. Starting with an empty set $T_0=\emptyset$ (with trivial overlap index of $\nu_{R,T0}=0$), the overlap index is maximized by sequentially including object t_{ki} for each next i into set T_i

$$T_i = \bigcup_{j=1}^{i} t_{k_j} \tag{10}$$

until no improvement in $\nu_{R,Ti}$ is achieved anymore:

$$0 = \nu_{R,T_0} \leq \nu_{R,T_1} \leq \nu_{R,T_2} \leq .. \leq \nu_{R,T_i} \; [> \nu_{R,T_{i+1}}] \tag{11}$$

or until all objects have been included into the mask (i=n). It can be proved that the set T_i, thus obtained, has a higher overlap index with R than any other combination of image objects t_k.

Mask Propagation. Our current implementation only propagates the brain mask from the center slice through the data in both directions (Figure 10a). The initial reference mask in the center slice is the largest candidate brain region. Although our results using this method are good, it is clear that errors, introduced in a slice, can propagate through the remaining slices unhindered. This could be avoided, or at least detected, if the mask propagation were performed from multiple equidistant reference mask slices (Figure 10b). This issue remains under investigation.

3. RESULTS.

From a large base of clinical scan data, 23 scans were selected for testing the algorithm. Selection was random, except for a visual screening for severe motion artifacts. 10 of these scans were acquired on a 1.5T imager, and 13 on an older 0.5T system. RF uniformity correction was

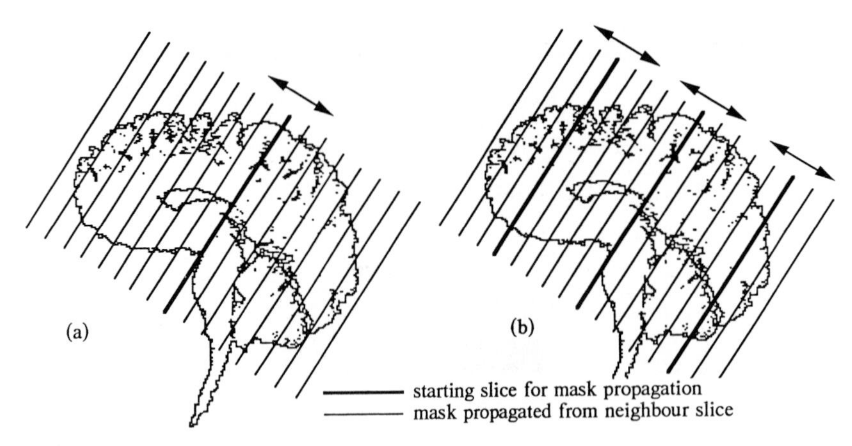

Figure 10: (a) Brain mask propagation starting from a single central reference slice location.
(b) Brain mask propagation from multiple reference slices (see text).

available for the 1.5T scans. No correction was available for the 0.5T data.

The results of these test runs are encouraging. It is important to note that no parameter settings were changed between different scans. The total processing time for a 40-slice scan on a Pixar/Sun 3-180 system is 3-4 minutes. Results for some typical images are shown in Figure 11.

The thresholds detected by both curve fitting algorithms were consistently adequate. The morphological processing steps in the algorithm successfully remove erroneous candidates near the skin. The object disconnection algorithm is very effective in removing undesired bridges between brain and other tissues. The overlap test procedure effectively guides the brain mask through complex shape changes such as traversing the brain stem or forking into the temporal lobes. In all 23 cases the brain contour was detected correctly in a large majority of slices. In the following paragraphs we summarize problems encountered in the test runs.

Incidental thresholding problems were invariably related to imperfections in, or absence of the RF-uniformity correction. In one of the 0.5T data sets the thresholding missed the lower side of the cerebellum due to intensity attenuation.

Frequently, brain regions in the most anterior and posterior slices are discarded erroneously by the distance thresholding procedure. The 3-D distance transformation considers the end slices to be the outer surface of the head, and thus treats end slices as closer to the surface of the head than they truly are. Severe partial volume effects also complicate segmentation in these slices.

It is not possible to stop eye movement during the scan, and hence difficult to avoid a certain motion blurring of the data in this region of the image. In combination with the complex anatomy near the eye, with small structures of often dramatic contrast, this often causes a certain jaggedness of brain contours near the eyes.

The overlap test is not always capable of distinguishing the tips of the temporal lobes from adjacent facial muscles or peri-ocular tissues. Sometimes the tips of the temporal lobes are propagated erroneously through 2 or 3 slices in frontal direction.

Other errors are encountered incidentally, such as the inclusion of a bone marrow region or the pituitary gland, or failure of the disconnection algorithm, e.g. when patient motion has degenerated the contrast between the temporal lobes and the nasal mucosa and muscles below.

4. DISCUSSION.

The algorithm presented in this paper can in its present form be used for certain applications. An interactive editing tool for correcting segmentation errors would be an adequate

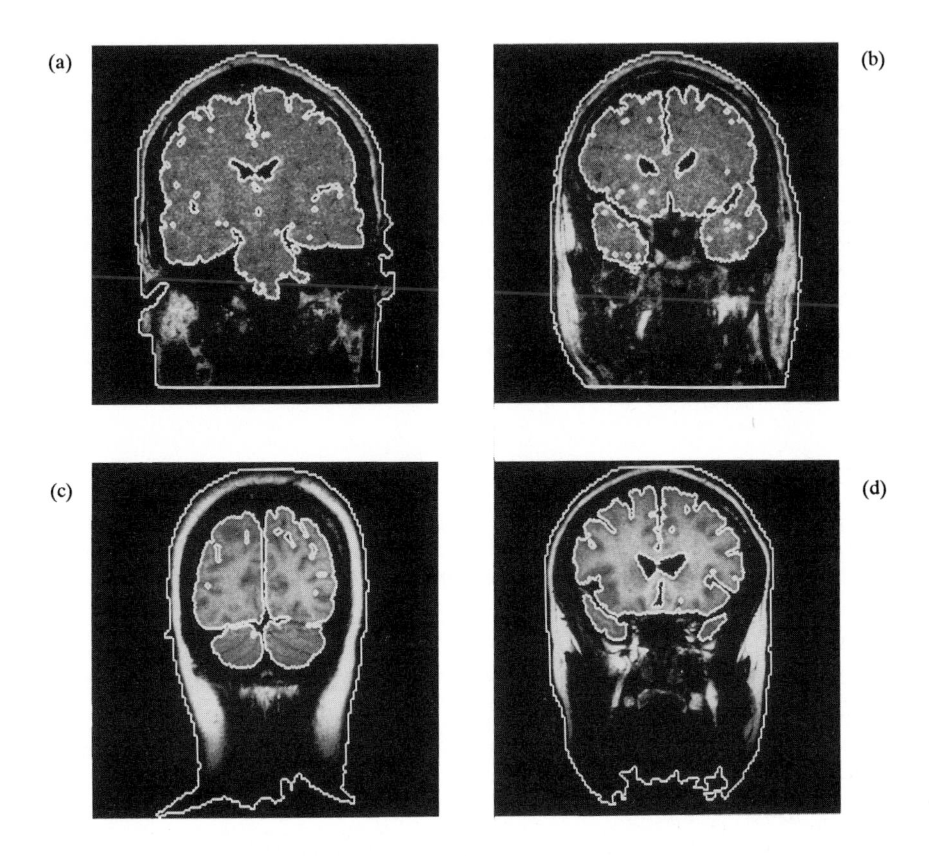

Figure 11: (a,b) Final contours of brain and head for the slice images used in examples. (c,d) Final contours for two other typical coronal slices.

complement for use with some 3-D rendering techniques, and offer substantial savings in operator time compared to completely interactive techniques. Most of the more than incidental segmentation errors discussed in the previous section can likely be eliminated by refinements of the algorithm, such as improved uniformity correction, and provisions for correction of end slices and for better definition of the temporal lobes. Another topic of interest for future research is the development of computationally efficient reliability tests within the algorithm.

Although the algorithm as described is specific to one problem, the key concepts, presented in this paper, can be extended to many other problems. Greylevel-based initial segmentation steps in MRI, whether based on thresholding, region-growing, or other principles, are likely to encounter the same problems of undesired connections of regions of interest to clutter. The disconnection algorithm presented here, based on the label image dilation operator, is an effective tool for dealing with these problems. The use of the fast distance transformation, a computationally inexpensive tool, to obtain geometric feature information, relative to known structures, has potential for automatic segmentation of other structures. More generally, the approach of direct encoding of knowledge about a structures of interest, relative to locations of already known structures, can result in computationally efficient automated detection algorithms.

We wish to conclude with some remarks about the correctness and accuracy of the detected contours. Visual inspection by an anatomy expert is an adequate method to determine the correctness of the detected contours. Quantitative statements about the accuracy of the contours, however, require a definition of ideal contours. Concentrating on a simple case of contours which represent a boundary between two tissues, ideal contours may be defined to follow the lines of averaged intensity, or equal partial volumes. The contours found by our algorithm are generated by global intensity thesholds. The techniques used for automatic threshold generation do not have information about the nominal image intensity of the structures directly adjacent to a structure of interest, but generate a threshold that works for any image intensity for which the histogram shows evidence. The contour locations will therefore not be optimal in the sense described above. If optimal contours are desired, a contour refinement pass must follow the algorithm as described here. Such a refinement pass could use information from either local intensity analysis, or from optimal edge operators (Bomans et al. 1990). Research on reliable algorithms for modification of complex contours to better correspond with optimal edge locations is at the present time still in a beginning stage.

Acknowledgements.

This work was supported in part by grant no. MH-44151 from the NIMH. The authors would like to thank Dr. Wido Menhardt for his valuable help and comments, and Dr. James C. Hoffman, for coming to their aid in with expertise in neuro-anatomy.

References.

Bomans M, Hoehne KH, Tiede U and Riemer M (1990), 3-D Segmentation of MR Images of the Head for 3-D Display, IEEE TMI, Vol 9-2, 177-183.

Borgefors G (1986), Distance Transformations in Digital Images, CVG&IP, Vol 34, pp 344-371.

Brummer ME, Van Est A and Menhardt W (1989), The Accuracy of Volume Measurements From MR Imaging Data, Proc. 8th Ann. Meeting SMRM, p. 610, Amsterdam, 1989.

Brummer M (1991), Intensity Correction of MR Image Data Acquired with Rigid RF Coils, Submitted to 10th Ann. Meeting SMRM, San Francisco.

Groen FCA (1978), Local Transformations, in: Course on Pattern Recognition and Image Processing 1978, Verhagen CJDM (ed), Pattern Recognition Group, Applied Physics Department, Delft University of Technology, Delft, The Netherlands.

Haralick RM, Sternberg SR and Zhuang X (1987), Image Analysis Using Mathematical Morphology, IEEE Trans. PAMI, Vol 9, pp 532-550.

Hoehne KH, Bomans M, Pommert A, Riemer M, Tiede U and Wiebecke G (1990), Rendering Tomographic Volume Data: Adequacy of Methods for Different Modalities and Organs, In: 3D Imaging in Medicine: Algorithms, Systems, Applications, Hoehne KH, Fuchs H and Pizer SM (eds), Springer Verlag, Berlin, pp 197-215.

Levin DN, Hu X, Tan KK and Galhotra S (1989): Surface of the Brain: Three-dimensional MR Images created with Volume Rendering, Radiology, Vol 171, pp 277-280.

Levoy M, Fuchs H, Pizer SM, Rosenman J, Chaney EL, Sherouse GW, Interrante V, Kiel J (1990): Volume Rendering in Radiation Treatment Planning, Proc. 1st Conference on Visualization in Biomedical Computing, Atlanta, pp 4-10.

Lewine RRJ, Gulley L, Risch S. Jewart R and Houpt J (1990), Sexual Dimorphism, Brain Morphology and Schizophrenia, Schizophrenia Bulletin, vol 16, 195-204.

Menhardt W (1990), Unscharfe Mengen (Fuzzy Sets) zur Behandlung von Unsicherheit in der Bildanalyse, PhD thesis, University of Hamburg, Hamburg.

Papoulis A (1984), Probability, Random Variables, and Stochastic Processes, 2nd Ed., McGraw Hill, New York.

Pizer SM, Cullip TJ and Fredericksen RE (1990), Toward Interactive Object Definition in 3D Scalar Images, In: 3D Imaging in Medicine: Algorithms, Systems, Applications, Hoehne KH, Fuchs H and Pizer SM (eds), Springer Verlag, Berlin, pp 83-105.

Raya SP (1990), Low-Level Segmentation of 3-D Magnetic Resonance Brain Images-A Rule-Based System, IEEE TMI, Vol 9-3, pp 327-337.

Rosenfeld A (1970), Connectivity in Digital Pictures, J. ACM, Vol 17, pp 146-160.

Rosenfeld A and Kak A (1982), Digital Picture Processing, 2nd ed., Academic Press, New York.

Windham JP, Abd-Allah MA, Reimann DA, Froelich JW, and Haggar AM (1988), Eigenimage Filtering in MR Imaging, J-CAT, Vol 12-1, pp 1-9.

TOWARDS AUTOMATED ANALYSIS IN 3D CARDIAC MR IMAGING

M Bister[1], J Cornelis[1] and Y Taeymans[2]

[1]IRIS Research Group, VUB, Pleinlaan 2, 1050 Brussels, Belgium

[2]Dept. of Cardiology, 5K12IE, UZ-RUG, De Pintelaan 185, 9000 Gent, Belgium

Abstract

The main purpose of the work described in this paper is to make a first step towards the automatization of the quantification of the ventricular volume in systole and diastole using MR Images.

To achieve this result, we pursue three partial objectives:

1. Obtain objective image segmentation. Manual organ delineations vary from physician to physician. An automatic segmentation, taking into account the typical nature of cardiac MR Images, should produce objective and reproducible results and take less time than manual segmentation.

2. Obtain reliable segment labeling. A computer system, which takes into account descriptions of the organs (scene knowledge), has to be developed to assist the physician in labeling the segments produced by the automatic segmentation. Interactive tools should be provided to show the results of segmentation and labeling to the physician and ask him for confirmation or corrections.

3. Obtain accurate volume measurement. Volume measurements will allow the evaluation and partial validation of the results obtained by the previous parts of the system.

The paper describes a prototype of a complete system for cardiac volume estimation. Detailed descriptions of the individual segmentation and labeling modules have been published previously. In this paper the emphasis lies on the interaction between these modules, their performances in the system, their 3D generalisation, and their evaluation based on cardiac volume estimations.

Keywords

Image analysis; segmentation; labeling; multiresolution pyramids; distance transforms.

1. INTRODUCTION

For each of the three partial objectives discussed in the abstract, we have analyzed one or more solutions.

1. Obtain objective image segmentation. A first tentative solution for this problem consisted of multiresolution pyramid image segmentation schemes (Bister, 1990d). Since these methods have serious drawbacks (Bister et al., 1990a) – which is briefly addressed in § 2.1 – we had to design another segmentation scheme to detect the heart cavities, called the Cavity Detector (CD) (Bister et al., 1989), (Bister et al., 1990c). This algorithm is discussed in § 2.2.

2. Obtain reliable segment labeling. A three-phased system based on a Probability Distribution Map (PDM) was developed, which allows not only to use part of the physician's knowledge about the organs, but also to acquire it in a user-friendly way (Bister et al., 1990b). This system is discussed in § 2.3.

We do not believe it is possible to automate the analysis of cardiac MR Images completely. Human interaction in this field is too complex to be replaced by computing machines. Moreover, human beings have an ability to tackle unexpected situations, exceptions, etc. which is hard to implement efficiently on a machine. Therefore, we have developed a system which is intended to *assist* the physician rather than to *replace* him. Our goal is to achieve more objectivity and reproducibility in the estimation of heart volumes and in the calculus of the stroke volume in comparison to the current manual procedures.

The system proposes a segmentation, which is submitted to a "control and labeling system" that can be refined gradually. In the beginning control and labeling is done mainly by the physician himself and gradually it is automated.

3. Obtain accurate volume measurement. In § 2.4 we discuss the pseudo-3D generalisation of the segmentation and labeling – the way to take into account the information in adjacent slices – and the algorithm used to compute the heart volumes. In § 3 results are shown, in § 4 they are discussed, and in § 5 conclusions are drawn.

2. METHODS

2.1. Pyramid Segmentation

At first view multiresolution schemes seem to be well adapted to medical image interpretation, where one is often confronted with the problem of detecting very small regions (heart valves, tumours,..) in the presence of much larger objects s.a. heart cavities, lungs, etc....

Although the literature reports that pyramid segmentation algorithms are successfully applied in a large number of fields (Rosenfeld, 1984) some of their basic properties have not been systematically investigated, such as shift-, scale- and rotation-invariance.

We studied the pyramid image segmentation algorithm thoroughly and developed several generalisations (Bister, 1990d). Some fundamental limitations of the pyramids such as their limited number of segments and the constraints on the shape of the segments were avoided by the development of a relaxation pyramid which supports brother-brother interaction on each level (Bister et al., 1990a).

Nevertheless, in the current state of the art, pyramids have to be rejected as general-purpose image segmentation algorithms, due to several severe problems such as shift-, rotation- and scale-variance, which are caused by:

- the sub-sampling introduced at higher levels of the pyramid to reduce computational cost and to transform global processing at full resolution into local processing at reduced resolution;
- the rigidity of the pyramid structure (Bister et al., 1990a).

2.2. The Cavity Detector

2.2.1. Introduction

The CD is especially designed to segment images of objects which are not completely surrounded by walls. This often occurs in medical images because of imaging artefacts (e.g.: the right ventricular wall is sometimes only partly visible because of movement during the MRI read-out gradient or due to partial volume effects), or because of the physical interruption of the cavity walls (e.g.: blood vessels connect the left atrial cavity and the lungs, and the ventricles and atria are connected when the valves are open). A segmentation scheme for cardiac MR Images should take care of this typical problem.

The only assumptions on which the CD is based are:

- segmentation in compact regions is required;
- the images have a high Signal-to-Noise ratio, a sufficient contrast between objects and background and sharp edges delimiting the objects.

Several generalisations have been implemented without changing the general principles to ensure good segmentation in the presence of elongated regions (§ 2.2.5), smooth edges (§ 2.2.6), spatial variations in gray value and/or contrast (§ 2.2.7). In § 2.2.8, the direct interaction of the CD with some pre- and postprocessing image transformations is discussed. Although the main subject of this paragraph is the CD as a tool for image segmentation, § 2.2.9 describes some other functions of the CD.

The CD is applicable in many other imaging domains such as X-ray CT scans of the brain, cardiac ultrasound images and even images of industrial scenes under structured illumination. In (Bister et al., 1990c), the generic nature of the CD is discussed.

2.2.2. General Principles

The Cavity Detector is a heuristic segmentation scheme which is based on the (supervised) sequential application of four operations, to which we attribute the following mnemonics:

1) the *Discriminator*, which makes a rough difference between the regions representing objects and the background;
2) the *Detector*, which determines a kind of "centres of gravity" of the regions, which are used as "roots" for the segments – each "root" producing a different segment;
3) the *Separator*, which delineates the segments;
4) the *Accurator*, which improves the borders of the segments (although logically the *Accurator* occurs last in the processing sequence, in the implementation it can be used either as a pre- or a postprocessing step – see § 2.2.6).

The typical problem handled by the CD is illustrated in Figure 1.

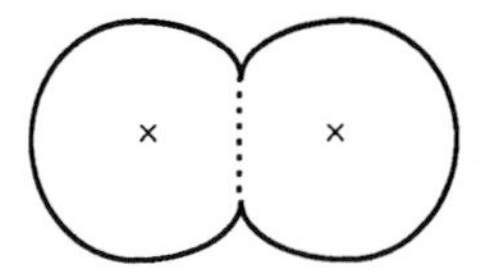

Figure 1. Illustration of the role of the Cavity Detector.

Both cavities are only partly separated by a wall. The objects and background are already separated (*Discriminator* – full line). The centres of gravity of both cavities are detected (*Detector* – crosses). Finally, both cavities are separated (*Separator* – dotted line). The *Accurator* step is not illustrated since in this example the input figure is binary and the borders of the regions are smooth.

2.2.3. Implementation

The *Discriminator* is realized by a global thresholding technique[1] followed by a logical filter which eliminates small objects, taking into account their gray value, size, shape and position in the image[2].

The *Detector* is realized by taking the local maxima of the Distance Transform[3] (DT) of the output image of the *Discriminator*. These are points which are far away from all walls. Even if a wall separating two cavities is only partially imaged (or if it is physically interrupted), the DT will have two local maxima, and hence it will produce two "roots". Since the DT is sensitive to noise which produces irregularities in the borders of the regions, one cavity can contain many local maxima lying very close to one another. In order to eliminate these spurious maxima, a second logical filter merges the maxima for which the sum of the heights is much larger than the geometrical distance between them.

The *Separator* consists of a conditional growth of the roots, i.e. the segments are grown along the steepest paths in the distance transformed image, hence the valley between two local maxima becomes the border of two adjacent segments.

The *Accurator* refines the borders between segments or between a segment and the background. Several techniques were tested (e.g. edge detection in the neighbourhood of the borders of the segments). In the case of high contrast imaging modalities (such as T1-weighted cardiac MRI), the use of the *Accurator* can be omitted.

All our software is developed in the C-language and under UNIX. Currently the analysis of a single 256 x 256 image requires approximately 1 minute on a SUN 3/60. Processing time is almost independent of the number of detected segments.

2.2.4. Results

A typical result is shown in Figure 2. We observe over-segmentation: some (elongated) organs are split into several segments. On the other hand, adjacent organs are almost never merged into one segment. In the case of over-segmentation, the different segments forming the complete organ have to be merged, either by the user, either by the control & labeling scheme. How the CD detects cavities is illustrated with the left atrium in this image (the elongated, horizontal cavity in the middle of the image): after the *Discriminator* step (middle left) this cavity is connected to both lungs (large cavities left and right of the left atrium), while after the *Separator* (down right) these three cavities are disconnected.

The robustness of the segmentation was tested with respect to changes in the parameters of the algorithm. When varying the mask for computing the DT, the threshold for the logical filter of the *Detector* and/or the *Discriminator* threshold, changes are usually small, and the automated choice of the threshold value is usually good (Bister et al., 1989), (Bister et al., 1990c).

---

[1] The thresholding technique is a modified Minimum Error Method (Kittler and Illingworth, 1986). We choose the threshold so as to minimize the mean square error between the input gray value and the average gray value over the classes.

[2] This logical filter is a modified salt-and-pepper filter: if a segment in the binary image has an area lower than a threshold, its value is switched (from black to white or from white to black). This condition is relaxed by allowing the threshold to be higher for elongated segments and for segments which are closer to the centre of the image, since these segments are more likely to be meaningful. Moreover, different thresholds can be defined for bright and for dark segments.

[3] The Distance Transform attributes to each pixel the value of (a linear function of) the shortest distance from that pixel to a "border" in the input image - typically, such a "border" is a value transition in a binary image. The most well-known DT is called "chamfer distance" or "weighted distance" (Borgefors, 1986).

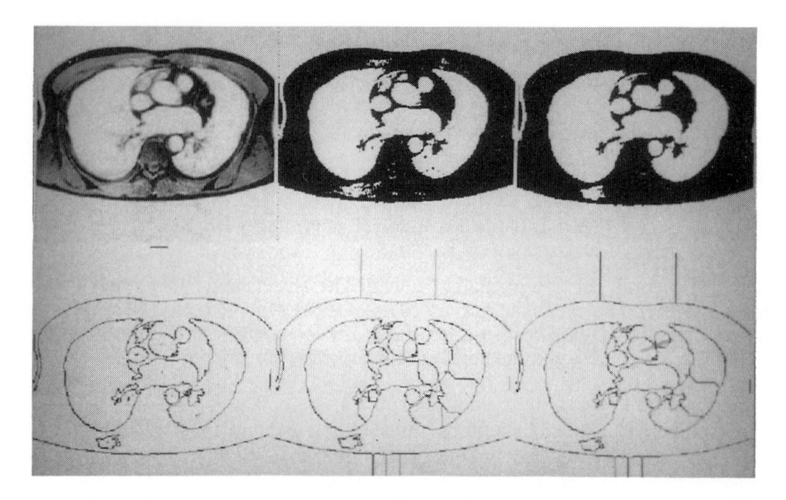

Figure 2. Top left: cardiac MR Image (input image); Top middle: result of the *Discriminator* before logical filtering; Top lef: result of the *Discriminator* after logical filtering; Bottom left: result of the *Detector* without *Detector* filtering; Bottom middle: result of the *Separator* without *Detector* filtering; Bottom right: result of the *Separator* with *Detector* filtering.

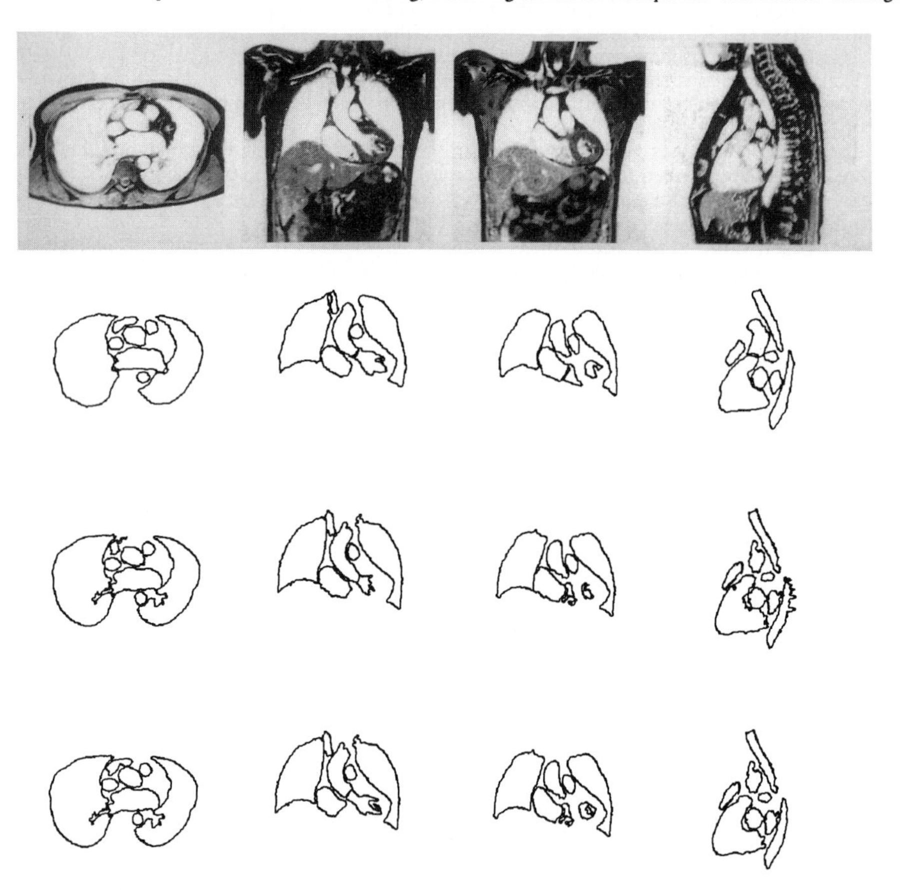

Figure 3. Validation of the CD in cardiac MR Images. First row: original images. Second row: manual segmentations; Third row: automated segmentations; Fourth row: manually corrected automated segmentations

For the validation of the CD, we compared manual segmentations by the cardiologist to automated segmentations. The results for four typical images are shown in Figure 3. In almost all the cases, the automated segmentation is superior to the pure manual one. Usually no organs are merged into one segment and some organs are split into several small segments. These segments have to be merged by a control & labeling scheme based on models or knowledge of the application area. The main corrections applied by the cardiologist to the automated segmentation are simplifications of the borders of the segments and some knowledge-based corrections (e.g.: due to trabeculae, the gray value threshold of the *Discriminator* should be put somewhat higher inside both ventricles; the limit between left ventricle and aorta is not exactly at the narrowest point in the structure, but somewhat higher towards the aorta). Smoothing of the borders of the segments produced by the CD will be discused in § 2.2.8. The corrections of the segments based on semantic information have to be achieved by the control & labeling scheme. Up till now, our processing chain ends with the labeling of the segments. However, it is intended to implement an automatic feed-back from the labeling to the segmentation via a rule based control algorithm which will take into account the knowledge-based corrections which are now accomplished by the cardiologist.

2.2.5. Generalisation for Elongated Objects

The detection of elongated regions (e.g. blood vessels in transverse direction) by the CD requires a modification in the second step of the algorithm (the *Detector*). In the *Detector*, a compact object is characterized by a point which is a local maximum of the DT. If an object is elongated, the extremum degenerates into a curve of constant height (a ridge). When we detect this ridge, we can also detect the elongated object.

A more detailed description of this generalisation is given in (Bister et al., 1990c).

2.2.6. Generalisation for Smooth Edges

In this paragraph, an *Accurator* for preprocessing is described. It consists of a steepest path growing algorithm applied to the reverse of the "edge image" (i.e. an image of the square of the gradient magnitude of the input (gray scale) image). Local maxima in the reverse of the edge image correspond to regions where the gradient has a local minimum, hence to homogeneous regions. We define these maxima as roots.

To each root, we attribute the gray value of the corresponding point in the original (gray scale) image. We then link pixels to one another along the steepest path in the reverse edge image, until a pixel is linked to a root. This procedure defines a (micro)segment containing all pixels which are linked to the same root. To all pixels belonging to the same (micro)segment, we attribute the gray value of its root.

The *Accurator* is illustrated for a 1D signal in Figure 4.

The *Accurator* has two properties – one positive and one negative – which are worth mentioning. The positive one is that, for regions with a linear variation of the gray value (which are very common in natural scenes), the second derivative is zero and hence s (x) is also zero. For such regions, pixels are linking to themselves, which means they are their own roots and getting their own gray values: regions having a linear variation of the gray value are not destroyed by the *Accurator* (see region "xxx" in Figure 4).

The negative property occurs when we have a roof-like edge profile, in which case the first derivative has a discontinuity but will not become zero, hence we will not have a root at this place and this edge will be destroyed by the algorithm (see region "yyy" in Figure 4). One way to avoid this is by forcing continuity, e.g. by applying preprocessing such as Gaussian smoothing. Another, more elegant way, is to link pixels only along paths with continuity of the gradient sign – i.e. if two opposite neighbours of a pixel have both a higher gray value or both a lower gray value than the pixel under consideration, this pixel will not link to either of both neighbours. In this way pixels on top of roof-like edge profiles will not link along these profiles. This means that details of 1 pixel thick will be preserved. Hence we call this procedure the "detail" option or "d-option".

Let us now consider how the *Accurator* fits in the framework of the CD. Since we apply the *Accurator* to the input image before the *Discriminator*, we can be sure that the borders of the *Discriminator* will always correspond to points where the gradient has a local extremum. Hence the *Accurator* realises the function for which it has been designed.

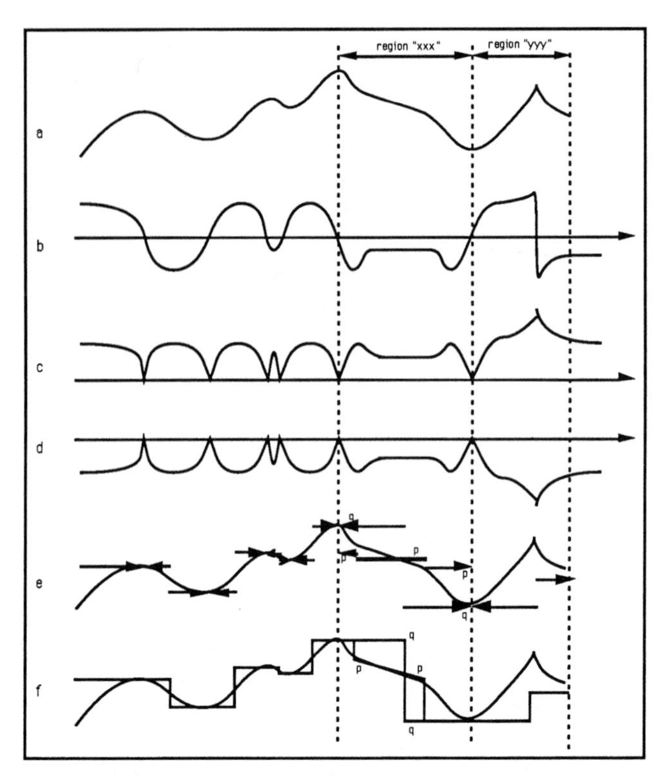

Figure 4. Different steps of a steepest path growing algorithm along the reverse of the edge transform to a 1-D signal (region "xxx" illustrates the constant derivative situation; region "yyy" illustrates the problems with a roof-like profile = discontinuity in first derivative). a) Input signal; b) derivative of the input signal; c) edge signal = square of the derivative; d) reverse of the edge signal; e) growing direction along the input signal (region "xxx": p and q are two alternative approaches: take each pixel in zero-derivative region as root (p), or grow along most nearby non-zero derivative (q)); f) output signal compared to input signal.

2.2.7. Generalisation for Spatial Variations in Image Quality

To cope with spatial variations of mean gray value and contrast in the image, several generalisations of the implementation of the *Discriminator* (§ 2.2.3) are possible, taking into account that only the borders of the resulting regions are being used by the subsequent subsystem. As *Discriminator* one can use: multilevel thresholding, edge detection, local thresholding, or a *Discriminator* adapted to the properties of the *Accurator* described in § 2.2.6.

Let us describe this last option in more detail. After the application of the *Accurator* (§ 2.2.6), we have a huge number of (micro)segments. Of course, these (micro)segments cannot be considered as valuable segments, since adjacent segments have sometimes gray values which are only slightly different, while logically the borders of the *Discriminator* should lie on transitions between (micro)segments with large differences in gray value. The *Discriminator* can be realised by merging adjacent (micro)segments with gray value differences lower than an a priori defined edge- or merge-threshold. The segments obtained at this point are either submitted to a global thresholding, or directly used as the input to the *Detector*.

2.2.8. Generalisation for Pre- and Post-Processing

For the cardiologist the borders of the segments obtained by the CD are too ragged (see § 2.2.4). To avoid this, we propose two types of approaches:

1) postprocessing, to wipe out the spurious details in the resulting segmentation – e.g. median filtering of the segmented image, or polynomial smoothing of the borders (e.g.: splines, as in (Eiho et al., 1987));

2) preprocessing, to reduce the amount of unnecessary details and noise in the input image – e.g. Gaussian blurring (Koenderink, 1984). The CD including *Accurator* gives both smooth and reliable borders when applied on images which have been preprocessed by a Gaussian smoother (Bister et al., 1990c).

2.2.9. Algorithms Derived from the CD

Until now, the CD was described as a tool for image segmentation. In this paragraph, we describe other functions for the CD, applicable in different fields of image processing.

2.2.9.1. Edgmentation

The *Accurator* (§ 2.2.6) – which grows regions taking into account edge information – , followed by the generalized *Discriminator* (last part of § 2.2.7) – which merges adjacent (micro)segments according to the difference in gray value between them – can be considered as a new segmentation algorithm on its own, comparable to the well-known split-and-merge algorithm (Horowitz and Pavlidis, 1976). The *Accurator* does the splitting operation, the *Discriminator* does the merging. Compared with the classical split-and-merge algorithm (which has the disadvantage of unnatural squarish borders of segments due to the typical splitting into quadrants (Haralick and Shapiro, 1985)), this new algorithm splits the image according to edge information, giving much more natural segments. In fact, we can consider this technique as a combined segmentation and edge detection/enhancement algorithm – hence the name "*Edgmentation*" for this new stand-alone algorithm.

Edgmentation has been implemented with an optional preprocessing step (Gaussian blurring), and can be tailored to specific applications by five distinct parameters: the sigma value for the Gaussian blurring, the criterion for root selection (strict or local extrema), the type of edge detector (Sobel, Prewitt, ...), the way roof-like profiles are handled, i.e. the "detail" option (see § 2.2.6), and the threshold for merging adjacent segments.

The results of the *Edgmentation* operator, applied on various types of medical images, are shown in (Bister et al., 1990c). One example for cardiac MRI is shown in Figure 5.

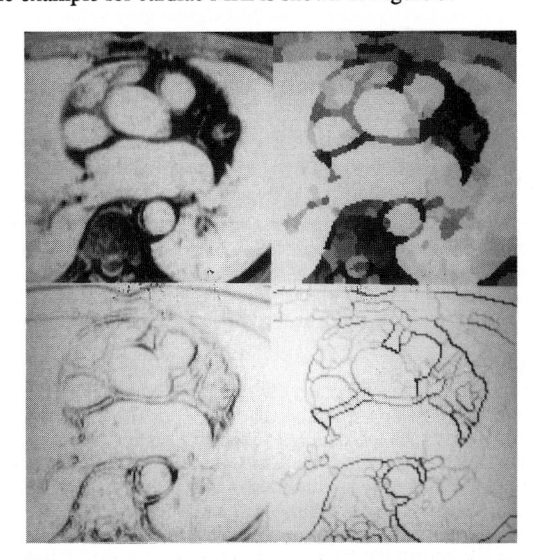

Figure 5. *Edgmentation* applied on a cardiac MR Image, with settings: sigma = 1, local extrema, 2x2 edge detection, detail option, merge threshold = 5. Top left: input image; top right: image after *Edgmentation*; bottom row: edge transform of the top row.

2.2.9.2. Region Classification

We can consider the *Accurator* alone (i.e. the *Edgmentation* with edge threshold equal to zero) as a generator of "atomic regions", which can be the basis for several region classification algorithms, such as statistical classifiers, AI labeling algorithms, etc. An example of an AI labeling algorithm is the PPL system (Harwood et al., 1988), which makes use of the SNF filtering technique (Harwood et al., 1984) to generate such atomic regions. Hence, we compared our *Accurator* with these SNF filters (Bister, 1990d).

The advantage of *Edgmentation* is that no iteration process is required. The *Edgmentation* process and the SNF filters have comparable results, although the approach is quite different.

2.2.10. Conclusions

The above considerations led us to believe that the CD might be a well-suited solution to the detection of heart cavities in MR Images. Moreover, several generalisations are possible, making this same technique useful for the segmentation of different types of medical images. The results of the CD in the case of cardiac MRI were validated by comparison with manual segmentation by a medical expert, and were often superior.

The Cavity Detector (CD) is a time-efficient algorithm for the segmentation of various types of biomedical images, including cardiac MRI. It is based on classical image processing techniques and it is particularly useful in solving often encountered problems in medical images, namely the bad or incomplete representation of walls and the presence of objects which are not completely surrounded by walls.

A new stand-alone segmentation and edge detection/enhancement algorithm was developed within the CD, namely "*Edgmentation*". It is basically a split-and-merge algorithm, taking into account edge information instead of gray value statistics. It can be used to enhance edges, as an *Accurator/Discriminator* in the CD, or to produce "atomic regions" for subsequent region classification by statistical algorithms or expert systems.

2.3. A Generic Labeling Scheme

2.3.1. Introduction

For the labeling of medical images, physicians use large amounts of knowledge. To transfer even a small fraction of this knowledge to machines is a very difficult task, because most of it is used implicitly and its explicit formulation requires important efforts. We have adopted an approach where part of the knowledge is extracted from the measurement data in an interactive learning phase, and part of it is explicitly formulated in rules (Bister et al., 1990b).

The labeling scheme introduced in this paragraph is applied on segmented images (e.g. images obtained after the application of the Cavity Detector to the original gray value images).

The labeling scheme consists of three steps:

1. The first step standardizes the image position (§ 2.3.2);
2. The second step checks the standardized image against spatial knowledge about the positions of anatomical objects (§ 2.3.3);
3. The third step generates a proposal for the labeling (§ 2.3.4).

2.3.2. Position Standardisation

On the basis of simple rules, implemented in C and involving position, relative position, size, gray value and adjacency of segments, the background, the lungs, the spine, the aorta descendens (AoD), the heart muscle and the heart cavities are identified. The centroid of the AoD is computed, as well as an estimation of the moments of inertia of the heart with respect to this centroid. The image is linearly transformed into a standardized position (shift of the origin to the centroid of the AoD, rotation of the axes according to the main axes of inertia, and scaling according to the polar momentum of inertia).

2.3.3. Probability Distribution Map

The segments are matched with a Probability Distribution Map (PDM), i.e. a map reflecting in each pixel the probability for each organ to be located in that pixel position. The overlap between the segment and the PDM for different organs defines a Labeling Probability Matrix (LPM).

Several different segments can get the same label, insofar as they all correspond to positions in the image with high PDM values for that organ. This means that the over-segmentation, observed when for example the Cavity Detector is used, is not a problem and that labeling of over-segmented organs is achieved without the classical combinatorial explosion problem.

The PDM can be simply computed by manual labeling of a set of images (learning group), and computing local probabilities from these labeled images. We developed a user-interface which allows easy manual labeling, and also fast correction of automatic labeling, so that newly labeled images can be added to the learning group.

As said previously, after matching the segmented image with the PDM, each segment gets a vector of probabilities to belong to each organ. This defines a Labeling Probability Matrix (LPM). Of course, this LPM takes only into account geometric model information. If we want to take into account other types of information, such as the number of occurrences of the organs (e.g. the aorta has two occurrences in transversal images, namely the aorta ascendens – the AoA – and the AoD), we must process the LPM in a more or less intelligent way.

2.3.4. Labeling Probability Matrix

The most simple way to analyze the LPM is of course to assign to each segment the label for which it has the highest probability.

We are currently experimenting with a procedure for analyzing the LPM which takes into account the cardinality of each label. The PDM itself contains implicitly information about the number of occurrences of the organs. If an organ occurs more than once, there will be two local maxima in the PDM for this organ. We can take advantage of this by detecting the number of local maxima for each organ, hence deducing the cardinality of the organs from the PDM itself.

2.3.5. Performances

The results of the rule-based labeling for the standardization of the position (the correct identification of the AoD, and the approximate identification and delineation of the heart region) are sufficient to align the images (Figure 6).

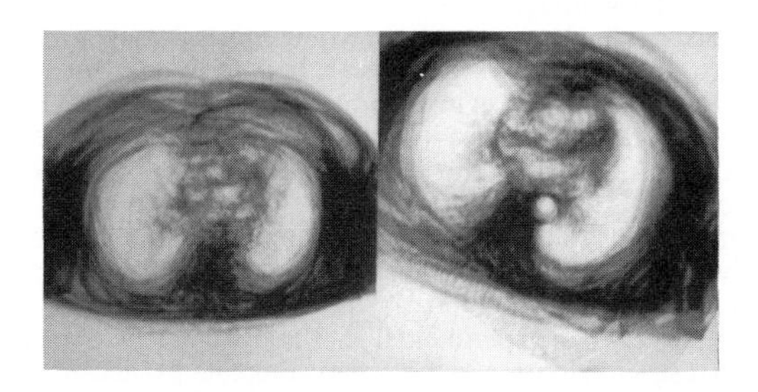

Figure 6. Image of the mean gray value calculated from 11 images with the same anatomical position in 11 different patients, without (left) and with (right) standardization. of the position.

The results of the most simple processing of the LPM (Figure 7: Colour Plate, p. 4) are generally in agreement with the manual labeling. Some improvements can be expected by taking into account the number of occurrences of the organs. Nevertheless, the matches are satisfactory, taking into account the large variability in thorax geometries.

As a conclusion, we can say that the results of the PDM labeling are promising, but will have to be improved in the future.

Although not using typical AI tools, the system makes use of some AI concepts such as trainability and separation between model (PDM) and "reasoning engine".

2.4. Pseudo-3D Generalisations

The three elementary steps of our system for the segmentation and labeling of MR Images have been described previously: preprocessing (with the *Edgmentation* program – i.e. the *Accurator* step of the CD), segmentation (with the CD) and labeling (with the PDM). The programs are developed in 2D; more-D extension is straightforward. In this section, we will discuss the pseudo-3D extensions of *Edgmentation*, the CD and the PDM. With "pseudo-3D" we mean a set of 2D slices out of a 3D object - i.e. a 3D image with coarse resolution along one dimension.

2.4.1. *Edgmentation*

Edgmentation implies the application of 5 successive operators on the image, namely: 1) Gaussian blurring, 2) edge detection, 3) extrema detection, 4) steepest path region growing, 5) merging.

Only steps 1, 2 and 4 use the between-pixel distance. These three steps were modified to make a distinction between within- and between-slice adjacency. The distance between 4-connected "within-slice neighbours" is equal to 1, while the distance between 4-connected "between-slice neighbours" is equal to the within-to-between sampling rate ratio R.

The results in (Bister, 1990d) show that the consistency of the results from one slice to another is improved. Pseudo-3D *Edgmentation* can be used for stand-alone image enhancement, especially adapted to the problem of pseudo-3D images with poor consistency and strongly variable image quality from one slice to the other.

2.4.2. The CD

To take into account the ratio R in the case of pseudo-3D images, remember that the CD consists of 6 successive operators (not taking into account the *Accurator*, which was discussed in previous paragraph):

1) thresholding, 2) filtering of small connected components, 3) Distance Transform, 4) extrema detection, 5) filtering and merging of adjacent extrema, 6) steepest path region growing.

Provision was made to make a difference between within- and between-slice adjacency for steps 3 and 6. For the pseudo-3D DT, we followed the same reasoning as in (Vossepoel, 1988).

Results of the pseudo-3D CD show consistency from one slice to the other. On the other hand, there is much more over-segmentation in pseudo-3D than in strict 2D (Bister, 1990d).

2.4.3. The PDM

In the labeling process, the first step is to standardize the image position. For this standardisation, we only need one transversal image located anatomically just above the LV. On the basis of this single slice, the complete set can be standardized in position. The pseudo-3D set itself can be a transversal, frontal or sagittal set, as long as the geometric parameters are suitably recorded and passed to the appropriate programs. In our programs, provision was made to read these informations automatically in the headers which are associated to each of these images.

After standardisation, each slice can be matched individually to the PDM, and evidence is accumulated from one slice to the other for each segment. In other words, one global LPM is computed for the whole set of slices, although each slice is matched individually with the PDM. The processing of the LPM doesn't take any supplementary geometric information into account, since the LPM itself contains all this information, hence the third step in the labeling process (§ 2.3.1) does not change with the dimensionality of the image set.

It is once again advantageous to use a pseudo-3D approach to the labeling whenever possible, since evidence is accumulated over several geometric levels, hence the labeling automatically takes into account the information in adjacent slices.

Two causes for the relatively bad experimental results (Bister, 1990d) can be pointed out:

1) the learning group is too restricted (currently, only two complete pseudo-3D image sets);
2) the rules for processing of the LPM are too restricted (see § 2.3.4).

The software discussed in § 2.4.3 is only a first step towards an automatic pseudo-3D labeling algorithm for cardiac MR Images. Further work will have to be done.

2.4.4. Volume Estimation

To estimate a volume from a set of (manually or automatically defined) segments in a set of slices, the most simple algorithm is to estimate the area of the segment in each slice (e.g. by simple pixel-counting), and to add them according to Simpson's rule. This is the procedure we applied.

We validate the results of this automatic volume estimation as follows

1) for each test person, we estimate the volume of LV and RV at several moments during the cardiac cycle; these volumes should vary smoothly in time;
2) for each test person, we estimate the volume of LV and RV in both diastole and systole, and compute the stroke volume for both LV and RV – they should be equal;
3) for two test persons, we did four different complete image acquisitions, at intervals of one week; for each test person the four results should be comparable since no major physiological changes occurred;
4) we asked a cardiologist to analyze manually the images of each test person in systole and diastole (manual tracing and labeling, and volume estimation with Simpson's rule), and compared this to the results of the automated analysis.

Results of these validation procedures are given in § 3 and discussed in § 4.

3. RESULTS

To test the consistency of the automatic results of the heart volume estimations in time, we estimated the volumes of LV and RV at several moments during the heart cycle[1]. We did this on recordings of the same healthy person at a week interval: the four curves should be comparable. Typical results for LV and RV are shown in Figure 8.

---

[1] Since automatic 3D labeling is not yet completely reliable, we decided to label the automatically defined segments manually.

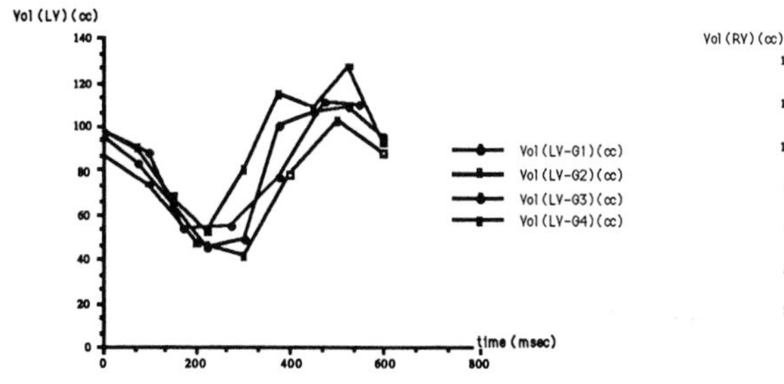

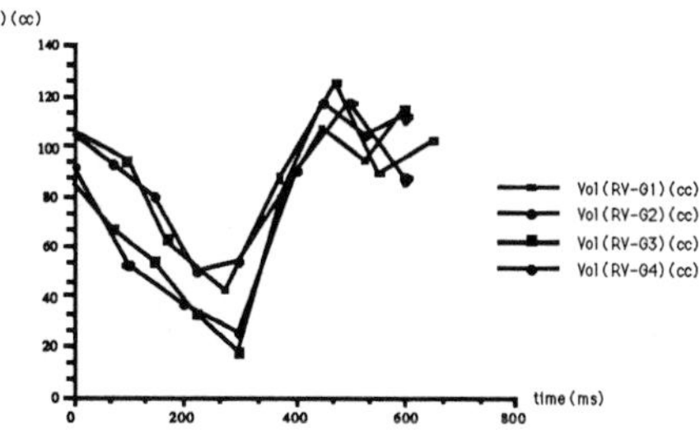

Figure 8. Volume of the LV (left) and RV (right) in function of time during the heart cycle for four recordings (G1,...G4) of the same (healthy) person at a week interval.

Numerical results for Diastolic and Systolic Volumes (V_D and V_S) and Stroke Volume (SV) were computed for LV and RV, as well by manual as by automatic segmentation[1]. Results are given in Table 1.

Table 1. Volumes of LV and RV in systole and diastole and Stroke Volume, as well for manual as for automatic segmentation, for four different recordings of the same test person.

	LV						RV					
	V_D (cc)		V_S (cc)		SV (cc)		V_D (cc)		V_S (cc)		SV (cc)	
	man.	aut.	man.	aut.	man.	aut.	man.	aut.	man.	aut.	man.	aut.
G1	88	95	40	49	48	46	69	86	17	17	52	69
G2	73	87	28	42	45	45	62	91	17	25	45	66
G3	75	98	30	54	45	44	62	105	24	43	38	62
G4		98		52		46		103		50		53

4. DISCUSSION

Time consistency of the results is quite good: within the heart cycle, the curves are quite smooth, except towards the end of the cardiac cycle; from one week to another, the same remark holds: good consistency during the first part of the cycle, less consistency towards the end of the cycle.

This is in accordance to what can be expected from cardiac MRI:

1) Image acquisition is triggered. This means that, at the R-peak in the ECG, the image acquisition is started, and images are acquired at regular time intervals from the trigger pulse on (e.g. every 100 ms). If the heart rate is regular (for example the R-R interval is equal to 900 ms), then the first image (TD = Time Delay between the trigger pulse and the image acquisition ≈ 0 [2]) is recorded at the beginning of the heart contraction, the fourth image (TD = 300 ms) at one third of the cardiac cycle, the seventh image (TD = 600 ms) at two third of the cardiac cycle, and the tenth image (TD = 900 ms) would be exactly at the same position in the heart cycle as the first. However, the heart rate varies during image acquisition, and even from one beat to another. The first image is always synchronised with the heart movement due to the triggering. But the images in the last part of the cycle will be recorded at the same absolute time delay from the trigger pulse on, but the position of the heart at that moment will be different. For example, if the R-R interval changes from 900 ms to 1200 ms (which is not uncommon, even in the absence of arrythmia), the first image (TD = 0) will still be recorded at the beginning of the heart contraction, the fourth image (TD = 300 ms) will now be recorded at one fourth of the cardiac cycle, the

[1] Manual segmentation was only performed on the Systolic and Diastolic images, so that a comparison of V_D, V_S and SV can be done. However, since all of the images in the cardiac cycle were not manually segmented, the manual results could not be plotted in function of time as in Figure 8.

[2] In reality, the time delay is at least equal to TE (Echo Time); this is visible in Fig. 6.1 and 6.2: the first image is taken at least 50 ms after diastoly.

seventh image (TD = 600 ms) at the half of the cardiac cycle, and the tenth image (TD = 900 ms) will only be at three fourth of the cycle.
2) Since images are acquired one line per heart cycle, this means that, for the images recorded late in the cardiac cycle, each line will be recorded at a slightly different position of the heart. Since image reconstruction, starting from the recorded data, requires the application of a Fourier transform, these irregularities will degrade the overall image quality as well as the exact time-information of the images, resulting in a degradation of the reliability of the estimated volumes.

Comparison between results of manual and automatic segmentation (Table 1) are quite good for the LV, and the most important physiological parameter – the Stroke Volume – is in excellent accordance from one week to another, as well as between manual and automatic segmentation. The results of the automatic segmentation for the individual volumes are usually somewhat higher than the manual results. This is due to the fact that the automatic segmentation takes into account all the trabeculae, and includes the inter-trabeculae volumes in the calculation of the ventricular volumes, while the physician has a tendency to neglect these volumes. They do not change much to the volume differences (the Stroke Volume), as was confirmed by the results.

For the RV however, the results are worse, and especially there is no agreement between the SV of the LV and of the RV. It should be noted that, although the manual results for the RV are better than the automatic results, they are still worse than both the manual and the automatic results for the LV. This is due to the fact that we used frontal images for this validation work. This orientation is not ideal for the analysis of the RV, because the tricuspid valve can not be easily seen in these images – while it is this valve that delimits the RV. Even the physician has many difficulties in delineating exactly the interface between RV and arteria pulmonalis – which is evident from the variability in the SV from one analysis to the other. It is logical that an automatic system – which does not use the large amount of knowledge which the physician uses to "guess" the position of the tricuspid valve – performs worse.

To the question as to why we did not use another orientation for the recordings, the answer is simple: to keep the acquisition time reasonable. We did transversal recordings because the back-to-front axis is the one which requires the less number of images to cover the LV and the RV completely. The better transversal orientation would require approximately twice as much slices, hence doubling the acquisition time, which was already approximately 2 hours[1].

It can be noted from the literature[2] that RV analysis is always much harder than LV analysis, hence we are not amazed, though disappointed, that this same rule holds for cardiac MR Imaging.

Even with manual labeling, it was hard to make the distinction between RV and RA. This is due to the fact that the interface between the RA and the RV – the tricuspid valve – is small (typically a couple of mm thick), moving, and parallel to the slices. Hence it disappears completely due to movement artefacts and partial volume effects. When looking, for example, to the left slice of Figure 9, we see clearly the 2D narrowing between RA and the RV. In the right slice also, we see this narrowing. The position of the interface between RV and RA is completely different in the two slices. In the middle slices we expect to see a whole interface region (Figure 9 – region "??"). This did not happen due to partial volume effects, hence the CD considers the lower part of the RA in the left slice and the upper part of the RV in the right slice as one segment.

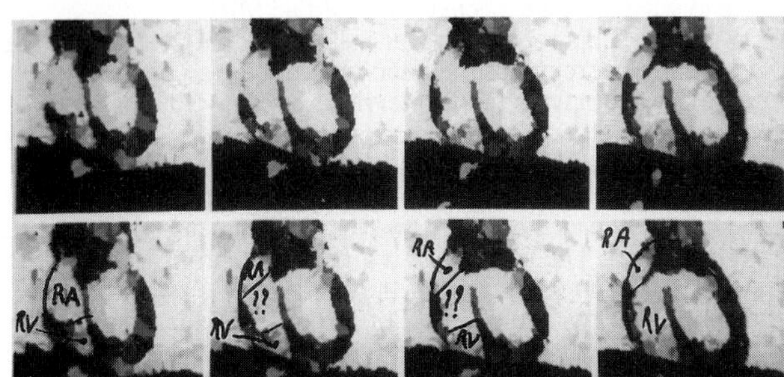

Figure 9. Four slices from a transversal series, illustrating the problem of missing the interface between RA and RV due to partial volume effects. Top row: original image; bottom row: labeled image. Four successive slices are visualised. The region labeled "??" should correspond to the interface between RV and RA, but this is not visible in the image itself.

[1] We used a rotated gated acquisition of T1 weighted MR Images, resulting in multi-slice, multi-phase images (Bister, 1990d).

[2] see Chapter 6 in (Yang et al., 1978) and Chapter 11 in (Weyman, 1982).

Partial volume effects are quite severe since the thickness of the slices is 8 mm (with a spacing of 4 mm between two adjacent slices). Smaller slices tend to be more noisy. Hence the only solution would be to take a better orientation. This is impossible due to limitations on the recording time.

Anyway, we can say that, for LV analysis, our automatic results are in good agreement with manual segmentation. Results of RV analysis however are not very reliable, mainly due to the image acquisition (slice orientation, partial volume effects, prolonged acquisition time).

5. CONCLUSIONS

While reading §§ 3 and 4, the reader might get the impression that the results for quantitative analysis of heart volumes and calculation of the stroke volume are rather negative. However, they have to be compared to what is available from other techniques. Typically, most of the volume computations in cardiology are done using one single, though ideally oriented slice, and several volume computations formulae. Their shortcomings are mainly evident when a deformation of the geometric model of the heart occurs due to an infarction (Weyman, 1982). Only the Simpson formula – which we used – is really satisfactory. Its application however requires an image quality which can nowadays only be achieved by MRI.

The advantages of MRI in cardiology are: excellent tissue contrast, spatial resolution, pseudo-3D capacity. Important disadvantages are the intrinsic dynamic limitations of MRI resulting in prolonged acquisition time and unavoidable partial volume effects due to the trade-off between slice thickness and signal-to-noise ratio.

Though MRI is more reliable for cardiac volume computations than other methods, it can only be useful as a reference method. Routine work with MRI will not be possible due to the prolonged acquisition times for images of diagnostic quality.

The main positive conclusion of this work is that computer analysis of cardiac MR Images is successful for LV stroke volume computations. For RV computations, results are less satisfactory (like in all other medical imaging methods). Further experiments can be done by using other orientations (which will cause longer acquisition times), and by using more model information during the segmentation (or re-segmentation) phase (interaction between segmentation and labeling modules). Future research will also be conducted towards the localisation and the quantification of infarctions.

Acknowledgements

The work was supported by FGWO grant nº 3.0054.88, IWONL grant nº 870424 and by the Belgian Ministry of Science Policy (RFO/AI/11). It was realized with the active support of PRIMIS. Many thanks to Prof. A. Rosenfeld and to Dr. N. Friedland (University of Maryland) for their many suggestions for the elaboration of the CD.

References

Bister M, Taeymans Y and Cornelis J (1989). Automatic segmentation of cardiac MR Images. In: Proc. Comp. in Card. '89. Ripley KL (ed), IEEE Computer Society, Los Alamitos, pp. 215-218.
Bister M, Cornelis J and Rosenfeld A (1990a). A critical view of pyramid segmentation algorithms. Patt. Recogn. Lett. 11:605-617.
Bister M, Cornelis J, Taeymans Y and De Cuyper B (1990b). A generic labeling scheme for segmented cardiac MR Images. In: Proc. Comp. in Card. '90. Ripley KL (ed), IEEE Computer Society, Los Alamitos.
Bister M, Schnall D, Deklerck R, Cornelis J and Taeymans Y (1990c). The Cavity Detector: a generic image segmentation algorithm. In: Proc. North Sea Conf. on Biomed. Eng. Cornelis J and Peeters S (eds), TI - K.VIV, Antwerp, topic 2.
Bister M (1990d). Computer analysis of cardiac MR Images. PhD Thesis, IRIS, VUB, Brussels.
Borgefors G (1986). Distance transformations in digital images. Comp. Vis. Graph. Im. Proc. 34:344-371.
Eiho S, Kuwahara M, Fujita Y, Matsuda T, Sakurai T and Kawai C (1987). 3-D Reconstruction of the left ventricle from Magnetic Resonance Images. In: Proc. Comp. in Card. '87. Ripley KL (ed), IEEE Computer Society, Los Alamitos, pp. 51-56.
Haralick RH and Shapiro LG (1985). Survey – Image segmentation techniques. Comp. Vis. Graph. Im. Proc. 29:100-132.
Harwood D, Subbarao M, Hakalahti H and Davis LS (1984). A new class of edge-preserving smoothing filters. CAR-TR-59, CVL, Univ. Maryland.
Harwood D, Prasannappa R and Davis LS (1988). Preliminary design of a Programmed Picture Logic. CAR-TR-364, CVL, Univ. Maryland.
Horowitz SL and Pavlidis T (1976). Picture segmentation by a tree traversal algorithm. J. ACM. 23:368-388.
Kittler J and Illingworth J (1986). Minimum error thresholding. Patt. Recogn. 19:41-47.
Koenderink JJ (1984). The structure of images. Biol. Cybern. 50:363-370.
Rosenfeld A (1984). Multiresolution image processing and analysis. Springer-Verlag, Berlin.
Vossepoel AM (1988). A note on distance transformations in digital images. Comp. Vis. Graph. Im. Proc. 43:88-97.
Weyman AE (1982). Cross-sectional echocardiography. Lea & Febiger, Philadelphia.
Yang SS, Bentivoglio LG, Maranhão V and Goldberg H (1978). From cardiac catheterization data to hemodynamic parameters. F.A. Davis Company, Philadelphia.

SEGMENTATION OF MAGNETIC RESONANCE IMAGES USING MEAN FIELD ANNEALING

W Snyder, [1] A Logenthiran, [2] P Santago, [1]
K Link, [1] G Bilbro, [3] , S Rajala, [3]

[1]Bowman Gray School of Medicine, Winston-Salem, NC
[2]AT&T Bell Laboratories, Medical Diagnostic Systems, Winston-Salem, NC
[3]Center for Communications and Signal Processing, North Carolina State University

Abstract

The problem of segmentation of Magnetic Resonance images into regions of uniform tissue density is posed as an optimization problem. A new objective function is defined and the resulting minimization problem is solved using Mean Field Annealing, a new technique which usually finds global minima in non-convex optimization problems, and performs particularly well on images. Noise sensitivity is evaluated by tests on synthetic images, and the technique is then applied to clinical images of a brain and a knee. The technique shows considerable promise as a method of quantitative change monitoring.

Key words:

Simulated annealing, restoration, reconstruction, optimization

1. INTRODUCTION

"Segmentation" is defined here as the process of assigning labels to pixels in such a way that each point is distinctly labeled according to the region to which that pixel belongs. Since segmentation is considered by many to be an essential component of any image analysis system, it is a problem which has received a great deal of attention in the literature; so much attention that any attempt to completely survey the literature in this paper would be too space-consuming. Instead, we focus on only one aspect of the problem- magnetic resonance images (MRI) of human anatomy.

In this paper, we pose segmentation as an optimization problem in which an objective function is defined whose minimum would represent a good segmentation. We find a good, possibly global, minimum using a previously reported technique known as "Mean Field Annealing", and we demonstrate the performance of the technique using both synthetic and real medical images.

2. BACKGROUND

Vision systems built to analyze scenes of man-made objects are successful to a large extent because most objects in the scene can be easily modeled to some degree of accuracy (Baker DC and Aggarwal JK, 1988). When it comes to natural objects, where the models of expected structure and their variations is enormous, conventional edge-based segmentation algorithms (Haralick R and Shapiro G, 1985) present serious inadequacies. A wide variety of natural imagery examples where segmentation poses some serious problems are found in magnetic resonance images (Shemlon S and Dunn SM, 1990).

Tissue classification based on a single property for each pixel typically does not generate good results (Ehricke HH, 1990). However, techniques that incorporate neighborhood interaction along with tissue density properties work well (Vannier M et al, 1989) because, in medical imagery, pixels within a particular tissue types have a tendency to be spatially related and connected. In other words, there is a high degree of local correlation among neighboring pixels in regions of a single tissue type (Lee RH and Leahy R, 1990). While our method exploits this correlation, it does not address the partial volume effect which can cause misclassifications on the edges of the region. For our current investigation, this effect has not been significant. Our method is in contrast to a maximum likelihood classifier which is based solely on density and does not exploit the fact that the subject may be considered as consisting of regions where each region is of a single tissue type. By introducing a "prior" distribution which models the spatial interaction between pixels in the tissue map, one can use *maximum a posteriori* (MAP) classification to obtain a smoother, more reliable segmentation (Geman S and Geman D, 1984). Hansen and Elliot (Hansen FR and Elliott H, 1982) use this approach to segment binary images.

3. AN OBJECTIVE FUNCTION FOR SEGMENTATION

We wish to maximize the *a-posteriori* probability $p(f|g)$ of the unknown correct image f given measured image g. Using Bayes' rule, we have the proportionality

$$p(f|g) \propto p(g|f)\,p(f) \qquad f = [f_1, f_2 \ldots f_N]^T. \tag{1}$$

For purposes of this paper, one can ignore the normalizing constant of proportionality and maximize the right hand side of (1). For the conditional density, a stationary Gaussian noise model allows us to write (Geman S and Geman D, 1984; Bilbro G and Snyder W, 1989; Hiriyannaiah H et al, 1989; Wolberg G and Pavlidis T, 1985)

$$p(g|f) = \frac{1}{\sqrt{2\pi}\sigma} exp \frac{-\sum_i (g_i - f_i)^2}{\sigma^2} \tag{2}$$

where the index i ranges over all the pixels in the image.

A number of researchers have noted that an image is appropriately represented by a two- or three-dimensional Markov field which allows (Besag J, 1974; Geman S and Geman D, 1984; Kashyap RL and Chellappa, 1983) the representation of the *a-priori* probability of given pixel value by a Gibbs distribution

$$p(f_i) = exp\left(\frac{\sum_j V_j}{T}\right). \tag{3}$$

The sum is taken over "cliques", connected neighborhoods, in our case, indexed by the pixel in question. T is an adjustable width parameter, and V are potentials which are, in general, functions of the pixels in the clique.

Forming the product of Equations (2) and (3) as indicated in Equation (1) and eliminating the constant term, we take natural logarithms thereby converting the problem from maximizing a probability to minimizing an objective function.

$$H(f) = \left(\sum_i \frac{(g_i - f_i)^2}{\sigma^2}\right) + \sum_i V_i \tag{4}$$

The dependence of H on the observation g has been suppressed for notational simplicity.

We will refer to the first, conditional, term of Equation (4) as the "noise" term (Geman S and Geman D, 1984) and to the second as the "prior". This gives the following form:

$$H(f) = H_n + H_p. \tag{5}$$

The prior term, H_p must reflect two properties, which we will initially address separately:

- Regions in the image are locally homogeneous. This assumption is reasonable for MRI density images of biological specimens and states that all the pixels within a given organ will have uniform density. As an example of using this assumption, striations of fat lying within muscle tissue would be segmented as distinct from the muscle regions.
- All pixels in the image have a density equal to one of a set of K known values $\{\mu_k\}$, $k=1,...,K$. Implementation of any technique based on this assumption might require that radiometric correction be performed prior to segmentation to correct for inhomogeneities in the magnetic field. However, the head and knee images acquired on our scanners (1.5T GE and Picker) show sufficient uniformity as to obviate this need. In addition, we found it unnecessary in this work to consider geometric distortions (Chang H and Fitzpatrick MJ, 1990).

We incorporate these two desires into the prior by writing the prior as a sum:

$$H_p(f) = H_h(f) + H_s(f). \tag{6}$$

Our experience in other image processing problems (Bilbro G and Snyder W,; Hiriyannaiah GL, 1989), suggests that the following form for H_h provides an objective function which, if minimized, results in a minimizing image $\hat{f}$ which will contain regions of uniform intensity separated by step boundaries

$$H_h = \sum_i -\frac{b_h}{\sqrt{2\pi\tau_h}} \sum_{j \in \aleph_i} exp(\frac{(\nabla f)_j}{\tau_h}) \quad (7)$$

where ∇ is a scalar operator defined for each pixel in $\aleph_i$, the neighborhood of pixel i, and having maximum value of zero whenever that neighborhood is uniform. τ_h is thus a measure of how much variation we permit/expect in a local neighborhood. The Gaussian form for (7) is chosen for two reasons: first, it effectively captures our intuition that neighborhoods should be uniform and provides us with a parameter, τ_h, which quantifies the strength of that intuition. Second, functions involving the Gaussian are sometimes exactly integrable allowing us, occasionally (Bilbro G and Snyder W), to find exact solutions to the mean field equations which we will discuss in the next section.

To meet the segmentation objective, we need a function whose value will be maximal whenever the intensity is equal to one of the K known values. Again, our previous success with Gaussian forms suggests

$$H_s = \sum_i \left(-\frac{b_s}{\sqrt{2\pi\tau_s}}\right) exp\left(-\frac{(f_i-\mu_1)^2 (f_i-\mu_2)^2 \ldots (f_i-\mu_K)^2}{\tau_s}\right). \quad (8)$$

Clearly, the value of (8) is maximal whenever f_i is equal to any one of the known brightness values μ_k. We thus have an objective function which captures three desires: 1) that the restored image, f, should resemble the data, g, 2) that the restored image should be locally homogeneous, and 3) that the restored image should have only the brightness values which were pre-specified. We accomplish these desires by solving

$$min_f \, H(f) . \quad (9)$$

We will accomplish this minimization using Mean Field Annealing.

3.1. Mean Field Annealing

Mean Field Annealing (MFA) is a technique for finding the minimum of complex functions which typically have many minima (Bilbro G et al, 1989). Here, we present only a brief description of the technique and refer the reader to (Bilbro G and Snyder W) for more detail. Assume we have a function $H(f)$, dependent on some vector f, and the knowledge that H is probably non-convex and possibly even non-differentiable. We approximate $H(f)$ with some other function $H_0(f,x)$ which is in some sense "easier"; that is, we may easily find its minimum. Here, x is some vector which parameterizes H_0. MFA theory allows an arbitrary choice of H_0, however, experience (Bilbro G and Snyder W; Bilbro G and Snyder W) indicates the choice of

$$H_0(f,x) = -\sum_i (f_i - x_i)^2 \quad (10)$$

for imaging problems with continuously-valued pixels. We thus must choose the parameter vector x in such a way that the function $H_o(f,x)$ most resembles $H(f)$. MFA accomplishes this by solving

$$min_x [\langle H(f) - H_o(f,x)\rangle - T \ln Z_0] \quad (11)$$

where the expectation operator uses a density

$$p_0(f) = Z_0^{-1} exp(\frac{H_0(f,x)}{T}) \quad (12)$$

with normalization Z_0. The parameter T which appears in the density is used in MFA in a manner analogous to that of the "temperature" parameter of Simulated Annealing (van Laarhoven JM and Aarts EHL, 1989). Proper choice of H_o, in particular, the choice of Equation (10), can be shown to reduce the problem of expression (11) to

$$\min_x \langle H(f) \rangle . \quad (13)$$

To implement MFA, we first determine an analytic form for $\langle H(f) \rangle$ by evaluating the expectation integral. We find that, in most cases, the simplification

$$\langle H(f) \rangle \approx H(x) \quad (14)$$

is either exactly (Hiriyannaiah G et al, 1989) or approximately (Bilbro G and Snyder W) true. Thus, the MFA algorithm becomes:

1..Set $T=T_{initial}$ (a problem-dependent parameter).
2..Using gradient descent or some other minimization technique, solve expression (13).
3..Reduce T.
4..If $T>T_{final}$, go to step 2.

In Figure 1, we illustrate these ideas with a one-dimensional example. The function *H(f)* is modeled by an approximating, convex function $H_o(f,x)$ centered at x. At high T, this function models the average behavior of *H(f)*. At low T, however, the algorithm has centered the mean x at the global minimum.

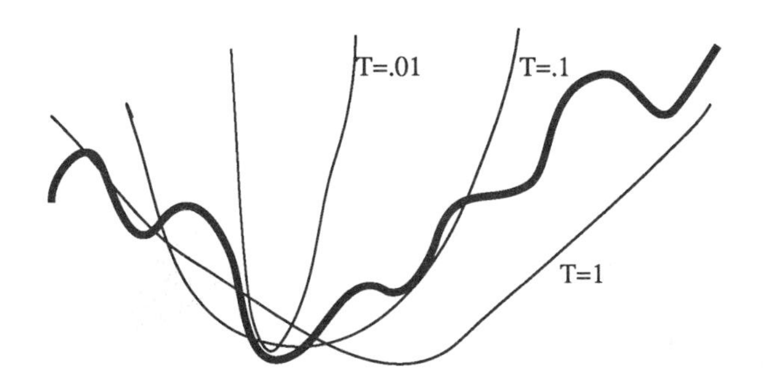

Figure 1 A non-convex energy function and several convex approximations

3.2. Application of MFA to *H(f)*

The strategy of MFA described in Equation (13) requires that we solve

$$min_x \langle H(f) \rangle . \quad (15)$$

The expectation of the noise term is straightforward to evaluate, however, for the prior term we have

$$\langle H_p(f)\rangle = \int df_1 \int df_2 \ldots \int df_N (H(f)) \frac{\exp\left(-\frac{H_0(f,x)}{T}\right)}{Z_0}. \tag{16}$$

Z_0 involves the nested integrals as well, and in some instances, this integral may be solved exactly, however, for the H of Equation (8), an analytic solution does not appear possible. Experience (Hiriyannaiah H, 1989; Bilbro G and Snyder W, 1989; Bilbro G and Snyder W), however, has shown that we should use the approximation of Equation (14), replace τ by $\tau + T$, and use the algorithm described in section 3.1.. The gradient descent step requires the analytic solution of $\frac{\partial H}{\partial x}$, which, although tedious, is neither difficult to evaluate nor difficult to program. We could use this result in a simple descent program to find the minimum of H at a given T, and then anneal on T as described above. An additional complication occurs, however, due to the presence of two different τ parameters.

Defining $t_h = \tau_h + T$ and $t_s = \tau_s + T$, the process of reducing T from a high value to zero changes into reducing t_h and t_s from some initial to some final (nonzero) value. We choose initial value for t_h using the same temperature estimates provided in (Hiriyannaiah H, 1990)

$$t_{hinitial} = 2\sigma^2 \tag{17}$$

For t_s, we estimate the initial value by noting that t_s should be of the same order as the numerator $\prod_{k=1}^{K} (f_i - \mu_k)^2$, whose variance is of order $(\sigma^2)^K$. We thus choose $t_{sinitial} = 2(\sigma^2)^K$.

For single temperature MFA image restoration problems, a geometric cooling schedule has been successful (Bilbro G and Snyder W, 1989; Bilbro G and Snyder W; Hiriyannaiah H et al, 1989). Here, we recognize, however, that the clustering of pixels into groups (homogeneity objective) should occur more rapidly than the choosing of particular values for those clusters (segmentation objective). We thus choose the update rule

$$t_h^{l+1} = \alpha_h t_h^l \text{ and } t_s^{l+1} = \alpha_s t_s^l \tag{18}$$

where l denotes the iteration number, and $\alpha_h < \alpha_s$. We determined experimentally that $\alpha_h = 0.95$ and $\alpha_s = 0.97$ resulted in good performance.

4. Experimental results

4.1. Noise Sensitivity

To determine a quantitative measure of the performance of the segmentation algorithm, a synthetic image was constructed similar to the familiar Shepp-Logan CT phantom (Kak AC and Slaney M, 1988). Noise was added to this image in varying amounts, and the segmentation algorithm was then applied. In an original, clean image, with K modes in its histogram, we define

$\Delta\mu_{min}^2 = \min_{i,j} (\mu_i - \mu_j)^2$, $i,j = 1,\ldots,K$. That is, $\Delta\mu_{min}$ represents the difference in intensities between those two regions whose densities are most similar, i.e., it is the height of the smallest possible step edge. With this definition, we may now define

$$SNR = \left(\frac{\Delta\mu_{min}}{\sigma}\right)^2 \tag{19}$$

with σ is the standard deviation of the additive Gaussian noise.

Since the image is synthetic, we know the exact geometric extent of each region. After noise is added and the region segmented by our algorithm, we would expect some pixels, particularly those near the boundaries of the object, to be incorrectly assigned--either to the background or to adjacent regions. Thus, for a particular region, we may define a segmentation error

$$E = \frac{\text{Number of misclassified pixels}}{\text{Area of region}}. \tag{20}$$

Figure 2 and Figure 3 show the noise-free phantom and histogram

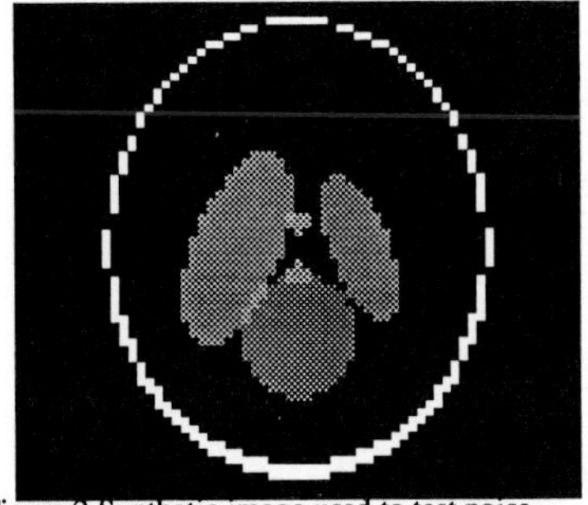
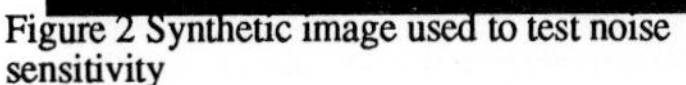
Figure 2 Synthetic image used to test noise sensitivity

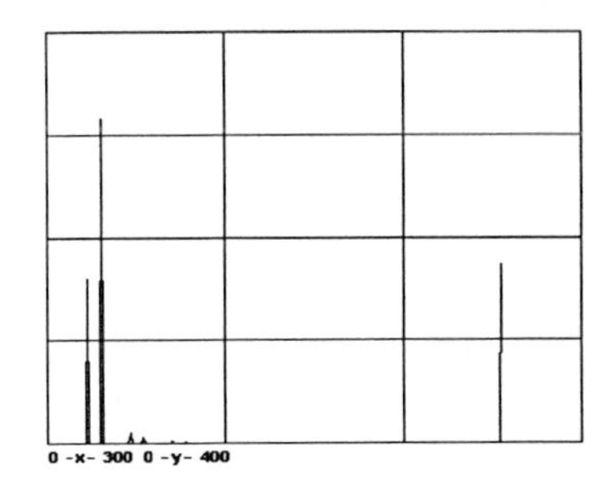

Figure 3 Histogram of Figure 2

Noise was added in varying amounts, and the segmentation algorithm was applied. The error performance is shown in Figure 4. In that figure, the segmentation performance is compared with that of Hiriyannaiah et al.(Hiriyannaiah H et al, 1989), which segments using only a homogeneity requirement. Clearly the addition of the segmentation term significantly improves performance.

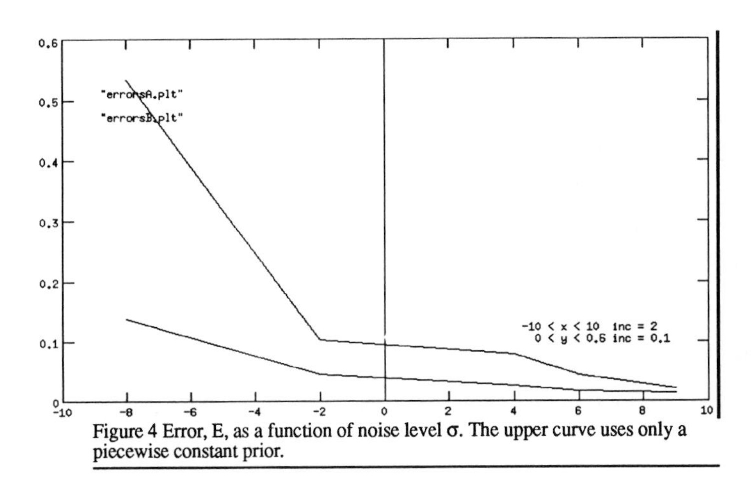

Figure 4 Error, E, as a function of noise level σ. The upper curve uses only a piecewise constant prior.

4.2. Segmentation of Brain Images

The algorithm was applied to an axial MR image of a brain obtained using a multi-echo pulse sequence with spin echo time, TE, and constant repetition time, TR, of (TE_1, TE_2, TR) = (20,100,3000) with all times in ms. The original image and its histogram are presented in Figure 5 and Figure 6.

Three intensity modes were chosen, corresponding to CSF, white matter, and gray matter. Using a single temperature gradient, holding $t_h=t_s$, the algorithm produced the results illustrated in Figure 7 and Figure 8. Running the same experiment with four modes and with two temperature gradients, as described in Equation (18) gives superior results as illustrated in Figures 9 and 10.

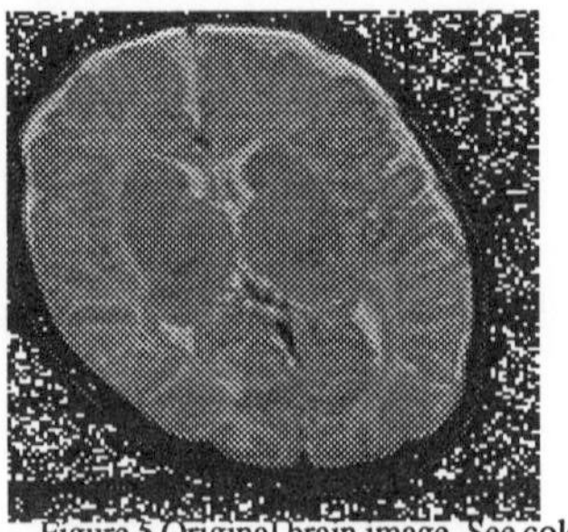

Figure 5 Original brain image. See color plate, p. 5

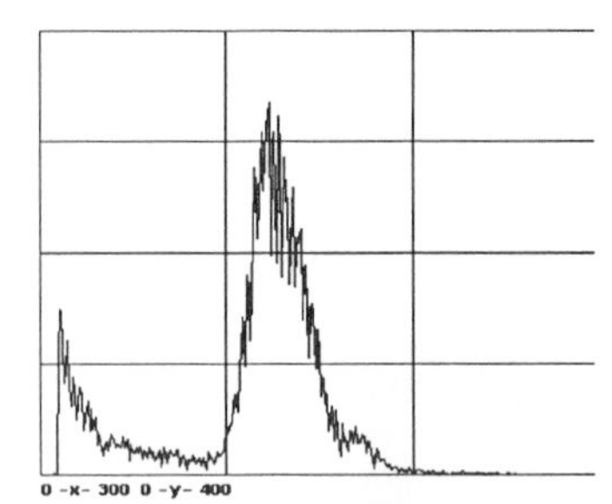

Figure 6 Histogram of Figure 5

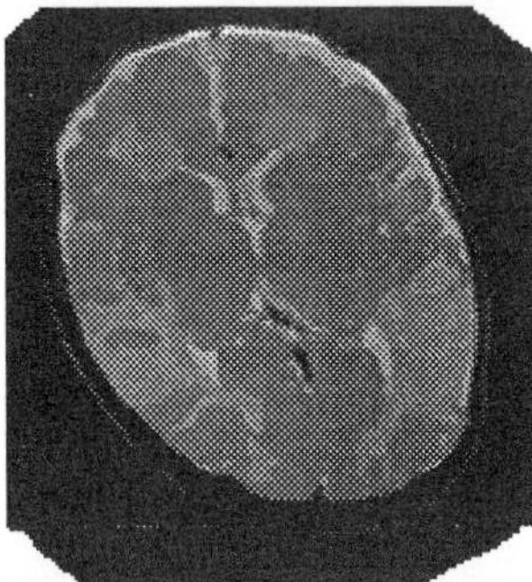

Figure 7 brain image segmented using one temperature. See color plate, p. 5

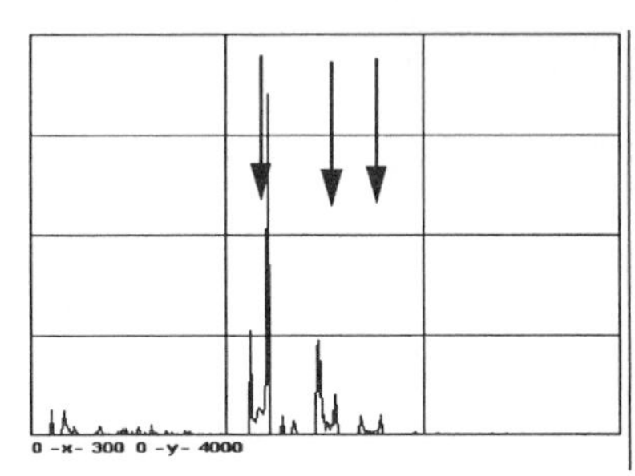

Figure 8 Histogram of Figure 5. Compare with Figure 6-- the three modes are clearly selected.

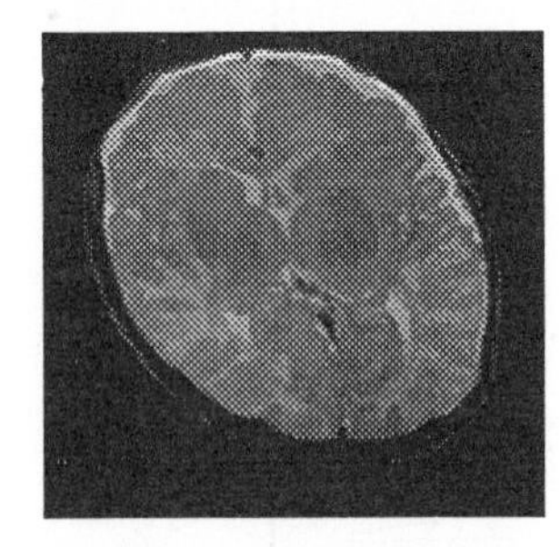

Figure 9 Brain image segmented using two temperatures. See color plate, p. 5

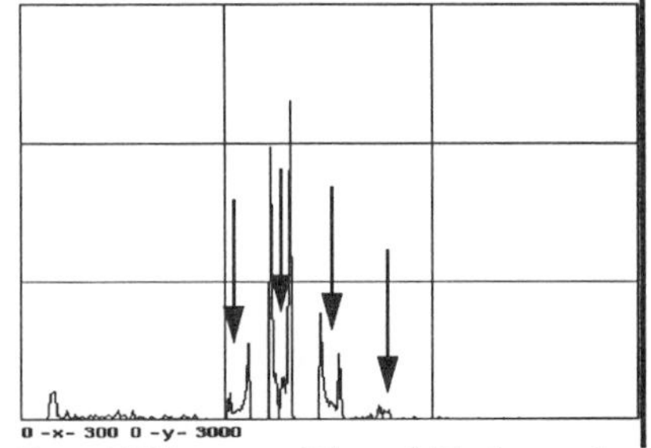

Figure 10 Histogram of Figure 5. The four modes are well differentiated. The smallest mode is CSF, and was represented by only a few pixels.

4.3. Variation of Relative Control Parameters

In order to demonstrate the effects of varying the bond strengths b_h and b_s (Equations (7) and (8)), the following experiment was performed. Figure 11 illustrates the result of varying the control parameters for segmenting the same brain image used above.

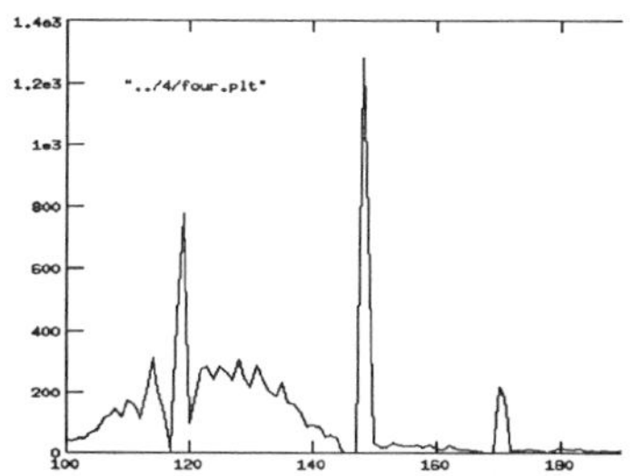

Figure 11 Histogram illustrating classification ambiguity resulting from low bond strength.

Although three modes were chosen, and are apparent in the histogram, the classification of many pixels remains ambiguous because the relative weights for the segmentation were chosen so that $b_h >> b_s$, leading to a restoration emphasizing the homogeneity objective at the expense of the segmentation objective. Another example of this effect is illustrated in the knee image and its histogram shown in Figure 12 and Figure 13. The large blocks of uniform brightness demonstrate that the b parameters may be used to control the relative strength of the two prior objectives.

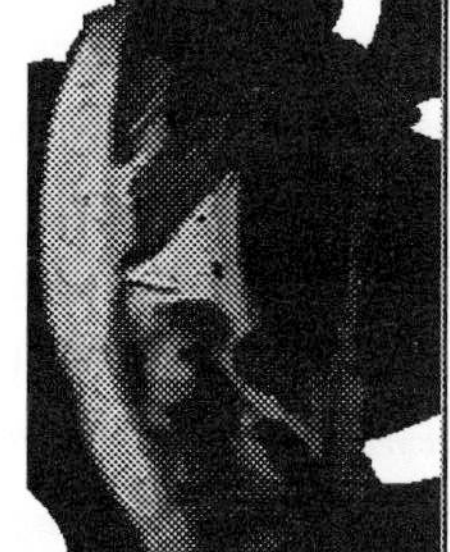
Figure 12 Segmented Knee image. See color plate, p. 5

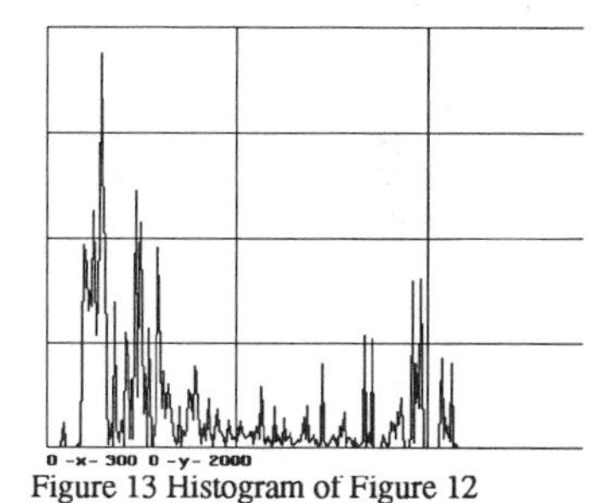

Figure 13 Histogram of Figure 12

5. CIINICAL EVALUATION

This technique is intended to serve as an adjunct to standard MR imaging. Its purpose is to serve in the delineation of subtle contrast differences between cerebral spinal fluid, white matter, and grey matter. Although standard MR imaging has contrast resolution superior to other medical imaging modalities, it has been our clinical experience that subtle lesions may be missed because they are not detectable to the human eye with standard techniques. By clearly delineating the different density structures within the brain, one in turn can more clearly identify abnormalities. In addition, one can quantify such abnormalities whether they be a mass within the brain or an abnormal accumulation of cerebral spinal fluid. The later is of particular importance as it may serve as a key in understanding the pathophysiologic sequela of nonobstructed versus obstructed hydrencephalus. Quantification of mass lesions will allow precise evaluation of the effects of radiation and/or chemotherapy.

To date, this technique has been applied in the knee and has proved extremely useful in studying the cartilaginous (meniscus) structures. We have found this technique to allow much more accurate assessment of meniscal injuries. This is important clinical information in that degenerative changes are abnormalities which are treated conservatively whereas frank tears require surgical intervention. In addition to improving our diagnostic accuracy in detecting tears, this technique has opened the door to reevaluating the pathophysiologic mechanisms involved in degenerative changes and their advancement to frank tears. Our initial data has indicated new subtle findings that were not previously identified with standard MR techniques which may change the way we study the knee with MRI.

6. CONCLUSION

We have shown that by using an appropriate statistical model and cost function and presenting segmentation as a minimization problem, a good segmentation can be found using optimization methods.

The use of MFA provides a straightforward and fast mechanism for solving the minimization problem, producing a segmented image as a solution. Such a segmentation is shown to provide good quantification of tissue type while also performing noise removal and edge retention. Diagnostic assistance is also possible using the color presentation of the segmented image.

As was seen in the knee segmentation, the method is sensitive to the parameters chosen for controlling the algorithm. Optimal parameter selection, such as for b_s and b_h, needs further investigation.

An obvious yet not so simple extension to the procedure is to have less operator input. By this we mean that the algorithm should decide on the number of regions and the means to use for the optimization. At this time, it seems that specifying the number of regions only would be a good first step. Perhaps indicating a seed point in each region could be used as well.

Finally using *a priori* information about the anatomy could assist greatly in the segmentation by providing bounds or constraints on the problem. Some work has been done in this regard, (Kapouleas I and Kulikowski C, 1988) and we are investigating the possibility of incorporating similar methods into our algorithm.

References

Baker David C and Aggarwal JK (1988). Geometry guided segmentation of outdoor scenes. SPIE Applications of Artificial Intelligence VI:576-583.

Besag J (1974). Spatial interaction and the statistical analysis of lattice systems. J Royal Stat. Soc., Series B, 36:192-326.

Bilbro G, Mann R, Miller T, Snyder W, Van den Bout D and White M (1989). Optimization by mean field annealing. In: Advances in Neural Information Processing Systems. Morgan-Kauffman, San Mateo.

Bilbro G and Snyder W (1989). Range image restoration using mean field annealing. In: Advances in Neural Network Information Processing Systems. Morgan-Kauffman, San Mateo.

Bilbro G and Snyder W. Mean field annealing, an application to image noise removal. Journal of Neural Network Computing.

Bilbro G and Snyder W. Mean field from relative entropy. Journal of the Optical Society of America A.

Chang Hsuan and Fitzpatrick Michael J (1990). Geometrical image transformation to compensate for MRI distortions. SPIE Vol. 1233 Medical Imaging IV, pp 116-127, February.

Derin H and Elliot H (1987).Modeling and segmentation of noisy and textured images using Gibbs random fields, IEEE Trans PAMI, vol. PAMI 9, pp39-55.

Ehricke HH (1990). Problems and approaches for tissue segmentation in 3D-MR imaging. Proceedings of the SPIE Vol. 1233 Medical Imaging IV: Image Processing, pp 128-137, February.

Geman S and Geman D (1984). Stochastic relaxation, Gibbs distributions, and Bayesian restoration of images, IEEE PAMI, vol 6, no 6.

Hansen FR and Elliott H (1982). Image segmentation using simple Markov field models. Computer Graphics and Image Processing, 20:101-132.

Haralick Rand Shapiro G (1985). Image segmentation techniques. Computer Vision, Graphics, and Image Processing 29:100-132.

Hiriyannaiah H (1990). Signal reconstruction using mean field annealing. PhD Thesis, North Carolina State University, Raleigh.

Hiriyannaiah H, Bilbro GL, Snyder WE and Mann RC (1989). Restoration of piecewise constant images via mean field annealing. Journal of the Optical Society of America A, December:1901-1912.

Kak AC and Slaney M (1988). Principles of computerized tomographic imaging. IEEE Press, New York.

Kapouleas I and Kulikowski C (1988). A model-based system for interpretation of MR human brain scans. Proceedings of the SPIE, Volume 914, Medical Imaging II, Feb.

Kashyap RL and Chellappa R (1983). Estimation and choice of neighbors in spatial-interaction model of images. IEEE Trans. Information Theory, IT-29, pp. 60-72, January.

Lee Rae Hand Leahy Richard (1990). Multi-spectral tissue classification of MR images using sensor fusion approaches. Proceedings of the SPIE Vol. 1233 Medical Imaging IV: Image Processing,pp 149-157, February.

Morris PG (1986). Nuclear magnetic resonance imaging in medicine and biology. In:, Clarendon Press, Oxford.

Nelson Thomas R (1990). Propagation characteristics of a fractal network: applications to the His-Purkinje conduction system. Proceedings of the SPIE, Newport Beach, California, 1233 Medical Imaging IV: Image Processing, pp 23-32, February.

Shemlon Stephen and Dunn Stanley M (1990). Rule-based interpretation with models of expected structure. In SPIE Vol. 1233 Medical Imaging IV, February, pp 33-44.

van Laarhoven JM and Aarts EHL (1988). Simulated annealing: theory and applications. D. Reidel, Norwell, Mass.

Vannier M et al (1989). Multispectral analysis of magnetic resonance images. Radiology, 221-224.

Wolberg G and Pavlidis T (1985). Restoration of binary images using stochastic relaxation with annealing. Pattern Recognition Letters, 3, 6:375-388.

A STOCHASTIC MODEL FOR AUTOMATED DETECTION OF CALCIFICATIONS IN DIGITAL MAMMOGRAMS

N Karssemeijer

Department of Radiology, University of Nijmegen

The Netherlands

Abstract

A stochastic model is developed to enable pattern classification in mammograms based on Bayesian decision theory. Labeling of the image is performed by a deterministic relaxation scheme in which both image data and prior beliefs are weighted simultaneously. The image data is represented by two parameter images representing local contrast and shape. Involving shape is necessary to distinguish thin patches of connective tissue from microcalcifications. A random field models contextual relations between pixel labels. Long range interaction is introduced to express the fact that calcifications do occur in clusters. This ensures that faint spots are only interpreted as calcifications if they are in the neighborhood of others.

Keywords

pattern recognition; image analysis; segmentation; mammography

1. INTRODUCTION

Application of digital image analysis in mammography is investigated to increase the quality and efficiency of breast cancer screening programs. One of the most important radiological signs for early detection of breast cancer is the presence and appearance of microcalcifications. These are small bright spots against the varying background density of the X-ray image. On conventional screen/film recorded mammograms microcalcifications as small 0.1 mm may be imaged. Much smaller spots can be identified histologically. Only clusters of calcifications are suspicious. Especially in screening programs, where more than 99% of the images are normal, automated detection of microcalcifications may improve the radiologist's performance to pick out abnormalities. The results of a detection algorithm can be used, for instance, to mark suspicious regions. Also display of a detected calcification pattern is found to be useful, because it makes interpretation of the spatial distribution of microcalcifications easier. This distribution is an important cue for distinguishing different types.

Procedures based on local thresholding to detect calcifications have been described by various authors (Fam et al, 1988, Chan et al, 1988, Davies and Dance, 1990). Using local thresholding accurate segmentation is possible to some extent. However, the performance reduces when parameters are set in such a way that a high sensitivity is obtained, in order not to miss very subtle calcification clusters. The false positive rate increases strongly then, due to background structure and noise. Especially background structure is difficult to deal with. An example is shown in Figure 1, which is a detail of a about 5 cm in cross section of a mammogram digitized with 12 bits

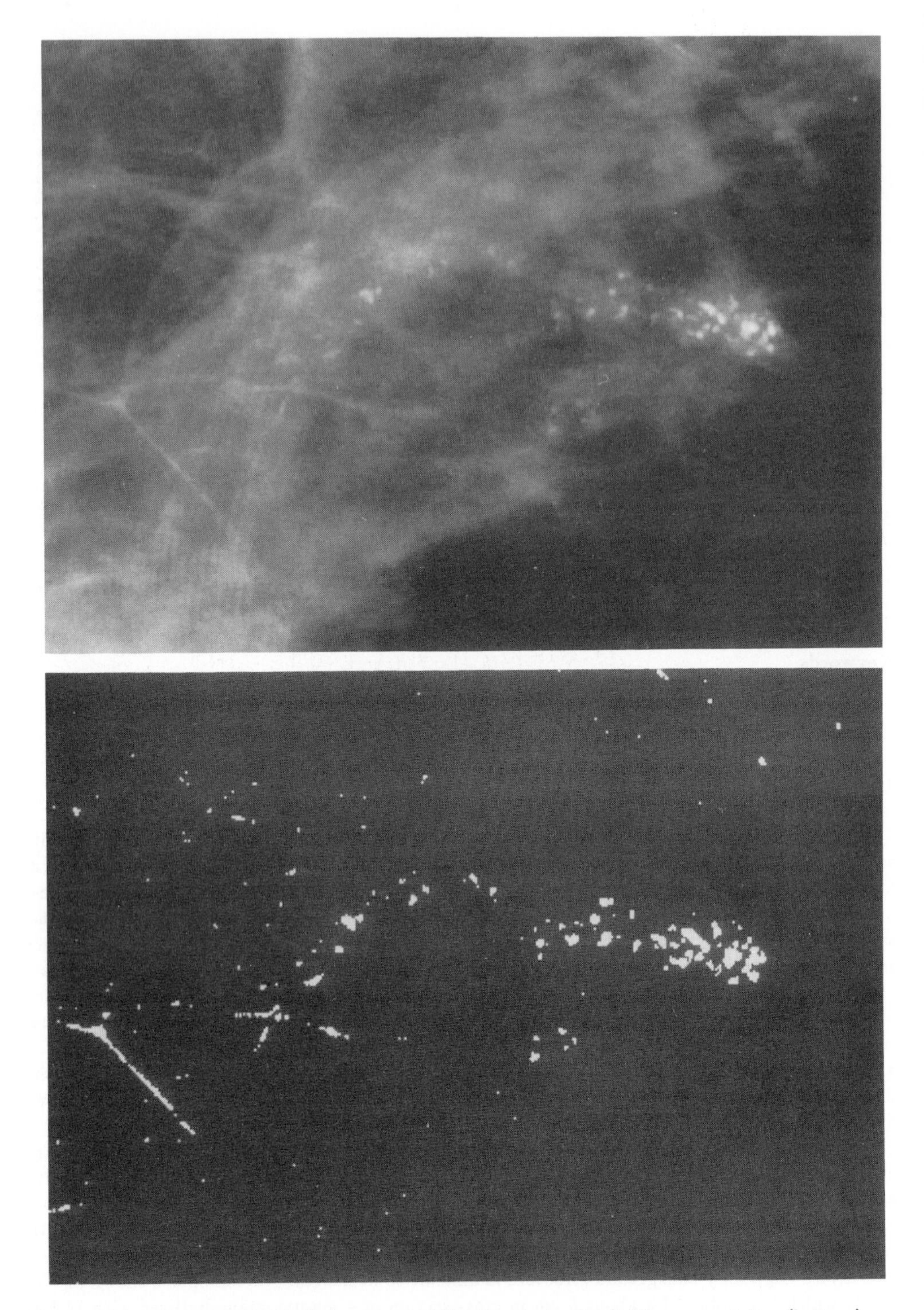

Figure 1. A mammographic detail showing a calcification cluster (top). The segmentation (bottom) was obtained by local thresholding. Thin layers of connective tissue cause significant errors.

CCD camera and a 0.1 mm pixel size. The line shaped microstructure, which represents connective tissue partitioning the breast in different compartments, is wrongly labeled. The segmentation result was obtained by applying a fixed threshold on the image of relative pixel values y'_i defined by

$$y'_i = y_i - \sum_{j \in \partial i} y_j, \tag{1}$$

with y_i denoting the original image data at site i. The neighborhood ∂i was chosen as a square area of nine pixels in diameter.

To optimize the performance of the algorithm the pixel values were rescaled first using an iso-precision criterion (Karssemeijer, 1991). On this scale the absolute error of the pixel values due to film noise and digitization noise is equal over the whole range of pixel values. The use of such a scale may improve the performance of feature detection algorithms significantly. It ensures that a constant sensitivity over the whole range of pixel values can be obtained with a parameter setting which is independent of the grey level. With respect to detection of calcification in mammograms there is a strong reason to avoid density dependent errors, because such errors tend to come up in clusters due to the background intensity pattern. This will hamper subsequent analysis aimed at distinguishing calcification clusters from noise.

Although the influence of the pixel value scale on feature detection algorithms may be considerable, the importance of of proper scaling is seldom recognized in the literature. For instance, when a logarithmic transform is applied on images digitized from film, as is reported by Chan et al (1988) and Davies and Dance (1990), the noise level in high density parts of the image is much higher than in brighter parts. This probably is the reason for the fact that all false positive clusters detected by the procedure described by Davies and Dance (1990) were located at low densities. In the method of Chan et al (1988) a noise dependent detection algorithm is applied at suspicious sites determined by local thresholding. The local noise level is estimated from the image data itself, around the site to be inspected. A disadvantage of this approach is that the estimation procedure will be sensitive to errors caused by normal image structure. By applying an iso-precision scale transform before feature detection local dependence on the noise level is also achieved, using more reliable noise estimates, however, based on prior knowledge of film and digitization noise.

Automated calcification detection procedures mostly consist of a second stage, in which it is attempted to improve the result of thresholding. True calcification clusters are classified on basis of various criteria like shape, grey level, size, or cluster size. Such a sequential step-wise approach is disadvantageous for various reasons. The main drawback is that the rigorous reduction of the image data that takes place in the first step is not justified by a model concerning the nature of the structures to be detected. Difficulties in later stages of the detection process do arise because image interpretation is to be based on unreliably segmented areas. The aim here is to avoid sequential mechanisms by using a statistical model of the image and iterative optimization.

2. METHODS

The introduction of stochastic models and Bayesian methods in image analysis has provided a general framework to model image data and to express prior knowledge. Particularly random field models have shown to be appropriate tools for modeling spatial context, while their use leads to segmentation algorithms which are feasible in practice. These algorithms are basically iterative rules for updating pixel labels in a process of local competition and coöperation. In this approach a segmentation X of a scene S is estimated by iteratively assigning pixel labels x_i with maximum a posteriori probability given the data Y. Two alternative choices are maximum a posteriori (MAP) estimation in which $p(X|Y)$ is maximized and maximization of posterior marginals (MPM) which aims at maximizing $p(x_i|Y)$ at each individual site i of S. Given a prior distribution $p(X)$ the posterior distribution of X given Y is

$$p(X|Y) \propto p(Y|X)p(X). \tag{2}$$

In practice suitable models for both the likelihood function and the prior $p(X)$ have to be adopted in relation to existing procedures for maximizing the posterior probability of X. Various details on the application of random field models and Bayesian image analysis are provided by Besag (1986). A more recent review of the subject is given by Dubes and Jain (1989).

Generally, a Markov random field (MRF) is used as a model for $p(X)$. A MRF is defined in terms of local dependence of the individual pixel labels x_i on the labels $x_{\partial i}$ in their neighborhoods. The probability distribution corresponding to a MRF has the following form:

$$p(X) = \frac{1}{Z}\exp[-U(X)], \tag{3}$$

where $U(X)$ is called the energy function and Z is a constant for normalization. Here we will restrict ourselves to the class of pairwise interaction models. In that case the energy function is given by

$$U(X) = \sum_{i=1}^{N} A_i(x_i) + \sum_{<ij>} B_{ij}(x_i, x_j), \tag{4}$$

in which $< ij >$ denotes summation over all neighboring pairs of sites in S and N is the total number of sites. A neighbor pair $< i, j >$ is by definition a pair for which $j \in \partial i$. In the following the interaction parameters B_{ij} are assumed to be independent of the positions of the sites i and j in S. The external field $A_i(x_i)$ is set to zero. For the conditional probability of a label to occur at a particular site it can be derived that

$$p(x_i = k|x_{\partial i}) \propto \exp\left[-\sum_{n=1}^{K} B_i(k,n)g_i(n)\right], \tag{5}$$

by comparing the energy functions of segmentations only differing at site i. The number of neighbors labeled as n is denoted by $g_i(n)$ and K is the number of classes.

The local probabilities defined in Eq. (4) are used in iterative updating schemes for estimation of the 'true' segmentation X^*. In methods based on sampling estimates are obtained by simulation of a Markov chain (e.g. Geman and Geman, 1984). Here a much faster method will be applied termed 'Iterative Conditional Modes' (ICM) suggested by Besag (1986). In this scheme pixel labels are updated by each time choosing the most probable label x_i' given the current estimate of the rest of the labeling $\hat{X}_{S\setminus i}$ and the data Y:

$$x_i' = \max_k \left[p(x_i = k|Y, \hat{X}_{S\setminus i})\right]. \tag{6}$$

Visiting each site in turn or at random the iterative process will converge to a local maximum of the posterior distribution $P(X|Y)$. Usually only few iteration cycles are needed for convergence. Using Bayes' relation and the Markov property the probability to be maximized can be written as

$$p(x_i = k|Y, \hat{X}_{S\setminus i}) \propto p(Y|x_i = k, \hat{X}_{S\setminus i})p(x_i = k|\hat{x}_{\partial i}). \tag{7}$$

In practice the logarithm of the expression is evaluated, which requires less computation. The ICM method can only be applied when a good initial estimate of the segmentation is available. This can often be obtained by maximum likelihood (ML), which maximizes the data term in Eq. (7) without considering neighboring pixel labels.

3. THE IMAGE MODEL

To apply statistical methods as described in the previous section both the appearance of microcalcifications and background have to be modeled. It is assumed that the background can be described by a global intensity pattern with independent additive noise superimposed on it. Thin bright patches of connective tissue like in Figure 1 do not fit in such a description. Therefore a third class is introduced to model line-like structures. To isolate calcifications from the noisy background local contrast is important. Therefore relative pixel values y_i' are to be used rather than absolute pixel values. These were calculated using Eq. (1) with the neighborhood ∂i chosen as above.

A distinction between faint microcalcifications and connective tissue can only be made on the basis of shape, as these two structures do have a similar contrast. The most obvious approach to modeling this shape difference would be through the use of $P(X)$, because it reflects geometric properties of the imaged tissue, independent of the imaging process. Having in mind, however, that iterative optimization is to be used to label the scene, this approach does not seem very attractive. As it restricts shape modeling to a description of contextual relations between neighboring labels, it rules out the possibility to obtain a reasonable initial estimate by maximum likelihood. As a consequence fast local optimization by ICM would not be possible. Another disadvantage is that shape clues modeled in $P(X)$ can only enter the labeling process through slow propagation of labels during iteration. From the point of view of computation this is rather inefficient. Information to decide whether a pixel belongs to a line or dot structure can be extracted much more straightforwardly from the image data. Therefore, in a rule for iterative local updating of pixel labels using all current information at hand, which is the basic idea behind the ICM algorithm, the local data around the site to be updated should not be ignored. For that reason the image data is involved here in modeling local shape characteristics.

Starting point is the use of ICM and relation Eq. (7). Conditional independence of the records y_i is assumed, and the imaging process is supposed to be without blur. Consequently, the following relation holds

$$p(Y|X) = \prod_{i=1}^{N} p(y_i|x_i). \tag{8}$$

To proceed, the standard approach is to replace the current labeling $\hat{X}_{S\backslash i}$ in Eq. (7) by the true segmentation. With assumption (8) this reduces the data term in Eq. (7) to $p(y_i|x_i)$. Clearly, this does not allow any freedom for making use of the local data $y_{\partial i}$ for updating the label at i. Therefore, another approach is chosen which does permit involvement of the local data. Instead of taking the labels surrounding site i for true they are supposed to be that uncertain that it is better to omit them from the conditioning set of the data term. This restricts influence of the current neighbor labels to the interaction term, which is defined by Eq. (5). Regarding the remaining expression modeling the likelihood of the data the following is assumed:

$$f(Y|x_i) = f(y_i', \theta_i|x_i), \tag{9}$$

where y_i' is the relative pixel value calculated as in Eq. (1) and θ_i is a shape parameter extracted from the image data in a small neighborhood of i. This shape parameter indicates whether a line

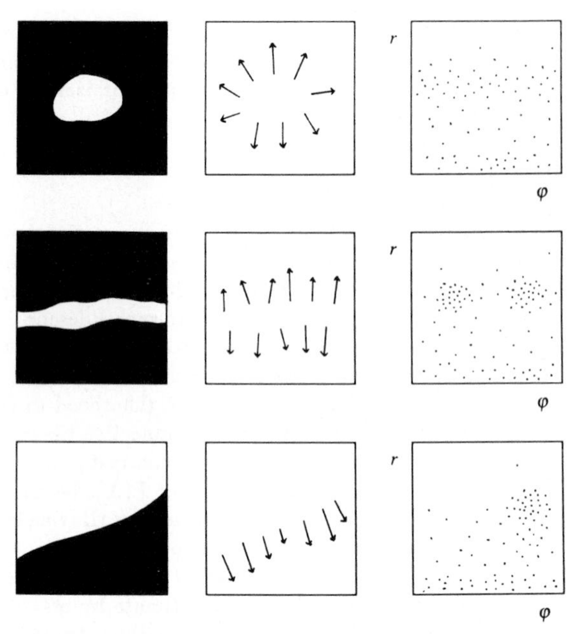

Figure 2. Calculation of the shape parameter θ_i from local data. The left column shows schematic diagrams of possible grey level configurations within a small neighborhood of i. The gradient is represented in the middle. In the right column this data is transformed to a (r, φ) space in which classification of local grey level patterns is more easy (see text).

or dot feature at site i is likely or not.

Calculation of the shape parameter θ_i from the local data is explained in Figure 2. The method resembles the Hough transform, which is used in image analysis for detection of straight lines (e.g. Ballard, 1984, Illingworth and Kittler, 1988). At each site in a small neighborhood of i the direction φ and magnitude r of the gradient are calculated. This data is accumulated in a (r, φ) table. The density $f_i(\varphi)$ of the gradient directions in the neighborhood of i is determined from this table for $r > T$, with T a fixed threshold applied to reduce the influence of noise. To normalize the φ-scale a coördinate transform is performed, defined by a shift of the least occurring gradient direction to $\varphi = 0$ followed by a translation to put the mean gradient $\bar{\varphi}$ direction at $\varphi = \pi$. The mean $\bar{\varphi}$ is calculated in the interval $[0, 2\pi)$. The shape parameter is subsequently calculated by

$$\theta_i^2 = \int_0^{2\pi} (\varphi - \varphi_b)^2 f_i(\varphi) d\varphi, \tag{10}$$

with $\varphi_b = \frac{1}{2}\pi$ in the interval $[0, \pi)$ and $\varphi_b = \frac{3}{2}\pi$ for $\varphi \in [\pi, 2\pi)$. Low values of θ_i indicate that a line structure at i is likely. An example of a shape parameter image is shown in Figure 3. A window of 9×9 pixels was used. The parameter values were scaled to arbitrary integer units and inverted for display purposes with $\theta' = 255(1 - \theta/4)$. These converted values of θ_i were used for further calculations.

Spatial continuity of the classes to be labeled is assumed to some extent. Due to the small size of the structures to be labeled constraints entering the labeling process through nearest neighbor interaction should be limited. Only the value of the parameter $B(k, n)$ modeling interaction between

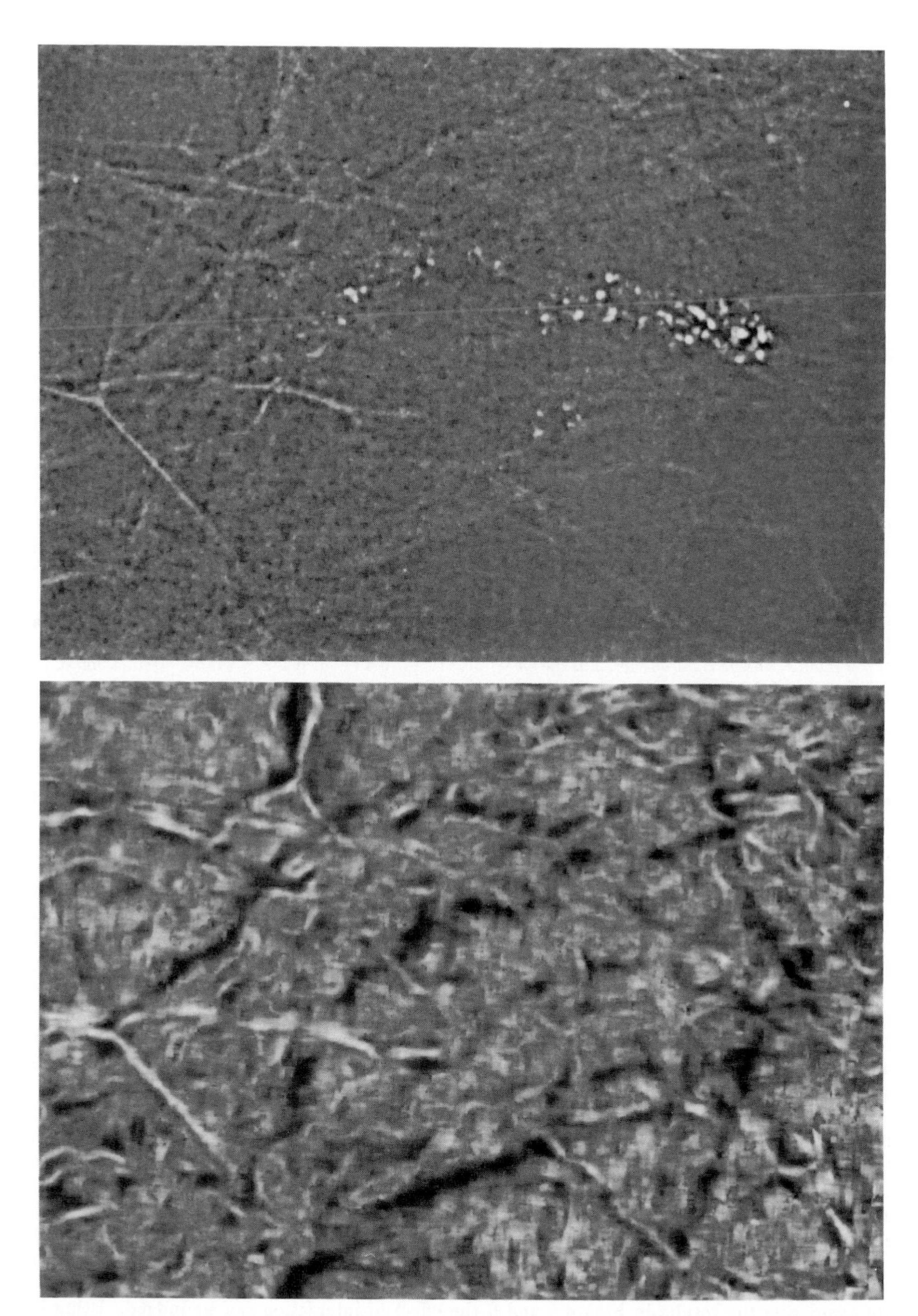

Figure 3. Parameter images representing local contrast (top) and shape (bottom) of the mammographic detail shown in Figure 1. Bright patches in the shape parameter image correspond to areas where connective tissue structures are more likely.

calcification and line labels must be chosen high, to disallow patches consisting of a mixture of these two classes. Interaction with the background class has to be chosen small or zero, to allow calcifications consisting of only one or a few pixels to survive if they are bright enough. A second order neighborhood is used to model nearest neighbor interaction. Apart from this, long range interaction is modeled between calcification and background. This permits bringing in the fact that calcifications occur in clusters. Faint isolated spots should not be interpreted as calcifications. The following model is used

$$p(x_i = k|x_{\partial i}) \propto \exp\left[-\sum_{n=1}^{3} B(k,n)g_i(n) - B'(k,n)h_i(n)\right], \tag{11}$$

in which $h_i(n)$ is the number of pixels labeled as n in a large neighborhood of i. In the following all sites within 30 pixels distance are considered, making up a total of about 2800 distant neighbors. This model may seem to induce an excessive amount of computation to evaluate the long range neighborhood at each site to be updated. Fast implementation is possible, however, because the number of calcifications is only small and because the only pairs of labels for which $B'(k,n)$ differs from zero are those for which k or n corresponds to calcification. Other pairs of labels do not contribute to the interaction sum. This means that only the number of calcification pixels in the neighborhood of i has to be known in order to update x_i. This can be computed efficiently, after each iteration cycle, by a convolution of the current calcification mask with the indicator function of the neighborhood. A space dependent field results, which strength relates to the local number of calcifications. Of course, this field can also be updated during labeling instead of at the end of each cycle. In order to reduce the computational load the local number of calcifications could also be determined at a lower resolution.

4. RESULTS

All digitized mammograms being used were scaled on an iso-precision scale. Digitization was performed with a 12 bits CCD camera (Eikonix 1412) using a 0.1 mm pixel size and a 0.05 mm aperture. The exposure was adjusted in such a way that the maximum output level of the camera was reached at film base density.

To obtain the preliminary result presented here conditional independence of the two parameters y_i' and θ_i describing the local data has been assumed. For a given set of labeled training examples this appeared to be the case in first approximation. Both the densities $f(y_i'|x_i)$ and $f(\theta_i|x_i)$ were estimated from these examples. They are shown in Figure 4. The following values for the interaction parameters were determined experimentally: $B(1,2) = 0.25$, $B(1,3) = 1.5$, $B(2,3) = 5.0$, and $B'(2,1) = B'(2,3) = 3/N_l$, $B'(2,2) = -120/N_l$, with the labels 1 corresponding to background, 2 to calcification, and 3 to line shaped structures. The parameter N_l denotes the number of pixels in the long range neighborhood. Other values follow from symmetry conditions, or else were put to zero. To prevent unrestrained expansion of calcifications in dense clusters it appeared to be necessary to limit the strength of the field representing the number of nearby calcification pixels. A cut-off at two percent of the maximum value N_l was applied.

An example of the results obtained using the procedure described above is shown in Figure 5. Comparison of this segmentation to the result of local thresholding in Figure 1 shows a large improvement. Calcifications are adequately distinguished from connective tissue, although the labeling of the latter class itself is not very good. This may be explained from the learning stage, which has not yet been optimized. Only one example of an image containing labeled line shaped structures was used. The segmentation in figure 6 was obtained by using nearest neighbor interaction only (e.g. $B'(2,n) = 0$, $n = 1,2,3$). To reduce the number of false positives the parameter value for interaction between calcifications and background was increased to $B(1,2) = 0.65$. By comparing the results in Figure 5 and 6 the effect of interaction can be judged. Using long range interaction the sensitivity inside the cluster is high enough to detect very subtle calcifications

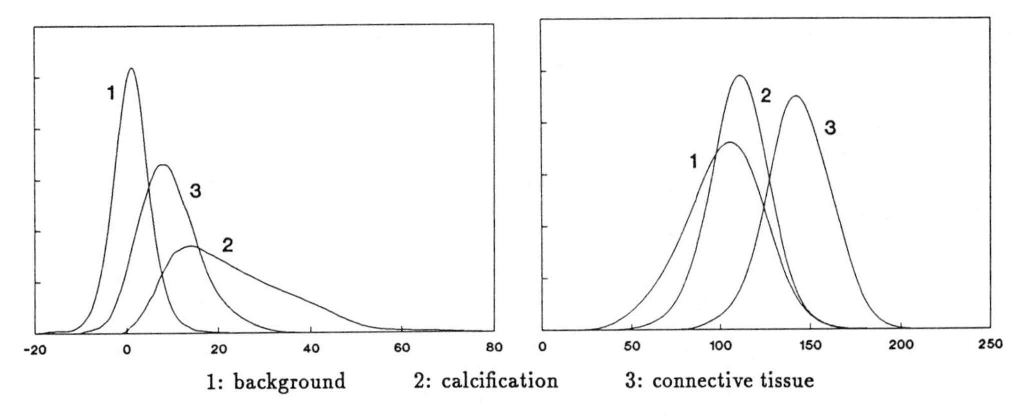

Figure 4. Estimated conditional densities of local contrast $f(y_i'|x_i)$ (left) and shape $f(\theta_i|x_i)$ (right).

as small as one pixel in size, whereas outside the cluster false positive detection is avoided. The two bright isolated spots near the upper right and left corners are film emulsion defects. In Figure 5 these are not being mistaken for calcifications because of their isolated positions and because of their very high local contrast, which is unlikely for both calcifications and background.

5. DISCUSSION AND CONCLUSIONS

The results of this study demonstrate the advantages of Bayesian image analysis in an application to computer aided mammography. A basically very simple iterative updating scheme appears to have enough potential to perform a complex recognition task. To be able to deal with planar projection images as mammograms, in which each pixel represents in fact a mixture of tissue classes, a modified version of the ICM algorithm was used for updating. This modification enables the use of parameter images describing local image features to represent the image data. Except for using relative grey levels this also allows modeling of local shape characteristics. This appeared to lead to a very efficient way of distinguishing long linear connective tissue structures from faint calcifications. In a recent paper Astley and Taylor (1990) also suggested combination of different cues derived from the local image data to improve the specificity and robustness of calcification detection algorithms. They used two cues generated by a top hat transform (local contrast) and an edge detector. Evidence from both cues is also combined in a Bayesian setting, with a nonspatial prior, however, which acts as a threshold to adjust the sensitivity of the algorithm.

To increase the likelihood of calcifications to occur in clusters long range interaction between pixel labels was modeled in the prior distribution $p(X)$, in addition to nearest neighbor interaction stimulating spatial continuity. In a fast implementation of the updating algorithm long range interaction appears as a dynamic space dependent field indicating whether there are other calcifications nearby. This resembles the use of prior probability functions to model spatial structure, as applied earlier for recognition of tissues in X-ray CT images (Karssemeijer, 1989).

By iterative optimization of the pixel labels classification is performed by combining all evidence at hand. In this process even calcifications as small as one pixel may survive provided the evidence for their existence is high enough. This is in contrast to the methods by Chan et al (1988) and Davies and Dance (1990), in which all candidates smaller than ca. 0.2 mm are rejected. With the current quality of mammographic techniques, calcifications as small as 0.1 mm may be perceptible indeed. They should not be ignored, because the presence of these fine granular calcifications may be a sign for intraductal carcinoma (Holland et al, 1990).

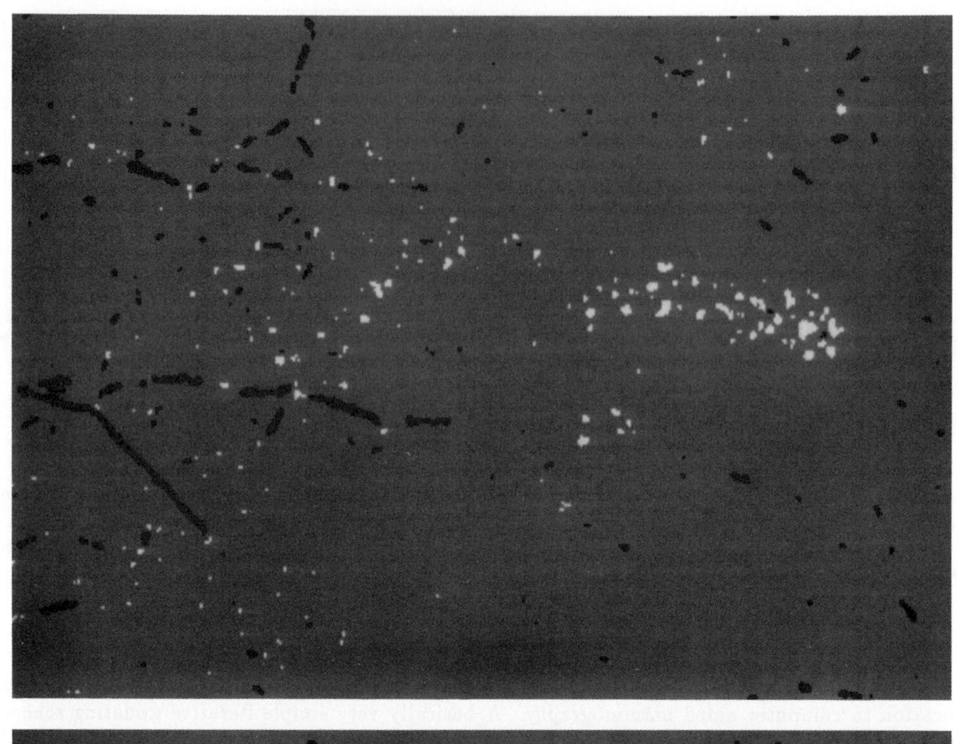

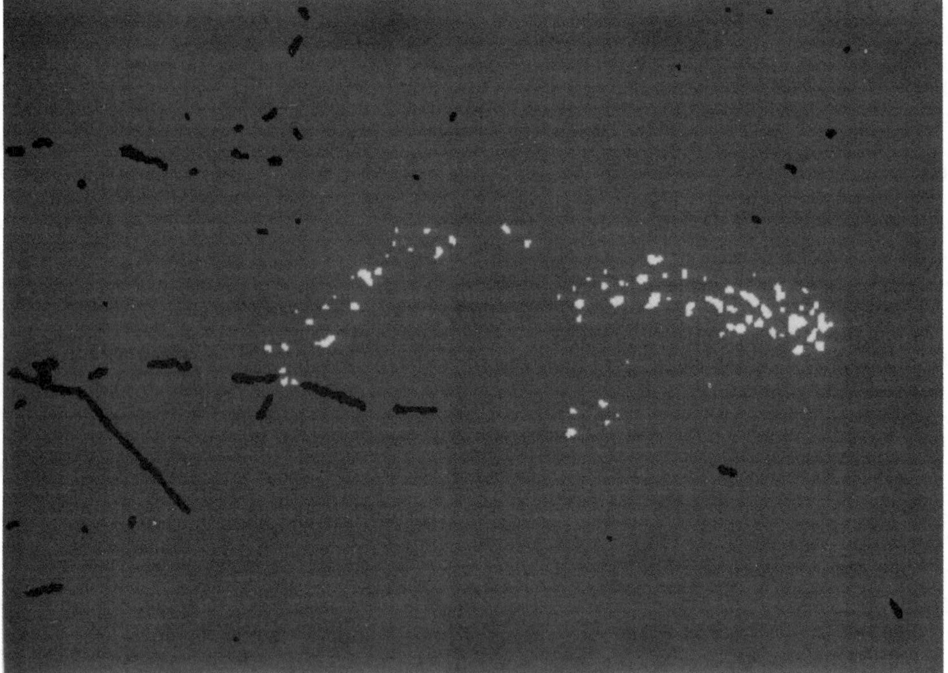

Figure 5. Calcification detection by iterative optimization applied to the image shown in Figure 1. The upper figure shows the segmentation after one iteration cycle. White represents calcifications while black stands for line-like structures. The result presented in lower figure was reached after 5 cycles.

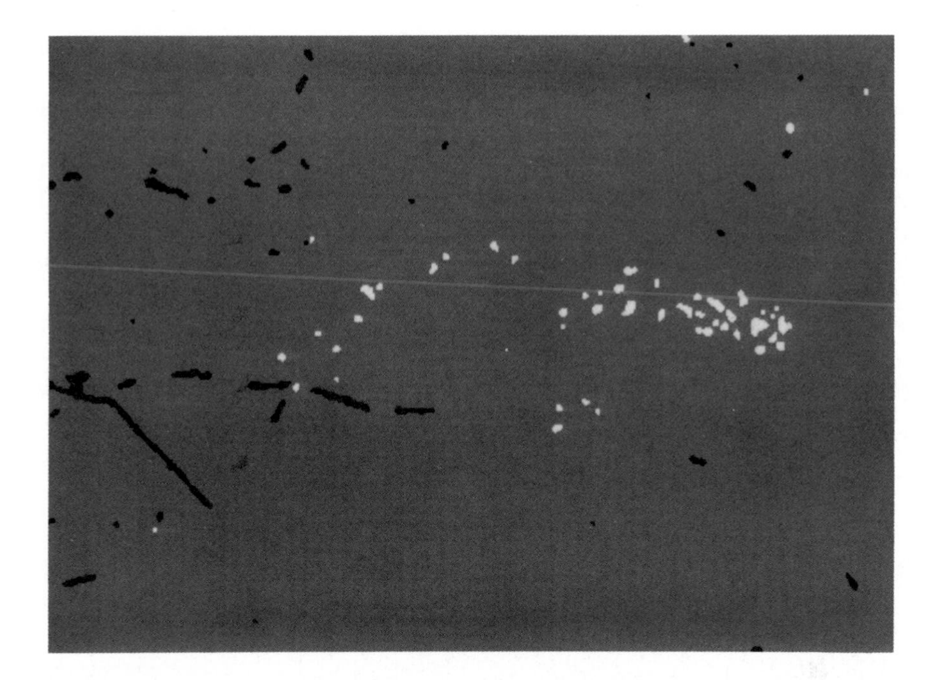

Figure 6. Segmentation result obtained by using nearest neighbour interaction only (five cycles). By comparison to the result presented in Figure 5 the influence of long range interaction can be judged.

Future work on calcification detection will include experiments applying more realistic models of long range interaction. Instead of being constant within a certain range the interaction should decrease more gradually. Also the use of proportionality to the number of calcification pixels nearby probably isn't the best choice for definition of interaction. Large calcifications do have too much influence in that way. For instance, the number of calcifications could be used instead. This can be implemented fairly easy by performing a restricted erosion of the calcification mask before calculating the field, leaving only one pixel per object.

Acknowledgements

This research was financially supported by a grant of the Dutch Praeventiefonds.

References

Astley S M and Taylor C J (1990) Combining cues for mammographic abnormalities. Proc. 1st. Brit. Mach. Vision Conf., 253-258.

Ballard D H (1984) Parameter nets. Artificial Intelligence 22:235-267

Besag J E (1986) On the statistical analysis of dirty pictures. J. Royal. Statist. Soc., Ser. B 48:259-302

Chan H P, Doi K, Vyborny C J, Lam K L and Schmidt R A (1988) Computer-Aided detection of microcalcifications in mammograms. Invest. Radiol. 23 (9): 664-671.

Davies D H and Dance D R (1990) Automatic computer detection of clustered calcifications in digital mammograms. Phys. Med. Biol. 35 (8): 1111-1118.

Dubes R C , Jain A K (1989) Random field models in image analysis. Journal of Applied Statistics 16 (2):131-164

Fam B H, Olson S L, Winter P F and Scholz F J (1988) Algorithm for the detection of fine clustered calcifications on film mammograms. Radiology 169 (2):333-337

Geman S, Geman D (1984) Stochastic relaxation, Gibbs distributions and the Bayesian restoration of images. IEEE Trans. Pattern. Anal. Machine. Intell. 6:721-741

Holland R, Hendriks J H C L, Verbeek A L M, Mravunac M, and Schuurmans J H (1990) Extent, distribution, and mammographic/histological correlations of breast ductal carcinoma in situ. Lancet, 335:519-522.

Illingworth J and Kittler J (1988) A survey of the Hough Transform. Comp. Vision, Graph. and Im. Proc. 44:87-116

Karssemeijer N (1989) A statistical method for automatic labeling of tissues in medical images. Mach. Vision and Appl. 3:75-86

Karssemeijer N and van Erning L J Th O (1991) Iso-precision scaling of digitized mammograms to facilitate image analysis. SPIE Med. Im. V, Image Processing (in press)

SCALE SPACE: ITS NATURAL OPERATORS AND DIFFERENTIAL INVARIANTS

Bart M. ter Haar Romeny Luc M. J. Florack Jan J. Koenderink
Max A. Viergever

Computer Vision Research Group, Utrecht University Hospital
Heidelberglaan 100, 3584 CX Utrecht, The Netherlands
email: luc@cv.ruu.nl, bart@cv.ruu.nl

Abstract

Why and how one should study a scale-space is prescribed by the universal physical law of scale invariance, expressed by the so-called Pi-theorem. The fact that any image is a physical observable with an inner and outer scale bound, necessarily gives rise to a 'scale-space representation', in which a given image is represented by a one-dimensional family of images representing that image on various levels of inner spatial scale. An early vision system is completely ignorant of the geometry of its input. Its primary task is to establish this geometry at any available scale. The absence of geometrical knowledge poses additional constraints on the construction of a scale-space, notably linearity, spatial shift invariance and isotropy, thereby defining a complete hierarchical family of scaled partial differential operators: the Gaussian kernel (the lowest order, rescaling operator) and its linear partial derivatives. They enable local image analysis in a robust way, while at the same time capturing global features through the extra scale degree of freedom. The operations of scaling and differentiation cannot be separated. This framework permits us to construct in a systematic way multiscale, orthogonal differential invariants, i.e. true image descriptors that exhibit manifest invariance with respect to a change of cartesian coordinates. The scale-space operators closely resemble the receptive field profiles in the mammalian front-end visual system.

Keywords

Scale-space; gaussian kernel; gaussian derivatives; differential invariants.

1. Introduction

Over the last few years there has been an increasing tendency in the image analysis literature towards a multiscale approach. A historical contribution to such an approach was the introduction of the **pyramid** (Burt et al. 1981). Though being based on a rather ad hoc method of averaging neighbouring pixels, this first model did capture the crucial observation of the inherently multiscale character of image structure.

For some time there has been discussion on the question of how to generate a **scale-space**, the continuous analogue of the pyramid, in a unique way, as there seemed to exist no clear way to choose among the many possible scale-space filters (Brunnström et al. 1990, Korn 1988, Lindeberg 1990, Babaud et al. 1986, Mallot 1989). One obviously needed a set of natural, a priori scale-space constraints.

A fundamental approach was adopted by Koenderink (1984), Witkin (1983), Yuille and Poggio (1986), who formulated an a priori constraint in the form of a causality requirement: no 'spurious detail' should be generated when increasing scale. This, together with some symmetry constraints, unambiguously established the Gaussian kernel (i.e. the **Green's function** of the **isotropic diffusion equation**) as the unique scale-space filter. Its width σ can be identified with spatial scale.

One can model an image as a scalar field on a finite-dimensional manifold and apply fundamental mathematical operations, like differentiations, to reveal local image structure. There exist many useful and rather well-established mathematical disciplines, notably differential geometry, tensor calculus, invariants theory, all of which have an increasing impact on nowadays image structure analysis.

In this paper we thoroughly discuss the fundamental concept of scaling as well as some natural constraints of a front-end visual system and show that a complete hierarchical set of **scaled differential operators** follows from these considerations. The lowest order kernel is the isotropic Gaussian. The higher order kernels are the scaled Gaussian directional derivatives, which form the natural, scaled differential operators.

With this set we can study local image geometry to any desired order. To this end we will introduce the concept of a **local jet** of order N, $J^N(L(P))$, also called N-jet (Poston and Steward 1978), defined as the equivalence class of functions L which share the same N-truncated Taylor expansion at a given point P. In other words, all images in a given N-jet are locally indistinguishable modulo higher order differences. Such a local N-jet can be represented with respect to a cartesian coordinate system by the set of partial derivatives up to N-th order, evaluated at the point P, so:

$$J^N(L(P)) \hat{=} \{L_{i_1 \dots i_n}(P)\}_{n=0}^{N} \tag{1}$$

The lower spatial indices attached to L all have values within the range $1 \dots D$, where D is the dimension of the image, and denote differentiation with respect to the associated spatial variable.

Derivatives of arbitrary order are generally well-defined and robust provided they can be calculated on a sufficiently high scale (relative to pixel scale and noise correlation width), and provided we have a sufficient resolution of intensity values (dynamic resolution, noise). We will not present a detailed discussion on these trade-offs here. In this paper we will restrict ourselves to $N \leq 3$.

This approach is valid in D dimensions, whereas much of the literature is limited to 1 or 2 dimensions (Witkin 1983, Burt et al. 1981, Lindeberg 1990, Brunnström et al. 1990). The operators show excellent performance in noisy imagery.

2. Theory

2.1. Physical versus mathematical operators

The only way to obtain structural information about a physical scene is to extract **observables** (i.c. images) with the help of some measuring apparatus. We inevitably have to face the problem of fixing the proper scale, because observables are always characterized by an intrinsic, finite scale range. Its lower bound is determined by the sampling characteristics of the device (**inner scale**), whereas the upper bound is limited by the scope of the field of view (**outer scale**).

The very fact that an image is a physical observable makes it subject to an extra constraint imposed by the **universal law of scale invariance**, which governs all laws of physics. There is no such scaling constraint on a mathematical, i.e. a dimensionless scalar field, defined on a dimensionless manifold, but it is instructive to observe how mathematicians alternatively constrain it by imposing convenient regularity conditions: a mathematical function is typically assumed to be 'sufficiently smooth', say a $C^N(\Omega)$-function on a D-dimensional domain Ω, with N sufficiently large to justify the operations performed on it. For a physical observable we cannot pose such smoothness constraints.

Clearly, it makes no sense to define a derivative of a sampled image in the strict mathematical sense (this would require the existence of an infinitesimal neighbourhood). For a digitized image, for example, this would yield the meaningless result of zero inside a pixel and would generally be undefined on the pixel's boundary. One usually circumvents this problem by considering neighbouring **pixels** instead of infinitesimally neighbouring mathematical **points** in the definition of a derivative. A well-known example of this is the 5-point Laplacean kernel (Jain 1989). This is, however, a non-robust and rather **ad hoc** solution that crucially relies on imaging conditions, like the size and shape of the pixels and other sampling characteristics. Using this operational Laplacean amounts to the implicit assumption that the relevant structures of interest have a spatial extent close to pixel scale or sampling width.

Disregarding the intrinsic dimensionality of an image or, in other words, its scaling degree of freedom, is the cause of the failure of naively applying differential geometry in image analysis. Dimensional analysis is a well recognised concept in physics, and its precise mathematical formulation will be used to argue for the necessity of a multiscale approach and to derive the unique scale-space operators for arbitrary dimensions $D > 1$.

2.2. Basic front-end vision constraints

Many interpretations of a front-end vision system are possible. We assume that its sole task is the ability to **see**, not **understand** things (cf. the 'sensorium' in Koenderink and Van Doorn 1987). By definition, a front-end vision system is assumed to be completely ignorant of any geometry of its input stimulus. Its task is to **establish** this geometry in a format that can be easily read out by various interpreting routines, which are themselves not considered to be part of the front-end. This lack of a priori geometrical knowledge argues for an a priori symmetric sampling and preprocessing of its input. Hence it is quite natural to define a front-end vision system by formulating a set of plausible symmetries. We propose the following set[1]:

- **linearity**: allowing for superposition of input stimuli.
- **spatial shift invariance**: implied by the absence of a preferred location.
- **isotropy**: implied by the absence of a preferred direction.
- **scale invariance**: implied by the absence of a preferred scale.

These basic symmetry requirements are rather weak, because we do not want the front-end system to commit itself to any specific task.

2.3. Scale Invariance

Let $F(x_1, \ldots, x_D)$ be some physical observable, e.g. the image luminance as a function of spatial coordinates, time, ..., etc. From a pure mathematical point of view there is no restriction whatsoever on the form of the function F. But because we are dealing with a **physical** entity, the requirement of **scale invariance** imposes a restriction on the form of F: only those functions are allowed that 'scale properly'. The precise meaning of this statement is expressed by the following law (Cooper and West 1988):

Universal Law of Scale Invariance: *Physical laws must be independent of the choice of fundamental parameters.* This is equivalent to:

Dimensional Analysis: *A function relating physical observables must be independent of the choice of dimensional units.*

Important are the quantities that do not change under the given scalings. They are called **dimensionless**. It is a necessary requirement to be able to express a physical relation in a unit free form.

Accordingly, simple dimensional analysis will reveal scale invariance. Remember, though, that it is the very notion of scale, in relation to the law of scale invariance, that justifies the method of dimensional analysis. The rigorous way of formulating dimensional analysis is through the **Pi-theorem** (see for a detailed discussion Olver 1986, pages 218–221).

---

[1]There may be asymmetries in the external environment the system has to operate in, like gravity, etc., which might argue for a less symmetric front-end vision model. On the other hand, this does not really limit the usefulness of these symmetries, since they can always be broken in a postprocessing stage. In addition, more restrictive (usually task-specific) symmetries may be imposed a posteriori as well.

2.4. Inner and Outer Scale

An image is just another physical observable, with an **inner scale** determined by the resolution of the sampling device, and an **outer scale** limited by its field of view.

In image analysis, there is a widespread concern with discretization effects. Strictly speaking, when we are interested in local image structure on the sampling device's inner scale, we are facing an apparent undersampling problem, from which there is only one escape: zooming in on the scene or resorting to a higher resolution acquisition.

Once having fixed the inner scale, all smaller scale image geometry has been destroyed and can by no means be reconstructed: 'there is no structure within a single pixel'. This 'catastrophical' destruction is of an intrinsically irreversible nature. One cannot expect things to be geometrically correct at the limiting lower scale boundary, where we have 'spurious inner scale detail'.

The scales of interest should preferably be relatively large compared to the imaging device's sampling width. Local image analysis on those scales is then of a continuous rather than a discrete nature (for discretization issues, see also Lindeberg 1990).

It pays off to study the human front-end visual system, being an astonishingly well performing device, having evolved over many millions of years (Hubel and Wiesel 1962). The front-end focusses on scales which are considerably larger than the eye's true inner scale: the scale of a typical rod or cone. It is not the output of individual rods and cones that is transferred, but only a **weighted** sum over typically several hundreds of them, making up a **receptive field** (RF). The profile of such a RF takes care of the small-scale 'spurious detail' generated by the individual rods and cones by scaling up to a larger inner scale in a very specific way. Only these larger scales are subject to further analysis. Indeed, numerous physiological measurements support the theory that RF profiles can be modelled by Gaussian filters of various widths (or their partial derivatives, see Young 1986a,b), which, as we will proof, is just the complete solution to our front-end requirements.

We also often encounter problems related to the device's limited outer scale. Finiteness of the image domain poses restrictions on the largest scales that are represented in each point of the image, depending on its position relative to the boundary. The further away from the nearest boundary, the larger the largest scale that is locally represented. On the boundary itself there is no scale information left. Reliable local geometry on a given spatial scale can only be found a minimal distance proportional to that scale away from the boundaries. The proportionality constant increases with the reliability threshold and also with the order of the differential geometric property you wish to study at that locus (because this requires a comparison of several neighbouring points).

Like with the inner scale problem, we may resort to the physiology of our own front-end to get a hint of how to deal with the outer scale problem. Multiple scaling is essentially achieved already at the acquisition stage (due to the many RF sizes), rather than by a postprocessing of a fixed-scale sampled image (the output of individual rods and cones is ignored as such). Thus the boundary problem **does not exist** in our front-end visual system.

These simple though important scale observations should suffice to support the claim that **a multiscale description of image structure is an indisputable necessity in front-end image analysis.**

We will introduce a continuous scale parameter σ which we allow to vary to account for the spatial scaling freedom. It has the dimension of a length and is used to define the notion of an 'immediate neighbourhood' of a point P on scale σ as the 'fuzzy' set of points within a sphere of radius $r(\sigma) \propto \sigma$ centered at P, i.e. the smallest spatial 'volume' (a length in 1D, an area in 2D) within which the image structure at that scale varies 'neither too much nor too little'.

Usually, of course, if σ is larger than the pixel size, the image structure **does** vary significantly over a distance σ, because of irrelevant small scale details. So then σ cannot denote inner scale. To reveal the 'pure' σ-scale structure of the image, we have to suppress those irrelevant details. This is most easily done in the Fourier domain by suppressing 'high' spatial frequencies: when interested in an inner scale of order σ we need a cut off frequency of order $\omega(\sigma) \approx 1/\sigma$. The question then arises of **how** to do the cut off. Danielsson and Seger (1990) use 'constant plateau' filters, but note their presumption that the sampled signal is already **bandlimited.** Without this presumption, as Witkin (1983) and Koenderink (1984) showed, there is essentially only one way to do it. However, their derivation relies on an assumption that can be phrased as 'prohibition of spurious detail'. We will not use this argument, but show show how the simple front-end symmetries set the stage for local image analysis in D dimensions.

2.5. The natural scaling operator in scale-space

Linear shift invariance implies that the rescaled image is a convolution of the input image with some kernel $G(\vec{x};\sigma)$:

$$L(\vec{x};\sigma) = \{L_0 * G(\,.\,;\sigma)\}\,(\vec{x};\sigma) \tag{2}$$

It is especially attractive to consider this property in the Fourier domain, in which the kernel becomes diagonal:

$$\mathcal{L}(\vec{\omega};\sigma) = \mathcal{L}_0(\vec{\omega})\mathcal{G}(\vec{\omega};\sigma) \tag{3}$$

The Pi-theorem states that there are only two independent dimensionless variables in this case. We may take these to be $\mathcal{L}/\mathcal{L}_0$ and $\vec{\Omega} \stackrel{\text{def}}{=} \sigma\,\vec{\omega}$. Let us therefore define

Natural Coordinates: Natural frequency (spatial) coordinates are defined as the dimensionless numbers $\vec{\Omega}$ ($\vec{X}$) associated with the coordinates $\vec{\omega}$ ($\vec{x}$) at scale-space level $\sigma > 0$ through

$$\vec{\Omega} = \sigma\,\vec{\omega} \qquad \textit{and} \qquad \vec{X} = \frac{\vec{x}}{\sigma} \tag{4}$$

We are able to derive the unique scale-space kernel from the constraints given in the previous section, and based on the consistency requirement that a **concatenation** of several rescaling operators should yield the same effect as a single operator associated with the resulting effective rescaling. In the appendix a proof is given that the **Gaussian kernel** is the unique scale-space kernel.

Having defined natural length units we also have a **natural distance measure** for the separation of two points $\vec{X}_1$ and $\vec{X}_2$ on a given level σ:

$$\delta_{1,2} \stackrel{\text{def}}{=} \|\vec{X}_2 - \vec{X}_1\| = \frac{\|\vec{x}_2 - \vec{x}_1\|}{\sigma} \tag{5}$$

Note its singularity at the highest (fictitious) resolution $\sigma = 0$. Two distinct points at $\sigma = 0$ are always infinitely far apart. This is quite natural indeed, for there can be an arbitrarily large amount of structure inbetween them. When measuring structures in terms of this natural distance concept, a typical multiscale ('fractal-like') image will not gain or loose a significant amount of structure per natural volume element.

Significant changes due to rescaling will occur only when we increase scale by an **order of magnitude** rather than by some absolute amount. Hence it is more natural to reparametrize our scale parameter, thus removing the artificial singularity at $\sigma = 0$:

Natural Scale Parameter: A natural, dimensionless scale parameter τ is obtained by the following reparametrization of σ:

$$\sigma = \varepsilon\exp\{\tau\} \qquad \text{or} \qquad \tau = \ln\{\sigma/\varepsilon\} \qquad \tau \in (-\infty, +\infty) \tag{6}$$

Note that we are forced to introduce, on dimensional grounds, a 'hidden scale' ε, which carries the dimension of a length. It is a property of the image, not of the universal scale-space kernel. An intrinsic scale inherent to any imaging device that presents itself is the sampling width or pixel width. Now we have a dimensionless scale parameter τ that indicates in a continuous manner the order of magnitude of scale relative to ε and that can take on, at least in theory, any real value. If we take ε to be the sampling width, then $\tau = 0$ corresponds to a resolution where the width of the blurring kernel is of the same order of magnitude as the pixel width ε, i.e. the inner scale of the original image. This sets a practical lower limit to the kernel widths, at which discretization effects will start to contribute to a significant degree. The range $\tau \in (-\infty, 0)$ corresponds to subpixel scales that are not represented in the image and in which all structure has been averaged out. When building up a scale-space it is most natural to use an equidistant sampling of τ, because it is this parameter that precisely formalises the physical notion of scale (Lindeberg 1990). An equidistant sampling of absolute scale σ would not be scale-independent.

In the next section we will show that the Gaussian scale-space kernel is merely the lowest order member of a complete, hierarchically ordered family of scale-space filters, all of which are compatible with our front-end vision requirements.

2.6. A complete hierarchical family of higher order operators

Although in principle the one-parameter Gaussian kernel is all one needs to generate a scale-space, it is highly insufficient for a **complete, local** description of image structure. In fact, this filter is the physical counterpart of the trivial mathematical identity operator in the sense that it extracts a **scaled copy** of a given input, representing the same scene merely on a different inner scale.

In this section we will show that the front-end vision requirements sec admit many more scale-space operations beyond mere scaling. We will derive a **complete, hierarchically ordered family of n-th order tensorial scale-space filters** $\{\gamma_{i_1 \ldots i_n}(\sigma)\}_{n=0}^{\infty}$ (in both spatial and Fourier representation) and discuss their role in front-end image analysis. The previously established Gaussian scale-space kernel naturally fits into this family as just the zeroth order, **scalar** member.

Since allowable kernels are diagonal in the Fourier domain, it is easiest to consider their Fourier representations. It is a common misconception to think that rotational invariance of the kernels implies that they only depend on the length $\|\vec{\omega}\|$ of the vector $\vec{\omega}$. This only holds for **scalar** kernels. It is easy to construct other, **tensorial** kernels within the isotropy constraint. In fact, any tensorial kernel must be proportional to a tensor product containing n factors $\vec{\omega}$, with $n = 0, 1, 2 \ldots$, since $\vec{\omega}$ is the only independent vector available. The proportionality constant must be a scalar. Putting in a scalar multiplier $\mathcal{G}(\vec{\omega}; \sigma)$ to account for proper scale fixing we can formulate the following claim:
Claim: A complete, hierarchically ordered family of multiplicative scale-space kernels is given in the Fourier representation by the set:

$$\{\mathcal{G}_{i_1 \ldots i_n}(\vec{\omega}; \sigma) = i\omega_{i_1} \ldots i\omega_{i_n} \mathcal{G}(\vec{\omega}; \sigma)\}_{n=0}^{\infty} \tag{7}$$

Alternatively, in the spatial representation, by the set of convolution filters:

$$\{G_{i_1 \ldots i_n}(\vec{x}; \sigma) = \partial_{i_1 \ldots i_n} G(\vec{x}; \sigma)\}_{n=0}^{\infty} \tag{8}$$

Note that the zeroth order kernel G_0 is the only scalar kernel. All higher order kernels are tensorial quantities. For example, the first order kernel, i.e. the **gradient**, is a vector.

The proof of this claim is given below, where we show that this **cartesian family** is sufficient for a complete determination of local image structure. Consider a given image $L(\vec{x}; \sigma)$ at a fixed scale. If we are interested in the geometrical structure of this image in the neighbourhood of some fixed point $\vec{x} \in \mathbf{R}^D$ we may consider its Taylor approximation up to a sufficient order N:

$$L(\vec{x} + \delta\vec{x}; \sigma) = \sum_{n=0}^{N} \frac{1}{n!} L_{i_1 \ldots i_n}(\vec{x}; \sigma) \delta x_{i_1} \ldots \delta x_{i_n} + \mathcal{O}(\delta\vec{x}^{N+1}) \tag{9}$$

Note that a scaled image $L(\,.\,; \sigma) \stackrel{\text{def}}{=} L_0 * G(\,.\,; \sigma)$ is a **smooth** function for all $\sigma > 0$, no matter how 'dirty' the initial condition L_0 may be (within certain very weak restrictions). Digitized images are, by the very fact of being digitized, always 'dirty' in the differential geometric sense (even in the

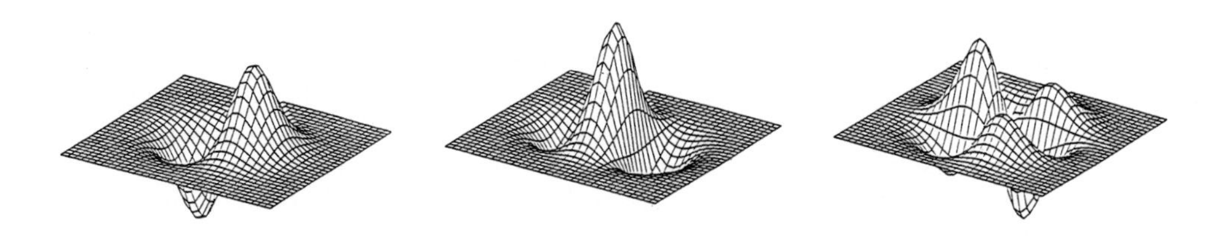

Figure 1: Some gaussian derivative profiles: G_x, G_{xx} and G_{xyy}.

absence of noise). In theory, L_0 may even be everywhere discontinuous. Scaled differentiation, as oposed to ordinary ('unscaled') differentiation, is well-posed by nature. Note the following identity:

$$L_{i_1 \ldots i_n}(\vec{x};\sigma) = L_0 * G_{i_1 \ldots i_n}(\vec{x};\sigma) \tag{10}$$

In other words, we have obtained the following important result: **In order to obtain the Cartesian partial derivatives of order n of a rescaled image $L(\vec{x};\sigma)$ one only needs to convolve the original image $L_0(\vec{x})$ with the corresponding partial derivatives of the zeroth-order Gaussian $G(\vec{x};\sigma)$.**

Since (9) represents the image's local geometry at $\vec{x}$ and at scale σ up to any desired order of precision N, we have proven the completeness of the constructed Cartesian kernel family. Each essential kernel component in the family corresponds to an independent degree of freedom. Thus it is also a minimal set. The zeroth order member of the Cartesian family represents the scaled identity operator, the higher order members constitute the physical, scaled counterpart of a **complete family of mathematical linear differential operators**. In Figure 1 the spatial profiles of a number of operators are shown.

An alternative, but equivalent way of looking at the completeness of this set of filters has been given by J. Koenderink and A. Van Doorn (1987, 1990), who took the isotropic diffusion equation as a fundamental starting point for the derivation of the complete family of scale-space filters or **local neighbourhood operators**, since this equation uniquely prohibits the generation of 'spurious detail' in scale-space (Koenderink 1984).

We end this section by noting that, although **in theory** n-th order derivatives of a **scaled** image are all well-defined, there is only one operational way of calculating them, viz. by a convolution of a lower scale image with their corresponding tensor components $G_{i_1 \ldots i_n}(\vec{x};\sigma)$ (cf. (10)). This brings us to another important result: **The operations of scaling and differentiation are essentially intimately related.**

The calculation of derivatives is most easily done in the Fourier domain, in which this filter family is diagonal:

$$L_{i_1 \ldots i_n}(\vec{x};\sigma) = \mathcal{F}^{-1}\left\{i\omega_{i_1} \ldots i\omega_{i_n} L(\vec{\omega};\sigma)\right\} = \mathcal{F}^{-1}\left\{L_0(\vec{\omega})\mathcal{G}_{i_1 \ldots i_n}(\vec{\omega};\sigma)\right\} \tag{11}$$

We have the important result that, because of the one-to-one correspondence between the Gaussian kernels $\gamma_{i_1 \ldots i_n}(\sigma)$ (on a fixed scale: 'horizontal image structure') and the Cartesian partial differential operators, it is straightforward to invoke the powerful machinery of well-established mathematical disciplines, like **differential geometry**, **tensor calculus** and **invariants theory** in a robust way. This enables us to study the visual system as a 'geometry engine' (Koenderink 1990).

2.7. Differential Invariants in Scale-Space

Once we have calculated the N-jet, we are provided with all partial derivatives of the image up to and including N-th order. However, one such derivative, L_x say, does not represent any geometrically meaningful property, since the choice of the coordinate axes is completely arbitrary. If we restrict ourselves to an orthonormal basis, we still have the possibility of rotating a given coordinate frame over any angle. Clearly, such a choice does not have anything to do with the image. On the other hand, it is also clear that any property that is invariant under such a coordinate transformation, must be connected to the image 'itself' and therefore can be given a geometric interpretation. The reverse is also true: every image property can be expressed through an invariant function (or 'invariant', in short). This shows that there is an intimate relation between **invariants theory** and **differential geometry**.

Although it is possible to give a coordinate-free description of image geometry in principle, we do need coordinates in actual calculations. To this end we simply choose any allowable coordinate system, but at the same time assure that the functions of interest are independent of that particular choice. Note that the term 'invariant' always implies the existence of a **group** of allowable transformations. In our case this is a rather 'minimal' group, viz. the product group of $SO(D)$, the Special Orthogonal group of all rotations in D dimensions (sometimes extended to $O(D)$, the full orthogonal group, by admitting reflections) and $T(D)$, the translation group. This is a very basic group, which we believe is especially important in medical imaging.

Those special combinations of image derivatives that exhibit such an invariance under cartesian coordinate transformations are called **cartesian differential invariants**. To get a basic understanding of the theory of cartesian invariants, it is necessary to understand the basics of **cartesian tensor calculus** (Kay 1988). In this section we will briefly outline how to construct functions describing true image properties.

It turns out that we can in fact construct an infinite number of invariants in each point of the image, but we will argue that there is only a small number of independent ones among these. This is to say that, to a given order N, we can build any geometrically meaningful quantity as a function of those (typically very few) independent or **irreducible invariants**.

2.7.1. Manifest Invariant Index Notation

It is clear that we cannot form an invariant out of a single derivative like L_x, whose value always crucially relies on the choice of the x-axis and thus varies among coordinate systems. It is the subject of tensor calculus to describe the transformation behaviour of quantities like L_x, called **tensor components**. A 'closed set' of tensor components, although given with respect to some arbitrarily chosen coordinate system, does however constitute a coordinate independent object, called a **tensor**. The meaning of the word 'closed' in this context is that, after a change of coordinates, each tensor component acquires a new value that can be expressed as some function of the old tensor components. This function only depends on the transformation parameters involved, i.c. a set of rotation angles and translation components (such a smoothly parametrized group of transformations is called a **Lie group** (Olver 1986)). For example, in 2D the partial derivative L_x changes, after a rotation over an angle α, according to the following rule:

$$L_{x'} = \cos\alpha L_x + \sin\alpha L_y \tag{12}$$

This shows that L_x cannot be the single component of a tensor. We should at least add the component L_y to it. Indeed, this suffices to obtain a 2-component tensor $\{L_x, L_y\}$, since these two components do transform in a closed way. Their transformations can be written in matrix form:

$$\begin{pmatrix} L_{x'} \\ L_{y'} \end{pmatrix} = \begin{pmatrix} \cos\alpha & \sin\alpha \\ -\sin\alpha & \cos\alpha \end{pmatrix} \begin{pmatrix} L_x \\ L_y \end{pmatrix} \tag{13}$$

It is clear that also the coordinates $\{x, y\}$, taken as a pair relative to some fixed origin, constitute a tensor. Let R_{ij} stand for the (i,j)-th element of the transformation matrix, then the abovementioned transformation can be conveniently written in condensed form as:

$$L'_i = R_{ij} L_j \tag{14}$$

Because L_i has only one free index it is called a 1-tensor or **vector**. But we can also consider tensors with more free indices (**n-tensors**). An example of a 2-tensor is the **Hessian**, i.e. the set of all second order partial derivatives: its transformation is given by:

$$L'_{ij} = R_{ik} R_{jl} L_{kl} \tag{15}$$

It has exactly three **essential components**, viz. $L_{xx}, L_{xy} = L_{yx}, L_{yy}$. This means that, because of the symmetry of the tensor, we cannot choose all its components independently.

By now it is obvious that all partial derivatives of a given order n form the components of an n-tensor. For each of its free indices its transformation law contains a transformation matrix with one free and one contracted index:

$$L'_{i_1 \dots i_n} = R_{i_1 j_1} \dots R_{i_n j_n} L_{j_1 \dots j_n} \tag{16}$$

These derivative tensors share the additional property of being **symmetric**, i.e. we can freely interchange indices without any effect, e.g. $L_{ij} = L_{ji}$. Thus there is a significant reduction of essential components.

Of great importance are the following two **constant tensors**: the symmetric **Kronecker tensor** δ_{ij}, which is always a 2-tensor, and the **Lévi-Civita** tensor $\varepsilon_{i_1 \dots i_D}$, which is a D-tensor in D dimensions. These tensors have invariant components, independent of the choice of the coordinate axes. They are defined as follows:

$$\delta_{ij} = \begin{cases} 1 & \text{if } i = j \\ 0 & \text{otherwise} \end{cases} \quad \text{and} \quad \varepsilon_{i_1 \dots i_D} = \begin{cases} 1 & \text{if } (i_1 \dots i_D) \text{ is even} \\ -1 & \text{if } (i_1 \dots i_D) \text{ is odd} \\ 0 & \text{otherwise} \end{cases} \tag{17}$$

The ε-tensor is **antisymmetric**, it changes sign upon an interchange of indices.

When including reflections into the transformation group, the ε-'tensor' is not a true tensor in the above sense anymore. Its significance still remains as a so called **relative** or **pseudo-tensor**. Its transformation law is then slightly modified so as to render its components invariant again:

$$\varepsilon'_{i_1 \dots i_D} \stackrel{\text{def}}{=} (\det R)^{-1} R_{i_1 j_1} \dots R_{i_D j_D} \varepsilon_{j_1 \dots j_D} = \varepsilon_{i_1 \dots i_D} \qquad \text{with} \qquad \det R = \pm 1 \tag{18}$$

Once we understand the transformation behaviour of the derivatives, we can try and combine them into (absolute or pseudo-) invariant combinations. This is in fact very easy. Given a set of tensors, the way to form an invariant is by means of **full contractions and alternations of indices** in a tensor product. A contraction is a procedure that pairwise reduces the number of free indices in a tensor by performing a restricted summation over them. More precisely, a contraction of i, j in L_{ij} by definition yields $L_{ii} = L_{ij} \delta_{ji}$.

An alternation of D tensor indices is defined as a full contraction of these indices onto the D indices of the ε-tensor, as e.g. (in 2D) in:

$$\varepsilon_{ij} L_i L_k L_{jk} = (L_x^2 - L_y^2) L_{xy} + L_x L_y (L_{yy} - L_{xx}). \tag{19}$$

Of course, functions of several invariants are themselves invariants.

If we consider reflections as well as rotations (after all, these also respect orthonormality of the coordinate basis), then the ε-tensor becomes a relative tensor and invariants containing an odd number of these become **relative invariants**, i.e. quantities that are invariant up to a possible minus sign (which shows up only when the orientation of the coordinate basis is reversed). In fact we can always write a relative invariant using exactly one ε factor. This follows from the fact that any tensor product of an even number of ε-tensors can be written in terms of δ-tensors. In 2D:

$$\varepsilon_{ij} \varepsilon_{kl} = \delta_{ik} \delta_{jl} - \delta_{il} \delta_{jk} \tag{20}$$

Similar relations hold in arbitrary dimensions.

Relative invariants are related to **oriented** geometrical objects. We will often speak of invariants, but silently admit relative invariants, too. The term **absolute invariant** is used to explicitly exclude relative ones.

2.7.2. Gauge Coordinates

In the index notation one refrains from choosing any particular coordinate frame. The invariance of a function then manifests itself through a full contraction of indices in the tensor products that make up the function. For this reason we speak of **manifest invariance** when using this notation.

Another way of forming manifest invariants is by singling out one particular, geometrically meaningful coordinate frame and using directional derivatives along its axes. There are several ways to set up such a coordinate frame. One useful way is to require, in each point of the image separately, one axis to coincide with the image gradient direction (called the w-axis henceforth). The other axis (v-axis) is then automatically directed tangentially along the isophote. By its very definition, L_v vanishes identically. This is precisely the motivation for this particular gauge. Because of the rotation freedom, this kind of requirement is always allowed, provided the image gradient does not vanish[2]. We call such a requirement a **gauge condition**, and the resulting coordinates **gauge coordinates**. It should always be checked whether a gauge condition is **admissible**, i.e. realizable through a suitable transformation provided by the transformation group at hand.

The directional derivative operators applied to the image will yield invariants. Invariance becomes manifest by writing differential invariants using these invariant differential operators.

To illustrate the use of differential geometry and at the same time show the power of gauge coordinates that are tuned to a particular problem, let us derive an expression for the **isophote**[3] **curvature** κ. The meaning of curvature of a planar curve may be intuitively clear; it is a measure for the local deviation from its tangent line (Spivak 1970). A useful definition is the following one: put a coordinate frame with its origin in the point P of interest on the curve. The x-axis should be tangent to the curve. The curve can then locally be described by a function $y(x)$ on an open interval around $x = 0$. In this system the curvature in the origin is defined as the second derivative $y''(0)$. In the (v, w)-system centered at P we have, by definition, $\kappa = w''(0)$, in which $w(v)$ denotes the function describing the isophote locally near $P(v=0, w=0)$. Now the isophote passing through the point P is implicitly given by the equation $L = L_P$. Taking first and second implicit derivatives of this equation with respect to v yields:

$$L_v + L_w w' = 0 \quad \text{and} \quad L_{vv} + 2L_{vw}w' + L_{ww}w'^2 + L_w w'' = 0 \, . \tag{21}$$

In P we have, by our suitable choice of gauge, $L_v(0) = 0$, hence also $w'(0) = 0$. So in P we have $\kappa \stackrel{\text{def}}{=} w''(0) = -L_{vv}/L_w$ (see also Clark 1989): In a similar way one may proof that the **flow line curvature** μ, i.e. the curvature of the integral curves of the gradient vector field (the orthogonal trajectories of the isophotes), is given by the formula $\mu = -L_{vw}/L_w$. So we have the following result:

$$\kappa = -\frac{L_{vv}}{L_w} \quad \text{and} \quad \mu = -\frac{L_{vw}}{L_w} \tag{22}$$

Because these invariants are closely related to simple isophote properties, they look simplest when written in this particular gauge.

It may come as a surprise to learn that, although we have calculated the isophote curvature in a **simplifying** coordinate system, it takes hardly any effort to arrive at the general expression in **arbitrary** coordinate systems. The method goes as follows: write down an invariant in manifest index notation that reduces to the simplified expression evaluated in gauge coordinates. By invariance, the two expressions are guaranteed to represent the same geometrical property. The following index notation for κ and μ (22) can be easily justified this way:

$$\kappa = \frac{L_i \varepsilon_{ij} L_{jk} \varepsilon_{kl} L_l}{(L_m L_m)^{3/2}} \quad \text{and} \quad \mu = \frac{L_i \varepsilon_{ij} L_{jk} \delta_{kl} L_l}{(L_m L_m)^{3/2}} \tag{23}$$

---

[2]The (v, w)-gauge is ill-defined in points with a vanishing gradient, but these points form a countable set, at least in **generic images**, images that are topologically stable against local distortions. Blurring a nontrivial image to a certain level of resolution always yields a generic image (which may, however, become obscured in a computer implementation by truncations due to the finite precision of L-values).

[3]An isophote is a contour of constant image values.

To see how this works, consider the isophote curvature expression in (23). Its denominator is the third power of the gradient magnitude, i.e. $L_w^3 = (L_m L_m)^{3/2}$, while its numerator can be translated from index into gauge notation as follows:

$$\kappa L_w^3 \quad = \quad L_i \varepsilon_{ij} L_{jk} \varepsilon_{kl} L_l \text{ (arbitrary system)} \quad = \quad -L_{vv} L_w^2 \text{ (gauge system)} \tag{24}$$

(the last equality follows from $L_v = 0$, $\varepsilon_{vw} = -\varepsilon_{wv} = 1$ and $\varepsilon_{vv} = \varepsilon_{ww} = 0$, so that the only nontrivial term is the one with indices $i = l = w$ and $j = k = v$). Similar arguments hold for μ in (23).

Another example of a manifest invariant is given by the well-known Laplacean:

$$\triangle L = L_{ii} = L_{vv} + L_{ww} = L_{ww} - \kappa L_w \tag{25}$$

in which we have used the expression for isophote curvature (22). This example shows that, in general, invariants can be interrelated. More specifically, the (v, w)-gauge nicely reveals the shortcomings of egde detection methods based on Laplacean zero crossings often encountered in the literature (Marr and Hildredt 1980, Hildredt 1980, Torre and Poggio 1986). The term L_{ww} is the second order image derivative along the gradient direction, i.e. normal to the isophote. If we define an edge as the locus of points of maximum gradient magnitude, which seems a quite natural choice, then the zero crossings of $\triangle L$ can only accurately describe edges if the isophotes are sufficiently straight, so that the curvature term can be ignored. It is well-known that this condition ceases to hold near corners and this is one deficiency of this zero crossings method. Another deficiency is the detection of **phantom edges** (Clark 1989), i.e. non-edge points detected by this zero crossings method. Even if the isophotes are straight, the L_{ww} zero crossings detect not only local maxima of L_w (**true edges**), but also local minima, which are the least likely candidate edge-points of all.

2.7.3. Complete Sets of Differential Invariants

It may be evident that we can construct an infinite number of invariants from any finite set of tensors by means of tensor multiplications and contractions. But it is also clear that the N-jet in a given point only has a finite number of independent degrees of freedom. For example, in 2 dimensions local image structure up to second order is completely determined by 5 independent 2-jet components, which is 1 less than the number of essential components because of the gauge degree of freedom. In the (v, w)-gauge in which $L_v = 0$ these correspond to the set $\{L, L_w, L_{vv}, L_{vw}, L_{ww}\}$. Therefore we might foresee the existence of a finite number of so called **irreducible polynomial invariants**, i.e. a set of basic polynomial invariants in terms of which all other invariants can be expressed. However plausible this argument may seem, the proof of it in the general case is far from trivial. A proof of existence was established by Hilbert, although the mathematical literature does not seem to provide an algorithm for the actual construction of such an irreducible set. In the simple case of the 2-jet, however, such an irreducible set can readily be given (Kanatani 1990). In 2D:

$$\mathcal{S} = \{L, L_i L_i, L_i L_{ij} L_j, L_{ii}, L_{ij} L_{ji}\} \tag{26}$$

An example of reducibility is given by the following identity:

$$L_{ij} L_{jk} L_{ki} = \left(\frac{3}{2} L_{ij} L_{ji} - \frac{1}{2} L_{ii} L_{jj}\right) L_{kk} \tag{27}$$

which can be verified most easily in the (p, q)-gauge, defined by the gauge condition $L_{pq} = 0$, i.e. the coordinate system in which the Hessian matrix of all second order derivatives is diagonal. This gauge is admissible, since it can always be realized by a suitable rotation. Because of invariance, this reducibility property holds in arbitrary coordinate systems.

In the next section we will take one or two examples for each of the lowest order jets ($N = 0 \ldots 3$). For the sake of presentation, we will consider the 2D case only, but it must be stressed that the dimensionality does not pose a fundamental restriction to the concepts introduced. In fact, in much of the previous theory we refrained from specifying the dimension of space explicitly wherever this was irrelevant.

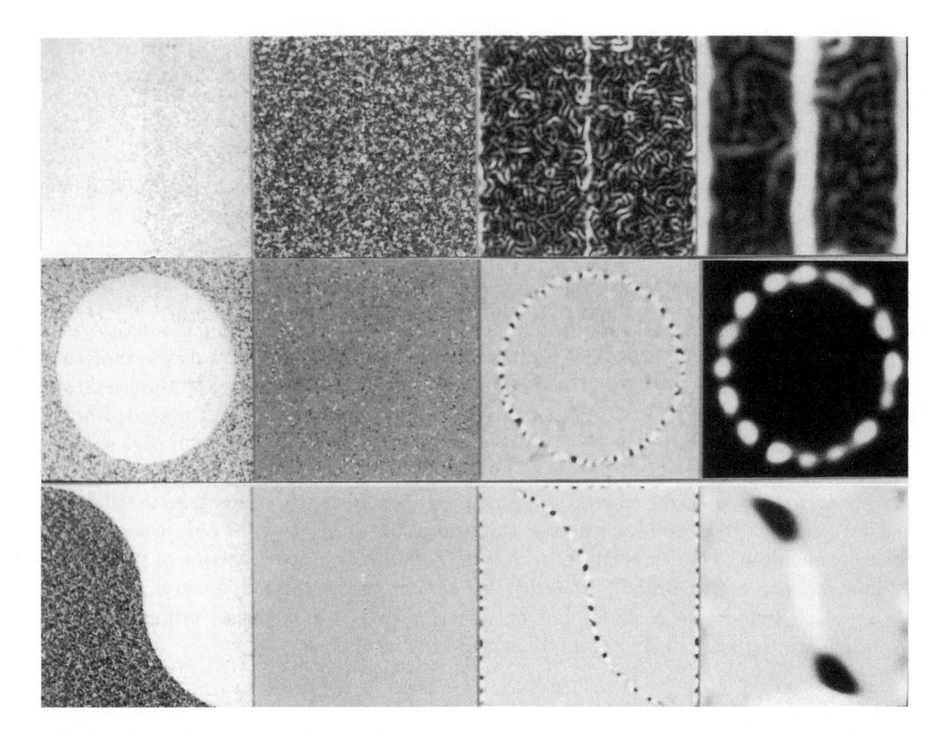

Figure 2: Some simple examples of invariants calculated for noisy test images on various scales. The additive gaussian noise imposed serves to illustrate the robustness of scale-space differentiation; **a**: 1-st order invariant L_w for noisy straight edge: 'edgeness', **b**: 2-nd order invariant $L_{vv}L_w^2$ for noisy polygon (16 corners): 'cornerness', **c**: 3-rd order invariant $L_{vvv}L_w^5 - 3L_{vv}L_{vw}L_w^4$ for noisy inflexion: 'bendedness'.

3. Applications

A trivial 0-jet example of a differential invariant is L, the local image intensity (in an implicitly given point P and on an implicitly given scale σ). A simple 1-jet example is $\sqrt{L_iL_i} = \|\nabla L\|$, the image gradient magnitude. It is most pronounced on edges, where there is a relatively strong change of intensity values over a relatively short distance. Note that this is just the Canny edge detector (Canny 1987, see also DeMicheli et al. 1989). A simple 2-jet example is the familiar Laplacean $L_{ii} = \triangle L$.

Figure 2 shows some differential invariants as they were calculated for noisy test images on several scales.

We already pointed out that there are basically only two independent, pure second order 'irreducibles', which can be taken as L_{ii} and $L_{ij}L_{ji}$ (26). Any other pure second order property can be expressed as some combination of these two. From a geometrical point of view this is clear, since a second order image property is always related to its 'deviation from flatness'. The local image intensity profile can deviate from its first order behaviour in 2 directions independently. There are 2 **principal directions**, corresponding to the coordinate axes of a system in which the mixed derivatives vanish (the (p,q)-gauge). The invariants L_{pp} and L_{qq} can then be regarded as measures for the deviation of flatness in these principal directions. This way the Laplacean $L_{ii} = L_{pp} + L_{qq}$ turns out to be twice the mean deviation from flatness, whereas the square root of $L_{ij}L_{ji} = L_{pp}^2 + L_{qq}^2$ is an absolute measure for the total deviation from flatness.

Another geometrical property closely related to this 'deviation from flatness' property is the

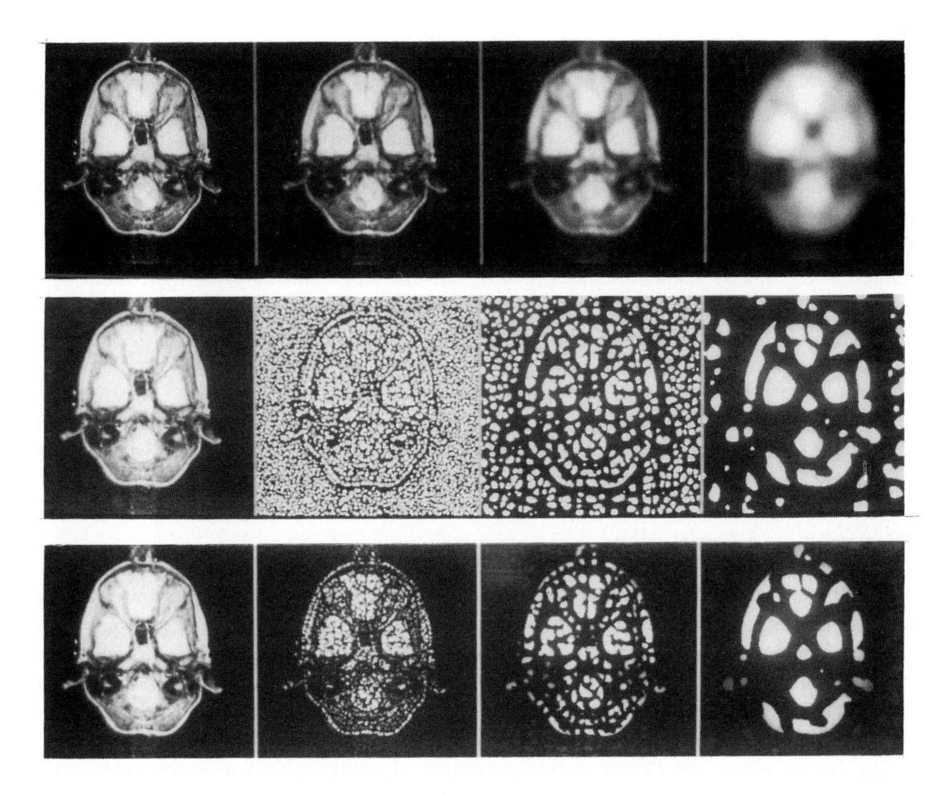

Figure 3: Patch classification based on 2-jet differential invariants calculated for an NMR images on various scales; **a**: scaled original image; **b**: light blobs, i.e. patches with positive umbilicity and negative Laplacean; note that these binary invariants are by their very nature unstable in (nearly) flat regions, since a small perturbation δL of L may easily cause them to flip. This is seen to occur in particular in the flat background; **c**: as in **b**, but now the patches have been weighted by the zeroth order images from **a** (on corresponding scales) so as to obtain continuous invariants again. This trick preserves the patches and may be used to obtain a hierarchically labelled patch classification by assigning a priority number to each patch corresponding to the relative ranking of its average intensity value (this ranking is not shown here). The patch with the highest average intensity value will then be the 'most pronounced blob' at that scale, etc. It is clear that the unstable patches surrounding the skull will acquire a low priority label in this way.

notion of light and dark 'blobs' in the image. These 'blobs' can be given an exact meaning by looking at the sign of the following invariant, called **umbilicity** (U):

$$U = \frac{\varepsilon_{ij}\varepsilon_{kl}L_{ik}L_{jl}}{L_{mn}L_{nm}} = \frac{2L_{pp}L_{qq}}{L_{pp}^2 + L_{qq}^2} \tag{28}$$

Note that we have normalized U such that $-1 \leq U \leq +1$. Dark and light blobs (or 'hills' and 'dales') now correspond to patches with equally signed principal deviations, i.e. with positive U, whereas the complementary, indifferent ('saddle-like') patches have negative U. The blobs are separated from the indifferent regions by the zero-crossings of U. We can single out only the light blobs by looking at the sign of the Laplacean in addition: blobs with negative (positive) Laplacean are light (dark) blobs.

Figure 3 shows the light blobs of an NMR image on various scales. The complementary dark blobs and indifferent patches have been suppressed.

The ordered light blob patches are reasonable primitives for a **fixed-scale segmentation**, and an across-scale linkage algorithm might be set up using these ordered fixed-scale segments to define linkage criteria for a more realistic, multiscale image segmentation.

4. Discussion and Conclusions

In this paper we have shown that the fundamental motivation for the construction of a scale-space is given by the physical nature of images and the universal law of scale invariance. Constraints arising from the lack of a priori geometrical knowledge naturally lead to the Gaussian kernel and its derivatives in $D > 1$ dimensions. The operations of scaling and differentiation are essentially intimately related. The study of invariants under a certain group of image transformations gives a robust mathematical basis for the study of image structure (Kanatani 1990).

The theory allows for the good understanding of many current available feature detection mechanisms, e.g. the Canny edge detector, Laplacean zero crossings, isophote curvature, etc. and puts these in the perspective of a broad class of differential invariants up to some order. The theory is applicable in many areas of computer vision: shape measures, shape from X, etc.

This theory must be further exploited by a closer look at the irreducible invariant 'building blocks', by inclusion of the temporal domain, stereo, and also by studying the '**deep structure**' in scale-space, incorporating our local theory into a global model.

The resemblance between the complete family of scale-space kernels and the mammalian receptive field profiles known from numerous neurophysiological data is encouraging: it suggests that our theory of differential invariants in scale-space is a promising attempt towards a robust simulation of some of the successful geometric routines actually working in the human visual system.

This theory may have an important impact on various topics in medical imaging, notably image segmentation, classification and pattern recognition.

A Derivation of the Gaussian as the unique scale-space kernel

One may write the (dimensionless) kernel $\mathcal{G}(\vec{\omega};\sigma)$ as:

$$\mathcal{G}(\vec{\omega};\sigma) = \mathcal{L}/\mathcal{L}_0 \stackrel{\text{def}}{=} \mathcal{G}(\vec{\Omega}) \tag{29}$$

For a **scalar** function, spatial isotropy implies that $\mathcal{G}$ depends only on the magnitude of the vector $\vec{\Omega}$:

$$\mathcal{G}(\vec{\Omega}) = \mathcal{G}(\Omega) \qquad \text{with } \Omega \stackrel{\text{def}}{=} \|\vec{\Omega}\|. \tag{30}$$

Let us choose σ to be such that, for fixed ω, the lower scale limit $\sigma \downarrow 0$ will leave the initial image unscaled, so:

$$\mathcal{G}(\Omega) \to 1 \qquad \text{as } \Omega \downarrow 0 \tag{31}$$

Also, the upper scale limit $\sigma \to \infty$ should lead to a complete averaging of the initial image:

$$\mathcal{G}(\Omega) \downarrow 0 \qquad \text{as } \Omega \to \infty \tag{32}$$

Performing several rescalings in succession should be consistent with performing a single, effective rescaling. More specifically, if σ_1, σ_2 are the scale parameters associated with two rescalings $\mathcal{G}(\Omega_1), \mathcal{G}(\Omega_2)$ respectively, then the concatenation of these should be a rescaling $\mathcal{G}(\Omega_3)$ corresponding to an effective scale parameter $\sigma_3 = \sigma_2 \oplus \sigma_1$, in which the abelian operator '$\oplus$' relates the effective scale parameter σ_3 to the parameters σ_1, σ_2. It is important to note that '$\oplus$' need not coincide with the familiar additive operator '+'. All that is required by consistency is that the set $\{\mathbf{R}_0^+; \oplus\}$ constitutes a **commutative semigroup** isomorphic to the commutative semigroup of image rescalings (with concatenation '$\circ$' as the semigroup operation, i.e. multiplication in the frequency domain and convolution in the spatial domain). This means that the following defining relations are applicable to $\{\mathbf{R}_0^+; \oplus\}$:

- **Semigroup operation:** $\forall \sigma_1, \sigma_2 \quad \exists \sigma_3 \stackrel{\text{def}}{=} \sigma_2 \oplus \sigma_1$
- **Associativity:** $\forall \sigma_1, \sigma_2, \sigma_3 \quad \sigma_3 \oplus (\sigma_2 \oplus \sigma_1) = (\sigma_3 \oplus \sigma_2) \oplus \sigma_1$
- **Null element:** $\exists \sigma_0 \ \forall \sigma \quad \sigma \oplus \sigma_0 = \sigma_0 \oplus \sigma = \sigma$
- **Commutativity:** $\forall \sigma_1, \sigma_2 \quad \sigma_1 \oplus \sigma_2 = \sigma_2 \oplus \sigma_1$

Note that we have already decided on the null element $\sigma_0 = 0$ by our limiting requirement (31). Note also the absence of inverse elements. Similar relations hold for the continuous semigroup $\{\gamma(\sigma), \circ\}$ of scale-space kernels in arbitrary representations (notably $\{G(\vec{x}; s), *\}$ and $\{\mathcal{G}(\vec{\omega}; \sigma), \cdot\}$).

The consistency requirement that there is a one-to-one correspondence $\gamma(\sigma) \leftrightarrow \sigma$ manifests itself mathematically by the existence of an isomorphism $\{\gamma(\sigma), \circ\} \simeq \{\mathbf{R}_0^+; \oplus\}$, i.e. a one-to-one map between these two semigroups preserving the semigroup structure:

$$\forall \sigma_1, \sigma_2 \qquad \gamma(\sigma_1) \circ \gamma(\sigma_2) = \gamma(\sigma_1 \oplus \sigma_2) \tag{33}$$

This isomorphism poses a very strong constraint on the form of the scale-space kernels. In the frequency domain, it reads:

$$\forall \vec{\omega}, \sigma_1, \sigma_2 \qquad \mathcal{G}(\vec{\omega}; \sigma_1)\mathcal{G}(\vec{\omega}; \sigma_2) = \mathcal{G}(\vec{\omega}; \sigma_1 \oplus \sigma_2) \tag{34}$$

There must also exist an isomorphism $\mathcal{P} : \{\mathbf{R}_0^+; \oplus\} \to \{\mathbf{R}_0^+; +\}$ that links the semigroup $\{\mathbf{R}_0^+; \oplus\}$ to $\{\mathbf{R}_0^+; +\}$, the familiar semigroup of non-negative real numbers equipped with ordinary addition.

On dimensional grounds (manifest scale invariance!), an allowable reparametrization of σ that preserves the limiting values must have the following form:

$$\mathcal{P} : \{\mathbf{R}_0^+; \oplus\} \to \{\mathbf{R}_0^+; +\} : \sigma \mapsto \lambda \sigma^p \tag{35}$$

for some dimensionless parameters $\lambda, p > 0$. Its inverse is given by:

$$\mathcal{P}^{-1} : \{\mathbf{R}_0^+; +\} \to \{\mathbf{R}_0^+; \oplus\} : \sigma \mapsto \left(\frac{\sigma}{\lambda}\right)^{\frac{1}{p}} \tag{36}$$

Because the factor λ merely rescales the scale parameter there is no loss of generality in putting $\lambda = 1$. Since the ordinary addition applies to $\{\mathbf{R}_0^+; +\}$ the following identity holds:

$$\gamma(\sigma_1) \circ \gamma(\sigma_2) = \gamma(\sigma_1 \oplus \sigma_2) = \gamma(\mathcal{P}^{-1}(\mathcal{P}\sigma_1 + \mathcal{P}\sigma_2)) \tag{37}$$

It is convenient to consider the frequency representation. If we define:

$$\tilde{\mathcal{G}}(\Omega) \stackrel{\text{def}}{=} \mathcal{G}(\Omega^p) \tag{38}$$

we get from (34):

$$\tilde{\mathcal{G}}(\Omega_1)\tilde{\mathcal{G}}(\Omega_2) = \tilde{\mathcal{G}}(\Omega_1 + \Omega_2) \tag{39}$$

The general solution to this constraint is a normalized exponential function:

$$\tilde{\mathcal{G}}(\Omega) = \exp(\alpha\Omega) \qquad \text{or} \qquad \mathcal{G}(\Omega) = \exp\left(\alpha(\Omega)^{1/p}\right) \tag{40}$$

in which α is an arbitrary constant.

To single out a unique scale-space kernel, we need two final, 'proper scaling' constraints for fixing the parameters α and p. One of them is **separability**:

$$\mathcal{G}(\Omega) = \prod_{i=1}^{D} \mathcal{G}(\Omega_i) \tag{41}$$

stating that an isotropic rescaling can be obtained either directly through $\mathcal{G}(\Omega)$ or by a concatenation of rescalings $\mathcal{G}(\Omega_i)$ by the same amount in each of the **independent** spatial directions $i = 1 \dots D$ separately. This fixes $p = 1/2$, so $s \stackrel{\text{def}}{=} \sigma^2$, not σ itself, is the 'additive' parameter:

$$\sigma_1 \oplus \sigma_2 = \sqrt{\sigma_1^2 + \sigma_2^2} \tag{42}$$

which is consistent with (31).

The second constraint is given by the limiting condition (32), which requires α to be negative. Its absolute value can be taken at will, since this merely defines the proportionality constant between scale σ and Gaussian width σ_{Gauss}. A convenient choice is to let them coincide, so that $\alpha = -1$.

So we have finally established the unique scale-space kernel. In the Fourier domain it is given by:

$$\mathcal{G}(\vec{\Omega}) = \exp\left(-\frac{1}{2}\Omega^2\right) \qquad \text{or} \qquad \mathcal{G}(\vec{\omega};\sigma) = \exp\left(-\frac{1}{2}\sigma^2\omega^2\right) \tag{43}$$

Acknowledgements

This work was supported by the Dutch Ministry of Economic Affairs Grant [VS-3DM]-50249/89-01.

References

Babaud, J., Witkin, A. P., Baudin, M., and Duda, R. O. Uniqueness of the gaussian kernel for scale-space filtering. **IEEE Trans. Pattern Anal. Machine Intell. 8**, 1 (1986), 26–33.

Brunnström, K., Eklundh, O., and Lindeberg, T. On scale and resolution in the analysis of local image structure. In **First European Conf. on Computer Vision, ECCV 90** (Antibes, Fr., 1990), pp. 3–13.

Burt, P., Hong, T. H., and Rosenfeld, A. Segmentation and estimation of image region properties through cooperative hierarchical computation. **IEEE Tr. SMC 11** (1981), 802–825.

Canny, J. A computational approach to edge detection. **IEEE Trans. Pattern Anal. Machine Intell. 8**, 6 (1987), 679–698.

Clark, J. J. Authenticating edges produced by zero-crossing algorithms. **IEEE Trans. Pattern Anal. Machine Intell. 11**, 1 (1989), 43–57.

Cooper, N. G., and West, G. B., Eds. **Particle physics**. Cambridge University Press, Los Alamos National Laboratory, 1988, ch. Scale and Dimension, pp. 4–21.

Danielsson, P. E., and Seger, O. Generalized and separable sobel operators. In **Machine Vision for Three Dimensional Scenes**, H. Freeman, Ed. Academic Press Inc., London, 1990.

De Micheli, E., Caprile, B., Ottonello, P., and Torre, V. Localization and noise in edge detection. **IEEE Trans. Pattern Anal. Machine Intell. 10**, 11 (1989), 1106–1117.

Hildreth, E. C. Implementation of a theory of edge detection. A. i. memo 579, Massachusetts Inst. Technol., Cambridge, 1980.

Hubel, D. H., and Wiesel, T. N. Receptive fields, binocular interaction, and functional architecture in the cat's visual cortex. **J. Physiol. 160** (1962), 106–154.

Jain, A. K. **Fundamentals of digital image processing**. Prentice Hall, Englewood Cliffs, NJ, 1989, ch. 9.

Kanatani, K. **Group-theoretical methods in image understanding**, vol. 20 of **Series in Information Sciences**. Springer-Verlag, 1990.

Kay, D. C. **Tensor calculus**. Schaum's Outline Series. McGraw-Hill Book Campany, New York, 1988.

Koenderink, J. J. The structure of images. **Biol. Cybern. 50** (1984), 363–370.

Koenderink, J. J., and van Doorn, A. J. Representation of local geometry in the visual system. **Biol. Cybern. 55** (1987), 367–375.

Koenderink, J. J., and van Doorn, A. J. Receptive field families. **Biol. Cybern. 63** (1990), 291–298.

Korn, A. F. Toward a symbolic representation of intensity changes in images. **IEEE Trans. Pattern Anal. Machine Intell. 10**, 5 (1988), 610–625.

Lindeberg, T. Scale-space for discrete signals. **IEEE Trans. Pattern Anal. Machine Intell. 12**, 3 (1990), 234–245.

Mallat, S. G. A theory for multiresolution signal decomposition: the wavelet representation. **IEEE Trans. Pattern Anal. Machine Intell. 11**, 7 (1989), 674–694.

Marr, D. C., and Hildreth, E. C. Theory of edge detection. **Proc. Roy. Soc. London B 207** (1980), 187–217.

Olver, P. J. **Applications of Lie groups to differential equations**, vol. 107 of **Graduate texts in mathematics**. Springer-Verlag, 1986.

Poston, T., and Steward, I. **Catastrophe theory and its applications**. Pitman, London, 1978.

Spivak, M. **A comprehensive introduction to differential geometry**, Publish or Perish Inc., Berkeley, 1970.

ter Haar Romeny, B. M., and Florack, L. M. J. A multiscale geometric model of human vision. In **Perception of visual information**, B. Hendee and P. N. T. Wells, Eds. Springer-Verlag, Berlin, 1991. In press.

Torre, V., and Poggio, T. A. On edge detection. **IEEE Trans. Pattern Anal. Machine Intell. 8**, 2 (1986), 147–163.

Witkin, A. P. Scale space filtering. In **Proc. IJCAI** (Karlsruhe, W. Germany, 1983), pp. 1019–1023.

Young, R. A. The gaussian derivative model for machine vision: visual cortex simulation. **J. Opt. Soc. Amer.** (July 1986).

Young, R. A. Simulation of human retinal function with the gaussian derivative model. In **Proc. IEEE CVPR** (Miami, Fla., 1986), pp. 564–569.

Yuille, A. L., and Poggio, T. A. Scaling theorems for zero-crossings. **IEEE Trans. Pattern Anal. Machine Intell. 8** (1986), 15–25.

SCALE AND SEGMENTATION OF GREY-LEVEL IMAGES USING MAXIMUM GRADIENT PATHS

L D Griffin, A C F Colchester and G P Robinson

Department of Neurology,
UMDS, Guy's Hospital, London SE1 9RT, England

Abstract

We present a technique for the construction of multi-scale representations of grey-level images. Unlike conventional representations the scales are discrete as opposed to continuous and their level is solely determined by the data. The technique is based upon connecting singular points in the image with maximum gradient paths. We also describe two segmentation methods which use the maximum gradient paths generated during the construction of the multi-scale representation. In both segmentation techniques the paths are used to determine significant ridges and troughs. The first technique operates directly on the image, while the second technique uses the magnitude of the image derivative.

Keywords

morphology; ridge; trough; saddle-point; grey-level skeleton; multi-scale representation.

1. INTRODUCTION

The need for a hierarchy of scales in bottom-up image analysis is widely accepted (Koenderink 1990). Gaussian blurring is usually used to create a scale-space (Witkin 1983; Koenderink and van Doorn 1984). This technique of producing a scale-space has many attractive properties. Many segmentation techniques employ the zero-crossings of the 2nd derivative of the different levels of the Gaussian scale-space e.g. Hummel (1986). It has been proved that the blurring does not create any new zero-crossings as the scale increases and that these zero-crossings uniquely determine the image up to a scaling constant (Yuille and Poggio 1986). Despite these nice properties this technique has an inherent difficulty in that it forms a continuous hierarchy of different scales, requiring further processing, normally user-intervention, to discover the significant levels. Psychophysical experiments have shown that the human visual system seems unable to discriminate between spatial information in scales closer than 1/8 of an octave in scale-space (Caelli et al 1983), but even the reduction of a continuous range of scales to a finite number still leaves a large number of scales to be examined or combined. There are several possible solutions to this problem: the discovery of certain domain-determined blurring levels; using a fixed selection of levels; or designing a segmentation technique which combines information from all scales (Bischof and Caelli 1988).

The first section of the paper describes the technique we use to generate the hierarchy of different scales. We have presented previously a preliminary implementation of the technique when used at a single scale. We first detect singular points (maxima, minima and saddle points) and then connect them using maximum gradient paths (MGPs) grown from saddle points (Rosin et al 1990).

Previous work on grey-level image segmentation can be considered to have taken two paths, namely edge-based segmentation and region-based segmentation. Edge-based methods use the concept of discontinuity and region-based methods use the concept of similarity (Fu and Mui 1981).

Segmentation techniques can be assessed under two criteria: detection and localisation. Edge detection techniques with good localisation such as Canny (1986) typically require further processing steps such as the explicit detection of corners, the use of edge-linking algorithms and thresholding based on edge strength and direction. Edge-based methods such as Marr-Hildreth (1980) overcome the problem of disconnected edge sections but have poor localisation and are unable to deal with junctions or sharp changes of direction. Although region-based techniques provide distinctly segmented areas and can deal with junctions, their localisation of edges is poor. Current methods for combining established edge- and region-based techniques (Nazif and Levine 1984; Hawkes et al 1990; Brelstaff et al 1990) have not proved satisfactory, and there remains a need for a more integrated approach to combining similarity and discontinuity measures.

In the second section we present two segmentation techniques that use the MGPs generated by the scale-space algorithm. The first technique uses the undifferentiated image and the second uses the differentiated image. Both segmentation techniques, like the smoothing, require no user intervention or setting of image-dependent constants. Also both produce network data structures to describe the edges, with associated confidences in each edge segment. Such a data structure we regard as essential to the higher level processing that follows an initial segmentation. In the final section of the paper we present some results of both the smoothing and segmentation algorithms.

2. METHODS

2.1. Scale

To outline the technique, we shall first explain how it is applied in the 1D case, and then describe the changes needed in order to move to 2D.

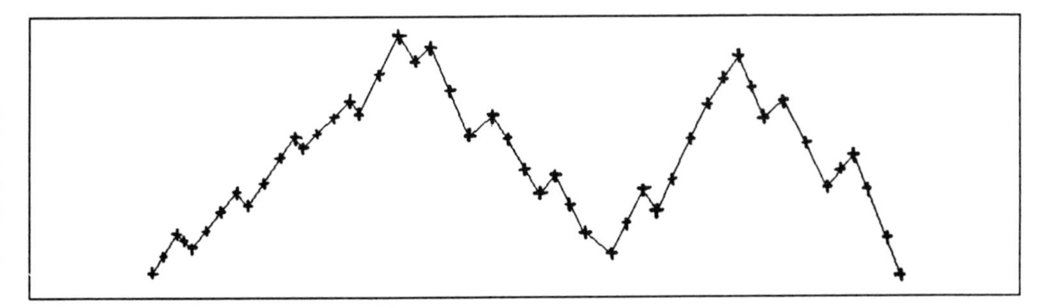

Figure 1 - A typical 1D signal. The crosses mark the original data points.

In Figure 1 we have a 1D signal. The original data points are marked with crosses; these have been joined with straight line segments to form a continuous everywhere-defined function. Our next step is to identify the maxima and minima of this function, and these are shown in Figure 2.

It can be seen that the maxima and minima partition the signal into distinct regions (we shall call them districts). Each district is bounded by a maximum and a minimum and is monotonic. To produce the first smoothing we calculate the centroid of each district. This gives us a new set of points which we again join with straight line segments (Colchester et al 1988) as shown in Figure 3. As can be seen we have jumped directly to the scale where the two gross peaks are present.

To use this technique in 2D we have four major steps. Firstly we need to create a continuous function from our input data points (which need not be grid-like). Secondly we detect maxima and minima in our continuous function. It is here that we find our first major change from the 1D case: we also detect saddle points. In the 1D case, since the maxima and minima alternate, it is obvious for any non-singular point in the input data which maximum and minimum it is associated with. In the 2D case the situation is more complex. We resolve the difficultly by growing maximum gradient paths (MGPs) both uphill and downhill from the saddle points (Rosin et al 1990). Regarding intensity

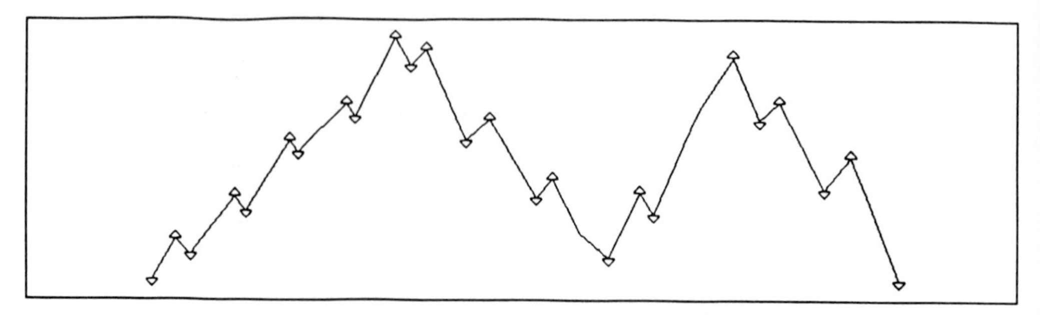

Figure 2 - The maxima and minima of the signal shown in Figure 1. The upright triangles mark maxima and the inverted triangles minima.

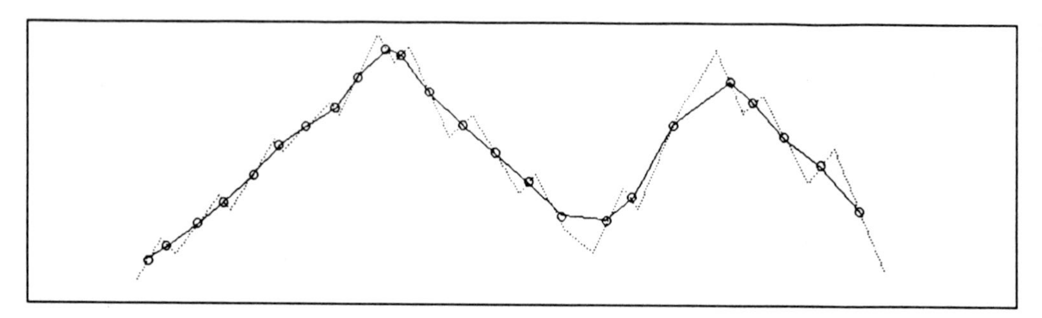

Figure 3 - The original signal is marked with a dotted line, the district centroids are marked with circles and the smoothed signal is marked with a solid line.

as elevation, a downhill MGP is the path that water would take, and an uphill MGP is the converse of it. Downhill MGPs terminate at minima and uphill MGPs at maxima. This gives us a graph whose nodes are maxima, minima and saddle points, and whose arcs are MGPs. It should be noted that in this paper 'MGP' refers to a maximum gradient path grown from a saddle. This is to be distinguished from MGPs grown from every point in the image which was the basis of our earlier method described in Colchester et al (1990). We refer to the regions bounded by the MGPs grown from saddles as districts (recall the 1D case). When smoothing we wish to preserve the local mean grey of the image whilst requiring that the excursions of the signal to local maxima and minima are lessened. To do this we constrain our smoothed signal to pass through the mean grey of each district at its centroid. The centroids of each district form a new set of points and are the result of the first smoothing. From these we create a continuous function and proceed as before.

This informal description of the algorithm is formalized in the following sections.

2.1.1. Creation of the Intensity Function

The first stage of the technique is to replace the discrete data points of the input image with a continuous everywhere-defined function (see Figures 4, 5 and 6 which are from a transverse cranial MR image). In the 1D case this is a trivial problem i.e. adjacent points are connected by straight line segments. The 2D case is resolved by using a Delaunay triangulation (Delaunay 1934; Lee and Schachter 1980). This is the triangulation of a discrete set of points such that the circumcircle of each triangle contains no data points in its interior. This definition has a unique solution except in the case of co-circular points. This ambiguity is highlighted in the case of grid-like points, which is typically the format in which image data is input. In our previous implementation (Rosin et al 1990) the problem was solved by interpolating an extra point in the centre of each group of four. However, this is an expensive solution, as it effectively doubles the initial number of data points.

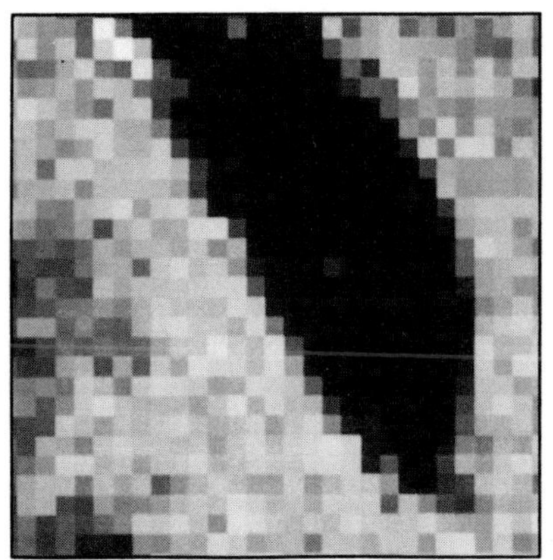

Figure 4 - A 40x40 ROI from an MR image. (i.e. The 0th generation data points).

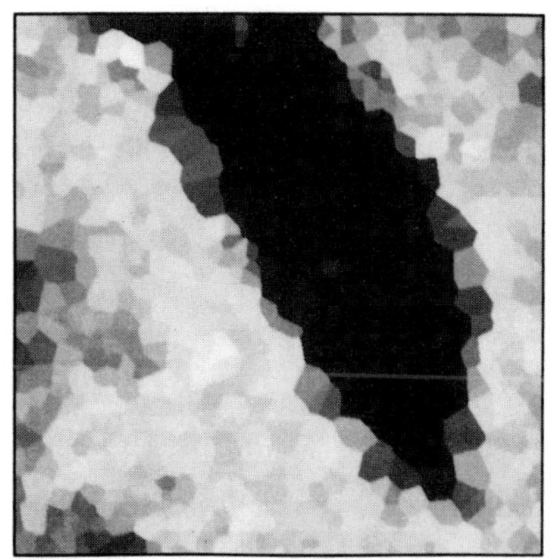

Figure 15 - The generation 1 data points displayed with 'Voronoi pixels'.

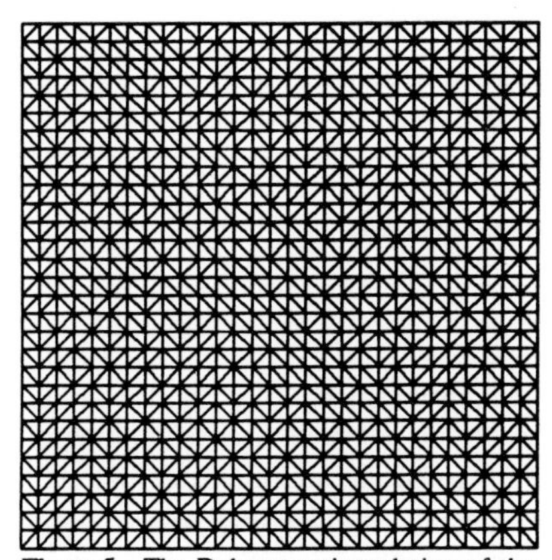

Figure 5 - The Delaunay triangulation of the data points of the image shown in Figure 4.

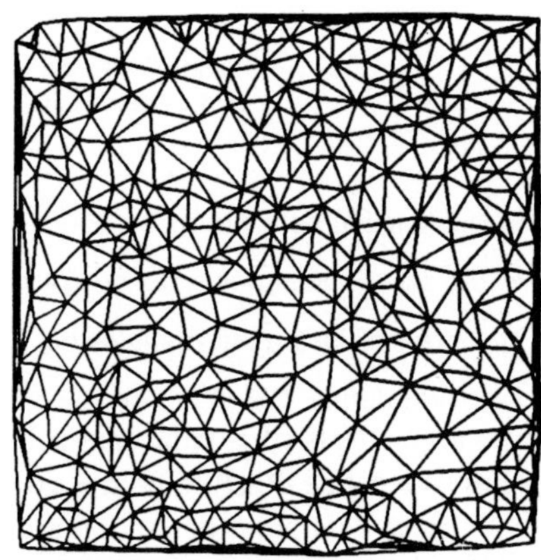

Figure 16 - The Delaunay triangulation of the generation 1 data points.

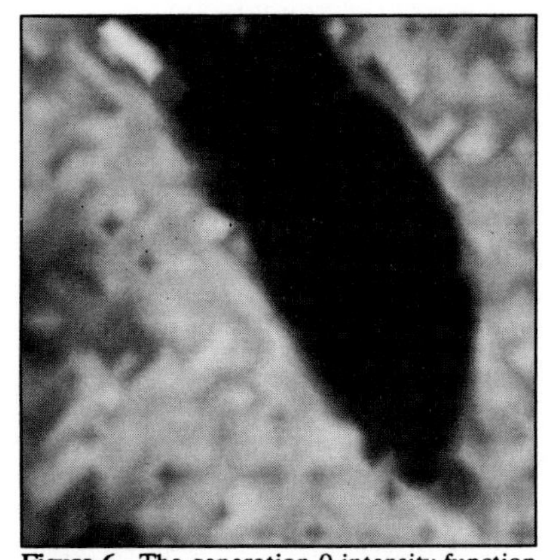

Figure 6 - The generation 0 intensity function derived from the image in Figure 4.

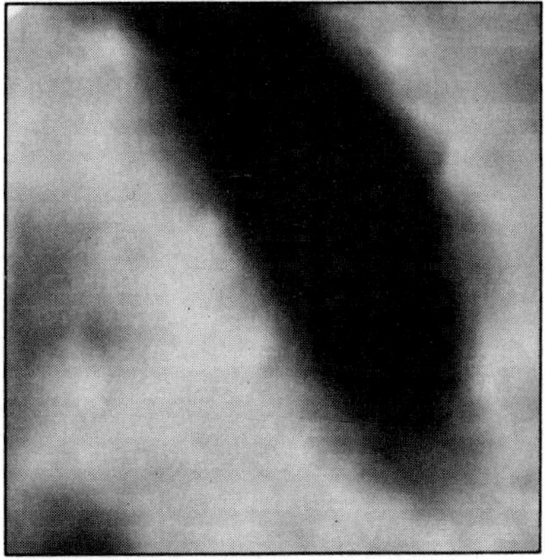

Figure 17 - The generation 1 intensity function derived from the image in Figure 4.

In the current implementation (see Figure 7), if $|v_1\text{-}v_4|<|v_2\text{-}v_3|$, then we will create the side $<v_1,v_4>$ otherwise we will create $<v_2,v_3>$. This technique has no significant effect on the performance of the algorithm as a whole. It should be noted that, except for this special case, the intensity values of the data points are not used in the construction of the triangulation. The triangle sides define which points are treated as neighbours.

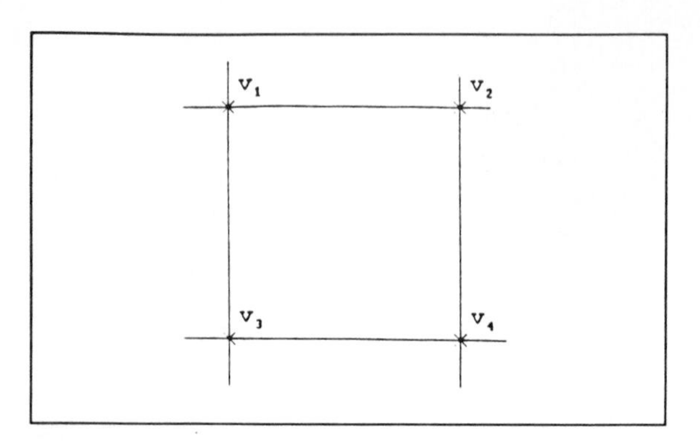

Figure 7 - 4 pixel data points arranged in a square. The numbers v_1, v_2, v_3 and v_4 are the intensities of the data points.

Let the original data points be $\boldsymbol{P}=\{p_1,\ldots,p_{N_P}\}$ where $p_i=<x_i,y_i,z_i>$, $x_i,y_i,z_i\in\mathbf{R}$.

Let the set of Delaunay sides be $\boldsymbol{S}=\{s_1,\ldots s_{N_S}\}\subseteq\boldsymbol{P}^2$

Let the set of Delaunay triangles be $\boldsymbol{T}=\{t_1,\ldots,t_{N_T}\}\subseteq\boldsymbol{P}^3$.

We define $\boldsymbol{CH}\subseteq\mathbf{R}^2$ to be the convex hull of $\mathbf{P}$ (ignoring intensity values). We shall also refer to points on the border of the convex hull; these shall be called exterior points as opposed to interior points. We define $\boldsymbol{P_E}\subseteq\boldsymbol{P}$ to be the exterior data points and $\boldsymbol{P_I}\subseteq\boldsymbol{P}$ to be the interior data points.

Each data point has an intensity value which we treat as a z co-ordinate and since 3 points in $\mathbf{R}^3$ define a plane, for each triangle $t_i\in\boldsymbol{T}$ we can find $a_i,b_i,c_i\in\mathbf{R}$ such that $z=a_ix+b_iy+c_i$ is the plane containing the 3 vertices of the triangle t_i.

We define the intensity function by first constructing the function $M{:}CH\rightarrow[1,\ldots,N_T]$ which maps any point in the convex hull to its containing triangle i.e. $<x,y>$ is within $t_{M(x,y)}$ and then the intensity function $I{:}CH\rightarrow\mathbf{R}$ is given by $I(x,y)=a_{M(x,y)}x+b_{M(x,y)}y+c_{M(x,y)}$. The function I is continuous and defined everywhere on the border and interior of the convex hull of the original data set.

2.1.2. Detection of Singular Points

The intensity function I that has been defined is composed of planar triangular facets. To simplify the following steps, we locate any connected data points that have the same intensity value and perturb the intensity of one of them. Formally, if $\exists p_i,p_j\in\boldsymbol{P} : <p_i,p_j>\in\boldsymbol{S},\ z_i=z_j$ then either z_i or z_j is changed by 1/1000th of a grey level to remove this situation, and any affected a_i's, b_i's and c_i's are recomputed. Effectively we are removing any plateaus from the intensity function I. Once all plateaus have been removed from our intensity function I : maxima, minima and saddle points will only occur at single original data points (see Figure 8; Colour Plate, p.6).

For each interior point $p_i\in\boldsymbol{P_I}$ we examine the function $g_i{:}[0,2\pi)\rightarrow\mathbf{R}$ which is the gradient going away from p_i in a radial direction (i.e. the radial gradient function). We can then define:-

$\boldsymbol{MAXIMA_I}=\{p_i\in\boldsymbol{P_I} : \forall\theta\in[0,2\pi)\ g_i(\theta)<0\}$

i.e. the set of interior points higher than all of their neighbours.

$\boldsymbol{MINIMA_I}=\{p_i \in \boldsymbol{P} : \forall\theta\in[0,2\pi)\ g_i(\theta)>0\}$

i.e. the set of interior points lower than all of their neighbours.

$\boldsymbol{SADDLES_I}=\{p_i \in \boldsymbol{P} : \exists\alpha,\beta,\gamma,\delta\in[0,2\pi)\ g_i(\alpha)=g_i(\beta)=g_i(\gamma)=g_i(\delta)=0\}$

i.e. the set of interior points with 4+ zero-crossings in their radial gradient function.

For a point $p_i \in \boldsymbol{P_E}$ on the exterior of the convex hull g_i is not defined for all $\theta\in[0,2\pi)$ but if we identify 0 and 2π we can say without loss of generality that g_i is defined on an interval $[\alpha_i,\beta_i]\subseteq[0,2\pi)$. In the case of an interior point $\alpha_i=0$, $\beta_i=2\pi$. We can now define:-

$\boldsymbol{MAXIMA_E}=\{p_i \in \boldsymbol{P_E} : \forall\theta\in[\alpha_i,\beta_i]\ g_i(\theta)<0\}$

i.e. the set of exterior points higher than all of their neighbours.

$\boldsymbol{MINIMA_E}=\{p_i \in \boldsymbol{P_E} : \forall\theta\in[\alpha_i,\beta_i]\ g_i(\theta)>0\}$

i.e. the set of exterior points lower than all of their neighbours.

$\boldsymbol{SADDLES_E}=\{p_i \in \boldsymbol{P_E} : \exists\theta,\psi\in[\alpha_i,\beta_i]\ g_i(\theta)=g_i(\psi)=0\}$

i.e. the set of exterior points with 2+ zero-crossings in their radial gradient function.

Thus for interior data points, 4+ zero-crossings are required in order to be a saddle while for an exterior saddle, only 2+ zero-crossings are required. Finally we define:-

$\boldsymbol{MAXIMA}=\boldsymbol{MAXIMA_I} \cup \boldsymbol{MAXIMA_E}$

$\boldsymbol{MINIMA}=\boldsymbol{MINIMA_I} \cup \boldsymbol{MINIMA_E}$

$\boldsymbol{SADDLES}=\boldsymbol{SADDLES_I} \cup \boldsymbol{SADDLES_E}$

2.1.3. Construction of MGPs

The next stage in processing is the construction of maximum gradient paths from the saddle points (see Figure 9; Colour Plate, p. 6). Regarding intensity as elevation, a downhill MGP can be thought of as the path that water would flow down, and an uphill path as the path water would take if the grey values of the image were negated. Since the intensity function I is composed of planar triangular facets the paths consist of sets of straight line segments. Considering a radial gradient plot of a saddle point (Figure 10) at least one MGP is grown between each pair of zero-crossings. For an exterior saddle, an MGP is also grown between the border of the convex hull and each zero-crossing adjacent to it (figure 11).

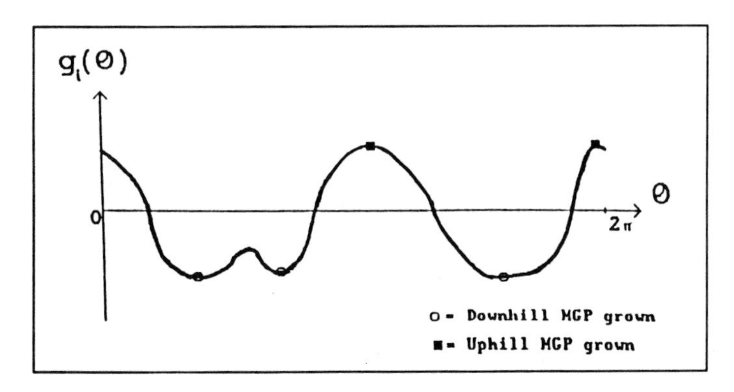

Figure 10 - Radial gradient plot of a typical Interior Saddle ($p_i \in \boldsymbol{SADDLES_I}$)

From Figures 10 and 11 it should be clear that for a saddle (interior or exterior) with n zero-crossings we grow at least n/2 uphill and n/2 downhill paths. To define this formally we must define $g_i'(\theta)$, the derivative of the radial gradient plot at p_i. This is shown in Figures 12 and 13.

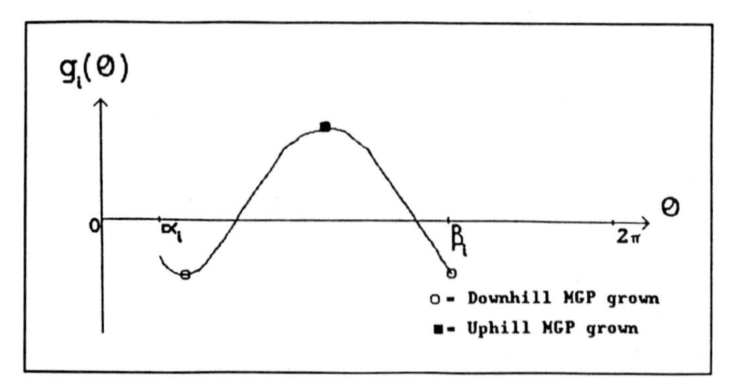

Figure 11 - Radial gradient plot of a typical Exterior Saddle *($p_i \in$ **SADDLES**$_E$)*. α_i and β_i are the directions along the convex hull.

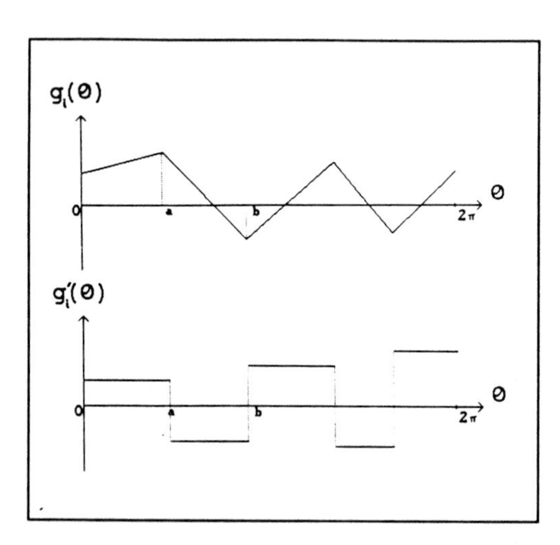

Figure 12 - Radial gradient plot and radial gradient derivative of a typical Interior Saddle *($p_i \in$ **SADDLES**$_I$)*

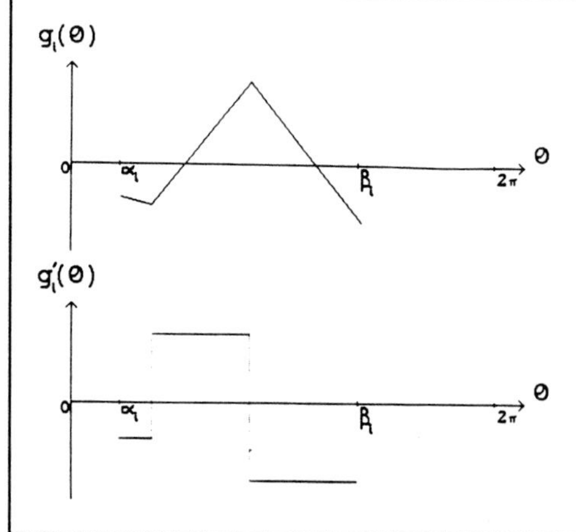

Figure 13 - Radial gradient plot and radial gradient derivative of a typical Exterior Saddle *($p_i \in$ **SADDLES**$_E$)*. α_i and β_i are the directions along the convex hull.

In Figure 12 $g_i'(a)$ and $g_i'(b)$ are not defined, so we must augment our standard definition of g_i' by saying that, if $g_i(x)$ is a local maximum or a local minimum, then $g_i'(x)=0$.

In the case of exterior saddles (Figure 13) we must also take account of the image border. We say that $g_i'(\alpha_i)=0$ iff

$g_i(\alpha_i)>0$ and $\exists\varepsilon>0\ \forall\alpha\ (\alpha-\alpha_i<\varepsilon \rightarrow g_i(\alpha)<g_i(\alpha_i))$

i.e. $g_i(\alpha_i)$ is positive and a maximum from the right.

or $g_i(\alpha_i)<0$ and $\exists\varepsilon>0\ \forall\alpha\ (\alpha-\alpha_i<\varepsilon \rightarrow g_i(\alpha)>g_i(\alpha_i))$

i.e. $g_i(\alpha_i)$ is negative and a minimum from the right.

and similarly for $g_i'(\beta_i)$. So in the above example (Figure 13) $g_i'(\beta_i)=0$ but $g_i'(\alpha_i)$ is undefined.

So, formally, from a saddle at point $p_i \in \boldsymbol{P}$ we grow uphill MGPs in each direction

$\theta : \theta \in [\alpha_i, \beta_i],\ g_i'(\theta)=0,\ g_i(\theta)>0$

i.e. all positive maxima in the radial gradient plot.

and downhill paths in each direction

$\theta : \theta\epsilon[\alpha_i, \beta_i], g_i'(\theta)=0, g_i(\theta)<0.$

i.e. all negative minima in the radial gradient plot.

The downhill MGP from the saddle at point $p_i\epsilon$***SADDLES*** in direction θ shall be denoted by

$D_{i,\theta} : \mathbf{N}\rightarrow \boldsymbol{P}$ where $D_{i,\theta}(0)=p_i$.

Similarly we denote the uphill path by

$U_{i,\theta} : \mathbf{N}\rightarrow \boldsymbol{P}$ where $U_{i,\theta}(0)=p_i$.

If $D_{i,\theta}(n)=p_j$ we can say that the MGP will continue in direction

$\theta_j\epsilon[\alpha_j,\beta_j] : g_j(\theta_j)<0, \forall\psi\epsilon[\alpha_j,\beta_j]\ \psi\neq\theta_j\ g_j(\theta_i)<g_j(\psi).$

i.e. θ_j is the global minimum of the radial gradient plot. Note that θ_j is not defined if $p_j\epsilon$***MINIMA*** which is correct as downhill MGPs terminate at minima.

When

$D_{i,\theta}(n)=p_j$ and $\exists p_k\epsilon\boldsymbol{P} : \langle p_j,p_k\rangle\epsilon\boldsymbol{S}, ((y_k-y_j)/(x_k-x_j))=\tan(\theta_j)$

i.e. the MGP runs along the side of a triangular facet.

then

$D_{i,\theta}(n+1)=p_k.$

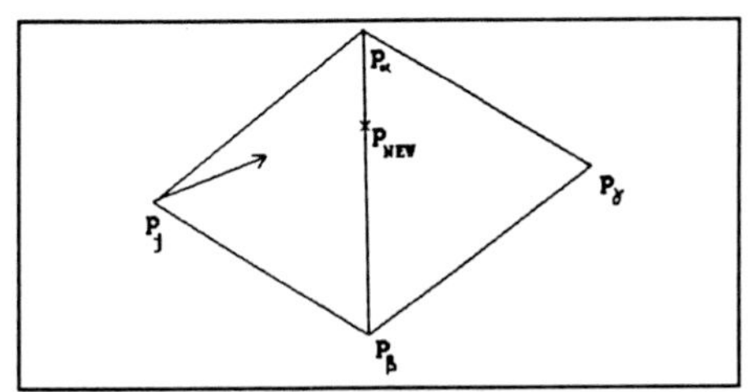

Figure 14 - An MGP running across a triangular facet.

However much of the time an MGP runs across a triangle rather than along its edge. This can be accommodated by adding additional points, sides and triangles to **P**, **S** and **T** so that a side does exist where an MGP wishes to run. Referring to Figure 14, if the maximum downhill direction from p_j is across the triangle $\langle p_j,p_\alpha,p_\beta\rangle$ then the co-ordinates $\langle x',y'\rangle$ of the point p_{NEW} can be used to calculate a new point $\langle x',y',I(x'y')\rangle$. This new point p_{NEW} is added to **P**. The side $\langle p_\alpha,p_\beta\rangle$ is removed from **S** and the sides $\langle p_\alpha,p_{NEW}\rangle$, $\langle p_\beta,p_{NEW}\rangle$, $\langle p_j,p_{NEW}\rangle$, $\langle p_\gamma,p_{NEW}\rangle$ are added to **S**. Also the triangles $\langle p_j,p_\alpha,p_\beta\rangle$, $\langle p_\gamma,p_\alpha,p_\beta\rangle$ are removed from **T** and the triangles $\langle p_j,p_\alpha,p_{NEW}\rangle$, $\langle p_j,p_\beta,p_{NEW}\rangle$, $\langle p_\alpha,p_\gamma,p_{NEW}\rangle$, $\langle p_\beta,p_\gamma,p_{NEW}\rangle$ are added to **T**.

So $D_{i,\theta}(n+1)=p_{NEW}$.

Finally

$D_{i,\theta}(n)=p_j, p_j\epsilon\boldsymbol{MINIMA} \rightarrow D_{i,\theta}(n+1)=UNDEFINED$

and

$D_{i,\theta}(n)=UNDEFINED \rightarrow D_{i,\theta}(n+1)=UNDEFINED.$

The construction of the uphill MGPs is analogous to this.

At the end of this process we have:-

$$\boldsymbol{DOWNS}=\{D_{i,\theta}\in\mathbb{N}\times(P\cup\{UNDEFINED\}) : p_i\in\boldsymbol{SADDLES},\ \theta\in[\alpha_i,\beta_i],\ g_i'(\theta)=0,\ g_i(\theta)<0\}$$

$$\boldsymbol{UPS}=\{U_{i,\theta}\in\mathbb{N}\times(P\cup\{UNDEFINED\}) : p_i\in\boldsymbol{SADDLES},\ \theta\in[\alpha_i,\beta_i],\ g_i'(\theta)=0,\ g_i(\theta)>0\}$$

This completes the construction of the MGPs.

2.1.4. Formation of Districts

We wish to partition **CH** into subsets in such a way that any two points in the same subset can be connected by a path completely contained in **CH** that does not cross an MGP. These subsets will be called districts. Since all points within the same triangle $t_i\in\boldsymbol{T}$ will be in the same district, recalling that due to modification of **P**, **S** and **T** MGPs only run along triangle sides, we only need to partition **T**. We define:-

Dt_it_j iff t_i and t_j are in the same district.

We formally define D by saying that, if the triangle t_i and t_j have some side in common, say $<\alpha,\beta>\in\boldsymbol{S}$ then

$$\neg\exists m\in\boldsymbol{DOWNS}\cup\boldsymbol{UPS}\ \exists n\ (m(n)=\alpha\wedge m(n+1)=\beta)\vee(m(n)=\beta\wedge m(n+1)=\alpha)\ \Rightarrow\ Dt_it_j$$

i.e. If there is no MGP running along the side between two adjacent triangles then they are in the same district.

From this it can be shown that D partitions **T**.

2.1.5. Construction of the next generation

To construct the next level in the scale-space (referred to as the next generation) we must construct $\boldsymbol{P'}$, the data points for the next scale. Our strategy is to form a new point for each district (see Figure 9; Colour Plate, p. 6).

If $\boldsymbol{D}_i\subseteq\boldsymbol{T}$ is a district, then associated with $\boldsymbol{D}_i$ is $\boldsymbol{A}_i\subseteq\boldsymbol{CH}$ the portion of the plane covered by the triangles in $\boldsymbol{D}_i$. $\boldsymbol{A}_i$ is finite and connected, and we can calculate $x_i', y_i'\in\mathbb{R}$ the co-ordinates of the centre of area of $\boldsymbol{A}_i$. The intensity of the new point is given by

$$z_i'=\frac{1}{|\boldsymbol{A}_i|}\int_{\boldsymbol{A}_i} I \qquad \text{i.e. the mean grey of the district.}$$

So we construct a p_i' for each district $\boldsymbol{D}_i$. With the new data set $\boldsymbol{P'}$ we can repeat the process of triangulation, detection of singular points, etc. (see Figures 15, 16 and 17, mounted beside Figures 4, 5 and 6). In Figure 15, the Voronoi polygon (Voronoi 1908) constructed around the centroid of each 0th generation district is shaded with the mean grey value of that district. In the caption we refer to these polygons as 'Voronoi pixels'. We continue this process until a newly constructed $\boldsymbol{P'}$ has only one point. This will occur if the previous generation has only a single district. We refer to the original data set as the 0th generation, and after that the 1st, 2nd,... generations.

2.2. Segmentation

We have developed two novel segmentation techniques which use the MGPs detailed in section 2.1. Both techniques are based upon the detection of ridges and troughs. The ridges and troughs are not detected with a filter but are constructed directly from the morphology of the image. Candidate ridges and troughs are first found by growing paths of maximum gradient from saddle points (see section 2.1.3 and Figure 9; Colour Plate, p. 6). Our hypothesis is that these paths form a superset of the perceptual ridges and troughs. The next stage of processing is to choose the correct subset. This is simplified when one realises that the controversial points which might belong to a ridge or to a trough are saddle points. Every section of candidate ridge has an associated section of candidate trough which crosses it at a saddle point. The problem then is to choose at each saddle whether to keep the ridge or the trough. At present we have a simple criterion for making the choice at a saddle point. For each ridge or trough an index of 'ridgeness' or 'troughness' (section 2.2.1) is calculated from the local morphology. The higher scoring of the ridge or the trough crossing at a saddle point is then selected. This means that the criterion uses no pre-determined thresholds and is local, where local is defined adaptively by the image morphology.

In the following sections (2.2.1, 2.2.2 and 2.2.3) we shall not distinguish between generations in the scale-space since the process is run identically on all generations.

2.2.1. Location of Ridges and Troughs

Uphill MGPs run along convexities and downhill MGPs along concavities, so it is a reasonable thesis that the saddle MGPs form a superset of what we would regard as perceptual ridges and troughs. Below we present a very simple criterion for extracting this subset. The criterion functions effectively in the vast majority of cases (see Figure 18; Colour Plate, p. 6).

Our strategy is to make a decision at each saddle as to whether the uphill MGPs (candidate ridges) or the downhill MGPs (candidate troughs) should be chosen as significant. A simple saddle is shown in Figure 19. $u_1,u_2,d_1,d_2>0$ refer to the grey difference between each peak or pit and the saddle and l^u_1,l^u_2,l^d_1,l^d_2 are the lengths of the MGPs.

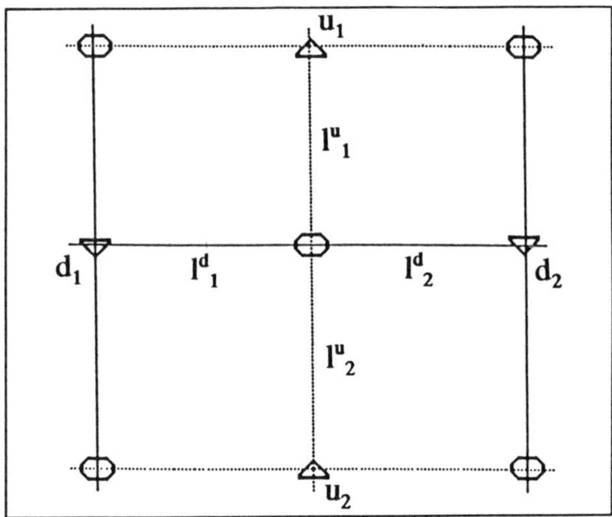

Figure 19 - A schematic diagram of a typical saddle and its associated MGPs. The hexagon marks the saddle, the upright triangle the peaks and the upside triangle the pits. The solid lines are downhill MGPs and the dotted lines uphill MGPs.

If $(u_1+u_2)(l^d_1+l^d_2) \leq (d_1+d_2)(l^u_1+l^u_2)$ then the uphill MGPs 'dominate' and a ridge is perceived; the two uphill MGPs are linked and the downhill MGPs are treated as minor concavities in the side of the ridge and are ignored. Alternatively if $(d_1+d_2)(l^u_1+l^u_2) < (u_1+u_2)(l^d_1+l^d_2)$ then a trough is perceived with the uphill MGPs being treated as minor convexities on the sides of the trough which can be ignored. At each saddle we define:-

$$\text{Ridge confidence} = \frac{(d_1+d_2)(l^u_1+l^u_2)}{(l^d_1+l^d_2)(u_1+u_2)+(l^u_1+l^u_2)(d_1+d_2)}$$

$$\text{Trough confidence} = \frac{(u_1+u_2)(l^d_1+l^d_2)}{(l^d_1+l^d_2)(u_1+u_2)+(l^u_1+l^u_2)(d_1+d_2)} = 1 - \text{ridge confidence.}$$

These confidences are used to optimize the segmentation within certain constraints (see Section 2.2.3) and can also be used to guide higher-level processing in the refinement of the initial segmentation.

2.2.2. Location of Edges via dynamic mid-grey thresholding

The ridges and troughs partition ***CH*** into what we term edge support regions (c.f. Burns et al 1986). Each edge support region is bounded by a single ridge and a single trough (and also possibly by the border of the image) and consists of several districts. Each edge support region is assumed to contain a single continuous edge.

Our edge location algorithm is based upon the concept of giving equal weight (at this early stage of processing) to the ridge at the top of the slope and the trough at the bottom. In the 1D analog this would be equivalent to choosing the mid-grey point on the slope. In many edge models this is a good approximation to the point of highest gradient which is the most commonly used edge location. To extend this edge definition into 2D we define an edge as the locus of points such that for each point the grey difference from it (its intensity being given by the intensity function I) to the nearest point (in $\mathbf{R}^2$) on the ridge is equal to the grey difference to the nearest point on the trough. This is like a dynamic mid-grey threshold between the ridge and the trough (see Figure 20; Colour Plate, p.6).

2.2.3. Location of Edges using ridges in the differentiated image

Many segmentation techniques make use of the gradient of an image (e.g. Kovasznay and Joseph 1955). Various methods have been proposed for the calculation of this derivative image such as using pairs of convolutions (Roberts 1965), differences of local averages (Rosenfeld and Thurston 1972), and surface fitting (Morgenthaler and Rosenfeld 1981). As described in section 2.1.1 we have already fitted a planar faceted function to our image. By calculating the magnitude of the maximum gradient vector of each planar facet we can produce an effective derivative image.

We grow MGPs on this derivative magnitude image and calculate ridge and trough confidences as before. Downhill MGPs run along level areas of the original image (low derivative magnitude) and uphill MGPs along steep areas (high derivative magnitude). The uphill MGPs are candidate edges and the downhill MGPs are candidate grey-level skeletons. A decision at a derivative saddle can be regarded as a choice between linking two regions into one (if the trough is chosen) or separating the two regions with edges (if the ridge is chosen). Making the ridge/trough decisions at each saddle independently (as done in section 2.2.2) produces a segmentation in which terminations of edges are possible. Instead we choose the ridges and troughs in such a way as to maximise the product of the confidences of the chosen ridges and troughs but with the extra condition that there should be no peaks in the derivative image with a single ridge terminating at them i.e. there are no terminations of edges. The optimization is currently done as follows. Initially we choose ridges and troughs at each saddle independently. We then detect any terminations of edges (i.e. ridges in the derivative image) and chose the more confident of (i) extending the edge so that it meets another edge or (ii) removing the whole edge. After optimization, the remaining uphill MGPs in the derivative image are treated as edges whilst the remaining downhill MGPs in the derivative image are the grey-level skeletons of the objects defined by the edges. As a consequence of this constrained optimization, boundary dropout is dealt with effectively and isolated edge segments are not chosen as edges.

3. RESULTS

The results of the scale-space technique on an MR image (Figures 21-23) and a SPECT image (Figures 24-26) are shown on the following pages. The MR example (Figures 21-23) is an 80x80 region of interest from a T1-weighted image. The curved ridge-like features on the left of the image are the scalp and skull, and the pale branching structure is white matter of the cerebral cortex. The anterior horn of a lateral ventricle is seen as a dark wedge-shaped structure in the bottom right hand quadrant of the image. The speckled dark area between the white matter and the skull is the cerebral grey matter. The speckled appearance of the white matter in the 0th generation image (Figure 21, left) has largely disappeared in the 1st generation image (Figure 22, left). In the 2nd generation image the white matter is an almost uninterrupted ridge (Figure 23, left). The skull and the scalp are preserved as separate ridges in the 1st generation (Figure 22, left) but have a much smoother outline. By the 2nd generation image the two ridges have almost completely coalesced (Figure 23, left). Producing further generations of this image terminates at generation 6 which is only a single point.

The central portions of Figures 21-23 show the edges calculated using the mid-grey method and the right hand portions show Canny edges for comparison. The values of σ for the Canny edges were chosen to correspond approximately to our generated scales. The mid-grey edges form closed loops around features. A large proportion of the perceived edges are successfully followed by computer edges. At a few specific locations obvious errors occur; the majority of these are accounted for by incorrect saddle decisions. Numerous small features detected in the 0th generation disappear by the 1st generation, and by the second generation the complex branching structure of the cerebral white matter is outlined almost as a single entity.

Figure 24 (left part) is a 60x60 HMPAO SPECT image of the brain showing a dark area of hypoperfusion (due to a stroke) on the right of the image. This set of images illustrates how ascending the generations simplifies the shape of complex objects. The pale area of high perfusion has moderately convoluted folds in its interior in the 0th generation image (figure 24, left). In the 1st generation image (Figure 25, left) the pale area has become a simple 'C' shape. In the 2nd generation image (Figure 26, left) the pale area has become a simple blob. Producing further generations of this image terminates at generation 5 which is only a single point.

The edges shown in the central portions of Figures 24-26 have been constructed using the mid-grey segmentation technique. The segmented regions correspond well to the perceived objects. However, close analysis shows that the localisation of edges on this type of image (with broad edge support regions) is less reliable. The right hand portions of Figures 24-26 show Canny edges for comparison. The values of σ for these edges were chosen to correspond to the scales that our technique has generated for this image. It is interesting to note that whereas the MR image produced scales approximately equivalent to Gaussian blurs of size 0.5, 2.0 and 4.0 the SPECT image produced scales equivalent to blurs of size 0.5, 3.0 and 9.0.

Some results of the segmentation technique which uses the magnitude of the image derivative are shown in Figures 27-32 (Colour Plate, p. 7). The technique has been applied to a 30x30 region of interest (Figure 27) from a larger digital subtraction cerebral angiogram on the 0th generation (finest scale) only. Figure 28 shows the complete set of differential MGPs. The uphill MGPs (red) appear to form a superset of the perceptual edges and the downhill MGPs (blue) a superset of the ridges and troughs (in the normal, undifferentiated image). Figure 29 shows the differential ridges and troughs when the selection is made at each saddle with no other considerations. In comparison, Figure 30 shows the selection which optimizes the confidence of the segmentation while preventing any terminations of edges. It can be seen that this constrained optimization improves the segmentation considerably. In Figure 30 the differential troughs are a type of grey-level skeleton of the segmented objects. Comparing Figure 31, which just shows the edges, to the Canny edges (σ chosen by us to produce the best segmentation) in Figure 32 shows the success of the technique in dealing with junctions and sharp changes of direction. It should also be noted that the localisation of the edges shows no obvious faults.

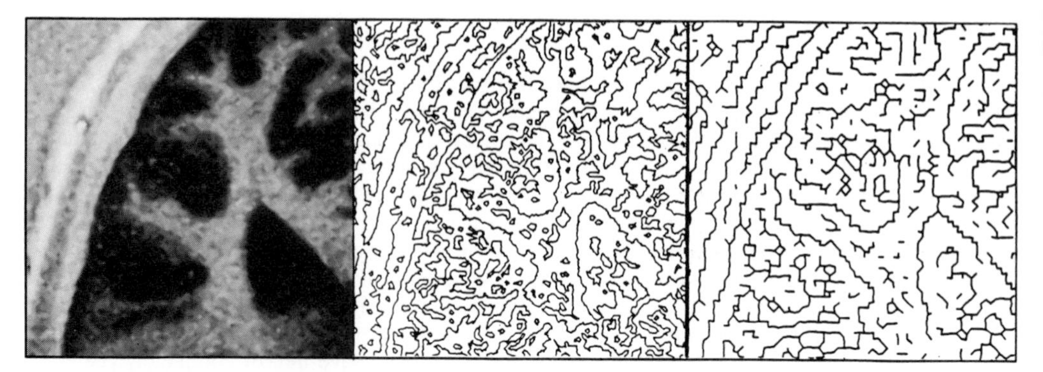

Figure 21 - Left: generation 0 intensity function of an 80x80 ROI from a transverse MR slice. Centre: edges found by 'dynamic mid-grey thresholding' around the ridges and troughs (section 2.2.2). Right: Canny edges (σ=0.5).

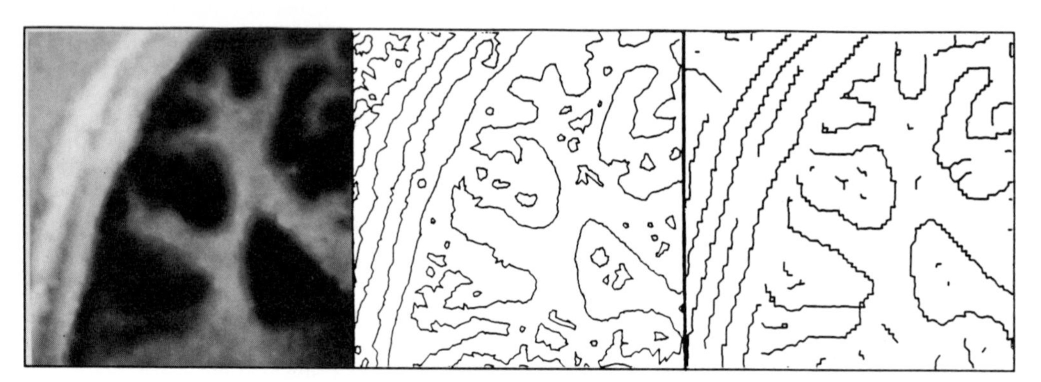

Figure 22 - Left: generation 1 intensity function. Centre: edges found by 'dynamic mid-grey thresholding' around the ridges and troughs (section 2.2.2). Right: Canny edges (σ=2.0).

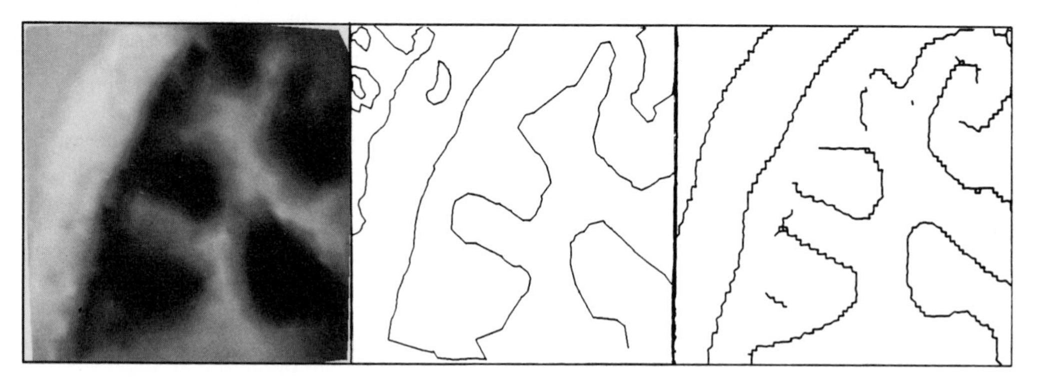

Figure 23 - Left: the generation 2 intensity function. Centre: edges found by 'dynamic mid-grey thresholding' around the ridges and troughs (section 2.2.2). Right: Canny edges (σ=4.0).

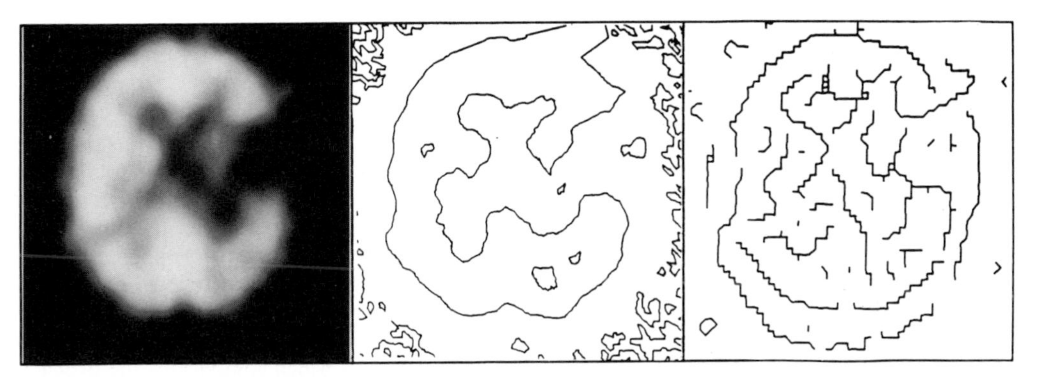

Figure 24 - Left: generation 0 intensity function of a 60x60 SPECT image. Centre: edges found by 'dynamic mid-grey thresholding' around the ridges and troughs (section 2.2.2). Right: Canny edges (σ=0.5).

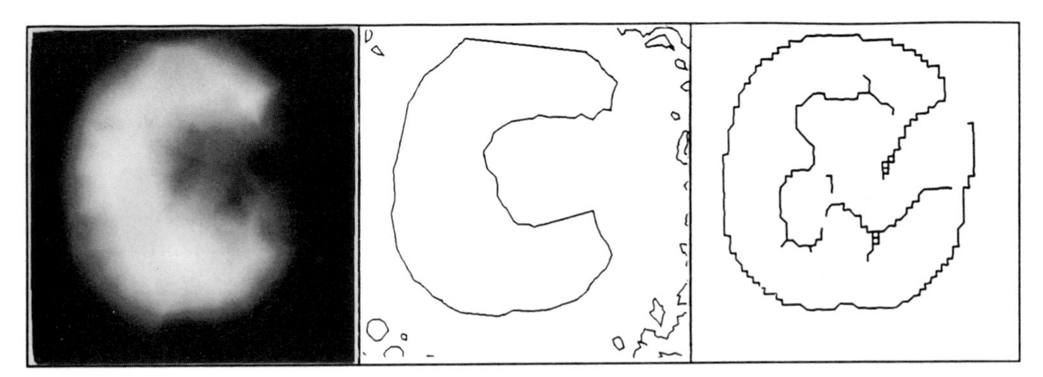

Figure 25 - Left: generation 1 intensity function. Centre: edges found by 'dynamic mid-grey thresholding' around the ridges and troughs (section 2.2.2). Right: Canny edges (σ=3.0).

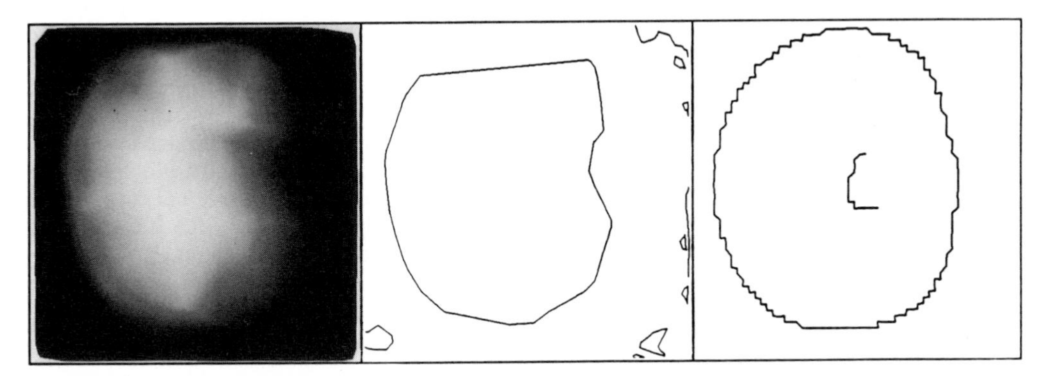

Figure 26 - Left: the generation 2 intensity function. Centre: edges found by 'dynamic mid-grey thresholding' around the ridges and troughs (section 2.2.2). Right: Canny edges (σ=9.0).

4. DISCUSSION

4.1. Scale

Gaussian blurring produces a continuum (or at least a large number) of scales for consideration. Only a small number of these scales seem useful to work with and the problem of selecting these scales is a significant one. Our technique is successful in only producing a small number of scales, and these scales do seem perceptually significant ones. However, there are many situations which the technique deals with inadequately. Figure 33 shows what can happen in the 1D case and equivalent phenomena are common when the technique is applied to 2D images. The problem is that whilst Gaussian blurring removes structure in order of absolute size our technique always removes the locally smallest level of structure. This means that structures of a similar size do not necessarily have a common scale at which they are all present which complicates the problem of multi-scale representation.

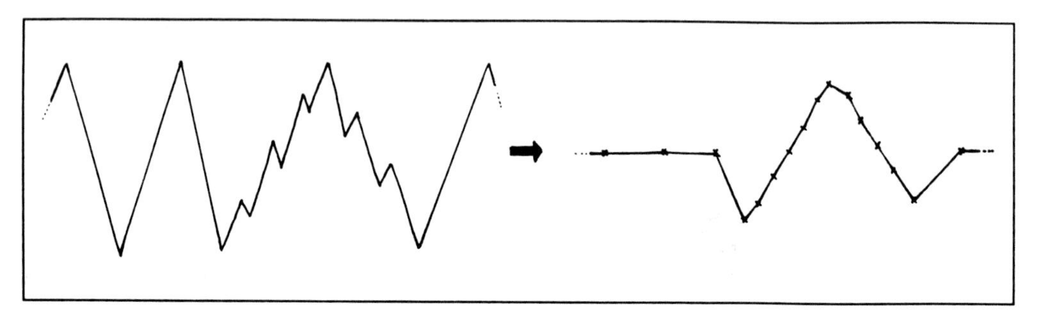

Figure 33 - Left: is the original 1D signal (0th generation). Right: the signal after one smoothing (1st generation).

4.2. Segmentation by dynamic mid-grey threshold

The first of the segmentation techniques (section 2.2.2) can be regarded as a type of region-based technique. The existence and approximate location of homogenous regions is determined by detecting ridges and troughs. Each ridge or trough is assumed to correspond to a perceptual entity (although this is only approximately true) and then the borders between them are constructed using a dynamic mid-grey threshold. The segmentation produced, like the Marr-Hildreth operator, consists of closed loops with no T-junctions or terminations of edges. There are three significant problems. Firstly, the criterion for choosing between ridges and troughs is not in total agreement with a human judge, although it is expected that this criterion can be improved. Secondly, the lack of T-junctions causes a very restricted segmentation topology. The third problem is the localisation of edges. The use of the dynamic mid-grey threshold is surprisingly effective but its weakness is shown when it encounters a 'staircase-type' set of edges. Only one edge is constructed and it is unlikely to correspond accurately to any of the perceived edges. Although this guarantees a single edges response between a ridge and a trough this is not always desirable.

4.3. Segmentation using the derivative magnitude image

The second technique operates on the magnitude of the image derivative. If one differentiates the image then the ridges in the differential image correspond to edges in the undifferentiated image. The troughs in the differential image are a type of grey-level skeleton. These grey-level skeletons differ in definition from those of Gauch and Pizer (1988) and Bailes (1990) but are a reasonable analogue to the binary skeleton as defined by Blum (1973). When we are dealing with the differential image we regard our 'ridgeness' score as a measure of local discontinuity and our 'troughness' score as a measure of local similarity. This technique is like a combined region-/edge-based approach where local (defined adaptively) measures of similarity and discontinuity are compared.

Unlike the dynamic mid-grey threshold technique this technique is unrestricted in the topology of its segmentation. The edge localisation of this technique appears to be excellent except for a few cases which seem to require the use of MGPs on the 2nd derivative of the image. These MGPs run along lines of high curvature (uphill MGPs) and low curvature (downhill MGPs). Such paths do seem occasionally to be perceptually significant. However, they are difficult to work with due to the amplification of noise when working with the 2nd derivative.

The occasional failing of the ridge/trough criterion mentioned in section 4.2 still applies to this technique, although its inadequacies seem to have less effect. The primary reason for this is the use of the constrained optimisation. Since most edges consist of several ridges in the differential image the inaccuracies of the individual ridge and trough confidences tend to even out. Suppose the edge between two posited objects consists of three candidate ridges with confidences 0.9, 0.1 and 0.8. If the selection of ridges and troughs was done at each saddle independently then the first two candidate ridges would be chosen and the third would not. This would be an unacceptable segmentation to the constrained optimisation and so would be rejected. The only acceptable segmentations would be with all three ridges selected or none selected. All three ridges being kept would have a confidence of 0.072 ($=0.9\times0.1\times0.8$), whereas rejecting all three would have a confidence of 0.018 ($=(1.0-0.9)\times(1.0-0.1)\times(1.0-0.8)$). Therefore all three candidate ridges would be kept as this is the more confident segmentation. This means that the technique deals with boundary dropout well and in an equivalent way does not respond to isolated edge segments.

5. CONCLUSIONS

To have a scale-space consisting of a small number of discrete levels which reflect genuine levels of structure would greatly aid later stages of the image understanding process such as entity matching. However the problem for our technique which is noted in section 4.1. is too serious too ignore. Nevertheless the partial success of the technique is illuminating and development of the ideas in order to reduce the number of Gaussian blurring levels which need to be considered would be a worthwhile enterprise.

Both segmentation techniques are independent from user-intervention and require no domain-dependent thresholds. In the network data structures produced by these techniques each edge or skeleton segment has an associated confidence (not a strength) making it easy to evaluate the 'difficulty' of splitting an object further or merging it with an adjacent object. Of the two techniques the derivative magnitude produces a more satisfactory segmentation and is superior to Canny. Improvement of the ridge/trough criterion should improve the technique further.

Acknowledgements

This research has been funded as part of the Alvey project MMI 134 "Model-based processing of radiological images". The authors would like to acknowledge Dr. D J Hawkes from the Department of Radiological Sciences, Guys Hospital (UMDS) and other members of the MMI 134 Consortium for their help.

References

Bailes DR (1990) The use of the grey-level SAT to find the salient cavities in echo-cardiograms. In: Proc. Brit. Machine Vison Conf. 1990, Sheffield University Press, pp151-156.

Bischof WF and Caelli TM (1988) Parsing scale-space and spatial stability analysis. Comp. Vision Graph. Image Process. 42:192-205.

Blum H (1973) Biological shape and visual science (part 1). Internat. J. Theo. Biol. 38:205-287.

Brelstaff JG, Ibison MC and Elliott PJ (1990) Edge-region integration for segmentation of MR images. In: Proc. Brit. Machine Vision Conf. 1990, Sheffield University Press, pp139-144.

Burns JB, Hanson AR and Riseman EM (1986) Extracting straight lines. IEEE Trans. Patt. Anal. and Machine Intell. PAMI-8:425-455.

Caelli TM, Brettel H, Rentschler I and Hilz R (1983) Discrimination thresholds in the two-dimensional spatial frequency domain. Vision Res. 23:129-133.

Canny J (1986) A computational approach to edge detection. IEEE Trans. Patt. Anal. and Machine Intell. PAMI-8:679-698.

Colchester ACF (1990) Network representation of 2D and 3D images. In: 3D Imaging in Medicine. Hoehne KH, Fuchs H, Pizer SM (eds), Springer-Verlag, Berlin, pp45-62.

Colchester ACF, Ritchings RT and Kodikara ND (1988) A method for multi-scale representation of data sets based on maximum gradient profiles: initial results on angiographic images. In: Proc. NATO ASI meeting on The Formation, Handling and Evaluation of Medical Images, Sep. 1988. Todd-Pokropek A and Viergever M (eds).

Colchester ACF, Ritchings RT and Kodikara ND (1990) A new approach to image segmentation using maximum gradient profiles orthogonal to edges. Image and Vision Computing. 8:211-217.

Delaunay B (1943) Sur la sphere vide. Bull. Acad. Sci. USSR(VII), Classe Sci. Nat. pp793-800.

Fu KS and Mui JK (1981) A survey on image segmentation. Patt. Recog. 3:3-16.

Gauch JM and Pizer SM (1988) Image description via the multi-resolution intensity axis of symmetry. In: Proc. of 2nd ICCV 1988, pp269-274.

Hawkes DJ, Hill DLG, Lehmann ED, Robinson GP, Maisey MN, Colchester ACF (1990) Preliminary work on the interpretation of SPECT images with the aid of registered MR images and an MR derived 3D neuro-anatomical atlas. In: 3D Imaging in Medicine, Hoehne KH, Fuchs H, Pizer SM (eds), Springer-Verlag, Berlin pp241-252.

Hummel RA (1986) Representations based on zero-crossings in scale-space. In: Proc. IEEE Conf. on Comp. Vision and Patt. Recog. pp204-209.

Koenderink JJ, van Doorn AJ (1984) The structure of images. Trans. Biol. Cyb. 50:363-370.

Koenderink JJ (1990) Solid Shape. MIT Press, Cambridge MA.

Kovasznay LS and Joseph HM (1955) Image processing. In: Proc. IRE 43:560-570.

Lee DT and Schachter BJ (1980) Two algorithms for constructing a Delaunay triangulation. Internat. J. Comp. Info. Sci. 9:219-242.

Marr DC and Hildreth E (1980) Theory of edge detection. Proc. Roy. Soc. B B-207:187-217.

Morgenthaler DC and Rosenfeld A (1981) Multidimensional edge detection by hypersurface fitting. IEEE Trans. Patt. Anal. Machine Intell. PAMI-3:482-486.

Nazif AM and Levine MD (1984) Low level image segmentation: an expert system. IEEE Trans. Patt. Anal. and Machine Intell. PAMI-6:555-577.

Rosin PL, Colchester ACF and Hawkes DJ (1990) Early visual representations using regions defined by maximum gradient profiles between singular points. In: Information Processing in Medical Imaging, Ortendahl D and Llacer J (eds), Wiley-Liss, New York. pp369-388.

Rosenfeld A and Thurston M (1972) Edge and curve detection for visual scene analysis. IEEE Trans. Comput. 21:562-569.

Voronoi G (1908) Nouvelles applications des parametres a la theorie des formes quadratiques. Deuxieme Memoire: Recherches sur la paralleloedres. Deuexieme reiene angnew. Math. 134, pp198-287.

Witkin AP (1983) Scale-space filtering. In: Proc. 7th Internat. Joint Conf. Art. Intell. pp1019-1022.

Yuille AL and Poggio TA (1986) Scaling theorems for zero-crossings. IEEE Trans. Pat. Anal. Machine Intell. PAMI-8:15-25.

USING UNCERTAINTY TO LINK 3D EDGE DETECTION AND LOCAL SURFACE MODELLING

O Monga, N Ayache, P Sander

INRIA Domaine de Voluceau-Rocquencourt - B.P. 105
78153 Le Chesnay Cedex, France

Abstract

We establish a theoretical link between the 3D edge detection and the local surface approximation using uncertainty. As a practical application of the theory, we present a method for computing typical curvature features from 3D medical images. We use the uncertainties inherent in edge (and surface) detection in 2- and 3-dimensional images determined by quantitatively analyzing the uncertainty in edge position, orientation and magnitude produced by the multidimensional (2-D and 3-D) versions of the Monga-Deriche-Canny recursive separable edge-detector. These uncertainties allow to compute local geometric models (quadric surface patches) of the surface, which are suitable for reliably estimating local surface characteristics, for example, Gaussian and Mean curvature. We demonstrate the effectiveness of our methods compared to previous techniques. These curvatures are then used to obtain more structured features such as curvature extrema and lines of curvature extrema. The final goal is to extract robust geometric features on which registration and/or tracking procedures can rely.

key words

Typical surface features, local curvature extrema, mean and Gaussian curvature, local surface modelling, uncertainty, 3D edge detection.

1 Introduction

Modern medical imaging techniques, such as Magnetic Resonance Imaging (MRI) or X-ray computed tomography provide three dimensional (3D) images of internal structures of the body, usually by means of a stack of tomographic images. In many applications, the physician asks for a segmentation of these 3D images into regions of interest he wants to manipulate, display, and characterize by objective measurements [AB+90]. The first stage in the automatic analysis of such data is 3D edge detection [ZH81, MD89, MDR91] which provide points corresponding to the boundaries of the surfaces forming the 3D structure. The next stage is to characterize the local geometry of these surfaces in order to extract points or lines on which registration and/or tracking procedures can rely [Koe90].

Sander and Zucker have proposed to compute surface singularities by the calculation of curvatures using local approximation and by iterative refinement of the curvature field [SZ87, SZ90]. In this paper we present a pipeline of processes which define a hierarchical description of the second order differential characteristics of the surfaces. We focus on the theoretical coherence of these levels of representation. Our levels of representation of the local geometry of the surfaces are :

- 3D edge points,
- Mean and Gaussian curvature, principal curvature directions,
- Local images of the curvatures,
- Characteristic points : curvature extrema, parabolic points, umbilic points ...

- Characteristic lines : Lines of curvature extrema, parabolic lines, Umbilic points ...

Three-dimensional edge detection is performed using recursive separable filters approximating the gradient or Laplacian as described in [MD89, MDR91, MDMC]. From these edge points we build an adjacency graph with position and gradient vector attached to each edge point.

To compute curvatures from this graph we fit a local model at the neighbourhood of each point. The local model is a quadratic surface and the fitting method is a Kalman filter. Our approximation scheme uses the locations of the edge points and also the gradient direction which approximates the normal to the surface. Using uncertainty we establish a theoretical link between the edge detection and the local surface approximation.

Our statistical results are then used as a solid theoretical foundation on which to base subsequent computations, such as, for example, the determination of local surface curvature using local geometric models or for surface segmentation [BJ88].

We tie the results of the analysis of the uncertainties involved in edge detection to the estimation of local geometric surfaces by a Kalman filtering technique. While the instantiation of these local models is similar to the method of Sander and Zucker [SZ90], the utilization of uncertainties yields improved results. In addition, Kalman filtering permits incremental and selective incorporation of new data, thus ensuring that the local models are fit to, and only to, relevant data points. We expect that this will permit us to effectively deal with the problem of discontinuities where the local surface smoothness assumptions break down.

From the local fitting, we calculate for each edge point a mean curvature, a gaussian curvature, and principal curvature directions, and covariance matrices defining the uncertainty. We define local curvature images by projecting at each point the curvatures of the neighbours onto the tangent plane. This yields the local aspect of the curvatures at a point.

From the curvature field we extract typical points such as curvature extrema, parabolic points, umbilic points ... This is done by using the local image of the curvatures defined at each (edge) point. For instance to select the extrema of the maximun curvature in the maximum curvature direction, we employ an algorithm very similar to the extraction of the extrema of the gradient magnitude in the gradient direction used in many 2D edge detection algorithms [Der87, Can86].

From these typical points, we extract characteristic lines such as line of mean curvature extrema, parabolic lines .... To obtain lines of mean curvature extrema, we perform a 3D hysteresis thresholding on the extrema curvature points using the mean curvature. Here again the algorithm is very similar to the one used to threshold the local gradient extrema described in [MDR91].

This work is described with more details in [MAS91].

2 Parametric local surface model

2.1 Problem formulation

In this section, we set up the local parametric surface models, describe briefly how to compute surface curvatures from the models, and present the problem of determining the parameters of the models from the data derived from the 3-D gradient operator. We thus assume that we are given the locations of estimated surface points P and their estimated surface normals $\mathbf{N}$ corresponding to the vector gradient, (see [MD89, MDR91] on how to compute them). Using $\mathbf{N}$ we can establish a *tangent plane coordinate system* at P, which we denote $(\mathbf{P}, \mathbf{Q}, \mathbf{N})$. Note that the basis $(\mathbf{P}, \mathbf{Q})$ of the tangent plane at P is arbitrary — the only constraint is that the coordinate system be right-handed and orthonormal.

In the following, P is the point at which the surface patch is being fit, and Q_i are neighbouring estimated surface points with associated normals $\mathbf{n}_i$, with both given in P's tangent plane coordinates (the development is simpler in these coordinates; we map everything into the actual image coordinates in §3.4.1).

2.2 Local geometric model

We assume that the data returned by the gradient operator represent noisy estimates of points and normals from a (smooth) surface $\mathcal{S}$. We treat a surface as a differentiable manifold and build local charts (parametrizations) at all the estimated surface points. Thus, at $P \in \mathcal{S}$ we assume that the local chart (ψ, U)

$$\psi : U \subset \mathcal{S} \to \mathbb{R}^2$$

(ψ a diffeomorphism, $P \in$ open set U) is such that $\psi(P) = (0,0)$ and its imbedding

$$\phi = i \circ \psi^{-1} : \psi(U) \subset \mathbb{R}^2 \to \mathbb{R}^3$$

in $\mathbb{R}^3$ (based on P's tangent plane coordinates) is the graph of some function $h : \psi(U) \to \mathbb{R}$ with

$$\begin{aligned} h(0,0) &= 0, \\ h_p(0,0) = \left.\frac{\partial h}{\partial p}\right|_{(0,0)} &= 0 \\ h_q(0,0) = \left.\frac{\partial h}{\partial q}\right|_{(0,0)} &= 0 \end{aligned}$$

(this is always true in some local chart). The Taylor expansion of h about the origin is

$$h(p,q) = \frac{1}{2}\left(h_{pp}p^2 + 2h_{pq}pq + h_{qq}q^2\right) + R.$$

Since our ultimate goal is computation of curvatures and related information, we take the simplest local chart which is appropriate, i.e., where

$$h = \frac{1}{2}ep^2 + fpq + \frac{1}{2}gq^2, \tag{1}$$

where we write

$$e = h_{pp}(0,0), \quad f = h_{pq}(0,0), \quad g = h_{qq}(0,0)$$

(known as a *parabolic quadric*).

2.3 Surface curvature

The curvature of the surface $\mathcal{S}$ at P can be computed from its local parametrization $\phi : \psi(U) \to \mathbb{R}^3$ in P's tangent plane coordinates. The surface normal at P is expressed as

$$\mathbf{N}(0,0) = \left.\frac{\phi_p \times \phi_q}{\|\phi_p \times \phi_q\|}\right|_{(0,0)}.$$

(In the following, we take it as understood that derivatives are evaluated at $(0,0)$, i.e., the $|_{(0,0)}$ is implicit.) The matrices

$$\begin{aligned} F_1 &= \begin{pmatrix} \langle\phi_p,\phi_p\rangle & \langle\phi_p,\phi_q\rangle \\ \langle\phi_q,\phi_p\rangle & \langle\phi_q,\phi_q\rangle \end{pmatrix}, \\ F_2 &= \begin{pmatrix} -\langle\phi_p,\mathbf{N}_p\rangle & -\langle\phi_p,\mathbf{N}_q\rangle \\ -\langle\phi_q,\mathbf{N}_p\rangle & -\langle\phi_q,\mathbf{N}_q\rangle \end{pmatrix}, \end{aligned}$$

are determined from the first and second fundamental forms respectively of the surface ($\langle\bullet,\bullet\rangle$ denotes inner product). The principal curvatures κ_1, κ_2 of ϕ at P in $(\mathbf{P}, \mathbf{Q}, \mathbf{N})$ coordinates are the two eigenvalues of the matrix $F_2F_1^{-1}$, and the Gaussian and mean curvatures are

$$\begin{aligned} \kappa_g &= \kappa_1\kappa_2, \\ \kappa_m &= \frac{\kappa_1+\kappa_2}{2} \end{aligned}$$

respectively. We show how to compute uncertainties in the curvatures in §3.4.2.

3 Recursive estimation of surface parameters

3.1 Instantiating the model

Now, we wish to determine the local quadric surface passing through point P which "best" (in a sense made precise below) fits neighbouring points $Q_i = (p_i, q_i, n_i)^t$ and their normals $\mathbf{n}_i = (\alpha'_i, \beta'_i, \gamma'_i)^t$. In P's tangent plane coordinate system, the equation of the quadric gives us a first measurement equation

$$E_1(e, f, g) = p_i^2 e + 2p_i q_i f + q_i^2 g - 2n_i = 0 \tag{2}$$

between the position of Q_i and the parameters e, f, g which are to be determined from the data.

In addition to estimated surface point locations, the 3-D gradient operator provides an estimate of gradient direction and we can use the measured normal $\mathbf{n}_i$ at point Q_i to further constrain the quadric surface parameters. We know that, in the tangent plane coordinates, the quadric's normal at point Q_i is

$$\mathbf{n}_i = \begin{pmatrix} -p_i e - q_i f \\ -p_i f - q_i g \\ 1 \end{pmatrix}.$$

Denoting the scaled normal *measured* at point Q_i by $(\alpha_i, \beta_i, 1)^t = (\alpha'_i/\gamma'_i, \beta'_i\gamma'_i, \gamma'_i/\gamma'_i)^t$, we obtain two more measurement equations

$$E_2(e, f, g) = p_i e + q_i f + \alpha_i = 0, \tag{3}$$
$$E_3(e, f, g) = p_i f + q_i g + \beta_i = 0. \tag{4}$$

Equations (E_1–E_3) are the three measurement equations which constrain the determination of the parameters e, f, g of the quadric at point P. These equations should be compared to the four equations E'_1–E'_4 of [SZ90]: Eq. E_1 is the same, but Eqs. (E'_2 – E'_4) there were based on unit normals and involved a non-linear combination of e, f, g. The only restriction on our equations here is the assumption that the normal $\mathbf{n}_i$ of Q_i measured in the tangent plane coordinates of P has a nonzero third component, which is reasonable if we assume that Q_i lies in the neighborhood of point P. (In fact, if the surface is regular, such a neighborhood exists [dC76, p.164], at least before discretization[1]). When this component vanishes, the local parametrization of the quadric in these coordinates is no longer valid, and point Q_i should not be taken into account.

Denoting

$$\mathbf{A}_i = \begin{pmatrix} p_i^2 & 2p_i q_i & q_i^2 \\ p_i & q_i & 0 \\ 0 & p_i & q_i \end{pmatrix}, \quad \mathbf{b}_i = \begin{pmatrix} 2n_i \\ -\alpha_i \\ -\beta_i \end{pmatrix}, \quad \mathbf{x} = \begin{pmatrix} e \\ f \\ g \end{pmatrix},$$

the measurement Eqs. (E_1–E_3) at P can be put in matrix form

$$\mathbf{A}_i \mathbf{x} = \mathbf{b}_i.$$

3.2 Non-recursive minimum variance least-squares solution

We wish to weight the measurement equations by the uncertainty of our measured parameters, i.e., the coordinates of points Q_i and attached normals $\mathbf{n}_i$. Once this is done (cf. next sections), we end up with a matrix $\mathbf{W}_i$ which is the covariance of $\mathbf{A}_i\mathbf{x} - \mathbf{b}_i$

$$\mathbf{W}_i = E\left[(\mathbf{A}_i\mathbf{x} - \mathbf{b}_i)(\mathbf{A}_i\mathbf{x} - \mathbf{b}_i)^t\right].$$

Then a weighted least-squares solution $\mathbf{x}$ to our problem at P using all $Q_i, i = 1, \ldots, n$ in some neighbourhood will therefore minimize

$$C = \sum_i (\mathbf{A}_i\mathbf{x} - \mathbf{b}_i)^t \mathbf{W}_i^{-1} (\mathbf{A}_i\mathbf{x} - \mathbf{b}_i),$$

[1] A common enough assumption throughout computer vision.

and is given by

$$\mathbf{x} = \left(\mathbf{A}^t\mathbf{W}^{-1}\mathbf{A}\right)^{-1}\mathbf{A}^t\mathbf{W}^{-1}\mathbf{b},$$

where

$$\mathbf{A} = \begin{pmatrix} \mathbf{A}_1 \\ \vdots \\ \mathbf{A}_n \end{pmatrix}, \qquad \mathbf{b} = \begin{pmatrix} \mathbf{b}_1 \\ \vdots \\ \mathbf{b}_n \end{pmatrix}, \qquad \text{and} \quad \mathbf{W} = \begin{pmatrix} \mathbf{W}_1 & & \\ & \ddots & \\ & & \mathbf{W}_n \end{pmatrix}.$$

3.3 Recursive solution

In fact, we implement a recursive solution to this problem, better known as a *Kalman filter* [Lue69, Aya91]. By this method, each time a new measurement Q_i is given, it is only necessary to compute $\mathbf{A}_i$ and $\mathbf{b}_i$ *for that point* and to update the current solution $(\mathbf{x}_i, \mathbf{S}_i)$ using the recursive equations

$$\begin{cases} \mathbf{x}_i &= \mathbf{x}_{i-1} + \mathbf{K}_i(\mathbf{b}_i - \mathbf{A}_i\mathbf{x}_{i-1}), \\ \mathbf{K}_i &= \mathbf{S}_{i-1}\mathbf{A}_i^t(\mathbf{W}_i + \mathbf{A}_i\mathbf{S}_{i-1}\mathbf{A}_i^t)^{-1}, \\ \mathbf{S}_i &= (\mathbf{I} - \mathbf{K}_i\mathbf{A}_i)\mathbf{S}_{i-1}. \end{cases}$$

$\mathbf{S}_i$ is the *parameter* covariance matrix

$$\mathbf{S}_i = E\left[(\mathbf{x}_i - \mathbf{x})(\mathbf{x}_i - \mathbf{x})^t\right]$$

relating the current estimate $\mathbf{x}_i$ and the ideal value $\mathbf{x}$ of the parameter vector. This is a measure of the quality of our estimate — a small covariance means that the computed estimate $\mathbf{x}_i$ is expected to lie close to the "actual" parameters $\mathbf{x}$. It is necessary to initialize the filter with $(\mathbf{x}_0, \mathbf{S}_0)$, which can be taken as $\mathbf{x}_0 = \mathbf{0}, \mathbf{S}_0 = \infty\mathbf{I}$ when *no a priori* information is available about any of the parameters.

3.4 Practical details

For simplicity, the preceding development was presented with all data assumed to be in P's tangent plane coordinate system. We now show how to transform from the actual data points Q_i and normal vectors $\mathbf{n}_i$, each with their respective associated covariance matrices $\mathbf{W}_Q$ and $\mathbf{W}_\mathbf{n}$, measured in a *global* coordinate system, i.e., the coordinate system of the image, to variables $\mathbf{v} = (p_i, q_i, n_i, \alpha_i, \beta_i)^t$ and associated covariance matrix $\mathbf{W}_i$ in P's tangent plane coordinates. Thus we can apply the above theory directly to the image data.

3.4.1 Computing parameters v

We now assume that $Q_i = (x_i, y_i, z_i)^t$ and $\mathbf{n}_i = (n_{x_i}, n_{y_i}, n_{z_i})^t$ are given in a global coordinate system $(\mathbf{X}, \mathbf{Y}, \mathbf{Z})$. To express them in P's tangent plane coordinates $(\mathbf{P}, \mathbf{Q}, \mathbf{N})$ at point $P = (x, y, z)^t$, we compute

$$\begin{pmatrix} p_i \\ q_i \\ n_i \end{pmatrix} = \mathbf{R}\begin{pmatrix} x_i - x \\ y_i - y \\ z_i - z \end{pmatrix}, \qquad \begin{pmatrix} \alpha_i' \\ \beta_i' \\ \gamma_i' \end{pmatrix} = \mathbf{R}\begin{pmatrix} n_{x_i} \\ n_{y_i} \\ n_{z_i} \end{pmatrix},$$

and

$$\alpha_i = \frac{\alpha_i'}{\gamma_i'}, \qquad \beta_i = \frac{\beta_i'}{\gamma_i'},$$

for

$$\mathbf{R} = \begin{pmatrix} P_x & P_y & P_z \\ Q_x & Q_y & Q_z \\ N_x & N_y & N_z \end{pmatrix}.$$

The coordinates of the tangent plane basis vectors at P expressed in the global image coordinates $(\mathbf{X}, \mathbf{Y}, \mathbf{Z})$ are

$$\mathbf{P} = \begin{pmatrix} P_x \\ P_y \\ P_z \end{pmatrix}, \qquad \mathbf{Q} = \begin{pmatrix} Q_x \\ Q_y \\ Q_z \end{pmatrix}, \qquad \mathbf{N} = \begin{pmatrix} N_x \\ N_y \\ N_z \end{pmatrix}.$$

3.4.2 Computing covariances $\mathbf{W}_i$

We assume that the covariance of point Q_i and its normal $\mathbf{n}_i$ are given in the *global coordinate system* $(\mathbf{X}, \mathbf{Y}, \mathbf{Z})$ by $\mathbf{W}'_{Q_i}$ and $\mathbf{W}'_{\mathbf{n}_i}$ respectively. Since, for any affine transformation of a random variable $\mathbf{v}_i \rightarrow \mathbf{w}_i = \mathbf{M}(\mathbf{v}_i - \mathbf{v})$ we have

$$E\left[(\mathbf{w}_i - \mathbf{w})(\mathbf{w}_i - \mathbf{w})^t\right] = \mathbf{M}\, E\left[(\mathbf{v}_i - \mathbf{v})(\mathbf{v}_i - \mathbf{v})^t\right] \mathbf{M}^t,$$

the corresponding covariance matrices expressed in P's tangent plane coordinate system $(\mathbf{P}, \mathbf{Q}, \mathbf{N})$ are

$$\mathbf{W}_{Q_i} = \mathbf{R}\mathbf{W}'_{Q_i}\mathbf{R}^t, \qquad \mathbf{W}_{\mathbf{n}_i} = \mathbf{R}\mathbf{W}'_{\mathbf{n}_i}\mathbf{R}^t,$$

which are 3×3 matrices.

In fact, if we express $\mathbf{n}_i = (\alpha'_i, \beta'_i, \gamma'_i)^t$ in the tangent plane coordinates, we must compute the covariance of $\alpha_i = \alpha'_i/\gamma'_i$ and $\beta_i = \beta'_i/\gamma'_i$. As a first order approximation, we compute the 2×2 matrix $\widetilde{\mathbf{W}}_{\mathbf{n}_i}$

$$\widetilde{\mathbf{W}}_{\mathbf{n}_i} = \mathbf{J}_1 \mathbf{W}_{\mathbf{n}_i} \mathbf{J}_1^t$$

where $\mathbf{J}_1$ is the Jacobian matrix

$$\mathbf{J}_1 = \begin{pmatrix} \frac{1}{\gamma'_i} & 0 & \frac{-\alpha'_i}{\gamma'^2_i} \\ 0 & \frac{1}{\gamma'_i} & \frac{-\beta'_i}{\gamma'^2_i} \end{pmatrix}$$

of the change of variables. Therefore the 5×5 matrix

$$\mathbf{W}_{Q,\mathbf{n}} = \begin{pmatrix} \mathbf{W}_Q & \mathbf{0}_{3\times 2} \\ \mathbf{0}_{2\times 3} & \widetilde{\mathbf{W}}_{\mathbf{n}_i} \end{pmatrix}$$

is the covariance of our measurement vector $(p_i, q_i, n_i, \alpha_i, \beta_i)$.

The 3×3 covariance matrix $\mathbf{W}_i$ is computed as a first order approximation by

$$\mathbf{W}_i = \mathbf{J}_2 \mathbf{W}_{Q,\mathbf{n}_i} \mathbf{J}_2^t,$$

where $\mathbf{J}_2$ is the Jacobian matrix

$$\mathbf{J}_2 = \begin{pmatrix} 2p_i e + 2q_i f & 2p_i f + 2q_i g & -2 & 0 & 0 \\ e & f & 0 & 1 & 0 \\ 0 & f & g & 0 & 1 \end{pmatrix}.$$

The 2×2 covariance matrix of curvatures is determined similarly,

$$\mathbf{W}_c = \mathbf{J}_3 \mathbf{W}_{Q_i} \mathbf{J}_3^t,$$

with the Jacobian

$$\mathbf{J}_3 = \begin{pmatrix} c & -2b & a \\ \frac{1}{2} & 0 & \frac{1}{2} \end{pmatrix}.$$

4 Estimating error in edge detection

In the reference [MDMC, MAS91], we determine the uncertainty inherent in edge detection in digital images by considering the 3-D (modified Canny) edge detector of Monga and Deriche [MD89, MDR91]. We first deal with uncertainty in edge position, and then with the error in edge direction and magnitude. We determine precisely the covariance matrices needed for the local quadric surface fitting described above, and derive the very interesting result that the uncertainty in edge position and magnitude is highly dependent on the orientation of the edge with respect to the image coordinate axes.

5 From curvatures to typical features

5.1 Introduction

For each edge point, the previous section determines the Gaussian and mean curvatures, principal curvature directions, and the corresponding covariance matrices. Note that the scale is defined by the size of the neighbourhood used to fit the local geometric model. In this section we deal with the extraction of more global curvature features from the local curvature information.

5.2 Local curvature maps

A practical way of characterizing the behaviour of the curvature in the neighbourhood of a point is to define local curvature maps (technically the pullback of the field onto the tangent plane as used in [SZ] for the computation of the direction field index). Given a point P and its tangent plane defined by $(\mathbf{P}, \mathbf{Q}, \mathbf{N})$. Let V be the intersection of a sphere whose center is P and radius r with the set of the edge points (this defines a neighbourhood of P), and let W be the orthogonal projection of the points of V onto the tangent plane. At each point of W we attach the curvatures of its corresponding points in V. The size of V could be determined using the distance of the points to the tangent plane and the angle between the gradient at a point and the gradient at point P. We thus define a map characterizing the behaviour of the curvatures around P.

5.3 Extracting lines of curvature extrema

From the local curvature images we can extract, for instance, the maxima of the maximum curvature in the maximum curvature direction . This may be done similarly to the classical extraction of the extrema of the gradient magnitude in the gradient direction in 2D edge detection methods [Der87, Can86].

Let $C(P)$ be the local maximum curvature map attached to point P, and let $G(P)$ be the maximum curvature direction. We compare the value of the maximum curvature along the straight line defined by P and $G(P)$ and retain P if its maximum curvature is a local extrenum along this direction. Thus we obtain local curvature extrema candidates. To remove false extrema, we perform a 3D hysteresis thresholding using the mean curvature already determined. This is done analogously to the thresholding of the local gradient extrema in the edge detection scheme described in [MDR91].

6 Results

6.1 Curvatures for synthetic data

We built a synthetic volume whose implicit equation is given by

$$z \leq ax^2 + by^2 + cx + dy + e$$

with $a = 1/20$, $b = 0.25$, $c = -4$, $d = -12$, $e = 224$. This was done by creating a 3-D digital image with points of maximum (resp. minimum) intensity within (resp. outside of) the volume. We extracted 3-D edges with the algorithm described in [MDR91]. These edge points correspond to the surface of an ideal elliptic paraboloid.

First, note that there exists a single surface point T such that the equation of the *entire* surface takes the reduced form

$$z' = \frac{1}{2}ex'^2 + fx'y' + \frac{1}{2}gy'^2$$

when (x',y',z') are expressed in the local tangent plane coordinate system attached to point T. This point is the vertex of the paraboloid and its coordinates are given by

$$T = (\frac{-c}{2a}, \frac{-d}{2b}, \frac{-c^2}{4a} - \frac{d^2}{4b} + e).$$

Point coordinates	Ideal	Positions	+ Normals	+ Uncertainty	Predicted σ
	$e = 0.1$	– 0.0037	0.0172	0.102	0.011
(40,24,1)	$f = 0$	0.0000	0.000	– 0.003	0.008
86 Neighbors	$g = 0.5$	0.324	0.360	0.494	0.02

Table 1: Estimation of the parameters (e, f, g) of the local quadrics

Point coordinates	Ideal	Positions	+ Normals	+ Uncertainty	Predicted σ
(40,24,1)	$C_g = 0.05$	– 0.0012	0.0062	0.0505	0.014
86 Neighbors	$C_m = 0.3$	0.16	0.16	0.298	0.023
(47,24,2)	$C_g = 0.022$	0.089	0.063	0.0157	0.008
46 Neighbors	$C_m = 0.23$	0.333	0.297	0.2024	0.030
(45,26,3)	$C_g = 0.010$	– 0.001	– 0.0003	0.018	0.009
54 Neighbors	$C_m = 0.12$	0.053	0.107	0.143	0.018
(36,22,2)	$C_g = 0.010$	0.013	0.018	0.012	0.006
50 Neighbors	$C_m = 0.122$	0.133	0.172	0.127	0.015
(39,23,1)	$C_g = 0.031$	0.011	0.023	0.017	0.023
28 Neighbors	$C_m = 0.22$	0.10	0.25	0.21	0.062
(38,28,5)	$C_g = 0.004$	0.009	0.010	0.006	0.003
50 Neighbors	$C_m = 0.07$	0.105	0.11	0.09	0.01

Table 2: Estimation of gaussian and mean curvatures C_g and C_m

For this particular point, the local quadric approximation is a *global* one, and the ideal parameters are $e = 2a$, $f = 0, g = 2b$. It is therefore possible to estimate these parameters with all the detected surface points (about 350 points). In this case the convergence is excellent towards the exact values. We show in table 1 the results obtained with a smaller but still rather large number of edge points (86). We show successively the results obtained for a least squares estimation with points only (measurement equation E1), then adding normals (measurement equations E2–E3). Finally, we show the results obtained when uncertainty on positions and normals is taken into account, following the computations of the previous sections. It is easy to check that not only is the estimate obtained much more accurate, but also that the computed standard deviation σ on the error estimation is perfectly coherent with the observed error.

At other points P on the surface, we applied our local quadric approximation with smaller neighborhoods (containing about 50 points). The size of the neighborhood is both controlled by limiting the angle between the neighbors' normal and P's normal, and also the distance between neighbors and P. We used the local approximation to compute locally the Gaussian and mean curvatures (C_g and C_m). In table 2, we show the results obtained for this computation, with the previous three methods. Here again, the results are much more accurate with the last approach, and the output covariance agrees almost perfectly well with the observed errors.

6.2 Typical curvature features for synthetic and real data

Figures 1 to 5 present some results for the determination of the extrema of the maximum curvature in the maximum curvature direction for a 3D MRI image of the face. We notice that our local approximation scheme provides a continuous maximum curvature field, allowing to detect reliably and accurately the curvature extrema. Regarding our experiences, this continuity is mainly due to the smoothness of the orientation of the gradient used for the local approximation.

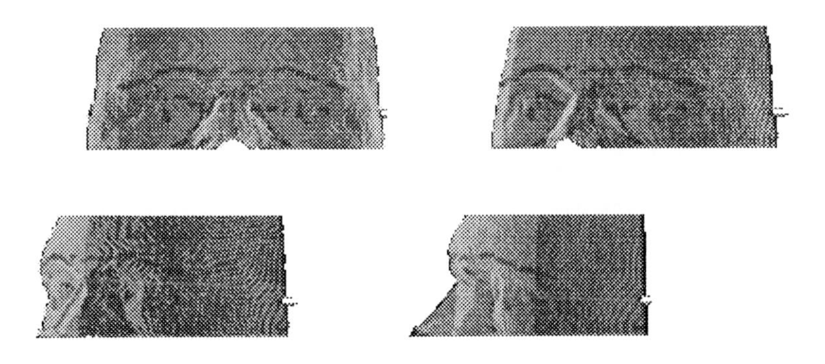

Figure 1: Perspective views of the 3D edges matched with the extrema of the maximum curvature in the maximum curvature direction colored in dark (the ratio defining the weighting point/normal in the least mean squares is about 1/40)

Using covariance matrices in our least mean square criterion introduces a ponderation between the equations taking into account the position of the points and the normal orientations. For the original data corresponding to figures 1 to 5 the step edges have a very strong amplitude. This implies that the localization criterion is over-estimated due to the first order approximation (the localization is inversely proportional to the step amplitude), and therefore the weight put on the position equations is too high. We also remark that the gradient coordinates are real values but that the point coordinates are integer values which could induce false discontinuity. Given that each point produces one measurement equation using its position and two measurement equations using the orientation of its gradient, we can evaluate the ratio between point information and normal information. If we apply exactly the theoretical calculus presented before for these data, we obtain a ratio of 1/12 (1 for point and 12 for normal). This allows to obtain rather good results but where some false discontinuities still remain. Experimentally a ratio of 1/40 yields a good trade-off between the smoothness and the preservation of the singularities. The distorsion of the theoretical optimum and the experimental one is due to the reasons we reported here. We also perform some experiences with a ratio of 1/4 and we obtain a bad continuity for the maximum curvature field.

The main practical conclusion of our experiences is that the gradient orientation (approximating the orientation of the normal to the surface) is a strong regularization criteria for the local approximation. This illustrate the applicability of our theoretical developments although its direct applicability are spoiled by first order approximations and by discretization.

7 Conclusion

Our main objective was to develop robust and reliable tools useful for modeling and analyzing surfaces of 3-D objects. In this paper we showed the importance of a careful quantitative analysis of the various sources of uncertainty for computing second order derivative features (mean and Gaussian curvatures) on a discrete surface.

We use a quantitative estimation of the uncertainty in edge position, orientation and magnitude produced by the multidimensional (2-D and 3-D) versions of the Monga-Deriche-Canny recursive separable edge-detector.

Figure 2: Projection of the extrema of the maximum curvature in the maximum curvature direction corresponding to the previous figure (the ratio defining the weighting point/normal in the least mean squares is about 1/40)

Figure 3: Projection of the extrema of the maximum curvature in the maximum curvature direction (the ratio defining the weighting point/normal in the least mean squares is about 1/40) ; the algorithm providing the extrema is slightly different.

Figure 4: Projection of the extrema of the maximum curvature in the maximum curvature direction (the ratio defining the weighting point/normal in the least mean squares is about 1/320)

Figure 5: Projection of the extrema of the maximum curvature in the maximum curvature direction (the ratio defining the weighting point/normal in the least mean squares is about 1/5)

Then, we revisited the algorithm initially proposed by Sander and Zucker for locally estimating the curvature of a discrete 3-D surface, and we modified the original measurement equations and proposed an optimal estimation scheme to account for the previously computed uncertainties and corrections.

We tested the corrected edge detector on 2-D and 3-D medical images and showed the importance of the corrected edge magnitude for edge detection. We also tested the surface modeling algorithm on discrete 3-D objects — not only are the results obtained more accurate, but the computed measure of uncertainty attached to the results agrees extremely well with the true one.

We also show how to use these curvatures to determine typical curvature features on which registration and/or tracking procedures can robustly rely.

Acknowledgements

We thank Serge Benayoun for the implementation of the extraction of the local curvature extrema. Nathalie Gaudechoux provided a substantial help in the preparation of this manuscript.

We thank Pr. Bittoun of the Kremlin Bicetre Hospital in Paris and Pr. J.L. Coatrieux in Rennes which provided the medical images.

This work has been partially supported by Ge-Cgr and some of the presented images were acquired in collaboration with the Advanced Image Processing Group of GE-CGR in Buc, France. We shall give the details of this acquisition process, and a thorough description of the results very soon in a forthcoming joint paper Inria-Ge-Cgr. This work was also partially supported by Digital Equipment Corporation and European AIM (Advanced Informatics in Medicine) Project Murim.

References

[AB+90] N. Ayache, J.D. Boissonnat, , L. Cohen, , B. Geiger, J. Levy-Vehel, O. Monga, and P. Sander. Steps toward the automatic interpretation of 3d images. In *Proceedings of the NATO Advanced Research Workshop on 3D Imaging in Medicine*, Travemünde, June 1990. NATO ASI Series, Springer-Verlag.

[Aya91] N. Ayache. *Artificial Vision for Mobile Robots – Stereo-Vision and Multisensory Perception.* MIT Press, Boston, 1991.

[BJ88] Paul J. Besl and Ramesh C. Jain. Segmentation through Variable-Order surface fitting. *IEEE Transactions on Pattern Analysis and Machine Intelligence*, PAMI-10(2):167–192, March 1988.

[Can86] John Canny. A computational approach to edge detection. *IEEE Transactions on Pattern Analysis and Machine Intelligence*, PAMI-8(6):679–698, November 1986.

[dC76] Manfredo P. do Carmo. *Differential Geometry of Curves and Surfaces.* Prentice-Hall, Englewood Cliffs, 1976.

[Der87] Rachid Deriche. Using Canny's criteria to derive a recursively implemented optimal edge detector. *International Journal of Computer Vision*, pages 167–187, 1987.

[Koe90] Jan J. Koenderink. *Solid Shape.* MIT Press, Boston, 1990.

[Lue69] David G. Luenberger. *Optimization by Vector Space Methods.* Wiley, New York, 1969.

[MAS91] Olivier Monga, Nicholas Ayache, and Peter Sander. From voxel to curvature. Technical report, INRIA, 1991. No 1356.

[MD89] Olivier Monga and Rachid Deriche. 3d edge detection using recursive filtering. In *Conference on Vision and Patern Recognition*, San Diego, June 1989. IEEE.

[MDMC] Olivier Monga, Rachid Deriche, Gregoire Malandain, and Jean-Pierre Cocquerez. Recursive filtering and edge closing: two primary tools for 3d edge detection.

[MDR91] Olivier Monga, Rachid Deriche, and Jean-Marie Rocchisani. 3d edge detection using recursive filtering: Application to scanner images. *Computer Vision Graphic and Image Processing, Vol. 53, No 1, pp. 76-87*, January 1991.

[SZ] Peter T. Sander and Steven W. Zucker. Singularities of principal direction fields from 3-D images. *IEEE Transactions on Pattern Analysis and Machine Intelligence.* To appear. Available as Technical Report CIM-88-7, McGill Research Center for Intelligent Machines, McGill University, Montréal.

[SZ87] Peter T. Sander and Steven W. Zucker. Tracing surfaces for surfacing traces. In *Proceedings of the First International Conference on Computer Vision*, pages 241–249, London, June 1987.

[SZ90] Peter T. Sander and Steven W. Zucker. Inferring surface trace and differential structure from 3-D images. *IEEE Transactions on Pattern Analysis and Machine Intelligence*, 12(9), September 1990.

[ZH81] S.W. Zucker and R.M. Hummel. A three-dimensional edge operator. *IEEE Transactions on Pattern Analysis and Machine Intelligence*, PAMI-3(3):324–331, May 1981.

BOUNDARY ESTIMATION IN ULTRASOUND IMAGES

W J Lin[1], S M Pizer[1,2,3], and V E Johnson[4]

Departments of [1]Computer Science,
[2]Radiology, and [3]Radiation Oncology
University of North Carolina
Chapel Hill, NC 27599-3175

[4]Institute of Decision Science and Statistics
Duke University
Durham, NC 27706

Abstract

Surface definition, a process of defining three dimensional surface from volume data, is essential in three dimensional volume data rendering. The traditional method applies a three dimensional gradient operator to the volume data to estimate the strength and orientation of surface present. Applying this method to ultrasound volume data does not produce satisfactory results due to noisy nature of the images and the sensitivity of certain signals to the direction of insonation. We propose a Bayesian approach to the surface definition problem of ultrasound images, and study this approach in two dimensions. We formulate the problem as the estimation of posterior means and standard deviations of Gibbs distributions for boundary believability and normal direction. A set of filters of directional derivatives of Gaussians are used to measure the edge strength and orientation at multiple scales. The likelihood function is based on the measurement at the smallest scale. The prior distribution reflects shape properties at multiple scales. It uses a pyramid algorithm for contour analysis where the lengths of contours are computed and contour gaps are closed at multiple scales. The outcome of the pyramid algorithm is the length and weight global attributes for each pixel. These attribute values are incorporated into the Gibbs prior using a data augmentation scheme. The design and implementation of such an approach are the subject of this paper.

Keywords

Volume rendering; surface estimation; Bayesian; likelihood function; prior; posterior; Markov random field; Gibbs distribution; data augmentation.

1. INTRODUCTION

Three dimensional arrays of digital data are being generated in many areas of medical imaging in ever increasing number. Multiple 2D slices of computed tomography (CT), magnetic resonance (MR), and single-photon emission computed tomography (SPECT) create volume data. A research project currently conducted in the Duke/UNC Engineering Research Center on Emerging Cardiovascular Technologies includes building a new generation transducer that can capture a three dimensional volume of ultrasound data in real time (von Ramm et al, 1988), (Shattuck et al, 1984). These volume data represent complex anatomy or functional process under study. Effective visualization of these volume data helps physicians in diagnostic interpretation or treatment planning.

Volume rendering, a method of direct rendering of volume data, has been successfully applied in visualizing volume data of CT, MRI, and PET images. The operational principle of volume rendering is to render the volume data directly instead of fitting geometric primitives and then rendering the primitives. This direct rendering is done by compositing images from the results of two separate and parallel processes. The first process performs surface classification to obtain a partial opacity for every voxel. The second process performs surface shading at every voxel of volume data with a locally computed surface normal. Non-binary classification increases the likelihood that small or poorly defined features are perserved (Levoy 1988).

Successful application of volume rendering depends heavily on the estimation of the local surface normal and surface classification. Usually one tries to take advantage of knowledge about the relationship of voxel values with surfaces and their normals. For example, in CT image studies one can make the assumption that CT numbers represent the percentages of material contained in voxels; hence the surface normal and surface classification can be obtained from the local gradient of an approximate percentage measure. Unfortunately this simple technique can not be satisfactorily applied to ultrasound volume data since ultrasound images suffer from serious speckle phenomena due to the coherent radiation source. Common speckle phenomena include random speckles spots from within soft tissues and broken contours on organ boundaries. Applying the above simple classification technique serves to pick up boundaries of random speckle spots as surfaces and to miss boundaries at contour gaps. The locally computed surface normals tend to be incorrect.

In this paper we study the surface classification and normal estimation problem from a Bayesian perspective in two dimensions. We formulate the problem as the estimation of posterior means and standard deviations of Gibbs distributions for boundary believability and normal direction. We show that the Gibbs distribution can be extended to model global structures by using a data augmentation scheme. We apply this method on ultrasound images in two dimensions and show the results. The remainder of this paper is organized as follows. The next section presents the Bayesian approach. The following two sections describe our filter design for producing edge-related measurements of ultrasound images to which Gibbs-compatible likelihood functions pertain. Then the design of Gibbs priors reflecting shape-related knowledge about object boundaries is presented, followed by a summary of the complete algorithm. We then present results and close with a discussion.

2. METHODS

2.1 Approach

Given an observed image we can apply measurements on the voxel values to determine whether a given voxel is on a boundary and, if it is, the associated normal direction. A representation of the target boundaries and normal directions on a grid can then be determined from the measurements. However, this representation will not always be the true representation of the boundaries and normal directions of underlying targets due to the image noise. Instead, there is a certain strength of conviction or believability associated with this representation.

The believability is obviously dependent on the measurements and is generally different in different parts of an image. We usually have strong conviction as to the representation of a part of an image when the measurements on that part show relatively high values and weak conviction when the measurement values are low. For example, the output of an edge detector affects our conviction about the presence of an edge — the higher the output value the stronger the conviction. This conviction is also affected by the measurements in the neighboring voxels. For example, the believability concerning the existence of a boundary at two neighboring voxels increases when their normal direction measurements are aligned and conform to the assumed boundary orientation but decreases when the normal directions are not aligned or contradict to the assumed boundary orientation.

Putting this dependency of believability on measurements and consistency into a mathematical form produces a Bayesian formulation for the conviction. In other words, the believability is really the posterior odds of a representation given measurements. It depends on a likelihood function, which is a distribution of measurements conditioned on a boundary representation, and a prior, which models the consistency in a representation, that is, the geometry of the situation.

Under the Bayesian framework we can define a random field with the same dimensions as the input image. Each site of the random field has two components: a *boundary value*, whose value is either zero or one, representing whether the corresponding voxel is on a boundary or not, and a *normal* which is the normal vector associated with the boundary. A representation of object boundary and normal direction is thus a sample drawn from the random field. In regard to this representation there is an associated posterior odds between zero and one, and the value is monotonic with the believability of this representation.

We compute the posterior means and standard deviations of boundary value and normal directions for each voxel. This produces a summary representation of the ensemble of all possible representations. This summary representation provides information useful for volume-rendering. For example, the mean of the normal direction of a voxel can be used as the surface normal for shading. The standard deviation of normal direction can be used to make the mean normal direction fuzzy before it is used as the surface normal. As a result a shiny surface might show a high confidence about the local surface normals while a dull surface show a low confidence. The mean of the boundary value can be used to modulate the opacity of the voxel and the standard deviation of the boundary value can be used to modulate the color. Here the color of a surface would show the relative confidence about our judgement of the presence of a local surface. Ideally this creates an image portraying the target shapes with additional visual information of confidence about the shapes.

Later we will show that under certain assumptions this random field has the Markovian property. By the equivalence theorem of Markov random fields and Gibbs distributions due to Geman and Geman (1984), the posterior distributions of this random field is a Gibbs distribution. The Gibbs sampler can then be used to draw samples from the random field repeatedly. We thus can compute the ensemble posterior mean and standard deviation.

2.2 Filter Design for Boundary Measurement

We take as our likelihood function the distributions of edge measurements conditioned on the presence or absence of boundaries and, in the case of boundary presence, the true normal direction. The edge measurements, including edge strength and edge orientation, are outputs of applying a edge fileter to the observed image. They should have well separated edge stength distributions between the cases of boundary presence and boundary absence, with the edge orientation distribution peaking at the true boundary normal direction.

The design of the edge measurement filter is application dependent. In this paper we shall concentrate on ultrasound images in two dimensions. We shall discuss briefly the properties of ultrasound image noise and then introduce a set of filters for measuring boundary and associated normal directions based on those properties. We shall then present our measured distribution of the filter outputs and approximations to the filter output distributions.

Ultrasound imaging is an unusual imaging modality in that it uses a coherent radiation source. A sound beam is produced at a transducer and directed into targets. Echoes are generated when the incident beam hits reflectors along the path of sonic transmission. Basically there are two types of reflection, the *specular reflection* and the *non-specular reflection.* The specular reflection is responsible for major organ outlines seen in diagnostic ultrasound examinations. These reflections usually appear to be very bright in the image and they form mostly continuous boundaries. We shall call this kind of boundary the *specular boundary.* Specular boundaries tend to be broken where the boundary orientation is parallel to the beam direction and the reflection is directed away from the transducer, a phenomenon called *echo drop-outs.* Specular boundaries also appear to be broken where the random scatterers around the boundary happen to produce a destructive component echo which is out of phase with the specular reflection.

Non-specular reflections originate from random inhomogeneous continuum. They appear in the image as a texture of random bright spots, called *speckles.* These echoes are mostly due to density and elasticity variations in soft tissues (Waag et al, 1989), (Nassiri and Hill, 1986). The intensity distribution of these echoes usually approximates Rician distributions with different mean intensities for different tissue types. The interface of different tissues also forms a boundary and sometimes can be visually detected in the image. We shall call this kind of boundary the *texture boundary.* We are interested in both specular boundaries and texture boundaries.

We wish to design a set of filters which provide accurate edge response and orientation for both specular and texture boundaries. We have selected directional derivatives of anisotropic Gaussians as the filter kernels (Korn, 1988), (Canny, 1986). The rationale for this is as follows. Suppose we know the local boundary orientation. We can compute the average intensities on the two sides of the hypothesized boundary. The absolute value of the difference of the two averages then is a good measure of edge strength. We would like to get as many samples as possible on either side to get reliable statistical averages. But at the same time we would like to avoid the problem of mixing intensities from other nearby image structures. The directional derivatives of anisotropic Gaussians with the elongated side oriented along boundaries satisfy these constraints. In our implementation we use such filters of fixed widths but increasing lengths at multiple scales. The purpose of more elongated filters, which will be detailed in the section describing the prior specificaton, is to aid closing broken boundaries of different gap sizes. The shape of such a filter is shown in Figure 1.

Notice that these filters work for specular boundaries as well as texture boundaries. The edge measurements from the smallest size filter is the basis of the likelihood function. The size of the smallest filter is chosen such that 2σ's is roughly the speckle size. That gives roughly two speckle spots on either side and should be large enough for good statistical averages. For specular boundaries smaller filters can be used to get better localization of edges. In current implementation we use the same filter sizes for both kinds of boundaries. (Strictly speaking, the filter size should change with the orientations of edges since the speckles are themselves anisotropic and always lie perpendicularly

Figure 1: The shape of our filters for measuring edge strengths and orientations. The filters are directional derivatives of Gaussians. Multiple such filters of fixed widths and increasing lengths are applied at every pixel.

to the beam direction.)

We apply the filters in several orientations and averaging the responses from all orientations. In the current implementation we use four orientations at 0, 45, 90, and 135 degrees. The response from each filter is taken to form a vector in the polar coordinate system with its corresponding angle. These vectors are transformed to a new vector by doubling their angles, averaging by vector summation, and then halving the angle of resulting vector. This angle doubling and halving before and after the vector summation averaging is necessary because we do not distinguish inner normals from outer normals for shading's purpose.

2.3 Empirical Approximation of the Likelihood Function

The likelihood function that we need to estimate is $p(f_b(i,j), f_n(i,j)|B(i,j), N(i,j))$, i.e. the filter response at voxel (i,j) given the true boundary situation at voxel (i,j). f_b and f_n are the magnitude and phase of the smallest scale filter response, $B(i,j)$ is either 1 or 0 representing boundary presence or not, respectively, and $N(i,j)$ is the normal direction in the case of boundary presence. Both f_n and $N(i,j)$ are modulated by 180 degrees, since inner normals are not distinguished from outer normals for shading's purpose, and quantized to 16 discrete orientations for computational efficiency. Here We shall assume that $f_b(i,j)$ is independent of $N(i,j)$ to simplify the mathematical treatment, although We recognize its limitation on purely specular reflections. It then follows that we need only estimate $p(f_b(i,j)|B(i,j))$ and $p(f_n(i,j)|f_b(i,j), N(i,j))$ since $p(f_b(i,j), f_n(i,j)|B(i,j), N(i,j)) = p(f_n(i,j)|f_b(i,j), B(i,j), N(i,j))p(f_b(i,j)|B(i,j), N(i,j))$ and $p(f_b(i,j)|B(i,j), N(i,j)) = p(f_b(i,j)|B(i,j))$ due to the assumption.

Three test images were used as the training set for determining these conditional distributions. These three images are a slice through a cone phantom simulating a diffuse target, a slice through a liver, and a slice of a baby doll hanging in a water tank, as shown in Figure 2. Contours in these images have been manually drawn as reference ground truth.

The histograms of f_b on the contours are computed and plotted in Figure 3. Regions in the images containing no boundaries were randomly selected to gather samples for computing the histograms of f_b in the absence of boundaries. These regions are marked in Figure 2 and the histograms are plotted in Figure 3. We have empirically fitted these histograms by normal distributions after a cube root transform. Figure 4 shows the transformed histograms and the fitting normal distributions. In these plots f_b is linearly scaled to between zero and one by normalizing by the maximal f_b in each image.

$P(f_n(i,j)|N(i,j), f_b(i,j))$ is estimated by the error in $f_n(i,j)$ with respect to $N(i,j)$, i.e., $p(h(f_n(i,j), N(i,j))|f_b(i,j))$, where $h(\cdot,\cdot)$ is a distance measure of normal angles. N(i,j) is computed from the drawn boundaries by fitting a straight line through five neighboring boundary voxels with least-square error. Figure 5 shows the histograms of $h(f_n(i,j), N(i,j))$. These histograms all peak at zero, indicating that our filters give approximately unbiased normal directions. These histograms also suggest normal distribution approximations.

Figure 6 shows scatter-plots of $h(f_n(i,j), N(i,j))$ versus $f_b(i,j)$. It appears that there is an exponential dependency relationship. We have approximated this dependency with an exponential function using a least-square-fit. Such an exponential function is shown in Figure 6 for the liver data set. The circular phantom does not show such a strong correlation between $h(f_n(i,j), N(i,j))$ and $f_b(i,j)$. This could be due to the facts that this image consists of speckles only and that there are relatively fewer sample points for this image. In summary, we have approximated $p(h(f_n(i,j), N(i,j))|f_b(i,j))$

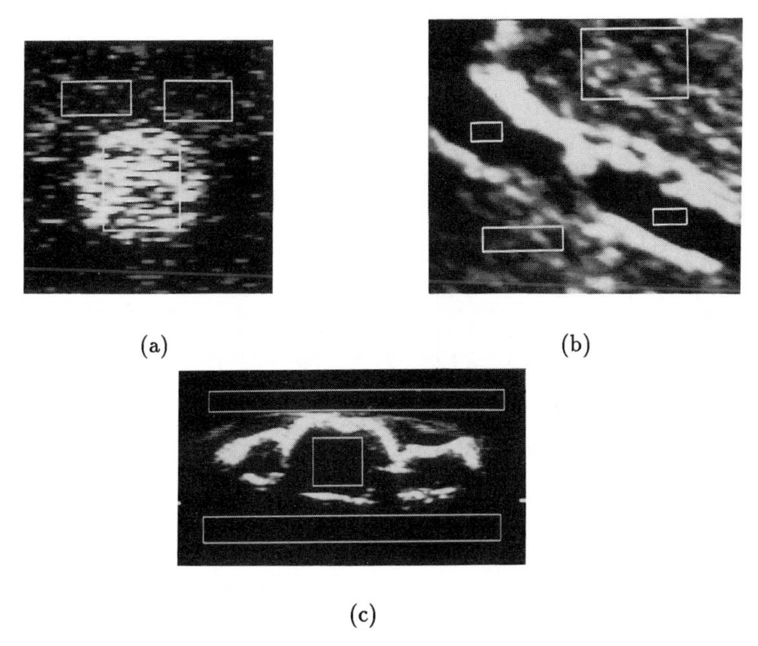

Figure 2: Test images: (a) Cone phantom, (b) Liver, and (c) Baby doll.

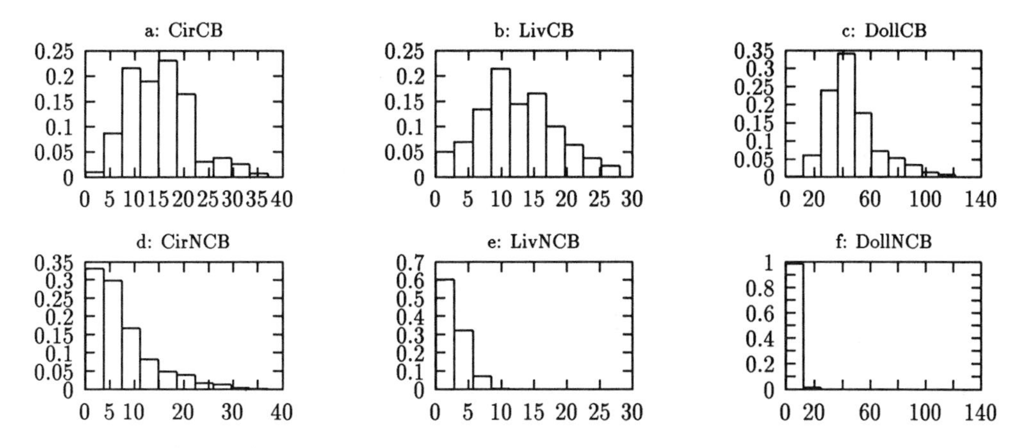

Figure 3: (a), (b), and (c): Histograms of f_b measured on boundaries of cone phantom, liver, and baby doll. (d), (e), and (f): Histograms of f_b measured on no boundary regions of cone phantom, liver, and baby doll.

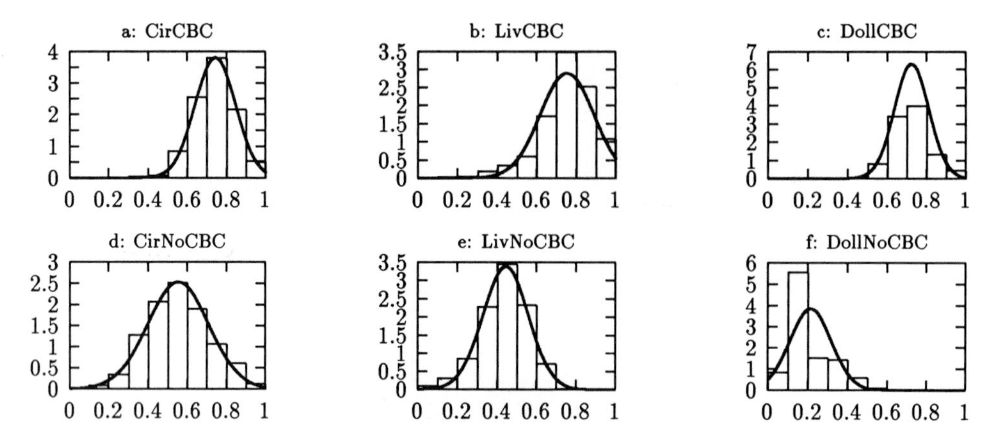

Figure 4: Histograms of f_b after cube root transform and normalization and the fitting normal distributions. Graph order is the same as in Figure 3.

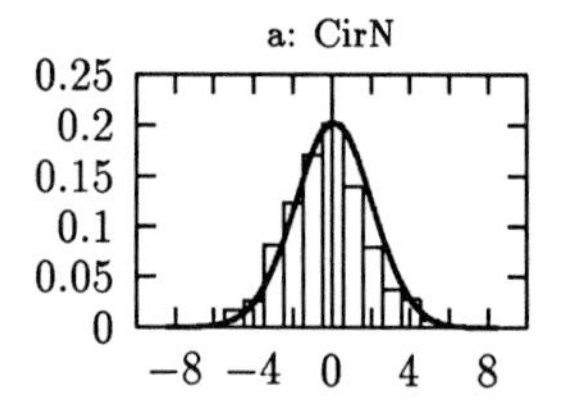

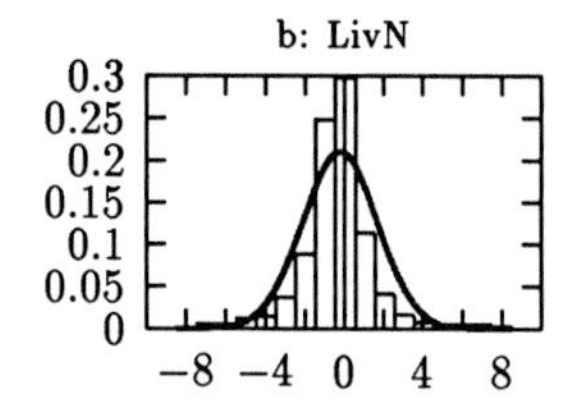

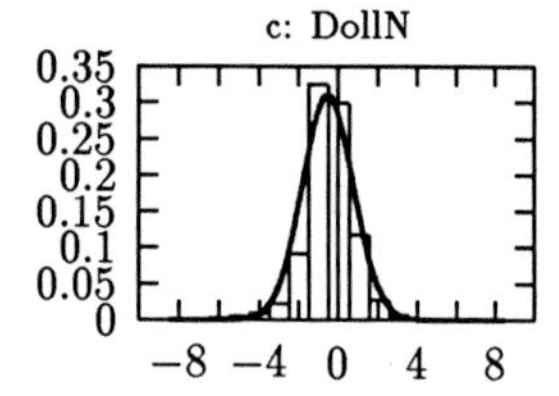

Figure 5: Histograms of f_n.

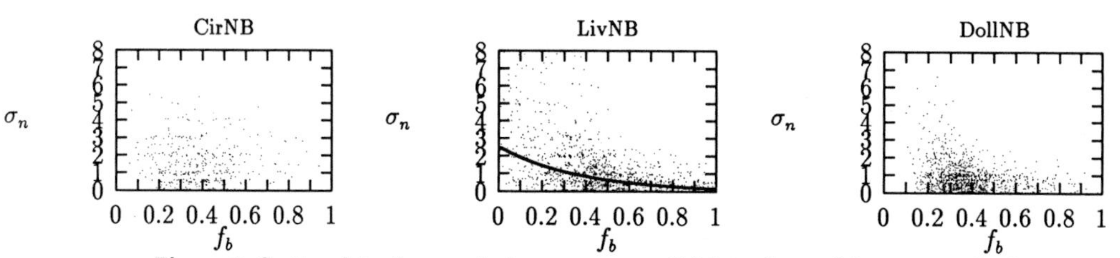

Figure 6: Scatter-plots of σ_n vs. f_b shows an exponential dependency. A least-square-fit gives $\sigma_n = 1.8 * \exp(-f_b/0.41)$, where f_b is the normalized edge strength measurement, for the liver data set.

with zero mean normal distributions with the variances depending exponentially on $f_b(i,j)$.

Finally, we have

$$p(f_b(i,j), f_n(i,j)|B(i,j), N(i,j)) =$$

$$\begin{cases} \frac{1}{2\pi\sigma_{b1}\sigma_n} e^{-V_{c1}} & \text{where } V_{c1} = (f_b - \mu_{b1})^2/2\sigma_{b1}^2 \\ & \quad +h(f_n(i,j), N(i,j))^2/2\sigma_n^2 \quad \text{if } B(i,j) = 1 \\ \frac{1}{(2\pi)^{1/2}\sigma_{b2}} e^{-V_{c1}} & \text{where } V_{c1} = (f_b - \mu_{b2})^2/2\sigma_{b2}^2 \quad \text{if } B(i,j) = 0 \end{cases} \tag{1}$$

where $\mu_{b1}, \mu_{b2}, \sigma_{b1}$, and σ_{b2} are all estimated from the test data set. $\sigma_n = \sigma_n(f_b(i,j))$ is computed from the approximating exponential function.

2.4 Prior Specification

The prior defines the distribution of object shapes. Due to the enormous possibilities of object shapes projecting onto the random field it is difficult to directly specify the joint distribution of the large number of random components in the random field. However, since local object shapes do not have absolute correlation with distant parts of global shapes, it seems reasonable to assume a Markovian property for the random field. In other words, we assume that the values of a given site depend only on values of nearby sites.

Random fields having Markovian property are called Markov Random Fields (MRFs). Geman and Geman (1984) have shown that for a MRF an equivalent Gibbs distribution can be defined on the field. Let $S = \{s_1, s_2, \cdots, s_N\}$ be the set of grid points on a MRF. A *neighborhood system*, $\mathcal{G} = \{\mathcal{G}_s, s \in S\}$ for S, is any collection of subsets of S for which 1) $s \notin \mathcal{G}_s$ and 2) $s \in \mathcal{G}_r \Leftrightarrow r \in \mathcal{G}_s$, where $\mathcal{G}_s$ is the set of neighbors of s. A subset $C \subseteq S$ is a *clique* if every pair of distinct sites in C are neighbors. A *Gibbs distribution* relative to $\{S, \mathcal{G}\}$ is a probability measure π on Ω, the state space of the random field, with the following representation:

$$\pi(\omega) = \frac{1}{Z} e^{-U(\omega)} \tag{2}$$

where ω is an outcome of the random field, Z is a constant and U, called the *energy function*, is of the form

$$U(\omega) = \sum_{C \in \mathcal{C}} V_C(\omega) \tag{3}$$

where $\mathcal{C}$ is the set of cliques for $\mathcal{G}$. Each V_C is a function on Ω with the property that $V_C(\omega)$ depends only on those components x_s of ω for which $s \in C$. Such a family $\{V_C, C \in \mathcal{C}\}$ is called a *potential*. Z is the normalizing constant:

$$Z \triangleq \int_\omega e^{-U(\omega)} \tag{4}$$

and is called a *partition function*.

We used a nearest-pair neighborhood system in modeling the local shape probability. There are four kinds of cliques in this neighborhood system: $(s_{i,j}, s_{i+1,j})$, $(s_{i,j}, s_{i,j+1})$, $(s_{i,j}, s_{i+1,j+1})$, and $(s_{i,j}, s_{i+1,j-1})$. For each kind of pair there are three combinations for the boundary values on the two grid points: both on boundaries, exactly one on a boundary, or neither on a boundary. For each case there are constraints on the normal directions for the points on boundaries: if both points are on boundaries the normal direction should change smoothly, while if only one point is on a boundary its normal direction should be oriented such that the assumed boundary direction does not point to the other point. The latter condition effectively penalizes blindly ending boundaries. Specifically, we have the following potentials for horizontal cliques:

$$\begin{aligned} V_{c2} &= B(i,j) * B(i+1,j) * h(N(i,j), N(i+1,j))^2/\sigma_{h1}^2 \\ &+ (1 - B(i,j)) * B(i+1,j) * h(N(i+1,j), 0)^2/\sigma_{h2}^2 \\ &+ B(i,j) * (1 - B(i+1,j)) * h(N(i,j), 0)^2/\sigma_{h2}^2 \end{aligned} \tag{5}$$

where $B(i,j)$ and $N(i,j)$ are the boundary value and normal direction at location (i,j), and $h(\cdot,\cdot)$ is the distance measure of two angles modulated by 180 degrees. σ_{h1} and σ_{h2} are scale parameters, which control how strongly the object shapes conform to the corresponding shape constraints. Potential functions for the other three cliques are similarly defined.

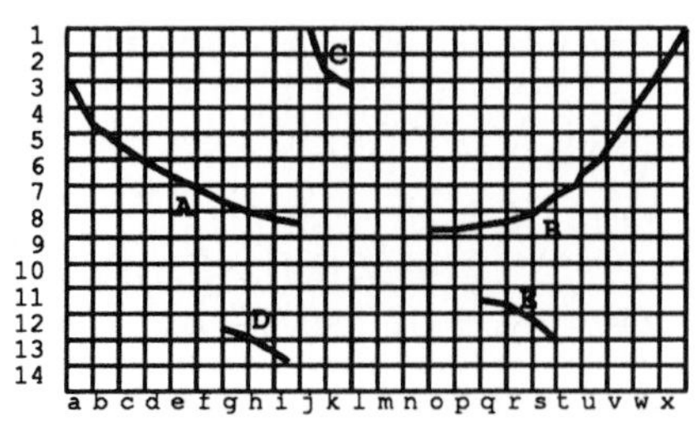

Figure 7: An example of current boundary estimation and the global attribute values. Please see text for the values of the global attributes.

This set of cliques can effectively capture the smooth property of surface normals and the unlikeliness of blindly ending boundaries. Nevertheless, it is insufficient for distinguishing true object surfaces from ones induced from erroneous filter response of random speckles. Furthermore, the broken surface speckle phenomenon may cause the filter to miss boundaries due to destructive echoes received at transducers. In either case we need to look beyond local neighboring properties to correctly determine the presence of true boundaries.

These problems can be overcome by analyzing object shapes in multiple scales to find support from increasing neighborhoods. Local boundary points suggested by the smallest scale filter are not to be regarded as true boundaries unless enough support is found from extended neighborhoods. Local non-boundary points are not to be regarded as non-boundaries if the global shape structures in some extended neighborhood strongly suggest that they are on broken boundaries.

The prior is designed with a component to capture global shape structures in multiple scales. It uses a parallel algorithm for the multiscale analysis and infers two global shape attributes, *length* and *weight*, for all points. The length attribute is the length of the boundary fragment of a boundary point. Boundary points induced from random speckles should have rather short length attribute values while those from true object boundaries should have longer length attribute values.

The weight attribute is designed to address the destructive echo problem on continuous surfaces. The weight of a point suggests the chances that point is actually on a boundary, and its value is between zero and one. The weight values are one for current boundary points and zero for non-boundary points not on boundary gaps. The weights of those points on boundary gaps are determined in two steps. First, a stochastic decision of whether to close the gap is made based on the configurations of the two matching ends and the filter measurements inbetween the matching ends. Then, if the decision is to close the gap, all points on the trajectory of the gap receive the same weight which is inversely proportional to the gap size. However, if the decision is not to close the gap, all points on the gap receive weight zero. The choice of a stochastic decision to close gaps, instead of a deterministic one, provides the capability to capture different closing alternatives when there are multiple possibilities.

Figure 7 gives an example of the length and weight attribute values. Curves A and B are part of a continuous curve with a gap while curves C, D, and E are random boundaries from speckles. The length attribute values are about 11 for all pixels covered by A, 13 for those pixels covered by B, 3 for those pixels covered by C, D, and E, and 0 for all other pixels. All pixels covered by any curve have a weight value 1. All other pixels have weight value 0, except those on the gap between curves A and B. For those pixels on the gap trajectory ((8,j)...(8,n)), they will have the same weight value between 0 and 1. Our current implementation uses $1/(log(s/2)+1.7)$ to compute this value where s is the gap size.

There are many choices for the gap-closing decision function and trajectory-computing algorithm. We propose a method to address the gap-closing and trajectory-computing problems in one unified approach. The idea is based on the active contour model where an energy-minimizing contour is sought given the image and internal constraints (Kass et al, 1988). For each potential gap, a spline is born to connect the two matching ends. The optimal positions of the spline is governed by the *image energy*, i.e., the filter measurements at the scale corresponding to the gap size and the internal bending and stretching energies of the spline itself. The filter measurements constrain the spline such that the spline is pulled towards pixels where the edge strength measurements are strong and the normal direction measurements conform to the local spline normal directions and is pushed away from pixels where the edge strength measurements are strong but the normal direction measurements contradict to the local spline normal directions.

This algorithm not only finds the gap trajectory, but also provides the energy at stable state as the basis of gap-closing decision. Whenever the algorithm detects a potential gap a spline is generated and the optimal positions of the spline is computed. The total energy of the spline at the optimal positions is taken as a score, s. Then a random value, r, between zero and one is uniformly drawn and compared to e^{-s}. If $r < e^{-s}$, the gap is closed; otherwise the gap is not closed. When there are multiple possible matching ends for an end, each will have a chance to be closed with the chances decrease with increasing scores. This seems reasonable since the score inversely reflects the conviction of a boundary present at a gap where the likelihood function shows relative low confidence. Notice that this choice of the total energy is in favor of closing small gaps over large gaps when there are multiple choices.

The algorithm computes length and weight attribute values by building a pyramid of cells of 6 by 6 overlapping receptive fields (Shneier, 1981), (Meer et al, 1990). At the base of the pyramid is a "thinned" version of the current boundary estimations of the Gibbs random field. A thinning procedure is applied to reduce the boundary thickness to one pixel in order to correctly compute length and weight values in the pyramid operation. At each level in the pyramid, boundary fragments in child cells are joined together in the parent cell and their lengths are summed. Potential gaps between a pair of boundary fragments within the same receptive field are closed according to the active contour algorithm. A gap is closed at a level corresponding to its gap size. This algorithm then makes a downsweep of the pyramid, tracing down the exact boundary locations while propagating the length and weight attributes at the same time. Please see (Lin, 1991) for details of the thinning, gap-closing and pyramid algorithms.

With the values of the length and weight attributes, $L(i,j)$ and $W(i,j)$, we define the following potential function:

$$V_{c6} = B(i,j) * \left(\frac{A_{L1}}{L(i,j) + C_L} + \frac{A_{W1}}{W(i,j) + C_W} \right) + (1 - B(i,j)) * \left(\frac{L(i,j)}{A_{L2}} + \frac{W(i,j)}{A_{W2}} \right) \quad (6)$$

where $A_{L1}, A_{W1}, A_{L2}, A_{W2}$ are scaling factors, and C_W, C_L are positive constants introduced to prevent infinite potentials when the attribute values are zeros. Although this potential function looks like a single element clique, it is actually derived from extended neighborhoods. The L and W in this potential are global shape attributes computed from the extended neighborhood, and serve as latent data in our data augmentation scheme to be described in detail in the following section. Without computing the L and W explicitly and viewing them as latent data, it would have been difficult to model the long range correlation in the Gibbs prior. By incorporating global attributes in a data augmentation scheme we have extended the use of Gibbs distribution to modeling global structures.

The parameters in the prior are determined empirically. There are two parameters in the local shape potential functions, i.e., σ_{h1} and σ_{h2}, and four parameters in the global attribute potential functions, i.e., A_{L1}, A_{L2}, A_{W1}, and A_{W2}. Their values are determined by assuming fixed energies for some typical configurations and are further fine-tuned by evaluating the results.

2.5 Architecture of the Solution Algorithm

Let X be the random field $X = \{x_{ij} = (B(i,j), N(i,j)), \forall i,j\}$. Let Y be the output of the smallest filter from the input image, i.e. $Y = \{y_{ij} = (f_b(i,j), f_n(i,j)), \forall i,j\}$. We shall assume the conditional distribution $p(y_{ij}|X)$ of observing filter output y_{ij} at voxel (i,j) given X depends only on x_{ij} and that the values of y_{ij} are independent of each other. We have made these assumptions to simplify the model. It follows from our prior specification and likelihood function that the data augmented posterior $p(X|Y, L, W)$ is a Gibbs distribution with the energy function:

$$U(\omega) = \sum_{C_i \in C} \sum_{i=1,6} V_{c_i}(\omega) \quad (7)$$

where V_{c1} comes from the likelihood function, $V_{c2}, V_{c3}, V_{c4}, V_{c5}$ come from the local shape constraints, and V_{c6} comes from the global shape attributes. The L and W act as latent data. They are not observed but have a distribution under the observed data Y. This distribution, $P(L, W|Y)$, is called a *predictive distribution.*

What we are really interested in is $p(X|Y)$. We use a data augmentation scheme, which is similar to an EM algorithm (Dempster et al, 1977), to iteratively calculate $p(X|Y)$ from the augmented posterior $p(X|Y, L, W)$ and the predictive distribution $p(L, W|Y)$ by alternatively sampling these two distributions (Tanner and Wang, 1987), (Gelfand and Smith, 1990). The idea is as follows. First, we note the following relationship between the posterior density and the predictive density:

$$p(X|Y) = \int_{\Omega_{L,W}} p(X|Y,L,W)p(L,W|Y)dL,W \tag{8}$$

$$p(L,W|Y) = \int_{\Omega_X} p(L,W|X,Y)p(X|Y)dX \tag{9}$$

where $\Omega_{L,W}$ is the sample space of the latent data, L and W, and Ω_X is the sample space of X. Substituting Eq. (9) into Eq. (8) and interchanging the order of integration, we obtain the following relationship for $p(X|Y) = g(X)$:

$$g(X) = \int_{\Omega_X} K(X,X')g(X')dX', \quad \text{where}$$

$$K(X,X') = \int_{\Omega_{L,W}} p(X|Y,L,W)p(L,W|X',Y)dL,W. \tag{10}$$

This is an integral transformation and can be solved by the iterative method: start with any initial approximation $g_0(X)$ to $p(X|Y)$, and successively calculate

$$g_{i+1}(X) = (Tg_i)(X) \tag{11}$$

using Eq. (10).

The integral transform in Eq. (11) is usually difficult to calculated analytically. Monte Carlo method can be used here to perform the integration. Putting all the above together we obtain the following iterative scheme: given the current approximation g_i to $p(X|Y)$,

(a) Generate a sample $(L,W)^{(1)},\cdots,(L,W)^{(m)}$ from the current approximation to the predictive density $p(L,W|Y)$.
(b) Update the current approximation to $p(X|Y)$ to be the mixture of conditional densities of X given the augmented data patterns generated in (a), i.e.,

$$g_{i+1}(X) = m^{-1}\sum_{j=1}^{m} p(X|(L,W)^{(j)},Y). \tag{12}$$

In step (a), the sample (L,W) values can be generated in two steps:

(a1) Generate X from $g_i(X)$.
(a2) Generate (L,W) from $p(L,W|X',Y)$, where X' is the value obtained in (a1).

When m is large, the two steps (a) and (b) in the above iterative algorithm provide a close approximation to one iteration of Eq. (11). However, even when m is as small as 1, the iteration is still "in the right direction" in the sense that the average of $p(X|Y,L,W)$ over the augmented data patterns generated across iterations will converge to the true $p(X|Y)$.

In a typical Markov Random Field only the interactions among the neighboring sites are explicitly formulated. Remote sites have influence among each other only through a chain of intermediate sites. The interaction among remote sites can not be readily apprehended from a chain of local interactions. Thus it is difficult to model a desired global interaction through the specification of local interactions.

The data augmentation scheme provides a mechanism to employ Markov Random Field with global information. The global interaction among remote sites can be specified through an intermediate representation which encodes the global information from the observed data. This intermediate representation serves as the latent data and facilitates the analysis of the estimated quantities.

The operation of the complete algorithm is shown in Figure 8. The current boundary values and normal directions to be estimated are shown in the shaded area. The initial boundary values are obtained by applying a threshold on $f_b(i,j)$, and the initial values of normal directions are $f_n(i,j)$. The algorithm then makes repeated updates on the whole field. Each iteration includes successive updates to each voxel. A sample from $p(L,W|Y)$ is drawn by performing the pyramid operation on the the current configuration. The Gibbs sampler is used to produce a new sample x_{ij} for each voxel. It involves first computing the augmented marginal posterior $p(x_{ij}|X_{S/ij},Y,L,W)$, where $X_{S/ij}$

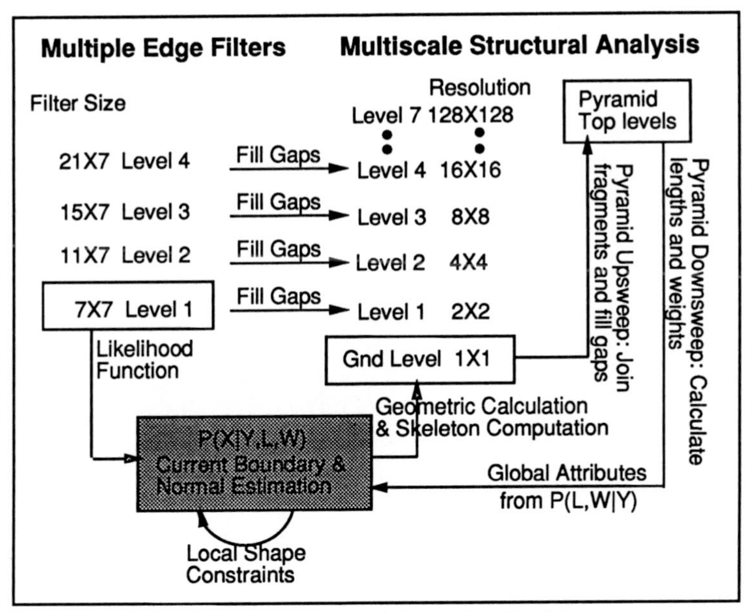

Figure 8: Operation of the Solution Algorithm.

denotes the neighborhood of x_{ij}. Each $p(x_{ij}|X_{S/ij}, Y, L, W)$ is computed from Eqs. (2), (4), and (7) with only those cliques including voxel (i, j). A sample is then randomly drawn accordingly and replaces the old x_{ij}.

In the current implementation we compute the pyramid only up to the sixth level. At this level each cell covers a 32 by 32 pixel region. This size is considered large enough to compute reliable global attributes to reveal false boundaries. The gap-closing operation is only performed up to the fourth level. At this level, the maximal gap size that can be closed is roughly 45 pixels. The values of $B(i, j), N(i, j)$ and squared values of them are accumulated during each iteration. They serve to compute the means and variances of the boundary and normal estimations.

3. RESULTS

Figure 9 shows the magnitude output from the smallest scale filter, contrasted with those from a Sobel filter, which is a typical filter used in volume rendering. These pictures show our filter gives more accurate boundary measurements on the boundaries of the diffuse phantom and smooths out broken boundaries of the baby doll image.

Figure 10 shows the three test images and their corresponding mean boundary values displayed as images. The mean boundary values are obtained by averaging over 150 samples of binary boundary values. These samples are generated by the Gibbs sampler with the data augmentation scheme described in the previous section. The original images are displayed on the left and the mean boundary values on the right.

The boundaries in these result images seem to be quite thick. This is due to two factors. First, the filter is designed to produce correct boundary measurements on texture boundaries. It has a large support and produces significant magnitudes near boundaries. Second, these significant magnitudes near boundaries have higher magnitude than those measured on random speckles. The likelihood function thus tends to suggest those pixels as on true boundaries. These thick boundaries can be thinned by a thinning algorithm.

The result of the baby doll image (Figure 10a) shows enhancement of some weak boundaries which can not be clearly seen in the original images. Unfortunately, these boundaries are mostly acoustic artifacts. They can be removed by selecting a higher mean magnitude values in the likelihood

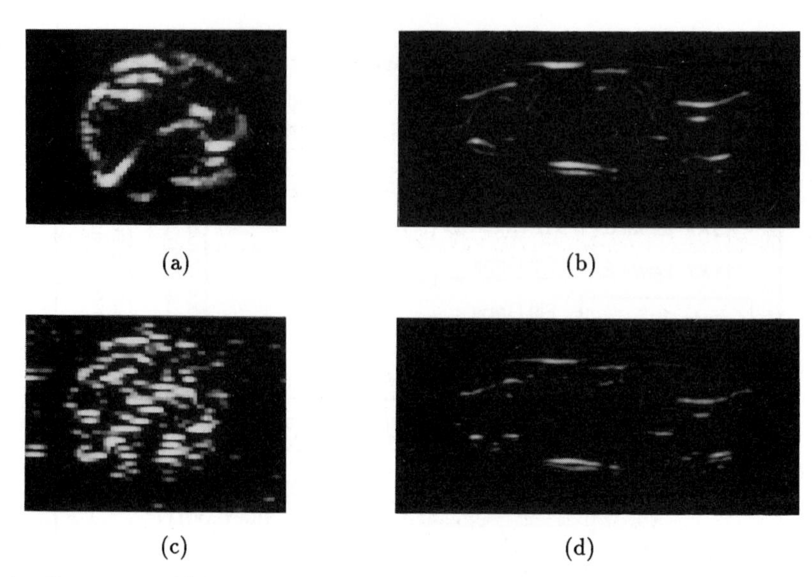

Figure 9: Our filter outputs: (a) Cone phantom, (b) Baby doll. Sobel filter outputs: (c) Cone phantom, (d) Baby doll.

image	μ_{b1}	μ_{b2}	σ_{b1}	σ_{b2}	σ_{h1}	σ_{h2}	A_{L1}	A_{L2}	A_{W1}	A_{W2}	C_L	C_W
doll	0.70	0.18	0.137	0.118	2	3	3	4	0.4	0.3	0.001	0.001
cone	0.73	0.51	0.137	0.118	2	3	3	4	0.4	0.3	0.001	0.001
liver	0.70	0.40	0.137	0.118	2	3	3	4	0.4	0.3	0.001	0.001

Table 1: The parameter values used to generate the results in Figure 10.

function. Figure 10b shows that the algorithm detects gaps (pointed to by the white arrows) and tries to close them. The algorithm also detects the gap in the liver image (Figure 10c), but the stable state energy of the snake associated with that gap is so high that it is closed with a very low probability. Thus only a few samples out of the 150 have this gap closed and the averaging result does not show this clearly. This suggests that the parameters in the gap-closing algorithm are not optimal and further fine-tuning is necessary. The values of the parameters used in producing these results are shown in Table 1.

4. DISCUSSION

Due to the large size of our filter, the localization of boundaries is not optimal. Particularly, the filter can not produce correct boundary strength on corners. As a results, small features or corners tend to be missed. In the future reasearch we wish to use a multiscale likelihood function in which filter outputs of different scales are weighted by types of boundary and boundary curvatures. This will allow us to use a smaller filter or filters specially designed to measure corners and should help produce more accurate measurements.

The local shape constraints seem to be the least important factors, compared with the global shape attributes and the likelihood function, in determining the final results. The results from the algorithm without the local constraints do not appear different. The results are sensitive to the estimated mean magnitudes of the filter measurement. An over-estimated μ_{b1} results in low boundary probabilities for weak boundaries and an under-estimated μ_{b2} results in high boundary probabilities for strong speckles.

A related problem is that the standard deviations of the filter measurements, σ_{b1} and σ_{b2}, might be fairly large and the means, μ_{b1} and μ_{b2}, might not be well separated. The likelihood function can not resolve ambiguous features and the results tend to have more errors even though the estimations of these parameters are correct. This happens when there are low contrast targets

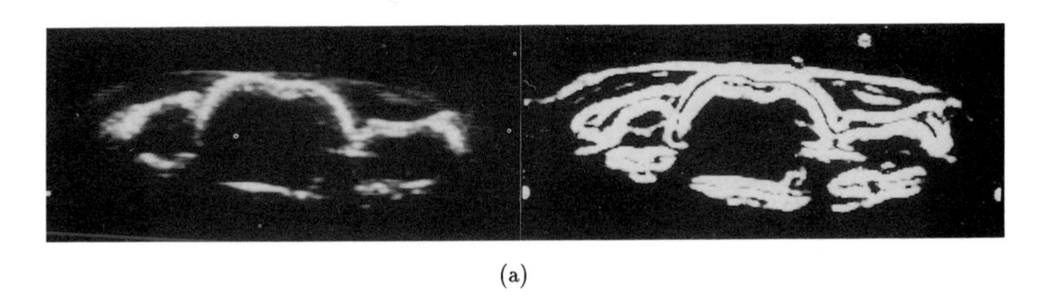

(a)

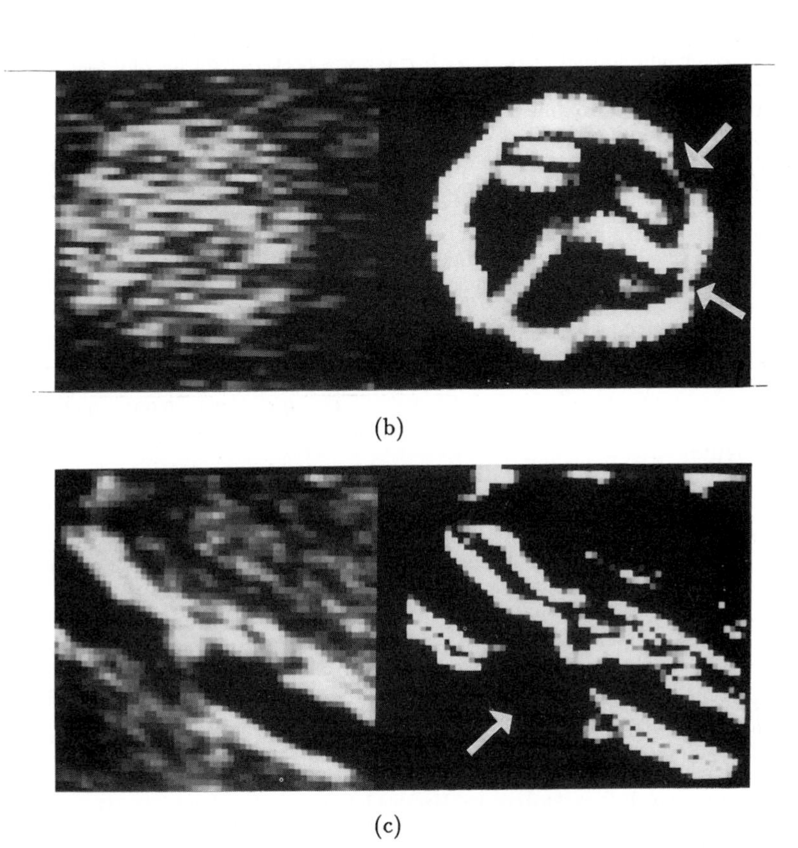

(b)

(c)

Figure 10: Test images and the mean boundary values displayed as images: (a) Baby doll, (b) Cone phantom, (c) Liver. The mean boundary values are obtained by averaging over 150 samples.

in the scene.

Current implementation takes about 25 seconds to produce the initial filter measurements at four scales and 4 seconds to produce one sample for a 128 by 128 images on a DEC3100 workstation with unoptimized code. Approximately 50 percent of the running time is spent on sampling, which can be optimized since most pixels are not on boundaries and have very small boundary strength measurement. About 15 percent of the running time is spent on gap-closing procedures and 35 percent on the pyramid algorithm and the rest of the algorithm.

5. CONCLUSIONS

We have described a Bayesian framework for estimating surfaces in noisy images. This framework generates samples of ensemble scenes for an observed image in the form of a random field of binary boundary value and normal directions. It computes posterior means and standard deviations of the ensemble at each pixel and produces a reference image of surface representation which can be further mapped to opacity for volume rendering.

We have applied this approach to two dimensional ultrasound images and have shown its usefulness in producing meaningful boundary representations. This boundary representation not only can be used in visualization but also can be the basis of image segmentation. We believe segmenting via this boundary representation will produce better results than segmenting via the local edge measurements only. We have implemented an active contour model using Gibbs distributions and are ready to test this idea using this model.

The Gibbs prior has also been extended to incorporate important properties of global shapes via multiscale analysis. By incorporating the global attributes in a data augmentation scheme we have extended the use of Gibbs distribution on modeling global structures. Future research includes quantitative evaluation of the algorithm and results and extending this model to three dimensions.

Acknowledgements

The authors wish to thank Dr. Olaf von Ramm of the Biomedical Engineering Department of Duke University, Mr. Antoine Collet-Billon of Philips-LEP in France, and Dr. Marc Levoy of the Computer Science Department of Stanford University for helpful discussions. Mr. Ryutarou Ohbuchi provided the baby doll image, and Philips-LEP provided the liver and phantom images. This research is supported by NSF grant number CDR-8622201 and NIH grant number P01 CA47982.

References

Canny J (1986). A computational approach to edge detection. IEEE Transactions on Pattern Analysis and Machine Intelligence. PAMI-8(6):679–698.

Dempster AP, Laird NM and Rubin DB (1977). Maximum likelihood from incomplete data via the em algorithm. Journal of the Royal Statistical Society. Ser. B(39):1–38.

Geman S and Geman D (1984). Stochastic relaxation, gibbs distributions, and the bayesian restoration of images. IEEE transactions on Pattern Analysis and Machine Intelligence. PAMI-6(6):721–741.

Gelfand AE and Smith AFM (1990). Sampling-based approaches to calculating marginal densities. Journal of the American Statistical Association. 85(410):398–409.

Hykes D, Hedrick WR and Starchman DE (1985). Ultrasound Physics and Instrumentation. Churchhill Livingstone, New York, N.Y., 1985.

Kass M, Witkin A and Terzopoulos D (1988). Snake: Active Contour Models. International Journal of Computer Vision. 1:321–331.

Korn AF (1988). Toward a symbolic representation of intensity changes in images. IEEE Transactions on Pattern Analysis and Machine Intelligence. PAMI-10(5):610–625.

Levoy M (1988). Display of surfaces from volume data. IEEE Computer Graphics and Applications. 8(3):29–37.

Lin WJ (1991). Boundary Estimation in Ultrasound Images. PhD thesis, U. of North Carolina, Chapel Hill, NC, 1991. To appear.

Meer P, Sher CA and Rosenfeld A (1990). The chain pyramid: Hierarchical contour processing. IEEE Transactions on Pattern Analysis and Machine Intelligence. PAMI-12:363–376.

Nassiri DK and Hill CR (1986). The use of angular acoustic scattering measurements to eliminate structural parameters of human and animal tissues. The Journal of the Acoustical Society of America. 79(6):2048–2054.

Shneier M (1981). Two hierarchical linear feature representations: Edge pyramids and edge quadtrees. Computer Graphics and Image Processing. 17:211–224.

Shattuck DP, Weinshenker MD, Smith SW and von Ramm OT (1984). Explososcan: A parallel processing technique for high speed ultrasound imaging with linear phased arrays. Journal of Acoustic Society of America. 75(4):1273–1282.

Tanner MA and Wong WH (1987). The calculation of posterior distributions by data augmentation. Journal of the American Statistical Association. 82(398):528–541.

von Ramm OT, Smith SW, Sheikh KH, and Kisslo J (1988). Real-time, three-dimensional echocardiography. Circulation. 78. Supplement II.

Wagg RC, Dalecki D and Christopher PE (1989). Spectral power determinations of compressibility and density variations in model media and calf liver using ultrasound. The Journal of the Acoustical Society of Americ. 85(1):423–431.

TOPOLOGICAL CLASSIFICATION IN DIGITAL SPACE

G. Malandain*, G. Bertrand† and N. Ayache*

* INRIA , B.P. 105 - 78153 Le Chesnay Cédex - France
† ESIEE , Labo IAAI - Cité Descartes - 2, bd Blaise Pascal - 93162 Noisy-le-Grand Cédex - France

Abstract

We propose in this paper a new approach to segment a discrete 3D object into a structure of characteristic topological primitives with attached qualitative features. This structure can be seen itself as a *qualitative* description of the object, because

- it is intrinsic to the 3D object, which means it is stable to rigid transformations (rotations and translations).
- it is locally defined, and therefore stable to partial occlusions and local modifications of the object structure.
- it is robust to noise and small deformations, as confirmed by our experimental results.

Our approach concentrates on topological properties of discrete surfaces. These surfaces may correspond to the *external* surface of the objects extracted by a 3D edge detector, or to the *skeleton* surface obtained by a new thinning algorithm. Our labeling algorithm is based on very local computations, allowing massively parallel computations and real time computations.

We present a realistic experiment to characterize and locate spatially a complex 3D medical object using the proposed segmentation of its skeleton.

Keywords

Digital topology; segmentation; simple surfaces; image analysis.

1. INTRODUCTION

Three-dimensional (3-D) images may come from several fields, the most popular one being the medical field, where images are produced by X-ray Computed Tomography (CT), Magnetic Resonance Imaging (MRI), Positon Emitting Tomography (PET), and more recently by Ultrasound Echography.

Automating the *interpretation* of these images is an awkward but important task for many applications where extremely accurate quantitative results are required or/and a large volume of data must be processed. A large class of interpretation tasks involves a *matching* stage, an accurate geometric registration between two 3D images, or between a 3D image and a 3D geometric and semantic model [ABC+90].

Prior to matching, a preliminary *segmentation* stage is necessary to reduce the original image into a more compact highly structured representation useful for interpretation [ABB+89]. We propose in this paper a new segmentation process which transforms a discrete 3D object into a structure of characteristic topological primitives with attached qualitative features. This structure can be seen itself as a *qualitative* description of the object, because

- it is intrinsic to the 3D object, which means it is stable to rigid transformations (rotations and translations).
- it is locally defined, and therefore stable to partial occlusions and local modifications of the object structure.
- it is robust to noise and small deformations, as confirmed by our experimental results.

The idea of characterizing 3D shapes with qualitative primitives was introduced by [San89] who studied differential singularities of 3D object surfaces. Our approach is complementary and original because we concentrate on different topological properties of discrete surfaces instead of geometrical properties. Finally, the core of our approach is based on very local computations, allowing massively parallel computations and real time computations.

This paper is organized as follows : first we recall previous work done on digital topology. Second, we describe our approach which is twofolds, a) local labelling of object points using a classification tree with two local measures, and b) global extraction of simple surfaces (or surfaces patches), using neighborhood properties. Third, we propose a new thinning algorithm using our classification. Fourth, we present a new distance transformation algorithm, which can compute distance on simple surfaces. Fifth, we describe two applications of this work, one using our thinning algorithm and the other using our classification for the segmentation and the spatial localization of the skeleton of a complex 3D medical object. We conclude this paper by a study of the advantages of this approach and some short and medium term future research topics.

2. BASIC DEFINITIONS

We present some basic definitions of 3D digital topology (see [Ros80, KR89, NA85, TYYF82]).

A 3D digital image Σ is a subset of Z^3. We consider only cubic lattices so that, for each point $x = (i, j, k)$, two types of neighbor may be defined:

6-neighbors : $N_6^*(x) = \{y = (i', j', k'), \mid i - i' \mid + \mid j - j' \mid + \mid k - k' \mid = 1\}$

26-neighbors : $N_{26}^*(x) = \{y = (i', j', k'), \max(\mid i - i' \mid, \mid j - j' \mid, \mid k - k' \mid) = 1\}$

We note $N_6(x) = N_6^*(x) \cup \{x\}, N_{26}(x) = N_{26}^*(x) \cup \{x\}$.

Two subsets $X \subset \Sigma$ and $Y \subset \Sigma$ are said to be 6-adjacent (26-adjacent) if there exist $x \in X$ and $y \in Y$, x and y being 6-neighbors (26-neighbors).

A 6-path (26-path) is a sequence $\rho_0, \rho_1, \dots \rho_i, \dots \rho_n$ of points such that ρ_i is 6-adjacent (26-adjacent) to ρ_{i-1}, for $1 \leq i \leq n$. An object $X \subset \Sigma$ is 6-connected (26-connected) if a 6-path (26-path) lying in X can be found between each pair of points of X.

A 6 (26)-connected component of X is a set $Y \subset X$ which is 6-connected (26-connected) and which is maximal for this property.

As in the 2D-case opposite types of connectedness should be used for X and $\overline{X}$, the complement of X.

Hence any further definition has to be viewed as a double definition: a "6-definition" and a "26-definition".

Let us introduce the notions of hole and cavitie. The notion of hole is not simple to define (see [PR71, Mor80]). For example a torus has one hole. The notion of cavitie is intuitive : for example any stone fruit has one cavitie, its stone.

The genus of $X, G(X)$ is the number of X's connected components plus the number of its cavities minus the number of its holes.

We have the following relations [Mor80]:

$$G_{26}(\overline{X}) - G_6(X) = 1, \quad G_6(\overline{X}) - G_{26}(X) = 1$$

In [PR71] a method for obtaining the genus of a 3D-object by local pattern matching in the 6-connectedness case is given. In [Mor80] the method for the 26-connectedness case is proposed.

A point $x \in X$ is call simple if its removal does not change the topology of the image.

In [Mor81] a characterization of simple points is proposed. A point x is simple if :

$$NC[X \cap N_{26}(x)] = NC[X \cap N_{26}^{*}(x)] \quad (1)$$
$$NC[(\overline{X} \cap N_{26}^{*}(x)) \cup \{x\}] = NC[\overline{X} \cap N_{26}^{*}(x)] \quad (2)$$
$$NH[X \cap N_{26}(x)] = NH[X \cap N_{26}^{*}(x)] \quad (3)$$
$$NH[(\overline{X} \cap N_{26}^{*}(x)) \cup \{x\}] = NH[\overline{X} \cap N_{26}^{*}(x)] \quad (4)$$

where $NC(A)$ stands for the number of connected components of A and $NH(A)$ denotes the number of holes in A. $NH(A)$ may be computed with the genus.

3. CONNECTED COMPONENTS NUMBERS

In the continuous space $\Re^3$, we propose a simple way to characterize the points of an object X with local topological properties, which is to examine an arbitrary small neighborhood $V(x)$ of $x \in X$ and to count the number of connected components in $V(x)$ and in $V(x) \setminus \{x\}$ ($\setminus$ is the symbol of substraction between sets). Four connected components numbers may be used:

- $C = NC[X \cap V(x)]$
- $C^* = NC[(X \cap V(x)) \setminus \{x\}]$
- $\overline{C} = NC[\overline{X} \cap V(x)]$
- $\overline{C}^* = NC[(\overline{X} \cap V(x)) \cup \{x\}]$

For example, if x belongs to a simple surface X, there exists $V(x)$ such that $C = 1$, $C^* = 1$, $\overline{C} = 2$, $\overline{C}^* = 1$. In a complex surface X, the curves which are the intersection of several simple surfaces may be characterized by $C = 1$, $C^* = 1$, $\overline{C} \geq 3$, $\overline{C}^* = 1$. A simple closed curve X is such that there exists $V(x)$ and $C = 1$, $C^* = 2$, $\overline{C} = 1$, $\overline{C}^* = 1$.

4. A NEW TOPOLOGICAL CLASSIFICATION

We propose to use such connected components numbers to get some information about the local characteristics of points in a 3D digital image.

In the following we assume that 26-connectedness is used for X, while 6-connectedness is used for $\overline{X}$. Instead of considering arbitrary small neighborhoods $V(x)$ we use a fixed neighborhood $N_{26}(x)$. For that reason, if we want to get the digital equivalents of the four numbers, we have to consider in the definition of the numbers the components adjacent to x rather that the components being in the neighborhood. Hence, we can define the discrete connected components numbers:

- C = number of 26-components of $X \cap N_{26}(x)$ 26-adjacent to x
- C^* = number of 26-components of $[X \cap N_{26}(x)] \setminus \{x\}$ 26-adjacent to x
- $\overline{C}$ = number of 6-components of $[\overline{X} \cap N_{26}(x)]$ 6-adjacent to x
- $\overline{C}^*$ = number of 6-components of $[\overline{X} \cap N_{26}(x)] \cup \{x\}$ 6-adjacent to x

Note that we use 26-adjacency for X and 6-adjacency between X and $\overline{X}$.

We see that, if x belongs to X:

1. C is always equal to 1 since all points of $X \cap N_{26}(x)$ are 26-connected to x
2. C^* is equal to the number of components of $[X \cap N_{26}(x)] \setminus \{x\}$ for the same reason: $C^* = NC[(X \cap N_{26}(x)) \setminus \{x\}] = NC[X \cap N^*_{26}(x)]$
3. $\overline{C}^*$ is always equal to 1 since all 6-components involved in that number are linked by x.

It follows that only two numbers are significant: C^* and $\overline{C}$. We now give a classification of points belonging to X according to the values of C^* and $\overline{C}$:

Type A - interior point :	$\overline{C} = 0$
Type B - isolated point :	$\overline{C} = 1$, $C^* = 0$
Type C - edge point :	$\overline{C} = 1$, $C^* = 1$
Type D - curve point :	$\overline{C} = 1$, $C^* = 2$
Type E - curves junction :	$\overline{C} = 1$, $C^* > 2$
Type F - surface point :	$\overline{C} = 2$, $C^* = 1$
Type G - surface - curve junction :	$\overline{C} = 2$, $C^* \geq 2$
Type H - surfaces junction :	$\overline{C} > 2$, $C^* = 1$
Type I - surfaces - curve junction :	$\overline{C} > 2$, $C^* \geq 2$

1. It should be pointed out that we assign a class to every possible value of $\overline{C}$ and C^*. It means that there will not remain any unclassified point, whereas this was the case previously [MR80].
2. For type A points, C^* is always equal to 1. Like in the 2D case, interior points may appear at the intersection of several surfaces (see figure 1).

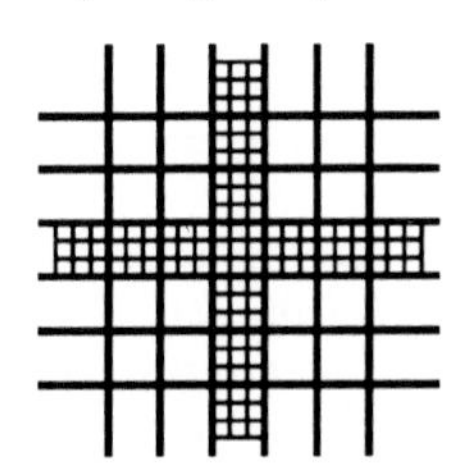

Figure 1: The horizontal trace of two intersecting vertical planes: the central point is an interior point.

3. Edge points may be of two types: simple points and non simple points. In fact it is clear that only edge points may be simple. The following property stands:

   a point $x \in X$ is simple iff:

$$\begin{aligned} &x \text{ is an edge point} && (5) \\ NH[X \cap N_{26}(x)] &= NH[X \cap N^*_{26}(x)] && (6) \\ NH[(\overline{X} \cap N^*_{26}(x)) \cup \{x\}] &= NH[\overline{X} \cap N^*_{26}(x)] && (7) \end{aligned}$$

   In effect, conditions (6) and (7) are identical to conditions (3) and (4) of the original definition of simple points [Mor81].

   Moreover, we have already seen that $NC(X \cap N_{26}(x)) = C = 1$, therefore condition (1) for simple points is equivalent to $C^* = 1$.

   Let b represent the number of connected components in $\overline{X} \cap N^*_{26}(x)$ not adjacent to x. We have $NC[\overline{X} \cap N^*_{26}(x)] = b + \overline{C}$ and $NC[(\overline{X} \cap N^*_{26}(x)) \cup \{x\}] = b + \overline{C}^* = b + 1$. Hence the condition (2) for simple points is equivalent to $\overline{C} = 1$. □

4. Simple surfaces may intersect without creating junction points. See for example figure 2 where two planes intersect : all points are classified as surface points. Connected components numbers are less powerful to describe junctions in discrete spaces than in continuous spaces. Nevertheless we present in section 5. a method to overcome this difficulty.

   The same problem still exists for junctions of curves. If we consider figure 2 as an intersection between several curves, we notice that all points are classified as curve points. We can overcome this difficulty by counting the neighbors of each curve points, if the number of neighbors is greater than 2, there is a junction between curves.

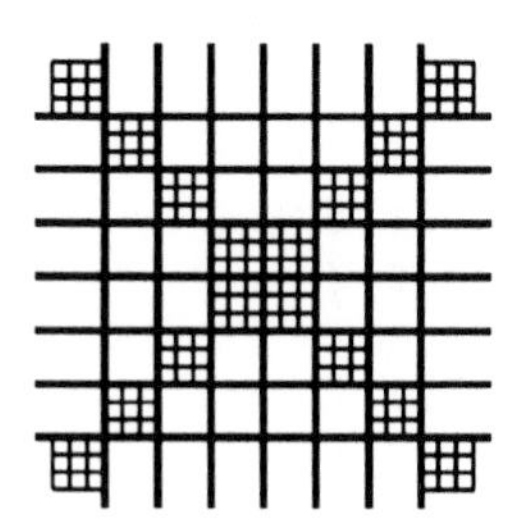

Figure 2: The horizontal trace of two intersecting vertical planes: all points are surface points.

5. TOPOLOGICAL SEGMENTATION INTO SIMPLE SURFACES

We call surface a connected set of points with a thickness of 1 (the deletion of any non-border point will create a hole). A surface can consist of several simple surfaces linked together by junctions (T-junctions, cross-junctions, ...).

As discussed above the connected components numbers do not allow, in a complex figure, to separate simple surfaces from each other (see figure 2) : if we delete all junction points of an image the remaining connected components are not necessarily simple surfaces.

Some work was done to characterize simple surfaces [MR80, Kim84, KF81]. Let us consider the definition of a simple surface proposed in [MR80] because a simple surface is considered here as a set of voxels rather than a set of faces like in [KF81]. We first compare this definition with our classification, then we show that this definition is not appropriate, finally we propose an algorithm for the complete extraction of simple surfaces.

A point $x \in X$ is a simple surface point, according to [MR80], if the following conditions are all satisfied :

1. $X \cap N_{26}^*(x)$ has exactly one component adjacent to x (in the X sense).

2. $\overline{X} \cap N_{26}^*(x)$ has exactly two components adjacent to x (in the $\overline{X}$ sense); call these components B_x and C_x.

3. for every $y \in X$ adjacent to x (in the X sense), y is adjacent (in the $\overline{X}$ sense) to some point in B_x and to some point in C_x.

If we use 26-connectedness for X, we can see that condition 1 is equivalent to $C^* = 1$, condition 2 is equivalent to $\overline{C} = 2$, hence these two conditions characterize what we call surface points. Let us consider now figure 2 : all points except the four central one's will be simple surface points. So this characterization succeeds in segmenting the complex surface into simple ones.

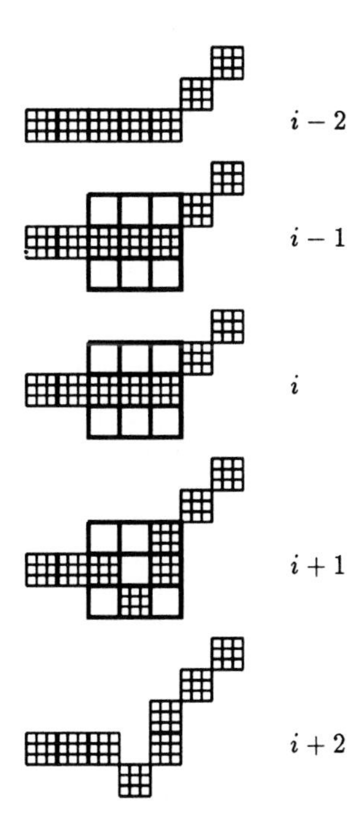

Figure 3: The point at the center of the window of the ith plane is not a simple surface point: the point at the upper right corner of the window in the $(i+1)$th plane does not satisfy condition 3.

However we did not retain this definition because it is not always appropriate. In figure 3 one can see a set of points which should normally be interpreted as a simple surface: the central point of the ith plane is not a simple surface point according to the above definition.

To overcome this difficulty, we propose to use an approach based on an equivalence relation.

Let x be a surface point. We call B_x and C_x the two connected components of $\overline{X} \cap N_{26}^*(x)$ 6-adjacent to x. We say that two surface points x and y are in relation if there is a 26-path $x_0, x_1, \ldots x_i, \ldots x_n$ with $x_0 = x$ and $x_n = y$ such as for $i \in [0 \ldots n-1]$:

- if $B_{x_i} \cap B_{x_{i+1}} \neq \emptyset$ and $C_{x_i} \cap C_{x_{i+1}} \neq \emptyset$
- or if $B_{x_i} \cap C_{x_{i+1}} \neq \emptyset$ and $C_{x_i} \cap B_{x_{i+1}} \neq \emptyset$.

It is easy to check that this relation is reflexive, symmetric and transitive, i.e. it is an equivalence relation.

We call simple surface any equivalence class of this equivalence relation. Let x belong to a simple surface $S \subset X$. If x has a 26-neighbor which does not belong to S and which is not an edge point, we say that x is a junction point of S.

We have now a topological segmentation of a 3D object into simple surfaces and surface junctions.

6. A NEW THINNING ALGORITHM

Several algorithms have been proposed for thinning 3D objects (see [Mor81, LVG80, TF82, HPJ84, GB90]. Two kinds of operations are considered : one kind reduces objects to surfaces and curves, the other one transforms them solely into curves. One basic idea is to remove in parallel simple points which satisfy some "geometrical" properties. Like in the 2D case, we have to use a cycle of several iterations in order to prevent some objects from vanishing (e.g. a two points thickness object). In the 3D case we can use a cycle of 6 iterations corresponding to the Up, Bottom, North, South, West, East directions.

We present now a new thinning algorithm whose the geometrical properties are based on connected components numbers.

Algorithm Do the following operations successively from the Up, Bottom, ... East sides of X until there is no changes:

1. detect all surface points and all curve points. Label them as end points.
2. delete all edge points from a given side of X, provided they are simple and are not labeled.

It should be pointed out that once a point is labeled, it remains labeled in further iterations. If it was not the case a simple surface will be shrunk by its edge. In fact one can see that the edge points of a simple surface will be deleted at the first cycle. All other points will be labeled at the first iteration so that no other points will be removed. Similarly a simple curve will have its end points removed. Other points will remain. It follows that the thinning algorithm preserves surfaces and curves except the very edge points and extremities. An example is shown in figure 4: a sharp corner generates a surface in the thinning process.

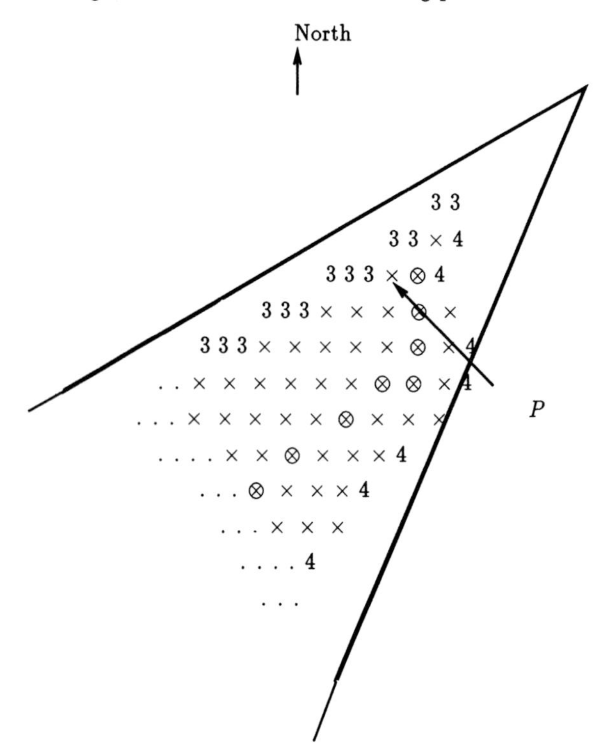

Figure 4: The horizontal trace of a "corner". All other planes are identical to this one. The point P is detected at the 5th iteration as surface point and generates a branch encircled.

Less sharp corners do not generate a surface. Hence this algorithm is well suited for describing objects which is composed of elongated parts. Other conditions may be used for end points in order to detect less sharp corners. This will not be discussed here.

We have to mention that the proposed algorithm needs two steps at each iteration. All points must be treated by the first step before beginning the second one. In fact it is easy to modify the algorithm in order to have independent steps. This can be done by using other labels. It allows to have an algorithm which can be implemented on a sequential computer more efficiently: each iteration can be done with a single scanning of the image.

7. DISTANCE TRANSFORMATION ON SIMPLE SURFACES

Distance transformation is a very important computation for many applications. It can be very useful for extracting dimensional characteristics of an object, or for estimating distance between several objects.

In previous work (see [Bor84] for example), distance transformations are proposed for binary pictures, consisting of *feature* and *nonfeature* elements. These transformations compute for each element an approximation of the distance to the nearest *feature* element.

We propose a new type of distance transformation for ternary pictures, consisting of *feature* elements, *nonfeature* elements and *background* elements. This new algorithm computes for each *nonfeature* element an approximation of the distance to the nearest *feature* element using a path consisting of *nonfeature* elements.

7.1. Classical algorithm

We present algorithms computing a 3-D distance transformation which can be easily extended to higher dimensions.

A classical algorithm consists of the successive application of two stages (forward and backward). Each stage consists in applying a specific mask (see figure 5) on a two-valued image : zero for *feature* elements and infinity otherwise (for *nonfeature* elements).

Forward

+d3	+d2	+d3	+d2	+d1	+d2	0	0	0
+d2	+d1	+d2	+d1	0	0	0	0	0
+d3	+d2	+d3	0	0	0	0	0	0

Backward

0	0	0	0	0	0	+d3	+d2	+d3
0	0	0	0	0	+d1	+d2	+d1	+d2
0	0	0	+d2	+d1	+d2	+d3	+d2	+d3

Figure 5: 3-D masks for the three-dimensional distance transformations

The forward mask is moved over the image from the left to the right, from the top to the bottom and from the front to the back. For each voxel, the sum of each 14-neighbors with the corresponding mask value is computed, and the new value of the voxel is the minimum of these sums. The backward mask is moved in the opposite way, and new values are computed identically.

Some values of $d1$, $d2$, and $d3$:

- Using $d1 = 1$ and $d2 = d3 = \infty$ we obtain the 6-neighbor distance or city block distance.
- Using $d1 = d2 = d3 = 1$ we obtain the 26-neighbor distance or chessboard distance.

7.2. Distance on simple surfaces

Computing distances on simple surfaces will be helpful for understanding their geometry. It makes it easy to erode a surface or to compute its skeleton.

We compute the distance on the surface from the its border. Using our topological classification, we know that the border of a simple surface consists of edge and junction points. These special points will be *feature* elements, simple surface points will be *nonfeature* elements (the background remains unchanged).

We transform an image into a ternary image which is three-valued : zero for *feature* elements, infinity for *nonfeature* elements and a negative number for *background* elements.

Our algorithm consists of two stages which will be iterated until stability. We use the same masks as the ones used by the classical algorithm (see figure 5).

The forward mask is moved over the image from the left to the right, from the top to the bottom and from the front to the back. For each *nonfeature* voxel, the sum of each *nonfeature* or *feature* 14-neighbors with the corresponding mask value is computed, and the new value of the voxel is the minimum of these sums.

The backward mask is moved in the opposite way, and new values are computed identically.

The speed of convergence depends on the shape of the simple surface.

8. EXPERIMENTAL RESULTS

We considered two X-Ray CT scanner 3-D images of a vertebra scanned in two different positions (see figures 6 and 8). These images come from the Rennes Hospital in France, and are a courtesy of Pr. J.L. Coatrieux. They contain about 256x256x50 voxels (each voxel is $0.5*0.5*1mm^3$) whose intensity is coded with 256 discrete values.

In effect, both vertebras can be registered by the application of a geometric transformation which is the combination of a translation, rotation and three different scalings along three orthogonal principal axes. This transformation was computed by matching globally the sets of data points : the rotation angle is 30 degrees along an axis which is tilted of 11 degrees with respect to the vertical axis and scaling is close to 0.9 in two directions.

8.1. Thinning edge detection results

We applied a 3-D edge detector due to Monga, Deriche and Canny [MDMC90] to the original data to extract the external surface trace of the vertebras. A result is shown in figure 7, where the detected edge points are thresholded local maximum of the gradient magnitude in the direction of the gradient. We applied the described thinning algorithm to the result to obtain a simple surface, By doing this, we suppressed more than 20% of the number of edge points, still preserving the topology of surfaces. We show in figure 7 the result.

8.2. Distance on simple surfaces

We applied our distance transformation on simple surface to a 3-D cylinder in which we removed a point. We used the chamfer distance for the computation, it takes 28 seconds of CPU time on a DEC 5000. We show in figure 9 the result.

8.3. Topological classification and segmentation results

We applied the thinning algorithm to the entire 3-D object, producing a skeleton shown in figure 6. These skeletons each contain about $14,000$ points, a figure which must be compared to the $170,000$ original data points in each 3-D-image. These skeletons, despite their instability to noise, are good examples for our algorithm because of the complexity of the surfaces they are made of.

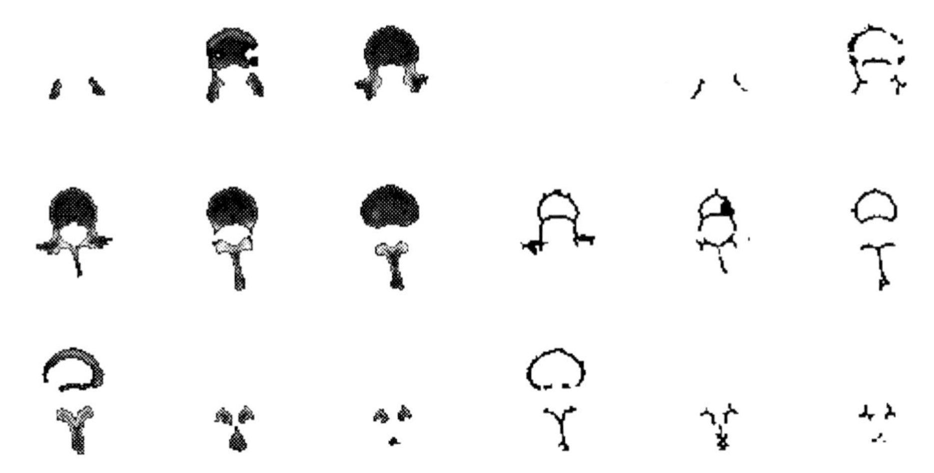

Figure 6: A few slices of the vertebra scanned in the first position and of its skeleton

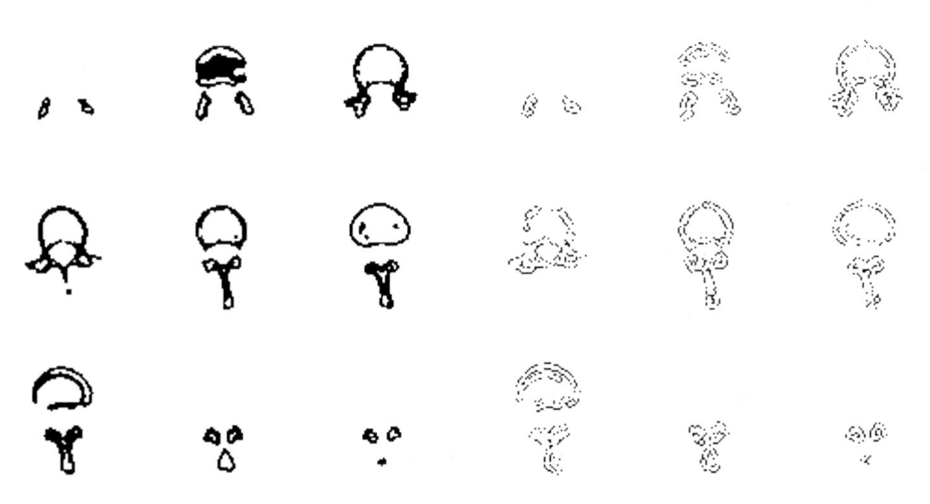

Figure 7: A few slices of the detected edges of the vertebra in both positions and of the edges points suppressed by the step of thinning

Figure 8: A 3D-representation of vertebra in both positions

Then, we applied our classification and simple surface extraction algorithm to label each skeleton point. This computation takes 40 seconds of CPU time on a DEC 5000. We show in figures 10 and 11 the main topological structures extracted for the two vertebras.

We extracted from these structures the junction points of both vertebras, then we applied the geometric transformation found between them (see figures 12 and 13). It is easy to check the accurate superposition of the extracted structures between the two vertebras. This shows the astonishing stability of the segmentation : a single quasi-rigid transformation can superimpose most of the structures within a tolerance of $\pm$ 1 voxel.

Instead of junctions, one can use extracted simple surfaces. In order to obtain a robust segmentation of our surfaces, we perform an erosion of order 2 of the simple surfaces from their border toward their interior. We use for that computation a thresholding of the result of our distance transformation on simple surfaces. We apply our characterisation and simple surface extraction algorithm to the result of erosion.

After this step of erosion, we apply a conditional dilatation of order 2 of the remaining simple surfaces with respect to the original skeleton. This is an opening of simple surfaces. We extract the simple surfaces of this result and we label them. The result of the step of labelling is shown in figures 14 and 15.

It is easy to check that we succeed in segmenting our complex surfaces to several simple surfaces which are stable (for more than 80% of them) from one object to the other. This will be used with profit in a forthcoming 3-D matching algorithm.

9. CONCLUSION

We have proposed a new approach to segment a discrete 3-D object into different classes of points which represent structures. The basic idea of this segmentation is the use of two local measures based on very local computations which allow massively parallel implementation. We obtain therefore a very fast segmentation of 3-D objects.

The proposed classification gives a very efficient set of structures, junctions or simple surfaces, which will allow more efficient registration of complex 3-D objects.

We applied our method to the segmentation into simple surfaces of the 3-D skeleton of a complex 3-D object, and the result appears to be remarkably accurate and stable with respect to quasi-rigid transformation, despite the relative instability of the skeleton itself.

Our method applies to the segmentation of 3-D edges, provided that the edge detector preserves the topology of junctions, which is not the case for known 3-D edges detector.

Our futur work includes a new 3-D edge detector, and the application of our segmenation method on 3-D edges, as well as its use to demonstrate robust matching of quasi rigid complex 3-D objects.

Acknowledgements

This work was partially supported by Digital Equipment Corporation and European AIM (Advanced Informatics in Medicine) Project Murim. The authors want to thank Olivier Monga who initialized this research and Pr. J.L. Coatrieux who provided us with the medical images.

References

[ABB+89] N. Ayache, J.D. Boissonnat, E. Brunet, L. Cohen, J.P. Chièze, B. Geiger, O. Monga, J.M. Rocchisani, and P. Sander. Building highly structured volume representations in 3d medical images. In *Computer Aided Radiology*, 1989. Berlin, West-Germany.

[ABC+90] N. Ayache, J.D. Boissonnat, L. Cohen, B. Geiger, J. Lévy Véhel, O. Monga, and P. Sander. Steps toward the automatic interpretation of 3d images. In *Workshop on 3D imaging in Medecine*, Travemunde, R.F.A., June 1990. NATO. Edited by Springer.

[Bor84] G. Borgefors. Distance transformations in arbitrary dimensions. *C. V. G. I. P.*, 27:321–345, 1984.

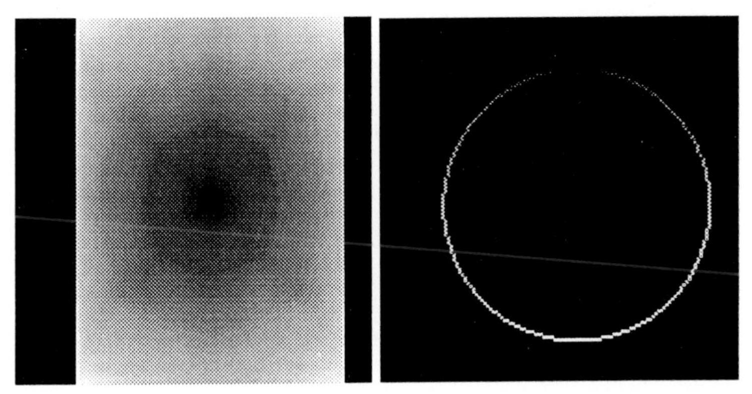

Figure 9: 3-D distance transformation on a cylinder. On the left (front view) the removed point is in front of the cylinder. On the right we show the slice with the removed point (at the top of the circle)

Figure 10: Upper view of the skeleton of the vertebra in both positions. Surface points are dark grey, junction points are light grey and edge points are medium grey

Figure 11: 3-D representations of the skeleton of the vertebra in both positions (with rescaling in the third dimension). Surface points are dark grey, jonction points are light grey and edge points are medium grey

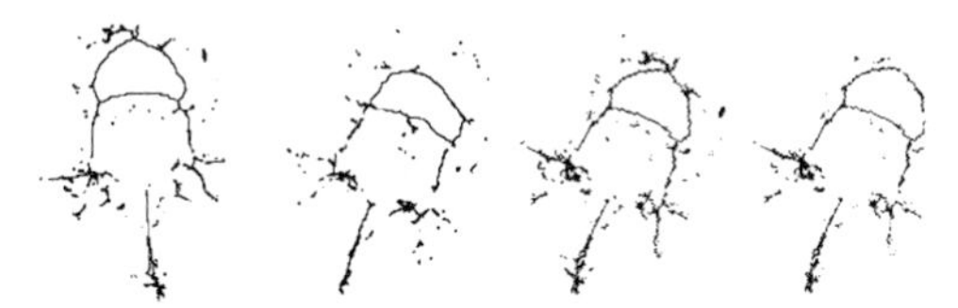

Figure 12: Projections on the XY-plane of the junction points of the skeleton of (from the left to the right) the first vertebra (1), the second vertebra (2), the first vertebra after the computed transformation (3), and the intersection of the dilated set of junction points of the second vertebra with the set of junction points of the first vertebra after transformation (4).

Figure 13: Same as figure 12 with projections on the YZ-plane.

Figure 14: Upper view of the skeleton of the vertebra in both positions after the labelling of simple surfaces remaining after an opening of order 2

Figure 15: Front view of the skeleton of the vertebra in both positions after the labelling of simple surfaces remaining after an opening of order 2 (without rescaling in the third dimension)

[GB90] W.X. Gong and G. Bertrand. A simple parallel 3d thinning algorithm. In *Proceedings 10th I. C. P. R.*, pages 188–190, June 17–21 1990. Atlantic City.

[HPJ84] K.J. Hafford and K. Preston Jr. Three dimensional skeletonization of elongated solids. *C. V. G. I. P.*, 27:78–91, 1984.

[KF81] N. Keskes and O. Faugeras. Surface simple dans z^3. In *3ème Congrès Reconnaissance des Formes et d'Intelligence Artificielle*, pages 718–729. AFCET, September 1981.

[Kim84] C.E. Kim. Three-dimensional digital planes. *IEEE Transactions on PAMI, PAMI 6*, 5:639–645, September 1984.

[KR89] T.Y. Kong and A. Rosenfeld. Digital to topology: introduction and survey. *C. V. G. I. P.*, 48:357–393, 1989.

[LVG80] S. Logbregt, P.W. Verbeck, and F.C. Groen. Three-dimensional skeletonization: principle and algorithm. *IEEE Transactions on PAMI, PAMI 2*, 1:75–77, January 1980.

[MBA91] G. Malandain, G. Bertrand, and N. Ayache. Topological segmentation of discrete surfaces. In *Proceedings of CVPR, Hawaii*, June 1991. also Research Report INRIA 1357.

[MDMC90] O. Monga, R. Deriche, G. Malandain, and J.P Cocquerez. Recursive filtering and edge closing : two primary tools for 3d edge detection. In *First European Conference on Computer Vision (ECCV), April 1990, Nice, France*, 1990. also Research Report INRIA 1103.

[Mor80] D.G. Morgenthaler. Three-dimensional digital topology: the genus. Tr–980, Computer Science Center, University of Maryland, College Park, MD 20742, U.S.A., November 1980.

[Mor81] D.G. Morgenthaler. Three-dimensional simple points: serial erosion, parallel thinning, and skeletonization. Tr–1005, Computer Science Center, University of Maryland, College Park, MD 20742, U.S.A., February 1981.

[MR80] D.G. Morgenthaler and A. Rosenfeld. Surfaces in three-dimensional digital images. Tr–940, Computer Science Center, University of Maryland, College Park, MD 20742, U.S.A., September 1980.

[NA85] A. Nakamura and K. Aizawa. On the recognition of properties of three-dimensional pictures. *IEEE Transactions on PAMI, PAMI 7*, 6:708–713, November 1985.

[PR71] C.M. Park and A. Rosenfeld. Connectivity and genus in three dimensions. Tr–156, Computer Science Center, University of Maryland, College Park, MD 20742, U.S.A., May 1971.

[Ros80] A. Rosenfeld. Three-dimensional digital topology. Tr–936, Computer Science Center, University of Maryland, College Park, MD 20742, U.S.A., September 1980.

[San89] P. Sander. Generic curvature features from 3-d images. *IEEE Transactions on Systems, Man, and Cybernetics*, November 1989. Special Issue on Computer Vision.

[TF82] Y.F. Tsao and K.S. Fu. A 3d parallel skeletonization thinning algorithm. *IEEE PRIP Conference*, pages 678–683, 1982.

[TYYF82] J.I. Toriwaki, S. Yokoi, T. Yonekusa, and T. Fukumura. Topological properties and topology-preserving transformation of a three-dimensional binary picture. In *Proceedings 6th I. C. P. R.*, pages 414–419, October 1982. Munich.

SHAPE-BASED INTERPOLATION USING A CHAMFER DISTANCE

G T Herman and C A Bucholtz

Medical Image Processing Group, Department of Radiology, University of Pennsylvania, Blockley Hall, Fourth Floor, 418 Service Drive, Philadelphia, PA 19104-6021, USA

Abstract

Shape-based interpolation is a methodology to estimate the locations of the picture elements (pixels) which would be contained in an organ of interest in non-existent slices through the human body from the locations of the pixels in the organ in slices that have been obtained by a tomographic imager. In this paper we motivate the need for shape-based interpolation and report on some quantitative experiments which were done to evaluate the relative performance of a number of interpolation methods for tomographic imaging of the human body. In particular, we introduce the new notion of shape-based interpolation using a chamfer distance and show that a statistically extremely significant improvement over previously proposed methods is achieved by this newly proposed interpolation method.

Keywords

Tomographic imaging; three-dimensional display; pixel classification; volume estimation; performance evaluation.

1. INTRODUCTION

In many medical imaging modalities which are tomographic in nature, such as computerized tomography (CT), magnetic resonance imaging (MRI), single photon or positron emission tomography (ET) and, for appropriately collected data, even ultrasound (US), the object of interest is reconstructed as a series of two-dimensional slices such that the separation between the central planes of these slices is typically much greater than the separation between the centres of the picture elements (pixels) which make up the slices (see, e.g., Herman, 1983). The situation is even more unsatisfactory in the case of serial imaging of a dynamic organ, such as the heart (see, e.g., Raichlen *et al.*, 1986), since typically considerable amount of motion, as compared to pixel size, takes place between successive images of the organ. For accurate quantitative analysis and/or display of information in such data (see, e.g., Udupa and Herman, 1990), some kind of interpolation between the slices seems to be desirable.

The classical way of doing that (see, e.g., Herman and Coin, 1980) is to do some type of interpolation (such as linear) of the gray values in the slices to estimate the gray values in the "missing" slices. Further processing (such as determination of the surfaces of specific organs of interest) is then done using this extended data set. There are two important objections that may be raised to this approach:

1. In many applications the objects of interest are represented in the images in such a way that automatic identification of their boundaries in the slices is not possible without higher order (expert) knowledge. At the current state of the art, such expert knowledge is typically provided by an expert user (such as a cardiologist) by manually tracing on the video screen (see, e.g., Raichlen *et al.*, 1986). To do this on the extended data set (which in the cardiac application may contain more than ten times as many slices as the original one; Raya and Udupa, 1990) puts an often unacceptable extra burden on the user.
2. If thresholding is the method used for identifying the objects of interest, as is often the case (see, e.g., Udupa and Herman, 1990 or Herman and Coin, 1980), a prior linear interpolation for estimating the

-100	-100	-100	-100	-100	-100	-100	+400	+400	+400
-100	-100	-100	+400	+400	+400	+400	+400	+400	+400

Figure 1. Values assigned to pixels in two one-dimensional "slices". Heavy line indicates the boundary of the object thresholded at 150.

-100	-100	-100	-100	-100	-100	-100	+400	+400	+400
-100	-100	-100	0	0	0	0	+400	+400	+400
-100	-100	-100	+100	+100	+100	+100	+400	+400	+400
-100	-100	-100	+200	+200	+200	+200	+400	+400	+400
-100	-100	-100	+300	+300	+300	+300	+400	+400	+400
-100	-100	-100	+400	+400	+400	+400	+400	+400	+400

Figure 2. Values assigned to pixels after linear interpolation of values in Figure 1. Heavy line indicates the boundary of the object thresholded at 150.

missing slices leads to an artifact which arises when the location of a boundary between two uniform regions (one with values above and the other with values below the threshold) shifts between two consecutive measured slices. Linear interpolation followed by thresholding results in the boundary in all intermediate slices being in the same place as in one of the measured slices, resulting in an abrupt, rather than smooth, change in boundary location as we move from one of the measured slices to the other. This idea is illustrated in Figures 1 and 2. The phenomenon can be beautifully observed in Figure 4 of (Hemmy and Tessier, 1985).

2. METHODS

2.1. Shape-Based Interpolation and Its Variants

Shape-based interpolation was recently proposed together with a demonstration of its usefulness in three- and four-dimensional medical image processing in (Raya and Udupa, 1990). A typical application is in CT, where slice separation is usually much greater than the pixel size within an individual slice. Assuming that a methodology exists to segment an organ of interest in the slice images, shape-based interpolation can be used to estimate the locations of the pixels which would be contained in that organ in (non-existent) intermediate slices. This is done by converting the segmented slice images into gray-

-195	-165	-135	-105	-75	-45	-15	+15	+45	+75
-75	-45	-15	+15	+45	+75	+105	+135	+165	+195

Figure 3. Distance-from-boundary values assigned to pixels in the two one-dimensional "slices" of Figure 1. (30 is taken to be the distance between pixel centers.) Heavy line indicates the boundary of the object thresholded at 0.

-195	-165	-135	-105	-75	-45	-15	+15	+45	+75
-171	-141	-111	-81	-51	-21	+9	+39	+69	+99
-147	-117	-87	-57	-27	+3	+33	+63	+93	+123
-123	-93	-63	-33	-3	+27	+57	+87	+117	+147
-99	-69	-39	-9	+21	+51	+81	+111	+141	+171
-75	-45	-15	+15	+45	+75	+105	+135	+165	+195

Figure 4. Values assigned to pixels after linear interpolation of values in Figure 3. Heavy line indicates the boundary of the object thresholded at 0.

level images, in which the grayness is the shortest distance (within the slice) of the pixel from the boundary of the organ (positive values for inside the organ and negative values for outside, see Figure 3). Segmented intermediate slices are estimated by interpolating the distance-representing gray-level slices and thresholding at zero, see Figure 4.

The distance function used in (Raya and Udupa, 1990) is a version of the so-called city-block distance. This distance can be calculated very efficiently, but it has the apparent disadvantage that it provides a relatively bad approximation to Euclidean distance (Borgefors, 1984, Borgefors, 1986). For this reason we decided to investigate whether or not a significant improvement can be observed in the performance of shape-based interpolation by using a distance function which approximates Euclidean distance more closely. We chose to study various chamfer distances as proposed by (Borgefors, 1984, Borgefors, 1986). These can be calculated by what is essentially a twice-repeated convolution process, with the convolving template having only a few nonzero elements. Thus calculation of such distances is not too much less efficient than the calculation of a city-block distance. (Important warning: we use the word "convolution" not in its customary technical sense. As will be seen, our convolutions are nonlinear. Nevertheless, they are inexpensive to implement due to the small sizes of the templates.)

Since the original chamfer distance method (Borgefors, 1984, Borgefors, 1986) calculates distances of non-object pixels from the nearest object pixel, for our application it would have to be used twice (once for the inside and then for the outside). However, we found a relatively simple adaptation which

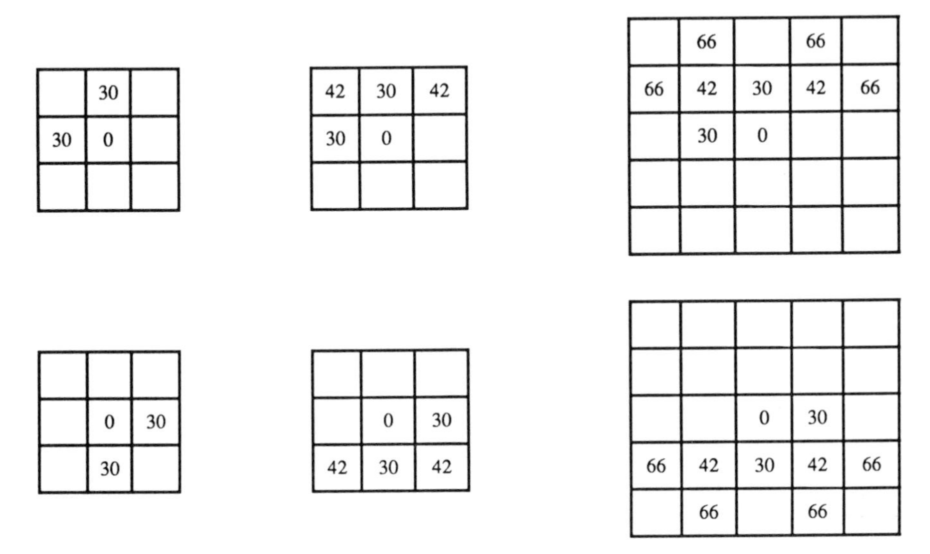

Figure 5. Templates for chamfer distance calculations. Left: city-block. Middle: near-optimal 3×3. Right: near-optimal 5×5. Top: template for first pass. Bottom: template for second pass.

allows us to calculate the two sets of distances simultaneously. Another difference between what seems to be appropriate here from the approach of (Borgefors, 1984, Borgefors, 1986) is that we are interested in the distance from the boundary. That is why we have assigned +15 and –15 to the pixels just inside and just outside the boundary in Figure 3. The assignment of (Borgefors, 1984, Borgefors, 1986) and, in fact, also of (Raya and Udupa, 1990) would be +30 and –30, respectively. We now describe precisely how the chamfer distances are defined and calculated.

In Figure 5 we show the templates to be used in the convolution processes. Prior to discussing exactly how they are to be applied, we have to discuss how the calculation is initialized. We do this first by the way of an example, which is appropriate for both the city-block and the near-optimal 3×3 templates. (The sense in which templates are "near-optimal" is explained by Borgefors, 1986, and will not be repeated here.)

Consider Figure 6. To initialize for the distance calculation, we replace all 0s by a very large negative number (in our case, –999) and we replace all 1s by a very large positive number (in our case +999), with the following exception: if a 0 shares an edge with a 1, then we replace it by –15 and if a 1 shares an edge with a 0, then we replace it by +15. (As will be explained below, further exceptions are to be used when initializing for the near-optimal 5×5 distance.) The result of this initialization is shown in Figure 7. (Note that in our example all the border pixels have 0s in them. In fact, in our implemented algorithms we pad the images with 0s prior to initialization and we remove these padding pixels from the final result of the chamfer process.)

We now explain the chamfer process which consists of two convolutions, the first using a top template in Figure 5 and the second using a bottom template in Figure 5.

The first convolution process updates the pixels row-by-row from the top to the bottom with a left-to-right ordering within the rows. Each pixel is updated as follows. A pixel which was assigned the value +15 or –15 during the initialization phase is left unaltered. Otherwise, the template is placed over the image in such a way that the central (0) entry in the template is over the pixel in question. What we do next depends on the sign of the of the content of the pixel to be updated. If it is positive (i.e., the pixel in question is inside the object), then for each image-pixel which is covered by a nonempty template-pixel we add together the contents of the image-pixel and the covering template-pixel. We

0	0	0	0	0	0	0	0	0	0	0	0
0	0	0	0	0	0	0	0	0	0	0	0
0	0	0	1	0	0	0	1	1	0	0	0
0	0	1	0	0	0	0	1	1	1	0	0
0	0	1	0	0	0	0	0	1	1	0	0
0	0	1	1	1	1	0	0	0	1	0	0
0	0	1	0	0	0	1	0	0	0	0	0
0	0	1	0	0	0	1	1	1	0	0	0
0	0	1	0	0	1	1	1	1	1	0	0
0	0	1	1	1	1	1	1	1	1	0	0
0	0	0	0	0	0	0	0	0	0	0	0
0	0	0	0	0	0	0	0	0	0	0	0

Figure 6. A binary array. Boundary is marked by a heavy line.

replace the value assigned to the pixel to be updated by the smallest of these sums. (Careful combination of the initialization and convolution processes insures that the new value is still positive.) If the current content of the pixel to be updated is negative (i.e., the pixel in question is outside the object), then for each image-pixel which is covered by a nonempty template-pixel we subtract the content of the covering template-pixel from the content of the image-pixel. We replace the value assigned to the pixel to be updated by the largest of these differences. (The new value is still negative.) The result of this first convolution process for the near-optimal 3×3 distance is shown in Figure 8.

The second convolution process updates the pixels row-by-row from the bottom to the top with a right-to-left ordering within the rows. The details of the process are identical to those in the first convolution, except that now the bottom template in Figure 5 (rather than the top one) is used. The result of this second convolution process for the near-optimal 3×3 distance is shown in Figure 9.

We see in Figure 9 that the value -27 is assigned to those pixels which are outside the object, but share a vertex (but not the edges at that vertex) with the object and the value $+27$ is assigned to those pixels inside the object which similarly share a vertex with the outside of the object. In order for the

-999	-999	-999	-999	-999	-999	-999	-999	-999	-999	-999	-999
-999	-999	-999	-15	-999	-999	-999	-15	-15	-999	-999	-999
-999	-999	-15	15	-15	-999	-15	15	15	-15	-999	-999
-999	-15	15	-15	-999	-999	-15	15	999	15	-15	-999
-999	-15	15	-15	-15	-15	-999	-15	15	15	-15	-999
-999	-15	15	15	15	15	-15	-999	-15	15	-15	-999
-999	-15	15	-15	-15	-15	15	-15	-15	-15	-999	-999
-999	-15	15	-15	-999	-15	15	15	15	-15	-999	-999
-999	-15	15	-15	-15	15	999	999	999	15	-15	-999
-999	-15	15	15	15	15	15	15	15	15	-15	-999
-999	-999	-15	-15	-15	-15	-15	-15	-15	-15	-999	-999
-999	-999	-999	-999	-999	-999	-999	-999	-999	-999	-999	-999

Figure 7. The result of initializing Figure 6 for either city-block or near-optimal 3×3 distance calculation.

near-optimal 5×5 chamfer templates (Figure 5, right) to achieve their desired effect, these values (±27) have to be assigned explicitly to the appropriate pixels during the initialization phase. (See Appendix for a detailed explanation.)

This completes our discussion of the various methodologies that we selected for comparison. Specifically, we compared the performance of six interpolation methods:

1. interpolation of gray values followed by thresholding;
2. shape-based interpolation as implemented in (Raya and Udupa, 1990);
3. shape-based interpolation with a city-block distance (Figure 5, left) in which pixels with an edge on the boundary are initialized to be half a pixel-step (i.e., 15) from the boundary, as illustrated in Figure 7 (in Raya and Udupa, 1990, such pixels are considered to be a pixel-step from the boundary);
4. shape-based interpolation with a city-block distance (Figure 5, left) in which pixels with either an edge or a vertex on the boundary are initially assigned their (approximately) Euclidean distances from the boundary (i.e., ±15 and ±27);
5. a near-optimal chamfer distance based on 3×3 templates (Figure 5, middle) with initial assignments

-999	-999	-999	-999	-999	-999	-999	-999	-999	-999	-999	-999
-999	-999	-999	-15	-45	-75	-105	-15	-15	-45	-75	-105
-999	-999	-15	15	-15	-45	-15	15	15	-15	-45	-75
-999	-15	15	-15	-27	-57	-15	15	27	15	-15	-45
-57	-15	15	-15	-15	-15	-27	-15	15	15	-15	-45
-57	-15	15	15	15	15	-15	-27	-15	15	-15	-45
-57	-15	15	-15	-15	-15	15	-15	-15	-15	-27	-57
-57	-15	15	-15	-45	-15	15	15	15	-15	-45	-69
-57	-15	15	-15	-15	15	27	45	27	15	-15	-45
-57	-15	15	15	15	15	15	15	15	15	-15	-45
-57	-27	-15	-15	-15	-15	-15	-15	-15	-15	-27	-57
-69	-57	-45	-45	-45	-45	-45	-45	-45	-45	-57	-69

Figure 8. The result of applying to Figure 7 the first convolution for the near-optimal 3x3 distance calculation. (See Figure 5, middle top template.)

to boundary pixels as illustrated in Figure 7; and

6. a near-optimal chamfer distance based on 5×5 templates (Figure 5, right) with initial assignments to pixels using both ±15 and ±27.

2.2. Methodology of Comparison

The methodology of comparison is illustrated by the following example.

We used a data set consisting of 61 contiguous CT slices of a patient's head, with 512×512 pixels in each. We segmented all the slices for bone by thresholding at the same level (CT value 200, on a scale in which water has value 0 and air has value -1024). For n from 2 to 60, we applied the six interpolation methods to slices $n-1$ and $n+1$ to estimate the segmented nth slice. We found, in particular, that the estimates provided by the third and fifth interpolation methods differed for a total of 21,960 pixels in the 59 estimated slices, with the chamfer distance providing a match of the actual segmented value

-99	-69	-57	-45	-57	-69	-57	-45	-45	-57	-69	-99
-69	-57	-27	-15	-27	-57	-27	-15	-15	-27	-57	-69
-57	-27	-15	15	-15	-45	-15	15	15	-15	-27	-57
-45	-15	15	-15	-27	-45	-15	15	27	15	-15	-45
-45	-15	15	-15	-15	-15	-27	-15	15	15	-15	-45
-45	-15	15	15	15	15	-15	-27	-15	15	-15	-45
-45	-15	15	-15	-15	-15	15	-15	-15	-15	-27	-57
-45	-15	15	-15	-27	-15	15	15	15	-15	-27	-57
-45	-15	15	-15	-15	15	27	45	27	15	-15	-45
-45	-15	15	15	15	15	15	15	15	15	-15	-45
-57	-27	-15	-15	-15	-15	-15	-15	-15	-15	-27	-57
-69	-57	-45	-45	-45	-45	-45	-45	-45	-45	-57	-69

Figure 9. The result of applying to Figure 8 the second convolution for the near-optimal 3×3 distance calculation. (See Figure 5, middle bottom template.)

12,479 times. Thus the null-hypothesis that there is no difference in performance due to the choice of the distance function (which would imply equal probability for being correct whenever the estimates are different) can be rejected in favor of the alternative that the method based on the near-optimal 3×3 chamfer distance performs better with just about absolute confidence.

The level of statistical significance is calculated as follows. Let Q be the total number of pixels which are estimated differently by two methods. Let q be the number of such pixels that have been correctly classified by the interpolation method which has greater accuracy for this particular data set. The null-hypothesis that the methods perform equally well implies that q is a random sample from a binomial distribution with total number of items Q and equal probabilities assigned to the two classes. The probability of randomly selecting an element from this distribution with value q or higher is the level of significance for rejecting the null-hypothesis. (For definitions of the statistical terminology see, e.g., Mould, 1989.)

The above discussion refers to the comparison of interpolation methods from the point of view of accuracy on a pixel-by-pixel basis. One of the major application of such methods is the estimation of volumes (Raichlen *et al.*, 1986, Kohn *et al.*, 1991). For this reason, we also compare the methods from the point of view of how well they estimate the total volume of the object in the intermediate slices.

3. RESULTS

The data sets used for comparing the methods were three different sets of CT slices of human heads and a set of Magnetic Resonance Imaging (MRI) slices of a brain specimen. In our initial experiments we found that the fifth method significantly outperformed the previous four methods. A consistent difference between the fifth and the sixth methods was not observed. Since the fifth method is computationally less expensive, it appears to be the method of choice. We decided that we shall compare the fifth method with the other five, but not the other five with each other.

Our clinical protocol for three-dimensional imaging of skulls from CT scans uses a threshold of 200 (Herman, 1988). For imaging skin, we considered thresholding either at value –200 or at value –100 reasonable. We therefore performed the experiment described in Methodology of Comparison for each of the three CT scans of heads at each of these three threshold levels. For one of the scans, we also performed the experiment at three higher threshold levels (561, 661, 761). By studying the resulting three-dimensional displays of our brain specimen (Kohn *et al.*, 1991), we see that numbers in the range 106–136 provide us with acceptable thresholds for segmenting brain in the MRI scan. However, the specimen was wrapped in a cloth whose MRI value is near the higher end of this range. We therefore performed our comparison on the MRI data with threshold 136. We did a second comparison at threshold 100, to see how well the methods perform when both the brain and the cloth wrapping it are considered to be part of the object of interest. Thus, we carried out altogether 14 comparison experiments.

In all cases the value of $Q/2$ (the mean of the binomial distribution) was well over 100 and so the normal approximation, with standard deviation $\sqrt{Q}/2$, is justified (Mould, 1989). With the exception of three instances (to be discussed below), we found that when comparing Method 5 with any of Methods 1–4, Method 5 was more accurate with a significance level of better than 0.00001. On the other hand Methods 5 and 6 did not differ significantly, sometimes the one sometimes the other being more accurate. Since Method 5 is computationally less expensive, it is clearly the interpolation method of choice among the six that we tested.

We now discuss our exceptional instances. A not particularly important exception is for the brain with threshold 136. Here we found that Method 5 is better than Method 4, but the significance level is worse than .2 (not significant). With the lower threshold for the brain (imaging the cloth) Method 4 is actually slightly more accurate than Method 5. The really interesting exception is the third one: for the skull thresholded at 761, Method 1 is more accurate than Method 5 at a significance level better than 0.00001. Thus, in this one experiment out of the fourteen, shape-based interpolation failed in comparison to the classical method (Herman and Coin, 1980). The reason is that at this clinically unreasonably high threshold, only the hard outer part of the skull is in the thresholded object. Thus the object consists of two thin parallel curving layers, which confounds the shape-based interpolation method.

Rather interestingly, our calculations of accuracy of overall volumes gave us much less clear-cut results. They are shown in Table 1, which we now discuss. First of all we note that, since errors are likely to occur near the boundaries of objects, the percentage error has much to do with the ratio of the area of the boundary to the volume of the object. Thus it tends to be small for the outer surface of the heads (l and m in Table 1), large for the skull (h in Table 1) and in-between these two for the brain. So these differences are due to the objects being imaged rather than to the interpolation method used. We also note that, in accordance with the previously stated results on accuracy of individual pixel classification, Methods 5 and 6 are very similar in performance. Also, with the exception of I.l (where the performances of the two methods are nearly identical), Method 5 always outperformed the classical Method 1 and half the time it is more than twice as accurate as Method 1. Similarly, for half the cases

Table 1. Percentage error in the estimates of the number of object pixels in the intermediate slices as compared to the actual number of object pixels. Rows refer to method number. I, J and K refer to the three CT data sets of heads thresholded at −300 (l), −200 (m) and +200 (h). MRI refers to the MRI data set of brain thresholded at 100.

	I.l	I.m	I.h	J.l	J.m	J.h	K.l	K.m	K.h	MRI
1	-0.23	-0.36	+9.03	-0.36	-0.77	+8.08	-0.65	-0.79	+9.95	-2.22
2	-0.35	-0.32	-8.36	-0.36	-0.17	-6.70	-0.57	-0.40	-16.10	-3.53
3	+0.21	+0.26	-2.92	-0.04	+0.16	-2.12	+0.26	+0.53	-9.75	+0.62
4	+0.25	+0.30	-2.44	+0.02	+0.21	-1.56	+0.32	+0.60	-8.89	+0.84
5	+0.26	+0.31	-2.39	+0.02	+0.22	-1.49	+0.33	+0.61	-8.75	+0.87
6	+0.25	+0.31	-2.40	+0.02	+0.22	-1.51	+0.33	+0.61	-8.77	+0.87

it is also very much more accurate than the previously proposed shape-based interpolation Method 2 and for the other half of the cases the two methods are rather similar. However, contrary to the results on the accuracy of individual pixel classification, the performances of Methods 3 and 4 are very similar to the performance of Method 5 for volume estimation. It appears that from the volume estimation point of view the most important aspect of our proposed modification to shape-based interpolation as proposed in (Raya and Udupa, 1990) is the initialization indicated in Figure 7.

Although this last result is surprising enough to warrant further investigation, it is not by itself indicative of a definite error in our experiments. A classifier which is more accurate on individuals may easily be less accurate on volumes. As an extreme example, consider a black and a white ball. The classifier which calls all balls black will be correct half the time, while a classifier which calls the black ball white and vice versa is never correct. Nevertheless, the latter will (accidently correctly) indicate that half the balls are black while the former will indicate (incorrectly) that all balls are black. Whether a phenomenon of this type is present in our interpolators (and what is to be done about it if it is) will form part of our future work on this topic.

4. DISCUSSION

In this paper we have proposed and evaluated some extensions of the shape-based interpolation method of (Raya and Udupa, 1990). In particular, we have suggested that pixels which share a boundary edge (one inside and the other outside the object) should both be considered to be at a distance from the boundary which is half the distance between adjacent pixel centers. Using such an initialization for distance calculations, we have developed a generalization of the chamfer distance calculation proposed in (Borgefors, 1984, Borgefors, 1986). This generalization allows us to simultaneously calculate distances within the object and its background by a double convolution process. We have evaluated the performance of a number of variants of our method. Based on our experiments, we can strongly recommend the interpolation method based on the near-optimal 3×3 chamfer distance (Figure 5, middle). Its accuracy seems to be as good as that of the method based on the near-optimal 5×5 chamfer distance (Figure 5, right), but it requires fewer calculations. From the point of view classifying individual pixels in the missing slices, we can say with extremely great statistical confidence that it is better than the classical method of gray value interpolation followed by thresholding (Herman and Coin, 1980) or any of the shape-based interpolation methods using the city-block distance (Figure 5, left) that we tried. From the point of view of volume estimation, the performance of our recommended method appears to be as good as any of the other methods that we tested and it is superior to the classical method (Herman and Coin, 1980) or the method which uses the version of the city-block distance proposed by (Raya and Udupa, 1990).

Regarding the last point, we point out that there is a computational price to be paid for the gain in accuracy. In (Raya and Udupa, 1990) there is a description of a very efficient implementation of their approach. Some of the ideas proposed in there do not carry over to the chamfer approach. While the double convolution used in the latter approach is quite acceptably efficient, it is not as efficient as the implementation described in (Raya and Udupa, 1990).

Another topic mentioned in (Raya and Udupa, 1990) which was not followed on in this paper is the use of cubic spline interpolation between the slices after distance calculation within the slices. In all our experiments we used linear interpolation between the slices. Since our main purpose was to compare the effects of different distance functions, this seemed to us an acceptable approach. However, it has been demonstrated in (Raya and Udupa, 1990) that for their distance function the use of cubic spline interpolation is preferable to linear interpolation. Cubic spline interpolation can be used with all the different distance functions discussed in this paper; we plan to evaluate in the future its usefulness in combination with chamfer distances.

A further topic that has been left for future research is an investigation of the one aspect in which shape-based interpolation clearly fails as compared to the classical method of gray level interpolation followed by thresholding. In the latter method it makes good sense to cover the object of interest by slices using additional slices at the ends. In an additional slice there is no object (i.e., if thresholded at the desired level it becomes blank). Nevertheless, the gray level information in it can be used to estimate gray levels in missing slices between it and the rest of the object. When these newly estimated slices are thresholded, they may very well be found to contain part of the object. As opposed to this, shape-based interpolation as described here does not allow us to go beyond the last slice in which some part of the object was identified.

5. CONCLUSIONS

We have demonstrated in this paper that shape-based interpolation using a chamfer distance has superior properties to previously proposed methods for estimating object locations in missing slices in tomographic radiology. We have outlined ideas by which even further improvements may be achieved.

Acknowledgments

The authors are grateful to Etemad Hossein for writing the first programs in our laboratory for comparing Methods 3 and 5 and to Jayaram K. Udupa for useful discussions. Our research is supported by NIH grants HL28438, MH42191 and CA50851. Carolyn A. Bucholtz is with the Division of Biotechnology of CSIRO, Australia; her visit to Philadelphia was made possible by a CSIRO Overseas Study Award.

References

Borgefors G (1984). Distance transformations in arbitrary dimensions. Comput. Vision Graph. Imag. Proc. 27:321–345.
Borgefors G (1986). Distance transformations in digital images. Comput. Vision Graph. Imag. Proc. 34:344–371.
Hemmy DC and Tessier PL (1985). CT of dry skulls with craniofacial deformities: accuracy of three-dimensional reconstruction. Radiol. 157:113–116.
Herman GT and Coin CG (1980). The use of three-dimensional computer display in the study of disk disease. J. Comput. Assist. Tomogr. 4:564–567.
Herman GT (1983). Special issue on computerized tomography. Proc. IEEE 71:291–435.
Herman GT (1988). Three-dimensional imaging on a CT or MR scanner. J. Comput. Assist. Tomogr. 12:450–458.
Kohn MI, Tanna NK, Herman GT, Resnick SM, Mozley PD, Gur RE, Alavi A, Zimmerman RA and Gur RC (1991). Analysis of brain and cerebrospinal fluid volumes with MR imaging. Part 1: methods, reliability, and validation. Radiol. 178:115–122.
Mould RF (1989). Introduction to Medical Statistics. Adam Hilger, Bristol, England.
Raichlen JS, Trivedi SS, Herman GT, St.JohnSutton MG and Reichek N (1986). Dynamic three-dimensional reconstruction of the left ventricle from two-dimensional echocardiograms. J. Amer. Coll. Card. 8:364–370.
Raya SP and Udupa JK (1990). Shape-based interpolation of multidimensional objects. IEEE Trans. Med. Imag. TMI9:32–42.
Udupa JK and Herman GT (1990). 3–D Imaging in Medicine. CRC Press, Boca Raton, FL.

-999	-999	-999	-999	-999	-999	-999	-999	-999	-999	-999	-999
-999	-999	-999	-15	-45	-75	-105	-15	-15	-45	-75	-105
-999	-81	-15	15	-15	-45	-15	15	15	-15	-45	-75
-999	-15	15	-15	-27	-51	-15	15	***21***	15	-15	-999
-999	-15	15	-15	-15	-15	-999	-15	15	15	-15	-999
-999	-15	15	15	15	15	-15	-999	-15	15	-15	-999
-999	-15	15	-15	-15	-15	15	-15	-15	-15	-999	-999
-999	-15	15	-15	-999	-15	15	15	15	-15	-999	-999
-999	-15	15	-15	-15	15	999	999	999	15	-15	-999
-999	-15	15	15	15	15	15	15	15	15	-15	-999
-999	-999	-15	-15	-15	-15	-15	-15	-15	-15	-999	-999
-999	-999	-999	-999	-999	-999	-999	-999	-999	-999	-999	-999

Figure 10. Partial result of applying to Figure 6 the first convolution for the near-optimal 5×5 distance calculation. (See Figure 5, right top template.)

Appendix

In this appendix we give a detailed illustration as to why one cannot simply use the initialization of Figure 7 for the 5×5 chamfer distance. Suppose that we attempt to do that and apply the template on the top-right of Figure 5 to the image in Figure 7. After a few steps we arrive at the situation indicated in Figure 10. The pixel that we reached is the one with ***21*** in it. We get 21 by adding 66 to –45, which is the content of the pixel two up and one to the right. The problem is that this –45 will be changed during the second convolution into a –27 and so the value 21 has been obtained based on the fallacious information that the pixel is 66 away from a pixel with value –45. In fact, the correct distance is 27, rather than 21, which is indeed the value that is explicitly assigned during the correct initialization phase for the 5×5 chamfer distance. This value is not altered during the first convolution because correct initialization also assigns the value –27 to the pixel containing –45 in Figure 10 and –27+66=39 is greater than 27.

Proper initialization insures that the final distance assignments in the whole image are correct. This means that for any pixel we can get to it from a boundary pixel (value ± 15) by a series of chamfer steps so that the final value assigned to the pixel is the sum of the value assigned to the starting pixel and the values assigned to the steps. The absolute value of the distance assigned to the pixel is as small as possible consistently with the condition of the previous sentence.

THIN-PLATE SPLINES
AND THE ATLAS PROBLEM FOR BIOMEDICAL IMAGES

Fred L. Bookstein

Center for Human Growth, University of Michigan, Ann Arbor, MI 48104 USA

Abstract

The **thin-plate spline**, a general-purpose interpolant for labelled point data, ties the geometry of image deformation to the classic biometric algebra of quadratic forms. The technique thus helps not only in the production of biomedical *atlases*—averaged or normative images of particular structures or configurations of parts—but also in the understanding of specimen-by-specimen variability around those atlases. This report summarizes the statistics of thirteen landmark points in midsagittal MRI images of nine normal adult human brains, produces and evaluates averaged images with the aid of those statistics, and pursues some implications for stereotaxy, for region-tracing, and for the understanding of averaged image structures.

Keywords

Image deformation; image warping; image averaging; landmark data; MRI; morphometrics; neuroanatomy; anatomical norms; brain shape.

1. INTRODUCTION

This is the fourth in a series of speculative essays for the IPMI community on morphometric methods for biomedical image analysis. My earlier contributions (Bookstein, 1986*b*, 1988, 1991*a*) have dealt with foundations of morphometrics for landmark data and with the role of landmark-based thin-plate splines in image relaxation to a template. Recent work in morphometrics has extended the spline techniques beyond computation and manipulation of averaged shapes to the study of variation around those averages. This paper attempts a preliminary extension of these newer methods from landmark data to archives of whole grey-scale images.

I will be directing your attention to some subtle problems at the foundation of the information content needed for atlases: information about the covariation of identifiable parts of the scene. New techniques from landmark-based morphometrics may be applied in this domain to generate both actual pictorial atlases (as sample mean images) and the statistics that guide their routine application to single clinical cases. The clinical context to be pursued here treats atlases of neuroanatomy as seen in MRI, and so the invocation for a single case has a name: the relatively ancient problem of *stereotaxy*. The idea of a landmark (see below) is not new to stereotaxy—you may already be familiar with the idea of the "AC-PC line," for instance—but it is perhaps unexpected that a single statistical method can extract all the information borne by landmarks in respect of both of stereotaxy's traditional goals, localization and diagnosis (the principal aims to which

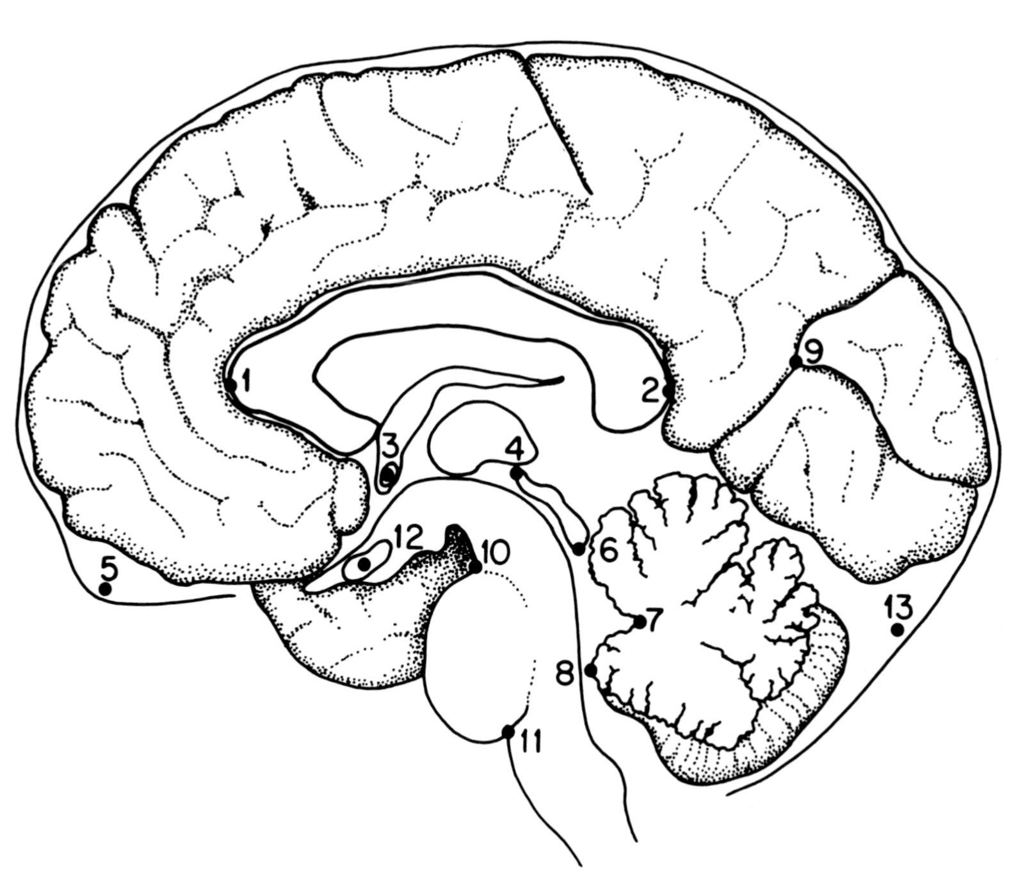

Figure 1. Thirteen landmarks on a schematic midsagittal MRI. They are named in the text.

atlases are put). The demonstration here uses a configuration of 13 landmarks from midsagittal MRI of nine UCLA medical students who were "otherwise normal." This set of landmarks (Figure 1) covers lower structures with reasonable completeness, but incorporates no information above the corpus callosum and no information about the curving outlines of the cortical surface. Using this set of 13 landmarks, I demonstrate a collection of techniques that have become standard in landmark-based morphometrics since 1985: shape averaging, matching of one shape to another by deformation, and description of shape variability by shape regressions and component analyses. Furthermore, because these landmark configurations were derived from pictorial data, I can also demonstrate a technique of directly visualizing the "average atlas" via pixel-by-pixel computations. Earlier appearances of these techniques in the neuroanatomical and the medical imaging literature, including the contributions to earlier IPMI proceedings and also Bookstein (1990*a*,*b*), have not explicitly considered the problems of atlases and their statistics.

The argument of this article is organized as follows. The next section introduces each of the morphometric techniques to be exploited: landmarks as data, the averaging of landmark sets via *shape coordinates*, the construction of a deformation relating any configuration of landmarks to any other via *thin-plate splines*, the extraction of statistically useful "components" or "factors" of these deformations, the application of the spline to deform individual clinical images into the geometry of the average shape, and the further use of these splines for stereotactic statistics by regression of the locations of some landmarks upon the locations of others. The third section carries out an exhaustive analysis of the demonstration data, explicitly constructing the average of these nine images and showing how these tools support sensible choices of stereotactic conventions. A more extensive analysis of some aspects of these data may be found in Bookstein, Mazziotta, and Valentino, 1991.

2. MORPHOMETRIC METHODS FOR LANDMARK DATA

The following subsections briefly review the several themes we need to borrow from the new morphometrics. For a survey of all these matters at monograph length, see Bookstein, 1991*b*.

2.1 Landmarks

For the description of sets of biological forms which differ substantially in shape, the most useful general approach begins with the location of *landmark points*. Landmarks, locations that have biological labels as well as geometric coordinates, are not intended to "represent" the individual image in any eidetic sense; rather, they directly sample the mathematical mapping that represents the "biological correspondence" or "homology" *between* any pair of images in the data set. The statistical methods for averaging sets of landmarks, for computing their correlations with various predictors (disease, age), and the like, are fully developed in both two dimensions and three.

Figure 1 shows the 13 landmarks of this demonstration. Some of these are fairly well-localized geometric points, while others are sensitive to subjective decisions about local outline curvatures or are characterized by a global orientation of coordinates. For a typology of landmark types, see Rohlf and Bookstein, 1990, or Bookstein, 1991*b*. Our 13 landmarks sample the midbrain and lower structures relatively well and take two experimental points (front and back) on the corpus callosum and the cortex as a whole.

The thirteen landmarks may be defined operationally as follows. **#1**: anteriormost point on genu of corpus callosum. **#2**: posteriormost point on splenium of corpus callosum. **#3**: anterior commissure. **#4**: posterior commissure. **#5**: anterior-inferior pole of cortex at the frontal sinus. **#6**: tip of inferior colliculus. **#7**: tip of fourth ventricle. **#8**: narrowing of cerebral aqueduct at obex. **#9**: junction of calcarine and parieto-occipital sulci. **#10**: top of pons at the peduncles. **#11**: ponto-medullary junction. **#12**: optic chiasm. **#13**: posterior pole of cortex at tentorium.

In any data set of landmark locations a few will prove missing or too blurred to locate with any accuracy in a few specimens. This data set began with a sample of fourteen images, but of those two were missing point **#1** (genu), four were missing **#3** (anterior commissure), and one was missing **#9**, the sulcal bifurcation. These incomplete cases were omitted. If this were not a methodological investigation but the actual empirical study of some hypothesis, the locations of the missing landmarks would have been estimated by regression on the others, and the computations continued using the full sample. The images and the landmark locations here were supplied by Dan Valentino under the direction of John Mazziotta of the University of California at Los Angeles.

2.2 Averaged Landmark Configurations

The morphometric consensus referred to above resolved a controversy among consumers of landmark data in regard to the computation of averages. Theorem 1 of Bookstein (1986*a*) showed that the statistics of all of the most reasonable alternatives for this averaging were *linearly equivalent*, so that for testing the significance of observed

Figure 2. Nine configurations of these thirteen landmarks, averaged and scattered according to two two-point registrations. The plus signs (+) represent the mean coordinates of the thirteen landmarks in Figure 1 to a registration on two of them. Digits are case numbers of actual registered landmark locations now viewed as residuals from these averages. (top) Registration to the AC-PC line, landmarks **#3** and **#4**, as in conventional stereotaxy. (bottom) Registration to the Chiasm-Colliculus line (**#12**−**#6**), which has the minimum over all pairs here of the mean-squared scatter of the digits about the + signs.

differences among groups of forms represented by landmarks, describing effects upon form or correlates of form, and similar goals, it does not matter how one computes the "mean form." The simplest of the available alternatives is the averaging by *shape coordinates*, Figure 2. In this approach, two landmarks are selected arbitrarily to be fixed in position, and the remaining locations are plotted with respect to these two fixed positions, by rotating, translating, and rescaling all the individual data sets as necessary. Figure 2 demonstrates two computations of the same average: one registered on landmarks **#12** (chiasm) and **#6** (colliculus), at the bottom, and the conventional registration on **#3** (anterior commissure) and **#4** (posterior commissure) above. The sensitivity of the

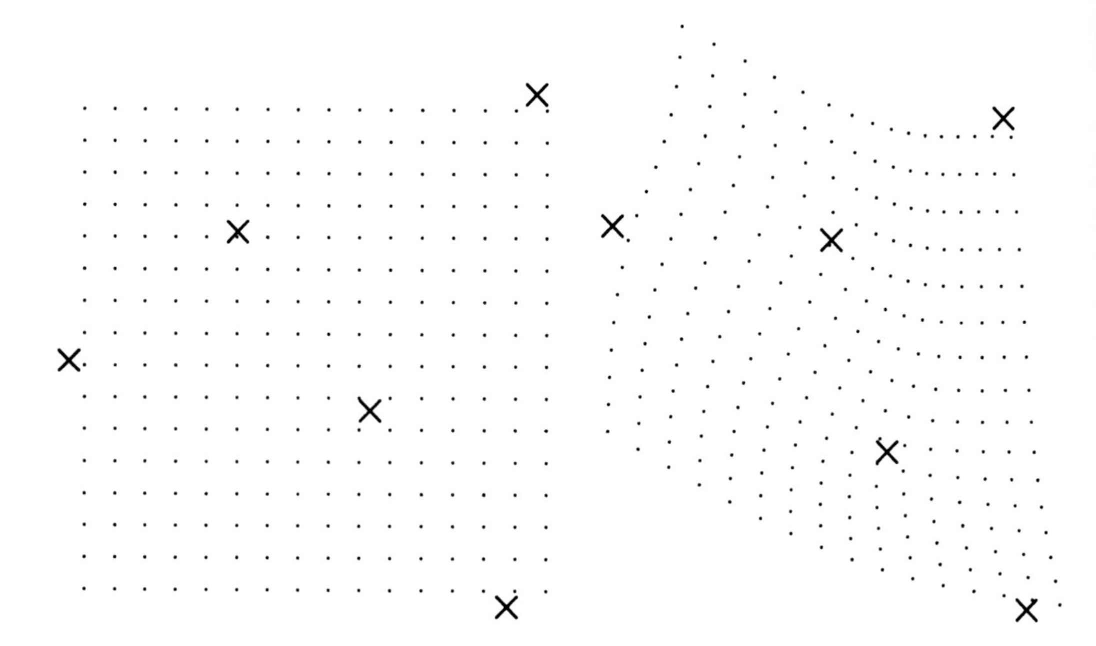

Figure 3. A single shape change of five landmarks and its interpolation by thin-plate spline. The mapping is shown D'Arcy-Thompson-style, as it deforms a square grid on the left into the exact interpolant of minimum integrated quadratic variation on the right.

average shape (the configuration of the + signs) to this decision is barely detectable. The scatters of digits about these means will be explained presently.

2.3 Deformation by Thin-Plate Splines

The construction of an atlas by image averaging, and its subsequent analysis, require a flexible data-driven functional that will map any subset of these landmarks from one form, or from the average, onto the same subset of landmarks in another form or in the average. An indefinitely wide variety of plane-to-plane mapping functions might serve, providing reasonable correspondences for the points in-between the landmarks, for which the homologues cannot be observed. From the family of possible interpolants, we need one that is optimal in some reasonable mathematical sense leading to useful quantifications of patterns.

The **thin-plate spline** was originally explored by Duchon and Meinguet in the 1970's for applications in interpolation theory; Terzopoulos (1983) introduced it to the literature of computer graphics, and Bookstein (1987, 1989, 1990*a*,*b*) brought it to the attention of the medical imaging community as a suitable general-purpose warping utility. For two-dimensional data the spline is a linear superposition of an affine term with the sum of 2-vector multiples of the function $r^2 \log r$, where r is the ordinary Euclidean distance of the point to be mapped from each of the starting landmark locations in turn. Such a summated function can be computed to map any set of three or more landmark locations onto any other set of the "same" points at different locations. The integral of the *quadratic variation*—summed squared second derivatives—is minimized over the class of functions consistent with the assigned pairing of landmarks. (The integral is taken over the whole plane.) While each spline is itself a nonlinear mapping, for a fixed starting form the coefficients of $r^2 \log r$ are linear functions of the Cartesian coordinates of the landmarks in the target configuration. If the shift of any Cartesian coordinate were taken to be "vertical" (upward, out of the picture plane), the map would represent the form of a thin steel plate adopting a configuration of minimal energy subject to being tacked at appropriate distances above the landmarks of the starting configuration. For changes from a fixed starting form, bending energy is a deficient quadratic form in the coordinates of the target form. There is not enough space here to publish the formulas for this interpolant, but after all IPMI has published them twice before. The complexity of the necessary computations is remarkably low, amounting to the inversion of a $(K+3)\times(K+3)$ matrix, where K is the count of landmarks. The entries of the matrix are 0's, 1's, original landmark coordinates, and the kernel function $r^2 \log r$ of their distances in pairs.

An example of how the spline serves to interpolate local nonlinearities of landmark rearrangement is shown in Figure 3. The data (X's in the figure) represent two "specimens" of an "outer triangle" of landmarks together with two interior points; the deformed grid on the right is the thin-plate spline (interpolation function) corresponding to these data. Figure 4 displays the decomposition of that spline into three *features of shape change* at decreasing geometric scale, as follows. At great distances, the spline expresses mainly the largest-scale change of form in the data, the contraction of that outer triangle of landmarks along an axis running northeast-southwest. This affine part is shown by itself in the top panel of the figure. It is the first of a series of three **partial warps**, which are vector multiples of the **principal warps**, eigenfunctions of the quadratic form which is the bending-energy of the equivalent "plate." (The affine part corresponds to the six-dimensional eigenspace of zero energy.) Interior to the triangle, the map bends smoothly to adjust for the relative reconfiguration of the interior points; and this bending (nonlinearity) itself has two further features. Approximately in the x-direction (Figure 4, bottom), the effective shift is *opposite* between the two interior points, as their relative displacement in the x-direction changes considerably from left to right. Approximately in the y-direction, Figure 4, middle, the interpolation is much smoother, as its principal feature is the *consistent* displacement of both interior landmarks with respect to the outside "corners." The y-feature is thus at larger scale than the x-feature; the inverse scales may be taken in the ratio of the eigenvalues (specific bending energies) of the principal warps along which they align, a ratio of about 2:1. Because the spline is linear in the coordinates of the target form, the displacement vectors involved in these partial warps add down the page to account exhaustively for the actual landmark shifts.

Figure 5 shows a typical application of the thin-plate spline to our little midsagittal data set. The top pair of panels shows the map taking one of the cases in this study to the "average shape" of Figure 2. The lower pair is the map taking that average shape back into the particular configuration of landmarks characterizing this case.

2.4 The Structure of Landmark Variability

The landmark-based analysis of image variability (Bookstein, 1991*b*, Chapter 7) proceeds by recourse to the partial warps, the features exemplified in Figure 4, as they apply to a sample of forms varying about their own mean. These features may be combined in a sort of morphometric principal-component analysis of their own, optimizing variance with respect to bending energy; it is as if each is weighted inversely by its geometric scale. (Remember that the spline is linear in the coordinates of the "target" landmarks, so the partial warps are a basis for the space of purely nonlinear transformations.) The resulting patterns of covarying landmark displacements, the **relative warps**, are geometrically and statistically orthogonal patterns of simultaneous

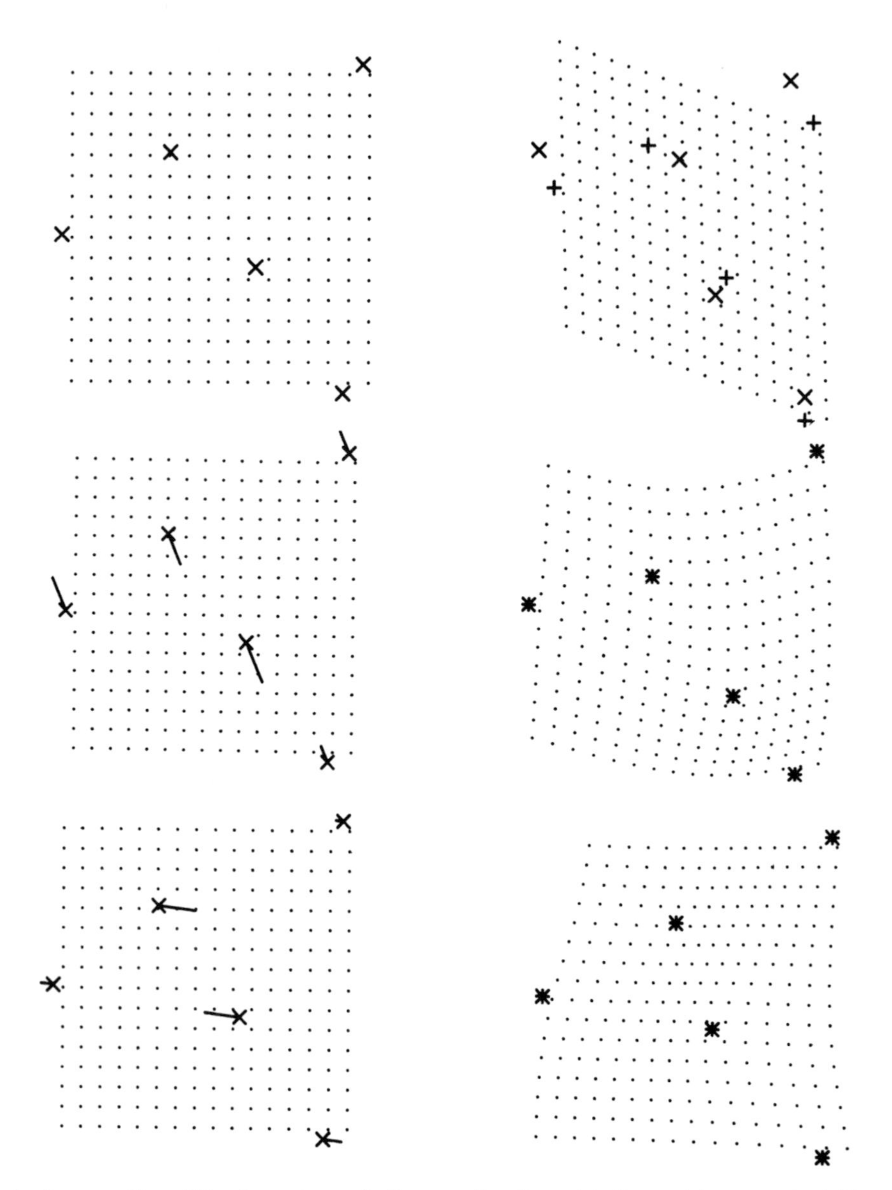

Figure 4. Decomposition of the shape change in the preceding figure into its partial warps. (top) The uniform component, which maps the left-hand landmarks to the positions marked by plus signs at the right, versus the positions they actually occupy (the *X*'s). (middle) Partial warp of largest scale (here, displacement of middle two landmarks in a consistent direction with respect to the outer three). Left, as displacements; right, as deformation. (bottom) Smallest-scale partial warp, displacement of middle landmarks opppositely with respect to the outer triangle. This is the most localized feature of deformation that this starting configuration can sustain.

landmark displacement that emerge in order of "variance per unit bending energy." The very largest-scale components emerge first: aspects of the uniform (homogeneous) change of shape all across the form. Then come, in general, multiples of the largest-scale

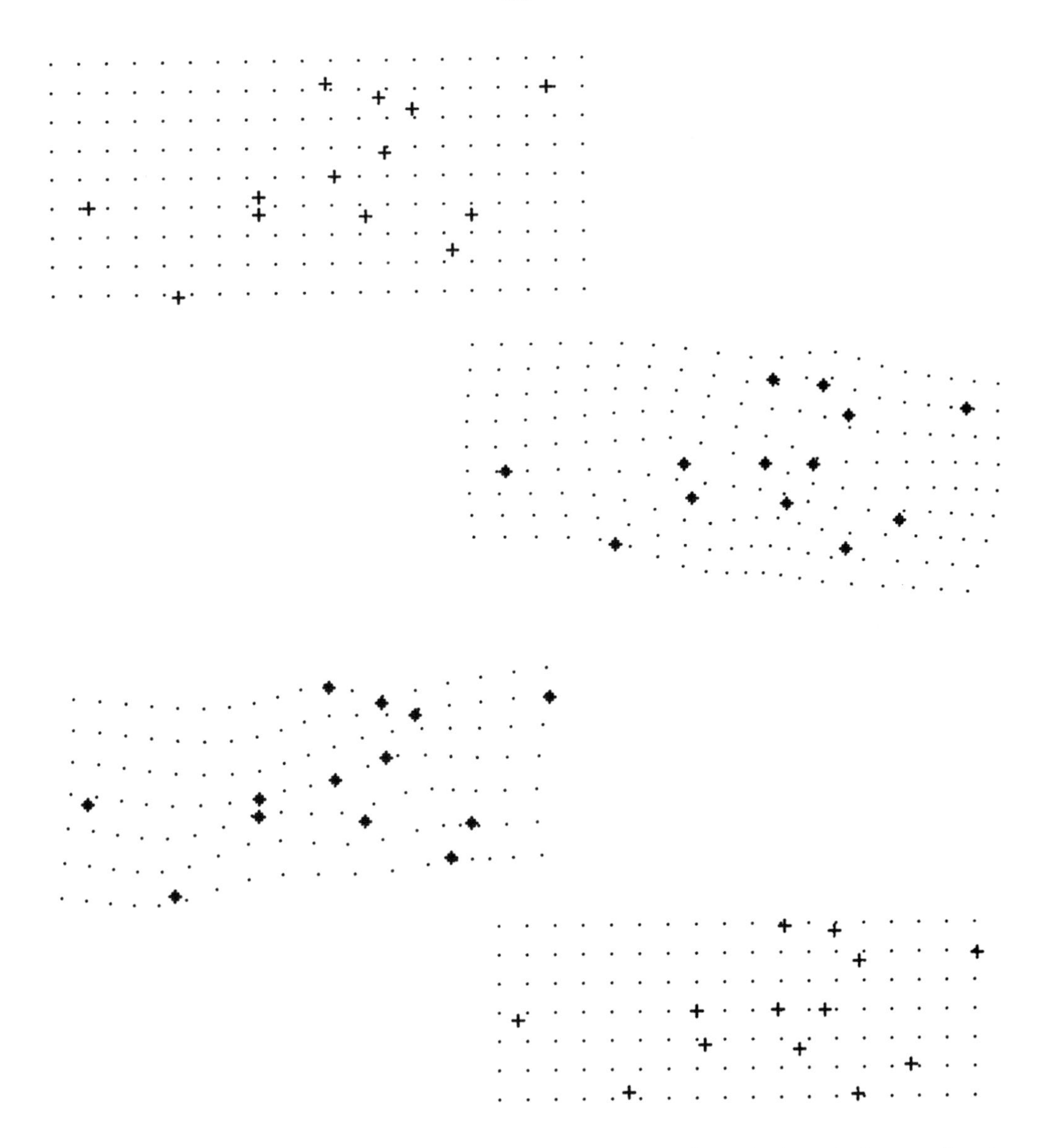

Figure 5. Thin-plate spline taking Case 2 to the averaged landmark configuration (top: the "unwarp") and the average to that of Case 2 (bottom: the "warp"). In both pairs the averaged configuration is the lower. The landmarks of the "starting" form of each spline are indicated with plus signs, those of the "target" form with stars. These mappings are very nearly inverses of each other.

principal warps, the least-localizable patterns of non-uniform shape change (for instance, even gradients across whole diameters of the form, or bends of a long axis); then the features at next-largest scale, and so on. This method is reviewed in detail in Section 7.6 of Bookstein, 1991*b*. The findings here—the two largest-scale patterns of spatial variation, and their scores case by case—are graphed in Figure 6. The first of these patterns represents the relative vertical displacement of the central landmarks with

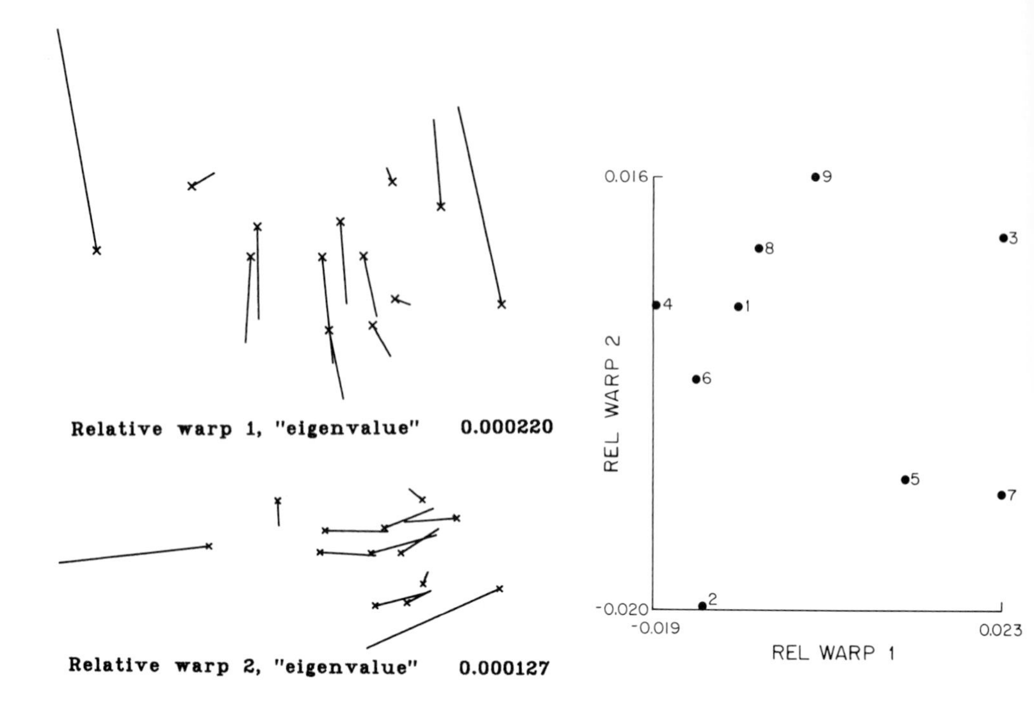

Figure 6. Relative warps for the full sample of nine cases. (left) First and second relative warps shown separately as displacements of "typical" cases from the mean landmark configuration. The first of these is the pattern of simultaneous landmark displacement from the average having the largest variance with respect to bending energy out of all those having no uniform component. The second is the pattern of largest variance uncorrelated with the first. Large-scale variability of a sample of forms is more efficiently described by its values on these two components (Figure 4, top), together with the uniform parts, than by any other combination of features. (right) The sample scatter of relative warp scores.

respect to the poles of the cortex; the second is their relative horizontal displacement; they vary in about a 2:1 variance ratio (which is not significant in a sample this small). An estimate of the "residual" shape variation remaining is shown in Figure 9.

2.5 Image Warping; Image Averaging

The thin-plate spline function does not merely produce grids; it can be used to warp any grey-scale image from one configuration of landmarks to another. As explained in Bookstein and Jaynes (1990), a computation at the pixel level relies upon the *inverse* warp, the spline taking the landmarks of the target configuration onto those of the starting form. To each pixel in the target configuration then corresponds an imputed location in the picture spanned by the original set of landmarks; it is the value of this pixel that will serve for grey-level in the "standardized" or "unwarped" image.[1] Should the resampling need to be to sub-pixel accuracy, the algebra of the thin-plate spline is

---

[1]Following a biometric rather than a computer-graphic approach, I imagine each case to have arisen as the warping of an average or modal "type specimen"; then an average is an "unwarping" of the idiosyncrasies of each case.

consistent with a popular interpolant on square grids of fiducials (the parametric cubic convolution). See Bookstein, 1990*b*.

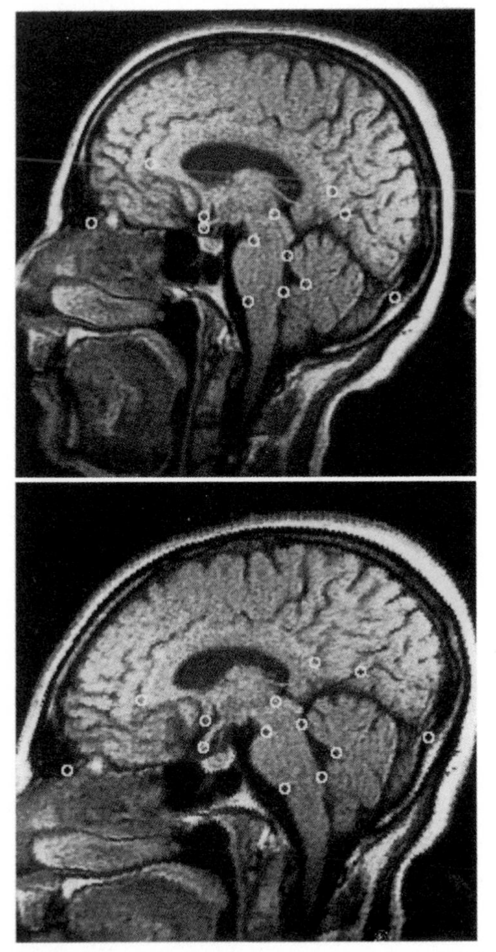

Figure 7. Application of the upper spline in Figure 2 to unwarp the entire MRI for case 2 onto the averaged landmark configuration.

Figure 7 presents the unwarping into the average landmark configuration for the case of which the landmarks have already been displayed in Figure 5. At every point of the "unwarped" picture, the original image data has been subjected to a resampled nonlinear transformation corresponding to the geometry of the corresponding region of the grid in the lower panel of Figure 5.

If the "unwarping" demonstrated in Figure 7 is applied to each of the original nine landmark configurations, there result nine grey-scale pictures of midsagittal images *all having the same biological coordinate system*, that is, all having their landmarks mapped to exactly the same locations. Averaging these pictures pixelwise, there results the image in Figure 8, the *morphometric average atlas* of all nine of the images with which we began. Such an average after nonlinear superposition may be imagined as if physically realized in a photography lab, as prints through a stack of overheated negatives just beginning to curl. Of course the technique is not limited to two-dimensional data.

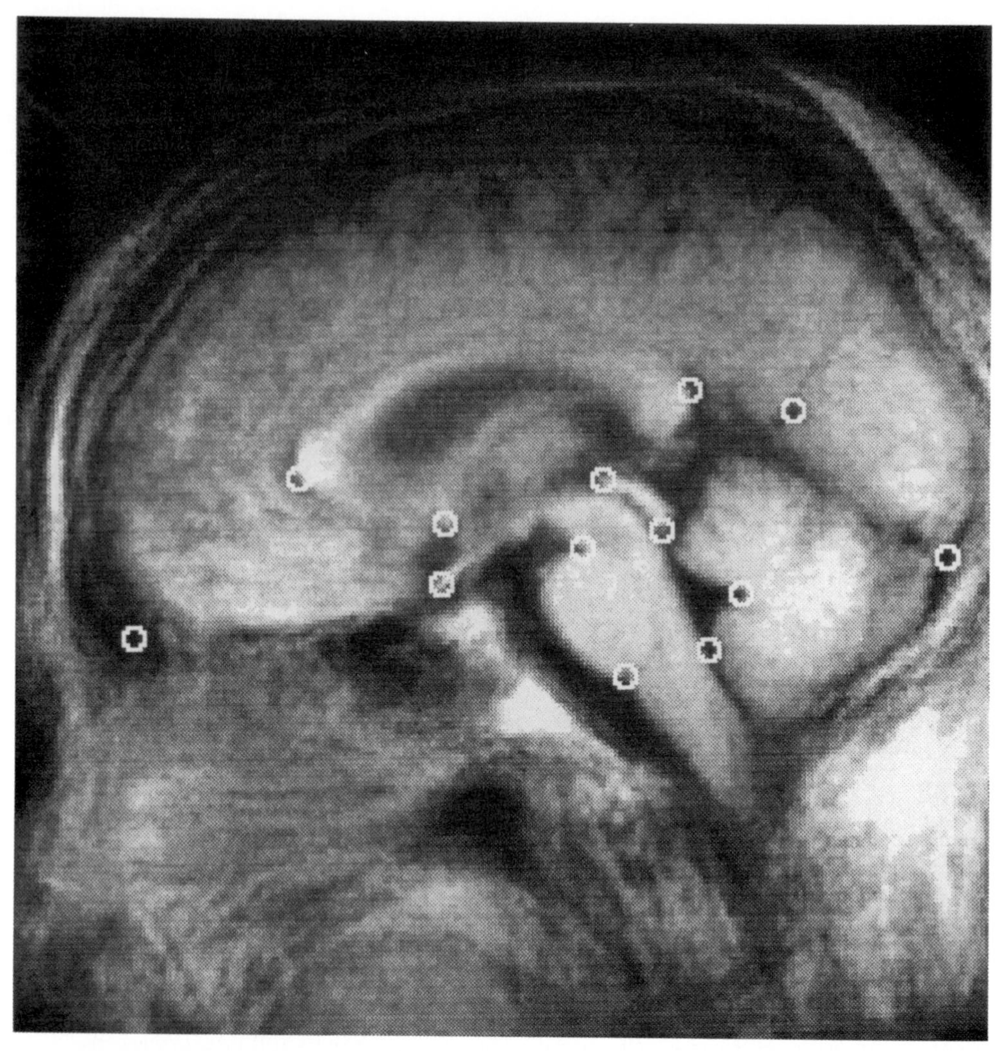

Figure 8. A "midsagittal MRI atlas": the pixelwise average of nine images like that of Figure 7. The text argues that atlases are more appropriately taken as such images than as images or tracings of any single form, however "typical." From Bookstein and Jaynes (1990).

2.6 Regressions of Landmark Locations from Atlas to Cases

With the aid of this flexible interpolation function, we are in a good position to explore the correlations of relative position among the 13 landmarks. In two dimensions, any subset of two landmarks or more drives a warping function, like that in the lower half of Figure 5, which takes the "averaged form" (atlas) onto each patient in turn. Interpolants based on two landmarks (underlying Figure 2, for instance) are simple Euclidean similarities. Those based on three are uniform shears; those for larger numbers of landmarks are thin-plate splines computed by the machinery already reviewed. Each of these mapping functions results in an interpolated or extrapolated location for every landmark that was not used in the spline. These interpolated landmarks take positions in the space of the case's original data that may be compared to the locations the landmarks actually *had* in that case's data. Such interpolants should be considered regression estimates, inasmuch as the spline is linear in the locations of the landmarks in the target image and itself takes the form of a mathematical projection.

Beginning from the same subset of the averaged landmark configuration in each instance, we can compute the corresponding "predicted values" of the remaining landmarks for each case. Because the deformations across this little "range of normal" are not too extensive, we can reflect the prediction error of each landmark as if it applied to the landmark's location in the *average configuration*. Once this is done we can display the pattern of landmark prediction errors for any selection of "regressors" (landmarks on which the spline is based) as a set of scatterplots, one per landmark, in approximately correct relative locations. Figure 2 includes these scatters for two midbrain baselines, Figure 9 for a quadrilateral which is "optimal" in a sense to be explained presently. Note that in all of these figures the landmarks appear to have considerably different "error variances" with respect to this sort of prediction. One would expect that predictions based on landmarks nearby would be likely to be substantially more accurate than those based on landmarks at a distance; likewise, predictions of noisy landmarks, and also prediction *by* noisy landmarks, ought to be less precise than predictions *of* relatively precisely locatable landmarks *by* others likewise precise. For instance, it appears that landmark **#10**, top of the pons, is both precisely locatable and precisely predictable from a variety of nearby baselines. It seems to be the most highly constrained location in the scene; also, it is one of the most ancient, evolutionarily speaking.

3. FINDINGS AND DISCUSSION

I will comment on the findings of this little study in three groups: the atlas (averaged image) that the set of all thirteen landmarks supports, the variability of the sample around the average landmark configuration, and the implications of that variability for the atlas and, more generally, for picture processing in samples of multiple images.

3.1 The Mean Landmark Configuration and the Mean Image

Figure 8 is the average of nine original MRI images pixelwise after unwarping to the shape of the mean landmark locations in Figure 2. Certainly there is strikingly more information in this figure than was put there explicitly in the course of its construction. While the computation referred only to the landmark positions, which are very discrete data, nevertheless several extended curves seem quite well-aligned also. The anterior outline of the pons (between landmarks **#10** and **#11**) and the outline of the fourth ventricle near landmark **#7** seem particularly consistent in the landmark registration, as seems, also, the entire outline of the cerebellum in this plane. More anteriorly, the stalk of the optic chiasm is clear in the average; and, above it, the whole arch of the corpus callosum is quite clear considerably inward from its two "ends."

Elsewhere, in regions of curving form at a considerable distance from these landmarks, this particular atlas is obviously blurred beyond any utility. At the upper right, the skull for one case soars out of the picture entirely, in clear misalignment with those of its classmates. At the forehead, the frontal bone, likewise, bulges to different extents in the different cases. (The curvatures seem to match well enough; it is only position that appears uncontrolled.) A landmark at the central sulcus (middle top of picture) would have helped, but the homologous location pertaining to that spot cannot be seen in this section (only by a shaded-surface visualization from above). At lower left, we are in the nose and the facial bones; the superposition here is obviously in chaos, as neuroanatomy has very little to do with shape below the cranial base.

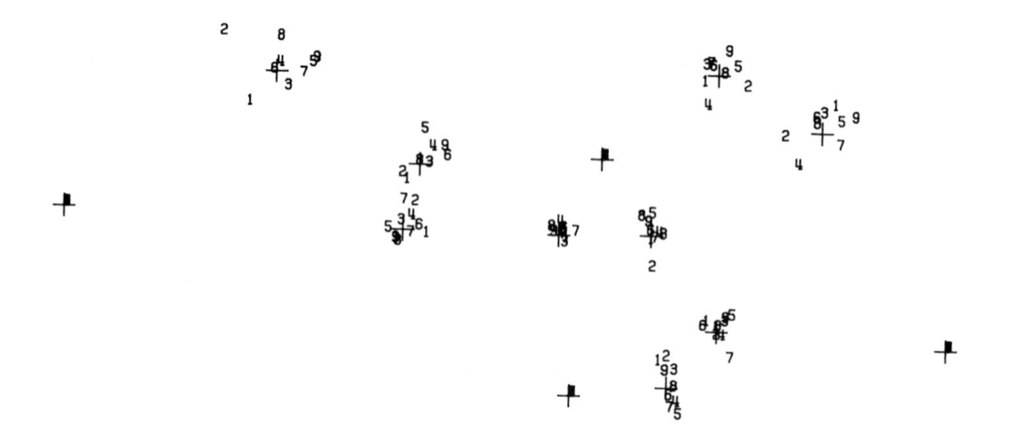

Figure 9. Scatter of "regression residuals" from the warping to the average using only a large quadrilateral of landmarks (**#4**, **#11**, **#5**, **#13**). Compare Figure 2. The analysis by relative warps indicates that most of the information about nonuniformity of shape change is encoded at this largest scale, so that four landmarks might be almost as effective as all thirteen for the purpose of describing relations of the individual case form to the average.

This averaged atlas on 13 points shows a great variety of contours at once clean and smooth. Individual variations in the curvature of specific boundary contours of organs seem to have been smoothed in the course of constructing the averaged atlas. Then the resulting image is substantially better-conditioned in respect of automatic edge-following algorithms (and their three-dimensional equivalents for surfaces). It would always have been preferable to construct the classic (hand-traced) atlases by tracing "many specimens at the same time," so as to archive only what they have in common. The technique of picture-averaging displayed here is the first potentially practical method I have seen for producing just such an "average tracing," by averaging *first*, after an appropriate landmark registration, and tracing *later*. In effect, the landmarks have "concentrated" the information about the edges upon which they are located. The sample average that they guide so strictly but gracefully constitutes the most appropriate "smoothing" of the picture, following which the tracing of edges can proceed in some much more robust way.

3.2 Variability around the Average Landmark Configuration

Let us inspect the panels of Figure 2 more closely. Each shows the "shape" of all nine landmark configurations as the location of each landmark (all the 1's, all the 2's, etc.) in a coordinate system registered on two in particular; recall that for statistical analysis it does not particularly matter which two. There are two important features to be noted in this pair of displays: the "error ellipses" are larger for the landmark scatters at the top, in general, than for the comparable scatters at the bottom; and the deviations of individual landmarks from the average are correlated among neighboring landmarks and, indeed, all the way across either diagram.

The first point is mainly ironic: the conventional AC-PC registration (commissure to commissure, as in Figure 2, top) is actually one of the *noisiest* possible for the sort of stereotaxy we are exploring, and the alternative shown at the bottom, from chiasm to colliculus, has the least r.m.s. scatter of any two-point registration for these cases. But the structure of correlation around these + signs strongly suggests that a registration on *more* than two points might represent a great improvement in the precision with which other landmarks can be statistically located. The conventional stereotactician indeed engages in just such an extension—fitting a "shoebox" to the head by scaling to the height of the vault over the commissures as well as their positions (cf. Fox and Kall, 1987)—but, as we shall see shortly, that adjustment misses most of the statistical signal that is actually present.

We can summarize the covariation of landmark positions around their means in this little sample by tracing two clusters of forms which "stay together" landmark by landmark. In the upper figure, these might be forms 1, 2, 7 and 5, 6, 9; in the lower figure, the cluster 1, 2, 7 remains, contrasted now against 5, 8, 9. There is a sort of "principal component" of shape variation indicated here in the way that the vector of separation between the two cluster means rotates across the figure. Still reading the lower frame, at landmark **#5** it points north by northeast; at landmark **#13**, south by southwest.

We can usefully divide this pattern of correlated variation into diverse features of different geometric scales, just as one single shape change was decomposed in Figure 4. At infinite scale is the variation of the "shoebox" into which these landmarks are fitted. We might ask, for instance, if points that deviate upward from the mean positions above the commissures correspond to points below the commissures that deviate downward (if, in other words, there is any systematic pattern of vertical stretching here). As one can see, configurations that are relatively high above the commissures, like 2 and 7 (at landmark **#1**), are not notably below the mean at the lower landmarks. Tests by the method of Bookstein, 1991*b*, Sec. 7.2, formally confirm this: for the 13 landmarks here, there is no substantial uniform factor.

There is, however, a factor of large-scale bending, corresponding to the end-to-end organization of the vectors contrasting those clusters of cases as they nutate about the form. The method of relative warps extracts these in order of "variance per unit bending," the relative eigenvalues of a pair of quadratic forms (the sample covariance and the bending energy of the spline as a function of the "target" configuration). Figure 6 has already showed the first two relative warps for this data set. They are a standard configuration, the contrast of the "middle" landmarks with those of the two "ends," first vertically and then, at half the variance, horizontally. The clearest contrasts here are thus directing our attention to the relative position of the midbrain landmarks as a whole with respect to the two cortical poles and their neighbors vertically and horizontally.

The dominance of the first pair of relative warps in this analysis indicates that a contrast of center with periphery is sufficient to archive most of the variability of this configuration. That is, these brains warp the average as if they were big, fuzzy quadrilaterals; further detail in the specification of these warps is not likely to lead to improvements in the sharpness of the atlas.

3.3 Implications for Medical Image Processing

If the largest-scale bending of this landmark configuration is derived from relocation of the middle of the form with respect to the "poles," then a registration (superposition rule) is likely to be more adequate to the extent it brings that anatomical variation under control. Figure 9 shows "regression residuals" of the nine images after they have been standardized to a quadrilateral of landmarks that rigidly fixes both the position and the orientation of the midbrain with respect to the cerebral poles in this way. The positions of the individual specimen landmarks (single digits) about these averages are now very nearly uncorrelated, indicating that little geometric information above the level of the landmarks individually remains beyond that at this largest scale. These residuals need not be "noise," of course—they might still have functional correlates—but they appear

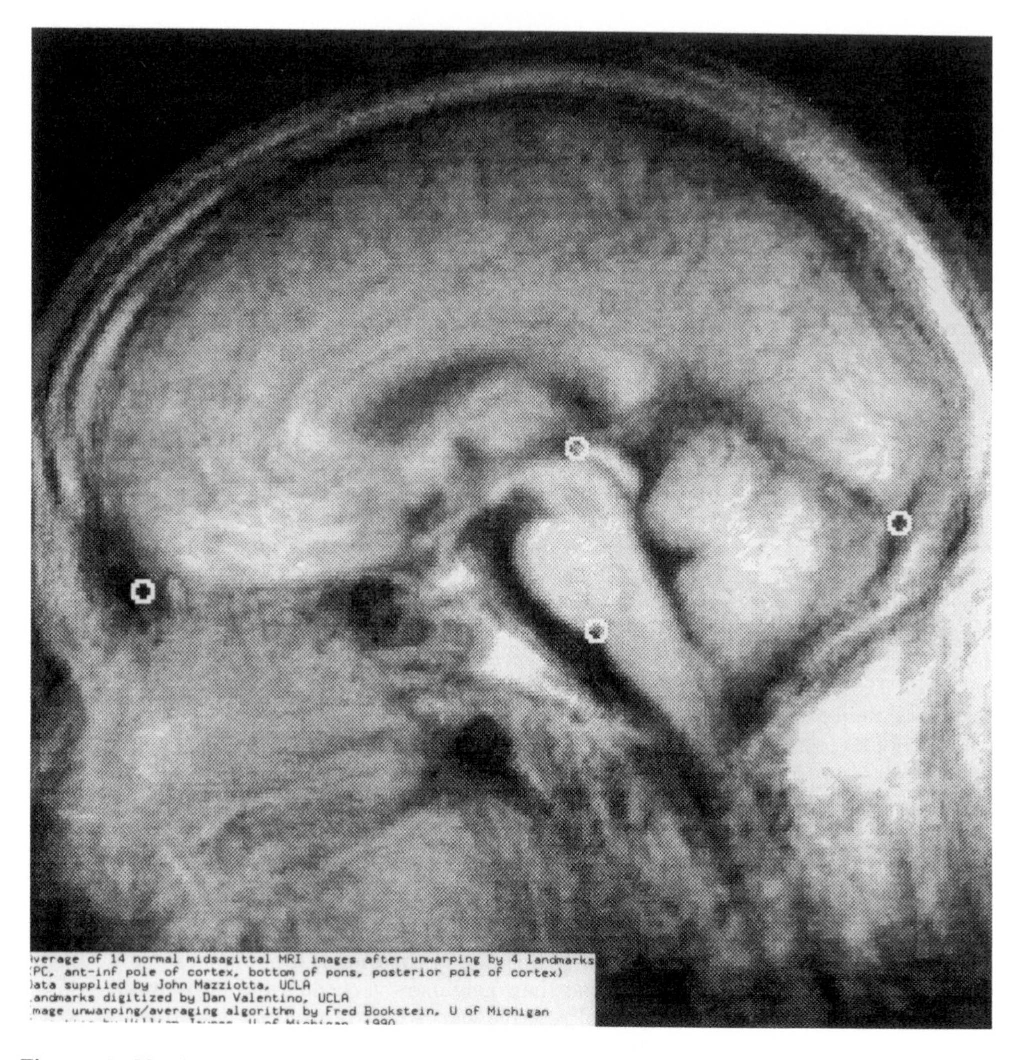

Figure 10. Morphometric atlas registering on only the four landmarks of the large quadrilateral in Figure 9.

geometrically independent. Up to this level of "noise," each configuration has been "measured" by the shape of this single quadrilateral (in effect, the "values" of its relative warp scores, as in Figure 6, right) together with its image as unwarped to this partially standardized coordinate system. The choice of this particular quadrilateral is sensible: it incorporates the longest available horizontal baseline, and thus the most precise estimate of that longitudinal bending which was the largest relative warp, crossed by a relatively long, relatively vertical segment locating midbrain structures and specifying any relative rotation. Then this set of four points conveys information about many aspects of the large-scale organization of this image at the same time: a fairly rich roster of parameters. In fact, this particular four-point registration here was not guessed at using such geometric arguments; of all possible four-landmark registrations, it has the smallest circles of error for the remaining "free landmarks." See Bookstein, Mazziotta, and Valentino, 1991. The averaged atlas it supports, Figure 10, is almost as sharply registered as that in Figure 8, which uses all 13 landmarks.

The discovery of this useful lozenge of landmarks for midsagittal stereotaxy strongly suggests that any automatic or semi-automatic algorithm for the analysis of scenes like these would do well to identify landmarks in the order for which the identification of the next is "most precise" in the sense of the residuals left by these partial splines. From landmarks **#4**, **#11**, **#5**, and **#13**, one can estimate the expected locations of all the others quite accurately; but each of these four contains enough unique information to require a large and careful search.

Tracing a large sample of edges in a landmark-rich scene, then, must involve some compromise between the smoothness of the sample edges individually and the variability of their location in the population undergoing averaging. One can imagine two different ways of using the unwarping based on the landmarks to drive a computation of the "average outline." In one approach, one would locate the edges separately case by case using the ordinary maneuvers of medical image processing, unwarp these edges onto the average image by image according to the superposition given by the landmarks there, and then average the positions of the nearly overlapping edges in the coordinate system of the mean shape into which every image is unwarped. This averaging of curves could be carried out by taking means of intercepts along nearly shared local normals. In another approach, one would average the pictures first, and then extract a consensual edge only once, in the averaged picture, where presumably its locus is a bit smoother or more generic. Under what realistic models for the distribution of location and sharpness of samples of edges is one of these protocols to be preferred to the other? The sharpness of the edge in the averaged picture (Figure 8) is a convolution of the sample distribution of edge locations and their separate localizations by the derivatives of the transverse grey-scale gradients. Is this sum of variances a reasonable statistic for the "net precision" of the location of that outline in the sample average? What role is played by the uncertainty of landmark locations parallel and perpendicular to nearby edges in calibrating this procedure? Under what circumstances, in particular, will the *omission* of landmarks from the registration procedure improve the sharpness of edges? A technology for the interaction of outlines with spline-based unwarping by modeling of the images that are their common source might provide the long-sought formal tie between the analysis of biological outlines and the analysis of landmarks (cf. Rohlf and Bookstein, 1990).

Acknowledgements

The work reported here was supported in part by NIH grants GM-37251 and NS-26529 to the University of Michigan (Fred L. Bookstein, Principal Investigator). The image manipulations were executed by William R. Jaynes.

References

Bookstein FL (1986*a*). Size and shape spaces for landmark data in two dimensions. Statistical Science **1**:181–242.

Bookstein FL (1986*b*). From medical imaging to the biometrics of form. In: Information Processing in Medical Imaging. Bacharach SL (ed), Martinus Nijhoff, Dordrecht, Netherlands, pp. 1–18.

Bookstein FL (1987). Morphometrics for functional imaging studies. In Mazziotta JC and Koslow SH, Assessment of Goals and Obstacles in Data Acquisition and Analysis from Emission Tomography: Report of a Series of International Workshops. Journal of Cerebral Blood Flow and Metabolism **7**:S23-S27.

Bookstein FL (1988). Toward a notion of feature extraction In: Information Processing in Medical Imaging. de Graaf CN and Viergever MA (eds), Plenum Press, New York, pp. 23–43.

Bookstein FL (1989). Principal warps: Thin-plate splines and the decomposition of deformations. I.E.E.E. Transactions on Pattern Analysis and Machine Intelligence **PAMI-11**:567–585.

Bookstein FL(1990*a*). Morphometrics. In: Toga, AW (ed), Raven Press, New York, pp. 167–188.
Bookstein FL (1990*b*). Distortion correction. In: Three-Dimensional Neuroimaging. Toga, AW (ed), Raven Press, New York, pp. 235–249.
Bookstein FL (1991*a*). Four metrics for image variation. In: Information Processing in Medical Imaging. Ortendahl DA and Llacer J (eds), Wiley-Liss, New York, pp. 227 – 240.
Bookstein FL (1991*b*). Morphometric Tools for Landmark Data. Cambridge University Press, New York, to appear.
Bookstein FL and Jaynes W (1990). Thin-plate splines and the analysis of biological shape. Video tape, 21 minutes.
Bookstein FL, Mazziotta JC and Valentino D (1991). Morphometrics for brain atlases: landmarks, deformations, and picture averaging. NeuroImage, submitted.
Fox PT and Kall B (1987). Stereotaxy as a means of anatomical localization in physiological brain images: Proposals for further validation. In Mazziotta JC and Koslow SH, Assessment of Goals and Obstacles in Data Acquisition and Analysis from Emission Tomography: Report of a Series of International Workshops. Journal of Cerebral Blood Flow and Metabolism **7**:S18-S20.
Rohlf FJ and Bookstein FL (1990). Proceedings of the Michigan Morphometrics Workshop. University of Michigan Museums, Ann Arbor, MI.
Terzopoulos D (1983). Multilevel computational processes for visual surface reconstruction. Computer Vision, Graphics, and Image Processing **24**:52–96.

NON-RIGID MOTION MODELS FOR TRACKING THE LEFT-VENTRICULAR WALL

A A Amini[†,*] R L Owen[†] P Anandan[‡] and J S Duncan[†,*]

Departments of Diagnsotic Radiology*, Electrical Engineering[†], and Computer Science[‡]
Yale University
New Haven, Connecticut 06510

Abstract

A unified framework for visual motion tracking of non-rigid objects with specific applications to the left ventricular endocardial wall motion is outlined. The theory considers both two dimensional contours and three dimensional surfaces and in each case uses an elastic model of the object with constraints on the types of motion allowed for tracking the movement. The basic theme in both two and three dimensional analysis is to match bending and stretching properties of shapes in consecutive time instances for deducing quantitative information about the motion of the LV wall. Several algorithms are presented, and applications to real and simulated data are included. At the end, future directions for research are discussed.

Keywords

optical flow; computational differential geometry; two-dimensional motion; three-dimensional motion; bending energy; conformal stretching

1. INTRODUCTION

Non-invasive techniques for measuring the dynamic behavior of the left ventricle (LV) can be an invaluable tool in the diagnosis of heart disease. The results gained from quantifying the motion can yield insight into the general health of the heart and provide clues regarding specific events affecting cardiac performance, such as infarct size or formation of aneurysm. This is a challenging problem mainly because motion of the LV is non-rigid. Rigid motion, as oppose to non-rigid motion, is better understood, and a plethora of techniques dealing with it have been proposed (Aggarwal, 1986). Motion of non-rigid objects, on the other hand, is a lot more complicated. This is predominantly because the *structure* of the object no longer remains the same, and in fact deforms with time (Huang, 1990).

The majority of efforts in the measurement of cardiac motion have dealt with two dimensional analysis, and for the purpose of discussion we divide these into two classes. In the first class of approaches, motion is measured by detecting changes in the intensity values of pixels in sequence of images. Included here are techniques which compute phase changes in the pixels across the temporal dimension (Links, 1980), those that find regional ejection fractions between end-systole (ES) and end-diastole (ED) from radial chords extending from the center of the ventricle (Areeda, 1982), and others that compute instantaneous velocities of points between successive frames (Mailloux, 1987). The latter techniques compute the motion at every spatial point in the image without attempting to extract the regional motion of points on the LV wall. This separation in general will be difficult.

A second group of approaches rely on initially extracting the boundary of the cardiac wall. Most of these approaches are semi-automatic in the sense that first an operator outlines the perceived cardiac wall in each image of the sequence. Following this stage, motion is computed at a sampled set of points on the boundary. There have been some attempts at automating the contour detection stage (Duncan, 1987c),(Gelberg, 1979). In either case, the algorithms are presented with a sequence of contours which specify the location of the cardiac wall in each image frame. The adopted schemes are mainly geometric such as the chord shortening method (Gelberg, 1979) which depends on accurately locating a reference point or line, and the centerline method (Sheehan, 1986) which measures the thickness of the region between the ED and ES contours and uses this in computing the motion. The inherent problem with this method is that it completely ignores shape cues in motion computation.

The alternate route to analysis is to artificially create distinct features that are visible in images of the LV. One method is to surgically implant markers on the LV wall, follow these through the wave of contractions and expansions, and infer the motion field for points that are close to the markers. Examples of such techniques are given in (Bookstein, 1989),(Ingels, 1975),(Slager 1986). Although useful to study characteristics of the motion of the LV for animals, invasive techniques do not extend well to the study of human hearts over large populations. (See (Ingels, 1975) who has used markers for studying dynamics of transplanted hearts). Moreover, these techniques usually require manual registration of the markers ((Bookstein, 1989) is an example).

A new, non-invasive technology that offers a different approach for tracking of points is referred to as MR tagging (Axel, 1989),(Zerhouni, 1988). The approach creates a magnetization grid which tags the underlying tissue, and uses the grid in following the tissue over a gated sequence. This is a viable solution to tracking 2D movement, however, this technique is still in the formative stages, and furthermore since the magnetic tags become weak over time, MR-tagging has limited ability in tracking points over the entire cardiac cycle. It is important to note that MR-tagging is really meant to aid the physician see where points have moved to, and little work has been done in attempting to automate the process. LV-wall motion analysis with MR tagging will rely on boundary-finding methods to decipher which of the tagged points lie on the LV wall. Finally, determining the motion in the tag plane will be a difficult task.

In comparison to two-dimensional analysis, three-dimensional analysis has not received a great deal of attention. Primarily, this has been due to the lack of available 4D (3-space and time) imaging methods. However, these methods are becoming increasingly available and popular. There are four major methods: Echochardiography (Raichlen, 1986), gated SPECT (Garcia, 1985), Cine-CT (Collins, 1986), and MRI (Lanzer, 1984). Among these methods, it seems that MRI and possibly Cine-CT are the most likely techniques to provide the highest resolution while permitting viewing of both the endocardial and epicardial walls, and thus are well suited for cardiac motion analysis. Use of true 3D information has advantages over 2D. Computing measures of wall motion or wall thickening can only be roughly estimated from two-dimensional images since the heart moves in and out of the plane of acquisition. Employing 3D information greatly enhances the accuracy of these measures.

In order to track LV motion in three dimensions, invasive markers and MR tags may be employed. However, with both of these techniques as noted previously, a correspondence problem exists. Once tags/markers have been placed and correspondence is made an important issue is how to compute a dense motion field from the sparse set of points with the available motion information. Although Bookstein's method (Bookstein, 1989) may be instrumental here, the approach that will be discussed in this paper, automatically chooses distinct shape properties as tags, optimally brings into correspondence the chosen tags, and uses stretching properties of the wall for computing a dense motion field over the totality of the surface. The methods that we will present for motion computation will provide a useful alternative to MR-tagging and invasive markers, while not depending on any given modality. In the case of MR-tagging, we believe that the presented methods can aid in solving the 3D correspondence problem noted above.

1.1. Shape-Based Tracking

For analyzing cardiac motion, we set out to find algorithms that (1) are fully automatic, (2) track the motion through-out the cardiac cycle – solely from image information and without requiring implanted markers, and (3) use physical models of the heart with specific allowable types of motion for adjacent points on the wall in interpreting the time-varying data. Points (2) and (3) require further explanation. Many of the current approaches to quantifying the LV wall motion use only information present in, or derived from, the ED and ES image frames. The LV actually goes through a temporal wave of contractions, with different types of deformations and movements occuring at each location on the heart wall. Each point follows its own unique trajectory in the three-dimensional space. Although there has been some work that has considered wall motion analysis using an entire sequence of images (Harris et al, 1974), (Areeda et al, 1982), these approaches are plagued by measuring motion as if it were emanating radially from a single point or a single control axis.

Physical modelling has become quite popular in the computer graphics literature. Terzopoulos (Terzopoulos, 1987), for example, uses elastically deformable models which are active and that respond in a natural way to applied forces, constraints, ambient media, and impenetrable obstacles. Physically-based approaches are interesting in that they incorporate flexible materials into the purely geometric models which were used traditionally in the computer graphics literature. It should be noted however that such approaches deal with the "forward" problem of creating animation sequences whereas we deal with the "inverse" problem of extracting the motion information.

The approach that we are taking to analyzing the motion of LV is based on physical elastic models. The physical modelling in two dimensions involves modelling the outlined contours of the LV as elastic rods; in 3D we use a thin-plate mechanical analog for the LV surface. This, we note, is different from Bookstein's method of representing the *warp* by thin-plate interpolants (Bookstein, 1989). The non-rigid motion can thus be summarized in determining the bending and stretching of the LV wall. There is an important theorem in differential geometry which assures us that finding these quantities is equivalent to unique determination of the planar-curve and the 3D-surface in a lapsed time (fundamental theorem of curves and surfaces) (Millman, 1977). This is important to recognize, since bending and stretching will be unique to the pair of geometric entities. For both two and three-dimensional analysis, we constrain the motion as well. Goldgof and Huang (1988) have used different stretching models, without any bending models, for describing non-rigid motion. They assume that point correspondences exist and subsequently use this information for motion classification. In fact, related methods measuring non-rigid motion invariably assume that correspondence of points is known. (See (Huang, 1990) for an overview of non-rigid motion in computer vision). We use bending and stretching models together, for obtaining point correspondences.

In order to arrive at point correspondences, in 2D, we assume that the motion field is smooth; thus requiring the stretchings to be similar for adjacent points on the contour. In 3D, we have formulated two stretching models. The first model assumes equal stretching of points all around the surface, similar to an inflating balloon. Another more general model assumes equal stretching around a given point, with the stretching factor now allowed to vary. The bending component of our motion model forces tracking of important shape tokens.

For analyzing motion, we assume that the boundary (for 2D), and the surface (in 3D) have been detected at the start of the sequence. The field of computer vision has increasingly realized the problems associated with image segmentation methods such as edge detection and region-growing. For example, the best edge detectors almost always introduce disconnected and spurious edges in images. For this reason, the emphasis of the vision community has turned towards methods that first assume a connected contour or surface and, via optimization strategies, deform the initial contour/surface in best segmenting the object of interest in the image. The data that the 2D motion detection algorithm will be illustrated with makes use of the parametrically deformable contours (Staib and Duncan, 1989). This algorithm is currently being extended to 3D (see (Staib, 1990)). There is also a class of inflating/deflating contours (Amini, 1990) that can be used in *weaving* a smooth surface from serial cross sections which can subsequently be used in tesselating a 3D surface (Ayache et al., 1989).

Organization of the rest of the paper is as follows: First we give a brief overview of token matching in computer vision. Section 3 reviews relevant concepts from differential geometry. Having laid the foundation, the two sections that follow discuss our 2D and 3D algorithms. At the end, we state future research directions.

2. TOKEN MATCHING

One method of tracking point-wise trajectories in images is to match identifiable elements - tokens - over time. These techniques (Ullman, 1979) compute image displacements between tokens in successive frames *generally in 2D* by matching each token in the first frame with a token in the second frame. In the computer vision literature, this is known as the correspondence problem. The strategies for solving this problem depend, to a considerable extent, on the level at which the matching process is performed. Matching can be established between simple tokens such as points or line segments, or between more complex tokens such as parts of structures or objects.

For example, if the tokens are simple dots then correspondence is determined primarily by distance, although other parameters may also affect the match. When the tokens are line segments, the primary correspondence parameters are distance, length, and orientation. Although parameters can also affect the match, a given line segment tends to match with one having similar length and orientation.

The level at which the correspondence is established determines the amount of processing required. Complex tokens require more processing, although their use can simplify the matching process since there are probably fewer such tokens in a given image. Simple tokens require little or no processing, although they usually have a greater number of possible matches. In essence, the correspondence problem can be simplified by using more complex tokens, usually at an increased computational cost. Motion is usually determined by matching edges or gray-level discontinuities. Nagel (Nagel, 1978) has discussed the approach in which region segmentation precedes motion extraction. In most studies matching is established between parts of object boundaries, as opposed to matching between complete objects. Pertinent to our work is (Hildreth, 1984) who uses zero-crossing contours in computing the *normal flow.*

We also assume that a set of tokens with distinct shape properties may be found. However, our assumption now carries over to surfaces in three dimensions. Points having distinct shape properties generally amount to high curvature points of curves in 2D and surfaces in 3D. With relative, small inter-frame motion in the time sequence, such tokens can be tracked reliably.

3. DIFFERENTIAL GEOMETRY FOR TRACKING SHAPE CHANGE

In this section, we review relevant concepts from differential geometry which will be used in the mathematical formulation of non-rigid motion computation. Computing non-rigid motion will include measurement of bending and stretching deformation of shape based on differential geometry.

Differential geometry is a branch of mathematics that deals with local description of smooth curves and surfaces. For curves, differential geometric properties that are of interest are the *curvature*, *speed*, and *torsion* of the curve measuring the *bending*, *stretching*, and *twisting* of the curves respectively. To state these on firm mathematical grounds, assume a representation of the curve $\alpha(s)$ where s is the arclength along the curve. If the tangent vector to each point on the curve is denoted by $\mathbf{T}(s)$, the rate of change of the tangent with respect to the arclength is a measure of bending of the curve and is by definition the curvature measure:

$$\kappa(s) = \frac{d\mathbf{T}(s)}{ds} \tag{1}$$

If the vector $\mathbf{B}(s)$ is the binormal of the curve, i.e., the vector perpendicular to both the tangent vector $\mathbf{T}(s)$ and the normal vector $\mathbf{N}(s)$, the torsion of the curve is expressed as:

$$\tau(s) = -\mathbf{B}'(s) \cdot \mathbf{N}(s) \tag{2}$$

Torsion measures the out of plane twisting of a 3D curve (for a planar curve this is zero.)

For surfaces, there are two basic mathematical entities that are considered in classical differential geometry. They are referred to as the first and second fundamental forms of a surface (Millman, 1977), (Eisenhart, 1955), (Kreyszig, 1975). The fundamental theorem of surfaces states that two surfaces have the same shape if these two quantities are the same as a function of parametrization of the two surfaces (Millman, 1977). A parametric representation of a surface $\mathbf{r}(u,v)$ is a 3-element vector function $[x(u,v)\ y(u,v)\ z(u,v)]^T$. The first fundamental form of a surface describes the amount of movement on a surface as a function of the movement in the parameter space. With $d\mathbf{r} = \mathbf{r}_u du + \mathbf{r}_v dv$, the first fundamental form is given by the quadratic form:

$$I(du, dv) = (d\mathbf{r}, d\mathbf{r}) = [du\ dv] \begin{bmatrix} E & F \\ F & G \end{bmatrix} \begin{bmatrix} du \\ dv \end{bmatrix} \tag{3}$$

where $E = (\mathbf{r}_u, \mathbf{r}_u)$, $F = (\mathbf{r}_u, \mathbf{r}_v)$, and $G = (\mathbf{r}_v, \mathbf{r}_v)$. The above equation can be written out as $I(du, dv) = Edu^2 + 2Fdudv + Gdv^2$. The symmetric matrix of first fundamental form coefficients is referred to as the metric of the surface and plays the role of "speed" for surfaces, which depending on the surface parametrization varies (for curves, arclength parametrization is the unit-speed parametrization).

The second fundamental form on the other hand measures the correlation between change in the normal vector $d\mathbf{n}$ and change in the surface position $d\mathbf{r}$ at (u,v) when moving by (du, dv) in the parameter space. With the normal vector expressed as:

$$\mathbf{n}(u,v) = \frac{\mathbf{r}_u \times \mathbf{r}_v}{||\mathbf{r}_u \times \mathbf{r}_v||} \tag{4}$$

the second fundamental form $II(du, dv)$ is computed as:

$$II(du, dv) = [du\ dv] \begin{bmatrix} L & M \\ M & N \end{bmatrix} \begin{bmatrix} du \\ dv \end{bmatrix} = -(d\mathbf{r}, d\mathbf{n}) \tag{5}$$

In the above equation, the symmetric matrix is referred to as the second fundamental form matrix and it can be shown to have the following components: $L = (\mathbf{r}_{uu}, \mathbf{n})$, $M = (\mathbf{r}_{uv}, \mathbf{n})$, and $N = (\mathbf{r}_{vv}, \mathbf{n})$. The first fundamental form and the metric are said to be intrinsic properties of the surface; that is to say, they do not depend on how the surface is embedded in 3-space. In contrast, the second fundamental form depends on embedding of the surface in 3-space, and therefore is referred to as an extrinsic property of the surface. An important observation about these quantities is that any property related to the first fundamental form is related to the stretching of the surface and properties related to the second fundamental form are related to the bending of the surface. By combining the first and second fundamental form matrices as

$$\beta = \begin{bmatrix} E & F \\ F & G \end{bmatrix}^{-1} \begin{bmatrix} L & M \\ M & N \end{bmatrix} \tag{6}$$

we arrive at the Weingarten mapping. This matrix in effect combines the first and second fundamental form matrices into one matrix. The Weingarten mapping, maps vectors in the tangent plane at a point on the surface to other vectors in the same plane. Since the metric tensor is a generalization of the speed of a curve, the Weingarten mapping may be viewed as the generalization of the curvature of a planar curve (Besl, 1986). The Gaussian curvature function K of a surface can be defined by

$$K(u,v) = det(\beta) = \frac{LN - M^2}{EG - F^2} \tag{7}$$

The Gaussian curvature is also an intrinsic property of the surface. The mean curvature which is an extrinsic property of the surface is defined as

$$H(u,v) = \frac{1}{2} tr(\beta) = \frac{EN + GL - 2FM}{2(EG - F^2)} \tag{8}$$

Other curvature properties of interest for a surface are obtained by considering the curvature of a line on the surface in the plane spanned by the normal to the surface at the point and the tangent vector in the direction of the line. The resulting curvature is referred to as the normal curvature at the point and may be parametrized as a function of the angle θ at the point:

$$\kappa_n(\theta) = \kappa_1 cos^2(\theta) + \kappa_2 sin^2(\theta) \tag{9}$$

the quantities κ_1, and κ_2 are referred to as the principal curvatures and are the minimum and maximum normal curvature at a point. They are the eigenvalues of the Weingarten mapping. The corresponding eigenvectors are the principal directions. It can be shown that the product of principal curvatures yields the Gaussian curvature and their average yields the mean curvature (Besl, 1986),(Millman, 1977). Our approach to measurement of non-rigid shape change of the LV draws from the concepts described in this section.

4. 2D ALGORITHMS

In this section, we discuss two-dimensional formulations for non-rigid motion computation of the LV wall. The formulation as described in (Owen et al, 1989) involves two separate stages. In the first stage, initial displacement vectors are estimated by matching points on pairs of successive contours by physically modelling the contour as an elastic rod and minimizing the bending energy for deformation. In the second stage, the initial velocity vectors are refined by smoothing of the initial displacement vectors. The constructed functional minimizes the linear combination of an approximation error term, maintaining adherance to the flow field computed by the bending term, and a smoothness error term maintaining consistency between flow vectors of neighboring points. By smoothing of the initial vector field, no guarantee is made however that the vectors emanating from the first contour will be constrained to specify a point on the second contour and presently we map the vectors by extending them to the second contour. Given a sequence of contours in successive time instances, the algorithms find flow vectors between each pair of contours in successive steps. Sampling the flow vectors and connecting them in consecutive steps allows for tracking motion of specific points over multiple frames.

4.1. Initial Displacement Vectors

Adopting the physical analogy of a thin elastic rod, for a planar contour, the potential energy of deformation for the rod is given by:

$$\int_{\Omega_1} \epsilon_{be}(s) + \epsilon_{st}(s) ds \tag{10}$$

where the first term measures the bending energy and the second term measures the stretching energy. As shown in (Landau and Lifschitz, 1986), the bending energy of the rod at a given point is:

$$\epsilon_{be}(s) = \frac{1}{2} EI\kappa^2(s) \tag{11}$$

The reader is referred to (Duncan, 1988) for a more in-depth discussion of bending energy and its computation from discrete data. In (11), E is the Young's modulus and I represents the inertia for the contour. This is the bending energy of deformation for a straight rod with circular cross section. To carry the analysis to a curved rod, the bending energy of deformation is simply related to the squared difference of curvature for the deformed and undeformed states of the rod. To compute initial match vectors numerically, the displacement is found at each point that minimizes the following penalty function:

$$e(\delta) = \frac{1}{2} EI \int_C [\kappa_g(s+\delta) - \kappa_f(s)]^2 ds \tag{12}$$

$f(s)$ represents the initial state of the contour and $g(s)$ its deformed state. The task is to find the δ at each point such that the above expression is minimized within a search window defined by varying δ. The physical constraint of the maximal movement of the LV wall that is possible within the sample time between two frames limits the region on the second contour over which the above match is computed for each point on the first frame. The precise definition of this region is not critical, but the window must be large enough to safely include the largest local displacement possible.

One way to compute a measure of stretching energy would be to fix the number of sample points on a contour and measure the increase in the average distance between points as a function of time (Duncan et al., 1991). The alternative route, as described in section 4.2., computes a stretching measure in the context of displacement computations for points on a contour.

4.2. Flow Field

The flow field computation minimizes the amount of stretching given the initial match vectors, and in doing so obtains a smooth flow field. The inherent assumption is that the displacement field for the LV wall is sufficiently smooth that any changes takes place in a gradual manner over time. We will discuss more strict motion models for the three dimensional formulation. The associated functional for motion estimation is given by the following expression:

$$\int_0^S C_1(s) \left\{ [u^\perp(s) - d^\perp(s)]^2 + C_2(s)[u^\top(s) - d^\top(s)]^2 \right\} + |\frac{\partial \mathbf{u}(s)}{\partial s}|^2 ds \tag{13}$$

In the above equation, $\mathbf{u}$ represents the displacement vector field that is to be estimated, and $\mathbf{d}$ represents the initial match vectors found from equation (12). The perpendicular and tangential components (to the contour) are each specified by superscripts $\perp$ and $\top$, and S is the length of the contour. We expect to obtain unique estimates of both components of the flow field if the curvature of the contour is high. At linear segments, the tangential component is locally indeterminate.

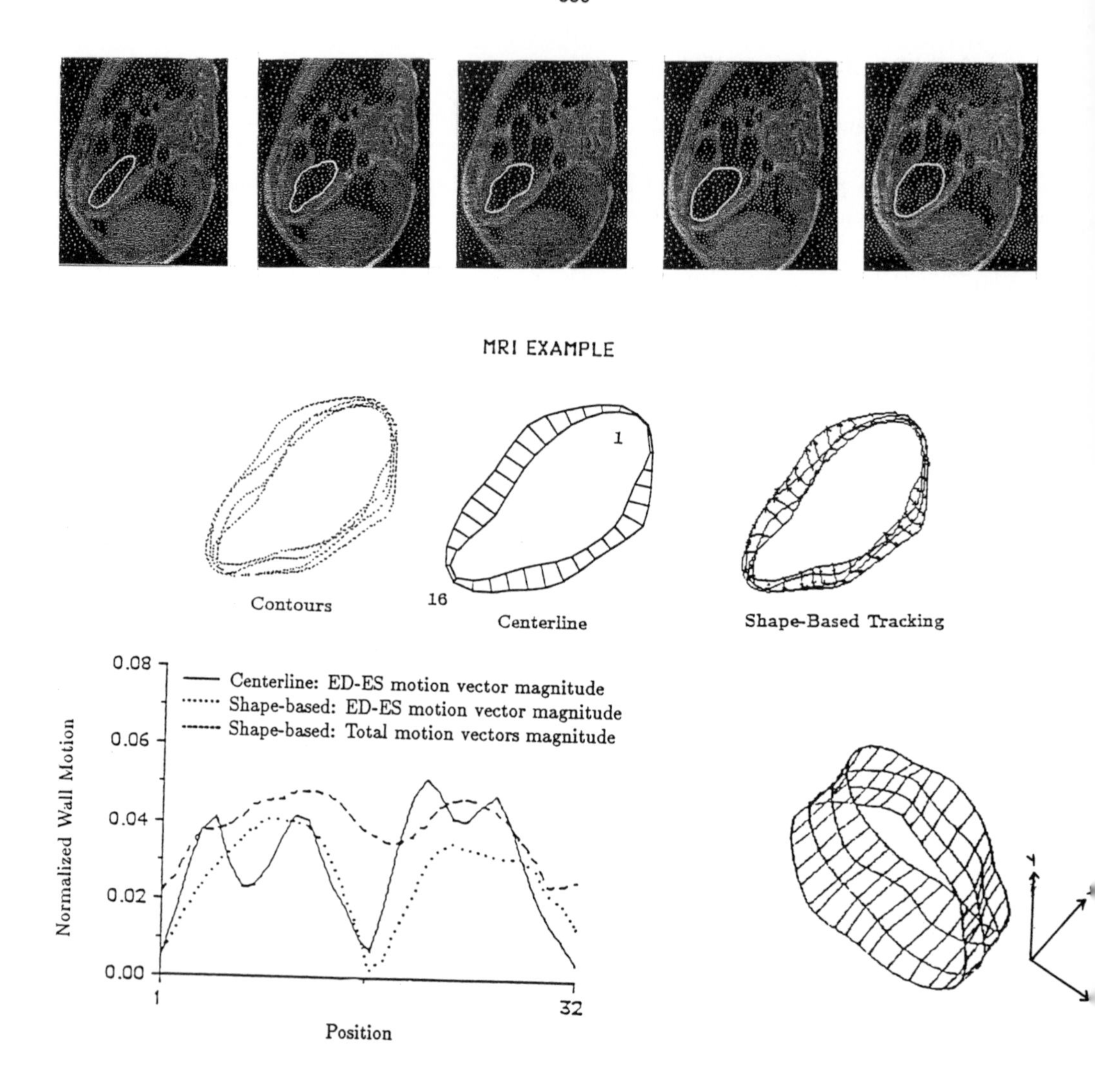

Figure 1: Temporal Sequence of gated MR oblique sagittal images and the detected contours is shown on top. Note that for each image in the sequence on top, the result of contour detection for each frame is used as the input to the algorithm for the subsequent frame. The overlayed contours and the computed flow vectors with both the centerline and the shape-based tracking methods are displayed next. In the bottom, comparison of the two methods and the results of applying shape-based tracking over multiple frames is illustrated. For the latter, the coordinate system is such that the x and y axes run left to right and top to bottom, and the time axis runs roughly parallel to the "flow-lines", into the plane of the paper.

$C_1(s)$ is inversely proportional to the minimum value of the error measure obtained in equation (12) corresponding to the best match, and $C_2(s)$ is a measure of sharpness of this measure around the match. In essence, the $C_1(s)$ confidence measure which weighs all terms in the integrand in equation (13) favors the points having a good match (i.e., uses these as anchors), whereas $C_2(s)$ which is a measure of variance of the error measure around the match favors projecting the flow vectors perpendicular to the contour in the regions that the contour is locally flat since at such points the error measure will have high variance (implying many possible candidate matches). It is interesting to note that irregularities on the LV contours actually help guide the algorithm's matching process, permitting the algorithm to track anatomically interesting features of interest.

4.3. Experimental Results

The minimization of the discrete form of the functional in (13) will yield a set of flow vectors that are displacement vectors extending from sampled points along the contour. Results of applying the gradient descent method for finding a local minimum of the above functional for a sequence of input contours found by automated boundary finding algorithm (Staib, 1989) in a sequence of MR images is given in Figure 1. Shown in this figure are quantitative comparison of the popular *centerline* algorithm (Sheehan, 1986) and shape-based tracking. In a capsule, the centerline algorithm fails to track points of high curvature such as corners correctly. Also shown is the result of shape-based tracking over multiple frames (in the spatio-temporal space). Due to the local smoothing, the current implementation does not guarantee that the vectors emanating from the first contour will touch the second contour, as a result, in a re-mapping stage, the smoothed vectors are mapped onto the second contour. An alternate approach is to realize the *variational* nature of (13), and use dynamic programming to explicitly constrain the vectors to go between successive contours (Amini, 1990).

Finally, in comparison with the clinically useful approaches such as the *centerline* algorithm, it is our opinion that the algorithm described here is more intuitively and physically meaningful with regard to measuring LV motion from 2-D contours.

5. 3D ALGORITHMS

Our three dimensional formulation is based also on a bending and stretching model. The physical model of the LV used now is a thin-plate loosely constrained to stretch in a predetermined way as will be shown later. The potential energy of an ideal-thin-flexible flat plate of elastic material which is a measure of the strain energy of the deformation is given in (Courant and Hilbert, 1957):

$$\begin{aligned} \epsilon_{be}(u,v) &= \frac{\kappa_1^2+\kappa_2^2}{2} \\ &= 2H^2(u,v) - K(u,v) \end{aligned} \tag{14}$$

The measure is invariant to 3D rotation and translation. The energy required to bend a curved plate to a new deformed state (assuming known matches between (u,v) and $\bar{u},\bar{v})$ is:

$$\epsilon_{be}(u,v) = \left\{\sqrt{(2H^2(u,v)-K(u,v))} - \sqrt{(2\bar{H}^2(\bar{u},\bar{v})-\bar{K}(\bar{u},\bar{v}))}\right\}^2 \tag{15}$$

Mean and Gaussian curvature of surface at time t_0 are given with no bars. The same quantities at $t_0+\Delta t$ are specified with bars.

A number of motion types can be described in terms of both the Gaussian and mean curvatures of the surface, and/or the first and second fundamental form coefficients. Rigid motion, which is the motion of an object without any bending or stretching and involving only translation and rotation, renders both H and K unaltered. Any human-made object moving in the environment typically exhibits rigid motion. Isometric motion is a more general motion type and can be stated as a motion that preserves distances along the surface as well as angles between curves. It can be shown that under isometric motion K remains the same while H changes. An example of this type of motion occurs when for example rolling a piece of paper into a cylinder or a cone. Clearly, this is non-rigid, and bending occurs, but note that no stretching of the surface takes place and thus distances remain unaltered. Conformal motion is more general and has the important property that it preserves angles between curves, but is not required to preserve distances between points. As a result of angle preservation under conformal motion, it can be shown that if a motion has the property that the stretching factor is uniform around a point, the motion necessarily is conformal (Eisenhart, 1955), (Kreyszig, 1975). When the stretching is constrained further such that it is not only the same around a point but is the same for all points around the surface, homothetic motion results. An example of homothetic motion occurs for example when a balloon is inflated.

There are important results in differential geometry that may be brought to bear in capturing these motion types, all related to change in coefficients of first fundamental form. We state the following general result. If there exists $\eta(u,v) > 0$ such that

$$\frac{E}{\bar{E}} = \frac{F}{\bar{F}} = \frac{G}{\bar{G}} = \eta(u,v) \tag{16}$$

the object's motion is conformal. Again, no bars specify values of quantities at time t_0, and bars specify the same quantities at the same position on the surface at a later time $t_0 + \Delta t$. η which may vary around the surface is called the stretching factor. If in (16), $\eta(u,v)$ is a constant, homothetic motion occurs, and yet if the constant is identity, isometric motion takes place.

In the stretching energy, we set out to model the general LV motion. Although one possible model is based on the homothetic assumption, we are currently considering the more general conformal motion. The objective function given below favors matching points that are close to the conformal model:

$$\epsilon_{st}(u,v) = \left(\frac{E}{\bar{E}} - \frac{F}{\bar{F}}\right)^2 + \left(\frac{F}{\bar{F}} - \frac{G}{\bar{G}}\right)^2 + \left(\frac{E}{\bar{E}} - \frac{G}{\bar{G}}\right)^2 \tag{17}$$

Combining equations (15) and (17), we arrive at an energy function that penalizes bending and deviation from conformal stretching:

$$\epsilon = \iint_{\Omega_2} \epsilon_{be}(u,v) + \epsilon_{st}(u,v)\,du\,dv \tag{18}$$

As can be seen, the stretching model formulated above favors the same value for $\eta(u,v)$ at each point on the surface. However, since this is sum of three quadratic error measures, the uniformity of $\eta(u,v)$ is not necessarily enforced. In order to force this uniformity in error calculations, we have also used the minimax procedure which as the name implies, minimizes the maximum of the three error terms. In statistical terms, minimax is a more *robust* error norm than sum of the quadratic errors.

5.1. Numerical Methods

The examples that will be given here are meant to model a portion of the smoothly curved LV wall with few relatively distinct locations of important shape information. However, we note that roughly the same computations will apply to a completely delineated surface. To compute the surface characteristics of a *Monge patch* in this case, the coordinates of points on the surface are given by:

$$\mathbf{r}(u,v) = [u \;\; v \;\; f(u,v)]^T \tag{19}$$

the components of metric, E, F, and G can be found as

$$\begin{aligned} E &= \mathbf{r}_u \cdot \mathbf{r}_u = 1 + f_u^2 \\ F &= \mathbf{r}_u \cdot \mathbf{r}_v = f_u f_v \\ G &= \mathbf{r}_v \cdot \mathbf{r}_v = 1 + f_v^2 \end{aligned} \tag{20}$$

Elements of the second fundamental form matrix L, M, and N are

$$\begin{aligned} L &= \mathbf{r}_{uu} \cdot \mathbf{n} = \frac{f_{uu}}{(1 + f_u^2 + f_v^2)^{1/2}} \\ M &= \mathbf{r}_{uv} \cdot \mathbf{n} = \frac{f_{uv}}{(1 + f_u^2 + f_v^2)^{1/2}} \\ N &= \mathbf{r}_{vv} \cdot \mathbf{n} = \frac{f_{vv}}{(1 + f_u^2 + f_v^2)^{1/2}} \end{aligned} \tag{21}$$

From the above quantities, Gaussian and Mean curvatures of the surface are obtained

$$K = \frac{f_{uu} f_{vv} - f_{uv}^2}{(1 + f_u^2 + f_v^2)^2} \tag{22}$$

$$H = \frac{1}{2} \frac{f_{uu} + f_{vv} + f_{uu} f_v^2 + f_{vv} f_u^2 - 2 f_u f_v f_{uv}}{(1 + f_u^2 + f_v^2)^{3/2}} \tag{23}$$

To estimate partial derivatives (in general from discrete data), the approach is to determine a continuously differentiable quadric surface (a linear combination of a set of orthonormal bases) that best fits the data on local patches of the surface. Subsequently, the derivatives are computed analytically at corresponding discrete points as a function of the coefficients of the basis functions. We will not describe the details of the method here, and refer the reader to (Besl, 1986). The advantages of the method however are in terms of speed, as the partial derivative computation can be done by convolution kernels separably, and robustness to noise due to the filtering effect of functional approximation. The disadvantage of this approach is that it smoothes over orientation and depth discontinuities.

Once the above quantities are found, we are in a position to find local displacement vectors $(\delta u(u,v), \delta v(u,v))$ at each point by finding the point on the deformed surface that minimizes the bending and the deviation from conformal stretching:

$$\begin{aligned} \iint_{\Omega_2} & \lambda_{be}((2H^2(u,v) - K(u,v))^{1/2} - \\ & (2\bar{H}^2(u+\delta u(u,v), v+\delta v(u,v)) - \bar{K}(u+\delta u(u,v), v+\delta v(u,v)))^{1/2})^2 + \\ & \lambda_{st}(\frac{E(u,v)}{\bar{E}(u+\delta u(u,v), v+\delta v(u,v))} - \frac{F(u,v)}{\bar{F}(u+\delta u(u,v), v+\delta v(u,v))})^2 + \\ & (\frac{F(u,v)}{\bar{F}(u+\delta u(u,v), v+\delta v(u,v))} - \frac{G(u,v)}{\bar{G}(u+\delta u(u,v), v+\delta v(u,v))})^2 + \\ & (\frac{E(u,v)}{\bar{E}(u+\delta u(u,v), v+\delta v(u,v))} - \frac{G(u,v)}{\bar{G}(u+\delta u(u,v), v+\delta v(u,v))})^2 du dv \end{aligned} \tag{24}$$

The above equation requires some explanation in relation with equation (18). Equation (18) assumed the corresponding points on the surface were known and arrived at a numerical value measuring the energy (with respect to our model) that was required to deform the surface. Now assume normalized coordinates both (u,v) and $(\bar{u},\bar{v})$, each of the coordinates ranging between zero and one. Then, by varying δu and δv for a fixed (u,v) as $(u+\delta u(u,v), v+\delta v(u,v))$, a disk is constructed in the parameter space corresponding to which an area on the second surface is specified. This defines the generalization of our search windows for the case of contours to surfaces. At each point on the first surface a number of possible matches on the second surface are hypothesized in this manner. The point leading to equation (24) being minimal is the found match. As in 2D, again the basic assumption is that important shape features only move within the resolution of the search window, but in addition, the search window now must accomodate for the largest possible stretching that can occur on the surface, although it would be possible to make the search window size a function of position on the first surface of the pair as suggested above. The weighting factors λ_{be} and λ_{st} weighing the bending and stretching terms specify the importance of each term. In practice, where there are important shape tokens, λ_{be} takes on a large value. In regions of the surface with less bending, the stretching term is given more importance. Assignment of values to weighting coefficients parallels the two-dimensional case. Regarding smoothing the flow vectors, we are currently investigating approaches to optimal smoothing of the 3D match vectors. A straightforward extension of the techniques given for the 2D case is certainly a possibility.

5.2. Experimental Results

Simulation results of the method are given in Figure 2. Two experiments are given here illustrating the efficacy of the algorithm in computing stretching and bending of surfaces over three frames of the four-dimensional volume. In the first experiment, bending of the surface was successfully followed by the algorithm. Note that although there were two different "tokens" present, in each case, the correct matching took place. Note however that due to the smoothing effect over the discontinuities, points close to bumps also get tracked. Surface matching based on bending energy requires an additional statement here. When there are no important landmarks that could be located, i.e., when there are ambiguities that might be present, the algorithm simply assumes minimal motion on the part of the ambiguous point, and hence finds the motion vector that has minimal length. This is the so-called *aperture* problem (Anandan, 1989) for surfaces. In the second experiment, a surface undergoes conformal stretchings. The algorithm reported an average error of 1.74 pixels from the conformal model between the first and second surfaces and 1.38 between the second and the third surfaces for the example on the left. In each case we associate these deviations from the conformal model with discretization errors.

6. CONCLUSIONS AND FUTURE RESEARCH

There are other extensions and alternative approaches to the the tracking of the LV wall that are of interest, such as tracking both endocardial and epicardial shape, and attempting to find both the bounding surfaces and their best interframe region matches simultaneously, also of interest to this work.

Validation of the method will be done with animal studies by comparing algorithm-derived motion to that found using implanted markers. Also, in order to get a handle on the general error properties of the 2-D algorithm with respect to the loss of a dimension, traced contours will be sent to the 2-D version of the shape-based flow algorithm described in the paper and a comparison will be made with the true 3D flow vectors detected following surface tesselation.

Once the regional motion has been captured as velocity vectors, the most important issue now is the question of how to optimally present this information for statistical analysis of patient data, and, ultimately, how to present the entire package of information to the clinical cardiologist for use in patient care.

At the more basic level, one avenue for further investigation will be in devising methods for incorporating non-elastic components of motion of the LV as part of the model. This is in accordance with known properties of the LV wall muscle fibers (Sagawa, 1973) which have elastic and non-elastic properties. The incorporation is ultimately necessary for improving the physically-based model presented here. Gaining a better understanding of the left-ventricular motion in terms

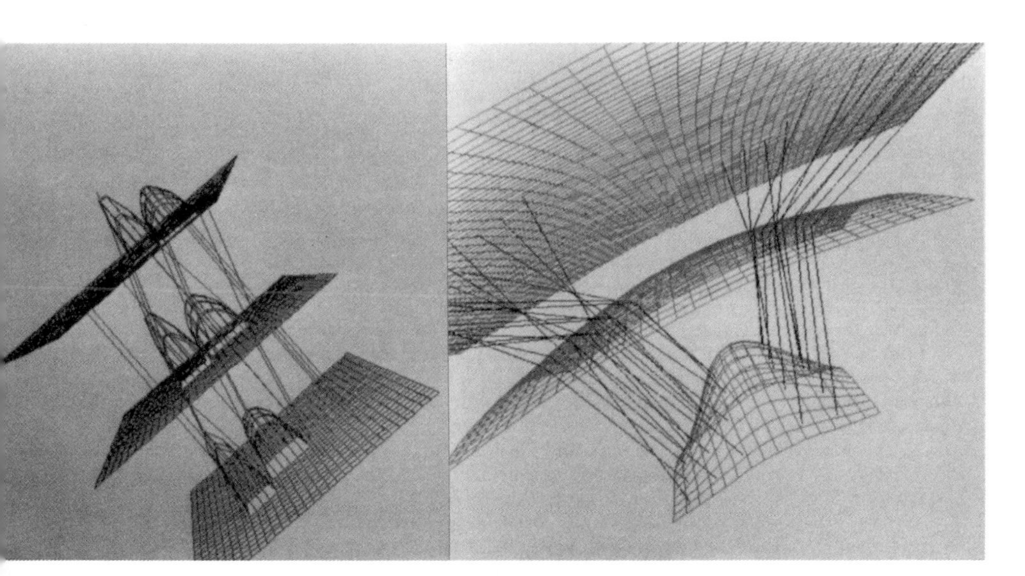

'**igure** 2: Motion computation using surfaces of a time-varying volume. Three such surfaces are shown here. Left: he algorithm's performance in detecting motion of moving bumps by minimizing the bending energy. Right: Motion f points on the surface is computed by minimizing deviations from conformal stretching. (See text).

of modelling and analysis can be both diagnostically important and at the same time provide a suitable source of information for constructing techniques useful for analyzing motion of non-rigid objects in general.

References

Aggarwal JK (1986). Motion and Time-Varying Imagery – An Overview. Proceedings of the Workshop on Visual Motion. IEEE Computer Society Press, Washington, D.C., pp. 1-6.

Amini AA, Weymouth TE and Jain R. (1990). Using Dynamic Programming for Solving Variational Problems in Vision. IEEE Transactions on Pattern Analysis and Machine Intelligence, PAMI-12:855-867.

Amini AA. (1990). Using Dynamic Programming for Solving Variational Problems in Vision: Applications Involving Deformable Models for Contours and Surfaces. PhD Thesis. The University of Michigan, Ann Arbor.

Anandan P. (1989). A Computational Framework and an Algorithm for the Measurement of Visual Motion. International Journal of Computer Vision, Vol-2:283-310.

Areeda J Garcia E Vantrain K Brown D Waxman A and Berman D. (1982). A Comprehensive Method for Automatic Analysis of Rest/Exercise Ventricular Function from Radionuclide Angiography. Digital Imaging: Clinical Advances in Nuclear Medicine. Society of Nuclear Medicine, pp. 241-256.

Axel L and Dougherty L (1989). MR Imaging of Motion with Spatial Modulation of Magnetization. Radiology, Vol-171:841-845.

Ayache N Boissonnat JD Brunet E Cohen L Chieze JP Geiger B Monga O Rocchisani JM and Sander P (1989). Building Highly Structured Volume Representations in 3D Medical Images. Proceedings of Computer-Assisted Radiology, Springer-Verlag, Berlin, pp. 765-772.

Besl P and Jain R (1986). Invariant Surface Characteristics for 3D Object Recognition. Computer Vision, Graphics, and Image Processing, Vol-30:33-80.

Bookstein F (1989). Principal Warps: Thin-Plate Splines and the Decomposition of Deformations. IEEE Transactions on Pattern Analysis and Machine Intelligence, PAMI-11:567-585.

Collins S and Skorton D (1986). Cardiac Imaging and Image Processing. McGraw Hill, New York.

Courant R and Hilbert D (1957). Methods of Mathematical Physics. Interscience, London.

Duncan JS (1987). Knowledge Directed Left Ventricular Boundary Detection in Equilibrium Radionuclide Angiocardiography. IEEE Transactions on Medical Imaging, MI-6:325-336.

Duncan JS Smeulders A Lee F and Zaret B (1988). Measurement of End Diastolic Shape Deformity Using Bending Energy. Computers in Cardiology, pp. 277-280.

Duncan JS Lee F Smeulders A and Zaret B (1991). A Bending Energy Model for Measurement of Cardiac Shape Deformity. Accepted to IEEE Transactions on Medical Imaging.

Eisenhart LP (1955). A treatize on Differential Geometry of Curves and Surfaces. Ginn and Company, Boston.

Garcia EV Van Train K Maddahi J Prigat F (1985). Quantification of Rotational Thallium-201 Myocardial Tomography. Journal of Nuclear Medicine, Vol-26:17-26.

Gelberg H Brundage B Glantz S and Parmley W (1979). Quantitative Left Ventricular Wall Motion Analysis: A Comparison of Area, Chord and Radial Methods. Circulation, Vol-59:991-1000.

Goldgof D Lee H and Huang T (1988). Motion Analysis of Nonrigid Surfaces. Proceedings of IEEE conference on Computer Vision and Pattern Recognition. IEEE Computer Society Press, Washington, D.C., pp. 375-380.

Harris L Clayton P Marshall H and Warner H (1974). A Technique for the Detection of Asynergistic Motion in the Left Ventricle. Computers and Biomedical Research, Vol-7:380-394.

Hildreth EC (1984). The Measurement of Visual Motion. MIT Press. Cambridge, Massachusetts.

Huang T (1990). Modeling, Analysis, and Visualization of Nonrigid Object Motion. International Conference on Pattern Recognition, Atlantic City, N.J., pp. 361-364.

Ingels N Daughters D Stinson E and Alderman E (1975). Measurement of Midwall Myocardial Dynamics in Intact Man by Radiography of Surgically Implanted Markers. Circulation, Vol-52:859-867.

Kreyszig E (1975). Introduction to Differential Geometry and Riemanian Geometry. University of Toronto Press, Toronto, Canada.

Landau L and Lifshitz EM (1986). Theory of Elasticity. Pergamon Press, Oxford, England.

Lanzer P Botvonick E Schiller N (1984). Cardiac Imaging Using Gated Magnetic Resonance. Radiology, Vol-150:121-127.

Links J Douglass K and Wagner H (1980). Patterns of Ventricular Emptying by Fourier Analysis of Gated Blood Pool Studies. Journal of Nuclear Medicine, Vol-21:978-982.

Mailloux GE Bleau A Bertrand M and Petitclerc R (1987). Computer Analysis of Heart Motion from Two Dimensional Echocardiograms. IEEE Transactions on Biomedical Engineering, BME-34:356-364.

Millman R and Parker G (1977). Elements of Differential Geometry. Prentice-Hall. Englewood Cliffs, New Jersey.

Nagel HH (1978). Formation of an Object Concept by analysis of Systematic Time Variation in the Optically Perciptible Environment. Computer Graphics and Image Processing, Vol-7:149-194.

Owen R Staib L Anandan P and Duncan J (1989). Measurement of Left Ventricular Wall Motion from Contour Shape Deformation. In: Information Processing in Medical Imaging. Ortendahl DA and Llacer J (eds.), Wiley-Liss,

Inc., New York, pp. 541-556.

Raichlen JS Trivedi SS Herman GT St. John Sutton MG and Reichek N (1986). Dynamic three-dimensional reconstruction of the left ventricle from two-dimensional echocardiograms. Journal of the American College of Cardiology, Vol-8:364-370.

Sagawa K (1973). The Heart as a Pump. In: Engineering Principles in Physiology, II. Brown and Gann (eds), Academic Press, New York. pp. 101-126.

Sheehan FH Bolson EL Dodge HT Mathey DG Schofer J and Woo HW (1986). Advantages and Applications of the Centerline Method for Characterizing Regional Ventricular Function. Circulation, Vol-74:293-305.

Slager C Hooghoudt T Serruys P Schuurbiers J Reiber J Meester G Verdouw P and Hugenholtz R (1986). Quantitative Assessment of Regional Left Ventricular Motion Using Endocardial Landmarks. JACC, Vol-7:317-326.

Staib L and Duncan JS (1989). Parametrically Deformable Contour Models. Computer Vision and Pattern Recognition. IEEE Computer Society Press, Washington, D.C. pp. 98-103.

Terzopoulos D (1987). Elastically Deformable Models. ACM Transactions on Computer Graphics, Vol-27:205-214.

Ullman (1979). Interpretation of Visual Motion. MIT Press, Cambridge, MA.

Zerhouni E Parish D Rogers W Yang A and Shapiro P (1988). Tagging of the Human Heart by Multiplanar Selective RF Saturation for the Analysis of Myocardial Contraction. Abstracts of the Annual Meeting of the Society of Magnetic Resonance in Imaging, San Francisco, California, page 10.

MODEL-BASED RECOGNITION OF MULTIPLE DEFORMABLE OBJECTS USING A GAME-THEORETIC FRAMEWORK

H. Işıl Bozma,* and James S. Duncan*,†

Departments of *Electrical Engineering and †Diagnostic Radiology,
Division of Imaging Science
Yale University, New Haven, CT 06510

Abstract

Image analysis systems aimed at robust segmentation via model-based recognition of multiple deformable objects require integration of a variety of modules. One approach to modularity and integration is based on decomposing the task into a set of coupled modules with their corresponding objectives, and then integrating them within a unifying framework. Although both theoretical and computational considerations indicate the importance of adhering to the multiple coexisting nature of the objectives, previous approaches to integration have been limited in this aspect. In (Bozma, 1990), an integration framework which overcomes this problem, is developed based on game-theoretic concepts. The contribution of this paper is to describe a segmentation system aimed at model-based recognition of multiple deformable objects, in which the modules are integrated using the game-theoretic approach. A secondary contribution is that the model used in object recognition, which is defined as a set of independent, yet coupled deformable objects, constitutes an extension of the previous models. Our experiments with several biomedical images using anatomical atlases as the basis for these models demonstrate the power of this system in medical applications.

Keywords

image processing systems; sensor fusion; nash equilibrium; parallel algorithms and architectures

1. INTRODUCTION

Accurate segmentation of a set of anatomical structures of interest is a common goal in medical image analysis since it can enable detailed study of the various properties of these structures. Segmentation has often been frustrated by the robustness of the final interpretations due to the underconstrained and complex nature of the problem. In order to overcome these difficulties, it has been suggested that a system aimed at segmentation should feature (Ballard and Brown, 1982):

- Segmentation via model-based recognition.
- Modularity and integration.

1.1. Segmentation via Model-Based Recognition

In order to overcome the problem of multiple interpretations - which result from the underconstrained nature of segmentation problem, it has been suggested that imperfect image data should be complemented by the a priori knowledge provided by a model (Ballard and Brown, 1982).

Model-based object recognition considers the segmentation problem as the matching of a model to the image-derived data such as edges via i.) constructing a model of objects based on a set of descriptors such as those of 2-D shape, ii.) defining a matching metric to assess the match of the model to the image-derived data by measuring how well the descriptors of the model correspond to the data, and finally iii.) modifying the descriptors as to optimize this matching metric. In order to deal with the diversity and irregularity of the anatomical shapes, this approach has been extended to segment deformable objects. Staib (1990) presents a model for deformable object recognition based on an object matching metric defined as the correlation of the deformable template with the image-derived data. Bajcsy and Kovacic (1989) apply a model of elastic matching to perform object matching of the brain atlas with CT brain images. A second extension to model-based matching has been to augment the match metric to incorporate relational (structural) matching as well. Fischler and Elschlager (1973) define their model as a set of rigid components held together by "springs", where the springs joining the rigid pieces constrain the relative movement of the components and measure the cost of the deformation required to match this model to the image-derived data.

1.2. Modularity and Integration

In order to i.) reduce the computational complexity of the image analysis required for segmentation, ii.) perform robustly in noisy images via redundancy of processing and iii.) better model the perceptual process, modularity and integration has proved to be essential (Hoff and Ahuja, 1989), (Aloimonos, 1989). Moreover, it has been suggested that such a system will exhibit synergistic behavior by overcoming the weaknesses of the individual modules. For example, in order to overcome the inherent limitation of each imaging modality to provide reliably only certain types of information, a system integrating different imaging modalities could compensate for the limitations of each individual imaging modality. Such a system could complement the measurement of certain functional information obtainable from positron emission tomography (PET), which is limited in its ability to provide accurate anatomical information, with the reliable anatomical information of magnetic resonance imaging (MRI) (Evans et al., 1988).

One approach to modularity and integration has been decomposing the task into several functional modules and then integrating them together within an unifying framework. Hence, the modular components are described systematically as follows (Bozma and Duncan, 1990): Each module $i \in 1, .., N$ is associated with an objective F_i, which is a function of both its output vector $\mathbf{p}_i$ and that of the other modules $\mathbf{p}_j, j \neq i$, which it needs to optimize with respect to its output $\mathbf{p}_i$. Once the task is decomposed into N functional modules and consequently N coupled objectives, integration of these modules can be achieved by prescribing how to link the objectives together. Examination of the previous approaches to integration has shown them to be limited in their ability to adhere to the *multiple coexisting nature* of the objectives of the system, which has several shortcomings. First, the objectives of such a system could potentially be different from the original multiple coupled objectives which will be discussed briefly later in this paper. A second disadvantage may be induced by computational considerations; namely it might be harder i.) to fully automate these approaches since they might need human monitoring with respect to coordination of decisions between the modules, or ii.) to map them to decentralized algorithms employing iterative techniques in order to reduce the computational complexity. Both of these limitations will have significant ramifications for robust, automated, and fast segmentation.

1.3. Our Approach to Segmentation

Our system incorporates both of the features: model-based recognition, and modularity and integration. In order to reduce the underconstrained nature of the problem, segmentation is based on model-based recognition. One of the contributions of this paper is to further extend model-based recognition by combining deformable object matching with relational matching, and using this model as a more versatile basis for guiding the process of matching to the image-derived data. This model - which will be referred to as the flexible model - is defined to be a set of independent deformable objects coupled together by loose (inequality) constraints. Even with an extended model such as the flexible model, the noise and ambiguity present in the image-derived data may not be easily interpretable or even reliable, and thus are potential causes of confusion and error to the matching process (Staib, 1990). In order to alleviate such problems, we follow the second principle and decompose the task into multiple modules and then integrate them together within an appropriate framework, which then corresponds to the image analysis system. In (Bozma and Duncan, 1990), a new approach to integration of vision modules which overcomes the limitations of the previous approaches and preserves the multiple coexisting nature of the objectives, is presented. This new approach is based on game-theoretic framework. In this model, a correspondence between a vision system and a N-player game is established, and then game-theoretic concepts are applied to solve the problem of multiple coexisting objectives. The primary contribution of this paper is to describe a vision system aimed at robust segmentation via model-based multiple deformable object recognition, in which the functional modules are integrated based on the game-theoretic approach. It is felt that this design is particularly applicable to biomedical images, where the objects (organs) have a continuum of shapes, have *a priori* models such as anatomical atlases and and the images are often noisy.

In the next section, the decomposition of the system into the two modules and their corresponding functional objectives are described. Section 3 examines the issues related to integration of the two modules, and presents both the previous approaches and the game-theoretic approach to solving the integration problem, which have been discussed extensively in (Bozma and Duncan, 1990). Finally, we demonstrate the power of the model-based deformable object system designed within the game-theoretic framework in a comparative experimental study with the previous approaches. Our applications, which involve a variety of nuclear medicine and MRI images, base their models on a variety of atlases of various structures, and hence illustrate that the robustness and reliability of the system make it attractive to serve as a tool for extracting both anatomical and structural information for diagnostic purposes.

2. MODEL-BASED RECOGNITION SYSTEM - MODULES

The object image to be segmented is obtained via a camera or other imaging techniques such as MRI, and input into the system. The system processes this signal or other signal-derived data in its various modules, which have been previously designed based on the decomposition of the task. In our system, the task of model-based recognition of multiple deformable objects is decomposed into two distinct modules: module 1 has the objective of inferring a discrete edge vector field from the two-dimensional image signal, alleviating any noise and other ambiguity causing effects and improving local curvilinearity, and module 2 has the objective of improving the shape descriptors provided by the model as to increase the match of the flexible model to the image-derived data. The flexible model provides only an initial estimate for the shape descriptors of the objects in the image. However, because the objects have deformable shapes and are only loosely coupled, there is still a continuum of possibilities near the initial shape descriptors provided by the model. Consequently, the model descriptors may not be correctly describing the objects of the image as is the case in Figure 2(right). Thus, module 2 needs to determine a new set of shape descriptors which in turn correctly depict the objects in the image. As the raw image signal is too complex to analyze directly, image-derived data such as edge vector field - which emphasizes specific features implicit in the raw signal and thus reduces the complexity of analysis - is deemed to be more appropriate to guide the segmentation process. However, due to the noise and various sources of ambiguity, even the image-derived data, which is the edge vector field in our case, will yield confusing information initially, and needs to be further processed. Hence, the objective of the first module is defined as improving the edge vector field.

Each module $i, i = 1, 2$ is then associated with an output and an objective F_i. In order to describe the modular components of the system precisely, the following must be specified: i.) the outputs of each module, ii.) the objective of each module, iii) the overall objective of the system. The output of each module are its decision variables. The objective of each module measures how well the output variables satisfy its a priori defined task. The overall objective of the system specify how the individual outputs and objectives relate to the overall task of the system. We will now discuss each for our 2-module system.

2.1. Outputs of Modules

The output of module 1 is an edge vector field, and the output of module 2 is a set of 2-D shape descriptors and a set of relations between pairwise shape descriptors. In the rest of this section, we will define each precisely.

The edge vector field represents the image data. Let $I_1 = \{(x_i, y_i) \mid i = 1, .., n_1\}$ be the discrete field corresponding to the image array. The incoming 2-D image signal $E(x_i, y_i)$ is passed through a series of filters which detect transition regions between two relatively uniform intensity areas. There are various available filters, ranging from simple differential high-pass operators to more sophisticated optimal filters such as noise compensating and predictive filters (Abdou and Pratt, 1979),(Canny, 1986), which are then applied to the 2-D image signal via discrete approximations. As a result of these operations, each point $i \in I_1$ is associated with a vector by $\mathbf{p}_{1_i} \in R^2$, which is an estimate of the brightness gradient $(\partial E/\partial x, \partial E/\partial y)$. As these transition regions commonly correspond to object boundaries, the vector $\mathbf{p}_{1_i} \in R^2$ specifies the magnitude and direction of an edge that is hypothesized to pass through that point and is referred to as the edge vector. The edge vector field $\mathbf{V} : I_1 \rightarrow R^2$ is formally defined as $\mathbf{V}(i) = (i, V(i))$ where $V(i) \triangleq \mathbf{p}_{1_i}$. Corresponding to each vector field, there exists a unique vector of edges $\mathbf{p}_1 \in R^{2n_1}$ which is formed by concatenating the vectors of all the points $i \in I_1$ as [1,2] $\mathbf{p}_1^T \triangleq \left[\mathbf{p}_{1_1}^T \cdots \mathbf{p}_{1_{n_1}}^T \right]$. For the the 78×64 image of Figure 1(right), the initial vector field corresponding to $\mathbf{p}_1^0 \in R^{2(78 \times 64)}$ is shown in Figure 2(left).

A set of 2-D shape descriptors represent the segmented objects. Let $I_2 = \{1, .., n_2\}$ be the set of objects in this model. Each object is defined by its 2-D closed contour, which is parametrically defined based on the truncated elliptic Fourier decomposition of the contour (Kuhl and Giardina, 1982), (Lin and Hwang, 1987). In this representation, the contour is represented by a parametrized curve $\alpha_{\mathbf{p}_{2_j}} : [0, 2\pi] \rightarrow R^2$, $j \in I_2$, consisting of two periodic functions as $\alpha_{\mathbf{p}_{2_j}}(s) = (x_j(s), y_j(s))$, which can in turn be represented by its Fourier expansion as:

$$\begin{bmatrix} x_j(s) \\ y_j(s) \end{bmatrix} = \begin{bmatrix} a_{j_0} \\ d_{j_0} \end{bmatrix} + \sum_{h=1}^{H} \begin{bmatrix} a_{j_h} & b_{j_h} \\ c_{j_h} & d_{j_h} \end{bmatrix} \begin{bmatrix} \cos hs \\ \sin hs \end{bmatrix}$$

The first two coefficients a_{j_0} and d_{j_0} are the mean values of $x(s)$ and $y(s)$ respectively, and indicate the position of the object.

$$a_{j_0} = \frac{1}{2\pi} \int_0^{2\pi} x_j(s) ds$$
$$d_{j_0} = \frac{1}{2\pi} \int_0^{2\pi} y_j(s) ds$$

The rest of the coefficients corresponding to the H harmonics represent its shape, where each har-

[1] In this paper, $\mathbf{p}_i$ refers to the output vector of the i th module. If $\mathbf{p}_i$ is formed by concatenating a set of vectors, $\mathbf{p}_{i_j}$ refers to the j th vector component of $\mathbf{p}_i$. Finally, $p_{i_{j_n}}$ refers to the n th component of $\mathbf{p}_{i_j}$.

[2] In v^T, superscript T indicates the transpose of the vector v.

monic is described by 4 coefficients:

$$
\begin{aligned}
a_{j_h} &= \frac{1}{\pi}\int_0^{2\pi} x_j(s)\cos(hs)ds \\
b_{j_h} &= \frac{1}{\pi}\int_0^{2\pi} x_j(s)\sin(hs)ds \\
c_{j_h} &= \frac{1}{\pi}\int_0^{2\pi} y_j(s)\cos(hs)ds \\
d_{j_h} &= \frac{1}{\pi}\int_0^{2\pi} y_j(s)\sin(hs)ds
\end{aligned}
$$

Thus, each 2-D object $j \in I_2$ can be represented by a vector of the elliptic Fourier descriptors - which correspond to the coefficients of the Fourier decomposition of the 2-D boundary. This vector of shape descriptors will be called the shape descriptor and denoted by $\mathbf{p}_{2_j} \in R^{4H+2}$ as defined in Eq. (1).

$$
\mathbf{p}_{2_j}^T = \left[\begin{array}{cccccccccccc} a_{j_0} & d_{j_0} & a_{j_1} & b_{j_1} & c_{j_1} & d_{j_1} & .. & .. & .. & a_{j_H} & b_{j_H} & c_{j_H} & d_{j_H} \end{array} \right] \quad (1)
$$

This model is described in detail in (Staib, 1990). Corresponding to the set of objects of the model, there is a unique vector of shape descriptors $\mathbf{p}_2 \in R^{n_2(4H+2)}$, which is obtained via concatenating the vectors $\mathbf{p}_{2_j}$ of all objects $j \in I_2$ as $\mathbf{p}_2^T \triangleq \left[\mathbf{p}_{2_1}^T \cdots \mathbf{p}_{2_{n_2}}^T \right]$. The initial vector of shape descriptors $\mathbf{p}_2^0$ is determined a priori based on general domain considerations. For the image of Figure 1(right), the model contains two objects ($n_2 = 2$) which are represented by 10 harmonics ($H = 10$), and the 2-D contours corresponding to the shape descriptors $\mathbf{p}_2^0 \in R^{2((4\times 10)+2)}$ of the model are shown in Figure 2(middle).

The relations among the objects correspond to pairwise coupling of the objects based on their relative geometric properties. In our system, three geometric properties- horizontal and vertical positioning and size - are considered as the basis for relating objects, but they could be easily expanded to include other properties as well. These geometric properties are then expressed in relative terms for any pair of objects $j \in I_2$ and $k \in I_2$ based on their respective shape descriptors $\mathbf{p}_{2_j}$ and $\mathbf{p}_{2_k}$ as follows:

$$
\begin{aligned}
\text{Relative horizontal positioning} &\triangleq p_{2_{j_1}} - p_{2_{k_1}} \\
\text{Relative vertical positioning} &\triangleq p_{2_{j_2}} - p_{2_{k_2}} \\
\text{Relative size}^3 &\triangleq \frac{\left(\sum_{h=1}^{H} h p_{2_{k_{4h+3}}} p_{2_{k_{4h+6}}} - h p_{2_{k_{4h+4}}} p_{2_{k_{4h+}}}\right)^2}{\left(\sum_{h=1}^{H} h p_{2_{j_{4h+3}}} p_{2_{j_{4h+6}}} - h p_{2_{j_{4h+4}}} p_{2_{j_{4h+5}}}\right)^2}
\end{aligned}
$$

2.2. Objectives of Modules

After each module i is associated with an output vector, it is then associated with a functional F_i, which mathematically represents the qualitative objective of the module. The qualitative objective of the first module is to improve the edge vector field by reinforcing local curvilinearity and removing noise. The qualitative objective of the second module is to find the shape descriptors which improve the match of the objects - whose shapes are represented parametrically by the shape descriptors - to the edge vector field.

2.2.1. Reinforcing Curvilinearity and Removing Noise in the Edge Vector Field

The initial edge vector field $\mathbf{p}_1^0$ provides only an initial estimate of where the edges are potentially located. Module 2 could proceed to correlate the model directly with the initial edge vector

[3] It can be shown that absolute value of the term in $(\cdot)$ of the numerator and the denominator corresponds to k th and j th objects' areas respectively.

field; i.e. is passed on directly to the module 2 without any attempts to improve it. However, due to the presence of noise and ambiguity that might be present in the imaging process as well as in the domain of application, a better idea is to improve the initial vector field (Zucker et al., 1988). One method of achieving this is to use an a priori model such as local curvilinearity, define a match metric to measure how well the initial edge vector field satisfies this model, and then modify the edge vector field in order to improve this match similar to (Hummel and Zucker, 1983), (Duncan and Birkholzer, 1989). The model of local curvilinearity requires that curves should be locally curvilinear, thin on average and isolated edges should be suppressed. The essential form of the match metric $f_1 : R^{2n_1} \rightarrow R$ measures the agreement between each edge vector $\mathbf{p}_{1_i}$ at each point i and its local context $\mathbf{s}_i(\mathbf{p}_1)$, which is a support function based on the model of local curvilinearity, as follows:

$$f_1(\mathbf{p}_1) \;=\; \sum_{i \in I_1} \mid \mathbf{p}_{1_i} - \mathbf{s}_i(\mathbf{p}_1) \mid^2$$

Let $\mathbf{p}_1^*$ represent the improved final vector of edges. The objective of module 1 is defined to find $\mathbf{p}_1^*$ as:

$$\mathbf{p}_1^* \;\in \arg\ \min_{\mathbf{p}_1} F_1(\mathbf{p}_1) \quad \text{where} \quad F_1(\mathbf{p}_1) \triangleq\; f_1(\mathbf{p}_1)$$

Since, in general, F_1 will be multimodal, the solution of interest will be dependent on the initial vector of edges $\mathbf{p}_1^0$.

2.2.2. Model-Based Segmentation of Multiple Deformable Objects

Model-based segmentation of multiple deformable objects is based on constructing a matching metric based on the flexible model which models the objects of segmentation as a set of independent deformable objects mutually coupled together by loose constraints. The flexible model is constructed based on domain-dependent considerations. For example, a flexible model of the heart can be constructed based on the the heart atlas which represents a 3-dimensional model of a healthy heart by providing geometrical, anatomical and functional information about each part of the heart as well as the relations among its parts. Such a flexible model is used to recognize the the right and left ventricles in the image of Figure 1. The matching metric, which is constructed based the flexible model comprises of two parts: i.) relational(structural) matching metric and ii.) a deformable object matching metric. This approach constitutes an extension of the two previous methods.

In order to evaluate the match of the the relations among a given set of shape descriptors with the existing relational constraints among the objects in the domain, a relational matching metric is used. The relations are specified in terms of geometric properties. As previously discussed, our model incorporates three relative geometric properties: horizontal and vertical positioning and size. Suppose that the domain considerations allow a range of possibilities, so that the geometric properties of each pair of objects are somewhere between the extremes of being totally unrelated and of being subject to fixed, hard constraints, and hence can be expressed as a set of inequality equations for each pair of objects j and k:

$$\begin{array}{llll}
\mu_{jk1} \leq & p_{2_{j_1}} - p_{2_{k_1}} & \leq \nu_{jk1} & \mu_{jk1}, \nu_{jk1} \in R \\
\mu_{jk2} \leq & p_{2_{j_2}} - p_{2_{k_2}} & \leq \nu_{jk2} & \mu_{jk2}, \nu_{jk2} \in R \\
\mu_{jk3} \leq & \dfrac{\left(\sum_{h=1}^{H} h(p_{2_{k_{4h+3}}} p_{2_{k_{4h+6}}} - p_{2_{k_{4h+4}}} p_{2_{k_{4h+}}})\right)^2}{\left(\sum_{h=1}^{H} h(p_{2_{j_{4h+3}}} p_{2_{j_{4h+6}}} - p_{2_{j_{4h+4}}} p_{2_{j_{4h+5}}})\right)^2} & \leq \nu_{jk3} & \mu_{jk3}, \nu_{jk3} \in R
\end{array}$$

The relations among the objects will arise from domain-dependent considerations. For example, in a flexible model of the heart, such relative geometric properties exist between the different anatomical parts of the heart.

Based on these relational constraints, one way of defining a relational matching metric is via measuring how well each inequality is satisfied. Thus, a function $c_{jkg} : R \rightarrow R$ is defined for each pair of objects $j \in I_2$ and $k \in I_2$ and for each geometric property $g, g = 1, 2, 3$, which penalizes deviations from the bounds of the inequality relations:

$$c_{jkg}(x) = \begin{cases} -(x-\mu_{jkg})^3 & \text{if } x \leq \mu_{jkg} \\ 0 & \text{if } \mu_{jkg} < x < \nu_{jkg} \\ (x-\nu_{jkg})^3 & \text{if } x \geq \nu_{jkg} \end{cases}$$

The relational matching metric, which is denoted by $f_2(\mathbf{p}_2) : R^{(4H+2)n_2} \rightarrow R$, is defined for all pairs of objects and geometric properties as:

$$f_2(\mathbf{p}_2) = \sum_{j,k \in I_2} c_{jk1}(p_{2_{j_1}} - p_{2_{k_1}}) + c_{jk2}(p_{2_{j_2}} - p_{2_{k_2}}) + c_{jk3}\left(\frac{\left(\sum_{h=1}^{H} h(p_{2_{k_{4h+3}}} p_{2_{k_{4h+6}}} - p_{2_{k_{4h+4}}} p_{2_{k_{4h+}}})\right)^2}{\left(\sum_{h=1}^{H} h(p_{2_{j_{4h+3}}} p_{2_{j_{4h+6}}} - p_{2_{j_{4h+4}}} p_{2_{j_{4h+5}}})\right)^2}\right)$$

In order to evaluate how well the set of the shape descriptors in the flexible model match those implicit in edge vector field, a deformable object matching metric is used. For example, in a flexible model of the heart, the segmented anatomical structures as represented by the shape descriptors need to match their counterparts which are implicit in the edge data information.

The deformable object metric $f_{12} : R^{2n_1 \times R^{(4H+2)n_2}} \rightarrow R$ is based on a measure of correlation between the 2-D contours corresponding to the model shape descriptors $\mathbf{p}_2$ and the vector field as specified by $\mathbf{p}_1$. Its definition is given in Eq. (2) where the summation is over all th objects $j \in I_2$, $\mathbf{p}_{2_j}$ denotes the subvector of $\mathbf{p}_2$ representing j th object's 2-D shape descriptors, the path of integration is defined by the parametrized curve $\alpha_{p_{2_j}}$ corresponding to each object's contour, $\mid V(\alpha_{p_{2_j}}(s)) \mid$ is the norm of the edge vector $\mathbf{p}_{1_i}$ associated with the point $i = \alpha_{p_{2_j}}(s)$ along the contour and $l \subset [0, 2\pi]$.

$$f_{12}(\mathbf{p}_1, \mathbf{p}_2) = \sum_{j \in I_2} \int_0^{2\pi} \mid V(\alpha_{p_{2_j}}(s)) \mid^2 + \int_{-l}^{0} \mid \alpha_{p_{2_j}}(s+l) - (\alpha_{p_{2_j}}(s) - V(\alpha_{p_{2_j}}(s))) \mid^2 dl + \int_0^{l} \mid \alpha_{p_{2_j}}(s+l) - (\alpha_{p_{2_j}}(s) + V(\alpha_{p_{2_j}}(s))) \mid^2 dl ds \quad (2)$$

Let us remark that the object matching metric, being a function of the outputs of both of the modules, accomplishes bottom-up interlevel coupling effectively.

The matching metric can then be defined based on the weighted averaging of the relational matching metric and the object matching. Let $\mathbf{p}_2^*$ represent the improved final vector of shape descriptors. The objective of module 2 is then defined as to find $\mathbf{p}_2^*$:

$$\mathbf{p}_2^* \in \arg\min_{\mathbf{p}_2} F_2(\mathbf{p}_1, \mathbf{p}_2) \quad \text{where} \quad F_2(\mathbf{p}_1, \mathbf{p}_2) \stackrel{\triangle}{=} [\gamma_2 f_2(\mathbf{p}_2) + \gamma_{12} f_{12}(\mathbf{p}_1, \mathbf{p}_2)]$$

Let us note that γ_1 and γ_1 and γ_{12} are *a priori* determined weights. In our system, they are taken to be $\gamma_1 = 0.5$ and $\gamma_2 = 0.5$ Since, in general, F_2 will be multimodal, the solution of interest will be dependent on the initial vector of edges $\mathbf{p}_1^0$ and the initial vector of shape descriptors $\mathbf{p}_2^0$.

2.3. Overall Objective of 2-Module System

The overall objective of the system is then qualitatively specified as follows: Given the edge vector field and the flexible model, improve the edge vector field by reinforcing local curvilinearity and removing noise, and find the shape descriptors which improve the match of the flexible model to the edge vector field. Hence, the system comprises of two vision modules and two *coexisting* objectives as defined in Eqs. (3).

$$\begin{aligned} &\text{Module 1:} \quad \mathbf{p}_1^* \in \arg\min_{\mathbf{p}_1} F_1(\mathbf{p}_1) \\ &\text{Module 2:} \quad \mathbf{p}_2^* \in \arg\min_{\mathbf{p}_2} F_2(\mathbf{p}_1, \mathbf{p}_2) \end{aligned} \quad (3)$$

Let us remark that, in general, module 1 could have a term including both $\mathbf{p}_1$ and $\mathbf{p}_2$ in its functional objective, which would correspond to the coupling from module 2.

3. INTEGRATION OF MODULES

Once modules are defined by functional objectives, integration is then mathematically equivalent to selecting an appropriate method for solving a set of coupled objectives. For ease of exposition, a 2-module system is considered. Also, our formulations are for the general case where each module's objective has coupling terms from other modules. One of the previous approaches to integration of vision modules is based combining multiple objectives into a single objective (Horn, 1989), (Aloimonos, 1989), (Gennert and Yuille, 1988). Let us refer to this approach as **System I**. This system can be mathematically described as:

$$\textbf{System I:} \qquad \mathbf{p}_1^*, \mathbf{p}_2^* \in \arg \min_{p_1, p_2} [\beta_1 F_1(\mathbf{p}_1, \mathbf{p}_2) + \beta_2 F_2(\mathbf{p}_1, \mathbf{p}_2)]$$

$\beta_1, \beta_2 \in R$ are the relative weights of each objective. A second approach assumes that the objectives of the modules are decoupled. In these systems, the flow of information among the vision modules has been sequential. Each vision module has to first complete its computation, resulting in a solution to its specified objective, before propagating this solution to all modules that will be subsequently using it (Binford, 1982), (Pratt, 1978). Let us refer to this system as **System II**. For example, in a strict bottom-up approach, module 1 finds the output $\mathbf{p}_1^*$ that best satisfies its objective F_1, and then this output is passed on to the next module which then tries to find the output vector $\mathbf{p}_2^*$. that best satisfies its objective F_2. Let us note that the objective F_1 of the first module is no longer a function of $\mathbf{p}_2$ since this is a strictly bottom-up approach. This system can be described mathematically as:

$$\textbf{System II:} \qquad \begin{aligned} \mathbf{p}_1^* &\in \arg \min_{p_1} F_1(\mathbf{p}_1) \\ \mathbf{p}_2^* &\in \arg \min_{p_2} F_2(\mathbf{p}_1^*, \mathbf{p}_2) \end{aligned}$$

The problem with these approaches is that they are based on restrictive assumptions such as that coexisting objectives can be subsumed under a single global objective or that they can be satisfied in a decoupled manner. They do not preserve the multiple coexisting nature of the objectives, and thus have several limitations as previously discussed. A new model of integration, which adheres to the multiple coexisting nature of the objectives and thus overcomes these limitations, is introduced in (Bozma and Duncan, 1990) as:

$$\textbf{System III: Game-Theoretic System} \qquad \begin{aligned} \mathbf{p}_1^* &\in \arg \min_{p_1} F_1(\mathbf{p}_1, \mathbf{p}_2) \\ \mathbf{p}_2^* &\in \arg \min_{p_2} F_2(\mathbf{p}_1, \mathbf{p}_2) \end{aligned}$$

This proposed model is based on game-theoretic setting in which a correspondence between a vision system consisting of N modules and a N-player game is established. After this correspondence, a N-module vision system can be described as a N-player game where each module is viewed as a single player and game-theoretic concepts become applicable for the N-module vision system. It is shown that a game-theoretic model of integration provides a powerful control structure by preserving the multiple and coupled nature of the objectives of the system as follows: Due to the coexistence of multiple objectives, a definition of the solution must be stated. Based on noncooperative game theory, the concept of a Nash equilibrium is a natural one for interconnected vision modules posed as a N-player game (Başar, 1987). Its mathematical definition is as follows (Nash,1958):
Nash Equilibrium: Let $F_i(\mathbf{p}_1, ..\mathbf{p}_N)$, $i = 1, .., N$ be the objectives for the N players respectively. If $\mathbf{p}^* = \{\mathbf{p}_1^*, .., \mathbf{p}_N^*\}$ is a Nash equilibrium of the game, then for each player $i = 1, .., N$

$$\mathbf{p}_i^* \in \arg \min_{p_i} F_i(\mathbf{p}_1^*, .., \mathbf{p}_{i-1}^*, \mathbf{p}_i, \mathbf{p}_{i+1}^*, .., ..\mathbf{p}_N^*)$$

where $\mathbf{p}_i$ is any admissible output for player i. Nash equilibrium is the optimal strategy for each of the players on the assumption that all of the other players are holding fast to their Nash outputs. Based on the successive iterations procedure previously developed, the computational description of the system is [4]:

$$\begin{aligned}
\mathbf{p}_1^t &= \mathbf{p}_1^{t-1} - \tau_1 \mathrm{D}_{p_1} F_1(\mathbf{p}_1, \mathbf{p}_2^{t-1})_{p_1^{t-1}} \\
&= \mathbf{p}_1^{t-1} - \tau_1 \mathrm{D}_{p_1} f_1(\mathbf{p}_1) \\
\mathbf{p}_2^t &= \mathbf{p}_2^{t-1} - \tau_2 \mathrm{D}_{p_2} F_2(\mathbf{p}_1^{t-1}, \mathbf{p}_2)_{p_2^{t-1}} \\
&= \mathbf{p}_2^{t-1} - \tau_2 \gamma_2 \mathrm{D}_{p_2} f_2(\mathbf{p}_2) - \tau_2 \gamma_{12} \mathrm{D}_{p_2} f_{12}(\mathbf{p}_1^{t-1}, \mathbf{p}_2)_{p_2^{t-1}}
\end{aligned}$$

τ_1, τ_2 are *a priori* determined step sizes with theoretical upperbounds. At each instant, each module i tries to find a new output $\mathbf{p}_i^t$ that best satisfy its objective F_i based on the previous set of outputs $\mathbf{p}^{t-1} = \{\mathbf{p}_1^{t-1}, \mathbf{p}_2^{t-1}\}$. It then updates its former output $\mathbf{p}_i^{t-1}$ to this new output $\mathbf{p}_i^t$ and broadcasts this to all other vision modules. The game finishes when all the vision modules have converged to their optimal strategies. The set of optimal strategies corresponds to the Nash solution, which is the best solution in the sense that it is secure against any attempts by any player to unilaterally alter its output.

4. EXPERIMENTS

In this section, we will present an comparative study that demonstrates experimentally some of the advantages of the game-theoretic approach to integration. The study was performed in the

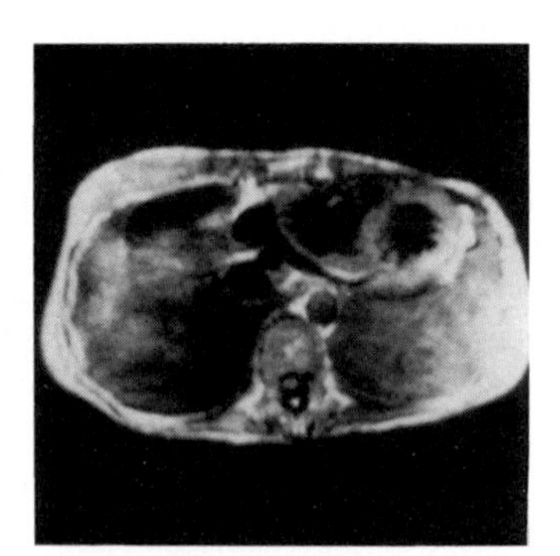

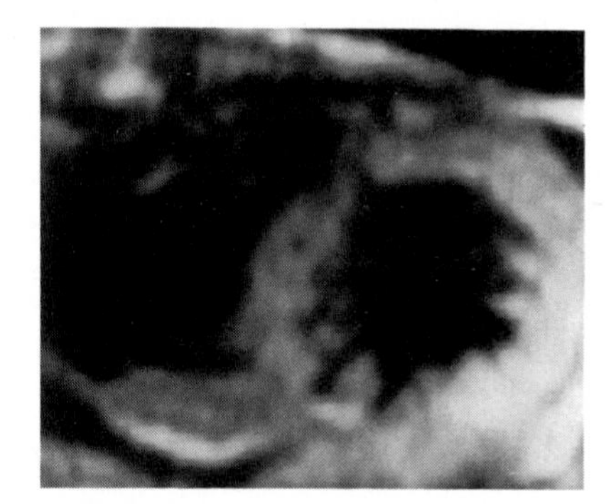

Figure 1: Magnetic Resonance Transaxial Cardiac Image Example; Left: 256 × 256 MRI Transaxial Cardiac Image; Right: 78 × 64 Subsection of the Cardiac Image.

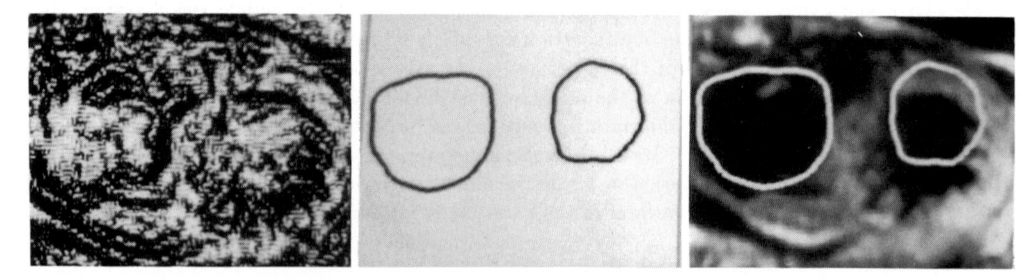

Figure 2: Magnetic Resonance Transaxial Cardiac Image Example: Initial; Left: Initial edge vector field; Middle: Initial 2-D Shapes; Right: Initial shapes on image.

[4] In this paper, $D_p(\cdot)_{p^t}$ denotes the gradient of $(\cdot)$ with respect to p evaluated at p^t.

context of *three different* 2-D deformable object recognition systems, all of which had the same task and consisted of identical modules, but differed in their integration mechanisms. These systems are respectively referred to as System I, System II and System III-Game-Theoretic System, as discussed in the preceeding section.

The first example is a 78×64 image of the left and right ventricle in Figure 1(right) from a 256×256 magnetic resonance transaxial cardiac image of Figure 1(left). The initial edge vector field, represented by $\mathbf{p_1}^0 \in R^{2(78\times64)}$ and computed using the Sobel edge operator, is seen in Figure 2(left). The initial 2-D shapes of the left and right ventricle corresponding to the initial estimate of shape

Figure 3: System I Results - Magnetic Resonance Transaxial Cardiac Image; Left: Final edge vector field; Middle: Final 2-D Shapes; Right: Final shapes on image.

Figure 4: System II Results - Magnetic Resonance Transaxial Cardiac Image; Left: Final edge vector field; Middle: Final 2-D Shapes; Right: Final shapes on image.

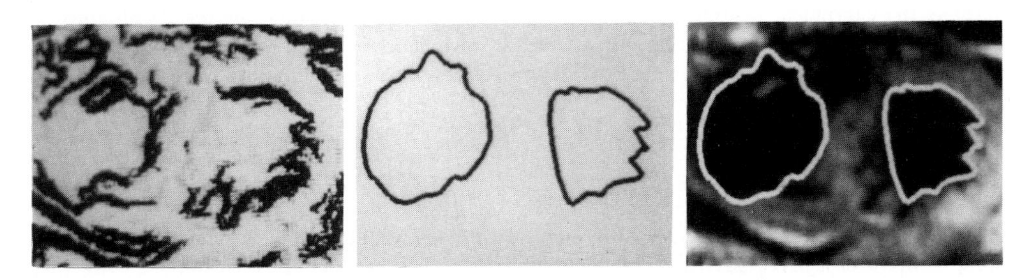

Figure 5: Game-Theoretic System Results - Magnetic Resonance Transaxial Cardiac Image; Left: Final edge vector field; Middle: Final 2-D Shapes; Right: Final shapes on image.

parameters $\mathbf{p_2}^0 \in R^{2(42)}$ as shown in Figure 2(middle) do not correctly represent the contours of actual two organs as can be observed in Figure 2(right). As seen in Figure 3 and Figure 4, both System I and System II fail completely at their tasks. Finally, Figure 5 show the results obtained by

Game-Theoretic System. We observe that the trabeculations in the left endocardial surface of the left ventricle are not picked up as accurately as the ones in the right endocardial surface of the same ventricle. This can be explained by computational considerations. Our experience has shown that higher harmonics are adjusted locally and the convergence is very slow as the the higher harmonics of the elliptic Fourier representation are computed.

The next example is that of 96×76 magnetic resonance sagital image of corpus callosum and midbrain shown in Figure 6(right). The initial edge vector field is shown in Figure 7(upper left) while the initial (model) contours are shown in Figure 7 (upper right). Again, the initial model objects do not correctly describe the underlying anatomical structures in the MRI image. As can be observed from Figure 8 and Figure 9, the accuracy of the System II in accurately delineating the corpus callosum and the midbrain is considerably less than that of the Game-Theoretic System. System II segments part of the fornix as being part of corpus callosum, while System III distinguishes between the two anatomical structures.

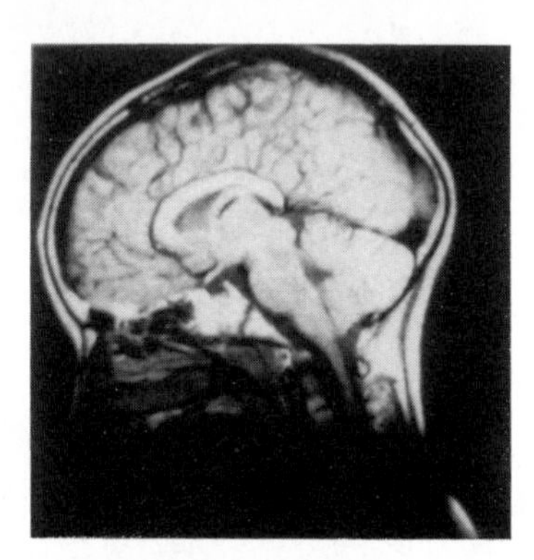

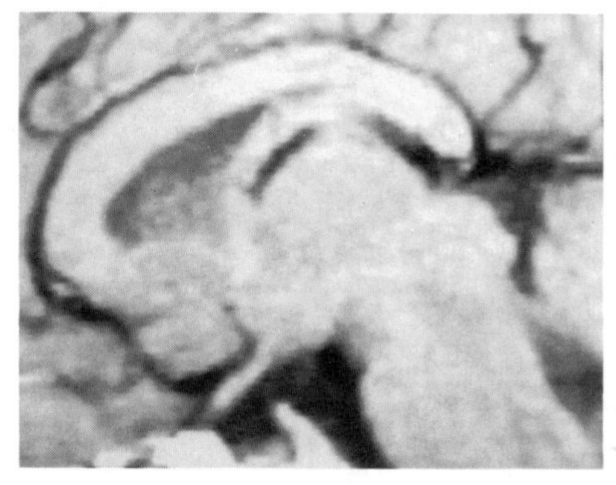

Figure 6: Magnetic Resonance Sagital Image of Brain; Left:256 × 256 MRI Sagital Image of Brain; Right:96 × 76 Subsection of the Brain Image.

The third example uses a 48 × 48 nuclear medicine image of the posterior view of kidneys as shown in Figure 10(upper left). The initial edge vector field, represented by $\mathbf{p_1}^0 \in R^{2(48\times48)}$ and computed based on the zerocrossings of the Laplacian operator, and is noisy as seen in Figure 10(upper right). The initial 2-D shapes of the two kidneys corresponding to the initial estimate of shape parameters $\mathbf{p_2}^0 \in R^{2(42)}$ in Figure 10(lower left) do not correctly represent the contours of actual two kidneys as can be observed in 10(lower right). As seen in Figure 11, System II fails to correctly determine the curvilinear contours of the kidneys. Finally, Figure 12 show the results obtained by Game-Theoretic System. Comparing our model to previous models, we see several advantages of preserving the multiple decoupled nature of the objectives of the system and having dynamic feedback between the modules.

5. DISCUSSION

Our experiments with several images have empirically demonstrated that the modules of a vision system: i.)have objectives that are not commensurable, ii.) have weak constraints among them that are not symmetric, and iii.) need partial information available from other modules. The game-theoretic integration is powerful because it addresses each of these issues effectively. In the context of 2-D deformable object recognition systems, we have seen that such an approach enables the system to be more robust by making it less sensitive to larger errors in the initial estimate of the shape variables.

A second advantage of the game-theoretic model is related to the scheduling of processing among the modules. In contrast to the previous methods where scheduling has been based on ex-

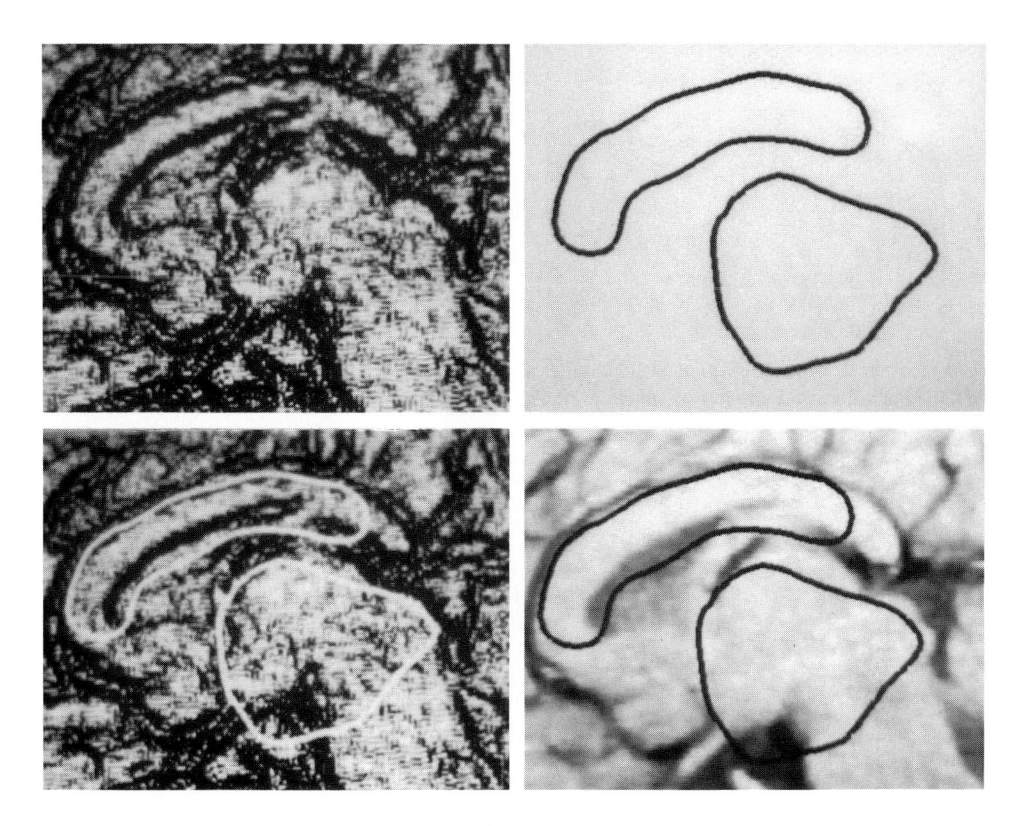

Figure 7: Magnetic Resonance Sagital Brain Image Example: Initial; Upper left: Initial edge vector field; Upper right: Initial 2-D shapes; Lower left: Initial shapes on initial edge vector field; Lower right: Initial shapes on image.

ternal criteria, the game-theoretic model has a control structure that automatically provides for the scheduling among the modules. The reason is that scheduling is naturally included in the definition of a N-player game via the specification of the decision making model. The decision making model defines which players will be playing at each stage of the game. Our experiments and theoretical results, which are to be published in a forthcoming paper, demonstrate the criticality of the timing of the moves to the system performance.

6. SUMMARY AND CONCLUSIONS

The main contribution of this paper is the development of a general deformable objects recognition system based on a game-theoretic framework. It is shown that a general deformable objects recognition system needs to integrate two modules. The first module improves the edge vector field of the original noisy image while the second module performs both object and relational matching of an *a priori* model (an anatomical atlas in the case of medical image analysis) with the edge vector field in order to accurately determine the objects in a given image. Consequently, the system's mathematical problem can be posed as finding optimal solutions for a a set of coupled coexisting functional objectives. It is then shown that the game-theoretic approach is appropriate for solving this problem because, in contrast to previous approaches to integration, game-theoretic framework adheres to the multiple nature of the objectives. We finally present several applications of this system aimed at specific clinically-related tasks.

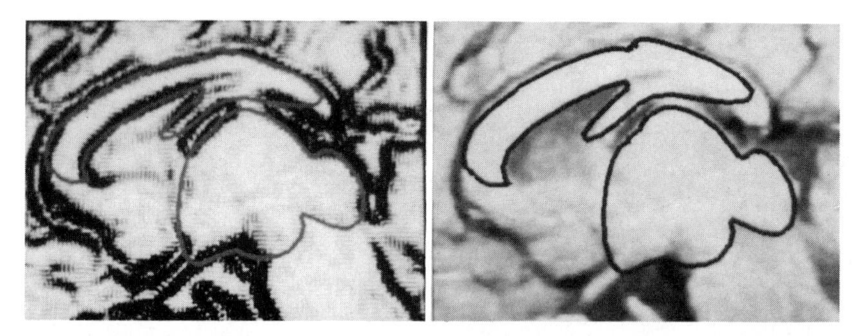

Figure 8: System II: Magnetic Resonance Sagital Brain Image Example; Left: Final Shapes on final edge vector field; Right: Final 2-D shapes on image.

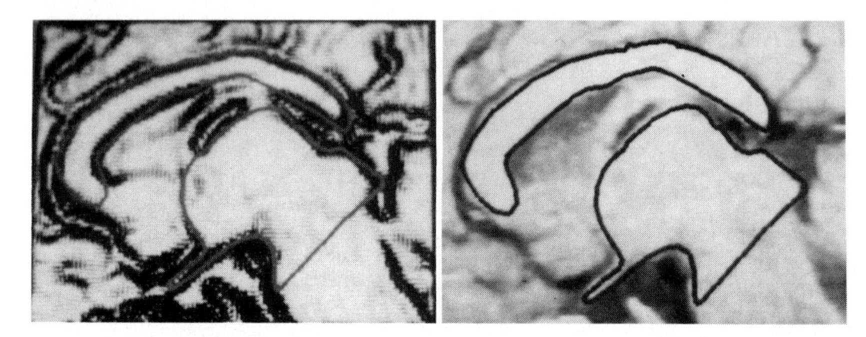

Figure 9: Game-theoretic System: Magnetic Resonance Sagital Brain Image Example; Left: Final Shapes on final edge vector field; Right: Final 2-D shapes on image.

Acknowledgements

We gratefully acknowledge the helpful discussions with Dan Koditschek and Kumpati Narendra, the technical and programming assistance of Ömür Bozma and Larry Staib, the assistance of Tom McCauley with the medical terminology and analysis, and George Zubal for providing nuclear medicine image data. This work was supported by NIH grant # R29-HL38333-02.

References

Abdou IE and Pratt WK (1979). Quantitative design and evaluation of enhancement/thresholding edge detectors. Proceedings of the IEEE, Vol-67:753-763.

Aloimonos JY (1989). Unification and integration of visual modules. In: Proceedings of Image Understanding Workshop (DARPA), pp. 507–551.

Ballard DH and Brown C (1982). Computer Vision. Prentice-Hall, Inc.

Basar T (1987). Relaxation techniques and asynchronous algorithms for on-line computation of non-cooperative equilibria. Journal of Economic Dynamics and Control, Vol-11:531–549.

T. Binford (1982). Survey of model-based image analysis systems. The Intern. Journal of Robotics Research, Vol-1:18–64.

Bozma HI and Duncan JS (1990). Integration of Vision Modules: A Game-Theoretic Approach. Center for Systems Science Technical Report 9016, Yale University.

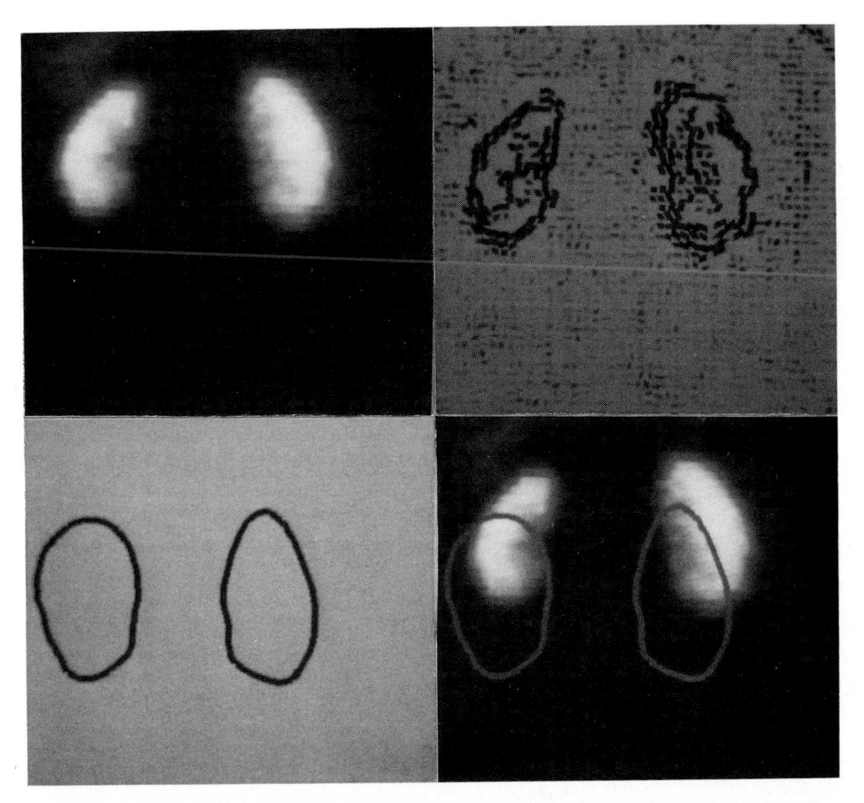

Figure 10: Nuclear Medicine Image of Kidneys Example: Initial; Upper left: 48 $\times$ 48 nuclear medicine image of kidneys; Upper right: Initial edge vector field; Lower left: Initial 2-D shapes; Lower right: Initial shapes on the image.

Duncan JS and Birkholzer T (1989). Reinforcement of linear structure using parametrized relaxation labeling. In: Information Processsing in Medical Imaging, pp:489–504.

Evans AC, Beil C, Marrett S, Thompson CJ, and Hakim A (1988). Anatomical-functional correlation using an adjustable MRI-based region of interest atlas with positron emission tomography. Journal of Cerebral Blood Flow and Metabolism, pp:489–504.

Fischler MA and Elschlager RA(1973). The representation and matching of pictorial structures. IEEE Trans. on Computers, Vol-22:67-92.

Gennert MA and Yuille AL (1988). Determining the optimal weights in multiple objective function optimization. In: Proceedings of Second International Conference on Computer Vision, pp:87–94.

Hoff W and Ahuja N (1989). Surfaces from stereo: Integrating feature matching, disparity estimation and contour detection. IEEE Transactions on Pattern Analysis and Machine Intelligence, Vol-11(2):121–136.

Horn BKP (1989). Height and gradient from shading. In Proceedings of Image Understanding Workshop (DARPA), pp:584–595. .

Canny J (1986). A computational approach to edge detection. IEEE Trans. on PAMI, Vol-8:679–697.

Kuhl FP and Giardina C (1982). Elliptic fourier features of a closed contour. Computer Graphics and Image Processing, Vol-18:236–258.

Lin CS and Hwang CL (1987). New forms of shape invariants from elliptic fourier descriptors. Pattern Recognition, Vol-20:535–545.

Nash J (1951). Non-cooperative games. Annals of Mathematics, Vol-54:286-295.

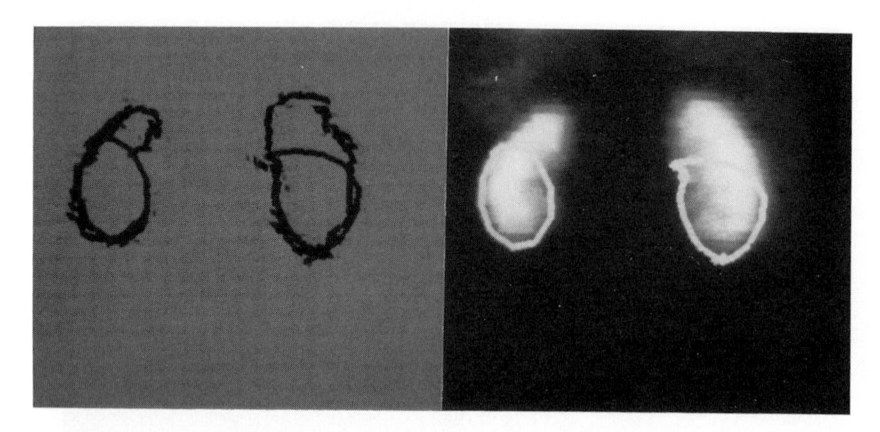

Figure 11: System II - Nuclear Medicine Image of Kidneys Example; Left: Final shapes on final edge vector field; Right: Final shapes on the image.

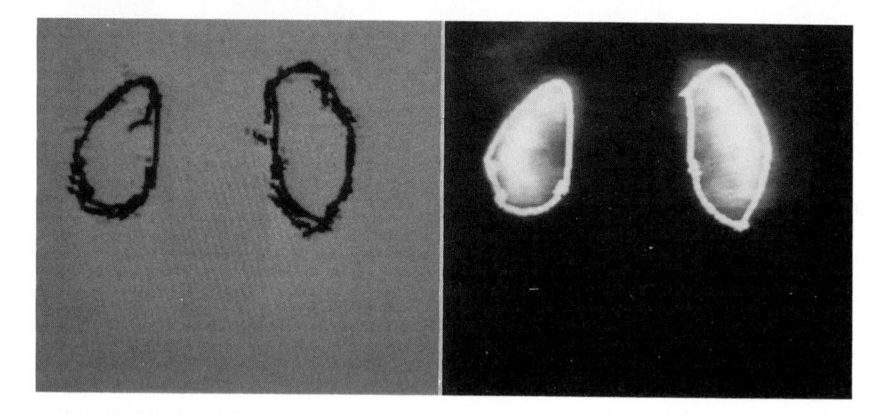

Figure 12: Game-Theoretic System - Nuclear Medicine Image of Kidneys Example; Left: Final shapes on final edge vector field; Right: Final shapes on the image.

Pratt W (1978). Dig. Image Processing. Wiley-Interscience.

Hummel RA and Zucker SW (1983). On the foundations of relaxation labeling processes. IEEE Trans. on Pattern Analysis and Machine Intelligence, PAMI-51:267–287.

Bajcsy R and Kovacic S(1989). Multiresolution elastic matching. Computer Vision, Graphics and Image Processing, Vol-46:267–287.

Staib LH (1990). Parametrically Deformable Contour Models for Image Analysis. PhD thesis, Yale University.

Zucker SW , David C , Dobbins A, and Iverson L, (1988). The organization of curve detection: coarse tangent fields and fine spline coverings. In: Proceedings of International Conference on Computer Vision, pp:568–577.

EXTRACTION OF BACKGROUND DISTRIBUTIONS FROM ABNORMAL DATA SETS: APPLICATION TO RADIOLABELLED LEUCOCYTE IMAGING

D C Barber and W B Tindale

Department of Medical Physics and Clinical Engineering
Royal Hallamshire Hospital
Sheffield S10 2JF
U.K.

Abstract

A new technique for the generation of images of the background distribution of tracer from a data set of abnormal images is described. The use of such an image in the analysis of images of labelled leucocytes in patients with Crohn's disease is outlined. Construction of the background image involves image registration, a weighted averaging technique to eliminate abnormal features in the mean image and an image normalisation technique based on mode rather than mean values. Clinical results show that the overall technique is capable of providing an index of disease activity.

Keywords

HMPAO; Image registration; background subtraction; Crohn's disease.

1. INTRODUCTION

Radionuclide studies allow quantification of physiological function. Frequently this involves measuring the amount of tracer taken up in some object seen in the radionuclide image. The principle is simple enough. If the object can be isolated, then the count rate measured over the complete object is proportional to the amount of tracer present. In practice the object cannot be isolated from the patient and this has two major consequences. The first problem is the problem of attenuation, which we shall not address here. The second is the background problem. Any measure of count rate over an object will be contaminated by activity within the surrounding tissues. In order to isolate the contribution from the object some estimate must be made of this background activity and this estimate subtracted from the measurement made over the object. This 'background problem' is important in many quantitative radionuclide applications and is often difficult to solve. No general solution exists.

The best way to estimate the background activity is from an image without the object of interest present. Since this image is rarely available it is necessary to model the background in some way. Often this modelling is implicit, involving unarticulated assumptions about the way the background varies between different parts of the image. For relatively complex backgrounds these assumptions may not be valid and will result in errors of estimation. Whether this matters or not depends on the application.

In the absence of a true background image one possible source of knowledge regarding the background distribution is an estimate from background images of the same type as the image being analysed. Barber (1976) and Barber and Sherriff (1978) showed how such an estimate might be constructed from such images using principal components analysis. It involved calculation of a mean image plus a set of orthogonal factor images from a population of background images. The mean image accounted for the grosser image structures in the data set; the factors for the variations between normal images. However the process was computationally expensive. We hypothesise that there are many situations where the mean image itself may provide most of the information needed.

Such an approach requires that a set of (normal) background images is available. In many studies collecting such a set may be ethically difficult. In this paper we shall describe a particular quantification problem which required the construction of a background image and how this image was assembled. The problem was made more difficult by the fact that a data set consisting purely of background images was not available. The underlying problem in this work therefore is the estimation of background activity from data which may contain some abnormal values. In the course of this paper two approaches to this problem will be presented.

1.1. The Clinical Problem

The objective assessment of disease activity in Crohn's disease is a difficult problem and there is currently no technique which is universally accepted as the gold standard. Whilst both radiology and colonoscopy play a primary role in the diagnosis of disease and the assessment of its extent, they have only a limited role in the determination of disease activity (Elliott et al, 1982; Gomes et al, 1986). The use of radio-labelled granulocytes, which migrate to sites of inflamed bowel in accordance with the severity of disease (Saverymuttu et al, 1982; Saverymuttu et al, 1986), has added a new dimension to this field and leucocyte imaging is now widely employed in the investigation of disease activity.

The recent introduction of a new radiopharmaceutical for white cell labelling, Tc-99m-hexamethyl-propyleneamineoxime (HMPAO), has enabled the selective labelling of granulocytes in a mixed white cell population (Peters et al, 1986). In patients without disease the tracer is distributed in the bone marrow and other organs of the reticuloendothelial system. In patients with disease there may be additional uptake of tracer in the gut, with the distribution and intensity of this uptake being dependent on the severity of the disease. Some examples of images from patients with active disease are given in Figure 1. This study seeks to quantitate disease activity.

2. METHODS

2.1. Clinical Groups

The study group consisted of 33 patients with a diagnosis of Crohn's disease. 9 of the patients had two white cell investigations, a minimum of ten days apart, and 6 were studied three times, giving a total of 54 investigations. Patients were classified as having active or inactive disease at the time of the scan, based on an overall clinical assessment which took into account the patient's symptoms and the results of endoscopy and blood tests. The scan results were not included in this assessment. In 34 cases the patient was classified as having active disease at the time of the scan, and in the remaining 20 the disease was considered clinically inactive. Following intravenous administration of the labelled cells anterior images of the abdomen were acquired at 2 hours.

2.2 Data Analysis

A number of semi-quantitative measures of disease activity have been derived, based on a visual assessment of the images. However these are essentially subjective and are also insensitive to small changes in bowel uptake and are therefore not ideally suited to longitudinal studies. The overall method of quantitation described in this paper can be simply summarised. A background image free of gut activity is estimated and then subtracted from the patient image. This reveals the abnormal areas associated with disease. The total counts collected in a fixed time is determined and used to derive a normalised numerical index of gut uptake.

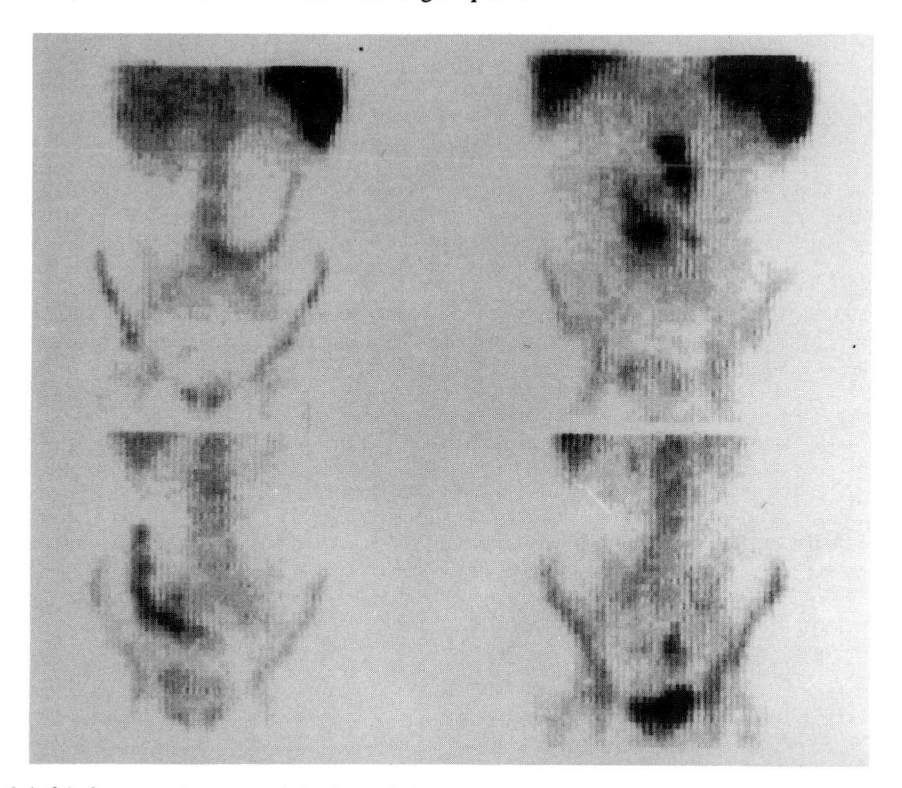

Figure 1. 4 of the images use to generate the background set.

2.2.1. Construction of the Background Image.

2.2.1.1. Image registration.

Barber (1976) and Barber and Sherriff (1978) noted that images taken from different patients will vary in size and position in a way which is irrelevant to the detection and analysis of abnormal areas. Before numerical analysis it is important to remove these variations from the image. In the present work, prior to the the construction of a background image, abdominal images taken from different patients were registered to a standard position and scaled to a standard size. Registration of one image from an image pair with the other involves the use of a coordinate transfer function (CTF), a function which determines which pixels in the first image correspond to which pixels in the second image (Gonzalez and Wintz, 1987). The work described in this paper uses a novel semi-automatic method for calculation of the CTF used to register two images, full details of which will appear elsewhere. An outline of the theory behind the method is given in the Appendix. In what follows it is assumed that all the images have been appropriately registered to a standard position.

2.2.1.2 Calculation of the weighted mean image.

The background image is calculated from a set of appropriately registered and scaled abdominal images. Although to a first approximation regions of abnormal gut uptake can be considered to be distributed randomly throughout the abdominal region, taking a simple pixel by pixel mean of all the images does not adequately eliminate the contributions from gut uptake, since a high proportion of the images may be abnormal. Since gut uptake represents increased activity a weighted average which biases against increased uptake should produce an image more representative of the underlying background activity.

In integral form a weighted average of a variable y is given by

$$a = \frac{\int_{-\infty}^{\infty} y w(y) P(y-m) dy}{\int_{-\infty}^{\infty} w(y) P(y-m) dy} \tag{1}$$

where P(y-m) is the probability distribution of y centred about the (unweighted) mean value and w(y) is the weighting factor given to each y. w(y) should not have any singularities within the range of integration. In the present case we wish to bias against large values of y. Since the only information available is the values themselves, one possibility is to set

$$w(y) = y^{-n} \tag{2}$$

i.e. an inverse power of y. The average then becomes

$$b = \frac{\int_{\alpha}^{\infty} y^{-(n-1)} P(y-m) dy}{\int_{\alpha} y^{-n} P(y-m) dy} \tag{3}$$

where the lower integral limit is some small positive value greater than zero to avoid the singularity at zero.

It can be shown that if m is the true mean of the distribution and μ_2 the variance then with some simplifications

$$b = m(1 - n\mu_2/m^2) \tag{4}$$

For the special case where the variance is equal to the mean value m, b is offset from m by the constant value n, so b is a biased estimate of m. In general the nature of this bias will depend on the distribution of pixel values in the background data set and on the value of n. The bias is likely to vary across the image. However simple simulations suggest that the change in the value of m at a point, caused by adding abnormal values of value greater than m to the distribution at that point, is greater than the change in b.

There is no obvious way of choosing n. Choice of large n will be most effective in reducing the effect of abnormal uptake, but with only a limited number of data values will become dependent on a few of these values only. In this paper n is chosen to be 2. Experimentally this seems to give good results.

In discrete form the estimate of the $(j,k)^{th}$ pixel in the background image is given by

$$Bg(j,k) = \frac{\sum_{i=1}^{N} 1/p_i(j,k)}{\sum_{i=1}^{N} 1/p_i(j,k)^2} \qquad (5)$$

where the summations are over the N images in the data set.

The abdominal images of the first 12 patients presenting to the study represented the data set used to construct the background image. Some of these images are shown in Figure 1. Since these images needed to be registered to a standard position, the best image to register to would be the background image itself. This was not available initially so one of the images in the data set was chosen to represent the standard position. All the images were registered to this image using the method outlined in the Appendix. Although scaling is included in the calculation of the CTF, this scaling factor was not included in the registration process itself and the images needed to be scaled to a standard amplitude. In a preliminary scaling, each image in the registered set was scaled to give a fixed value of one million counts in an abdominal region of interest, which included the pelvis, lumbar spine and, depending on the anatomy of the patient, some of the liver and spleen. A preliminary background image was calculated from the image data set using Eq. (5).

However the scaling factor for an image in the background set determined using the counts in an abdominal region of interest will be affected by the presence of abdominal abnormalities. The weighted mean analysis will be most effective in removing the contributions from abnormal regions if the distribution of underlying background values is as narrow as possible. In other words the amplitude scaling applied to an image in the data set must reflect as closely as possible the amplitude of the underlying background distribution in the image. Having calculated a preliminary background image each image in the background data set was then rescaled to this image using the method outlined in the next section.

2.2.1.3 Rescaling the images.

The relationship between an image G in the background set and the preliminary background image Bg is modelled as

$$G = m * Bg \qquad (6)$$

The scaling parameter m could be obtained from a least means squares fit between G and Bg. However if G contains an abnormality then the value of m will be affected by the presence of the abnormality. In particular m will be overestimated. Ideally the value of m which would be obtained if the abnormality were not present is required.

Figure 2a shows a plot of pixel values in Bg against pixel values in an image with marked gut uptake. The underlying structure of this plot is a narrow bivariate distribution with a 'tail' of values extending upwards in the plot. These represent pixels within the abnormal gut region in G. In the absence of an abnormality the image m * Bg (Eq. (6)) would plot as a line along the 'background' principal axis of this distribution. In the presence of an abnormality the principal axis is shifted away from the 'background' axis. In order to estimate this axis in the presence of abnormal pixels a method based on finding the mode line of this distribution is used since the mode line is likely to be a better estimate of the 'background' axis than estimates based on least mean squares methods.

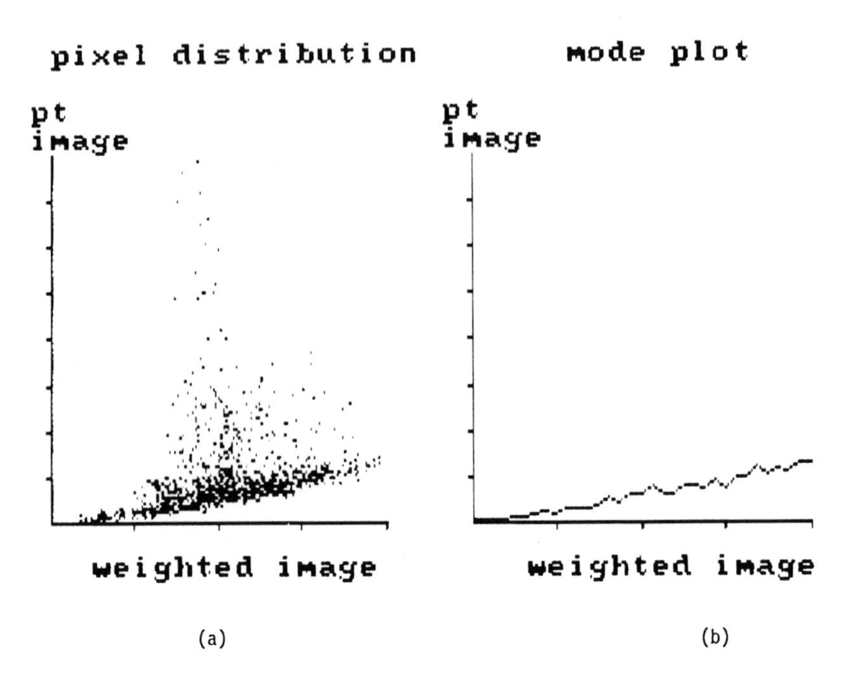

Figure 2. Use of the mode technique a) Plot of pixel values in the pelvic region of interest for a registered patient image vs background image; b) Mode plot corresponding to the data given in a).

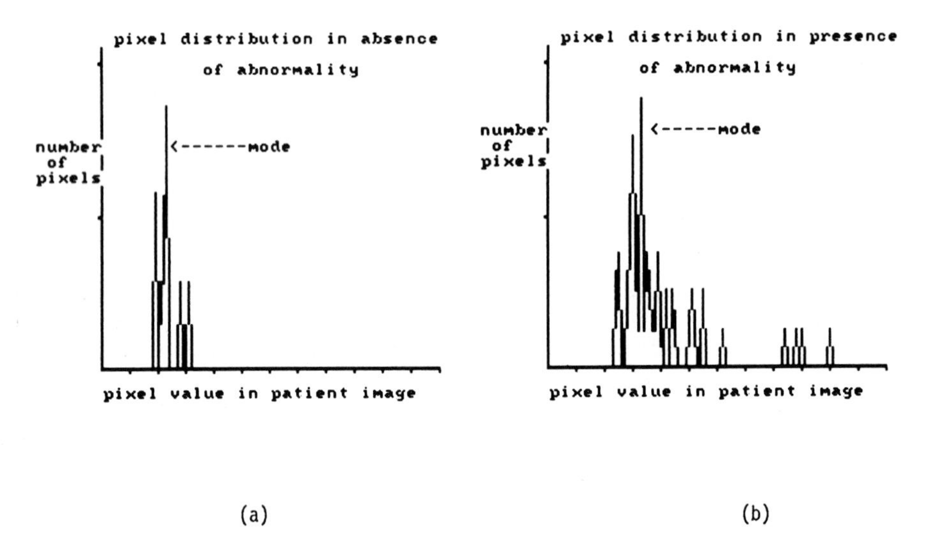

Figure 3. a): The pixel value distribution in an image without abnormal gut uptake for a particular grey level in the standard image. b): The pixel value distribution in an image with abnormal gut uptake for a particular grey level in the standard image.

2.2.1.4 Mode scaling - the mode line.

For any grey level taken from Bg a plot of values of the corresponding pixels in G would show a unimodal distribution in the absence of an abnormality (Figure 3a) and a bimodal distribution or a distribution with a tail when an abnormality is present (Figure 3b). Unlike the mean of the distribution of pixel values, the mode of the distribution of pixel values in G for a particular grey level in Bg will only be weakly affected by the presence of abnormal pixels.

The scaling parameter m between G and Bg is therefore obtained in the following way. The pixels in Bg were segmented into 32 grey level bands. For each of these grey levels in Bg, a histogram of the values of the corresponding pixels in the image G was produced and the mode value determined. Examples of such histograms are given in Figure 3. In total, 32 mode values were obtained from image G corresponding to the 32 grey levels bands of Bg. A plot of these values is shown in Figure 2b for the distribution in Figure 2a. They form a straight line along the principal axis of Figure 2a which constitutes the mode line for this distribution. As expected it shows no obvious deviations due to the presence of the abnormality. A weighted least squares fit of the mode line to a straight line through the origin, with weighting factors proportional to the number of pixels contributing to the peak of the histogram from which each mode value was calculated, provides an estimate of the required scaling parameter m between the two images, G and Bg.

As long as there are more pixels from 'normal' areas than pixels from abnormal areas contributing to a histogram, the mode will be insensitive to the presence of abnormalities. The scaling parameter m should therefore be insensitive to the presence of abnormalities in the image G, provided the number of pixels occupied by the abnormality is fairly small and these pixels are distributed across several of the histograms.

The mode was taken to be the maximum of the histogram. This measure is fairly sensitive to the presence of noise but because the scaling factor is determined from up to 32 values this is probably not as important as it at first appears.

After scaling all the images in the background set in this way a further weighted average was produced from this rescaled data set. This represents the final background image. This is shown in Figure 4.

In order to show the effects of using a weighted averaging compared to an unweighted averaging, an unweighted mean was computed from the same data set as the weighted mean. This image is shown in Figure 5 and the difference (mean minus weighted mean) between these images in Figure 6. This image shows uptake in the gut indicating that the weighted mean has suppressed uptake in the gut. It also shows uptake in the liver, spleen and bowel areas, as might be expected since the exact position of these organs and their uptake varies between subjects. The registration process did not include the liver and spleen. There is also some residual marrow activity in the pelvis, indicating some bias in the weighting process as suggested above. Several segments of the gut structure seen in Figure 6 can be identified with gut activity seen in the images of Figure 1.

2.2.2 Quantitation of bowel uptake.

To determine the gut activity in a new image the image was first registered to the background image and scaled using the mode technique described above. Inspection of Figure 2b suggests that the mode line does not pass though the origin, and in fact a more complete model of the background in an image is given by

$$G = m * Bg + c \quad (7)$$

In this phase of the analysis this model was used. Once m and c had been calculated for each patient image from the mode line, the background image for that patient image was then generated from Bg, using Eq. (7) and subtracted from the aligned image. The residue which remains after subtraction represents the difference between the two images and will consist predominantly of

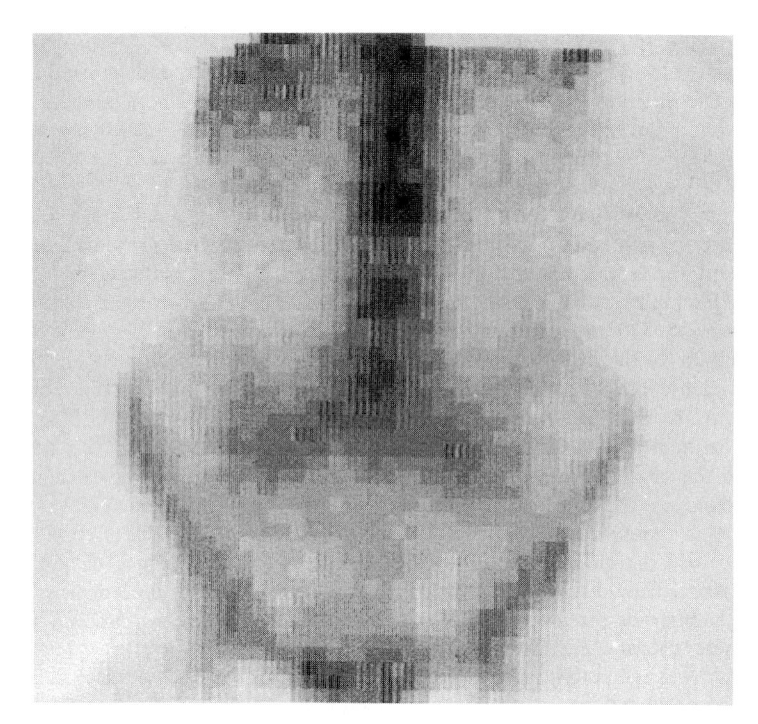

Figure 4. The weighted mean image.

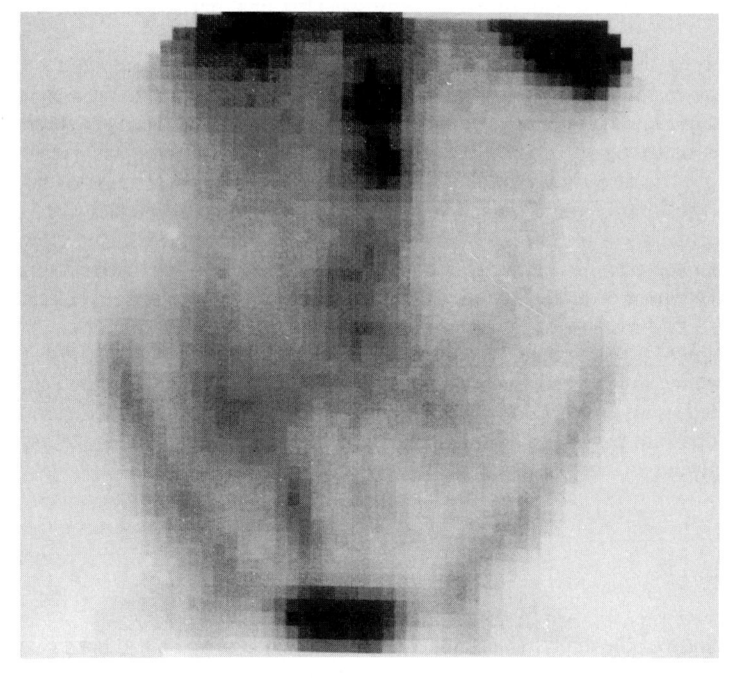

Figure 5. The unweighted mean image.

abnormal bowel. Figure 7 shows an image registered and subtracted in the above manner.

Since the standard image, even after registration, was not a perfect representation of the background some residue was present on the subtraction image. Estimation of the bowel uptake required further delineation of the bowel activity (which could be seen clearly on the subtraction image), although the accuracy of this delineation is not critical. Since this aspect of the analysis is not central to this paper and space is limited further details will not be presented here. We note simply that, since all the images were scaled to a standard background image, it was possible to normalise the estimated gut activity and produce an index, the scan score, which reflected the relative gut uptake between patients. It is independent of patient size, circulating white cell count and administered dose.

Scan scores were typically below 20 for patients with inactive disease and ranged up to 400 for patients with active disease. The median scan score for the active group was 67, compared with 12 for the inactive group. Analysis of these data indicates that maximum accuracy of the scan score for predicting disease activity is obtained with a threshold value of 20. A scan score of 20 was taken as the border between active and inactive disease.

3. RESULTS

On the basis of a scan score of 20 delineating between active and inactive disease, the scan score correctly identified 16 out of 20 cases with inactive disease and 31 out of 34 cases with active disease, giving an overall accuracy of 87%.

4. DISCUSSION

Determination of the amount of radiotracer uptake by some object in an image almost always requires the estimation of the background activity in the region of the object. One possible approach to estimating this activity, in the absence of an image of the true background without the object, is to produce an estimate from similar but object-free images taken from other patients. In practice sufficient 'normal' images may not be available. Formally the problem has become one of estimating a parameter associated with a distribution of normal values from a distribution which, though partly normal, contains some abnormal values.

Two methods have been described. The first uses a weighted average to bias against the abnormal values in the distribution. This gives an estimate of the background distribution which should be less affected by the presence of abnormal values in the data set than the simple mean. Experimentally this appears to be the case. For simplicity the choice of the weighting factor was limited to one which could be determined from each data value on its own. In this study, the reciprocal of the square of the pixel value was used. Higher powers could be used. As the (inverse) power of the pixel value is increased, the weighted mean value will be dominated by an increasingly small number of individual pixel values, which may lead to the creation of a noisy background image. Conversely, as the power of the pixel value tends to zero, the weighted mean image tends towards the simple mean image and abnormalities will not be removed.

The second method used, the mode technique, is also a powerful method of calculating distribution parameters in a way insensitive to the presence of abnormal outliers. It allowed scaling factors for image scaling and subtraction to be calculated in a way which did not depend on the abnormality present. The principal problem in the method is that, because the mode value computed is simply taken to be the position of the maximum value in the histogram of the distribution, it is rather sensitive to noise. This is not particularly problematic in the present case but needs to be addressed at some stage. A reliable algorithm for the determination of the mode of a distribution in the presence of noise is needed. In principle with a large enough data set the background image could have been calculated in a similar manner. However with only 12 data values per pixel, this approach would not have been feasible.

Given a sufficiently large population of patients such that a reasonable number of subjects

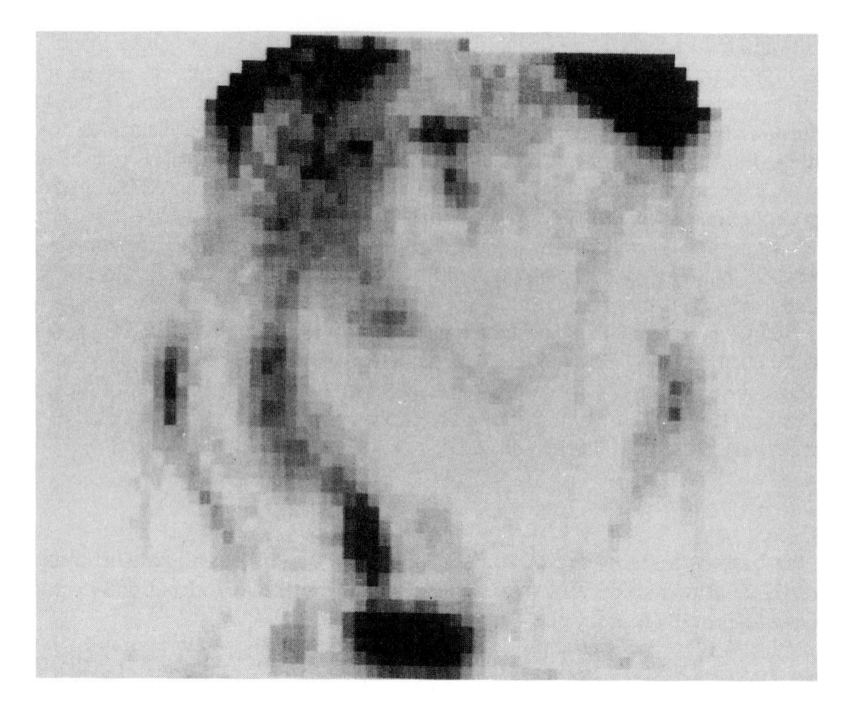

Figure 6. The difference image between Figures 4 and 5.

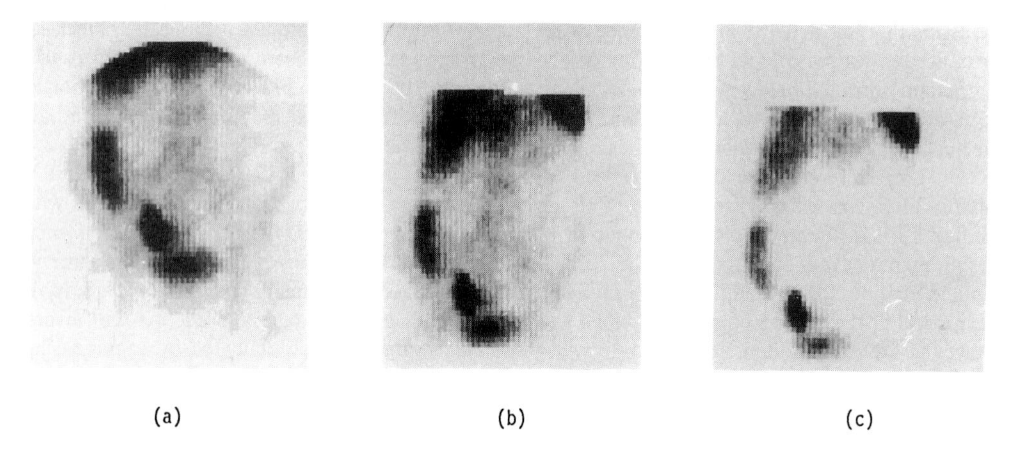

(a) (b) (c)

Figure 7. (a) Original patient image. (b) Patient Image after registration. (c) Subtraction image.

produced images without visible abnormalities it would be tempting to use these images to estimate the background. In this case it may not be necessary to use the weighted mean. We intend to explore this approach. However if the number of such images is limited it does introduce the problem of bias into the determination of the scan score ranges unless a second independent set of 'inactive' patients is available.

Of the 54 investigations comprising this study, the scan score correctly classified 16 out of 20 scans on patients with clinically inactive disease. Three of the scans which were misclassified were repeat studies on patients who had been given treatment for previously active disease. In two cases, the scan score had decreased from its previous level, although not to a value of less than 20. In the third case, the patient had a large abdominal mass which persisted throughout the period of treatment, and she suffered a relapse on the day of the scan, which was accompanied by a rise in scan score from its previous level. The fourth case of misclassification was a borderline result with a scan score of 22.

5. CONCLUSIONS

The work presented here has formed part of a project for the quantitation of disease activity in Crohn's disease using a tracer for which previously only subjective image assessment was possible. It allows the progress of disease or its remission to be followed in a quantitative manner. It shows that a useful estimate of the image background can be derived even if a full set of abnormality-free images is not available and that this estimate can be used to isolate abnormal areas for further quantitiation. In addition the mode technique of image subtraction may have other applications in image subtraction in the presence of abnormalities, for example, in thallium-technetium parathyroid imaging.

APPENDIX A

In order to align two images it is necessary to specify a coordinate transfer function (CTF). If the transformation is not known a priori then it is often modelled using a polynomial function in x and y

$$r = a_{11} + a_{12}x + a_{13}y + a_{14}xy + a_{15}x^2 + a_{16}y^2 \ldots \qquad \text{(A1)}$$
$$s = a_{21} + a_{22}x + a_{23}y + a_{24}xy + a_{25}x^2 + a_{26}y^2 \ldots.$$

A simpler form of this equation drops the quadratic terms in x and y to become the affine transform

$$r = a_{11} + a_{12}x + a_{13}y \qquad \text{(A2)}$$
$$s = a_{21} + a_{22}x + a_{23}y$$

If the values of r,s,x,y are known for a sufficient set of landmarks or control points in the two images then the resulting two sets of simultaneous equations may be solved for the a_{ij}. For low resolution images it may be difficult to specify landmarks with sufficient accuracy. A new approach to the problem of finding the coefficients of the CTF through a directed search is presented here. Suppose the images to be registered are called F and M. The aim of registration is to transform image M in such a way that it is registered with image F. Image F is assumed fixed. Image M can be considered to have moved relative to image F. One approach to the problem of detecting image motion analyses it in the following way (Horn and Schunck, 1981). Let the image function at time t be f(x,y,t) and the image function at time $t+\delta t$ be $f(x+\delta x, y+\delta y, t+\delta t)$. The image has moved in some way. Expanding the latter image function as a Taylor series to first order terms gives

$$f(x+\delta x, y+\delta y, t+\delta t) = f(x,y,t) + \delta x\frac{df}{dx} + \delta y\frac{df}{dy} + \delta t\frac{df}{dt} \quad \text{(A3)}$$

Since $f(x+\delta x,y+\delta y,t+\delta t)$ and $f(x,y,t)$ represent the same object point on the two images, by definition

$$f(x+\delta x, y+\delta y, t+\delta t) = f(x,y,t) \quad \text{(A4)}$$

so that

$$0 = \delta x\frac{df}{dx} + \delta y\frac{df}{dy} + \delta t\frac{df}{dt} \quad \text{(A5)}$$

or in simpler notation

$$0 = \delta x\ x' + \delta y\ y' + \delta t\ t' \quad \text{(A6)}$$

$\delta x,\delta y,\delta t$ is the movement which takes $f(x,y,t)$ to $f(x+\delta x,y+\delta y,t+\delta t)$. If $f(x,y,t)$ is image F then δx and δy are given by

$$\delta x = r - x \quad \text{(A7)}$$
$$\delta y = s - y$$

δt can without loss of generality be set equal to unity. Evoking the CTF given in Eq. (A1) we can write

$$r-x = a_{11} + (a_{12}-1)x + a_{13}y \quad \text{(A8)}$$
$$s-y = a_{21} + a_{22}x + (a_{23}-1)y$$

although all of the following is in principle applicable to the general polynomial of arbitrary order. Let x' and y' be the image gradients in the x and y direction. Substituting into equation A6 gives

$$-t' = (a_{11} + (a_{12}-1)x + a_{13}y)x' + (a_{21} + a_{22}x + (a_{23}-1)y)y' \quad \text{(A9)}$$

Consider a point x,y on $f(x,y,t)$. The term $-t'$ is simply

$$-t' = F(x,y) - M(x,y) \quad \text{(A10)}$$

and combining these last two equations gives

$$F(x,y) = M(x,y) + (a_{11} + (a_{12}-1)x + a_{13}y)x' - (a_{21} + a_{22}x + (a_{23}-1)y)y' \quad \text{(A11)}$$

For each pixel in image F, x,y and the values x' and y' are known. Eq. (A11) is therefore linear in the parameters a_{ij}. One such equation is available for each image pixel and therefore the set of simultaneous equations is overdetermined (assuming we are considering more than eight independent pixels), allowing a solution for the parameters of the coordinate transfer function.

In the present case, because the images are from different patients they will not have the same amplitude. In this case Eq. (A11) can be rewritten as

$$F(x,y) = a_{00}M(x,y) + (a_{11} + (a_{12}-1)x + a_{13}y)x' + (a_{21} + a_{22}x + (a_{23}-1)y)y' \qquad \text{(A12)}$$

where a_{00} is a scaling factor.

Not all pixels are of equal use in determining the a_{ij}. Pixels in image F which are in regions of low gradient cannot contribute to the registration process, since this is driven by gradients. If the images are non-zero at the edge of the image field (as is often the case) then there may be pixels in one image which do not have equivalent pixels in the other. Both these considerations suggest that selection of pixels to be used in the registration process using a region-of-interest may be useful or even necessary.

Image registration can often be improved by use of histogram equalised images (Gonzalez and Wintz 1988) and this means that the scaling factor a_{00} may not be appropriate for subsequent image subtraction.

References

Barber DC (1976). Digital computer processing of brain scans using Principal Components. Phys. Med. Biol. 21:792-803.

Barber DC and Sherriff SB (1978). Computer analysis of liver scans. In: Biosigma 78 Proc. of . Conf. Signals and Images in Medicine and Biology II, INSERM, Paris, pp1-5.

Elliott PR, Lennard-Jones JE, Bartram CI, Swarbrick ET, Williams CB, Dawson AM, Thomas BM and Morson BC (1982). Colonoscopic diagnosis of minimal change colitis in patients with a normal sigmoidoscopy and normal air-contrast barium enema. Lancet 1:650-652.

Gomes P, Du Boulay C, Smith CL and Holdstock G (1986). Relationship between disease activity indices and colonoscopic findings in patients with colonic inflammatory bowel disease. Gut 27:92-95.

Gonzalez RC and Wintz P (1987). Digital Image Processing. Addison Wesley, Reading, Mass, pp146-52.

Gonzalez RC and Wintz P (1987). Digital Image Processing. Addison Wesley, Reading, Mass, pp247-51.

Horn BKP and Schunck BG (1981). Determining optic flow. Artificial Intell. 17:185-203.

Peters AM, Danpure HJ, Osman S, Hawker RJ, Henderson BL, Hodgson HJ, Kelly JD, Neirinckx RD and Lavender JP (1986). Clinical experience with Tc-99m hexamethyl propylene amine oxime for labelling leucocytes and imaging inflammation. Lancet 2:946-949.

Saverymuttu S H, Peters A M, Hodgson H J, Cadwick V S and Lavender JP (1982). Indium-111 autologous leucocyte scanning : comparison with radiology for imaging the colon in inflammatory bowel disease. Br. J. Med. 285:255-257.

Saverymuttu SH, Camilleri M, Rees H, Lavender JP, Hodgson HJF and Chadwick VS (1986). Indium-111 granulocyte scanning in the assessment of disease extent and disease activity in inflammatory bowel disease. Gastroenterol. 90:1121-1128.

A QUANTITATIVE COMPARISON OF CURRENT METHODS OF FACTOR ANALYSIS OF DYNAMIC STRUCTURES (FADS) IN RENAL DYNAMIC STUDIES

A S Houston[1] and K S Nijran[2]

[1]Department of Nuclear Medicine, Royal Naval Hospital Haslar, Gosport PO12 2AA, England and
[2]Department of Nuclear Medicine, Charing Cross Hospital, London W6 8RF, England

Abstract

Five methods of performing factor analysis of dynamic structures are compared in this paper. These are apex-seeking; the intersection method; cluster analysis; spatial constraints; and simple structure. Variants of these have been implemented in a workstation environment. The methods were tested on sequential images of 40 individual kidneys obtained from 20 radionuclide dynamic studies. For each kidney, estimates of whole kidney transit time, parenchymal transit time, and glomerular filtration rate had been made using conventional region-of-interest techniques. Each method produced three curves (representing three structures) which, in almost every case, corresponded to parenchyma, collecting system, and blood background. These curves were normalised to represent the total counts per frame contributed from each structure, and whole kidney transit time, parenchymal transit time, and glomerular filtration rate estimated for each kidney. For whole kidney transit times, values obtained from factor methods, with the exception of apex-seeking, were in good agreement with those obtained conventionally. For parenchymal transit times, values obtained from factor analysis were higher than, and poorly correlated with, values obtained conventionally. For glomerular filtration rate, values obtained from factor analysis were generally higher than, and moderately correlated with, values obtained conventionally.

Keywords

Principal components analysis; apex-seeking; intersection method; cluster analysis; spatial constraints; simple structure; whole kidney transit time; parenchymal transit time; glomerular filtration rate.

1. INTRODUCTION

The use of factor analysis of dynamic structures (FADS) as a means of separating overlaying structures in nuclear medicine dynamic studies was suggested independently by Bazin et al. (1980) and Barber (1980). The technique involves using temporal differences in the behaviour of these structures to separate their contributions to the series of two-dimensional sequential images.

A principal components analysis (PCA) (e.g. Schmidlin, 1979) is first performed on the population of activity/time curves obtained for each pixel in the image. The study will then be defined by a few components in the temporal domain and corresponding sets of coefficients in the image domain. Each coefficient matrix may be expanded into a sequential set of images by multiplying, at each frame, by the appropriate element of its component vector. Summing these over all principal components will give an approximation to the original dynamic study data.

It has been shown (Barber, 1980) that a spatially invariant, homogeneous structure may be described by a matrix representing its spatial contribution to the image and a vector representing its temporal behaviour. If these are known, then, in a manner analogous to the last paragraph, a sequential set of images may be obtained, representing the structure's contribution to the dynamic study data. In cases where the study comprises a finite number of such structures, the dynamic study data will be the sum of their contributions. Since a clear analogy exists between the results of the PCA and the physiological representation, it remains to find the appropriate transformation of the PCA data.

Since principal components are orthogonal by definition, component vectors and coefficient matrices will contain negative elements. Since the number of photons detected in a nuclear medicine study is never negative, an oblique transformation is performed to eliminate, or reduce significantly, the negativity in these vectors and matrices. The resultant factors and factor images are said to be "physiological" (Bazin et al., 1980).

It was demonstrated (Houston, 1984), that the use of a non-negativity constraint alone did not yield a unique solution in most instances. As such, results obtained from FADS can rarely be used quantitatively. Attempts at quantification in the past have either been in first-pass studies where blood background is not a major problem (Villanueva-Meyer at al., 1986; Philippe et al., 1988), or, alternatively, have received critical attention (Murase et al., 1988; Šámal and Kárný, 1989). It is generally accepted that further constraints are usually required.

Attempts to provide these additional constraints may be called constrained FADS (or COFADS). Such attempts include the intersection method (Nijran and Barber, 1985; 1986), which involves adopting a physiological model for at least one of the structures; the use of cluster analysis (Houston, 1986), which requires a definition of the degree of overlay in the image; the method of spatial constraints (Nijran and Barber, 1988), which requires a definition of the extent of certain structures in the image; and the method of simple structure (Šámal et al. 1988), which requires the definition of a region of pure background.

In this paper, each of these methods are compared with the use of the non-negativity constraint alone. Quantitative parameters are derived from renal dynamic studies and compared with values obtained from conventional subjective region-of-interest (ROI) analysis.

2. METHODS

2.1. General Methodology

2.1.1. Patient Studies Used.

Twenty renal dynamic studies (40 kidneys), obtained from archives at the Royal Naval Hospital Haslar, were analyzed using constrained factor analysis. These were collected as 50 sequential frames, each of 20 seconds, following an intravenous injection of 150 MBq $^{99}Tc^{m}$-DTPA. The images were obtained as 64x64 matrices. The studies were selected to represent a range from normal to non-functioning kidneys, and, in all but a few cases, exhibited evidence of pathology.

2.1.2. Parameters Evaluated.

Three measurements had been obtained previously for each kidney: whole kidney transit time (WKTT); parenchymal transit time (PTT); and glomerular filtration rate (GFR). Transit times were calculated using deconvolution analysis according to the "matrix algorithm" method of Diffey et al. (1976). A region of blood background, drawn over the aorta above and clear of the ureters, was used as the input function. Appropriate regions were drawn to define whole kidney and parenchyma (Britton et al., 1979). The calculation of GFR was due to a method by Sampson et al. (1985) and involved the initial phase of the whole kidney activity/time curve and a dose-related standard value. Parenchymal curves were used to define the extent of the initial phase. These procedures were performed by technicians as part of a routine analysis and were therefore quite independent of the processing used to analyze the results obtained from factor analysis.

2.1.3. Method Variants Compared.

Variants of five different FADS methods were tested. These were apex-seeking (three variants); the intersection method; cluster analysis (two variants); spatial constraints (two variants); and the method of simple structure. Programs for apex-seeking (two variants), cluster analysis (both variants) and spatial constraints were written by one of the authors (ASH); while all other programs were written by the other author (KSN): in the case of the method of simple structure this amounted to an adaptation of code kindly provided by Martin Šámal of Charles University, Prague. It is important to make this distinction as the normalisation used in pre-processing the data was different for each author. For this reason, the treatments will be described separately.

2.2. Treatment one (ASH)

2.2.1. Principal Components Analysis.

Individual kidney images were obtained by appropriate zooming to form a sequence of 32x32 matrices in each case. A rectangular mask was then used to limit further the spatial extent of the region to be analyzed. PCA was applied to the data according to the method proposed by Schmidlin (1979). This involves a normalisation (Benzécri, 1973) which reduces the dimension of the problem by one. It is assumed at this point that a maximum of three structures is present in the image. Because of the normalisation used, two principal components are required to represent dynamic data for three structures. Each pixel will therefore be represented by two coefficients, which may be plotted as co-ordinates in feature space.

2.2.2. Apex-seeking Method.

Barber (1980) has shown that the "true" factors may be constructed from points lying in the feature space defined by the PCA. These points, which also define the transformation, will form the apices of a triangle. For non-negative factor images, the triangle must contain all the pixel points; for non-negative factors, the apices must lie within a domain which can be determined (Houston, 1984). FADS uses these constraints alone; COFADS attempts to isolate these apices by applying additional constraints. In practice, since noise is present in all dynamic studies, both methods allow some negativity to remain, usually in the factor images.

Although FADS uses no *a priori* information, the algorithm requires initial values for the co-ordinates of the apices. These were determined, in the first instance, by the method of Di Paola et al. (1982). In order to test the importance of these values, the routine was repeated using the initial approximation proposed by Barber (1980).

2.2.3. Cluster Analysis.

Both cluster analysis and the method of spatial constraints require a classification of pixels in the image. For cluster analysis, clustering takes place in the feature space defined by the PCA. Pixels are therefore classified automatically according to their temporal behaviour. On the other hand, the method of spatial constraints involves an operator-defined partitioning of the image. Both methods require *a priori* information.

For cluster analysis, this involves a knowledge of which structures overlay. It is not necessary to know where they overlay. For example, if we have three structures, A, B, and C, it is sufficient to know that there exist pixels in the image where, for example, A exists in isolation; A overlays B; A overlays C; or all three overlay. Sets of overlaying structures may be defined in the image in this way, e.g. {A}, {A,B}, {A,C} and {A,B,C} in the example. These sets are said to be explicitly defined. A list of all explicitly defined sets forms the required *a priori* information. New sets formed from union or intersection of explicitly defined sets are said to be implicitly defined.

Two clustering algorithms were compared, namely Gravitational Clustering (GC) (Wright, 1977) and Hierarchical Ascendant Classification (HAC) (Jambu, 1978). Previous attempts to apply these methods to factor analysis of dynamic studies have been documented: GC (Griffin et al., 1982; Houston, 1986; Houston, 1988); HAC (Hannequin et al., 1990). For both methods, co-ordinates were weighted according to Hannequin et al. (1990), i.e. by the inverse square-root of the corresponding eigenvalue, and by the total activity in each pixel. However, for display purposes and subsequent analysis, the latter weighting was not used.

In either case, the effect of clustering in both feature and image domains is displayed on a TV monitor when 15 or fewer clusters remain. At present, termination of the clustering program is operator controlled.

2.2.4. Spatial Constraints.

The method of spatial constraints requires *a priori* information about the extent of certain structures. This must equal or exceed the actual extent. The rationale behind this method is to isolate some pixels for which that structure is absent. In the example previously described, a region defining the extent of structure C will isolate pixels containing structures A and B only. Similarly, a region defining the extent of structure B will isolate pixels containing A and C only.

For renal studies, this involves the definition of regions limiting the extent of the parenchyma and collecting system. These were defined by the operator.

As with cluster analysis, the effects of this partitioning is displayed on a TV monitor in both feature and image domains. Since there is clear analogy with cluster analysis, pixels points classified together will be said to form a cluster in feature space.

2.2.5. Subsequent analysis for COFADS methods.

The operator is then asked to define the elements, i.e. apices, sides, of the required transformation. It was previously shown (Houston, 1986) that each explicitly defined set will define a transformation element of dimension equal to the number of structures in the set.

First of all, one or more clusters corresponding to a single structure are identified. The centroid of the cluster(s) is defined as an apex of the transformation. Clearly, up to three such apices may be defined. If three apices are defined the transformation is also defined; otherwise clusters corresponding to two overlaying structures are identified. A straight line with best perpendicular least squares fit to the points of the cluster(s) is defined as a side of the transformation. If one of these structures has its apex defined, the line is constrained to pass through it. Clearly, if both apices are defined, the side is pre-determined. Similarly, two such sides will intersect at an apex. (These examples correspond to implicitly defined sets). This procedure is continued until as many apices and sides as possible are defined.

In general, only one apex, corresponding to blood background, will be defined. In many cases, for part of the image, each structure will overlay blood background with no contribution from the third structure. If, in addition, all three structures overlay at some part of the image, the situation reflects that of the example previously described, with blood background corresponding to structure A.

It may assumed that this is the case in the region of a single kidney with structures B and C corresponding to parenchyma and collecting system. In such a case, the apex corresponding to blood background may be defined. Lines constrained to pass through this apex and corresponding to (i) parenchyma overlaying background; and (ii) collecting system overlaying background; may also be defined. The transformation will be complete if apices corresponding to parenchyma and collecting system can be defined on these lines. The position of these apices is further constrained by the fact that they must lie within the domain of non-negative factors.

In order to determine the remainder of the transformation (in this case the position of two apices on two lines) a background constraint is used. Part of the image contains pure background. The limits of this area can be defined either by the operator (as in the current study), or by using the appropriate cluster(s). For pixels outside this region, background is estimated using an adaptation of bi-linear interpolation (Goris, 1976). A search is made from each pixel in four directions: up; down; left; and right; and the first pixel defined as pure background is found in each direction. If four such pixels are found, conventional bi-linear interpolation is applied. An adjustment is made if less than four pixels are used. If, for any pixel, no background pixels are found, a second or subsequent pass is made using the newly-defined background estimates. Background structure may also be estimated as a function of the remaining transformation elements, which are chosen to give the best least-squares fit between the two estimates (Houston, 1986). Constrained optimisation ensures that the apices lie within the domain of non-negative factors (Houston, 1988).

2.2.6. Normalisation Required For Quantification.

From the resultant choice of apices, factor curves and images are obtained for each method. With appropriate normalisation, the curves will represent the total counts per frame corresponding to each structure. This involves reversing the initial Benzécri normalisation, and then scaling each factor by the total number of counts in the appropriate factor image.

2.3. Treatment two (KSN)

2.3.1. Principal Components Analysis.

PCA of dynamic image data results in principal components and coefficients images. The principal components are said to define a feature space (S space) in the temporal domain, in which the physiological factors plot. The coefficient images also define a feature space (Q space) in the spatial domain, in which the factor images plot. The methods described below apply constraints in either Q space or S space.

2.3.2. Apex-seeking Method.

This is the original (and most widely used) method. It uses the constraints of non-negativity of factor curves and factor images (Barber, 1980; Bazin et al., 1980). The problem may be interpreted in the following way for a three factor solution. A 3-dimensional plot of coefficients for each pixel curve in the study produces a distribution of points. The non-negativity criterion defines a pyramid whose apex is the origin of this space within the domain of non-negative factors and the pyramid contains all such points. These conditions ensure that factor curves contain no negative elements and that factor images will be non-negative everywhere. However, due to noise in the data, a small number of points will lie outside the pyramid.

2.3.3. Intersection Method.

This method attempts to isolate one or more factors by assuming a physiological model for a dynamic structure in a study. Feature spaces are obtained for both study and model and their intersection defines the space in which the true factors exist (Nijran and Barber, 1985). In the case of a renal study, a theory space (T) is obtained by applying PCA to a population of theoretical curves obtained from a mathematical model for the kidney. A study space (S) is defined by PCA applied to patient data. An intersection of T and S defines a 2-dimensional space in which two physiological factors for a kidney exist in their true form.

A modified apex-seeking procedure incorporating the constraint that two of the three factors lie in the intersection space is used to extract a complete solution for that study (Nijran and Barber, 1986).

2.3.4. Spatial Constraints.

Here the roles of pixels and image dependent elements are reversed. Constraints are applied in Q space. In most studies, it is possible to define the spatial extent of a dynamic structure using a mask defined by an ROI. Pixels lying outside the ROI will have zero values in the corresponding factor image. Such ROIs are delineated for as many structures as possible, and this condition used as a constraint. (Nijran and Barber, 1988). One factor structure which cannot be determined in this way is the blood background, which is present throughout the image. A linear model is used to estimate this factor image (assuming uniform blood background).

2.3.5. Simple structure.

This method also applies constraints in Q space. A varimax transformation yields factor images with enhanced factor specific areas and suppressed factor common areas. A reference area of background is selected where factors will have zero values. This area is used to index the level to which varimax factors should be balanced, and hence a more correct further transformation is performed (Šámal et al.,

1988). The algorithm used here has been provided by Martin Sámal, written on Fortran 4 on a PDP 11 machine. The sources have been modified to run on a Bartec Micas 5 Sun microsystem.

2.3.6. Normalisation Required For Quantification.

For each method, the factor curves are normalised to the total counts of the region of the study used in the analysis, i.e.

$$T(t) = a_1K_1(t) + a_2K_2(t) + a_3B(t), \quad (1)$$

where T(t) is the time-activity curve for the region of the study analyzed; $K_1(t)$ and $K_2(t)$ are the two factor curves for the kidney; B(t) is the background factor curve; and $a_{i=1,3}$, are the weights. A direct matrix solution is obtained for the above equation.

2.3.7. Operational Details.

In the case of treatment two, there was no pre-processing of the data prior to PCA. However, rectangular regions were defined limiting the extent of the area under investigation for all methods, and outlining a small area of pure background as required by the method of simple structure. Two regions, limiting the extent of parenchyma and collecting system, were defined, as required by the method of spatial constraints. Because of differences in processing involved in the two treatments, all regions had to be re-drawn. Regions which performed the same task in the two treatments were drawn by the same operator (ASH), within an interval of two weeks, in order to reduce problems of reproducibility. Similarly, although the techniques of treatment two are essentially those of one author (KSN), the analysis was performed by the other (ASH).

2.4. Quantitative Analysis

In every case the outcome of the analysis will be three curves and three images. These should correspond to parenchyma, collecting system and blood background, with curves normalised to total counts. A whole kidney curve may be defined as the sum of the curves corresponding to parenchyma and collecting system. The curves corresponding to whole kidney, parenchyma and blood background may then be used to determine estimates of WKTT, PTT and GFR, exactly as in the routine analysis performed using ROIs.

The analysis was performed on a Bartec Micas V system based on a Sun 3/180 computer with a Motorola 68020 processor and a 68881 arithmetic co-processor.

3. RESULTS

3.1. Qualitative Inspection

From a qualitative viewpoint, it was difficult to assess the results. For 36 of the 40 kidneys, curves corresponding to parenchyma, collecting system, and blood background could be identified for all methods. Three kidneys, where this was not always possible, had measured GFRs of less than 10 mls/min. In such cases the use of feature-oriented methods, such as cluster analysis, tended to produce a kidney curve, a blood background curve, and a curve (and image) which could be assigned to liver or spleen; while spatial-oriented methods, such as spatial constraints, produced two similar looking kidney curves, together with a blood background curve. For the remaining kidney, standard curves were produced by all methods except for simple structure, which gave the error message "Singular matrix" and exited without producing curves or images.

It was noticeable that the method of spatial constraints, used with no normalisation (treatment two), produced curves with a fair degree of negativity present. It is thought that these apices may not be constrained to lie within the domain of non-negative factors, although this is currently under investigation.

All other methods produced curves and images which, in the majority of cases, appeared similar on visual inspection.

3.2. Problems Arising From Quantification

On occasion, it was not possible to measure certain parameters from the factor curves.

It proved impossible to measure the GFR of one kidney. Since conventional means had estimated a GFR of 1 ml/min, this was not altogether surprising.

The algorithm for deconvolution requires the definition of a plateau for the initial part of the impulse response function. This is equivalent to the definition of a minimum transit time, which was taken as 100 seconds. When this was not possible, deconvolution was said to have failed. This happened, for at least one method, in the case of 13 kidneys used in determining WKTT, and for 16 kidneys used in determining PTT. It may be of interest that deconvolution failed only once (for both WKTT and PTT) using conventional ROIs. The method with the poorest record of deconvolution failure was spatial constraints (both variants), followed by cluster analysis (both variants).

In order that the results be comparable, these kidneys were omitted from the analysis of the appropriate parameter. It was also necessary to omit the single kidney which caused the method of simple structure to fail. This affected the GFR parameter alone, as it had already been rejected for the others. Results were obtained for all three parameters from 23 kidneys.

3.3. Whole Kidney Transit Times

Table 1 shows the results obtained for WKTTs using 27 kidneys. In general, the values are of the correct order of magnitude when compared with those obtained using the conventional ROI method and the correlations with this method are good. The poorest correlation is obtained for unnormalised apex-seeking, while the worst agreement between means is found for normalised apex-seeking.

Table 1. Means and standard deviations obtained for whole kidney transit times and correlation with measured values. Twenty-seven kidneys were used.

METHOD	Mean	SD	corr. coeff.
Measured from ROIs	249.3	112.6	----
Treatment one (ASH)			
Apex-seeking (Di Paola)	373.6	129.2	0.86
Apex-seeking (Barber)	364.7	124.1	0.85
Gravitational clustering	273.9	137.3	0.95
Cluster Analysis (HAC)	262.5	122.1	0.90
Spatial constraints	267.6	132.1	0.88
Treatment two (KSN)			
Apex-seeking	287.5	132.6	0.75
Intersection method	295.8	132.6	0.80
Spatial constraints	251.7	139.0	0.86
Simple structure	255.5	133.7	0.83

3.4. Parenchymal Transit Times

Table 2 shows the results obtained for PTTs using 24 kidneys. In every case the estimate of the mean is high when compared with the conventional ROI method, while the correlations, with the possible exception of normalised spatial constraints, are poor.

Table 2. Means and standard deviations obtained for parenchymal transit times and correlation with measured values. Twenty-four kidneys were used.

METHOD	Mean	SD	corr. coeff.
Measured from ROIs	159.5	43.0	----
Treatment one (ASH)			
Apex-seeking (Di Paola)	268.0	95.3	0.42
Apex-seeking (Barber)	251.1	75.3	0.37
Gravitational clustering	206.2	82.8	0.35
Cluster Analysis (HAC)	194.0	70.0	0.20
Spatial constraints	204.5	71.6	0.64
Treatment two (KSN)			
Apex-seeking	256.0	118.7	0.15
Intersection method	272.4	112.2	0.19
Spatial constraints	181.9	71.5	0.26
Simple structure	163.5	49.0	0.15

3.5. Glomerular Filtration Rates

Table 3 shows the results obtained for GFRs using 38 kidneys. With two exceptions, the estimates are high when compared with the conventional ROI method while the correlations are, at best, fair. Simple structure gives a somewhat reduced estimate with good correlation. It is interesting that the other exception is unnormalised spatial constraints, where the mean is below that obtained conventionally and the correlation is reasonably good. This is in stark contrast to the results obtained for normalised spatial constraints, where the mean is exceptionally high and the correlation very poor.

Table 3. Means and standard deviations obtained for glomerular filtration rate and correlation with measured values. Thirty-eight kidneys were used.

METHOD	Mean	SD	corr. coeff.
Measured from ROIs	45.8	22.7	----
Treatment one (ASH)			
Apex-seeking (Di Paola)	105.2	56.6	0.63
Apex-seeking (Barber)	100.8	58.2	0.56
Gravitational clustering	70.8	38.2	0.59
Cluster Analysis (HAC)	70.0	36.5	0.59
Spatial constraints	139.4	131.0	0.04
Treatment two (KSN)			
Apex-seeking	89.3	46.2	0.36
Intersection method	98.6	56.4	0.41
Spatial constraints	32.7	17.3	0.78
Simple structure	35.6	22.9	0.85

4. DISCUSSION

4.1 Quantitative Analysis

4.1.1. Whole Kidney Transit Times.

It is clear that WKTTs obtained from COFADS are comparable with those obtained conventionally. The requirements for the calculation of this parameter are a blood curve and a whole kidney curve. The magnitudes of these curves are not relevant to the process. Therefore, apart from the fact that the two kidney curves require to be combined in the correct proportion, the normalisation to total counts is not tested. From this result, it is reasonable to assume that COFADS is capable of defining the correct shapes for the blood background and kidney factors.

4.1.2. Parenchymal Transit Times.

For PTTs, the results are not nearly so convincing. This parameter requires that a parenchymal curve and a blood background be defined. This result implies that the separation of the two kidney factors was not as successful as seemed to be the case from visual inspection.

4.1.3. Glomerular Filtration Rates.

The results for GFRs are unspectacular. This parameter is calculated using differences in temporally adjacent count levels. It also uses the parenchymal curve to determine the onset of the clearance phase. As such, it is unclear whether the normalisation to total counts or the definition of the parenchymal curve is responsible.

The results obtained for simple structure give some hope that reasonable estimates of absolute values can be obtained. The under-estimation of the mean GFR is, however, in excess of 20%, a value which is clearly not acceptable.

A point of concern is that, after simple structure, unnormalised spatial constraints produced the best results. These curves were reported as having many negative elements, which, in themselves, would not affect the GFR estimation since count differentials, rather than actual counts, are involved. It was also noted that some COFADS methods which constrained the apices to lie within the domain of non-negative factors found these apices on or close to domain limits. The implication that, in these cases, the "true" curves might be formed from apices lying outside the domain of non-negative factors would, if real, enforce a major re-thinking of the rationale behind the background constraint.

It is worth stating, however, that the use of the conventional ROI method as a "gold standard" is, at best, questionable.

4.2. Ease of Use of Methods

As well as comparing the methods quantitatively, it is useful to consider the ease with which each can be used. The following describes the operator interaction involved, given that the region under investigation has been defined.

4.2.1. Apex-seeking Method.

In the case of FADS, no operator interaction is necessary. In contrast, COFADS methods require varying degrees of input from the operator.

4.2.2. Intersection Method.

The main difficulty for the intersection method is the choice of an appropriate model. Although this is a fairly major task, it needs to be done only once, i.e. at the outset. The only other operator requirement is a definition of which curve corresponds to each structure.

4.2.3. Cluster Analysis.

In the case of cluster analysis, the operator is currently responsible for (a) determining the limits of pure background; (b) terminating the clustering; and (c) combining the clusters. Criteria already exist for automation of (a) and (b). The cluster(s) used to determine the background may be used to define the extent of pure background. A problem which results is that the edge determined in this manner is often contaminated by the adjacent structure. Ways of overcoming this problem are currently being investigated. Termination of clustering depends on several factors, including the distance between the closest clusters, and the effect an additional step has on the image and feature domains. It should not prove too difficult to derive an appropriate stopping criterion.

It is likely that the combination of clusters will prove the most difficult task to automate. In this respect, HAC would appear to have distinct advantages over GC, as its later termination results in fewer clusters, and its spatial interpretation is much clearer.

4.2.4. Spatial Constraints.

For spatial constraints, the operator is involved only in the partitioning of the image domain. This task is inherent in the method.

4.2.5. Simple Structure.

In the case of simple structure, the operator need only define a region of pure background. Again, this task is inherent in the method.

4.3. Computation Time

It is also important to consider the time taken by each method.

4.3.1. Treatment One Methods.

For treatment one, the PCA phase takes less than two minutes per view, while FADS takes about 10 seconds. Of the clustering techniques, GC and HAC take from 2 - 15 minutes and from 5 - 15 minutes respectively, depending on the complexity of the image. This time includes operator interaction. The method of spatial constraints takes around 10 seconds, but first requires that the operator partition image space, a task which can take about one minute.

4.3.2. Treatment Two Methods.

For treatment two, FADS (including PCA) takes around 1¼ minutes; the intersection method requires about 1½ minutes; while spatial constraints (excluding partitioning) and simple structure each require about 3½ minutes.

5. CONCLUSION

Although COFADS represents an advance on FADS for the quantification of radionuclide dynamic studies, it does not appear to be suitable, at least in its present form, for the determination of absolute values, or for values whose calculation requires the reliable determination of more than two structures. It is hoped, however, that this work will identify problem areas and lead ultimately to the goal of objective quantification of these studies.

Acknowledgement

The authors are grateful to Martin Sámal of Charles University, Prague for providing the original code for the method of simple structure.

References

Barber DC (1980). The use of principal components in the quantitative analysis of gamma camera dynamic studies. Phys. Med.Biol. 25:283-292.
Bazin JP, Di Paola R, Gibaud B, Rougier P and Tubiana M (1980). Factor analysis of dynamic scintigraphic data as a modelling method. An application to the detection of metastases. In: Information Processing in Medical Imaging. Di Paola R and Kahn E (eds), INSERM, Paris, pp. 345-366.
Benzécri JP (1973). L'analyse des correspondances. Dunod, Paris.
Britton KE, Nimmon CC, Whitfield HN, Hendry WF and Wickham JEA (1979). Obstructive nephropathy; successful evaluation with radionuclides. Lancet i:905-907.
Diffey BL, Hall FM and Corfield JR (1976). J. Nucl. Med. 17:352-355.
Di Paola R, Bazin JP, Aubry F, Aurengo A, Cavailloles F, Herry JY and Kahn E (1982). Handling of dynamic sequences in nuclear medicine. IEEE Trans. Nucl. Sci. NS-29:1310-1321.
Goris ML, Daspit SG, McLaughlin P and Kriss JP (1976). Interpolated background subtraction. J. Nucl. Med. 17:744-747.
Griffin DW, Donovan IA, Harding LK, White CM and Chackett KF (1982). Application of gravitational clustering analysis to liquid gastric emptying. Phys. Med. Biol. 27:1263-1267.

Hannequin P, Liehn JC and Valeyre J (1990). Cluster analysis for automatic image segmentation in dynamic scintigraphies. Nucl. Med. Commun. 11:383-393.
Houston AS (1984). The effect of apex-finding errors on factor images obtained from factor analysis and oblique transformation. Phys. Med. Biol. 29:1109-1116.
Houston AS (1986). The use of set theory and cluster analysis to investigate the constraint problem in factor analysis in dynamic structures (FADS). In: Information Processing in Medical Imaging. Bacharach SL (ed), Martinus Nijhoff, Dordrecht, pp. 177-192.
Houston AS (1988). The use of cluster analysis and constrained optimisation techniques in factor analysis of dynamic structures. In: Mathematics and Computer Science in Medical Imaging. Viergever MA and Todd-Pokropek A (eds), NATO ASI Series F:39:491-503.
Jambu M (1978). Classification automatique pour l'analyse des données. 1: Méthodes et algorithmes. Dunod, Paris.
Murase K, Mochizuki T and Hamamoto K (1988). Quantitative aspect of two-factor analysis applied to radionuclide gated cardiac studies. Eur. J. Nucl. Med. 14:190-195.
Nijran KS and Barber DC (1985). Towards automatic analysis of dynamic radionuclide studies using principal-components factor analysis. Phys. Med. Biol. 30:1315-1325.
Nijran KS and Barber DC (1986). Factor analysis of dynamic function studies using *a priori* physiological information. Phys. Med. Biol. 31:1107-1117.
Nijran KS and Barber DC (1988). The importance of constraints in factor analysis of dynamic studies. In: Information Processing in Medical Imaging. de Graaf CN and Viergever MA (eds), Plenum, New York, pp. 521-529.
Philippe L, Mena I, Darcourt J and French WJ (1988). Evaluation of valvular regurgitation by factor analysis of first-pass angiography. J. Nucl. Med. 29:159-167.
Šámal M and Kárný M (1989). Quantitative aspects of factor analysis in nuclear cardiology. Eur. J. Nucl. Med. 15:165-166.
Šámal M, Sůrová H, Kárný M, Maříková E, Pěnička P and Dienstbier Z (1988). The reality and meaning of physiological factors. In: Information Processing in Medical Imaging. de Graaf CN and Viergever MA (eds), Plenum, New York, pp. 499-519.
Sampson WFD, Hamilton CI and Jenkins L (1985). Estimation of GFR and functioning renal parenchyma from ^{99m}Tc-DTPA(Sn) renogram. Eur. J. Nucl. Med. 11:A48 (abstract only).
Schmidlin P (1979). Quantitative evaluation and imaging of functions using pattern recognition methods. Phys. Med. Biol. 24:385-395.
Villanueva-Meyer J, Philippe L, Cordero S, Marcus CS and Mena I (1986). Use of factor analysis in the evaluation of left to right cardiac shunts. J. Nucl. Med. 27:1442-1448.
Wright WE (1977). Gravitational clustering. Pattern Recognition 9:151-166.

CONFIRMATORY ASPECTS IN FACTOR ANALYSIS OF IMAGE SEQUENCES

M Šámal[1], M Kárný[2] and D Zahálka[1]

[1]Charles University and [2]Czechoslovak Academy of Sciences,

Salmovská 3, 120 00 Prague 2, Czechoslovakia

Abstract

Confirmatory approach in factor analysis of image sequences is specified by an employment of considerable initial information in the processes of factor extraction and rotation and by the possibility to verify hypotheses assumed in advance. Confidence interval for factor contribution is introduced and its utility in an assessment of factor significance demonstrated. Based on a partial apriori knowledge of resulting factor image, the method for a multiple subtraction of images is derived and its noise-rejection properties demonstrated. Quantitative transformation of factor curves into the compartmental scheme is described and the method is applied to a dynamic radionuclide study of renal function.

Keywords

interval estimate, image subtraction, compartmental model, scatter correction, dynamic scintigraphy

1. INTRODUCTION

Factor analysis of experimental data is usually considered to be an exploratory method. The adjective "exploratory" emphasizes the minimum need for apriori knowledge of data structure and specifies the method in contrast to confirmatory procedures fitting the data to a hypothetical model.

As applied to a sequence of medical images, factor analysis provides a few artificial factor images with specific information content (DiPaola et al, 1982). Factors may be of interest for several reasons. Data compression and noise rejection are regularly achieved positive outcomes. The performance of the method in this regard has been admitted without much controversy, as the precision of original data reconstruction can be evaluated by standard statistical tests.

On the other hand, the feature-extraction function of factor analysis is more controversial. This is because the specific and unique solution of analysis depends on additional information or an apriori hypothesis which, in current solutions, usually reflects assumptions on intrinsic data structure rather than the problem-related hypothetical model.

As regards factor analysis of dynamic radionuclide studies, perhaps the most frequent criticism stated by potential clinical users is the lack of a physiological model which could help to predict the results and interpret the factors. Factor images and time-activity curves are thus evaluated and interpreted empirically, in the same way as their original counterparts (with an important difference of added difficulty in understanding the obtained result).

In our opinion, the lack of a physiological model does not necessarily invalidate the method completely. As many reports have demonstrated, factor analysis of image sequences is reliable enough to provide its minimum goal, i.e. effective data reduction and transformation which are useful per se. The method does not produce unacceptable artifacts because, in principle, it naturally reflects the true information content in the data. Nevertheless, as the clinical potential of the method becomes useful due to the availability of reliable algorithms and more rapid computers, the validity of feature extraction performance should be verified.

In factor analysis, the application of constraints derived from a data-independent model was introduced by Joereskog (1969). The method has been accepted and developed by some authors while others, giving priority to classic data-based analysis, refused the model-based approach as "Procrustean".

In factor analysis of image sequences, several successful attempts to develop a confirmatory approach have already been made by D.C.Barber and his group (Barber and Nijran 1982, Nijran and Barber 1986, 1988, Martel and Barber 1989). The aim of our contribution is to discuss some confirmatory aspects of factor analysis in a more general way. The precision of factor estimates is analysed briefly before the two confirmatory applications are reported. Since recently factor analysis has been used most frequently for the correction of scatter in scintigraphic images and the analysis of dynamic scintigraphic data, the two related problems are discussed in more detail.

2. METHODS

2.1. Factor Analysis Model

The model of factor analysis and different ways used to estimate its parameters are described specifically elsewhere (DiPaola et al, 1982, Houston, 1988, Šámal et al, 1988). Briefly, an image sequence of m images having n (n>m) pixels each is framed into a data matrix Y (m,n) of dimension (m,n) and expressed as the sum of matrix X (m,n) of data reproduced by factor analysis and error matrix E (m,n)

$$Y = X + E = C.V + E \tag{1}$$

where matrix X is modelled as the product of matrix C (m,k) of factor coefficients relating m original images to k (k<m) factors and matrix V (k,n) containing individual values of k factors on n objects. In dynamic scintigraphy, matrix C contains factor time-activity curves while matrix V contains corresponding factor images.

The aim of factor analysis is to extract matrices C and V from Y. As the first step, matrix Y is factorized into its principal components

$$Y = U.W + E \tag{2}$$

where matrix U (m,k) contains the coefficients and matrix W (k,n) the pixel values of the first k significant principal components. Unknown matrices C and V are achieved by the transformation

$$C = U.T^{-1} \tag{3}$$

$$V = T.W \tag{4}$$

employing the regular transformation matrix T (k,k) which is derived using assumptions on matrices C, V and E.

2.2. Precision of Parameter Estimation in the Factor Analysis Model

The factor images and curves resulting from factor analysis are contaminated due to an error produced both by a not fully relevant mathematical model and uncertainty of its parameter estimates. While the model error is expected to affect the results in a systematic way, the parameter estimation error is primarily of a stochastic nature.

Solutions to the problem of random error estimation in factor analysis have been reported by many authors investigating the precision of the original data reconstruction in order to determine the proper number of significant factors (Hannequin et al, 1989, Malinowski, 1977a,b, Harman, 1960). However, in practice, the question of error distribution among the extracted factors also is of importance.

Using a Bayesian approach and the assumptions of both confidence interval symmetry and maximum sharpness, the probability that the factor curve falls into a specified interval has been calculated. The derivation of a final equation is based on normal regression and the search for the confidence interval of multivariate Student's distribution. The details of the derivation exceed the scope of this contribution and will be published separately. The confidence intervals for factor contributions have been calculated by integration over the confidence interval of the respective factor curve.

2.3. Initial Information on Factor Structure

Some problems inherent in image subtraction (those due to motion, attenuation, scatter, signal-to-noise reduction, etc.) can be overcome when more than two images are subtracted, the specific distortion in each of them is of a different weight. The procedure best suited to the purpose is factor analysis. However, if a sufficient amount of apriori information is known in advance (as is usual in scatter correction or dual-isotope imaging), eigenvectors need not be extracted at all; yet the result is equivalent to complete confirmatory factor analysis including the rotation to oblique factors. The result can be evaluated statistically as is usual in regression experiments.

The method is based on the assumption that some values in factor images are known in advance and that the factor analysis model, as described by Eq. (1), can be read in the opposite direction. Factor images thus can be considered as linear combinations of original images and estimated by regression.

Each image of an analysed image sequence is considered to be the superposition of two partial images: the image of the structure of interest and the "remaining" image which contains the other structures (background, noise, motion artifacts) that should be rejected. In every original image, the two partial images of hypothetical structures are of different weights.

Using this idea, matrix X in Eq. (1) can be decomposed into two parts, X_i and X_r

$$Y = X_i + X_r + E \tag{5}$$

where (m,n) matrix X_i represents the image of interest multiplied by m different weights, and (m,n) matrix X_r contains the same arrangement of the remaining image weighted by another m different weights. It is not important whether the matrix X_r is regarded as being formed by several factors or by just one composed dynamic structure. As above,

$$Y = C_i.V_i + C_r.V_r + E \tag{6}$$

where C_i and C_r are matrices (m,1) and (m,k-1) containing coefficients of the factor of interest and the remaining factors, respectively, and V_i and V_r are corresponding factor images (1,n) and (k-1,n). After multiplying Eq. (6) by C_i' from the left

$$C_i'.Y = C_i'.C_i.V_i + C_i'.C_r.V_r + C_i'.E \tag{7}$$

the columns of matrix C_r can be selected to be orthogonal, both to each other and also to C_i. This selection is possible because it does not harm the calculation of the image of interest and because the structure of the remaining image and its coefficients is not important. Then

$$C_i'.Y = C_i'C_i.V_i + C_i'.E \tag{8}$$

$$V_i = (C_i'.C_i)^{-1}.C_i'.Y + E_i \tag{9}$$

where the error term

$$E_i = - (C_i'.C_i)^{-1}.C_i'.E \tag{10}$$

keeps the statistical characteristics of matrix E and makes the least-squares solution of equation (9) possible. After another substitution

$$b_i = (C_i'.C_i)^{-1}.C_i' \tag{11}$$

the least-squares estimate of regression coefficients b_i (1,m) can be performed

$$V_i = b_i.Y + E_i \tag{12}$$

provided that enough entries in vector V_i are known. Then, using the partial knowledge of V_i and the corresponding part of matrix Y, both the vector b_i and, consequently, the complete matrix V_i are estimated.

In many applications of imaging practice, the pixels can be found in an image matrix where the structure of interest is not present. For instance, when scintigraphic images are registered at different photon energy windows the region can be selected where the photopeak signal is expected to be minimal. In the image sequence of digital subtraction angiography a region can be selected where no useful contrast is expected, etc. At all the selected locations the entries of the vector V_i are assumed to be zero. If there is at least m (number of images in a sequence) of such locations, the subset of homogeneous linear equations can be derived from Eq. (12) and solved for m unknown entries of vector b_i while one of them is assigned an arbitrary value

$$V_{0,i} = b_i . Y_0 + E_{0,i} \tag{13}$$

where $V_{0,i}$ is the row vector (1,p>m) containing zero elements only selected as a subset of V_i, Y_0 is the submatrix (m,p) of matrix Y containing the identical subset of columns as the vector $V_{0,i}$, and $E_{0,i}$ is the same subset of E_i. The arbitrary scale in the resulting image V_i can be transformed to the proper one using the method developed for factor analysis (Šámal et al, 1988). However, due to the orthogonality of "factors" V_i and V_r, only the factor image of interest will be of real scale. If, under special circumstances, other than zero values are known in advance, the set of equations becomes heterogeneous and can be solved in proper scale without arbitrariness.

The Eq. (13) can provide a sensible estimate of b_i only if some additional algebraic requirements are satisfied. The most important one is that only those zero entries in V_i are useful which correspond to non-zero entries in identically located pixels of matrix V_r. In other words, the part of the structure which is going to be subtracted from the original image must be present in the region selected as the reference zero region of V_i. Otherwise, the method described above cannot be used and standard factor analysis should be applied instead.

If more information is available in advance, more than one structure of interest can be derived from the image sequence in the same way. The only condition is that the specific locations of zero entries in expected factor images have to be known. As the derived equations represent the standard regression equations, the corresponding statistical measures of estimate validity can be applied.

2.4. Initial Information on Factor Function

The ultimate goal of dynamic radionuclide study is to describe investigated metabolic or functional processes quantitatively in terms of an accepted physiological model. Both compartmental and deconvolution analyses are based on well established physiological models of organ function (Blaufox, 1989, Britton and Nimmon, 1989). Their results, however, are critically dependent on the quality of input data. Moreover, the models which are simple enough to be evaluated and related to measurable quantities do not correspond to real physiology, while more complicated models are difficult to verify as a sufficient number of variables usually cannot be measured in man.

Factor analysis, on the other hand, provides formal extraction of significant dynamic data components. Although not physiological in a true sense, they reflect the information content of the study and its intrinsic structure properly. If the acquisition scheme is designed well to detect some physiological process and the relevant information is involved in data then factor analysis necessarily grasps this information. It can be used then not only as a source of input data for a physiological model but also to evaluate and modify factor analysis results by feedback.

Factor analysis is designed to provide images and curves which are independent of projection superposition in scintigraphic images. They can be used to evaluate patient data in a similar way to that of compartmental analysis and invasive pharmacokinetic distribution studies. Indeed, there are reports supporting the reality of this expectation (Oster et al. 1987, 1989).

The compartmental model can be described by a set of first order differential equations

$$D = -C'.K \tag{14}$$

where D is an (m,k) matrix containing time derivations of activity in m time points and k compartments, C' is a (m,k) matrix of compartmental time-activity curves, and K is a (k,k) matrix of rate constants (fractional turnover and transfer rates).

The model is established if matrix K is known (for details see e.g. Rescigno et al, 1990, Wellner, 1986) . Corresponding compartmental time-activity curves then can be formed using the set of exponentials and coefficients derived as eigenvectors and eigenvalues of matrix K

$$C = M.N \tag{15}$$

where M is a (k,k) matrix of k eigenvectors of matrix K, and N is a (k,m) matrix of k exponentials sampled in m time intervals (Mueller-Schauenburg 1973, 1976, Wellner 1986). Exponents are equal to eigenvalues of matrix K.

The entries of matrix K can be estimated using known or assumed parameters of the model, and curves C can then be calculated and used as the coefficients in Eq. (1). The correspondence between the model and measured data thus can be evaluated immediately. On the other hand, the curves C achieved by factor analysis and their differences D can be substituted in equation (14) which then can be solved numerically for the unknown matrix K. The rate constants suggested by factor analysis can be compared with those estimated from a model or by independent methods.

Evaluating the compartmental model, factor analysis can help to visualize the spatial distribution of exponentials and also to evaluate the contribution of respective exponentials to individual compartmental time-activity curves

$$X = M.N.V \tag{16}$$

where the product N.V demonstrates the spatial distribution of exponentials.

Eq. (14) and (16) can be used either to formulate constraints for a compartmental model-related solution in factor analysis, or, alternatively, to verify the validity of results achieved by factor analysis. In a similar way, the procedures of deconvolution analysis can be applied to factor analysis output.

3. RESULTS

3.1. Interval Estimate of Factor Contribution

Interval estimates of factor contributions have been calculated in phantom and renal dynamic studies. The phantom consisted of two factors with significant overlap of their structures.

Table 1: Interval estimate of factor contributions in the phantom. Contributions are given in % of reproduced count rate. Relative error is the ratio between the confidence interval and the contribution in %.

noise level % maximum pixel value	95% confidence interval factor 1	factor 2	relative error factor 1	factor 2
7	19.6 - 20.1	19.6 - 19.9	2.4	1.6
14	16.1 - 16.5	16.2 - 16.9	2.9	4.0
21	7.5 - 7.8	9.0 - 9.5	3.7	5.0

The true contribution of each factor to the data reproduced by factor analysis was 21 % while the background was 58 %. Three different levels of random noise have been introduced with maximum amplitude of 7, 14, and 21 %, respectively, of maximum pixel value in the study. Results are given in Table 1.

Addition of noise to the data reduces the contrast of factor dynamic structures. Although the relative error of factor contribution increases with noise, even in data where the contrast has been reduced to approximately 30 % the error is about 5%.

Dynamic renal study consisted of 15 images of one kidney selected from a standard $^{99\text{-}m}$Tc DTPA study. Factor analysis has been performed firstly on rough data and secondly after repeated 9-point smoothing of images. Results are given in Table 2.

Table 2: Interval estimate of factor contributions in dynamic renal study. Contributions are given in % of reproduced count rate. Relative error is the ratio between the confidence interval and the contribution in %. Factor 1 corresponds to renal parenchyma, factor 2 to renal pelvis.

number of smoothing procedures	95% confidence interval		relative error	
	factor 1	factor 2	factor 1	factor 2
0	30.2 - 38.1	40.8 - 45.0	23.0	9.6
1	39.6 - 45.3	41.3 - 43.7	13.4	5.5
2	43.7 - 48.7	43.9 - 46.3	10.8	5.3
3	48.3 - 53.5	45.8 - 48.1	10.2	4.8

Smoothing of data increases the contrast of factor dynamic structures. Originally smaller contrast of greater structure increases faster. The improvement in relative error is significant with first smoothing only, while repeating the procedure has less effect.

3.2. Multiple Image Subtraction

The procedure derived in 2.3. has been verified with phantoms and a series of spin-echo NMR images (Šámal et al. 1990). In preliminary experiments, good suppression of complex motion artifacts has been achieved in the image sequences of digital subtraction angiography.

As an illustration, a simple phantom is presented consisting of 5 images in a 3x3 matrix. Images are composed of two components simulating the "photopeak" and the "scatter" count rate (Figure 1).

(a)			(b)		
2	0	0	7	8	6
0	10	0	8	10	8
0	0	2	6	8	7

Figure 1: Two images representing "photopeak" (a) and "scatter" (b) from which the phantom sequence has been formed. The ratio of (a) and (b) in respective images was 2:8, 3:7, 5:5, 7:3, and 8:2. Pixel value of background was 5, maximum noise amplitude +/- 5.

Additional background and random noise has been introduced. The result of factor-like multiple image subtraction has been compared with optimized subtraction of two images and the changes of contrast-to-noise ratio have been registered. Comparisons of contrast-to-noise ratio in phantom experiments are given in Table 3.

Table 3: Comparison of contrast and noise in subtracted and factor-like images. Noise is measured as the standard deviation of background values. Contrast is measured as a ratio between the central-pixel value and the background mean value in the subtracted image. Contrast-to-noise ratio is related to the best subtracted image.

	noise	contrast	contrast / noise
best subtracted image	0.603	8.36	1.00
factor-like image from sequence of 2	0.255	3.61	1.02
factor-like image from sequence of 4	0.293	5.42	1.33

The experiment demonstrates that with two images the contrast-to-noise ratio is the same in both the subtraction and factor-like images while with four images the contrast-to-noise ratio increases in the factor-like image.

3.3. Physiological Model of Renal Function

As a preliminary experiment with MAG_3 dynamic renal study, the factor curves and their differences have been introduced into Eq. (14) and the matrix K of rate constants calculated. The results of a child patient with single kidney, normal function of parenchyma but delayed outflow from the pelvis are given in Table 4.

Table 4: Rate constants (min^{-1}) in compartmental model of kidney derived from factor curves. Diagonal entries represent fractional turnover rates in the respective compartments. Off-diagonal entries represent fractional transfer rates in the row-to-column direction.

	blood	parenchyma	pelvis
blood	-0.117	0.147	-0.015
parenchyma	-0.627	0.318	0.273
pelvis	0.072	-0.306	0.228

Rate constant for renal plasma clearance is -0.117 min^{-1}. The reciprocal values of diagonal entries are equal to mean transit times. Thus, in the examined kidney, the parenchymal mean transit time is 3.14 min while the pelvic one is 4.39 min. The clearance is below and both the transit times above the mean values reported by Carlsen et al (1986) but still within the normal range.

Using the data in Table 4, the corresponding compartmental system is described in Figure 2.

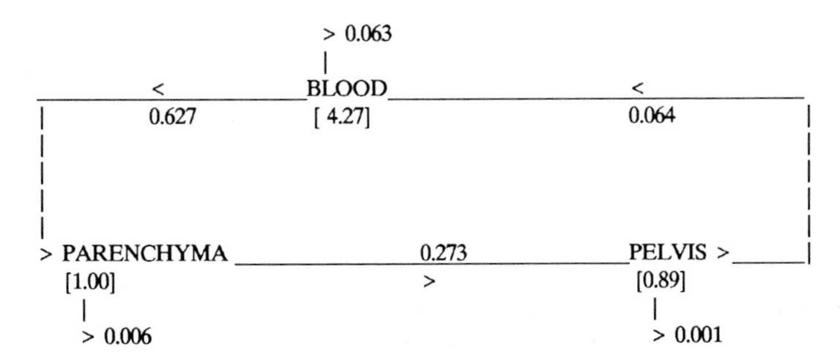

Figure 2: Compartmental system derived from Table 4. Rate constants are normalized for the kidney pool and given in reciprocal minutes. Pools (the numbers in brackets) are related to renal parenchyma. Arrows indicate the direction of flow.

The size of "blood pool" is limited to relatively small field of view. The flow from the pelvis to the blood represents the communication to outer environment because the bladder is outside the region selected for analysis.

The factor curves have been used for the reconstruction of the compartmental model directly without smoothing, fitting, or other preprocessing. Both the procedure of factor analysis itself and the use of least squares in the solution of Eq. (14) support the relative robustness of the procedure against noise.

4. DISCUSSION

The methods used to solve Eq. (1) can be classified from several points of view as working in space versus time (or energy) domain, using different ways to normalize input data or different procedures to subtract the background count-rate. In our discussion, the most relevant criterion is that of the amount of initial information supplied from external (data-independent) sources.

Provided a sufficient amount of information is available, there is no need for factor analysis. Eq. (1) is then solved directly as is usual when region-of-interest (ROI) dynamics are evaluated in functional scintigraphic studies. The specific part of matrix V is known and corresponding entries of matrix C are calculated. The method described by Martel and Barber (1989) is very close to this scheme as it is based on the assumption that the image of the specific dynamic structure is known. An alternative method reported by Barber's group (Barber and Nijran, 1982, Nijran and Barber, 1986) requires knowledge of characteristic time course in one or more columns of matrix C before the complete solution of Eq. (1) is found. With Barber's methods, very good results are achieved at the price of a great deal of apriori information. In practice, however, this information is not always available.

On the other hand, when nothing useful is known in advance, the solution of Eq. (1) relies on purely formal criteria like the orthogonality of components and the variance maximization. In fact, the analysis of principal components in Eq. (2) represents exactly such an extreme, and also the different types of orthogonal factor rotation belong to this category of solutions. Using formal criteria, the analysis of any data can always be performed. However, while providing useful description of the information contained in data matrix, the factors usually cannot be interpreted more than formally.

Although the majority of factor analysis solutions exist between the above extremes, the scale of additional information introduced in order to solve Eq. (1) still can be used to differentiate the methods. The minimum information (positivity criterion) is used by the classic apex-seeking procedure (Barber, 1980, DiPaola et al, 1982) which has been recalled by a recent contribution of Van Daele et al (1991). Methods demanding more detailed information of dynamic structures are represented by those of Nijran and Barber (1988) and Houston (1988) while the approach introduced by our group (Šámal et al, 1988, 1989) has been motivated by the search for a compromise assuming simultaneously both the positivity condition and superficial knowledge of factor image structure. In our opinion, the constraints introduced to guide extraction and rotation of factors should be tight enough to prevent ambiguity but, at the same time, sufficiently general in order to ensure the data-based derivation of factors.

Considering the stochastic error in factor estimation the results given in Table 1 and 2 confirm the relative robustness of the method with regard to noise. Different methods of factor analysis may differ significantly in this regard because the result depends on which product of data matrix is used for subsequent factorization. This point has been discussed in detail in our previous IPMI contribution (Šámal et al, 1988) and the derivation of interval estimate has only allowed the error to be measured in the factor-specific way.

Once a certain level of apriori information is available, it is possible to re-invert the factor analysis model in Eq. (1) and, consequently, the problem to find a specific structure can be solved by regression. The improvement in contrast-to-noise ratio (reported originally by Martel and Barber, 1989) increases with the processed number of images. A greater amount of useful information thus can be extracted from observed data and the result can be evaluated statistically.

In contrast to the method described by Martel and Barber (1989), much less additional information is necessary using multiple subtraction. In contrast to the method of Nijran and Barber (1988) and that reported by Todd-Pokropek and Gagnon (1989, 1990), there is no need for principal component analysis. Nevertheless, the result of multiple subtraction is equivalent to that of complete factor analysis including the rotation to oblique factors.

Besides performing image subtraction, the procedure described above is useful to interpret true factor images, too. In our opinion, the idea of image subtraction is more acceptable for clinical users than the not-so-friendly terms used in factor analysis. Moreover, the description of a factor image as a result of multiple subtraction of original images is not oversimplification. Factor image is not a suspicious construction but a linear combination (or weighted sum) of original images. Factor curves then represent the weights of factor images in the images of original sequence.

Transformation of dynamic data into one simple matrix of quantitative parameters as described in section 3.3. may provide a clinical user with the required numbers on organ function. Indeed, the factor analysis is the only method able to supply the reasonable input for compartmental experiment in man. Considerable advantage is the robustness against noise. In contrast to the method described by Nijran and Barber (1986), there is no need for apriori characterization of factor curves except of the acceptance of first-order kinetics in Eq. (14).

If the factor rotation is not "physiological", the numbers in the extracted matrix K become unreasonable. This is the case, for instance, when the curves of orthogonal factors are introduced into the Eq. (14). However, when pathology is present, the decision between the wrong rotation and function may appear difficult. In our opinion, the evaluation of factor images and curves together with the matrix K simplifies the problem to a very acceptable level.

In man, different compartmental models can be verified using factor analysis and Eq. (14). The fit between the matrix K derived from the factor curves and that derived from a physiological model can be evaluated statistically. Besides the verification of both methods, the purpose of comparative experiments is to examine the possibility of direct introduction of Eq. (14) as a constraint in factor analysis.

Using factor analysis in a quantitative way, two cautions about factors should be remembered. First, the resulting factors represent dynamic structures which do not always correspond to defined anatomical units. In kidney example, the borders between "parenchyma" and "pelvis" or between "body background" and "parenchyma" are strictly defined only if the dynamics of the respective dynamic structures are strictly separated. Otherwise, several different structures may be involved in one factor, however, the result of factor analysis still reflects the data consistently. A second caution is necessary if a diagnostically important dynamic structure is a linear combination of other prominent parts in the image. Such a structure is not extracted by factor analysis as a specific factor but usually is artificially divided among all the remaining factors. Thus the image of a spleen in equilibrium cardiac study is decomposed into two dynamic structures with atrium-like and ventricle-like dynamics because the spleen itself manifests exactly the same dynamics as the overlap between the atria and ventricles.

Validity of physiological constraints derived from the compartmental model remains to be verified clinically. In our opinion, however, the results of preliminary experiments fully justify a further research in the field.

5. CONCLUSIONS

In contrast to exploratory applications, the confirmatory approach in factor analysis of image sequences is specified by an employment of more initial information and by a possibility to verify the fit between the results of data analysis and the hypotheses assumed in advance. Different methods are classified according to the amount of external information introduced in calculation. The methods cover the wide range between the standard regression techniques and the classic factor analysis. The emphasis is given on both the availability of external information and the prevention of data deviation when the information supplied is not fully relevant.

Evaluation of random error in factor analysis has been extended on the level of individual factors. The significance of extracted factors thus can be evaluated using not only the factor contribution to reconstructed data but also its confidence limits.

The method for multiple subtraction of images has been derived on the ground of partial apriori knowledge of a resulting image. Using principles common in factor analysis and related methods, the procedure requires neither an analysis of principal components nor a detailed description of unknown structures. In comparison with standard subtraction, the method improves the contrast-to-noise ratio in subtracted images. It is intended to be used as a scatter correction technique in scintigraphy, however, its application is not limited to radionuclide imaging.

The method for a quantitative transformation of factor curves into the compartmental scheme has been derived and applied to renal radionuclide dynamic study. In contrast to standard methods of dynamic data processing, the procedure requires neither preprocessing of data nor preliminary hypotheses on compartmental model except the basic conception of the first-order kinetics. Resulting matrix contains quantitative description of all the compartments detected by factor analysis in a form related directly to comparable clinical measures.

Both methods provide the results which can be evaluated in terms of a hypothesis assumed before factor analysis experiment. The software for IBM compatible PC has been developed and the methods are verified experimentally.

Acknowledgements

The authors acknowledge with thanks the kind help of Ing. J. Zmrzlík with the preparation of camera-ready form of the contribution.

References

Barber DC (1980). The use of principal components in the quantitative analysis of gamma camera dynamic studies. Phys.Med.Biol. 25:283-292.

Barber DC and Nijran KS (1982). Factor analysis of dynamic radionuclide studies. In: Nuclear Medicine and Biology. Raynaud C (ed), Pergamon Press, Paris, pp. 31-34.

Blaufox MD (1989). Compartment analysis of the radiorenogram. In: Evaluation of renal function and disease with radionuclides. Blaufox MD (ed), Karger, Basel, pp. 98-107.

Britton KE and Nimmon CC (1989). The measurement of renal transit times by deconvolution analysis. In: Evaluation of renal function and disease with radionuclides. Blaufox MD (ed), Karger, Basel, pp. 108-116.

Carlsen O, Kvinesdal B and Nathan E (1986). Quantitative evaluation of iodine-123 hippuran gamma camera renography in normal children. J.Nucl.Med. 27:117-127.

DiPaola R, Bazin JP, Aubry F, Aurengo A, Cavailloles F, Herry JY and Kahn E. (1982). Handling of dynamic sequences in nuclear medicine. IEEE Trans.Nucl.Sci. NS-29:1310-1321.

Hannequin P, Liehn JC and Valeyre J (1989). The determination of the number of statistically significant factors in factor analysis of dynamic structures. Phys.Med.Biol. 34:1213-1227.

Harman HH (1960). Modern factor analysis. The University of Chicago Press,Chicago, IL.

Houston AS (1988). The use of cluster analysis and constrained optimisation techniques in factor analysis of dynamic structures. In: Mathematics and Computer Science. Viergever MA and Todd-Pokropek AE (eds), Springer - Verlag Berlin, Heidelberg, pp. 491-503.

Joereskog KG (1969). A general approach to confirmatory maximum likelihood factor analysis. Psychometrika 34:183-202.

Malinowski ER (1977a). The theory of error in factor analysis. Anal.Chem. 49:606-612.

Malinowski ER (1977b). Determination of the number of factors and the experimental error in a data matrix. Anal.Chem. 49:612-617.

Martel AL and Barber DC (1989). A new approach to dynamic study analysis. In: Information Processing in Medical Imaging. Proceedings of XI-th International Conference, June 19-23, 1989, Berkeley, CA.

Mueller-Schauenburg W (1973). A new method for multi-compartment pharmacokinetic analysis: the eigenvector decomposition principle. Eur.J.Clin.Pharmacol. 6:203-206.

Mueller-Schauenburg W (1976). On some applications of the eigenvector decomposition principle in pharmacokinetic analysis.In: Mathematical Models in Medicine. Berger J et al (eds), Springer-Verlag Berlin, Heidelberg, New York, pp.226-242.

Nijran KS and Barber DC (1986). Factor analysis of dynamic function studies using a priori physiological information. Phys.Med.Biol. 31:1107-1117.

Nijran KS and Barber DC (1988). The importance of constraints in factor analysis of dynamic structures. In: Information Processing in Medical Imaging. deGraaf CN and Viergever MA (eds), Plenum Press, New York, pp. 521-529.

Oster ZH, Som P, Bazin JP, DiPaola M, Raynaud C, Atkins HL and DiPaola R (1989). The role of factor analysis in the evaluation of new radiopharmaceuticals. Nucl.Med.Biol. 16:85-89.

Oster ZH, Som P, Raynaud C, Atkins HL, Bazin JP, DiPaola M and DiPaola R (1987). Factor analysis for in-vivo evaluation of radiopharmaceuticals. J.Nucl.Med. 28:A 519.

Rescigno A, Thakur AK, Brill AB and Mariani G (1990). Tracer kinetics: a proposal for unified symbols and nomenclature. Phys.Med.Biol. 35:449-465.
Šámal M, Kárný M, Sůrová H, Pěnička P, Maříková E and Dienstbier Z (1989). On the existence of an unambiguous solution in factor analysis of dynamic studies. Phys.Med.Biol. 34:223-228.
Šámal M, Kárný M, Sůrová H and Zajíček J (1990). Feature extraction from NMR images using factor analysis.In: Tissue Characterisation in MR imaging. Higer HP and Bielke G (eds), Springer-Verlag Berlin, Heidelberg, pp.161-164.
Šámal M, Sůrová H, Kárný M, Maříková E, Pěnička P and Dienstbier Z (1988). The reality and meaning of physiological factors. In: Information Processing in Medical Imaging. deGraaf CN and Viergever MA (eds). Plenum Press, New York, pp. 499-519.
Todd-Pokropek A and Gagnon D (1989). Image reconstruction using energy information: convergence criteria. Nucl.Med.Commun. 10:248.
Todd-Pokropek A and Gagnon D (1990). The use of principal component analysis methods for manipulating images acquired over a wide energy window. In:Radioactive Isotopes in Clinical Medicine and Research. Hoefer R and Bergmann H (eds), Schattauer Verlag, Stuttgart, in press.
VanDaele M, Joosten J, Devos P, Vandecruys A, Willems JL and DeRoo M (1991). A new vertex-finding algorithm for the oblique rotation step in factor analysis. Phys.Med.Biol. 36:77-85.
Wellner U (1986). Kinetic models of metabolism. Nucl.-Med. 25:138-141.

TRAINABLE MODELS FOR THE INTERPRETATION OF ECHOCARDIOGRAM IMAGES

Richard A Baldock[1]

MRC Human Genetics Unit, Western General Hospital
Crewe Road, Edinburgh EH2 2XU, UK.

19th December 1990

Abstract

In this paper we report on the use of explicit models in the interpretation of echocardiogram images. The problem is considered as an example of a general biomedical image interpretation task and the modelling techniques used can be applied to a wide range of problems. The models are built as a hierarchy of components and the parameters of each component are determined from training examples and prior "expert" knowledge. For each component the model encodes information about the average case and information about the expected distributions, for example sizes, shapes and positions. The models are used within the interpretation process to assess hypothesised matches and to guide further processing. Details of the design and implementation of the model components, the refinements and training techniques and the the results of application to the echocardiogram images are presented.

Keywords

Biomedical image interpretation; ultrasound; cardiac measurement; model-based methods; shape models.

1. INTRODUCTION

The problems associated with the interpretation of echocardiograms are typical of a wide variety of biomedical image analysis tasks. The images are noisy, contain many imaging artefacts (Kremkau and Taylor 1986) and are obtained from a domain that is highly variable and flexible. Some imaging modalities have better resolution and less noise but are correspondingly required to provide more detailed information and so the interetation task is always at the limit of what can be achieved with the equipment. The only way to extract the maximum amount of information with respect to a particular task is to use all available knowledge about the image content. In biomedical applications it is often the case that the structures present in the image, and their orientation, are known and the task is to find the best possible detailed match to allow feature mesurement and estimation. For this reason we employ model-based methods, eg. see Binford (1981) and Besl and Jain (1985), in which the prior information is encoded as a symbolic or iconic model (in our case the anatomy) and we need techniques to use this model to interpret the image. In this paper we present part of a model-based system for interpreting echocardiogram images.

Another consideration for the design of a system to analyse biomedical images is the fact that imaging equipment is always improving and the required task is always changing. This implies that

[1]Supported on an SERC grant as part of the Alvey project MMI/134

the system must be readily modified and be able to adapt to new conditions. To address this problem the models must allow convenient retraining or adaptation, ie. there must be mechanisms for easily building models and training the various parameters by example. Ideally the systems in routine use should be continuously adapting and optimising to their current usage.

So far the term model is used in a very general sense, typically "model" is used to describe an object, eg. its geometry, but also to describe a process, eg. the imaging mechanism, and in general any knowledge about anything can be termed a model. The only common theme between different uses of the term model is the notion of *prescription* - a model is something you can use to generate an instance. In its most trivial or iconic form the model is just a list of instances and the rule for generating an instance is simply "pick-one". In a more complex form the model may be a set of equations to be solved or a series of rules to be followed. In this discussion the models are principally of the iconic variety although not at the level of images but at the level of anatomical components. Procedural ana rule-type knowledge can be attached but may be implicit in general-purpose methods and rule-bases for controlling the matching. Imaging models will be implicit in terms of the procedures, cues, matching cost functions, geometric transforms and matching/feedback control. It is our belief that the models must be as *explicit* as possible to allow easy inspection and validation and so that the anatomical and physiological parts are clearly separated from the imaging and processing model. This should allow the models to be used with different modalities and the comparison of results at the anatomical level.

A basic requirement of the model system is that it must be able to encode biological shape *and* normal variation, and provide a mechanism by which the information can be employed during the matching process. For the application considered here, variation over time is also important. The model must also contain information about the imaging modality, for example appearance of structures and scale. The model is to be used to generate a confidence or probability of a match and so must encode all the required probability distributions, including information about the reliability of the techniques and software used to analyse the image. The required information about anatomy can be obtained independently but the modality and software reliability data can only be established by experiment and training. For this reason we believe that the modelling and model representation, training and matching are bound tightly together and must be considered simultaneously and developed as a whole.

The knowledge to be built into the system includes:

- anatomy - sizes, shapes etc.
- physiology - blood flow, time variation.
- radiology - specific modality and machine charcteristics, artefacts, specific view parameters etc.
- case history - age, sex, clinical condition etc.
- imaging model / histology - tissue response.
- image processing expertise
- inference and control mechanisms.
- clinical parameter estimation - numbers etc. required by the clinician

One final consideration for any system to build and use such models is that it must be possible to build from small components so that work at one level of detail can be used as a component of a larger structure. Thus it is clear that a convenient structure to adopt for building model descriptions is a hierarchy, large-to-small, general-to-particular. Models built in this way should be easier to understand and manipulate.

In this paper we describe the modelling part of a model-based system to interpret echocardiogram images. The matching algorithm we assume is to first identify some cues, use the model to predict or hypothesise interpretations of the cues, then use the model plus the image to refine and test the result. This strategy is termed *Cue-Hypothesis-Refine/Verify.* The primary cue finding-process is considered elsewhere (Bailes 1990), in this paper we discuss only the model representation and the

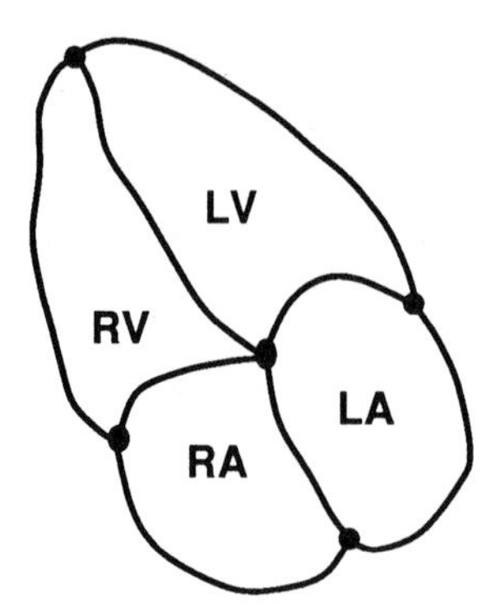

Figure 1: The four-chamber model of the the heart, eight polylines and five vertices.

associated refinement and training mechanisms. In section 2.1 we discuss the general requirements of the models for our particular task and in section 2.2 we present the details of the model component representation. In section 2.3 we show how a dynamic programming technique can be used to refine intial estimates of a structure using information encoded in the model and in section 2.4 the model training procedures. Section 3.1 describes some results of using these techniques on echocardiogram images and in section 4 we discuss the results and indicate how the full system may operate.

2. METHODS

2.1. Models

There are many ways in which the information for any particular task can be encoded and if each representation holds all the information required then in principle it is possible to converted from one form to another and all such representations are equivalent. In practice, however, certain operations are easier or more efficient in one model representation as opposed to another and it this consideration that determines the actual implementation. In the echocardiogram example the structures of interest are the boundaries of the heart chambers and are most appropriately represented by a number of linear components. This is one type of primitive required in a complete system but is sufficient to explore functional requirements of a modelling system to represent and match structures in the echocardiogram example.

In this paper we concentrate on the four-chamber view of the heart which provides sufficient complexity to test the modelling mechanisms. For our purpose we represent the four-chamber view of the heart as a number of (curved) line primitives, with relationships between lines that have a common end vertex. This model is shown schematically in figure 1 and is comprised of eight polylines and five vertices interconnected as shown. Each line component represents a cut through the corresponding muscle wall except for the valves which are expected to highly variable. The typical measurements required for the ventricles are of an internal boundary which we assume we can locate once the myocardium has been matched.

Figure 2 shows a highly simplified symbolic version of the same model where the detailed shapes of the iconic model are encoded within the corresponding node and the common vertices are represented as relationships (not depicted) between nodes on the graph.

Each component must hold information about the expected size, shape and distributions of the corresponding anatomical structure, and have associated imaging information, for example scale factors, image position and expected grey-level profile. The information encoded in the model must be accessible for a number of purposes. Firstly it should be possible to generate examples of the corresponding stuctures, with no other information this will be the average of the distributions represented but may be modified in the light of additional evidence, for example cues that have been

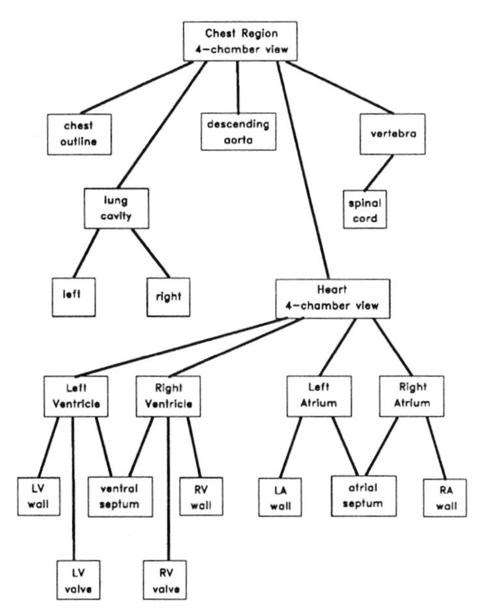

Figure 2: A possible model for the 4-chamber view of the heart and the surrounding region of the body. The links shown indicate **part-of** relationships. Each node would contain information about sizes, shapes, positions and there are many other relationships between nodes, in particular **next-to**. This model is certainly not complete but is just an example of how the data will be structured.

located in the image or information in the medical case-history. Secondly, given a hypothesis for a particular structure it must be possible to use the model to improve or refine the match of the hypothesis with respect to the image and model. Finally, the refined hypothesis or "interpretation", must have an associated probability or degree of belief that provides a measure of how well the image structure matches the model structure, ie the model must be capable of assessment. These three capabilities of prediction, refinement and assessment are central to the application of the model to a task. A further important capability in any complete system is to be able fit a given example or set of examples to generate a model instance. Implicit in this mechanism is the ability to update a given model with a further example.

2.2. Polyline Structure Representation

Each linear component of the model must encode information part of which is anatomical, eg size and shape, and part of which is the imaging model, eg position, scale and appearance. It is important that information arising from independent knowledge sources is only linked by the application so that models at the anatomical level can be reused with other images.

For this purpose the structures used to encode the model information are shown in figures 3-4. The polyline model shown in figure 3, describes the shape of the anatomical structure and its expected size and position within the model coordinate frame.

When applied to a particular imaging modality the various scale factors, position and angle are modified in the light of knowledge of the scanning machine. Additionally a profile model, also shown in figure 3, is attached to form an *image-polyline model*, figure 4, which holds sufficient information to allow a match to an actual image. This is now a component model specialised to a particular modality.

Each polyline has two special points which we refer to as fiducial points, by default at each end of the line, but in principle anywhere. The scale, angle and position of the line are determined by

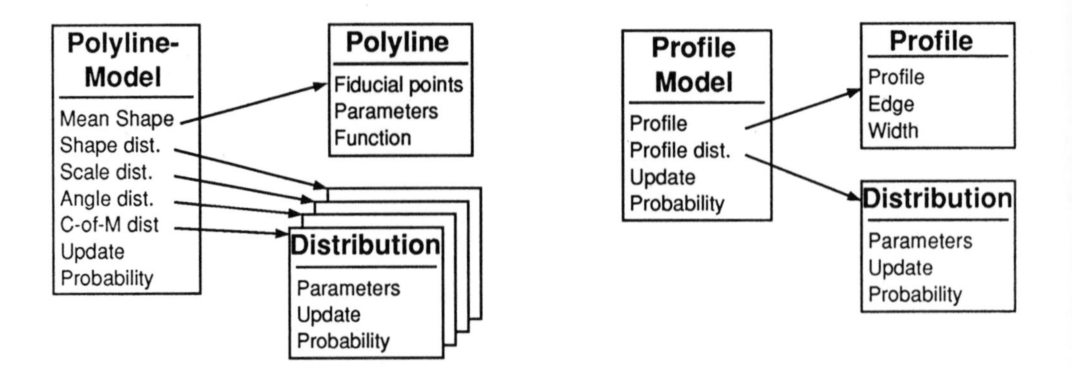

Figure 3: Structure used hold hold polyline and profile model data. Each polyline model has its mean shape encoded via a conformal transformation and distributions for the shape, scale, angle and centre-of-mass. The profile models encode the mean grey-level profile obtained from a training set and a probability density to encode the variability of the training set.

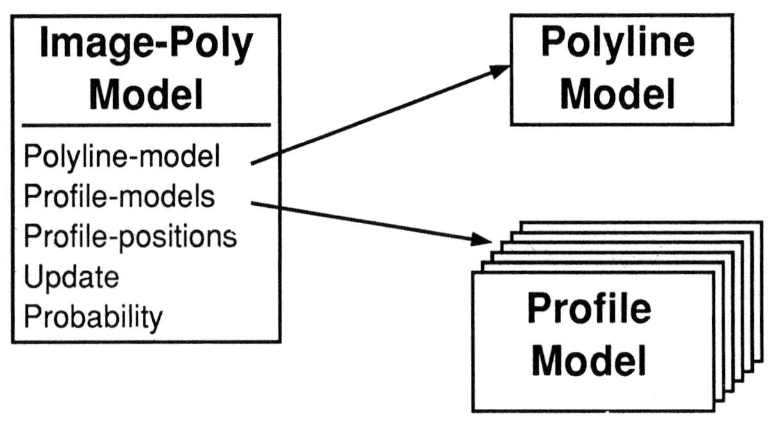

Figure 4: An image-polyline model is comprised of a polyline model and a number of profile models corresponding to the profiles detected at a number of points along the line

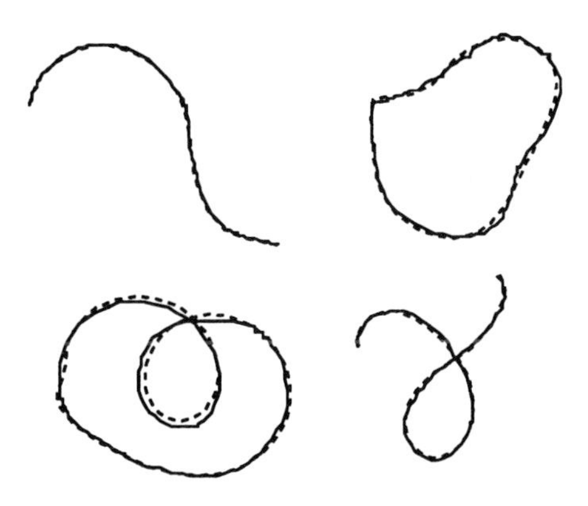

Figure 5: The solid lines indicate the original polyline and the dashed line the fitted polynomial curve. The polynomial order was ten in all cases.

these points and the shape is determined with respect to a normalised coordinate system defined by the two fiducial points. The shape is represented by a polynomial conformal transformation and the shape distribution is parametrised as a γ-distribution density of a distance function. The position, angle and scale distributions are assumed to be normal. The representation used to encode the shape is arbitrary provided a minimum descriptive power and functionality is provided, the choice of conformal transformations is mainly for convenience and the economy of the representation. Thus points on the polyline are mapped onto the real axis in a transformed space by

$$z = P(w) = \sum_{j=0}^{j=m} A_j w^j,$$

where $z = x+iy$, $w = u+iv$, $A_j = a_j+ib_j$ and $i = \sqrt{-1}$. Figure 5 shows the fit of this transformation to a number of curves. The fitting process is fast and the representation compact. It is straightforward to show how transformations of the original polyline can be expressed as transformations of the polynomial parameters and various feature values are easily computed. The fitting process is readily constrained to enforce equal endpoints and equal end derivatives.

The average shape of two input lines defined as the line which minimises the square of differences from the given lines is trivially determined by averaging the respective polynomial coefficients. This also provides a mechanism for describing smooth transformation from one line to another which we ultimately require to encode time dependencies. An example of sinusoidal time variation between two end point curves is shown in figure 6 where $1\frac{1}{2}$ cycles are displayed.

Profiles are represented as a series of grey-levels across the line. Each profile records its width and edge position, ie. that position that is to be regarded as the edge. The idea is that profiles are recorded at a number of points along the line and used to establish a profile model for the line. This is an average profile plus distribution parameters of how well example profiles match the model profile which allow the estimation of the probability that a test profile matches the stored model. The distance between two profiles is defined in terms of the correlation of the two grey-level sequences aligned at their respective edge points.

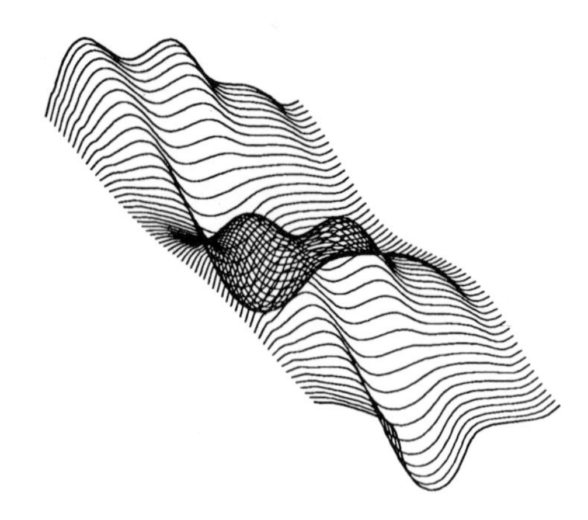

Figure 6: Time evolution of a curve using the interpolation formula given in the subsection on time variation. The diagram is a display showing one and a half cycles.

2.3. Refinement Techniques

The information stored in the model structures is sufficient to allow the calculation of a confidence or degree of belief of a given example with respect to the model. The basis of the model-based methods is that if all available information is employed then a better match and interpretation should be possible. To this end we want to use the model to drive a refinement process to improve any intitial hypothesis. For the lineal object we are considering the algorithm should search the image close to the given line for a new line that better matches the information on size, position, shape and importantly match to the profile model. A convenient and efficient mechanism is provided by the dynamic-programming techniques described by Cooper *et al* (1988). In this technique a number of perpendiculars to the given line are defined, see figure 7, and along each of these lines a match cost is calculated.

The match cost has a number of terms, some are local for example how well the profile matches the model, others are non-local, for example the expected curvature of the line. The refinement process finds the new line drawn through points on each perpendicular that minimises the global cost. The true global minimum is efficiently found by using dynamic programming methods.

In our case we consider four contributions to the cost function. Some relate to the polyline model and others arç associated with the expected accuracy of the initial hypothesis and each cost is calculated as the value of a normalised probability density function. The profile cost is determined from the correlation of the actual grey-level profile with the model profile. The correlation value densities are parametrised by a β-distribution and the probability that the test profile corresponds to the model profile is calculated using Bayes rule and assuming the prior model probability is constant. If the correlation value is c and the probability of obtaining this value, *assuming* the test profile was an example of the model, is denoted $Pr(c \mid model)$ then the probability that the test profile is an example of the model is

$$Pr(model \mid prof) = \frac{Pr(c \mid model)Pr(model)}{Pr(c)}.$$

The second term in the cost is also local and is calculated from the deviation distance from the initial line. This is calculated by assuming a normal distribution for this deviation with parameters dependent on the expected variation in shape of the component and the expected error of the initial hypothesis. In practice these parameters are determined during a training phase using the "true" lines input by an expert.

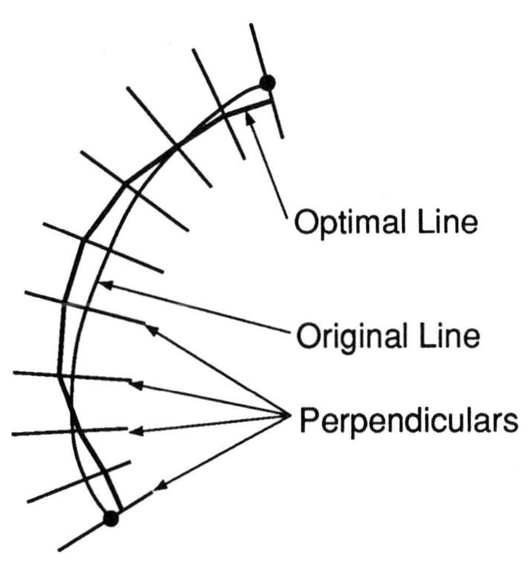

Figure 7: During the refinement process a new line marked optimal is located near to the original line such that the cost function is minimised. The perpendiculars are used to sample the space near the original line.

The third term and fourth terms are non-local and impose a cost for deviating in angle and curvature respectively from the initial line. The angle deviation from the initial line is approximated by the angle of the straight line from the previous point in the path to the current point, thus

$$tan(angle) = \frac{\delta distance}{\delta s},$$

where s is measured along the curve. The curvature is approximated by finding the change of angle at the previous point,

$$curvature = \frac{\delta angle}{2\delta s}.$$

The total probability of any point is the product of each of the contributing terms and the cost we wish to minimise is the negative log of the probability. This yields a sum of terms which are weighted to allow some additional flexibility and account to some extent for not knowing the true measurement errors and other ignored factors. Our refinement cost is therefore

$$\begin{aligned} Cost &= w_{prof}(-log(profile)) + w_{dist}(-log(Pr(distance)) \\ &+ w_{angle}(-log(Pr(angle)) + w_{curv}(-log(Pr(curvature)). \end{aligned}$$

These weights are estimated by training and in fact are all equal apart from the profile cost.

In addition to the costs arising from the model and the initial estimate, there may be further constraints arising from earlier processing or relationships from other components. These are easily included by modifiing the local cost to take account of any existing probability estimates of the position of the line within the image. In particular in our case we will want to impose constraints on the positions of the fiducial vertices which in general will be coincident with those of another component. We impose the constraint by adding a square well cost to the local cost function.

2.4. Model Building and Training

Any detailed representation of a complex object requires input of all the shapes sizes and parameters. To build up sufficient training data for the system to be robust implies considerable time spent by the radiologist or clinician. To avoid this the system has been designed to allow a "boot-strap" mode for building the model database. After designing the structure of the model the test images are presented in turn to the radiologist to interactively mark the required structures. These are recorded and used to build the model by updating the current version. Every structure must be delineated in the first image but for subsequent images the refinement process can be employed. Once a version of the required model exists then it may only be necessary to mark a number of key points on subsequent images and to confirm the result of the refinement process. In this way the model is gradually updated as each new image is considered. This is also the mechanism by which the model can be allowed to learn during routine use. Provided the operator is able to confirm that the results of the match process provide a good example of the model then that information can be used to update the model. This can allow gradual adaptation to new circumstances or machine settings.

After the test images have been marked with the structures of interest then these can be used to optimise the refinement cost parameters. In the system we have developed each "true" result is randomly perturbed in position a number of times and then used as initial estimates for the refinement process. A measure of success is calculated and summing these success values over all the test set yields a measure of how good the refinement process parameters are. This can be repeated many times to allow testing and optimising of the refinement cost parameters. By this means global information about the expected shape of each component is built into the refinement parameters.

3. RESULTS

3.1. Echocardiogram Example

Our test application for these ideas is the location of ventricles in 4-chamber view echocardiogram images and the tracing of the movement of the walls of the chambers through a sequence of images. The matching process will first locate a number of cues for the ventricle positions and then the model employed to predict the wall positions. The initial estimates are refined using the model, to achieve the best possible interpretation of a particular set of cues with respect to the model and the given image. In this paper we do not consider the cue finding mechanism (Bailes 90) but assume that it is capable of providing estimated postions for our vertices within 10-15 pixels of the true position. The hypothesis is then refined and the confidence of the match is determined from the confidence of the fit to the model and the overall cost of the refinement fit.

In figure 8 the component of the model corresponding to the ventricular septal wall has been interactively traced. This is used to generate an initial model.

In figure 9 this model has been used to locate the same structure in a new image from two approximate end points. The model building system requests confirmation of the match but in this case the new line is not acceptable and will need redrawing. After many images have been delineated the system is allowed to optimise the refinement function and figure 10 shows the result of applying this to a new image, again with only the end points given as cues. Now the result is acceptable to allow immediate updating of the model.

4. DISCUSSION AND CONCLUSIONS

In this paper we have described a model representation for the analysis of ultrasound images of the cardiac region. The model is trainable and entirely empirical but is sufficiently flexible to learn complex shapes and distributions, typical imaging transformations and imaging and matching characteristics sufficient to match in a difficult imaging modality. The training of the model is by example and is "boot-strapped" interactively. The operator delineates the required structures by hand for the first image. The system uses this knowledge to establish an initial guess of the parameters of the model. In subsequent images the operator will only need to mark a few cues on the image and the system will attempt to match the rest. After possible correction the operator need only confirm the correct match, for that information also to be included within the model. In this way after a number of training cycles all parameters within the model can be established with relatively little effort by the operator. A valuable spin-off of building in this training mechanism is that during routine use of the system the model can be continuously updated by new examples provided the operator can confirm or edit the match.

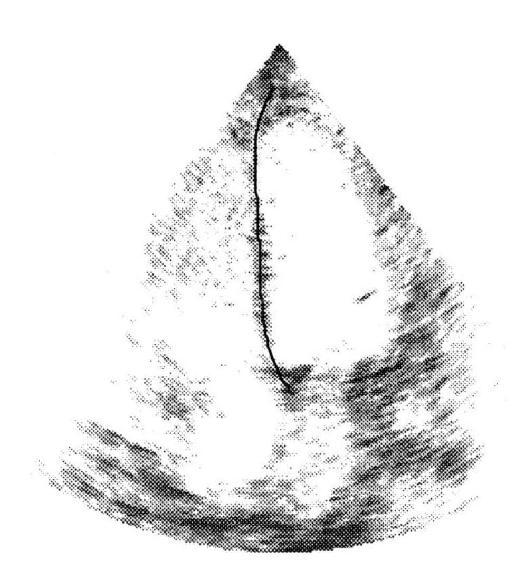

Figure 8: Test echocardiogram image with model component corresponding to the ventricular septum marked by the operator to be used to generate the initial model parameter set.

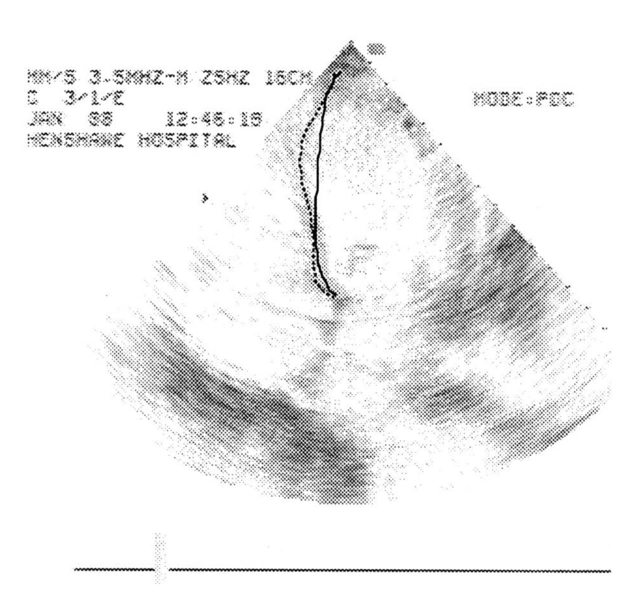

Figure 9: A different echocardiogram showing how the refinement method can use two starting points and the model derived from other images to find a better estimate. The solid line is the hypothesis and the dashed line the optimal line found using dynamic programming. In this instance the refined line would need editing before it could be used to update the model.

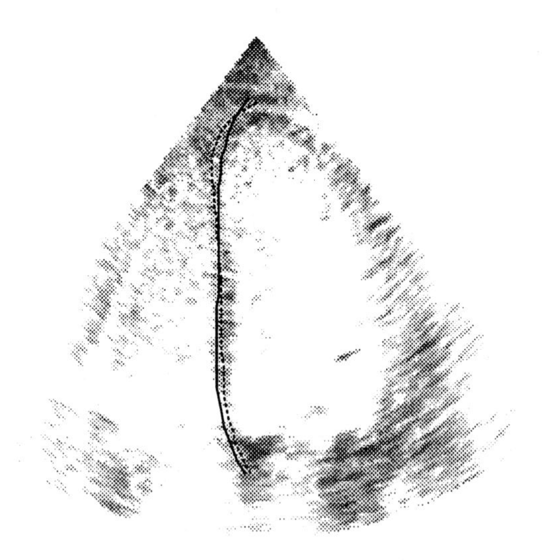

Figure 10: A third echocardiogram showing how the refinement method can use two starting points and the model derived from other images to find a better estimate. The solid line is the hypothesis and the dashed line the optimal line found using dynamic programming. In this instance the refined line is acceptable to update the model.

References

Bailes DR (1990). The Morphological Grey Level SAT. Alvey report: MOBPRIM/Mu/Rep3/900110.

Besl PJ and RC Jain (1985). Three-Dimensional Object Recognition. Computing Surveys 17.

Binford TO (1982). Survey of Model-Based Image Analysis Systems. The International Journal of Robotics Research, 1:18-64.

Cooper DH, N Bryson and CJ Taylor (1988). An Object Location Strategy Using Shape and Grey-Level Models. Proc 4th Alvey Vision Conference :65-71.

Kremkau FW, and KJW Taylor (1986). Artifacts in Ultrasound Imaging. J. of Ultrasound Medicine 5:227-237.

KIDS :
A DISTRIBUTED EXPERT SYSTEM
FOR BIOMEDICAL IMAGE INTERPRETATION

A Ovalle and C Garbay

Equipe de Reconnaissance des Formes et de Microscopie Quantitative
Laboratoire TIM3 - IMAG - Université Joseph Fourier
CERMO - BP 53X - 38041 GRENOBLE Cédex
Tél: 76 51 48 13 Télefax: 76 51 49 48
E-mail(s): Ovalle@Imag.Imag.Fr & Garbay@Imag.Imag.Fr

Abstract

This paper is about KIDS (Knowledge-based Image Diagnosis System), a knowledge-based system to assist the human diagnosis of cytological specimen. After a short state of the art, we describe the expertise involved in the exploration of cytological specimen. MAPS (Multi-Agent Problem Solver) is afterwards rapidly presented, as defining a generic environment for multi-agent system design. Modelling of medical expertise is then presented : the global architecture of KIDS is firstly described, followed by a precise description of the agent role, resources and behaviour. We finally discuss the present state of the system as well as the existing human-computer interaction facilities.

Keywords

High-level Modelling of Human Reasoning; Distributed Artificial Intelligence : Multi-Agent Design; Software Architecture for Computer-assisted Problem Solving; Human-Computer Interaction for Breast Cancer Cyto-Diagnosis;

1. INTRODUCTION

Expert System technology has been largely applied to the Medical Diagnosis field. This research domain has moreover contributed to the expansion of other research topics such as Knowledge Engineering, Cognitive Ergonomy or Reasoning Modelling. The purpose of this paper is to describe an approach to the modelling of human cognitive behaviour, more precisely in the field of cytological specimen investigation, and also to present the current development of a multi-agent system (KIDS) for biomedical image interpretation.

A survey of current knowledge-based system development is presented in part 2. The third part involves a description of the expertise in cytopathology, more specifically considering the domain of breast cytology. Modelling of cytopathologist expertise is presented in part four, based on a generic distributed programming environment called MAPS. Human-computer interaction facilities are finally presented, considering not only the possibilities offered by KIDS itself but also the ones offered by MAPS programming environment.

2. STATE OF THE ART

From the early work of Prewitt and Mendelsohn (1966), a large amount of research effort have been devoted to microscopic imagery (Koss, 1987). These research efforts have conducted to the design of automated systems which were very early commercialized (Preston, 1976).

As a consequence of this rapid growth of the field, the main limitations of the approach have rapidly emerged, including:

- procedural and predefined task scheduling, involving the successive image acquisition, segmentation, description and data analysis;
- unefficiency of segmentation process;
- complexity of the data analysis tools;

Knowledge-based systems have thus been designed, to gain a more flexible control of the various procedures, but also to allow the modelling of the pathologist way of understanding images. Such approach should moreover provide the ability to combine between qualitative and quantitative descriptors, to use human-based reasoning strategies and to provide the user with well documented explanation of its behaviour (Bartels and Weber, 1989a).

Rule-based approach still remains the most classical methodology for knowledge representation. In fact, this methodology suggests a rich and flexible framework for the task of structuring and specifying large amount of knowledge. It has particularly been used in USERS (Garbay and Pesty, 1988), an expert system for cytological diagnosis of bladder cancer. This system, based on interacting with an external observer, is able to propose or confirm a diagnosis hypothesis under control of a prediction/verification strategy.

A rule-based Pathology Expert Consultation System (PECS) has been developed for the assessment of premalignancies, malignancies and noncancerous conditions (Baak and Kurver, 1988). In addition to diagnosis and prognosis decision making support, this sytem permits the integration of data-base information, worksheet and text files with the rules and control structures related to the current application. Graphical, textual or combined explanations are provided upon user request, even by means of interrupting the system execution.

The system developed by (Dhawan, 1988), and devoted to the early detection of melanoma is also a rule-based system. This specimen diagnosis system broadly uses two main rule-based systems which develop low-level and high-level analysis and interpretation of images of the skin lesions for the detection of melanoma. In addition, a set of secondary modules such as "Knowledge About Local Features e.g. Skin Color" or "Melanoma Prognostic KB" gives more precise information about individual cases and assist both rule-based systems.

A similar approach has been adopted by (Wu et al., 1989) to develop an Expert System for Chromosome Analysis using a prediction/verification strategy. A low level module is responsible for metaphase segmentation and feature computation (centromere localisation, median axe determination...). These features are, in his turn, used by the high level module which is responsible for classifying/interpretating image data. While a classical baysian approach is used in the hypothesis generation phase, a “classical” inference process is used in the confirmation phase.

A complex system for Histological Image Understanding (HES) has been conceived by (Bartels et al., 1989b), which combines image analysis with diagnostic abilities by comprising two subsystems, namely, SES and DES. SES, Scene Segmentation Expert System, is devoted to identifying the image components, their features and relations. To reach such goal, SES is coupled to a Digital Image Processing Library, DIPL. SES is mainly a rule-based system, whose inference engine works according to a forward strategy. Rules are gathered into modules according to some goals to be reached (e.g. superposed nucleus decomposition) and govern the dynamic application of image analysis procedures. A model of the histological specimen is used to guide the segmentation process; such model is implemented as a frame-based semantic network.

DES, Diagnostic Expert System for histopathology, is designed as a shell to conceive diagnosis support systems; it has been written under ICON, a classical programming langage (C like) extended for list processing. Current version of this system is devoted to colon cancer diagnosis. Knowledge representation is particularly original : it is not based on a rule formalism but rather on “arrays” correlating “symptom” vs corresponding “diagnosis”. SES and DES communicate by means of an additional system, called IES (Interpretation Expert System) whose role is to interpret iconic criteria provided by SES in terms of abstract symbolic results, to be handled by DES.

A mixed object/production rule representation scheme has been used in a system devoted to the Histological Diagnosis of Breast Cancer (Garbay and Pesty, 1988). This system describes the histolo-gical image as a hierarchical arrangement of objects (ducts, lobules, acini, ...) involved at various organization levels (from general architecture to individual cells). The various image descriptors are gained by means of a human interaction process. The rule-base is comprised of several modules, whose activation depends on the current analysis level: the histological image is explored according to a top-down hierarchical mode, under control of a prediction/ verification strategy.

A commercial frame-based AI development system (ART 3.1) has been used to build a chromosome analysis and classification system (Pycock and Taylor, 1989). The representation of structural descriptors, generic control strategies and domain specific knowledge is made explicit into this system. A frame-based representation scheme is used to describe the chromosome model and a rule-based system controls the execution of the entire system.

A Knowledge-Based Chromosome Analysis System (Piper et al., 1989) devoted to chromosome image segmentation is currently being built under the context of SBS, a blackboard system shell.

Chromosome and cell models to which images are matched are made explicit by means of a frame-network. A set of SBS experts (or knowledge sources) have been conceived, each of which is responsible for a particular domain task in the process of image segmentation and cooperate for solving these tasks. Among the "main" experts (a dozen all in all) we can quote the "cell selection expert", "image enhancement expert" or "statistical chromosome classifier expert".

3. EXPERTISE IN CYTOPATHOLOGY

Clinical cytology concerns the study of cells that have been sampled, stained and fixed upon transparent slides, in order to be observed through a microscope (Gompel, 1982). Various cytological examination styles exist, which apply to a rich variety of organs. It is thus very difficult to exhibit general approach or rules that would apply to any specimen investigation. We have focussed on breast cytology, but hope to be able to examplify rather general reasoning mechanisms, that would apply to other cytological examinations.

Table 1: Formal representation of main tasks involved in cytological specimen exploration.

TASKS	INPUT	OUTPUT	OBJECTIVE	APPROACH	REMARKS
T1	Specimen	Specimen Descriptors	Material Inspection & Checking	Random Scanning	Not Diagnosis Oriented
T2	Specimen	Selected Fields	Selection of Informative Fields	Systematic Scanning	Heavily Context Dependent
T3	Selected Fields	Morphologic Descriptors	Morphologic Description	Low & High Level Examination	Low Res. may reveal Sufficient
T4	Morphologic Descriptors	Cell Types	Cell Type Identification	Prediction Verification	*ibid.*
T5	Cell Types	Diagnosis	Diagnosis Formulation	Evidence Gathering	Influence further Gathering
T6	Diagnosis & Patient Data	Diagnosis Confirmation	Diagnosis Validation	Consistency Checking	May imply Resuming the Analysis

Cytological specimen examination involves two main steps: a preliminary step consisting in material scanning and inspection, conducted by cyto-technicians, followed by a decision-making step assumed by cytopathologists. The preliminary step, assumed by cytotechnicians, consists in validating the specimen lisibility and significance (Task T1) as well as manually marking the significant fields of the slide (Task T2). The interpretation step, performed by the cytopatologist, must lead to the gathering of morphological evidences, allowing to identify the lesions that might appear on the specimen as benign or malign. This step implies the achievment of three different tasks, namely morphological analysis (T3), cell identification (T4) and finally diagnosis interpretation tasks (T5). The lastest task is finally aimed at validating the inferred diagnosis hypothesis (T6): it consists in exploiting additional clinical and administrative data related to the examined patient. Such process leads the pathologist to evaluate the proposed diagnosis hypothesis as fully coherent or not.

These tasks are formally described in Table 1 according to a structured schema involving the definition of input and output items, task objective and approach, as well as some remarks. The dependency between tasks may be represented by means of a task dependency graph (figure 1). A backward dependency, from tasks T4, T5 and T6 to task T2, may be observed.

3.1. Specimen Validation Task

Specimen validation task (T1) is aimed at acquiring global specimen features, such as cell sociology and distribution or staining quality. In fact, sampling and preparing conditions affects the global specimen aspect, its lisibility, as well as the significance of supported morphological evidences. A random scanning of the preparation is accomplished, that leads to the acceptance or rejection of the smear.

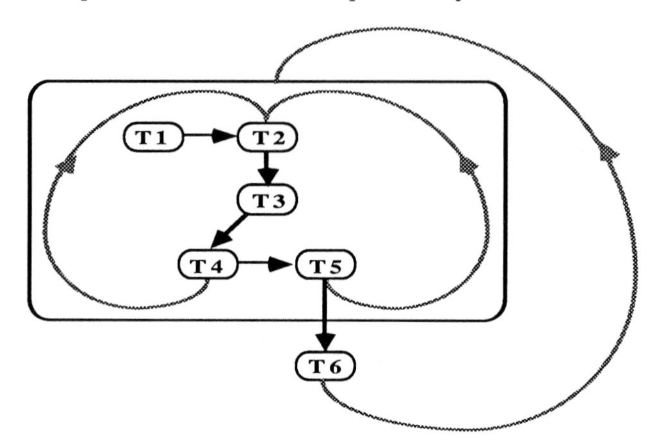

Figure 1 : Specimen Diagnosis : the task dependency graph. Bold arrows indicate the forward scheduling between tasks, dotted arrows indicate the backward dependency between tasks.

3.2. Field Selection Task

Field selection (T2) is performed by using a systematic specimen exploration strategy: it is aimed at selecting a set of fields carrying information that is considered as "a priori" relevant regarding the diagnosis formulation process. Selection criteria may involve specimen appearance features as well as diagnosis evidences: when the cell density is particularly low, for instance, any field containing cells would be selected; in the opposite case, if the cell density looks normal while current diagnosis hypothesis reveals a potential for malignancy, a field would not be selected except if displaying apparently "abnormal" cells.

3.3. Cell Description and Identification Task

The cell identification task is aimed at identifying cell types displayed by the explored field. Such identification is based on morphological indicators provided during the morphological analysis task (T3); a low magnification level is firstly used, which may be followed by a high magnification examination, if needed to confirm the type of the cell under examination (prediction/verification strategy). This last phase is not always performed: using low magnification may reveal enough for identifying adipocyte cells, for example.

Main morphological features can be found into Table 2 in association with correlated cell types, among which:

- Adenoma (AD)
- Fibroblast (FB)
- Benignant_Nude_Nucleus (BNN)
- Malignant_Nude_Nucleus (MNN)
- Apocrine_Metaplasia (AM)
- Spumous Cell (SC)
- Epithelial Cell (EC)
- Myoepithelial Cell(MC)
- Cancer Cell (CC)

3.4. Diagnosis Formulation Task

The set of observations provided from previous tasks and particularly cell identifiers conduct to formulate a diagnosis hypothesis (Taks 5), which is progressively documented, confirmed or rejected, in the course of a sequential and systematic scrutiny of remaining fields. During this interpretion step, the amount of cells observed for each different cell type is computed: such scores in fact provide essential support to the diagnosis formulation process.

Table 2 : Cross-correlation table tying cell morphological features to cell types (partial view).

	AD	FB	BNN	MNN	AM	SC	EC	MC	CC
Arrange-ment	Isol. Group	Isol.	Nucleus	Nucleus	Group Sup.	Isol.	Group Sup. Isol.	Isol.	Group Sup. Isol.
Size	Large	Small Med.	Small	Small	Med.	Large Med.	Small	Small	Large Med.
Shape	Round	Elong.	/	/	Round	Round	Round	Elong.	Round Irregul.
Cyto Aspect	Homo.	Homo.	/	/	Red Granul	Vacuol.	Homo.	Homo.	Homo. Vacuol.
N/C Ratio	Small	Med.	/	/	Med. Small	Small	Med.	Med.	Large Med.
Vac. Numb.	Rare	Absent	Absent	Absent	Absent	Nume.	Absent	Absent	Rare
Vac. Size	Large	/	/	/	/	Small	/	/	Large Med.

Table 3 displays a cross-correlation matrix which associates "cell types" to "diagnosis", among which:

- Epithelial Hyperplasia (HEP);
- Adenoma (AD);
- Fibroadenoma (FAD);
- Mastose (M);
- Cyst (K);
- Lipome (L);
- Cancer (C), to which grades 1, 2 or 3 are associated.

Table 3 : From cell types to pathologies (partial view).
The signe * indicates that cell type identification reveals difficult in that case.
The signe / indicates the absence of cells for this given type.

	HEP	AD	FAD	M	K	L	C
AD	> 0	> 0	> 0	> 0	> 0	> 100	> 0
FB	< 20	< 20	> 20	< 20	/	/	/
BNN	/	/	> 0	/	> 0	/	> 0
MNN	/	/	/	/	/	/	> 0
AM	/	/	/	> 0	> 0	/	/
SC	> 0	> 0	> 0	> 50	> 50	/	> 0
EC	< 5	> 5	> 5	< 20	< 5	/	> 0
MC	*	*	*	*	/	/	/
CC	/	/	/	/	/	/	> 0

3.5. Validation of Formulated Diagnosis Task

The patient medical file contains his name, age and sex, as well as all information about his medical background such as surgical interventions, or last or current treatments. In addition, reports from various medical examinations are also included.

Such patient file is consulted by the pathologist upon completion of specimen investigation. A "blinded" specimen investigation is, in fact, required to guarantee diagnosis objectivity. Once patient file has been consulted, the pathologist could resume the analysis, thus modifying the previously proposed diagnosis hypothesis.

4. MODELLING OF MEDICAL EXPERTISE

Main concepts of MAPS programming environment are firstly described. KIDS system is afterwards presented, by considering successively the agent role and ressources as well as the system main functionnalities.

4.1. MAPS Environment

MAPS (Multi-Agent Problem Solver) is defined as a generic environment dedicated to the design of distributed knowledge-based systems. It involves 2 classes of agents, called Knowledge Server (KS) and Knowledge Processor (KP), communicating by message sending.

These agents are respectively responsible for handling descriptive and operative knowledge elements : they thus entail a distribution of knowledge in terms of static and dynamic elements, that allows to reach a balance between problem modelling and problem solving abilities.

4.1.1. The concept of agent

We define an agent as a knowledge-based system communicating with other agents by message sending. Communication is based on a very simple message sending protocol: any request is sent in an explicit way to a specific agent, according to a "command" mode (Hautin and Vailly, 1986) in opposition to information sharing mechanism as in the blackboard approach (Hayes-Roth, 1985). For that purpose, any information is always denoted by a quadruplet of the form (Agent Object Attribute Value). Any agent moreover knows what are the kind of request that may be sent to the other ones : it is considered that no conflict may occur among the agents.

Each agent is furthermore conceived as an autonomous entity, i.e. it is provided the capacity to decide on its own the way to behave in front of external events: any agent is conceived as able to answer a limited set of requests, according to predefined resolution scheme. These scheme, as shortly described later on, define the actions to undertake, depending on the agent own ressources, but also on the ability of other agents to bring their support.

4.1.2. Knowledge Server

A Knowledge Server (KS) is an agent whose role is to represent, maintain and disseminate a set of descriptive knowledge elements. Such agent is comprised of a set of objects, describing some problem elements, a set of primary rules, describing potential inner propagation strategies, and finally 2 sets of exploratory rules, driving the selection of elements to be transmitted or requested outside (Propose and Wonder rules). These rules are applied according to a forward chaining scheme by a proprietary control structure.

Several internal methods are also defined. Get and Update, on one hand, control the "read/write" access to any information: in case an information is not available, it is requested outside, by sending the request Solve to a KP agent; the reception of a new information entails the knowledge base updating, followed by primary rules application. Propose and Wonder, on the other hand, drive the selection of informations to be transmitted or requested outside (request Process or Solve transmitted to a KP agent): their role is merely to control the application of the corresponding exploratory rules, through the inference engine.

Two external methods are finally defined, Receive and Supply, whose "algorithmic structure" is provided below. These methods are activated upon external request.

```
Receive                               Supply

    (And Update                          (Or Get
        (Or (And  Propose                    (And  Propose
                  (Send KP Process))               (Send KP Process))
            (And Wonder                      (And  Wonder
                  (Send KP Solve)                  (Send KP Solve)
                  (Send self Receive)))            (Send self Receive)))
```

As can be seen, in case of a "Receive" request, a first attempt is made to Update the knowledge base, then to search a valid information (Propose) and finally to send it outside (Send KP Process). In case of failure, at any of these three steps, the KS agent has the ability to search for a lacking information (Wonder), to request it to an outside KP agent (Send KP Solve), and finally to resume the whole Receive cycle: the basic role of a KS agent is the completion and dissemination of knowledge. In case of a Supply request, the element is first of all tentatively obtained (Get call), and then sent back to the caller. In case of failure, the KS agent has the ability to try and resume the current unsuccessful solving strategy, by entering a Propose/Wonder cycle, which is similar to the previous one.

4.1.3. Knowledge Processor

A Knowledge Processor (KP) is an agent whose role is to represent and handle a set of operative knowledge elements. Such agent involves a set of rules, which may describe any deductive or procedural analysis, and a set of meta-rules driving their selection. Their handling is controlled by a proprietary control structure, working according to forward or backward chaining strategies.

Two internal methods are also provided, called "Data-Drive" and "Goal-Drive", aimed at calling the inference engine to select and apply the rules, given respectively a data to process or a problem to solve. In case an information has to be requested or transmitted outside, the request Supply or Receive are sent to the concerned KS agents.

Two external methods are finally defined, Process and Solve, which in turn call the corresponding internal methods (respectively Data-Drive and Goal-Drive). The behaviour of a KP agent may thus be drawn as a very simple linear scheme: such agent may only conclude to the success or failure of its activities. The sole processing alternative it might propose in case of failure occurs at the very rule selection level.

4.1.4. From Agents to Architectures

A "minimal" generic architecture for problem solving may thus be defined as involving one KS agent communicating with one KP agent: the KS agent in this case involves both the data and results associated to the problem at hand, while the KP agent involves some problem solving elements. A more realistic scheme however implies 2 KS agents communicating with one KP agent, in order for the data and results to be distributed among 2 different agents (figure 2).

Such "minimal" generic architecture may in turn be specialized, to model complex problems: distributing static knowledge elements among several KS agents allows to differentiate them according to their conceptual levels, and thus to handle them in a dedicated way. Distributing dynamic knowledge elements among several KP agents allows to examplify successive processing steps, but also to point out the presence of high level processing alternatives.

From a dynamic point of view, finally, the functional as well as structural agent features that have been described give rise to powerful problem solving strategies (Manon et al., 1988), which appear to depend on the own ressources and competences of the various agents, but not on the individual decision of a centralized control structure.

4.2. KIDS (Knowledge-Based Image Diagnosis System)

Cytopathologic specimen investigation for breast diagnosis formulation demonstrates the need to model and handle two types of knowledge elements:

- descriptive knowledge elements such as : knowledge about specimen appearance, field significance, cell morphologies, ... ;
- operative knowledge elements such as : knowledge about specimen exploration strategies, image processing operators, cell type identification and diagnosis formulation rules...

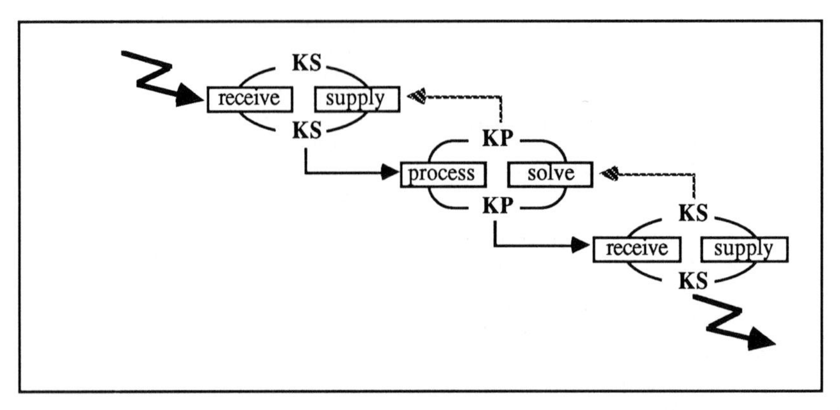

Figure 2 : A "minimal" multi-agent architecture: the reception of a Receive request (left-most server) provokes the development of a data-driven resolution strategy; a goal-driven resolution strategy is simultaneously developped (dotted arrows), due to potential request for lacking information.

A multi-agent approach appears as a way to handle such knowledge in an integrated and cooperative framework. KIDS (Knowledge based Images Diagnosis System) has thus been designed under MAPS as a specialized agent network (figure 3). Each KS or KP is in charge of particular knowledge sets and tasks.

4.2.1. Description of KIDS Agents

The network presented in figure 3 provides a high-level modelling of pathologist expertise : each KP-agent may easily be interpreted in terms of the various pathologist tasks (cf table 1), related KS-agents very simply encapsulates the knowledge elements appearing as "input" or "output" elements for these tasks. Each of these agents are briefly described below, in the form of synthetic tables.

Agent	**KS-Specimen**
role	acquire and transmit a view of the specimen
resources	--> objects : specimen image file and memory buffer descriptors, flags
	--> propagation rules : IF (1) specimen file name & (2) specimen image location & (3) specimen size are known THEN load and display specimen image
	--> proposition rules : IF (1) specimen image is loaded & (2) it has not ever been proposed for investigation THEN propose specimen image to agent KP-Explore

Agent	**KP-Explore**
role	explore the specimen and send each field, one after the other, for further analysis
resources	--> action rules : IF (1) specimen image descriptors are obtained from KS-Specimen THEN repeat from 1 to specimen field number : send current field to KS-Low-Resolution for further analysis

Agent	**KS-Low-Resolution**
role	send specimen field for further analysis or magnification, by requesting the user agreement
resources	--> objects : specimen field and memory buffer descriptors, flags
	--> proposition rules : IF (1) current field image is loaded & (2) user agrees to handle current field & *cc : user may consider current field as irrelevant and reject it* (3) user does not agree to use high magnification & (4) current field has not yet been proposed THEN propose current field to KP-Analyze for further investigation

	--> interrogation rules :
	IF (1) current field image is loaded &
	(2) user agreement to handle current field has not ever been requested
	THEN request the agreement of the user
	IF (1) user agreement to handle current field has been obtained &
	(2) user agreement to use high magnification has not ever been requested
	THEN request the agreement of the user

Agent	**KP-Analyze**
role	segment the field image and send each cell, one after the other, for further identification
resources	--> action rules : IF (1) field image descriptors are obtained from KS-Low-Resolution THEN (1) segment current field image (2) repeat for each detected cell in current field image : - highlight cell contour by a graphical procedure - send detected cell to KS-Cell-Morph for further morphological investigation
remarks	--> image segmentation procedure : - computation of thresholds by means of histogram analysis (Fisher classification); - image thresholding; - binary closure; - connected component labelling; - geometrical decomposition of cell aggregates.

Agent	**KS-Cell-Morphology**
role	send a cell together with morphological features for further identification
resources	--> objects : cell label & main morphological features, among which : cell_arrangement, cell_size, cell_shape, cytoplasm_texture, N/C ratio, number_of_vacuoles & vacuole_size
	--> proposition rules : IF (1) user agrees to analyze current cell & *cc : user may consider current cell as irrelevant and reject it* (2) its cell_arrangement, cell_size, cell_shape, cytoplasm_texture, N/C ratio, number_of_vacuoles & vacuole_size are known THEN propose current cell for further identification
remarks	cell morphological descriptors are currently obtained via the activation of interrogation rules which directly request the user. A new KP agent will be added to provide for their automatic computation.

Agent	**KP-Identify**
role	identify cell type, based on the cross-correlation table tying cell morphological features to cell types
resources	--> action rule example : IF (1) cell_arrangement = isolated & (2) cell_size = small or medium & (3) cell_shape = elongated & (4) cytoplasm_texture = homogeneous & (5) N/C ratio = medium THEN send "cell type = fibroblast" to KS-Cell-Type

Agent	**KS-Cell-Type**
role	store and update quantitative scores describing cell type frequency of appearance
resources	--> objects : cell type name and score, among which : cancer, adenoma, fibroblast, benign_nude_nucleus, malignant_nude_nucleus, metaplasia, spumous, epithelial, myoepithelial, debris
	--> propagation rules : IF (1) cell type = myoepithelial & (2) user agrees with proposed cell type THEN increase by 1 the myoepithelial cell type score
	--> interrogation rules : IF user agreement on proposed cell_type has not ever been requested THEN request the agreement of the user on proposed cell_type
remarks	proposition rules, into this agent, are intended to propose a cell type to KP-Interpret agent each time such cell type has been identified.

Agent	**KP-Interpret**
role	propose a diagnosis, based on the cross-correlation tabletying cell types to possible diagnosis
resources	--> action rules : IF (1) fibroblast cell score > N & (2) epithelial cell score < M THEN send "diagnosis = fibroadenoma" to KS-Diagnosis

Agent	**KS-Diagnosis**
role	store potential diagnosis hypothesis
resources	--> objects : Description of diagnosis hypothesis among which : Epithelial Hyperplasia, Adenoma, Fibroadenoma, Mastose, Cyst, Lipoma, Cancer
	--> interrogation rules : IF user agreement on proposed diagnosis has not ever been requested THEN request the agreement of the user on proposed cell_type

It should be noticed that agent implementation is still rudimentary : our main concern has been to give them enough resources to make the system exhibit some coherent behaviour : our objective is now to provide them with accurate knowledge elements and refined abilities. Among these agents, some have not yet been implemented : KP-Validate,whose role is to validate the quality of cytological specimen, KP-Magnify, whose role is to magnify current exploration field, and finally KS-High-Resolution, whose role is to store current magnified field.

4.2.3. KIDS Functional Behaviour

An image of the specimen is acquired by KS-Specimen agent and then transmitted to KP-Explore agent. A particular specimen exploration strategy is implemented in that agent, in the form of a loop (see circular arrow in figure 3) which controls the successive sending of fields to KS-Low-Resolution agent. After reception by KS-Low-Resolution, the user agreement is requested to know whether analyzing such field is relevant or not, or wether it would need high magnification analysis.

After reception by KP-Analyze agent, the field is segmented and a second loop is implemented, which controls the successive sending of cells to KS-Morphology agent. The latter is then responsible for collecting relevant morphological information and send it to KP-Identify agent for further identification. It should be noticed that user agreement is once more requested, to know whether analyzing current cell is relevant or not. The various cell descriptors are currently requested off the user : the contour of current cell is graphically highlighted, for that purpose.

Cell type identification is afterwards performed by KP-Identify agent by means of inference rules, based on a cross-correlated table in which cell morphologies are tied to cell types. Such identification may then be validated or not by the user (activation of KS-Cell-Type interrogation rules). In case of acceptance, a scoring procedure is activated as conclusion part of KS-Cell-Type propagation rule. In the opposite case, the next cell is simply proposed for analysis, i.e. control is sent back to loop 2. The cell type scoring directly assists the diagnosis formulation performed by KP-Interpret and based on a cross-correlated table tying cell types to pathologies.

Proposed diagnosis hypothesis is finally received by KS-Diagnosis agent; activating propagation rules then entails requesting the user to validate or not the proposed diagnosis. Control is sent back to loop 2 and then loop 1 anyhow, i.e. field / specimen exploration is pursued even if a satisfactory diagnosis has been obtained. Different behaviours could of course be programmed.

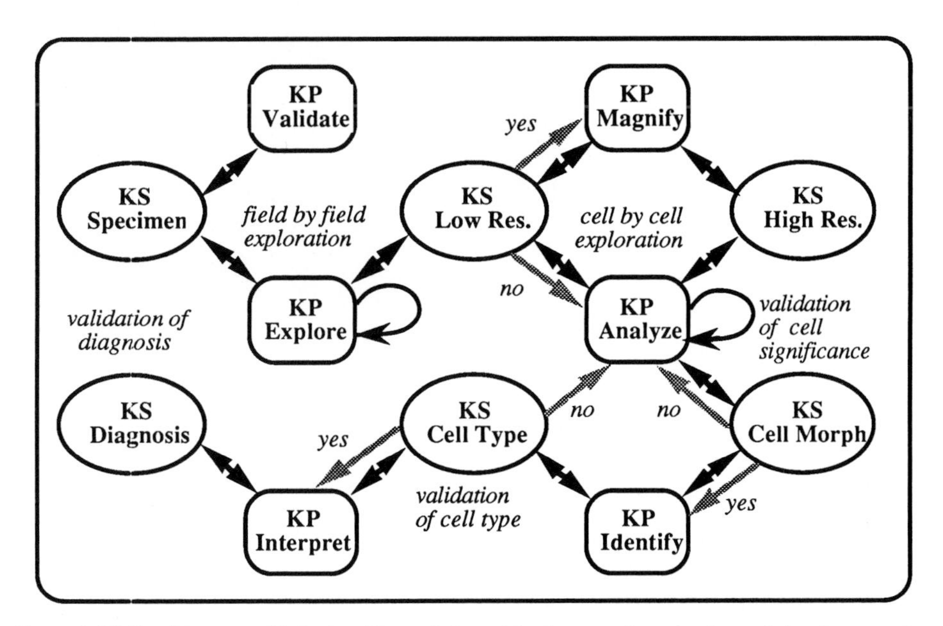

Figure 3: KIDS architecture and behaviour. Two main loops (circular arrows) may be observed, denoting respective field by field and cell by cell analysis strategies. Dark arrows indicate the flow of information between agents while grey ones indicate user-based decision flows.

5. HUMAN-COMPUTER INTERACTION

Human-computer interaction for applications designed under MAPS programming environment is performed through two independent processes which communicate according to a message sending protocol : one of them is dedicated to controlling any interaction occuring at the MAPS level, while the other controls any interaction occuring at the application level (figure 4). The interface of MAPS environment (leftmost interaction windows on figure 4) displays the entire system architecture, where each agent may be dynamically accessed for consultation, updating or request sending. Any application running under MAPS may thus be activated by triggering any of its agents; an execution trace may finally be obtained by simply selecting the desired trace level.

Four main parts may be distinguished in KIDS graphical interface (rightmost interaction windows on figure 4). The "Specimen Validation & Field Exploration" part involves KS-Specimen, KP-Explore, and KS-Low-Resolution agents; it provides low and high level views of the specimen; a superposed square grid displays the partitioning of the specimen into fields. The "Image Analysis" part involves KP-Analyze agent : it displays both thresholded and labelled field images, together with a grey level image of the field where the contour of the cell under interest is highlighted in red. The "Cell Type Identification" part involves KS-Cell-Morphology, KP-Identify and KS-Cell-Type agents : it displays a list of morphological criteria together with user-provided values. A list of possible values is provided for each in the neighbouring window. This window is otherwise dedicated to the "Findings Interpretation" part : it involves KP-Interpret and KS-Diagnosis agents and displays a list of score for each observed cell type together with the proposed diagnosis hypothesis.

6. DISCUSSION AND CONCLUSION

In this paper we have presented current progress on KIDS (Knowledge-based image diagnosis system) a distributed knowledge-based system to assist the human diagnosis of cytological images. KIDS

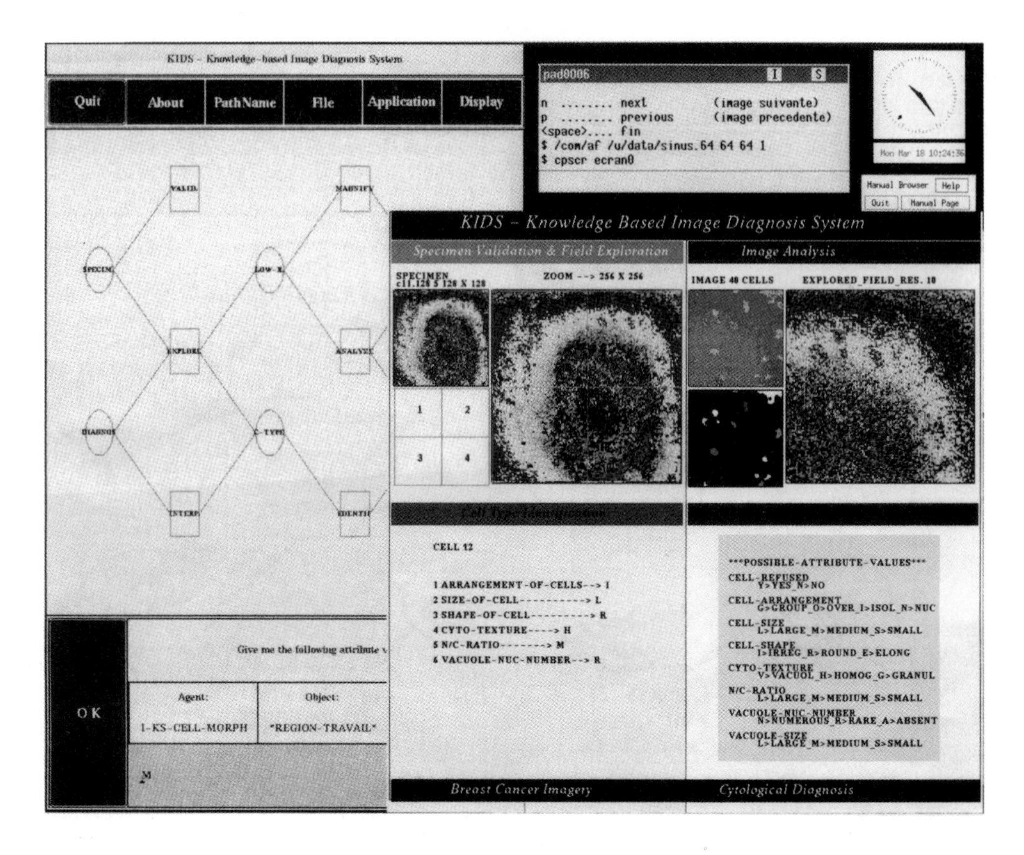

Figure 4: The KIDS graphical interface (rigthmost interaction window) interacting with the MAPS programming environment (leftmost interaction window).

has been implemented under MAPS, a generic programming environment for multi-agent system design. In fact, MAPS offers simultaneously a formalism for expertise modelling and a distributed programming environment . MAPS is implemented in C++, it runs within UNIX, under X WINDOWS environment, on an APOLLO workstation. KIDS is programmed under MAPS as a set of agents, objects and rules definition : any such “program” is compiled and then executed under MAPS.

Further improvements would concern the image analysis as well as diagnosis formulation parts. Concerning image analysis, a multi-agent computer vision system called KISS (Knowledge-based Image Segmentation System) and designed under MAPS, is currently under evaluation. It could provide a firm basis to improve KIDS segmentation performances, thus improving cell type identification on which diagnosis formulation accuracy is clearly dependent. In fact, one major feature of systems conceived under MAPS is their connectivity to others : several systems may be merged together, thus creating larger knowledge-based architectures showing enhanced functionalities and performances.

Regarding cell type identification and diagnosis formulation process, handling of hypotheses should be refined : a prediction / verification scheme should in particular be introduced at the morphological analysis level, and afford the possibility to scrutinize a field at a higher resolution level, in order to verify a pending cell type hypothesis.

Control of the main exploration loops should finally be refined, by introducing field and cell selection criterion that would depend not only on user agreement but rather on current diagnosis hypothesis and cell type scoring : selecting a field would for example depend on the gravity and confidence of current diagnosis hypothesis (a low grade diagnosis imposes a rigourous examination of a specimen, while a high grade may orient the investigation towards specific findings).

Acknowledgements

The authors wish to thank Pr. Seigneurin for his thorough contribution and patience during various interviews for the purpose of knowledge elicitation and medical expertise modelling.

References

Baak J.P.A. and Kurver P.H.J. (1988). Development and Use of a Rule-Based Pathology Expert Consultation System. Analyt Quant Cytol Histol, 10(3):214-218.

Bartels P.H. and Weber J.E. (1989a). Expert Systems in Histopathology. Analyt Quant Cytol Histol, 11(1):1-7.

Bartels P.H., Bibbo M., Graham A., Paplanus S., Shoemaker R.L. and Thompson D. (1989b). Image understanding system for histopathology. Analytical Cellular Pathology, 1:195-214.

Dhawan A.P. (1988). An Expert System for the Early Detection of Melanoma Using Knowledge-Based Image Analysis. Analyt Quant Cytol Histol, 10 (6):405-416.

Garbay C. and Pesty, S. (1988). Expert system for biomedical image interpretation. In "Artificial Intelligence & Cognitive Sciences ", pp. 323-345. J. Demongeot, T. Hervé, V. Rialle, C. Roche (eds), Manchester University Press.

Gompel, C. (1982). La cytologie mammaire. Atlas de Cytologie Clinique, pp.125-139. Maloine s.a. éditeur.

Hautin F. and Vailly A. (1986) La coopération entre systèmes experts. Actes des Journées nationales du PRC-GRECO "Intelligence Artificielle", Cepadues Editions.

Hayes-Roth, B. (1985). A blackboard model of control. Artificial Intelligence Vol 26, pp. 251-321.

Koss, L.G. (1987). Automated Cytology and Histology: a historical perspective. Analyt Quant Cytol Histol, 9(5):369-374.

Manon G., Pesty S.and Garbay C. (1988). KIDS (Knowledge-based Image Diagnosis System), a specialized architecture. Proc. of the 9th ICPR, pp. 995-997, IEEE Computer Society Press.

Pycock D. and Taylor C.J. (1989). Chromosome Classification in a General Purpose Frame-based Interpretation System. In press.

Piper, J., Baldock, R., Towers, S. and Rutovitz D. (1989). A Knowledge-Based Chro-mosome Analysis System. In "Automation of Cytology: Advances in Systems and Techniques ", J. Piper and C. Lundsteen (eds), Springer Verlang, Heidel-berg.

Preston, K. (1976). Digital Picture Analysis in Cytology. In "Digital Picture Analysis", pp.209-294, A. Rosenfeld (eds), Springer Verlag.

Prewitt, J. and Mendelsohn (1966). Analysis of cell images. Ann. N.Y. Acad. Sci. 128:1036-1053.

Hayes-Roth, B. (1985). A blackboard model of control. Artificial Intelligence Vol 26, pp. 251-321.

Wu Q., Suetens P. and Oosterlink A. (1987). Toward an expert system for chromosome analysis. Knowledge-Based Systems, 1(1):43-52.

LEARNING OF UNCERTAIN CLASSIFICATION RULES IN BIOMEDICAL IMAGES: THE CASE OF COLPOSCOPIC IMAGES

E Binaghi and A Rampini

Istituto di Fisica Cosmica e Tecnologie Relative - C.N.R. Milano
Via Ampere 56, 20133 Milano, Italy

Abstract

Knowledge acquisition is always a critical step in the development of a knowledge-based computing system. In the particular area of the interpretation of biomedical images, the assignement of meanings to image patterns is based on obscure and intrinsically vague criteria which are difficult to asses and transform into a suitable machine representation. Automatic learning tecniques may be a promising tool in addressing this problem. The paper illustrates a methodological procedure based on fuzzy set theory and using fuzzy logic for the automatic learning of classification rules for biomedical image interpretation systems. It also provides a detailed description of the application of the procedure in the development of a system for the automatic detection of preneoplastic and neoplastic lesions in colposcopic images. Plans to employ the system contemplate its use in educational applications, in diagnostic review for research purposes, and as an online support in clinical practice.

Keywords

Learning; classification; colposcopic images.

1.INTRODUCTION

Knowledge acquisition is a critical step in the development of a knowledge based computing system (Gaines B R, 1988), (Brenker J and Wielinga B, 1987). It involves eliciting the knowledge that human experts use to solve a particular problem and then transforming it into a suitable machine representation. Difficulties may arise in acquiring the overall knowledge directly from the experts. Even when efficient interview techniques are used to structure the elicitation process, these specialists are always forced to formulate their expertise explicitly, as rules or other abstractions which may not truly reflect their internal mental representation. The danger that the knowledge elicited may be an artefact created under the pressure of questioning must be always considered. When dealing with biomedical image interpretation, in particular, domain knowledge is strongly structured and presents a large number of items and complex semantic relations. Visual inspection of biomedical images is an immediate implicit cognitive process. The assignment of medical meanings to image patterns is based on obscure and complex criteria which may be difficult to assess and transform into a suitable machine representation.

Automatic learning tecniques may be useful in addressing this problem: while experts have difficulty formulating their knowledge abstractly, they find it easy to demonstrate their expertise in specific performance situations.

Consequently schemes for learning by examples offer domain experts a potential tool for the direct transfer of their specific exemplified knowledge, and provide inductive algorithms to induce knowledge into a general abstract form.

In this paper a method for learning by examples is proposed to support the acquisition of classification rules in knowledge-based systems for biomedical image interpretation systems, which use fuzzy reasoning techniques based on fuzzy production rules to manipulate certain and uncertain knowledge related to image contents and deduce interpretations (Binaghi E et al. 1990). The rules are organized in different sets. We focus our attention here on classification rules regarding the evaluation of features which characterize the structures of interest and the decision assigning these structures to a given class. Classification rules are then encoded in fuzzy production rules structured as evaluation-decision pairs. These define a fuzzy relation between linguistic description of characterizing features and decision classes.

The strength of implication in fuzzy production rules is approximated by a linguistic variable reflecting the intrinsic human variability and vagueness with which a conclusion may be inferred.

Difficulties may arise in eliciting fuzzy production rules directly from experts and in particular in expressing the different strengths of implication, which may vary from rule to rule.

An intuitive learning by examples technique is proposed which uses fuzzy reasoning operators and in the form of fuzzy induction in a fuzzy logic framework to automatically infer the strength of implication of classification rules, given a training set of already classified images.

The methodological procedure designed for the automatic learning of classification rules for biomedical image interpretation systems is illustrated in the following sections. A detailed description of its application to the development of a system for the automatic detection of preneoplastic and neoplastic lesions in colposcopic images is also provided. Plans to employ the system contemplate its use in educational applications, in diagnostic review for research purposes, and as an online support in clinical practice.

2. THE LEARNING OF FUZZY CLASSIFICATION RULES FOR BIOMEDICAL IMAGE INTERPRETATION.

2.1. Previous work in the field

Various methods have been proposed as ways of generating fuzzy production rules. The literature provides examples of rules generated in fuzzy logic controllers and fuzzy models, by trial and error approach, or by analyzing experts'experience (Tong R M, 1978).

How to obtain the relations between the variables from a collection of statistical data has been studied by Pedrycz (1984,1985), using the method of fuzzy relational equations. A stastistical method of learning the membership of fuzzy naming relations from imprecise descriptions of the input patterns has been reported by De Mori and Saitta (1980) and applied in the medical diagnosis field (Lesmo L, 1982).

Although the statistical approach guarantees the availability of efficient algorithms, it is difficult to monitor the sequence of operations involved and assign an intuitive meaningto each of them.

When dealing with complex and multivariate domain of medical diagnosis, and medical diagnosis supported by biomedical image interpretation in particular, it is important to adopt flexible and controllable learning method in a way which always allows the explanation and monitoring of intermediate results.

The learning from example method proposed here, is a direct method which makes use of fuzzy reasoning operators and automatically generates fuzzy production rules, given a predefined set of premises and diagnostic classes.

2.2. Fuzzy logic-based representation of classification rules in biomedical images

The knowledge representation language and the deductive scheme adopted to represent biomedical image interpretation processes are referred before illustrating the inductive learning technique.

A basic concept of our approach to biomedical image interpretation is that domain knowledge may be affected by an intrinsic uncertainty due to the presence of medical concepts that are intrinsically qualitative, and to the presence in the images of objects with fuzzy features and of noise introduced by the imaging process. Conventional inferential methods, based on binary logics and probability theory have proved to be insufficient for handling most of the ac-

tual situations present in the cognitive process employed by physicians in interpreting biomedical images. The problem can be addressed by using fuzzy set theory and fuzzy logic formulated by Zadeh.
The classification rules involved in biomedical image interpretation may be modelled in terms of fuzzy production rules having the following general form (Zadeh L A, 1981):

$$R_i\!: \; A_i \rightarrow^{w_i^k} D_k$$

where

- A_i is a fuzzy compound declarative proposition of the form S_1 *isA* $_{1,j1}$ *And* ... *And* S_n *is* $A_{n,jn}$, and the term S_h *is* $A_{h,jh}$ represents the linguistic description of the h- feature related to a given image structure (ex.: the contour is regular);
- $A_{h,jh}$ is a term belonging to the predefined term set of the linguistic variable S_h (ex: Sharpness = (well defined, medium, smoothed)). It denotes a fuzzy set in a given universe of discourse U_h characterized by a membership function $\mu_{A_{j,jh}}(u_h)$, $u_h \in U_h$. Applying the rule pertaining to composition (Zadeh L A, 1981) , the membership function of the global fuzzy set associated with the antecedent A_i of the rule R_i is given by: $\mu_{A_i}(u) = \min(\mu_{A_{1,j1}}(u_1) \ldots \mu_{A_{n,jn}}(u_n)),\ u \in U,\ U = U_1 \times \cdots \times U_n \quad (1)$.
- $D_k, 1 \leq k \leq K$ belongs to a predefined set of diagnostic classes (for example: (Definite Disease, Disease Excluded)
- w_i^k is a term expressing the strength of implication in the fuzzy production rule, i.e. the degree of satisfaction or certainty with which a conclusion may be inferred. It belongs to the term set of a linguistic variable W the values of which are the possible linguistic expression of the strength of implication for a given application (for example very high, high, medium etc.), and is therefore a fuzzy set defined in the interval [0,1]. It is usually, for the sake of simplicity, reduced to a fuzzy number by applying suitable defuzzification algorithms (ex.: very high = 0.9, high = 0.7, medium = 0.5 etc.) (Binaghi E, 1990).

The possibility value $\Pi(D_k)$ of assigning the diagnostic class D_k to a given imagestructure is computed according to Zadeh's possibility theory (Zadeh L A, 1981 rbracket:

$$\Pi(D_k) = \max_i(\min(\mu_{A_i}(u), w_i^k)), \quad 1 \leq i \leq R_k$$

where R_k is the cardinality of the set of available rules.

2.3. Learning of Fuzzy Production Rules

The problem of learning fuzzy classification rules is formulated in terms of how to automatically induce, from a training set of example T^k , the strengths of implication w_i^k for each rule R_i and for a given diagnostic class D_k. These examples are classified image structures, each of which may be formalized in the following form:

$$t^k = [u_t\,;\, j_t^k], \quad t^k \in T^k$$

where u_t is the vector of measurements obtained in correspondence with the selected set of features, j_t^k is the expert's judgement expressing the strength or degree of satisfaction with which the image structure represented by u_t may be assigned to class D_k.

Experts must use predefined linguistic labels, such as *very high*, *medium*, *very low*, to express their mental judjement; these labels are automatically translated into numbers in the interval [0,1].

Given the diagnostic class D_k, let:

- l,the cardinality of the set of all the possible antecedent (A_i)
- m, the cardinality of the training set T^k
- and p, the cardinality of the term set W containing the terms with which the strength of the implication may be expressed.

Matrices M_1^k and M_2^k are then defined:

$$M_1^k = \begin{bmatrix} \mu_{A_1}(u_1) \ldots \cdots \mu_{A_1}(u_p) \\ \ldots \\ \ldots \\ \mu_{A_l}(u_1) \ldots \cdots \mu_{A_l}(u_p) \end{bmatrix}$$

$$M_2^k = \begin{bmatrix} \mu_{w_1^k}(j_1^k) \ldots \mu_{w_m^k}(j_1^k) \\ \ldots \\ \ldots \\ \mu_{w_1^k}(j_p^k) \ldots \cdots \mu_{w_m^k}(j_p^k) \end{bmatrix}$$

In the case in which fuzzy sets $w_1^k \ldots w_m^k$ are defuzzified into numbers, matrix M_2^k is the result of a crisp comparison (by equality) between values $w_1^k \ldots w_m^k$ and values $j_1^k \ldots j_p^k$.

By performing the matrix product

$$G = M_1^k o M_2^k$$

using max-min operators, matrix G with dimension $l \times m$ is obtained. The element g_{ij} expresses the induced degree of certainty with which the rule

$$A_i \rightarrow^{w_j^k} D_k \quad (2)$$

may be generated.

A decision-making activity is performed to select the best strength value w_j^k which may be assigned to each A_i, D_k pair. The rule $A_i \rightarrow w_j^k D_k$ is created if:

$$g_{ij} = \max_h(g_{hj}) \quad 1 \leq h \leq m$$

3. COLPOSCOPIC IMAGE CLASSIFICATION

The formal procedure presented above has been applied to the problem of designing an automatic system to support the classification of colposcopic images for the identification of neoplastic and preneoplastic regions. This could be a valuable contribution to the generation of diagnostic rules in a sector where the intrinsic complexity of the scene (two images are involved for each subject), the presence of noise introduced during the imaging process, and the large amount of information the physician must correlate in his visual analysis make interview techniques inappropriate.

3.1. The Medical Problem

Recent studies of cervical intraepithelial neoplasia and invasive cancer of the uterine cervix show the need to integrate the traditional Papanicolaou Smear with visual inspection (colposcopy) (Darnell Jones D E et al., 1987), (Lowzosky et al., 1982). But colposcoy, in spite of its amply proved capability of reducing the percentage of wrong diagnoses ("false negatives"), is not widely employed basically because it is both expensive and time consuming (Morell N D et al., 1982).

In this situation, an automatic system supporting medical diagnosis, by selecting from a large number of subjects those showing some suspect alteration of uterine epithelial tissues, could substantially reduce (by about 60%) the population group for which further examination is recommanded.

These considerations have spurred studies in the application of automatic image interpretation techniques to digitized colposcopic pictures in order to identify patients at risk for cervical neoplasia. The first important result is the possibility of dividing subjects into two main groups identified on the basis of characteristic features of an epithelial region called the "transformation zone", where the anomalous alterations appear in the early stage. One group consists of subjects for which no further investigation is currently required (with a "Normal Transformation Zone": NTZ), while the second consists of subjects (with "Atypical Transformation Zone": ATZ) for which visual inspection or additional colposcopically directed biopsy is recommanded.

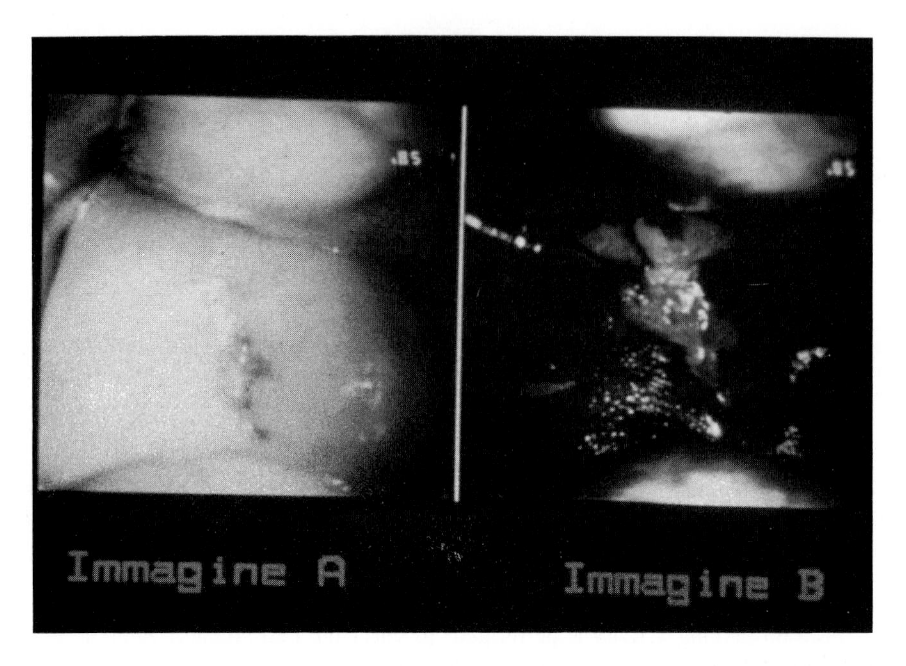

Figure 1 An example of a pair of colposcopic images with Atipical Transformation Zone

Epithelium	Pattern	Image A color	Image B color	Topology
MALPIGHIAN		Pink	Black	Outer
COLUMNAR		Red	Red-Yellow	Inner
IMMATURE		Light Pink	Red-Yellow	Transition zone
ATYPICAL		White-Light pink	Red-Yellow	Transition zone

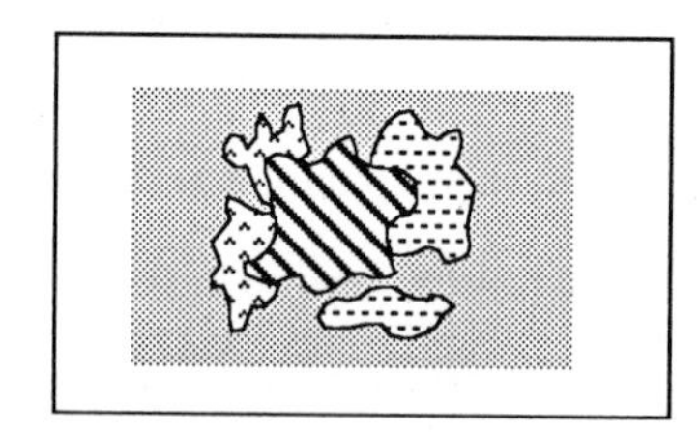

Figure 2 Possible types of tissues

3.2. The Image Model

The scene observed and the physical phenomenon involved are respectively the uterine cervix with its different types of epithelial tissues and possible processes of alteration of the squamocolumnar junction (SCJ). Each subject is observed after ordered and non-reversible supplies of two different reagents, respectively acetic acid and iodine solution, that give a selective colouring to the different types of tissues. The colposcopic pictures, normally used by the expert in his visual analysis, and recorded on standard film, are digitized using a microdensitometer with sampling interval of 50 micron and a dynamic range for each RGB component of 256 grey levels.

Two pictures ("imageA" and "imageB")are available for each subject. An example of a pair of these images can be seen in Figure 1.

The possible types of tissues and their spatial interrelations are illustrated in Figure 2: the malpighian (or squamous) epithelium is found from the outer region toward an approximate center of the scene, the junction (if visible) separates this tissue from the columnar (or ectopic) tissue. The transformation zone, formed where malpighian tissue encroaches columnar tissue is the locus where metaplastic or neoplastic lesions, known respectively as "immature" and "atypical" epithelia, generally appear.

The expert's visual analysis correlates the information contained in both images and collects elements relating to color, texture and relative position of the different tissues to reach a conclusive judgement. This judgement, if the SCJ is visible, ascertains the presence of any lesions and the need of a directed biopsy. If the junction is not visible (imageB is completely brown), the use of different diagnostic tools mey be recommanded.

3.3. Classification strategy

The automatic classification system has the goal of screening for subjects for whom further investigation is not necessary. The goal is reached with a complex interpretation strategy that consists mainly in selecting and performing those visual procedures which first focus analysis on significant regions in the scene and subsequently allow the extraction of discriminant features. Classification is performed on the basis of the selected features.

For the sake of brevity, we focus our attention on classification tasks here. A detailed description of the preprocessing and processing activities involved in the application may be found in (Della Ventura A et al., 1989).

Classification has been devided into two steps, preliminary classification and final classification, with subsequent complexity reduction limiting both the space of events and image analysis activities.

3.3.1. Preliminary Classification

Preliminary classification discerns two classesof subjects:

- C_1 , collecting normal subjects (N), in which only malpighian and columnar epithelia are present, and subjects with atypical transformation zone (ATZ)
- C_2 class, including subjects with immature epithelium.

To do so the physician considers the degree of definiteness of the SCJ, which is identified on image B as the border between IODO- and IODO+ zones.

Two linguistic variables are defined to represent contour features:

- Sharpness = (*well defined*, *medium*, *smoothed*)
- Morphology = (*regular*, *medium*, *irregular*)

Terms in the above term sets are fuzzy sets defined in the universes of discourse as

- $U_{Sharp.} = [0, 255]$
- $U_{Morph.} = [0, \pi/2]$

respectively.

To compensate for illumination effects, sharpness and mophology are measured by first performing a local linear stretching (Della Ventura A et al., 1989).
$U_{sharp.}$ contains values obtained by averaging gradient module values determined, with edge detector masks, for each contour pixel.

Values in $U_{morph.}$ are wiggliness measures. They are obtained by computing the differences of gradient direction on the values of image B.

To exemplify the definition of linguistic variables, Figure 3 shows the membership functions of fuzzy sets associated with labels of Sharpness and of Morphology.

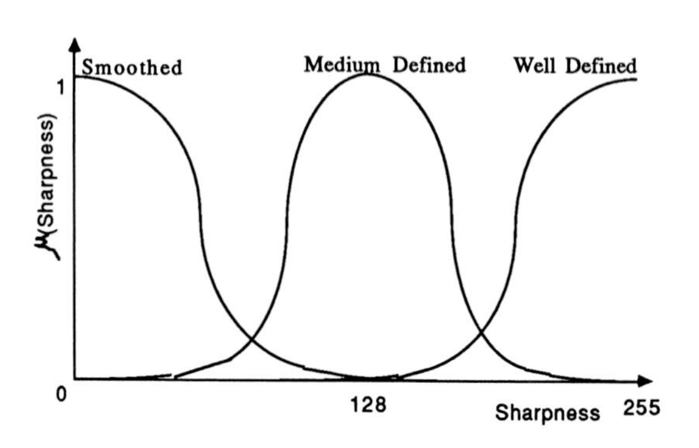

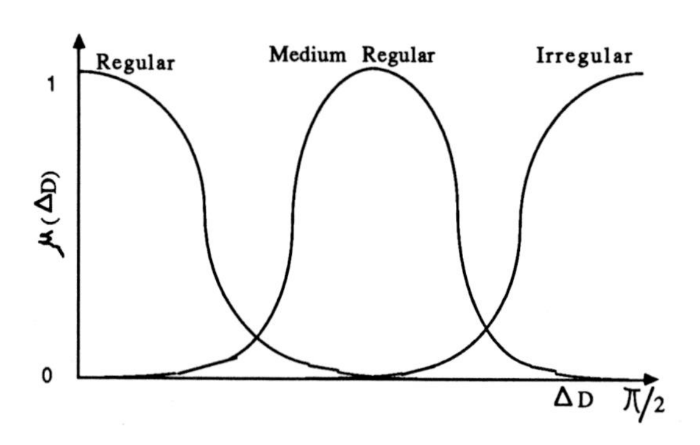

Figure 3Membership functions characterizing terms of Sharpness and Morphology

The degree of belonginess of image contour to one of the predefined class is linguistically expressed with a term set associated with the linguistic variable W:

$W = $ (*very low*, *medium*, *high*, *very high*) .

The training set is composed of 10 classified images $(I_i,[j_i^1 j_i^2])$ $1 \leq i \leq 10$. Images are classified by physicians as belonging to C_1 and C_2 with different values of belonginess j_i^1, j_i^2 choosen from W's values. The matrices M_1^1 and M_2^1 for class C_1 are defined according to the definition provided in section 2:

$$M_1^1 = \begin{bmatrix} \mu_{well\ def.,reg.}(I_1) \ldots \cdots \mu_{welldef.,\ reg.}(I_{10}) \\ \ldots \\ \ldots \\ \mu_{smooth.,\ irreg.}(I_1) \ldots \cdots \mu_{smooth.,irreg.}(I_{10}) \end{bmatrix}$$

$$M_2^1 = \begin{bmatrix} \mu_{verylow}(j_1^1) \ldots \mu_{veryhigh}(j_1^1) \\ \ldots \\ \ldots \\ \mu_{verylow}(j_{10}^1) \ldots \cdots \mu_{veryhigh}(j_{10}^1) \end{bmatrix}$$

The matrix product $M_1^1 o M_2^1$ gives us the matrix $G_{9\times5}$; strengths of classification rules are inferred by maximizing matrix G and performing suitable thresholding on the strength values accepted.

The same procedure is adopted for class C_2.

Two sets of rules (for C_1 and C_2 respectively) generated with the threshold of acceptance set at the value $t = 0.5$ are:

Sharpness	**Morphology**
well defined	*regular* $\rightarrow^{very\ high}$ C_1
well defined	*medium* $\rightarrow^{high}$ C_1
medium	*regular* $\rightarrow^{high}$ C_1
medium	*medium* $\rightarrow^{low}$ C_1
smoothed	*regular* $\rightarrow^{low}$ C_1
smoothed	*irregular* $\rightarrow^{very\ low}$ C_1
well defined	*regular* $\rightarrow^{very\ low}$ C_2
well defined	*medium* $\rightarrow^{low}$ C_2
medium	*regular* $\rightarrow^{low}$ C_2
medium	*medium* $\rightarrow^{medium}$ C_2
smoothed	*regular* $\rightarrow^{high}$ C_2
smoothed	*irregular* $\rightarrow^{very\ high}$ C_2

3.3.2. Final Classification

In the second classification step, the SCJ contours extracted in the previous phase as belonging to an atypical transformation zone (ATZ) or columnar transformation zone are now looked for in Image A. The SCJ in Image A is localized via generalized Hough Transform (HT).

The IODO-region extracted from Image B in the previous processing phase is now shifted and overlaid on image A (Figure 4) on the basis of the HT results.

Classification is based on color analysis; in particular, the selected feature is the difference in intensity, between internal and external regions with respect to the SCJ contour, measured using a component of the RGB space.

A linguistic variable is defined to linguistically qualify the differences in intensity together with their signs, passing from the internal region to external region:

- ΔI = (*negative*, *null*, *positive*)

Figure 5 shows the membership functions characterizing terms of the above term set. A preliminary set of classification rules has been generated for N and ATZ classes on the basis of a training set of 5 images.

The results of the two classification steps are then combined to obtain a definitive classification.

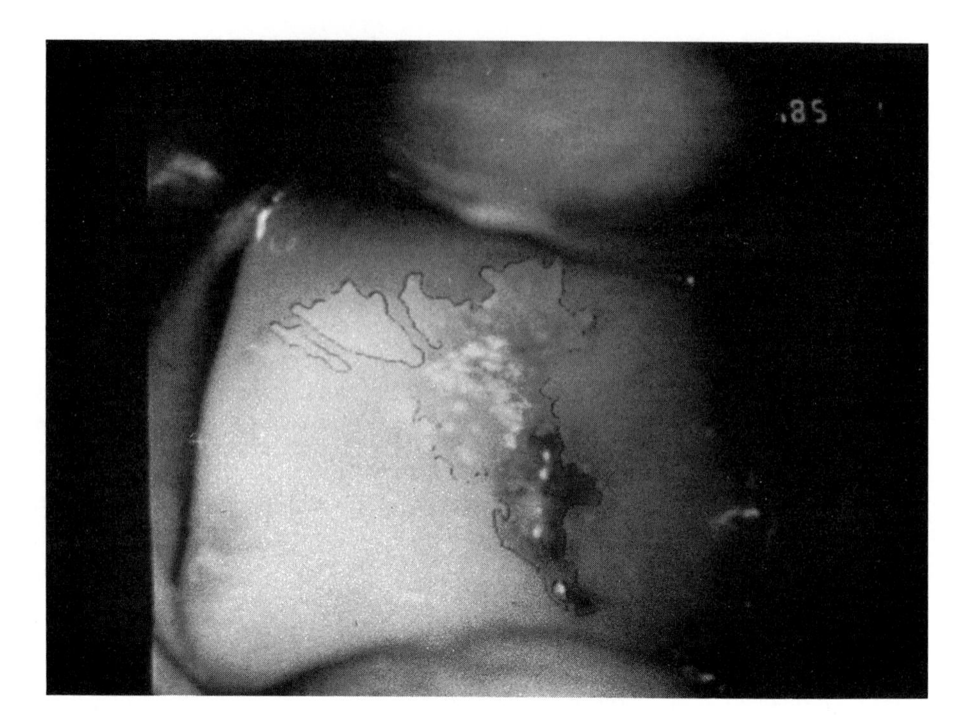

Figure 4 The contours extracted from image B superimposed on image A

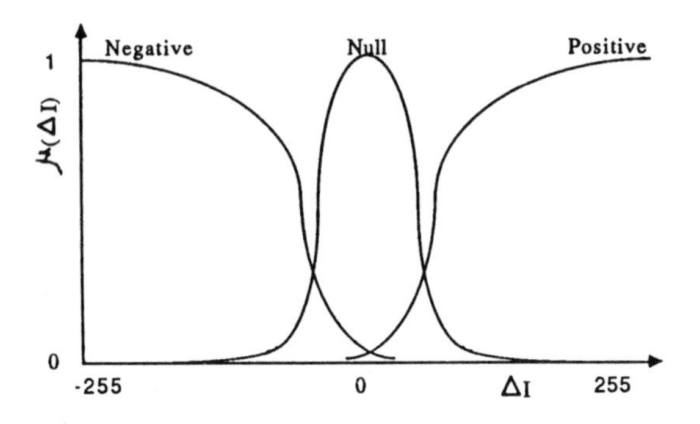

Figure 5 Membership functions carachterizing terms expressing the differences in intensity

4.PRELIMINARY RESULTS AND FUTURE WORK

The application of the learning technique to the case of colposcopic images has been developed with preliminary training sets of classified images but it has not yet been tested on representative number of patients. However, as a first assessment of its performance, the diagnostic results were compared with human diagnoses in 30 cases.

Plans have been made to extend the training set of classified images in order to improve system's accuracy.

Aknowledgements

The authors sincerely thank Dott. Anna Della Ventura, of the C.N.R. Istituto di Fisica Cosmica, Milan for her extremely helpful and highly qualified scientific supervision and Dott. M. Sideri of the Prima Clinica Ostetrica Ginecologica dell'Universita' di Milano for kindly offering his expert advice in the medical field.

References

Gaines B R (1988). An Overview of Knowledge-acquisition and transfer. In: Knowledge Acquisition for Knowledge-Based Systems. Gaines B R and Boose J H (eds), Academic Press, New York, pp. 3-22.

Breuker J and Wielinga B (1987). Use of Models in the Interpretation of Verbal Data. In: Knowledge Acquisition for Expert Systems, Kidd A L(eds), Plenum Press, New York, pp. 17-42.

Binaghi E, Della Ventura A, Rampini A, Schettini R (1990). A Fuzzy knowledge-Based System for Biomedical Image Interpretation. In proceedings of IPMU, Paris, pp.170-172.

Tong R M (1978). Synthesis of fuzzy models for industrial processes International Journal General Systems. 4: 143-162.

Pedrycz W(1984). An identification algorothm in fuzzy relational systems. Fuzzy Sets and System. 13:153-167.

Pedrycz W (1985). Applications of Fuzzy Relational Equations for methods of reasoning in presence of Fuzzy Data. Fuzzy Sets and Systems. 16: 163-175.

De Mori R and Saitta L (1980). Automatic learning of fuzzy naming relations over finite languages. Information Sciences. 21: 93-139.

Lesmo L, Saitta L, Torasso P (1982). Learning of Fuzzy Production Rules for Medical Diagnosis. In: Approximate Reasoning in Decision Analysis. Gupta M M and Sanchez E (eds), pp.249-260.

Zadeh L A (1981). PRUF - a meaning representation language for natural languages. In: Fuzzy Reasoning and its Applications, Mamdani E H and Gaines B R. (eds) Academic Press, London, pp. 1-58

Binaghi E (1990). A Fuzzy Logic Inference Model for a Rule-Based System. Medical Diagnosis, 7(3): 134-141.

Zadeh L A (1981). Fuzzy Sets as a Basis for a Theory of Possibility. Fuzzy Sets and Systems, 5: 3-28.

Darnell Jones D E , Creasman W T , Dombroski R A , Lentz S S , Waeltz J L (1987) Evaluation of the atypical Pap smear. Am. J. Obstet. Gynecol. pp.157-544.

Lozowsky M.S. (1982). The combined use of cytology and colposcopy in enhancing diagnostic accuracy in preclinical lesions of the uterine cervix. Acta Cytol., pp.26-285.

Morrell N D , Taylor J R , Snyder R N (1982). False Negative citology rates in patients in whom invasive cervical cancer subsequently developed. Obstet. Gynecol., pp.41-60.

Della Ventura A, Pennati G, Sideri M (1989) Computer Aided Screening of Subjects at Risk for Cervical Neoplasia. In: Recent Issues in Pattern Analysis and Recognition, Lecture Notes on Computer Science, Cantoni V, Creutzburg R, Levialdi S, Wolf W (eds), Springer-Verlag, Berlin, 399, pp. 338-350.

A FUZZY MODEL FOR THE PROCESSING AND RECOGNITION OF MR PATHOLOGICAL IMAGES

Dellepiane S , Venturi G , Vernazza G

Department of Biophysical and Electronic Engineering

University of Genoa

via Opera Pia 11a, I16145, Genova - Italy

ABSTRACT

The methodologies for the generation of model describing two echoes of MR pathological images of the head are presented. A vocabulary set has been chosen and formalized consisting of attributes and relations for the characterization of the organs and tissues contained in the image. The analitic study of a training set of images, associated with expert aid for medical aspects has permitted the creation of the model whose robustness has been proved on a set of test images. The most important and innovative characteristics of the model are the hierarchical subdivision between organ-father and sub-organs, together with the distinction between anatomical and acquisition-dependent properties of the image.

The results obtained by utilizing the model inside the system IBIS, devoted to image recognition, are presented. The same model has been used for various patients, thus proving its ability to face individual variabilities. For each patient the cerebrum, the pars ossea, the cutis, and the pathological area (where present) have been recognized with sufficient accuracy, while the ventricle has often been identified with lower precision.

KEYWORDS

Knowledge-based systems; propositional models; segmentation; model matching; visual data; symbolic data; numerical data.

1. INTRODUCTION

The final goal of automatic interpretation consists in the matching of models with data. In other words, this process aims to find a correspondence between an image (or image subparts) and predefined models of the world under analysis in order to obtain a description of what is represented in the input image.

For the purpose of automatic interpretation of real images, a knowledge-based (KB) approach seems to be the most appropriate, since it can manage different sources of information, address complexity and uncertainty problems, and take into account relational aspects (Hanson and Riseman, 1988). Moreover, the models used by a KB system can be very general, independent of the system's architecture and implementation, hence widely utilizable and flexible.

In this paper the methodologies applied for the development of a model of MR pathological images of the head (in the presence of lesions of the tumor type) will be presented. Special attention has been focused on the results obtained by utilizing the model inside the Knowledge-Based system IBIS (Interpretation of Biomedical, Industrial and remote Sensing images) (Vernazza et al, 1987) (Dellepiane et al, 1988) (Dellepiane et al, 1989).

In particular, the process of model generation is explained. It starts from a set of training images whose features are analyzed to find discriminant descriptions of single objects and object subparts and their mutual relationships .

Only one model for each scene type is derived from the training set images; it takes into account individual characteristics, deviations in image acquisition, and noise effects.

The model generation follows the subsequent steps:

- Selection of an appropriate vocabulary set to describe the chosen class of images. This vocabulary consists of a set of attributes and relations able to characterize the organs and tissues of the image;

- Formalization of the vocabulary in a mathematical way. For this step the fuzzy approach has been adopted to take into account uncertainty;

- Description of an image class. This step needs the selection of a set of training images and an analitic study of their properties. The aid of experts is needed for the medical aspects;

- Test of the model with a set of images able to verify its robustness.

The generated model aims to be general and applicable to other system architectures, as demonstrated for the iconic fuzzy-set approach proposed in a previous paper (Menhardt and Schmidt, 1988) and in a EEC project (COVIRA, 1990).

Results obtained by applying this model to a set of test images are reported; they demonstrate the applicability of one ideal model to various real scenes of the same type.

The specific images considered in this paper are transaxial slices acquired from the first and second echoes of multiecho sequences, at the superorbital level; each slice shows signs of a lesion of the tumor type. For each case, two input images are given to the system, and the model refers to the related acquisition sequences.

The model efficiency is thus tested by considering different patients and by evaluating the errors incurred.

2. METHOD DESCRIPTION

The approach followed by the system IBIS is the classical one, which subdivides the interpretation process into the low-, middle- and high-level processings (corresponding to the numerical, symbolic and semantic abstraction levels) driven by a mixed strategy which allows for feedback and backtracking mechanisms. The primitives extracted by the segmentation step, and used for the symbolic interpretation process, are of the region type. The final results consist in associating a group of regions with an object, and in assigning each single region a fuzzy membership value related to its degree of reliability.

The particular domain application knowledge is wholly contained in the model. As a consequence, it seems evident that the development of the model assumes a central role for the system performances.

The main model requirements are:

- efficiency and tolerance: the model must represent accurately both complex images affected by noise and variable data on the same class of objects;
- discriminating power: the model must be able to discriminate between various classes of objects.

The fuzzy approach has been adopted to easily represent the model, as it seems to be one of the most suitable methods for handling imprecise and uncertain situations.

The contemporaneous fuzzy representations of data and of the matching mechanism allow for a powerful use of uncertainty.

In addition, model complexity, individual characteristics of real objects, and noise effects are faced not only by exploiting the general constraints expected between and within scene elements but also by adding some redundancies to these constraints (Hanson and Riseman, 1988). By combining many partially redundant rules, the effect of a single unreliable rule is reduced by the effect of feature cooperation.

As a consequence, redundancies allow for the selection of less constrained linguistic values for a single feature, thus generating a more fault-tolerant description.

On the other hand, it is not advisable to utilize a description based on the complete set of features available to the system, as sometimes this description may be affected by errors, for instance, when an unreliable feature is used, even though it is not required.

Once the application domain has been selected, a vocabulary set is created, containing the most suitable properties and relations to be used in describing the model. They are chosen according to the domain considered, and constitute the features on which the model and the matching process are based. This phase is based on a set of images (training set) acquired by using a fixed set of acquisition parameters.

3. VISUAL DATA STRUCTURE

Visual data are made up of numerical and symbolic data. Various abstraction levels are represented in accordance with the organization of the processing levels involved in the recognition task.

At the beginning of the interpretation process, visual data consist of the original image (or images) only. Subsequently, they are increased by generating transformed images, by extracting symbols through the segmentation step, and by inferring new symbols related to the interpretation result.

At the numerical-representation level, original and filtered images are sets of pixels, each characterized by a feature vector referring to location and intensity attributes.

The segmentation process provides a new image: to each pixel a label is assigned, corresponding to the region the pixel belongs to.

Then, the same image, after the segmentation process (i.e., at a higher symbolic level) can be described as a set of elementary regions.

The segmentation process makes it possible to reduce the dimensions of the patterns to be classified, by passing from the number of pixels in the image to the number of grown regions. So far, the processing has been performed without using the domain knowledge; the segmentation process exploits the gray-level distribution in the original image (or images) to find groupings of similar pixels (Ronco et al, 1990).

Other data levels are generated, starting form this point up to the interpretation process, by comparing data with model elements.

The actual interpretation process, that is, the matching between model and data elements, is performed at the symbolic level by comparing regions and model descriptions, on the rules of fuzzy logic.

The final result of the interpretation process consists of a set of hypothesized scenes, that is a grouping of hypothesed objects, for which relationships are verified as in the model, and each characterized by a fuzzy membership value. Accordingly, the best hypothesis is supposed to be the one with the highest fuzziness value. However, possible alternatives are available to the user; they are particularly significant when their fuzzy membership values are quite similar.

4. MAGNETIC RESONANCE APPLICATION

4.1. The anatomical model structure

The anatomical model has to contain the descriptions of the organs and of the anatomical regions visible in the slices parallel to the Reid line in the superorbital part.

Such a description has to be derived from the training images, and has to characterize anatomical objects to allow for their recognition, without aiming at a complete descriptive model.

The level of detail was chosen following the advice of medical experts: only the descriptions of the anatomical regions that are of some interest and that can be detected in an MR slice by using the selected sequence parameters were included in the domain knowledge. However, when using the present approach, addition of more details or modifications to existing information are extremely easy.

Table 1 shows linguistic values that the attributes can assume in the context of the present medical application and the related fuzzy compatibility functions. The shape of fuzzy functions is, at present, predetermined once the application domain has been selected, and it does not take into account population distributions from the training images.

Finally, at the bottom of the table, the operators that can be applied to such linguistic values are given.

Relations are presented in Table 2: the first section gives the relations derived from the attributes;the second section gives specialistic relations which have to be computed directly on images. The linguistic values applicable to such relationships are given,too. Only the "not" operator is provided for relationships.

Attributes and relations, as well as their linguistic values and fuzzy operators, are collected in appropriate vocabularies and associated with the algorithms provided to the system and able to compute them starting from an original and a segmented image.

Some features refer to domain-independent attributes, which define densitometric, topological and geometric properties that are of major importance for any kind of digital real images. Another type of features describes the properties related to the specific application domain considered, e.g., centrality, which is computed on the basis of the elliptical geometry of transversal slices.

In terms of the content, features can be further divided into features referring to anatomical knowledge (Ronco et al, 1990), which is general for all biomedical images of supertentorial slices, and features describing the specific physical characteristics of an MR signal. So it is possible to utilize the same anatomical knowledge to process different biomedical images, by only changing the physical features.

Generally, algorithms for property measurements are not precise enough; the use of a fuzzy approach and the combination of various attributes allow one to reduce such uncertainty. In the future, a larger number of more precise tools will certainly yield more reliable interpretation results, even for very complex images.

An additional hierarchical level was used to describe the model of human anatomy, taking into account the intrinsic taxonomy of anatomical regions. Some anatomical objects can be subdivided into a number of anatomical subregions (called suborgans) that describe them in greater detail. For instance, gray and white matters are suborgans of the brain. This additional hierarchical level introduces a set of nodes linked to the father organs.

Independently of the model's hierarchical organization, in the present recognition system a specialized strategy for the search for suborgans is applied. Detection of a suborgan starts only after the identification of the father organ. The search area is thus restricted to the area of the father organ, thus increasing the probability of locating the suborgan correctly and reducing computation time. Of course, the failure to identify a suborgan does not invalidate the detection of the father organ.

Table 1. Attributes, linguistic values, and related fuzzy functions for the application domain of tomographic images. The last line contains the operators that can be applied to the linguistic values. (SIZE = dimension in pixels; GRAY = average-gray-level on both the first and second echoes; CCx, CCy = centroid coordinates; m = minimum coordinate, M = maximum coordinate; FIT = region_size/MBR_area; ELONG = Max_MBR_side/min_MBR_side; DIREC=direction of MBR).

ATTRIBUTES Ai	LINGUISTIC VALUES	FUZZY FUNCTION
SIZE:	large medium-size small	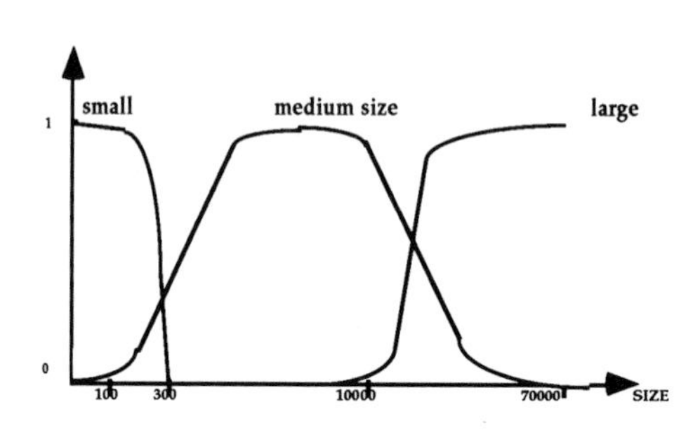
GRAY (first and second echo):	very-dark dark medium-bright bright very-bright	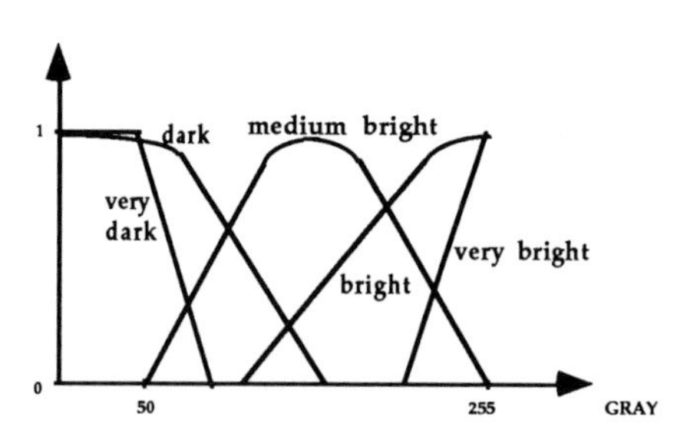
mx,my,Mx,My:	right medial left occipital parietal-temporal frontal	
ELONG:	elongated	
FIT&ELONG:	circular circular-and-elongated	
CENTRALITY:	peripheral central	
OPERATORS:	very,quite, not	

Up to now, the same segmented image has been used to detect both organs and suborgans; in the future, it will be extremely useful to base the main organ's recognition on a very rough segmented image, whereas a more detailed segmentation will be performed on the area of an organ to detect its suborgans, thus avoiding merging and recognition of single parts.

Table 2. Relationships and linguistic values for the application domain of tomographic images.

RELATIONS based on:	LINGUISTIC VALUES:
SIZE:	larger-than similar-size-as smaller-than
GRAY (first and second echo):	brighter same-gray darker
CCx,CCy mx,my,Mx,My:	frontal-of occipital-of right-of left-of homosagittal homocoronal
mx,my,Mx,My:	outside inside
CENTRALITY:	more-central more-peripheral

RELATIONS Ni:	LINGUISTIC VALUES:
relative position	in-contact-with

OPERATORS: not

4.2 Training on pathological slices

Three medical cases (denoted as Nos. 1451, 2397, 2065) were considered in order to generate the model for a class of slices showing lesions of the tumor type.

For each patient, the first and second echoes of a multiecho sequence were considered, for echo times TEs equal to 50 and 100, respectively, and for TR equal to 2000 (the original images of patients and 2397 are displayed in Figure 1).

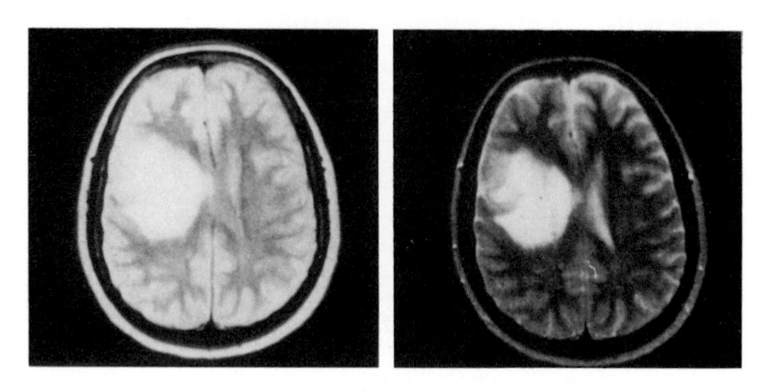

Figure 1. Original echoes of patient N. 2397.

The elementary regions of each training image were then manually assigned to the correct class; this resulted in labelling each region by the code of the identified class. In Figure 2, the manual segmentation is shown overlapped to the original second echo.

For the regions assigned to the same organ in the various training images, the histograms of feature distributions were analyzed to derive the discriminative properties of that organ and of its subparts with respect to the other organs in the scene.

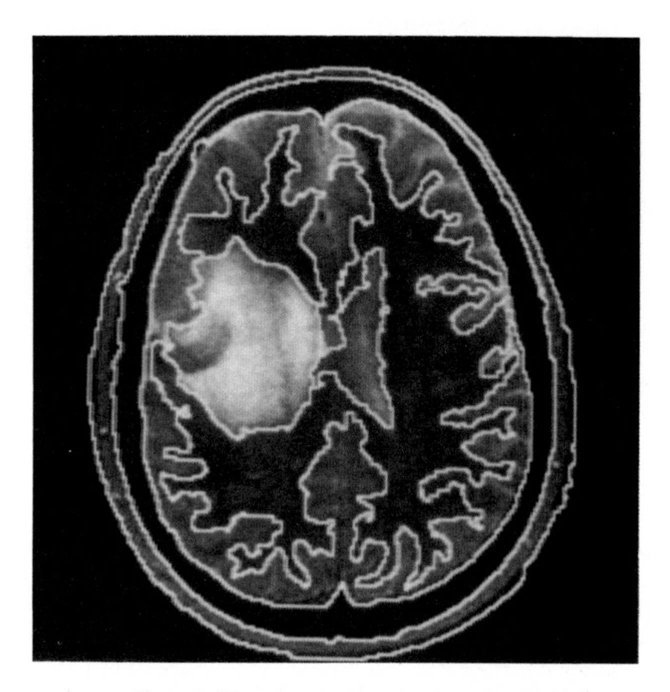

Figure 2. Manual recognition map for patient N. 2397.

After the manual region labelling, the most similar compatibility function was associated with each feature distribution.

Unary descriptors of organ properties were used as long as they made it possible to discriminate between classes.

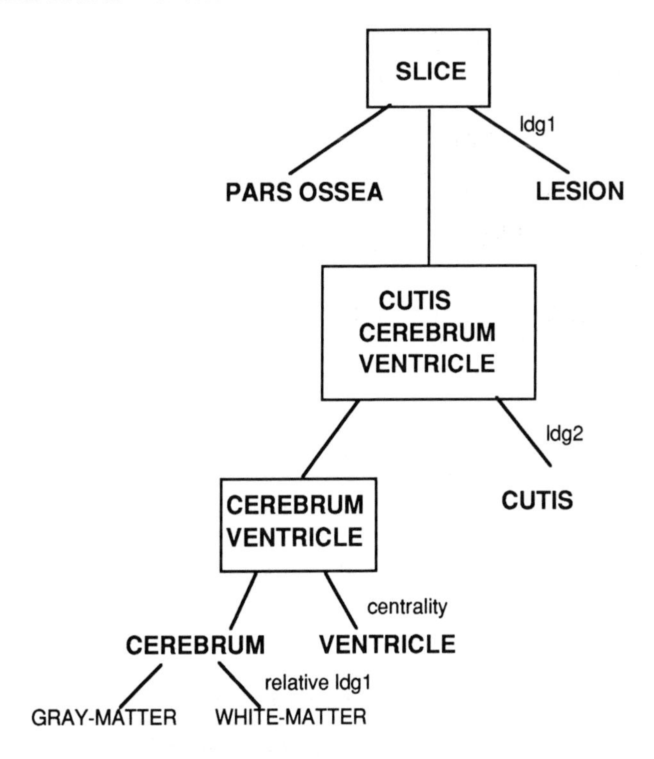

Figure 3. Feature-tree consistent with a progressive employment of features for classification.

The classes' distributions were analyzed in terms of a feature to define the discriminative power and the significance of that feature. It should be noted that a feature could be regarded as being one of the best for the description of a class, but not for the description of another class.

Distributions that were not significant were soon discarded, for example, if an attribute assumed a very wide range of values when computed on an object's subparts (uniform distribution).

A distribution was considered to be separate enough even when queues were partially overlapped. The fuzzy function associated with a feature distribution was selected to be larger than the distribution itself in order to take into account possible deviations in real cases. The combination of several functions designed over various features allowed us to discard those samples (belonging to other classes) which were contained in the queue parts, since they were supposed not to verify the other class descriptions.

From Figure 3 one can see that, by applying these rules, some organs (e.g., the pars ossea and the lesion) can be well separated just according to gray-level attributes, while some others, (e.g., the cutis, the ventricle and the cerebrum), need some additional features to be assigned to a given class.

As a consequence, in order to distinguish between these classes, the second-echo gray level was used, which is mainly able to separate the cutis from the area of the cerebrum and the ventricle. Then, by using centrality, the ventricle and the cerebrum were separated.

Finally, it was possible to discriminate between gray matter and white matter only by utilizing the relative information inside the area detected as the cerebrum, as, the separation of these anatomical regions requires a more detailed analysis than that performed in the previous steps.

The process of adding new features to describe object classes can be represented, in an iconic fashion, by a feature-tree, as shown in Figure 3. Starting from a global slice image, a feature is used to decompose it into some classes or class groupings. A feature is considered to be significant enough if further separation is accomplished by adding it to this tree. It should be noted that such separation is not complete, in that branches of the tree are associated not with fixed thresholds but with fuzzy functions that produce a fuzzy separation between classes.

Table 3. The anatomical model

ANATOMICAL REGION:	SUBPART ATTRIBUTES:	GLOBAL ATTRIBUTES and RELATIONS:
PARS OSSEA	very dark-ch1 very dark-ch2 peripheral	not in-contact-with background more-peripheral cerebrum more-central cutis in-contact-with cutis
LESION	very bright-ch1 very bright-ch2	not in-contact-with background
CUTIS	medium-bright-ch1 bright-ch2 peripheral	not in-contact-with cerebrum more-peripheral cerebrum in-contact-with background
CEREBRUM	medium-bright-ch1 medium-bright-ch2 peripheral	large-region not in-contact-with background
VENTRICLE	medium-bright-ch1 medium-bright-ch2 very central	not in-contact-with background
SUBSTANTIA GRISEA (suborgan of cerebrum)	brighter-than-ch1 cerebrum brighter-than-ch2 cerebrum more-peripheral cerebrum not in-contact-with ventricle	
SUBSTANTIA ALBA (suborgan of cerebrum)	darker-than-ch1 cerebrum darker-than-ch2 cerebrum more-central cerebrum	

When, after the insertion of all available features, terminal nodes of the tree correspond to single classes, the set of features utilized is considered to have sufficient discriminative power.

The features used to pass from the root node to the terminal node corresponding to a class constitute the best subset for that class.

To introduce some redundancies, additional features are considered together with the Global attributes are computed on whole objects by extracting the features of the complex region made up of the object parts. Only the significant ones are included in the model, mainly when they refer to properties that are not inherited by the subparts from the whole organ (e.g., dimension properties). For instance, the cerebrum has "large-region" as a global attribute, but this is not an attribute of its subparts, which are smaller.

At this point, the information derived may still be insufficient to discriminate between some objects. Relationships are then introduced, even when not necessary, in order to avoid errors in some uncertain situations that may occur when testing the consistency of object configurations.

The generated model is described in Table 3. It points out that the gray-level attributes and the positions of anatomical objects inside a global section represent the most useful information for the description of such objects. Features and relationships are given for an object if considered necessary, by the model generation phase, to the image recognition task. In such a way a simple model description is achieved, which does not aim at completely characterizing anatomical organs and regions, but which is sufficient to the present recognition goal.

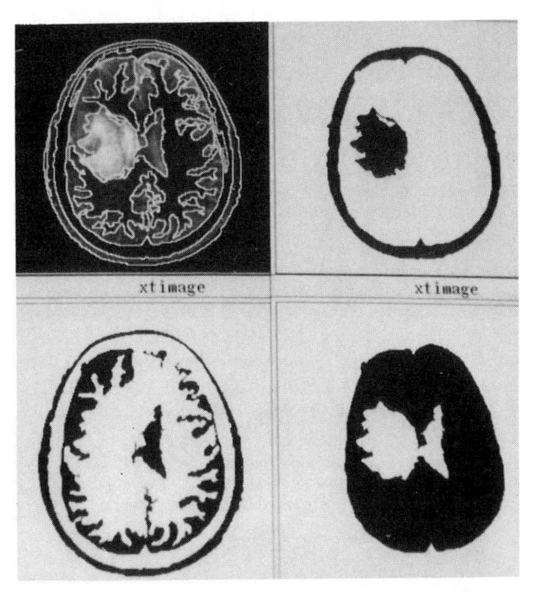

Figure 4. Automatic recognition map for patient N. 2397.

The application of this model during the interpretation process provided the results given in Figure 4 for the patient No.2397. The upper-left portion of the picture shows the contours of the recognized organs overlapped to the original second echo. The upper-right portion of the picture shows the regions recognized as ventricle, pathology and skull. The lower-left portion of the image shows the regions recognized as skin and gray-matter, the lower-right portion shows the regions recognized as brain.

As can be noticed from a visual comparison of the manual and automatic recognition results for this patient, very small discrepancies exist as regards principal organs, while greater differences arose between the sub-organs of white-matter and gray-matter. The reason is that differences in the description of these objects are so small that is often impossible to achieve an automatic classification with a sufficient reliability value. It is important to notice that, in such doubtful situations, the system assigns the regions to the father organ. In the present case a number of regions belonging to white and gray matter have been automatically labelled as brain.

4.3 Test on real pathological images.

Some test images were acquired by using the same parameters as for the training cases. They refer to a different set of patients, and not all of them include pathological areas.

The processing of a test case is described in the following; Figure 5 shows the original echoes.

The results of the automatic interpretation, obtained by using the model described in subsection 4.1, are presented in Figure 6.

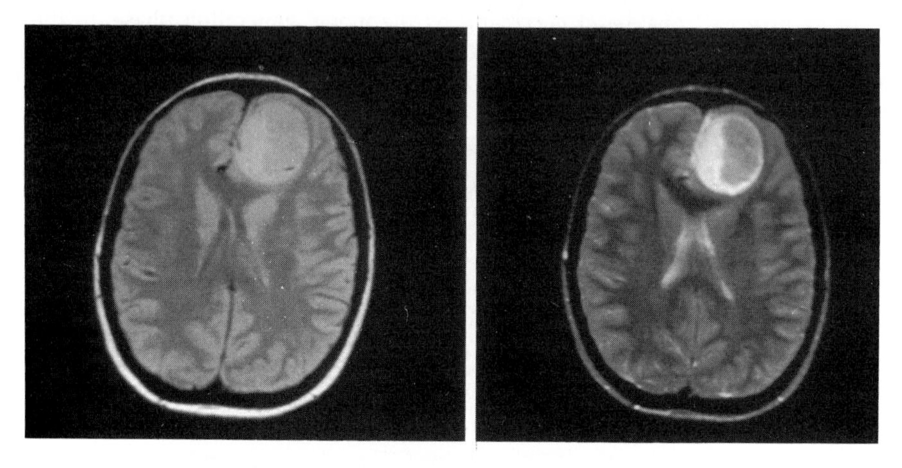

Figure 5. Two original echoes of the first test case.

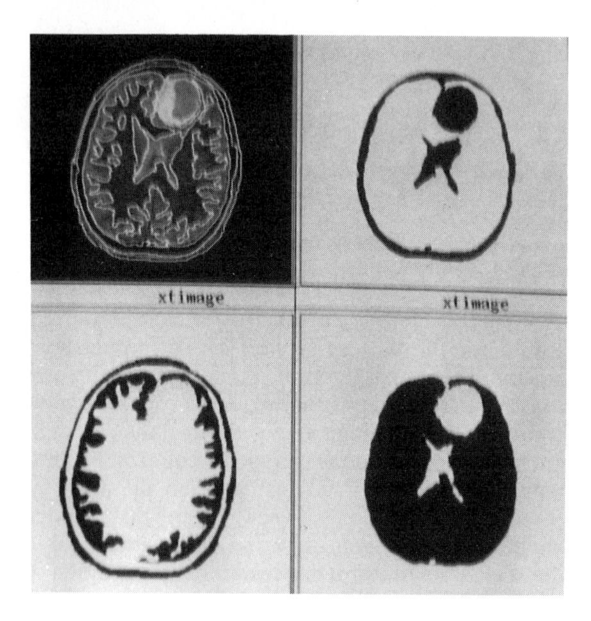

Figure 6. Automatic recognition results for the first test case.

All anatomical regions searched for were located in the analyzed images, and corresponded quite well to the real configurations. To evaluate the error introduced, a quantitative measure was obtained by comparing the automatic interpretation results with the manually assigned region labels. Deviations of the detected organs' contours from the real ones were not considered. This kind of error is referred to the segmentation process, which is required to obtain a good oversegmentation of the whole image.

The confusion matrix for the case shown in Figure 5 is presented in Figure 7, where insertion and substitution errors (i.e., the percent number of wrong pixels) are shown for each organ, anatomical object or tissue. Rows indicate how pixels belonging to an organ

		BACKGROUND	CEREBRUM	PARS-OSSEA	CUTIS	VENTRICULUS	LESION		SUBSTANTIA ALBA	SUBSTANTIA GRISEA		INSERTION ERROR
(40752)	BACKGROUND	40692			60							0.15
(15911)	CEREBRUM		16367									0
(3285)	PARS-OSSEA			3146	139							4.2
(2368)	CUTIS				2368							0
(1599)	VENTRICULUS		456			1143						28.5
(1544)	LESION						1544					0
(7960)	SUBSTANTIA ALBA		273						7654	33		3.8
(7069)	SUBSTANTIA GRISEA		1478						46	5545		21.5
	SUBSTITUTION ERROR	0	13.8	0	8	0	0		0.5	0.46		

Figure 7. Confusion matrix.

have been automatically interpreted; columns refer to the pixels assigned to an organ by the automatic process. The exact organ size is put in brackets before the name of each organ.

The last columns gives the percent insertion error, while the last row gives the percent substitution error.

Generally, the errors found in test cases range below 5 percent for the organs or anatomical objects. In the present case a higher error was incurred .pa between cerebrum and ventricle areas. In particular, this image is more sensitive than others to the lack in the model of the nucleus caudatus, which corresponds to the region responsible for the error. It is expected that the insertion in the model of this additional anatomical region would improve the recognition results, by decreasing error rates.

For suborgans or tissues, as, for instance, gray and white matter, the error seems higher; in this case, it reflects only an imprecise recognition, and not an error, since the confusion is only between an organ and its suborgans.

5. CONCLUSIONS

The validities of the approach and of the model were demonstrated by means of practical applications to various pathological tomograms. Results proved the sufficient efficiency and tolerance of the generated model, even if only three cases have been considered in the training set. Some innacuracies are still present, also due to the approximate definition of the manual recognized anatomical objects.

The use of a simple method for model generation was introduced. The repeatability and the quantitative characteristics of such a method suggest the possibility of easily making it automatic, thus allowing the knowledge engineer to avoid performing a time-consuming model acquisition process. The accurate selection and formalization of the vocabulary makes it possible an easy insertion of new organs knowledge (i.e., the nucleus caudatus) or the processing of images acquired at different slice levels.

In this paper MR images have been analyzed; however, there are valid reasons to believe that the same method may be generalized and extended to other types of images, obtained by different acquisition methodologies. The model is easily extendible to MR multislices (by extending the model and processing algorithms to 3D space), as demonstrated in (Dellepiane et al, 1988), and with few changes about the densitometric attributes to CT images (Dellepiane et al, 1990). Further developments of the present approach will allow the recognition of an organ and the extraction of accurate data on its properties, through the use of different information sources. In particular, very interesting advantages will be achieved by integrating different techniques (e.g. CT,MR and PET) in stereotaxy applications, where the exact localization of a pathology and a accurate analysis of its morphological aspects are of major importance for the planning of an appropriate therapy, as well as for didactical and training purposes.

To improve the model quality, it should be advisable a more precise definition of fuzzy functions behaviour, for instance by analyzing distributions and statistical characteristics of interesting populations, over a set of pre-recognized training images. In addition, in application domains that involve specialistic shape, as in biomedical images (especially in 3D space), it would be advisable to make use of analogical information, like an anatomical atlas. The latter could be employed together with the described propositional model to allow for more precise and more reliable interpretation results.

In addition to the above points, which constitute the goals of future research, a more extensive use of 3D models and 3D data would be of major importance.

ACKNOWLEDGEMENTS

The present work was partially developed within the framework of the project COVIRA (AIM-A1011), supported by the EEC, and of a national grant by MURST (40%). In particular, the training and test images were acquired with a Philips GYROSCAN T5 (0.5 Tesla) MR Tomograph.

REFERENCES

Dellepiane S, Regazzoni C, Serpico SB and Vernazza G (1988). Extension of IBIS for 3D organ recognition in NMR multislices. Pattern Recognition Letters Vo8:65- 72.

Dellepiane S, Regazzoni C, Serpico SB, Vernazza G (1989). An application-independent knowledge-based framework for complex image recognition. IAPR Positano 5th Int. Conf. Image Analysis and Processing Proc., World Scientific Pub., pp.309-316.

Dellepiane S, Leonardi M, Venturi G and Vernazza G (1990). Automatic recognition for 3D organ visualization from CT spatial sequences. Riv. di Neuroradiologia Vo3:81-93.

COVIRA- Computer Vision in Radiology, Project A1011 of the AIM programme (Advanced Informatics in Medicine) of the European Community. Consortium partners: Philips Medical Systems (Prime Contractor), IBM UK Scientific Center, DIBE University of Genova (Italy), University of Hamburg (FRG), G. Maranon General Hospital Madrid (Spain) (1990).

Hanson AM and Riseman ER (1988). Vision, Brain and Cooperative Computation. Academic Press, New York.

Menhardt W and Schmidt KH (1988). Computer Vision on Magnetic Resonance. Pattern Recognition Letters Vo8:73-85.

Ronco M, Vio R, Dellepiane S and Vernazza G (1990). Hierarchical image segmentation: a K-B system using fuzzy functions. Signal Processing V (Theories and Applications) Proc. of EUSIPCO-90, Elsevier Science Pub., 1735-1738.

Vernazza G, Serpico SB and Dellepiane SG (1987). A knowledge-based system for biomedical image processing and recognition. IEEE Trans. on Circuits & Systems, CAS-34:1399-1416.

LINEAR DISCRIMINANTS AND IMAGE QUALITY

H H Barrett[1,2,3], T Gooley[3], K Girodias[2],
J Rolland[1,2,4], T White[1,2] and J Yao[1,2]

[1] Optical Sciences Center, University of Arizona, Tucson, AZ 85721
[2] Department of Radiology, University of Arizona, Tucson AZ, 85724
[3] Program in Applied Mathematics, University of Arizona, Tucson, AZ 85721
[4] Dept. of Computer Science, University of North Carolina, Chapel Hill, NC

Abstract

The use of linear discriminant functions, and particularly a discriminant function derived from the work of Harold Hotelling, as a means of assessing image quality is reviewed. The relevant theory of ideal or Bayesian observers is briefly reviewed, and the circumstances under which this observer reduces to a linear discriminant are discussed. The Hotelling oberver is suggested as a linear discriminant in more general circumstances where the ideal observer is nonlinear and ususally very difficult to calculate. Methods of calculation of the Hotelling discriminant and the associated figure of merit, the Hotelling trace, are discussed. Psychophysical studies carried out at the University of Arizona to test the predictive value of the Hotelling observer are reviewed, and it is concluded that the Hotelling model is quite useful as a predictive tool unless there are high-pass noise correlations introduced by post-processing of the images. In that case, we suggest that the Hotelling observer be modified to include spatial-frequency-selective channels analogous to those in the visual system.

Keywords

Image quality; medical imaging; linear discriminant functions; ideal observer; Hotelling trace.

1. INTRODUCTION

A general definition of image quality has proven to be an elusive goal. Indeed, in the image-processing literature, image assessment is most often purely subjective, and no objective definition of quality is even attempted. The radiology literature is somewhat more sophisticated in this respect; image quality is usually defined there in terms of how well some observer can perform some task of diagnostic interest. The difficulty in that case is in choosing a task and an observer.

By far the most common observer of real radiographic images is the physician, though there is also considerable interest in automated or machine observers. For the human observer, task performance can be measured by psychophysical studies. If the task is binary (i.e., the observer has only two possible choices), the results of such studies can be analyzed by use of ROC (receiver operating characteristic) curves. A common figure of merit for image quality is thus the area under the ROC curve

(AUC) or the associated detectability index d' or d_a.

Though psychophysical studies and ROC analysis satisfy our requirement for a rigorous definition of image quality, there are still many problems in practice. The studies are time consuming and expensive, especially if the observers are physicians or if real clinical images are used. Moreover, the results are too specific to answer many questions of practical importance. An ROC study can give a definitive comparison of two imaging systems for one particular disease entity and one set of engineering parameters for each system, but it says nothing about how either system would perform with other parameters or for other diseases.

For these reasons, there is considerable interest in the use of model observers for which the performance indices such as AUC can be calculated rather than measured. If we had a model observer whose performance correlated well with that of the human, we could use it to study the effects of variation of task or system parameters. Such a tool would be extremely valuable for optimizing and effectively using radiographic imaging systems.

The most widely investigated model observer is the ideal or Bayesian observer, defined as one who has full statistical knowledge of the task and who makes best use of that knowledge to minimize a suitably defined risk. The strategy of the ideal observer for a binary task is to calculate a test statistic called the likelihood ratio and to compare it to a threshold in order to decide between the two alternatives; this strategy maximizes the AUC. The performance of the ideal observer sets an upper limit to the performance obtainable by any observer, including the human, and it might be hoped that a system optimized for the ideal observer would also be optimized for the human.

Though this approach seems reasonable, significant problems are encountered in practice. Most importantly, the likelihood ratio is only rarely calculable. Indeed, almost all investigations of the ideal observer have concentrated on detection of an exactly specified signal (or perhaps discrimination of two exactly known signals) superimposed on an exactly known background. We refer to such situations as SKE/BKE (signal known exactly, background known exactly). The SKE/BKE paradigm is obviously quite different from clinical radiology where, even for simple lesion-detection tasks, the background is cluttered with normal anatomic structures and the lesion to be detected is highly variable in size, location, shape and contrast.

The reason for the concentration on SKE/BKE tasks is that the likelihood ratio in that case can be calculated by simple linear filtering. For detection of a known signal on a flat background, where the only randomness is measurement noise that can be modeled as a stationary, white, Gaussian random process, the likelihood ratio is the output of a matched filter. If the noise is stationary and Gaussian but not white, the likelihood ratio is calculated by a so-called prewhitening matched filter.

Even in the SKE/BKE case, the performance of an ideal observer can be very different from that of a human observer. For example, Myers et al. (1985) found that human performance relative to the ideal was dramatically degraded by certain kinds of noise correlations. One interpretation of this result, and of similar results by other authors, is that the human observer is incapable of performing the prewhitening operation. This interpretation has led to the suggestion that the correct model for predicting human performance is the quasi-ideal or non-prewhitening (NPW) ideal observer who uses a simple matched filter, even in the presence of colored noise, to derive a test statistic. Though this test statistic is inferior to the optimum test statistic (the likelihood ratio), it does have the virtue of correctly predicting human performance in a range of SKE/BKE tasks.

Unfortunately, as we shall see in Section IV, the NPW model can yield very poor correlation with the human if there is inherent randomness in the task. Furthermore, the ideal observer is usually not an option except for SKE/BKE since the likelihood ratio is impossible to calculate. We must therefore look for other observer models that remain calculable for a wide variety of realistic tasks yet correlate well with the human observer.

These problems have led us to consider various linear discriminant functions, where the test statistic is a linear function of the data, as potential observer models. It is our hope that a suitable linear model will be found that will be computationally tractable for a wide variety of realistic tasks and will also be a good predictor of human performance as measured by ROC.

We have given particular attention to the optimum linear discriminant, which is often ascribed to Fisher (1936) but which had its origins in a classic paper by Hotelling (1931). We therefore refer to this observer model as the Hotelling observer.

It is the goal of this paper to survey efforts at the University of Arizona to determine the usefulness of linear discriminant models, and especially the Hotelling model, as tools for the assessment and optimization of imaging systems.

2. MATHEMATICAL BACKGROUND

2.1. Problem Statement

A digital image consisting of M pixels can be represented as an Mx1 column vector **g**. This vector is related to the object being imaged, denoted **f**, by a relation of the form

$$\mathbf{g} = \mathrm{H}\mathbf{f} + \mathbf{n} \ , \tag{1}$$

where **n** is a vector representing the measurement noise and H is an operator representing the imaging system, including any processing or reconstruction steps. If we consider only linear imaging systems and represent the object **f** in discrete form as an Nx1 column vector, then H is an MxN matrix. More generally, however, H can be a nonlinear operator, especially if a reconstruction algorithm is included, and **f** can represent a continuous object. There is no loss of generality in writing the noise as an additive term, even in the nonlinear case, provided the statistics of **n** take into account the statistics of **f** as well as the nature of H.

Note that **g** is a random vector, both because of the measurement noise and also because many different objects **f** will be imaged. We shall consider both sources of randomness in what follows, though only the measurement noise is present in SKE/BKE problems.

We assume that the task of interest is to observe a particular image **g** and use it to classify the corresponding **f** that produced the image into one of K classes. The simplest case is the binary task where K = 2 (e.g. normal vs. abnormal or lesion-present vs. lesion-absent). A general discriminant function for this binary task is a scalar test statistic $\lambda(\mathbf{g})$. The classification is performed by comparing this test statistic to a threshold λ_t; if $\lambda(\mathbf{g}) > \lambda_t$, **f** is said to belong to class 1, while otherwise it is classified into class 2. The theory of discriminant functions is concerned with finding the best functional form for $\lambda(\mathbf{g})$ and with assessing the accuracy of the classification procedure. If $\lambda(\mathbf{g})$ is a linear function of **g**, it is referred to as a linear discriminant.

2.2. Ideal Observers and Matched Filters

One theoretical route that yields a linear discriminant is to assume an ideal observer and to model the noise **n** as a Gaussian random process. In general, the test statistic used by the ideal observer is the likelihood ratio, defined by

$$\lambda_{ideal} = \frac{p(\mathbf{g}|\,1)}{p(\mathbf{g}|\,2)} \ , \tag{2}$$

where $p(\mathbf{g}|\,k)$ is the probability density of **g** given that it was produced by an object in class k (k = 1 or 2). This test statistic is usually very difficult to determine and a highly nonlinear function of **g**. In the special case where p(**g**| k) is a multivariate normal probability density function, with the same covariance matrix **K** for both classes, λ_{ideal} is given by the so-called prewhitening matched filter:

$$\lambda_{PW}(\mathbf{g}) = [\bar{\mathbf{g}}_2 - \bar{\mathbf{g}}_1]^t \mathbf{K}^{-1} \mathbf{g} \ , \tag{3}$$

where $\bar{\mathbf{g}}_k$ is the mean image for class k, and the superscript t denotes a matrix transpose. (Hence, if **a** and **b** are two vectors, $\mathbf{a}^t\mathbf{b}$ denotes their scalar product.) This test statistic λ_{PW}, which is clearly a linear function of **g**, can be calculated for each image if $\bar{\mathbf{g}}_1$ and $\bar{\mathbf{g}}_2$ and the common covariance matrix **K** are known. As an aside, if we model the noise as a Gaussian random process but with different covariance matrices for the two classes, the likelihood ratio turns out to be a quadratic function of **g**.

The model that led to Eq. (3) is very restrictive, and it is not obvious whether it is applicable to real radiographic images. One situation in which it is applicable is simple signal detection where the object consists of a flat background on which some weak signal can be superimposed. If the object is a gamma-ray emitter and we image it through a pinhole or collimator, the measurement noise is rigorously Poisson, but we can usually approximate the Poisson law by a Gaussian. If there is some linear post-processing filter, the Gaussian noise remains Gaussian but it becomes correlated. This is precisely the situation for which Eq. (3) describes the likelihood ratio, with $\bar{\mathbf{g}}_2 - \bar{\mathbf{g}}_1$ being the image of the signal to be detected, so in this case the prewhitening matched filter is indeed ideal.

Unfortunately, even in this simple situation, the prewhitening matched filter is not a good model for the human. There is considerable evidence, reviewed in Section IV, that shows that the human cannot perform the prewhitening operation. A better model for SKE/BKE signal-detection problems might be the non-prewhitening (NPW) matched filter where the test statistic is given by

$$\lambda_{NPW}(\mathbf{g}) = [\bar{\mathbf{g}}_2 - \bar{\mathbf{g}}_1]^t \mathbf{g} \ . \tag{4}$$

This form, which differs from Eq. (3) only by deletion of $\mathbf{K}^{-1}$, has a simple physical interpretation. For signal detection, the expected difference signal $\bar{\mathbf{g}}_2 - \bar{\mathbf{g}}_1$ is an image of the signal to be detected, so the observer simply lays a template of this signal image over the image **g** and integrates; the integral of the product of $\bar{\mathbf{g}}_2 - \bar{\mathbf{g}}_1$ and **g** is then the test statistic.

2.3. Scatter Matrices

If the object f is regarded as a random variable, the probability densities p(**g**| k) that enter into the likelihood ratio must take into account the variability of f as well as the measurement noise **n**. Even though Poisson noise can be modelled as Gaussian, it is highly unlikely that a multivariate Gaussian law would adequately describe realistic medical objects f. The true densities are very difficult to determine, and even if they could be found, the likelihood ratio would be a nonlinear function of **g**, making the performance of the ideal observer difficult to analyze.

For these reasons, we consider test statistics λ(**g**) that are constrained from the outset to be linear. To define these test statistics and analyze their performance, we describe the first- and second-order statistics of **g** by use of two "scatter matrices" $\mathbf{S}_1$ and $\mathbf{S}_2$. The interclass scatter matrix $\mathbf{S}_1$, which measures how far the class means for the data values deviate from their grand mean $\bar{\mathbf{g}}$, is defined as

$$S_1 \equiv \sum_{k=1}^{K} P_k(\bar{g}-\bar{g}_k)(\bar{g}-\bar{g}_k)^t , \tag{5}$$

where K is the number of classes (two in the binary problem), P_k is the probability of occurrence of class k, $\bar{g}_k$ is the class mean for the kth class (where the overbar denotes an ensemble average that accounts for the variability in **f** as well as **n**), and the grand mean is given by

$$\bar{g} = \sum_{k=1}^{K} P_k \bar{g}_k . \tag{6}$$

The intraclass scatter matrix $\mathbf{S}_2$ is the average covariance matrix, given by

$$S_2 \equiv \sum_{k=1}^{K} P_k K_k , \tag{7}$$

where the kth class covariance matrix is given by

$$K_k \equiv \langle (g-\bar{g}_k)(g-\bar{g}_k)^t \rangle_k , \tag{8}$$

where the angular brackets have the same meaning as the overbar, a full ensemble average over all objects **f** in class k and all realizations of **n**.

If there are M pixels in the image, both $\mathbf{S}_1$ and $\mathbf{S}_2$ are MxM matrices. Since $\mathbf{S}_2$ represents an ensemble covariance matrix, it will usually also have rank M. The rank of $\mathbf{S}_1$, on the other hand is much less than M, in fact just K-1, where K is the number of classes (Fiete et al., 1987). Thus, for a two-class problem, $\mathbf{S}_1$ has rank one, and it can be written as a single outer product:

$$S_1 = P_1 P_2 (\bar{g}_2-\bar{g}_1)(\bar{g}_2-\bar{g}_1)^t = \mathbf{x}\mathbf{x}^t , \tag{9}$$

where

$$\mathbf{x} \equiv \sqrt{P_1 P_2}(\bar{g}_2-\bar{g}_1) . \tag{10}$$

2.4. Optimum Linear Discriminants

The first step in using the scatter matrices $\mathbf{S}_1$ and $\mathbf{S}_2$ to form a linear discriminant is to solve the eigenvalue problem:

$$S_2^{-1} S_1 \, \mathbf{u}_p = \mu_p \mathbf{u}_p , \quad p = 1....M, \tag{11}$$

where $\mathbf{u}_p$ is an Mx1 column eigenvector and μ_p is its associated eigenvalue. Since $\mathbf{S}_1$ has rank K-1, as noted above, there are just K-1 nonzero values of μ_p. Each of the eigenvectors $\mathbf{u}_p$ corresponding to nonzero μ_p can then be used to form a linear feature given by

$$\lambda_p(g) = \mathbf{u}_p^t g , \; p = 1...K-1 \tag{12}$$

The hyperplanes $\lambda_p(g) = C_p$, where the C_p are constants, then partition the g space into K disjoint regions corresponding to the K classes. One common choice of the constants leads to the following decision rule: Choose the class k for which

$$\sum_{p=1}^{K-1} [u_p{}^t(g - \bar{g}_k)]^2 = \min . \tag{13}$$

In other words, the class chosen by this rule is the one for which the image being tested is closest to the class mean, where distance is measured in the eigenvector basis as indicated in Eq. (13). Other choices for the constants C_p lead to other decision rules, but all of them are based on partitioning the **g** space with some set of hyperplanes, so in this sense all are linear discriminants.

The situation is much simpler in the two-class problem, for which there is only one $\lambda_p(g)$, and we therefore drop the index p. In that case, an explicit solution to the eigenvalue equation is given by

$$\mathbf{u} = S_2{}^{-1}\mathbf{x} \tag{14}$$

$$\mu = \mathbf{x}^t S_2{}^{-1}\mathbf{x} . \tag{15}$$

Direct substitution of Eqs. (14) and (15) into Eq. (11) will verify that the **u** and μ given by these equations are indeed an eigenvector and eigenvalue, respectively, of $S_2{}^{-1}S_1$. The decision rule is then to compute $\lambda(g) = \mathbf{u}^t\mathbf{g}$ and compare it to a threshold C; class 1 is chosen if $\lambda(g) > C$ and class 2 if $\lambda(g) < C$.

It is interesting to compare the Hotelling test statistic $\mathbf{x}^t S_2{}^{-1}\mathbf{g}$ to the prewhitening matched filter given in Eq. (3), which can be written as $\mathbf{x}^t\mathbf{K}^{-1}\mathbf{g}$. The difference is that the noise covariance matrix **K** has been replaced by the more general weighted covariance matrix S_2. Thus if the only source of variability in **g** is additive Gaussian noise, the Hotelling observer is the same as the ideal observer. More generally, the ideal observer uses a test statistic that is nonlinear in **g**, and the Hotelling test statistic is a linear approximation to it.

2.5. Performance Measures

A common and important measure of task performance for binary decisions is the detectability index d_a, defined by

$$d_a{}^2 = \frac{[E(\lambda(g)|\,2) - E(\lambda(g)|\,1)]^2}{P_1\,\mathrm{var}(\lambda(g)|\,1) + P_2\,\mathrm{var}(\lambda(g)|\,2)} , \tag{16}$$

where $E(\lambda(g)|\,k)$ is the conditional mean of the test statistic $\lambda(g)$ given that **g** comes from class k, while $\mathrm{var}(\lambda(g)|\,k)$ is the corresponding conditional variance. It is well known that, if $\lambda(g)$ is Gaussian, d_a is related to the area under the ROC curve by

$$\mathrm{AUC} = \frac{1}{2} + \frac{1}{2}\,\mathrm{erf}\left[\frac{d_a}{2}\right] , \tag{17}$$

where erf( ) is the error function.

Although d_a is an accepted index of performance in binary classification tasks, it is not readily extended to tasks with more than two alternatives. For such tasks a possible performance metric is the Hotelling trace J, defined by

$$J = tr(S_2^{-1} S_1) \ , \tag{18}$$

where tr denotes the trace (sum of the diagonal elements) of the matrix. Since the trace is a scalar invariant, it can be calculated in a representation where the matrix is diagonal, and hence tr(A) is the sum of the eigenvalues of A. For the case at hand, we already know that $S_2^{-1} S_1$ has only K-1 nonzero eigenvalues, so the trace is a sum of K-1 terms. For K=2, $tr(S_2^{-1} S_1)$ is just the μ given in Eq. (15).

The Hotelling trace is an intuitively appealing figure of merit for classification performance. It is a single scalar, so it can be used for system optimization. It increases if the system is modified in such a way that the class means become more widely separated, increasing the norm of the vectors that constitute S_1, and it also increases if the variability in the image due to noise or other factors is decreased, since that corresponds to reducing the covariance terms that go into S_2 and hence in a sense increasing S_2^{-1}.

Moreover, in the two-class problem with Gaussian assumptions, J is given by

$$J = P_1 P_2 [d_a(Hot)]^2 \ , \tag{19}$$

so optimizing a system for maximum J is equivalent to optimizing the d_a or AUC for an observer that uses the Hotelling feature operator **u** to form a test statistic.

2.6. The NPW Observer

For later reference, we give the expression for the d_a index for the NPW observer in terms of S_1 and S_2. As shown by Barrett (1990), we have

$$[d_a(NPW)]^2 = \frac{1}{P_1 P_2} \frac{[tr\ (S_1)]^2}{tr(S_1 S_2)} \ . \tag{20}$$

3. COMPUTATIONAL METHODS

3.1. Image Modelling and Training Sets

One scheme we have used for implementing the Hotelling observer is to begin with a realistic three-dimensional mathematical model of some organ or organ system, allowing variability in both normal anatomy and in the nature and placement of lesions or other pathology (Cargill, 1989). This model is then used to create training sets of objects in two or more classes (normal and abnormal classes in the simplest case), and these simulated objects are used with an accurate model of the imaging system to create training sets of images. Ideally, these images would be indistinguishable from ones obtained with real clinical objects and physical imaging systems; the model developed by Cargill (1989) is very close to this ideal in the case of radiocolloid imaging of the reticuloendothelial system (liver, spleen and spinal bone marrow). Further work will be needed for other organ systems.

Given a training set of images created this way, a straightforward attempt to implement the Hotelling prescription would be to estimate the ensemble scatter matrices S_1 and S_2 from the training set, denoting the estimated matrices by $\mathcal{S}_1$ and $\mathcal{S}_2$, and then try to form $\mathcal{S}_2^{-1}\mathcal{S}_1$ by usual matrix manipulations; this method fails badly. One problem is that the matrices are huge. If the images are 64x64, then $\mathcal{S}_1$ and $\mathcal{S}_2$ are each 4096x4096. Furthermore, if the number of images in the training set is less than 4096, $\mathcal{S}_2^{-1}$ does not exist.

Fiete et al. (1987) have described a way to avoid the singularity of $\mathcal{S}_2$ in some cases. The trick is to take advantage of the fact that we can generate noise-free images in simulation studies. We can express S_2 rigorously as the sum of two matrices:

$$S_2 = S_2^{nf} + C_n \ , \tag{21}$$

where S_2^{nf} is the S_2 that would result from noise-free images, and C_n is the covariance matrix of $\mathbf{n}$. Using noise-free simulated images, we can get an estimate of S_2^{nf}, which we shall denote as $\mathcal{S}_2^{nf}$. Furthermore, C_n can often be modelled theoretically. For example, if we consider pinhole or collimator imaging, the noise is Poisson and C_n is a diagonal matrix with diagonal elements given by the mean values of each pixel in $\mathbf{g}$. These mean values can be estimated rather accurately from the noise-free training set, so we also have an estimate $\mathcal{C}_n$ of C_n. Combining these two estimates, we get our final estimate for S_2, namely

$$\mathcal{S}_2 = \mathcal{S}_2^{nf} + \mathcal{C}_n \ . \tag{22}$$

Since the second term in $\mathcal{S}_2$ is full rank and both terms are non-negative definite, the inverse now exists, but it is still may not be practical to calculate it directly. Instead, we now take advantage of the fact that we do not need to know $\mathcal{S}_2^{-1}\mathcal{S}_1$ completely; rather, we need only to find its dominant eigenvectors and eigenvalues. Furthermore, $\mathcal{S}_1$, like S_1, has rank K-1, where K is the number of classes, so for a two-class problem we have to find only a single eigenvector and eigenvalue. The eigenvector is just a 64x64 image in our example, so now we have a relatively routine image reconstruction problem; an iterative algorithm for its solution is given by Fiete et al. (1987). Once the eigenvector is found, its scalar product with each image generates the scalar test statistic. The Hotelling trace can then be readily calculated, and its variation with any number of engineering parameters can be efficiently studied.

3.1. Region of Interest

The approach of estimating S_1 and S_2 from a noise-free training set requires that we have a theoretical model for C_n, but this is not always possible. If we have a nonlinear processing step in the imaging procedure, we cannot specify C_n, even though we may know a great deal about the statistics of the noise prior to the nonlinearity. Since nonlinear reconstruction algorithms are very valuable in tomography, this is a severe limitation.

One way to ensure that $\mathcal{S}_2$ is full rank is to have more images in the training set than pixels in one image. While this is difficult for even 64x64 images, it may be possible to restrict the task in such a way that the observer needs only a subset of the pixels to perform it. For example, if we wish to detect a small lesion in a fixed location, then S_1 corresponds to a template that covers only a few pixels. According to Eqs. (12) and (14), the template is applied not to $\mathbf{g}$ but rather to $S_2^{-1}\mathbf{g}$, and the S_2^{-1} operation spreads information from the lesion location in $\mathbf{g}$ to surrounding pixels. The extent of this spread is difficult to predict, since it depends on the exact nature of S_2, but the spread should certainly be less than the full size of the image. Thus it may be possible to choose a small region that encompasses $S_2^{-1}\mathbf{g}$ yet contains fewer pixels than the number of training images. An experimental approach to choosing the region size would be to start with the very small region defined by S_1 and to gradually increase it, estimating J at each region size. If the estimate of J approaches a constant before the number of pixels in the region exceeds the number of images, that constant value can reasonably be taken as a good estimate of the ensemble J.

3.3. Effects of Location Uncertainty

One objection to the method just described might be that the use of a fixed lesion location is unrealistic. A simple calculation will show, however, that this restriction is really not very severe. Suppose that the task is to detect a small lesion that might be in one of L nonoverlapping locations. From Eq. (9), $\mathbf{S}_1$ is (within a constant) just the outer product of $\bar{g}_2-\bar{g}_1$ with itself. Let s_ℓ denote the mean difference signal $\bar{g}_2-\bar{g}_1$ when the lesion is in the ℓth location. Since the lesion locations do not overlap, we can write

$$\mathbf{S}_1 = \frac{P_1P_2}{L^2}\sum_{\ell=1}^{L} s_\ell s_\ell^t = \frac{1}{L^2}\sum_{\ell=1}^{L} \mathbf{S}_{1\ell}, \tag{23}$$

where $\mathbf{S}_{1\ell}$ is the $\mathbf{S}_1$ matrix that would result from a lesion in the ℓth location alone. If we assume that the lesion is weak and does not significantly affect the noise level in the image, the class covariance matrices $\mathbf{K}_k$ are approximately equal, and $\mathbf{S}_2 \cong \mathbf{K}_1 \cong \mathbf{K}_2$. Thus $\mathbf{S}_2$ is approximately independent of the lesion location, and we have

$$J = \mathrm{tr}(\mathbf{S}_2^{-1}\mathbf{S}_1) = \frac{1}{L^2}\sum_{\ell=1}^{L} \mathrm{tr}(\mathbf{S}_2^{-1}\mathbf{S}_{1\ell}), \tag{24}$$

which, if all locations are statistically equivalent, is just 1/L times the value of J for the fixed-location case. Thus, if we are comparing two imaging systems, the one that gives the higher J for detection of a lesion in a fixed location will also give the higher J for detection of the same lesion in one of L nonoverlapping locations; the absolute J value will be reduced by 1/L, but the rank ordering of the systems will be preserved.

3.4. Pseudoinverses

While $\mathcal{S}_2$ may not have an inverse, it always has a unique Moore-Penrose pseudoinverse, which we shall denote by $\mathcal{S}_2^+$. Thus a rather obvious way to estimate J is by

$$\mathcal{J} = \mathrm{tr}[\mathcal{S}_2^+\mathcal{S}_1]. \tag{25}$$

This estimate of J has several nice properties, though space does not permit a full exposition here. First, note that $\mathcal{S}_2$, being a weighted sum of covariance matrices, may be defined in one of two ways. The unbiased estimate of the ensemble $\mathbf{S}_2$ is given by

$$\mathcal{S}_2 = \frac{1}{N-1}\sum_{k=1}^{K} P_k \sum_{n=1}^{N} [\mathcal{g}_{kn} - \bar{\mathcal{g}}_k][\mathcal{g}_{kn} - \bar{\mathcal{g}}_k]^t, \tag{26}$$

where N is the number of training images per class, $\mathcal{g}_{kn}$ is the nth training image from the kth class, and $\bar{\mathcal{g}}_k$ is the sample mean image for the kth class. The alternative definition of $\mathcal{S}_2$ uses 1/N in place of 1/(N-1) in this equation, resulting in a small bias but yielding a maximum-likelihood estimate of $\mathbf{S}_2$. It will be shown in a subsequent paper that $\mathcal{J}$ using the 1/N definition of $\mathbf{S}_2$ is a maximum-likelihood estimate of J, while the 1/(N-1) definition leads to a minimum-variance estimate. In practice, N is relatively large, so these two estimates will not differ appreciably.

One might think that actual calculation of $\mathcal{S}_2^+$ would be an enormous computational task, since $\mathcal{S}_2$ is a 4096x4096 matrix for 64x64 images. In fact, however, it suffices to work with a KNxKN matrix. In most of our work, we take K=2 and N=32, so all that is required is a singular-value decomposition of a 64x64 matrix, an eminently feasible task. Again, details will be published separately.

3.5. Stationary Background Models

Another way to reduce the computational burden is to assume spatial stationarity for the image statistics. To see the advantage of this assumption, consider first one-dimensional images. In that case, the covariance matrices are Toeplitz and usually approximately circulant. This means that knowledge of one row or column of the matrix is sufficient to specify the full matrix, greatly reducing the number of independent parameters we must determine. Moreover, a circulant matrix can be diagonalized by means of a discrete Fourier transform, and the resulting diagonal elements are samples of the power spectral density. Thus knowledge of the power spectrum for each class yields the covariance matrices and hence $\mathbf{S}_2$ and $\mathbf{S}_2^{-1}$, with the latter guaranteed to exist unless all of the class power spectra vanish identically at some spatial frequency.

The situation in two spatial dimensions is more cumbersome, but the same basic conclusions hold. The covariance matrices are block-Toeplitz and approximately block-circulant, so they can be diagonalized by a 2D discrete Fourier transform. In both 1D and 2D, calculation of $\mathrm{tr}(\mathbf{S}_2^{-1}\mathbf{S}_1)$ reduces to performing an integral over the Fourier domain (Barrett et al., 1989; Myers et al., 1990).

4. EXPERIMENTAL RESULTS

4.1. Initial Studies Using Training Sets

As an initial test of the use of the Hotelling trace as a quality metric, we created a simple two-dimensional phantom of random, overlapping ellipses, roughly representing a liver (Fiete at al., 1987). The task was to detect a small cold lesion of random size, shape and contrast. The images were blurred with Gaussian blur functions of different widths, and Gaussian noise of various amplitudes was added; 32 normal and 32 abnormal images were generated for each combination of blur width and noise level. Each of these images was presented to 10 human observers, who were asked to use a six-point rating scale to specify how certain they were that a lesion was present. These data were analyzed to produce ROC curves from which values of the index d_a were computed for each value of noise and blur.

The Hotelling test statistic was calculated by use of Eq. (22), with $\mathcal{C}_n$ just being a constant times the unit matrix, and the Hotelling feature operator was found by an iterative search algorithm. This feature operator was then used to construct ROC curves for the Hotelling observer and to compute d_a. A remarkable correlation (r=0.99) between Hotelling and human d_a values was found.

4.2. Collimator Optimization

Later, a more realistic extension of this study was performed, with the objective of determining the optimal collimator to use in planar radiocolloid imaging of the liver (Fiete et al., 1987; White et al., 1989). In this study, three-dimensional mathematical phantoms (Cargill, 1989) of the reticuloendothelial system were generated to model a healthy class, while another group of mathematical phantoms with elliptical cold regions in the liver simulated a diseased class. Images of these objects through 24 parallel-hole collimators with various bore diameters (D_b = 1 mm to 7 mm) and bore lengths (L_b = 1 cm to 11 cm) were calculated, taking into account the attenuation and scatter in the body, spatial resolution due to the collimator and camera, and Poisson noise. The Hotelling trace J was calculated from these images for each collimator. As in the initial Fiete study, Eq. (22) formed the basis for the calculation, but this time $\mathcal{C}_n$ was derived from a Poisson noise model, so it was diagonal but not a multiple of the unit matrix.

Each liver was imaged through each collimator for two different lengths of time.

For a short imaging time, the best collimator had bore diameter of 1 mm and bore length of 1 cm. At a longer (and more reasonable) imaging time, the best collimator had a higher resolution and was slightly more efficient (D_b = 3 mm, L_b = 5 cm). Some typical images for the shorter time and the corresponding J values are given in Figure 1.

It is interesting to note that the ultra-high-resolution long-bore collimators performed poorest in both of these cases. An auxiliary study to chart the variations of J as a function of imaging time for a few of the collimators found that the long-bore collimators did indeed perform better than the others, but at such long imaging times (or high doses) as to be prohibitive for clinical studies. A psychophysical study to corroborate the conclusions of this theoretical investigation has also been performed, and again a good correlation between human and Hotelling performance was found (r=0.83).

4.3 Algorithm Optimization

Gooley has investigated a number of statistical image reconstruction algorithms for use with a coded-aperture cardiac SPECT imager. Among the methods studied were maximum likelihood techniques, including both the popular expectation-maximization (EM) algorithm and a Monte Carlo search routine. Other algorithms included various kinds of prior information through the specification of a prior probability density or hard constraints on the object (or class of objects) to be reconstructed.

The objects used in this study were ellipses representing a left-ventricular cross-section at end systole. The abnormalities to be detected were akinetic wall segments represented as small protrusions of the upper surface of the ellipses. The objects, both normal and abnormal, were taken to be binary, i.e., each pixel was constrained to take on one of two distinct values. This assumption is reasonable in first-pass cardiac studies where a pixel either contains a uniform amount of the injected radiopharmaceutical or it contains none of the radiopharmaceutical. Some algorithms made use of this information while others did not.

Gooley has completed a psychophysical study of the images from a total of 16 algorithms and obtained d_a values for the human observer for each algorithm. Efforts to compare these psychophysical data with d_a(Hot) and d_a(NPW) are in progress at this writing. Preliminary results indicate that the psychophysical results do not correlate well with d_a(NPW).

4.4. Effects of Noise Correlation in an SKE/BKE Task

Myers et al. (1985) studied the performance of human and ideal observers for detection of a nonrandom disk signal superimposed on a spatially uniform, nonrandom background. The object was imaged through an aperture having a point spread function $p_1(\mathbf{r})$, and Poisson noise was added to the resulting blurred image. A deblurring filter with PSF $p_2(\mathbf{r})$ was then used to partially compensate the blur due to the aperture.

The main variable in this study was the aperture PSF $p_1(\mathbf{r})$. The PSF of the deblurring filter was adjusted so that all imaging systems considered had the same final point spread function, $p_1(\mathbf{r})*p_2(\mathbf{r})$ (where $*$ denotes convolution), regardless of the form of $p_1(\mathbf{r})$. Four different functional forms were used for $p_1(\mathbf{r})$; each was a low-pass filter but there were different rates of falloff in the frequency domain. Therefore, the corresponding deblurring filters $p_2(\mathbf{r})$, required to hold $p_1(\mathbf{r})*p_2(\mathbf{r})$ constant, had four different rates of high-frequency boost. Normally, increased high-frequency boost would increase the noise level in the images, but in this study the exposure time was also varied in such a way as to keep the performance of the ideal observer constant. Thus all images had the same overall point spread function and the same ideal-observer detectability performance; the only difference was in the noise correlation

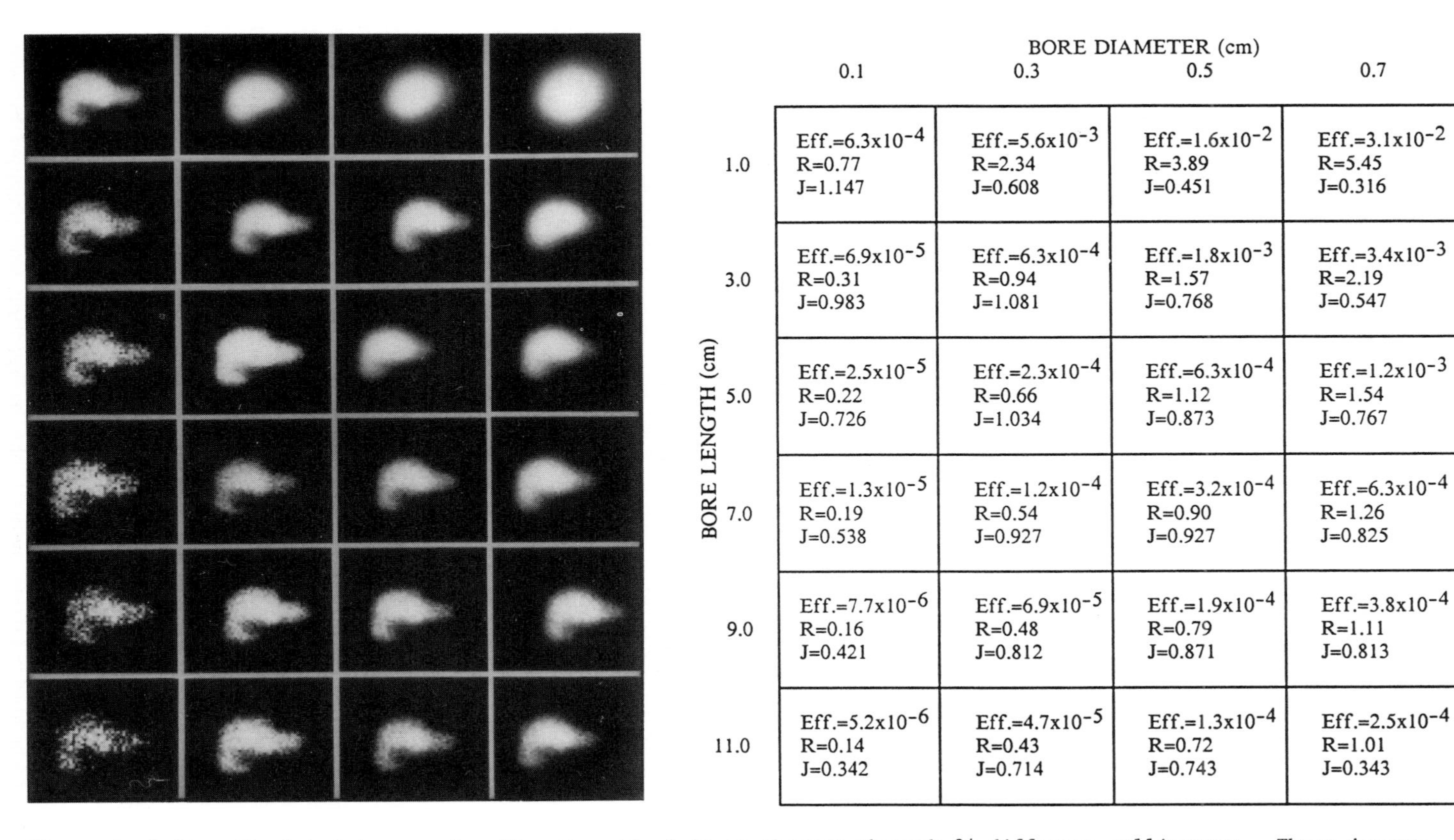

BORE LENGTH (cm) \ BORE DIAMETER (cm)	0.1	0.3	0.5	0.7
1.0	Eff.=6.3×10^{-4} R=0.77 J=1.147	Eff.=5.6×10^{-3} R=2.34 J=0.608	Eff.=1.6×10^{-2} R=3.89 J=0.451	Eff.=3.1×10^{-2} R=5.45 J=0.316
3.0	Eff.=6.9×10^{-5} R=0.31 J=0.983	Eff.=6.3×10^{-4} R=0.94 J=1.081	Eff.=1.8×10^{-3} R=1.57 J=0.768	Eff.=3.4×10^{-3} R=2.19 J=0.547
5.0	Eff.=2.5×10^{-5} R=0.22 J=0.726	Eff.=2.3×10^{-4} R=0.66 J=1.034	Eff.=6.3×10^{-4} R=1.12 J=0.873	Eff.=1.2×10^{-3} R=1.54 J=0.767
7.0	Eff.=1.3×10^{-5} R=0.19 J=0.538	Eff.=1.2×10^{-4} R=0.54 J=0.927	Eff.=3.2×10^{-4} R=0.90 J=0.927	Eff.=6.3×10^{-4} R=1.26 J=0.825
9.0	Eff.=7.7×10^{-6} R=0.16 J=0.421	Eff.=6.9×10^{-5} R=0.48 J=0.812	Eff.=1.9×10^{-4} R=0.79 J=0.871	Eff.=3.8×10^{-4} R=1.11 J=0.813
11.0	Eff.=5.2×10^{-6} R=0.14 J=0.342	Eff.=4.7×10^{-5} R=0.43 J=0.714	Eff.=1.3×10^{-4} R=0.72 J=0.743	Eff.=2.5×10^{-4} R=1.01 J=0.343

Figure 1 Left: Simulated images of a 3D mathematical liver phantom through 24 different collimators. These images are representative of a large set consisting of a total of 1,536 images (64 phantoms, 24 collimators) that we have generated. Half of the phantoms, including the one used in this figure, contained a cold lesion. Right: Parameters of the collimators used to generate the images at the left, arranged in the same order as the images. Eff. denotes efficiency, R denotes resolution in cm at a distance of 8 cm, and J denotes the calculated Hotelling trace.

structure.

Since the task was SKE/BKE, the ideal observer and the Hotelling observer for this study were identical, and both were implemented by means of the prewhitening matched filter as specified in Eq. (3). The psychophysical study, on the other hand, revealed that the performance of the human observer was greatly degraded when there was a strong high-pass character to the noise. The ideal/Hotelling observer failed completely to predict the influence of noise correlations on human performance. The NPW model, on the other hand, accurately described the psychophysical data.

A later theoretical investigation by Myers and Barrett (1987) found an alternative model that also explained the psychophysical data of Myers et al. (1985). These authors considered the effects of the spatial-frequency-selective channels in the visual system that have been found in a variety of psychophysical and neurophysiological experiments. The ideal-observer model was modified so that the observer had to pre-process the images through these channels, with a resulting loss of information, before calculating the likelihood ratio. The performance of this so-called channelized ideal observer was found by Myers and Barrett to be indistinguishable from that of the NPW observer for a wide range of tasks. The channels thus provide a plausible explanation of the human's inability to prewhiten.

4.5. Stationary Background Statistics

We have performed an extensive analysis of the effects of spatial inhomogeneity of the background on detection of a known lesion (Barrett et al. 1989; Rolland, 1990; Myers et al., 1990). The background was described as a stationary random process with a Gaussian autocorrelation function, and the signal to be detected was at a fixed location and had a Gaussian profile. The object, consisting of the background and, in half the images, the signal, was imaged through a pinhole aperture having either a hard-edged square profile or a smooth, Gaussian profile. Important variables included the width of the aperture (relative to the width of the signal) and the exposure time. The performance of three observers -- human, Hotelling and NPW -- were determined as a function of these variables.

The theoretical performance of the NPW and Hotelling observers was simple to determine since, as indicated in section 3.5, the $\mathbf{S}_2$ matrix in the case of a stationary background is diagonalized by a Fourier transform, so calculation of the traces in Eqs. (18) and (20) reduces to integration in the Fourier domain. The integrals were performed numerically, resulting in plots of $[d_a(\mathrm{Hot})]^2$ and $[d_a(\mathrm{NPW})]^2$ as a function of aperture width and exposure time (Myers et al., 1990).

As shown in Figure 2, there are striking differences in these plots for the two observers. The performance of the Hotelling observer increases steadily with increasing exposure time, while the NPW observer shows a saturation at a very short time. Not surprisingly, the NPW observer, which does not take into account any statistical properties of the background, is far more sensitive to the inhomogeneity than is the Hotelling observer. Finally, the predictions of the two observer models for the optimum aperture size to use is somewhat different.

The psychophysical studies performed by Rolland (1989) allow an unambiguous choice between the two models. As seen from Figure 2, the dependence of the human d_a on aperture size, exposure time, and degree of background lumpiness is very well predicted by the Hotelling model and not at all by the NPW model.

4.6. Effect of Higher-order Statistics

Rolland's psychophysical study described above was later extended by Yao (unpublished). Yao considered two different ways of generating the stationary random process for the background. In one method, the background was obtained by spatial filtering of a white, Gaussian random process, so that the grey-level probability densi-

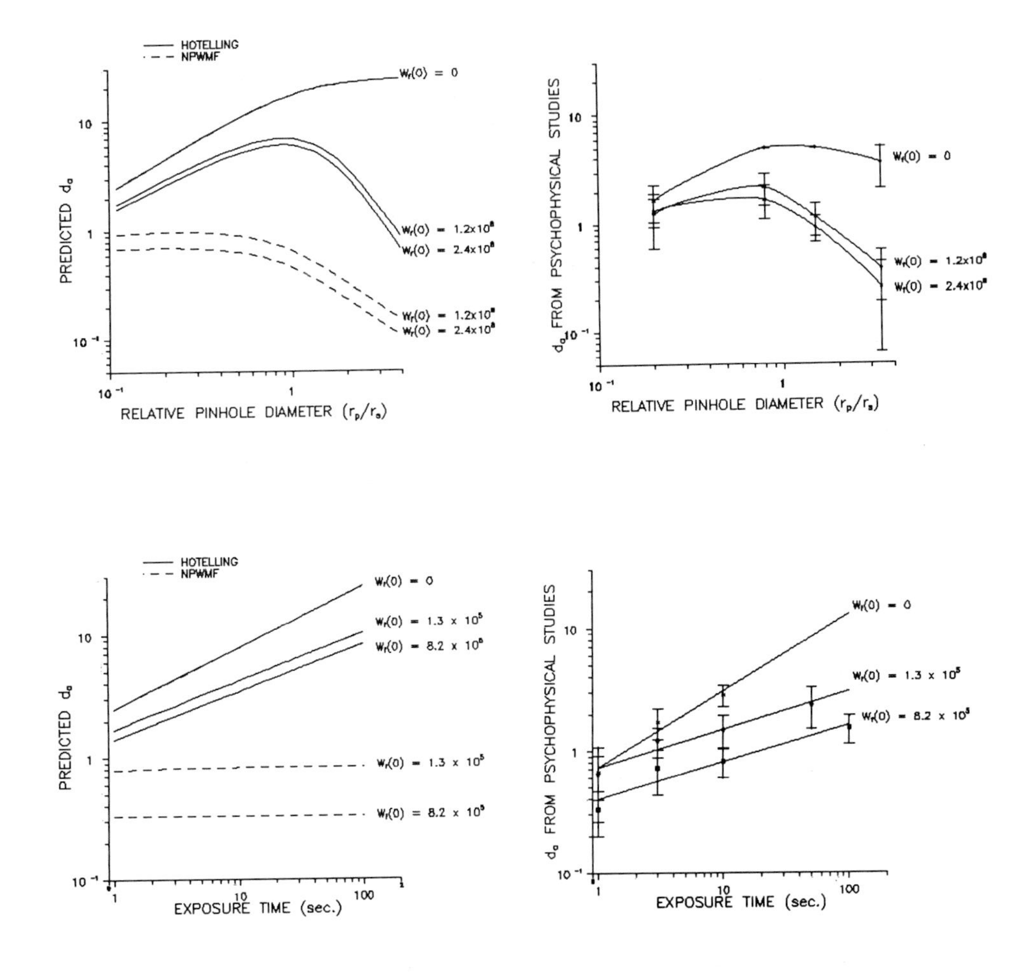

Figure 2 Results from Rolland (1990) on the detectability of a known lesion in an inhomogeneous background. Top left: Performance metrics (d_a) for the Hotelling and NPW observers as a function of aperture size. The parameter $W_f(0)$ specifies the degree of nonuniformity of the background. The upper curve, labelled $W_f(0) = 0$ is for the uniform background for which the Hotelling and NPW observers are identical. Top right: Performance of the human observer for the same parameters as in the theoretical curves at the left. Bottom left: Performance for the Hotelling and NPW observers as a function of exposure time. Bottom right: Performance of the human observer for the same parameters as in the theoretical curves at the left. Note that, except for a constant shift, the human curves agree very well with those for the Hotelling observer and not at all with those for the NPW observer.

ties remained Gaussian. The other method simply summed up N randomly placed Gaussian blobs. For large N this method also yields a Gaussian grey-level density, but for small N it is decidedly non-Gaussian.

Yao performed a psychophysical study using both methods for generating the background, and with large and small N in the second method. The parameters of the background were adjusted so that all three approaches yielded backgrounds with exactly the same mean, variance and autocorrelation function, but with different higher-order statistics as indicated by the grey-level histograms. The result of the study was that the human observer had the same detection performance for objects located in the three backgrounds. Thus the higher-order statistics do not seem to play a role, at least for this task. This result lends further support for considering the Hotelling observer, who has knowledge of only first- and second-order statistics.

4.7. Estimation Tasks

The Hotelling and ideal observers are both based on the assumption that the task of the imaging system is either detection of some abnormality or classification of the objects into two or more classes (differential diagnosis). In nuclear medicine, on the other hand, the task is often to extract some quantitative information from the image. In the medical literature, this task is called quantitation, while in the statistics literature it is known as estimation. Important examples of quantities to be estimated in nuclear medicine include the cardiac ejection fraction or the concentration of some receptor-specific tracer in a region of the brain.

Smith and Barrett (1986) demonstrated that the Hotelling trace could be used to select an optimum coded aperture, and later these same authors (Smith and Barrett, 1988) showed that the performance of various aperture codes for a detection task, as measured by the Hotelling trace, correlated well with the performance on an estimation task as measured by a mean-square error. This prompted us to examine theoretically the relationship between these two apparently very different tasks. We were able to derive a quite general set of mathematical relations between performance metrics for detection and estimation (Barrett, 1990). For detection performance, we considered either the Hotelling trace or a non-prewhitening matched filter; as estimation metrics we considered either the ensemble mean-squared error or the relative variance in a region-of-interest estimate.

In each case we found that the detection metric can be rigorously written as the estimation metric times a product of four factors. The first factor just accounts for the lesion size and contrast as in the Rose model, so we call it the Rose factor. The second factor accounts for the bias in the estimator, the third factor accounts for the complexity of the scene (or, equivalently, the conspicuity of the lesion). while the final factor accounts for the effects of noise correlation. We call these last three factors the bias, conspicuity and correlation factors, respectively.

Taken together, these factors provide a comprehensive picture of how different characteristics of the imaging system or the object affect performance on different tasks. General matrix expressions for all four factors have been derived, and specific forms have been worked out for several radiographic imaging modalities. The details of this theory are found in Barrett (1990).

5. DISCUSSION

The experimental results presented above lend considerable credence to the use of the Hotelling observer to predict the performance of the human, at least for the purpose of evaluating and optimizing imaging systems. The psychophysical studies performed by Fiete, White, Rolland and Yao all show unequivocally that the human performance correlates well with that of the Hotelling observer for the tasks considered. It is hoped that the results obtained by Gooley on comparison of algorithms will also fit this pattern, but computation of the Hotelling trace for these images is still in progress at this writing.

There is, however, one major study for which the Hotelling observer fails badly to predict human performance, and that is the work of Myers et al. on SKE/BKE detection in correlated noise. As noted above, the Hotelling and ideal observers are identical for this study, but the NPW observer is the one that predicts human performance. By contrast, in the Rolland study the Hotelling observer correctly predicted the effects of background inhomogeneity on human performance, while the NPW observer failed badly to do so.

One possible way to reconcile these apparently contradictory results is to include channels in the Hotelling model. We have already seen that this addition removes the discrepancy between the Hotelling (or ideal) model and the psychophysical results obtained by Myers. It might also be expected that a channelized Hotelling observer could also take proper account of background statistics and therefore correctly predict human performance in the Rolland study. Studies to examine this possibility are in progress.

Acknowledgements

The authors have benefitted greatly from discussions with many people, including Robert Wagner, Kyle Myers, Stephen Moore, Charles Byrne and John Denny. This work was supported by the National Cancer Institute under grants PO1 CA23417 and RO1 CA52643.

References

Barrett HH, Rolland JP, Wagner RF, and Myers KJ (1989). Detection of known signals in inhomogeneous, random backgrounds. Proc. SPIE, 1090:176-182.

Barrett HH (1990). Objective assessment of image quality: effect of object variability and quantum noise, J. Opt. Soc. Am. A. 7:1266-1278.

Cargill EB (1989). A mathematical liver model and its application to system optimization and texture analysis. Ph.D. Dissertation, University of Arizona.

Fiete RD, Barrett HH, Smith WE, and Myers KJ (1987). The Hotelling trace criterion and its correlation with human observer performance. J. Opt. Soc. Am. A, 4:945-953.

Fisher RA (1936). The use of multiple measurements in taxonomic problems. Ann. Eugenics 7 (part2):179-188.

Gooley TA (1990). Quantitative comparisons of statistical methods in image reconstruction. Ph. D. dissertation, University of Arizona.

Hotelling H (1931). The generalization of Student's ratio. Ann. Math. Stat. 2:360-378.

Myers KJ, Barrett HH, Borgstrom MC, Patton DD, and Seeley GW (1985). Effect of noise correlation on detectability of disk signals in medical imaging. J. Opt. Soc. Am. A, 2:1752-1759.

Myers KJ and Barrett HH (1987). Addition of a channel mechanism to the ideal-observer model. J. Opt. Soc. Am. A, 46:2447-2457.

Rolland JPY, (1990). Factors influencing lesion detection in medical imaging. Ph. D. dissertation, University of Arizona.

Smith WE and Barrett HH (1986). Hotelling trace criterion as a figure of merit for the optimization of imaging systems. J. Opt. Soc. Am. A, 3:717-725.

Smith WE and Barrett HH (1988). Linear estimation theory applied to the evaluation of *a priori* information and system optimization in coded-aperture imaging. J. Opt. Soc. Am. A,5:315-330.

White TA, Barrett HH, Cargill EB, Fiete RD, and Ker M (1989). The use of the Hotelling trace to optimize collimator performance. The Society of Nuclear Medicine 36th Annual Meeting, St. Louis, MO, (abstract).

EDGE-AFFECTED CONTEXT FOR ADAPTIVE CONTRAST ENHANCEMENT

R Cromartie[1] and S M Pizer[1,2]

[1]Department of Computer Science
[2]Departments of Radiology and Radiation Oncology
The University of North Carolina at Chapel Hill
Chapel Hill, N.C. 27599

Abstract

Contrast enhancement is a fundamental step in the display of digital images. The end result of display is the perceived brightness occurring in the human observer; design of effective contrast enhancement mappings therefore requires understanding of human brightness perception. Recent advances in this area have emphasized the importance of image structure in determining our perception of brightnesses, and consequently contrast enhancement methods which attempt to use structural information are being widely investigated. In this paper we present two promising methods we feel are strong competitors to presently-used techniques. We begin with a survey of contrast enhancement techniques for use with medical images. Classical adaptive algorithms use one or more statistics of the intensity distribution of local image areas to compute the displayed pixel values. More recently, techniques which attempt to take direct account of local structural information have been developed. The use of this structural information, in particular edge strengths, in defining contextual regions seems especially important. Two new methods based on this idea are presented and discussed, namely edge-affected unsharp masking followed by contrast-limited adaptive histogram equalization (AHE), and diffusive histogram equalization, a variant of AHE in which weighted contextual regions are calculated by edge-affected diffusion. Results on typical medical images are given.

Keywords

Edge-limited diffusion, human vision.

1. INTRODUCTION

Display of a digital image is the process by which the array of recorded intensities is presented to the human observer as a light image. For medical images the overall goal of display is the detection, localization and qualitative characterization of anatomical objects represented by the intensity variations in the recorded data. A necessary step in this process is the rule by which the recorded intensities of the original image are mapped to the display-driving intensities of the display device (Figure 1). When performed explicitly, this step is called contrast enhancement. After this mapping is performed, the image undergoes further transformations, first within the display device and then in the human visual system. The effective design of contrast enhancement mappings requires a thorough understanding of these transformations. It would be ideal if the display device/observer combination could be made linear, so that equal differences in display-driving intensity would be perceived as equally

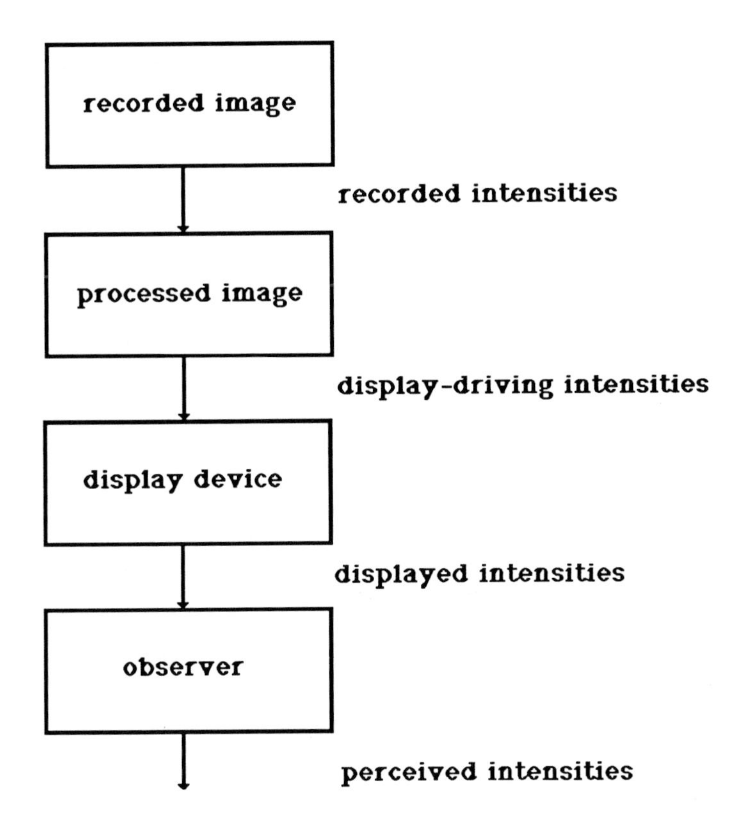

Figure 1. The steps of digital image display.

different. A description of acheiving this linearity given an appropriate model of brightness perception is given in (Pizer, 1981a). The difficulty lies in our incomplete understanding of human brightness perception. We know, for instance, that brightness perception is a function not only of luminance, but also of the spatial context of the stimulus. In particular, edges seem to play an important role in contrast perception. This growing understanding of human visual perception is reflected in recent research in contrast enhancement.

In this paper we first present a survey of contrast enhancement techniques, concentrating on recent developments in locally adaptive methods. Classical adaptive methods have centered on the calculation of various statistics of the local intensity distribution and the use of these to amplify local contrasts. More recently, methods have been developed which attempt to take explicit account of local image structure, especially edges. We have developed a pair of related techniques of this type. These methods are based on advances in our understanding of human visual perception. We will describe and show the results of some of these methods on typical medical images, and indicate directions for further research.

2. CONTRAST ENHANCEMENT APPROACHES IN THE LITERATURE

2.1. Global (Stationary) Contrast Enhancement

A global or stationary enhancement mapping is a grey-level transformation based solely on the intensity of each pixel:

$$I'(x,y) = f(I(x,y)).$$

The goal is to find a function which best utilizes the full range of display grey levels. Among these methods are intensity windowing, histogram equalization and histogram hyperbolization.

If we identify a subrange of image grey levels corresponding to features of interest this subrange can be expanded linearly to fill the full range of intensities. This technique is called intesity windowing. Pixels whose values fall outside the selected range are mapped to the minimum or maximum level. This technique is commonly used in the presentation of CT images. For example, in chest CT images, a "lung window" and a "mediastinum window" are chosen and applied, producing two images. These two images are then viewed side-by-side by the radiologist. This method has the advantage of being easily computed and in the case of CRT displays can be made interactive by an implementation which directly manipulates the lookup table of the display device. One difficulty is that objects occupying widely separated areas of the intensity range cannot be well presented in a single image. A perhaps more serious difficulty is that the perception of object boundary locations can depend critically on window selection.

Global histogram equalization is justified by the argument that for noise-free images it maximally transmits information as to scene intensity values (Cormack, 1981). In this method, a pixel's grey level is mapped to its rank in thegrey-level histogram of the entire image, scaled so that the output image fills the full range of intensities. The enhancement mapping is thus proportional to the cumulative distribution function of the image intensities. The result is that intensity values having greater numbers of pixels will be allocated a greater number of display levels, and the resulting histogram will be as flat as possible. There is however the compression of intensities that occur less frequently in the global histogram, which results in a loss of contrast in some areas of the image (Figure 2).

In histogram hyperbolization (Frei, 1977), a transformation of intensities is sought that results in a flat histogram of perceived intensities. Since the luminance response of the first stage of the human visual system is approximately logarithmic, it is argued that the shape of the histogram of displayed intensities should be approximately hyperbolic. Essentially, what is sought is histogram equalization after the effect of retinal processing. Thus a histogram-equalized image presented on a perceptually-linearized display should result in perceived brightnesses very close to those of a histogram-hyperbolized image displayed without linearization. This approach assumes a display device which is linear in its presentation of absolute luminances. Its main weakness is the strong dependence of our visual system on local context; brightness perception is not a simple function of absolute luminance.

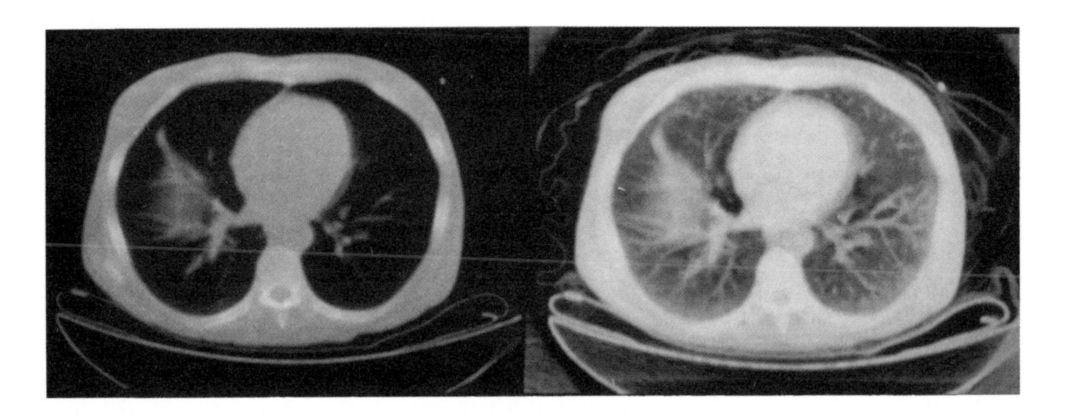

Figure 2. Chest CT scan -- original (left) and processed using global histogram equalization (right).

2.2. Adaptive Contrast Enhancement

An adaptive contrast enhancement mapping is one in which the new intensity value for a pixel is calculated from its original value and some further information derived from local image properties:

$$I'(x,y) = f(I(x,y), D_N(x,y)),$$

where N(x,y), the *contextual region*, is some spatial neighborhood of (x,y) in the image which includes the pixel of interest. For computational efficiency, it is most usual for N to be a square region centered on (x,y), but as we shall see, this need not be the case. Furthermore, the size and shape of the contextual region may itself vary throughout the image, based on either local statistics or local structural information.

A large number of adaptive contrast enhancement methods can be viewed as some variation of high-pass filtering. The oldest and most widely-used of these is *unsharp masking*. Known in its photographic form for at least sixty years, unsharp masking has also been applied to digital images. It can be defined as:

$$I'(x,y) = \gamma(I(x,y) - I^*_N(x,y)) + I^*_N(x,y),$$
$$= \gamma(I(x,y)) + (\gamma-1)(I^*_N(x,y)),$$

where $I^*_N(x,y)$ is a weighted average of intensities over the contextual region and γ is a constant gain factor. The term $(I(x,y) - I^*_N(x,y))$ is a high-frequency component sometimes referred to as the *detail image*. A γ between 0 and 1 results in a smoothing of the image, a γ greater than 1 results in emphasis of the high-frequency detail image. Unsharp masking has been applied and tested with varying success on a wide range of medical images (Loo *et al.*, 1985, Sorenson, 1987). It has a noticeable sharpening effect on edges, but when the gain factor is high enough to present very small details well, ringing artifacts are introduced across strong edges, and breakup of image objects can occur (Figure 3).

Unsharp masking can be generalized in a number of ways. One way is to replace the constant gain with separate weights for the high and low-frequency terms:

$$I'(x,y) = A(I(x,y) - I^*_N(x,y)) + B(I^*_N(x,y)).$$

An example of a method using this formulation is the statistical difference filter (Wallis, 1976, Harris, 1977). In this method, A is chosen so that the variance within the contextual region is made as nearly constant as possible, subject to a preset maximum gain to avoid over-enhancement of areas of very small standard deviation. B is a constant which serves to restore part of the low-frequency component. The method has been shown to produce objectionable artifacts and finding suitable values for the weighting factors, the maximum gain and the window size proves difficult. A related method (Gordon *et al.*, 1984, Dhawan *et al.*, 1986) is based on the definition of a measure of the contrast at a pixel:

$$C = |I(x,y) - I^*(x,y)| \;/\; (I(x,y) + I^*(x,y)),$$

which yields a value in the range 0-1. Enhancement consists of computing a new contrast C' and modifying the intensity of the pixel according to this new contrast as follows:

$$I'(x,y) = I^*(x,y)\,(1+C')\,/\,(1-C'), \quad \text{if } I(x,y) > I^*(x,y)$$

$$= I^*(x,y)\,(1-C')\,/\,(1+C'), \quad \text{if } I(x,y) < I^*(x,y).$$

An advantage of this method is that arbitrary enhancement functions can be easily applied. Results depend critically on window size, however, and

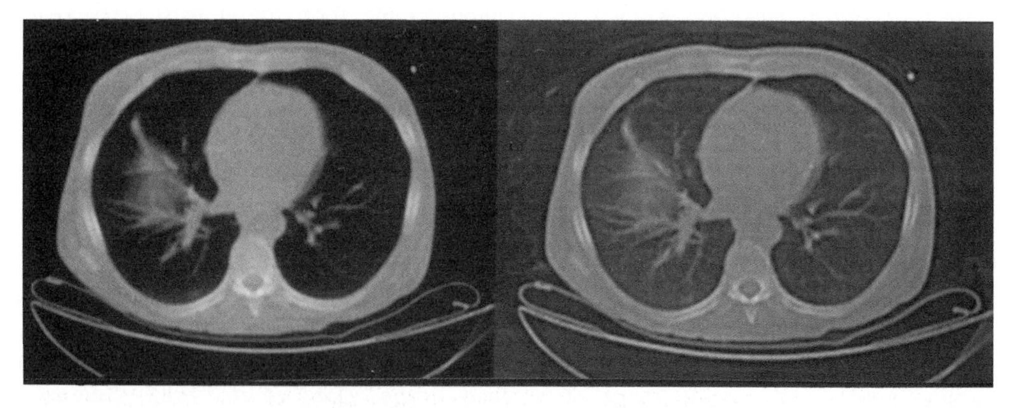

Figure 3. Unsharp masking applied to the same image as in Figure 2 with two different gain factors -- $\gamma = 2$ (left) and $\gamma = 5$ (right).

blurring of edges is a problem. Moreover, the need to rescale the range of I' to the original range defeats the attempt to actually acheive the contrasts C' that are desired.

In *adaptive histogram equalization (AHE)* (Pizer, 1981b, Zimmerman, 1985), the histogram is calculated for the contextual region of a pixel, and the transformation is that which equalizes this local histogram. It development is logical both from the point of view of the information theory basis of global histogram equalization and from our knowledge of the human visual system. We are very sensitive to local relative contrasts, but insensitive to both absolute luminance and widely-separated relative luminances. AHE provides a single displayed image in which contrasts in widely-varying recorded intensities can be easily perceived. AHE has demonstrated its effectiveness in the display of images from a wide range of imaging modalities, including CT, MRI and Radiotherapy portal fims.

While providing excellent enhancement of the signal component of the image, AHE also enhances noise. In addition, shadowing of strong edges can occur in certain types of images. This latter problem has been analyzed and a suggested remedy given in the context of high resolution digital chest radiographs in (Rehm *et al.*, 1990). In *contrast-limited adaptive histogram equalization* (CLAHE) (Pizer *et al.*, 1987), the enhancement calculation is modified by imposing a user-specified maximum on the the height of the local histogram, and thus on the slope of the cumulative histogram which defines the mapping. The enhancement is thereby reduced in very uniform areas of the image, which prevents over-enhancement of noise and reduces the edge-shadowing effect of unlimited AHE (Figure 4). Several investigators have examined the possibility of using unsharp masking as a pre-processing step for CLAHE (Blume, 1987, Rehm, 1990). More about this will be said later.

2.3. Methods Incorporating Structure

It has been recognized for some time that local image structure plays a crucial role in our perception of contrast, and enhancement techniques which incorporate local structural information are a logical result. There are two ways in which the above methods may be extended to include structural information.

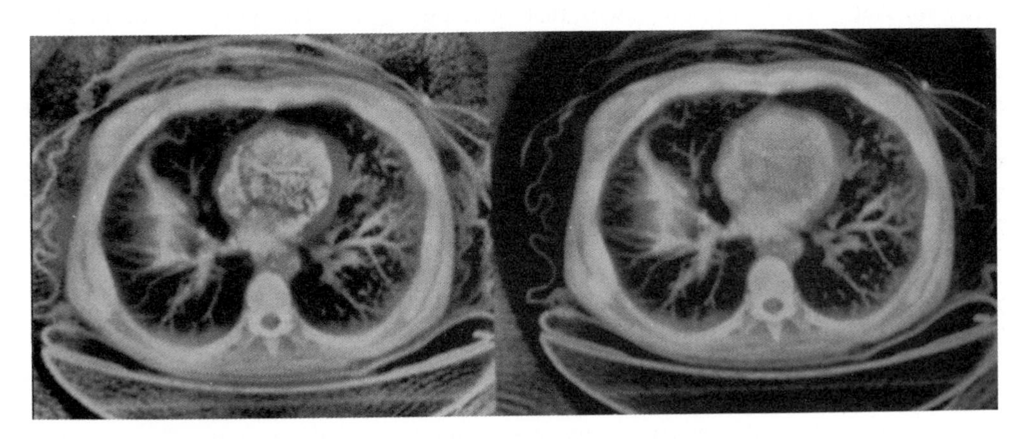

Figure 4. Images processed with AHE (left) and CLAHE (right).

One is to change the enhancement calculation itself; the other is to change the contextual region over which the calculations are done. Examples of each of these approaches are presented below.

An interesting extension of Gordon's technique (Beghdadi and Le Negrate, 1989) uses a modified contrast definition based on the detection of edges within the contextual region. In essence, the edge-grey-value of a pixel is its intensity weighted by the local edge strength at that pixel as computed by the Sobel, Laplacian or other edge operator. These edge-weighted values are then used in the calculation of the contrast measure as before. This method has an edge-enhancing effect when compared to the original formulation, but the exact effect depends strongly on the shape of the enhancement function and the choice of window size.

Several ways have been proposed of adjusting the contextual region over which the contrast enhancement is calculated. The idea is to adaptively restrict the local context to that which is relevant to perception of the pixel under consideration. Exactly what constitutes relevance in this sense is not entirely clear and depends to a large extent on the visual model we employ, but it is certain that perceived object boundaries are important in defining relevant context.

Gordon's method has also been extended by introducing a limited set of different window sizes, and choosing the appropriate size on a pixel-by-pixel basis throughout the image. This is done by analysing how the contrast function changes across these different window sizes. As the window size increases, the contrast of a central object will increase until the inner window just covers the object. This window is then used to calculate the enhancement as before. Even by restricting the available windows to a few possible sizes, the computational burden is large. Moreover, the use of square windows limits the ability to adapt to actual image structure, and as is the case with all the variants of this method, the rescaling problem remains.

Kim and Yaroslavskii (1986) use analysis of the local histogram to define subsets of the contextual region, and the enhancement mappings are applied over these subsets rather than the entire region. One method uses only a portion of the histogram centered on the pixel of interest, the includes only those pixels of the contextual region which fall within a certain intensity range surrounding the value of that pixel. To the extent that nearness in the histogram or nearness in absolute intensity corresponds to closeness within the image, this has the effect of restricting the calculations to within object boundaries. These measures are, however of doubtful validity -- either method may result in a contextual region of disconnected pixels. Moreover, while the contextual region does indeed change across the image, the overall window size remains fixed. To be entirely satisfactory, an adaptive neighborhood must both have some mechanism for responding to object boundaries and also not be limited by an imposed overall shape. Two methods which meet both these criteria are now examined.

3. NEW METHODS INCORPORATING EDGE STRUCTURE

While many of the techniques discussed thus far are quite useful, they have as their weakness that context is determined in a way that is at best indirectly related to the grouping schemes used by visual systems. Thus we seek some way of determining the relevant context, and we can use recent advances in understanding human vision to help us. We will apply this idea in extending two of the most attractive methods from those discussed earlier: unsharp masking followed by AHE, and CLAHE.

We know that an important early stage of human vision involves the calculation of an edge map. Best evidence seems to suggest that our perception of brightness is controlled by a sort of diffusion process in which the perceived contrast of these edges acts as an insulation strength that partially blocks the diffusion (Cohen and Grossberg, 1984).

Originally developed in the context of edge detection and the theory of scale-space, anisotropic diffusion (Perona and Malik, 1988) offers a way of producing truly variable contextual regions in a manner very like the above description of the calculations of the human visual system. Stationary blurring, which corresponds to the pixel averaging discussed above performed over unvarying contextual regions, can be modeled as a solution to the heat-conduction or diffusion equation:

$$I_t = c\Delta I = c(I_{xx} + I_{yy}),$$

where c is a constant controlling the rate of diffusion. If we let c vary according to local image features, we obtain

$$I_t = \mathrm{div}(c(x,y,t)\Delta I) = c(x,y,t)\nabla I + \Delta c \Delta I,$$

the *anisotropic diffusion* equation. Here ∇ is the gradient and Δ the Laplacian operator. If we knew for a given time t the location of object boundaries, we could set c to be 1 within those boundaries and 0 outside. In this way, we could entirely eliminate interaction across region boundaries. Since we do not know precisely the object boundaries, this is not possible; moreover this is not the result we want. We can chose c to be a monotonically decreasing function of the edge strength such that diffusion across edges is limited, but not eliminated. As an example, we use the function

$$c(|\nabla I|) = \exp(-|\nabla I|^2 / 2K^2),$$

where the parameter K can be viewed as selecting an edge strength up to which diffusion is allowed and beyond which the diffusion is strongly limited. Thus, diffusion using a large value of K blurs all but the very strongest edges, while selecting a very small K blocks diffusion except across very weak edges.

3.1 Unsharp Masking Using Edge-Limited Blurring

The context for unsharp masking is the blurring kernel. We have used an implementation of anisotropic diffusion (Gerig and de Moliner, 1989) in an variant of unsharp masking which can be viewed as the standard unsharp masking formula using a variable rather than fixed contextual region. The result is a relative increase in the blurring of low-contrast edges. This means, on amplification of the detail image, a relative increase in the enhancement of small details. A comparison of unsharp masking using isotropic Gaussian blurring and using edge-limited blurring is given in Figure 5. With edge-limited blurring, there is considerably more enhancement of the finest details of the image, and yet a noticeable lack of the ringing artifacts across strong edges that can be seen in the isotropically-blurred version. We have used this edge-limited unsharp masking as a pre-processing step before applying CLAHE, and the result is an image with excellent grey-scale without the edge artifacts of CLAHE used alone (Figure 6).

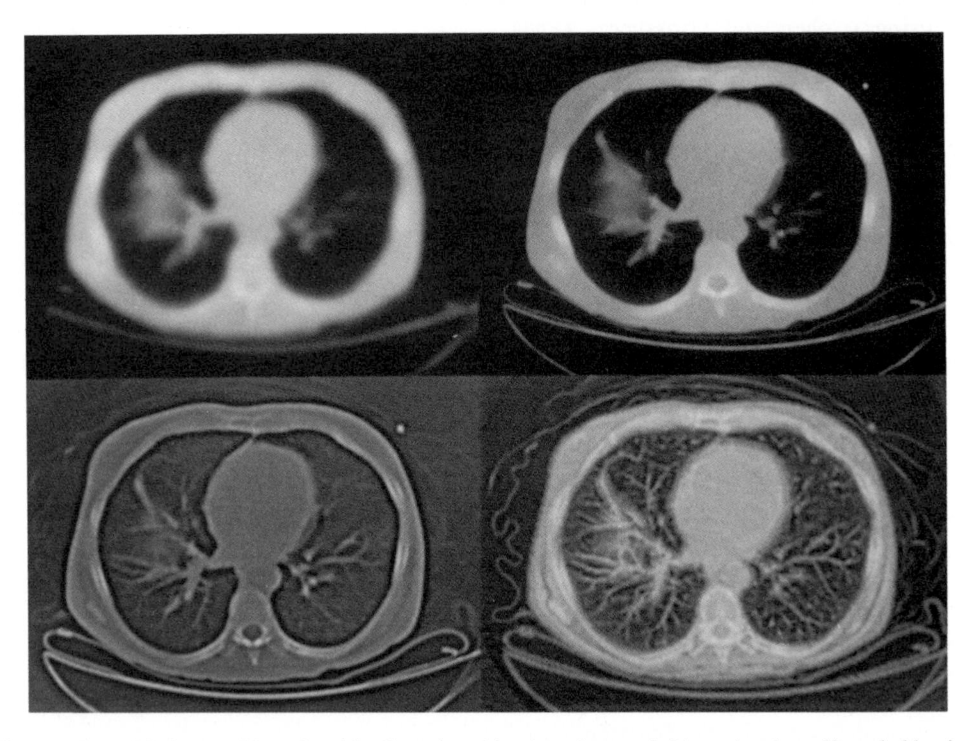

Figure 5. CT image blurred with Gaussian blurring (upper left) and edge-affected blurring (upper right). The resulting unsharp-masked enhancements are pictured below. The gain factor in both cases is $\gamma = 4$.

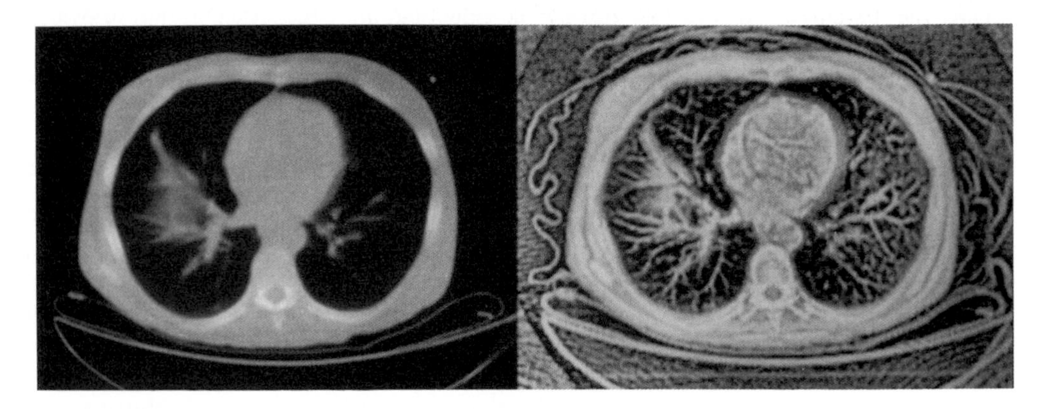

Figure 6. Original CT image (left) and enhancement using edge-affected unsharp masking followed by CLAHE (right).

3.2. Diffusive Histogram Equalization

For AHE, the context is all the pixels contributing to the histogram from which we calculate the output grey level. These contributions can have variable weights, so the weight values are also part of the definition of the contextual region. Here, too, we have used edge-affected blurring to achieve a variable-neighborhood adaptive histogram equalization algorithm we call *diffusive histogram equalization (DHE)*. In this method, we calculate the effect of all pixels of a given grey-level on the histograms of all pixels in the image in a single step. This is done by applying edge-limited blurring to an image made by placing a positive starting value in each pixel of the current grey level and zeros eveywhere else. The edges used to limit the diffusion are the edges of the original image. After diffusing, the histogram value for each pixel is contained in this image. Thus the influence of one pixel on another is limited by intervening edges. The contextual region is truly variable, not only in overall shape (potentially the entire image could be the contextual region) but also in the variable weighting of each pixel within the region. Figure 7 shows the contextual region for one pixel in the chest CT image, and the result of applying DHE to that image.

4. DISCUSSION

We have applied the two methods discussed above to a number of images from different modalities with encouraging results. We expect to conduct formal analysis of these methods in the near future. It is important in designing tests for evaluation of contrast enhancement methods to keep in mind the particular task being performed. Enhancement methods are often compared on the basis of their ability to increase detectablity of either standard test patterns or very subtle artificially-produced lesions imposed on real medical images. This detection task is certainly important for many imaging modalities, but may not be the most important in every case. Boundary localization, shape characterization and comparison of absolute luminances are some other viewing tasks which may be

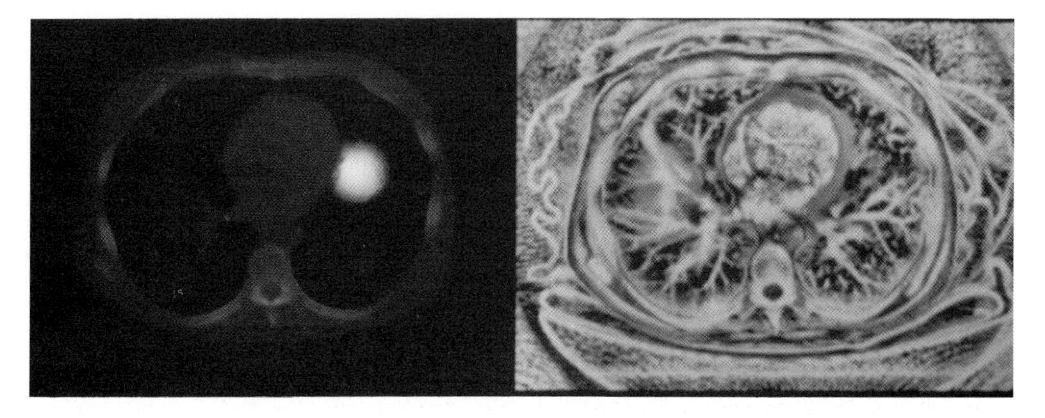

Figure 7. Diffusive histogram equalization -- edge-affected contextual region for one pixel superimposed on original image (left) and final enhanced image (right).

of importance. Diffusive histogram equalization is a particularly aggressive enhancement method, yielding images exceptionally rich in structural detail. We expect it might not be particularly well-suited for lesion detection. By comparison, the images produced by edge-affected unsharp masking followed by CLAHE are perhaps more suited to making the qualitative judgements required by lesion detection or appreciation of gross shape features. The ultimate test of any contrast enhancement method designed for use with medical images is whether or not it provides increased diagnostic accuracy or efficiency in a clinical setting. In choosing among contrast enhancement methods, we must generally be content with some approximation to this test. Moreover, it may take a considerable amount of training for the clinician to effectively use images processed by means of these enhancements.

Another task-related matter is how noise is treated. Noise is unwanted image detail, so its definition depends on what detail is wanted, as well as what is known about the properties of the image-formation process. With such a decision, the contrast enhancement must be chosen not simply to convey signal differences, but to convey them relative to noise. This problem is acute for DHE, which tends to bring out contrast at all image levels. This can be controlled to some extent by allowing the diffusion to continue relatively long, effectively increasing the size of the contextual region. It would certainly be possible to apply a contrast-limitation factor, as in CLAHE, which would result in images with less apparent noise.

We have discussed these techniques without paying particular attention to the cost of computing them. The anisotropic diffusion calculations are very expensive, both in processor time and memory requirements. Certainly, for a method to be usable in the real world, the implementation must be relatively fast, even real-time. Many of the methods discussed above, particularly the adaptive ones, are too computationally expensive to be clinically valuable. Many can be speeded up by using recently developed image processing computers or other specialized hardware.

5. CONCLUSION

Digital image processing is becoming more and more important in medical imaging as we move from film-based to computer-based imaging systems. An important part of this is the effective display of these digital images, and contrast enhancement is an essential step of the display process. We have tried to give an indication of the importance of an accurate visual model in the development of these techniques, and have outlined the development of two contrast enhancement methods, edge-affected unsharp masking followed by AHE and diffusive histogram equalization. Both of these methods are based on our knowledge of how the visual system determines context. As better models of human visual perception are formulated, we will be able to design contrast enhancement methods which more effectively complement our perceptual capabilities.

Acknowledgements

We wish to thank Dr. Julian Rosenman, leader of the project entitled "Adaptive Histogram Equalization for Radiotherapy" within the research program in Medical Image Presentation at The University of North Carolina at Chapel Hill.
This work is supported in part by NIH Grant #P01 CA47982.

References

Beghdadi A and Le Negrate A (1989). Contrast enhancement technique based on local detection of edges. Computer Vision, Graphics, and Image Processing 46: 162-174.

Blume H and Kamiya K (1987). Auto-ranging and normalization versus histogram modifications for automatic image processing of digital radiographs. Proc. S.P.I.E. 767 Medical Imaging: 371-386.

Cormack J and Hutton BF (1981). Quantitation and optimization of digitized scintigraphic display characteristics using information theory. Medical Image Processing: Proceedings of the VIIth International Meeting on Information Processing in Medical Imaging, Stanford University: 240-263.

Dhawan AT, Buelloni G and Gordon R (1986). Enhancement of mammographic features by optimal neighborhood image processing. IEEE Transactions on Medical Imaging MI-5 No. 1: 8-15.

Frei W (1977). Image enhancement by histogram hyperbolization. Computer Graphics and Image Processing 6: 286-294.

Gerig, G and de Moliner R (1989). Personal communication.

Gordon R (1986). Enhancement of mammographic features by optimal neighborhood image processing. IEEE Transactions on Medical Imaging MI-5 No. 1: 8-15.

Grossberg S (1984). Neural dynamics of brightness perception: features, boundaries, diffusion, and resonance. Perception and Psychophysics 36 (5): 428-456.

Harris, Jr. JL (1977). Constant variance enhancement: a digital processing technique. Applied Optics 16: 1268-1271.

Kim V and Yaroslavskii L (1986). Rank algorithms for picture processing. Computer Vision, Graphics, and Image Processing 35: 234-258.

Loo LD, Doi K and Metz C (1985). Investigation of basic imaging properties in digital radiography 4. Effect of unsharp masking on the detectability of simple patterns. Medical Physics 12: 209-214.

Perona P and Malik J (1988). Scale-space and edge detection using anisotropic diffusion. Report UCB/CSD 88/483, Computer Science Division University of California, Berkeley, CA.

Pizer SM (1981a). Intensity mappings to linearize display devices. Computer Graphics and Image Processing 17: 262-268.

Pizer SM (1981b). An automatic intensity mapping for the display of CT scans and other images. Medical Image Processing: Proceedings of the VIIth International Meeting on Information Processing in Medical Imaging, Stanford University: 276-309.

Pizer SM, Amburn EP, Austin JD, Cromartie R, Geselowitz A, ter Haar Romeny B, Zimmerman JB and Zuiderveld K (1987). Adaptive histogram equalization and its variations. Computer Vision, Graphics, and Image Processing 39: 355-368.

Rehm K, Seely GW, Dallas WJ, Ovitt TW and Seeger JF (1990). Design and testing of artifact-suppressed adaptive histogram equalization: A contrast-enhancement technique for the display of digital chest radiographs. Journal of Thoracic Imaging 5, No. 1: 85-91.

Sorenson J (1987). Effects of improved contrast on lung-nodule detection A clinical ROC study. Investigative Radiology 22: 772-780.

Wallis R (1976). An approach to the space variant restoration and enhancement of images. Proceedings of the Symposium on Current Mathematical Problems in Image Science, Monterey, California, Naval Postgraduate School.

Zimmerman JB (1985). Effectiveness of Adaptive Contrast Enhancement. Ph.D. dissertation, Department of Computer Science, The University of North Carolina at Chapel.

INCREMENTAL VOLUME RENDEREING ALGORITHM FOR INTERACTIVE 3D ULTRASOUND IMAGING

R Ohbuchi and H Fuchs

Department of Computer Science,
University of North Carolina at Chapel Hill,
Chapel Hill, NC 27599, U.S.A.

Abstract

This paper describes a medical 3D ultrasound imaging system that incrementally acquires and visualizes a 3D volume from a series of 2D images. The system acquires the image from a conventional B-mode 2D echography scanner, whose scanhead is attached to a mechanical tracking arm with three degrees of freedom. It reconstructs a stream of 2D images with their locations and orientations into a 3D array of regularly spaced samples, to be rendered by a modified front-to-back image-order volume rendering algorithm. Visualization is done so that each incoming 2D image slice promptly affects the rendering result. This paper concentrates on the incremental volume rendering algorithm that takes advantage of the incremental scanning to reduce image generation time per each input image slice. We describe a new fast ray-clipping scheme called *D-buffer* algorithm that is based on the Z-buffer algorithm. It is followed by another speedup scheme called *hierarchical ray caching,* and a method to efficiently integrate geometric objects with volume data in image space.

Keywords

3D ultrasound echography; Ray-casting; 3D imaging; Z-buffer algorithm; D-buffer algorithm; hierarchical ray-caching.

1. INTRODUCTION

We have been working toward a medical 3D ultrasound scanner system that will acquire and display a 3D volume in real-time. We call this system an 'ultimate' 3D echography (3DE) system. It will acquire, display, and manipulate a 3D volume image in real-time. Such 'real-time-ness' can be crucial for an application such as cardiac diagnosis, where a static 3D image or even a dynamic display of a series of images acquired over many cardiac cycles by gated acquisition might miss certain kinetic features. The 'ultimate' system has two major components: the image acquisition part and the image visualization part. We at UNC-Chapel Hill have been working on the real-time 3D visualization part of the system, while Dr. Olaf von Ramm's group at Duke University has been working on the real-time 3D image acquisition part of the system.

To study the issues involved in 'ultimate' 3D ultrasound visualization before the real-time acquisition subsystem becomes available, we have been developing an *incremental, interactive 3DE* scanner system. This system will acquire and display the 3D image incrementally at an interactive rate. It will use a state-of-the-art medical real-time 2D ultrasound echography scanner as an image input, where a user guided scanhead is tracked with 3 degrees of freedom. Using the geometric information acquired for each 2D image slice, the system incrementally reconstructs the 3D array of sample points spaced regularly on the Cartesian coordinate from a stream of 2D slices. These slices are located and oriented arbitrarily with 3 degrees of freedom. The system renders the reconstructed volume, again incrementally, to produce volume rendered 2D images. This visualization is done incrementally so that every new 2D slice acquired affects the final image promptly, thus offering a high degree of interactivity to the user.

We present here the brief introduction to the incremental, interactive 3DE system and the incremental volume rendering algorithm designed for a high degree of interactivity. We briefly describe the 'ultimate' system, then present a review of previous works on 3D ultrasound imaging. This is followed by the sketch of the incremental acquisition and visualization system we have been working on. We describe the basic incremental volume reconstruction and rendering algorithm in section 2. Section 3 and 4 introduces enhancements to the basic algorithm. In section 3, a fast ray-clipping algorithm called *D-buffer* algorithm is presented, which is followed by a proposed enhancement to the idea of ray-caching called *hierarchical ray-caching* for faster compositing. Then, in section 4, another algorithm is proposed which combines D-buffer with the hierarchical ray-caching to allow fast integrated rendering of polygons, polyhedron defined volumes for cutaways, etc., to provide fast interaction with the volume image.

1.1. The 'Ultimate' 3D Ultrasound Scanner

Among various medical imaging modalities, ultrasound echography is the closest to achieving 3D real-time acquisition, even though other modalities such as MRI are becoming faster. The new scanner being developed by Dr. Olaf von Ramm's group at Duke University (Shattuck, 1984) will enable the 'ultimate' 3DE system to acquire 3D images at real-time rate. Due to the velocity of sound limitation (about 1540 m/s in water), scanning a 3D volume with reasonable resolution (128 x 128 x 128 or more) in real-time (e.g., 30 3D-frames/s) requires parallel processing. The new scanner will use a single-transmit/multiple-receive scheme called *Explososcan* to increase data acquisition bandwidth. The first implementation will use a 16 x 16 2D-array transducer along with 16 x 16 x 64 digital delay-lines implemented in VLSI chips for 3D beam steering and focusing with 64-way multiple simultaneous reception.

Once the real-time 3DE data has been obtained, it has to be visualized. We at UNC-Chapel Hill have been working on this problem. One major effort is a display system based on a volume rendering technique for visualization that will display 3DE data in real-time. Such a display system has to cope with the challenge of very high data bandwidth. The real-time 3DE scanner above will produce about 2-4 Million points per frame, or 60-120 Million points per second. Visualizing a 3D volume data of this bandwidth in real-time requires very large computational power (on the order of Giga floating point operations per second) if a straightforward algorithm is used. We are approaching this issue through parallelism and algorithm efficiency.

Providing a visualization method that is effective to the user is another major issue. It involves standard problems associated with visualization of 3D data, such as obscuration. To look at the inside of the left ventricle, various objects in front, e.g., images of fat tissue and part of myocardium, must be removed to reveal the object behind. This usually requires extensive manipulation of the 3D image data. On top of those are the additional characteristics of ultrasound data that can hinder the visualization, e.g., speckle noise, gain variation, shadowing, specular artifact. Wei-jyh Lin at UNC-Chapel Hill has been working on an algorithm to estimate surfaces in 2D echography image under the presence of speckle noise using multi-scale filtering technique (Lin, 1991). In the future, this surface estimator will be added as a front-end to the display system described above.

1.2. Previous Work in 3D Ultrasound Imaging

Many works on 3D ultrasound imaging have been done in the past. Most of the studies known to the authors have tried to reconstruct 3D data out of imaging primitive data of lesser dimension (usually 2D image), instead of directly capturing 3D data. This is due to the obvious lack of real-time 3DE scanner.

To reconstruct 3D image from images of lesser dimension, the location and orientation of imaging primitives must be available. Coordinate values are either explicitly tracked, as in (Brinkley, 1978) (Ghosh, 1982) (Hottier, 1989) (Raichelen, 1986) (McCann, 1988) (Nikravesh, 1984) (Stickels, 1984) (Mills, 1990) using mechanical, acoustic, or optical tracking mechanisms; or it is controlled implicitly at the time of acquisition (Lalouche, 1989) (Nakamura, 1984) (Billion, 1990). The most interesting recent work in the latter category is on a near-real-time, automatic 3D scanner system (Billion, 1990). This system is the closest yet to the real-time 3D ultrasound scanner, and is being developed at Phillips Paris Research Laboratory. It is a follow up to the earlier work (Hottier, 1989) which, like our system, was a manually guided scanner with mechanical tracker. This near-real-time 3D scanhead is a 'double wobbler' mechanical sector scanner, where a conventional wobbler 2D sector scanhead with annular array transducer is rotated, or wobbled, in an additional axis by a stepping motor to provide 3D scanning. In a period of 3 to 5 seconds, 50 to 100 slices of 2D sector scan image can be acquired. The use of an annular array transducer in this system will give better spatial resolution than the real-time 3D phased array system mentioned above by von Ramm et al, but the latter will provide better temporal resolution.

For presenting the scanned result, there are two forms: non-visual, and visual. The latter can be classified further by the rendering primitives, which are either (geometric) graphic or image. The majority of the earlier studies (Brinkley, 1978) (Ghosh, 1982) (Raichelen, 1986) (Nikravesh, 1984) (Stickels, 1984) had non-invasive estimation of the volume of the heart chamber as their primary objective. Thus, often the reconstruction is only *geometric*. A typical process involved using a digitizer to manually trace

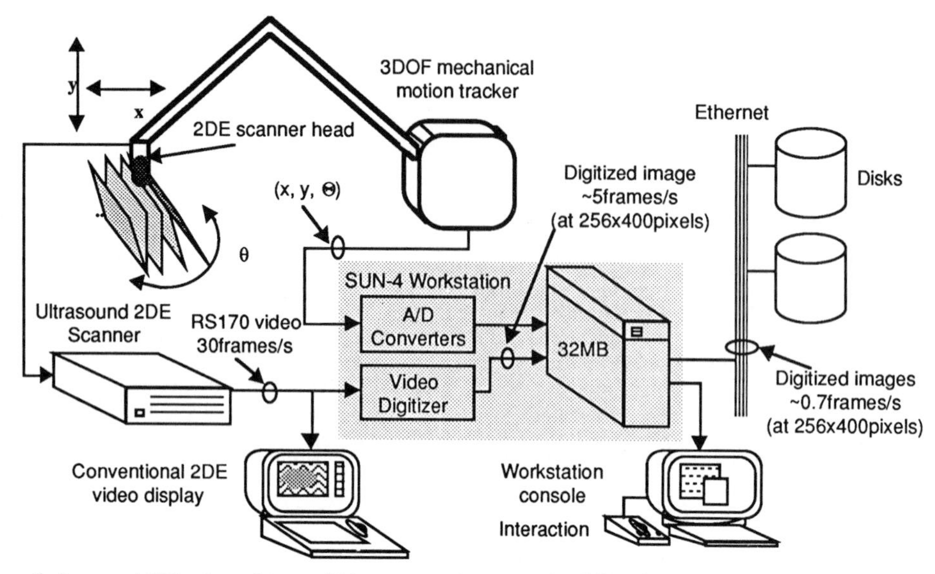

Figure 1. Incremental 3D volume data acquisition system using conventional 2D echography scanner

2DE images stored on video tape. Since the visual presentation is a secondary matter, wire frames or a stack of contours are used to render the geometrical reconstruction result

More recent studies (Nakamura, 1984) (Lalouche, 1989) (McCann, 1988) (Hottier, 1989) (Billion, 1990) actually reconstructed 3D grey level *images*, preserving the grey scale, which can be crucial to tissue characterization. (Lalouche, 1989) is a mammogram study using a special 2DE scanner which can acquire and store 45 consecutive parallel slices with 1 mm intervals. It is reconstructed by cubic-spline interpolation, and volume rendered. (McCann, 1988) performed the gated acquisition of a heart's image over a cardiac cycle. They volume rendered 2DE images stored on video tape. Upon reconstruction, 'repetitive low-pass filtering' fills the spaces between radial slices, which suppresses the aliasing artifact.

(Billion, 1990) uses two visualization techniques: re-slicing by an arbitrary plane, and volume rendering. The former allows faster but 2D viewing on a current workstation. The latter allows 3D viewing, but often involves cumbersome manual segmentation. The reconstruction algorithm is a straightforward low-pass filtering.

1.3. Incremental, Interactive 3D Acquisition and Visualization

To study the issues of real-time ultrasound 3D echography visualization before the real-time 3DE acquisition system becomes available, we have been studying an *incremental, interactive 3D echography (3DE) system*. In this system, a user-guided scanhead mounted on a 3 degrees of freedom (3 DOF) mechanical tracking apparatus will acquire a series of 2D image slices as well as the corresponding geometries, i.e., location and orientation, of each slice. Using these geometries, regular 3D volume data, where sample points are at uniform intervals in Cartesian coordinates, are reconstructed from a series of 2D images with irregular geometries. This reconstruction process and the following volume rendering process take place incrementally as each new 2D image slice arrives. Each new 2D image will affect the final rendered image without waiting for the rest of the slices to arrive. (Since scanning may continue for an indefinite period of time, possibly sweeping the same volume many times, waiting for the 'rest' loses its meaning.)

Figure 1 shows the image acquisition system for the incremental, interactive ultrasound scanner system (See Figure 2a for the picture). The 2DE scanner mounted on a mechanical tracking arm with 3 DOF acquires 2D image frames at a maximum rate of around 30 frames/sec. Each 2DE image slice from the ultrasound scanner is video-digitized in real-time by a Matrox MVP-S video digitizer board and copied into a SUN-4 workstation. Due to the MVP-S's frame buffer design, this copying process is relatively slow, running at 5 frames/s for 256 x 400 pixel images.

The potentiometers in the 3D tracking arm transduce the coordinates and orientation (x, y location and angle θ) of each of the image frames through a Data Translation DT-1401 A/D converter board. The

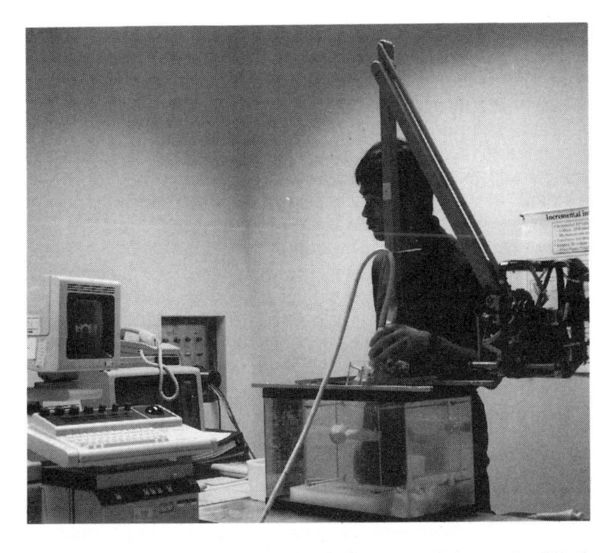

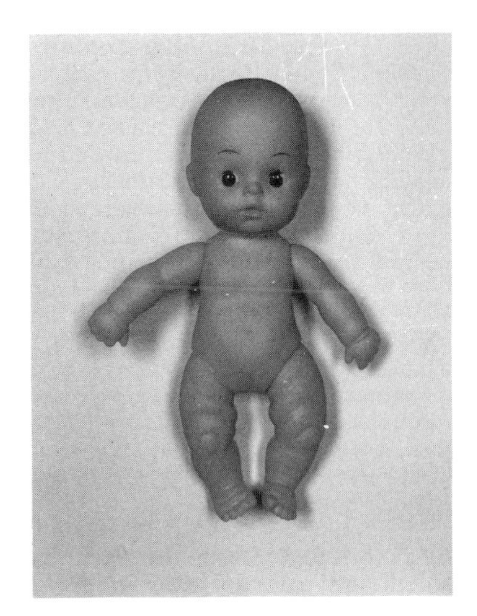

Figure 2a. (Left) Scanning setup for the incremental, interactive 3D ultrasound scanner system.
Figure 2b. (Right) The doll phantom. The height is about 17 cm.

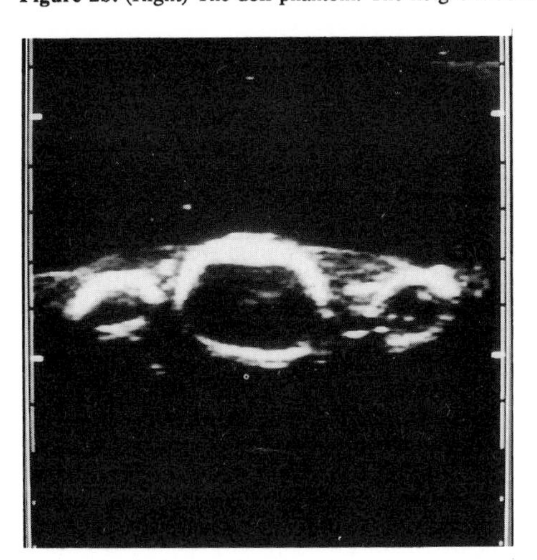

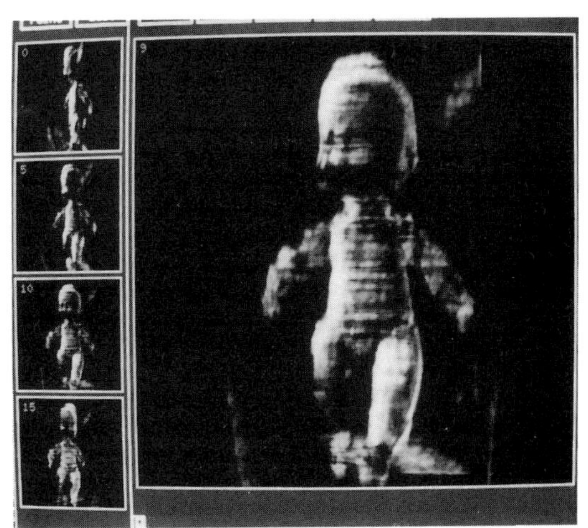

Figure 3a. (Left) A 2D echography image of the torso section of the doll phantom.
Figure 3b. (Right) Reconstructed and volume rendered image of the phantom.

tracking system has an accuracy of about 1 mm in x-y position and a maximum sampling rate of around 800 sample/s. The arm came from the previous generation 2D echography scanner Rohe ROHNAR 5580. ROHNAR 5580 used tracking to generate 2D images from a 1D scanhead. In our system, tracking is used to generate 3D volume from a state-of-the-art 2D scanner.

The system acquires image and geometry only if the change in geometry exceeds a certain preset threshold. This regulates to some degree the spatial sampling rate. Since the scanning of the 3D volume takes some time, the object must be stable enough. Examples of possible imaging targets are a liver or an immobilized woman's breast. We do not consider the scanning of a moving target, such as a beating heart, as an objective.

This acquisition setup is similar to the work at the Paris Phillips Research Laboratory by François Hottier et al (Hottier, 1989), which used a similar mechanical tracking arm. In their work, scanned 2D images and the scanhead tracking result are stored on a video tape and a PC's disk, respectively. Images and their matching geometry values are selected later off-line. We have used the real-time video digitizer and the tracking arm to acquire image and geometry at the same time.

Acquired 2DE image slices are incrementally reconstructed by local interpolation into regularly sampled 3D volume data by merging the new data with the existing volume data. As the slices accumulate into 3D volume, the rendering takes place, which shows the buildup of 3D volume. We will describe the details of the incremental volume rendering algorithm later in this paper. One thing of note is that the volume can be scanned repeatedly, to get better spatial sampling or to gain a a better acoustic window (e.g., to avoid bones). The reconstruction algorithm considers these modes of use, and provides reconstruction buffer update policies that include 'complete replace' and 'replace by weighted average'.

Currently, we store the images and coordinates into a disk of a file server connected by Ethernet for later reconstruction and rendering experiments on workstations such as DECStation 3100. In 1991, we hope to demonstrate a system that will use a powerful graphics multicomputer, Pixel-Planes 5 (Fuchs, 1989), for parallel reconstruction and rendering. In this parallel system, we expect the incremental visualization process to take place at an interactive rate of more than a frame per second without ever storing the image onto the disk. As mentioned, image read-out from the video digitizer is slow, and this might limit the processing speed of the entire system to about 5 frames/s acquisition rate. Assuming a 4 mm elevation resolution of the scanner image slice and a Nyquist sampling rate of 2 mm, a target volume of 20 cm thickness can be scanned as 100 slices of equally spaced parallel planes. Scanning the volume with 50-100 slices will take 1.4-3 s, assuming a 30 2D-frames/s acquisition rate, or 10-20 s, assuming a 5 2D-frames/s acquisition rate.

We have conducted a preliminary data acquisition and rendering experiment, which is reported in (Ohbuchi, 1990). We used an ATL Mark-4 scanner with a 3.5 MHz linear scanhead as the image input, and scanned a phantom (baby doll, in Figure 2b) and a human forearm in a water bath. Figure 3a shows the 2D echography image of the doll phantom. Figure 3b shows the reconstructed and rendered image of the doll from 90 slices of roughly parallel, roughly 2 mm interval 2D image slices. It is incrementally reconstructed using a forward-mapping octahedral kernel reconstruction algorithm, which will be explained later.

2. INCREMENTAL VOLUME VISUALIZATION

In this section, the algorithm for the incremental, interactive visualization of 3DE images from a sequence of 2DE slices is presented. We can divide the visualization stages of the incremental, interactive 3DE system into two major stages; *reconstruction* and *rendering*. The emphasis in this paper is on the incremental volume rendering algorithm.

Figure 4 shows the pipeline of the incremental volume visualization. Video digitized 2DE image slices from the scanner are incrementally *reconstructed* into a 3D scalar field sampled at uniformly spaced 3D grid points on a Cartesian coordinate system. Reconstruction is done into a reconstruction buffer, which resides in the world coordinates. It is then rendered using a modified front-to-back image-order volume rendering algorithm as developed by Levoy (Levoy, 1989). A reconstructed 3D scalar field in the reconstruction buffer is *classified* (non-binary classification) and *shaded* to produce a 3D shade buffer that also sits in the world coordinates. The next steps are the *ray-casting* process, where sample points in the world coordinates are transformed into 3D screen coordinates and then composed into 2D screen coordinates to yield a volume rendered image.

2.1. Incremental Volume Reconstruction

Several different approaches are possible to visualize irregularly sampled volume data. Theoretically, a stream of 2D image slices with variable geometry can be rendered directly by a certain rendering algorithm. Some algorithms to directly render irregular data samples have appeared recently (Garrity, 1990) (Wilhelms, 1990) (Miyazawa, 1991) (Neeman, 1990) (Shirley, 1990).

We have chosen to reconstruct regularly sampled 3D volume data in the *reconstruction buffer,* which is then rendered using more or less conventional volume rendering algorithms that expect volume data samples at regular 3D grid points. One reason for this is the speed of rendering. Once reconstructed, the volume data sampled on a regular 3D grid stored in 3D array with implicit geometry and connectivity offers easy access to the data. Another reason is that we wanted to allow multiple scanning sweeps of the target volume by the scanhead. 2D images from multiple passes are merged into a single image. Achieving this with direct rendering requires storing a large, unknown number of raw input image slices, and this is neither memory nor computationally efficient.

In designing the visualization scheme, we assumed that the target scalar field being sampled is continuous. Thus, if the target volume is sampled with a sufficient sampling rate by the 2D image slices, the resulting image of the target volume generated from these 2D slices must be continuous. For example, the resulting image should not look like a set of intersecting image planes with empty spaces in between. This means that the reconstruction requires some form of interpolated resampling. We have to be careful

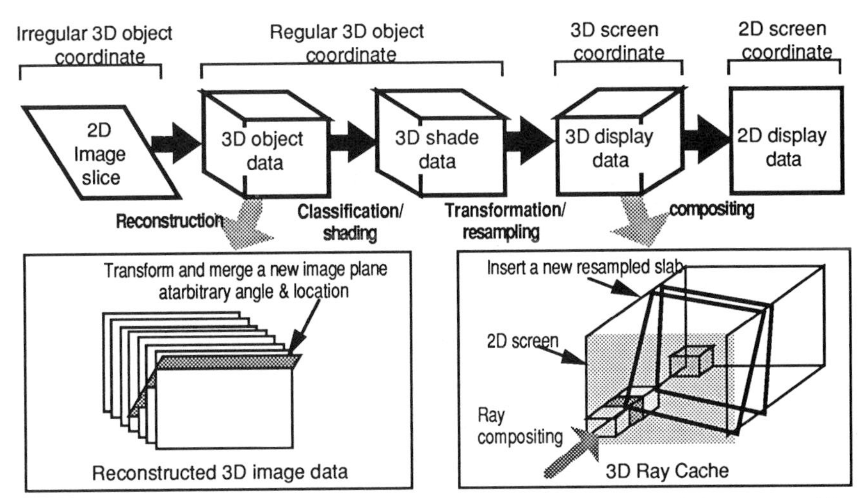

Figure 4. Incremental volume visualization processing stages. Irregular data slices are first reconstructed into regularly sampled 3D volume data. The volume data go through a standard volume rendering process, but the processing is confined to the inside of the slab formed by the last two image slices.

about interpolation. If the sampling rate is not sufficient, there is no way to recover information that was not sampled upon acquisition. The user should be made aware, by some means, of the undersampling..

We try to reconstruct a 3D array of sample points at regular intervals on Cartesian coordinates from a series of 2D image slices with 3 DOF as the imaging primitive. We assume that there is a rectangular volume in the real-world we are imaging, which will be sampled by points uniformly spaced on the 3D Cartesian coordinates. The system discards the sample points on 2D image slices that are outside this rectangular volume.

The next question regards what interpolation techniques to use. One important characteristic of our visualization scheme is that each input image must promptly be reflected in the final image. Thus, the interpolation method used for reconstruction can not be what Franke (Franke, 1982) calls the 'global' method of interpolation, where the interpolant is dependent on all data points. Both of the two reconstruction methods we have implemented so far are the 'local' kind.

Ideally, the interpolation function should depend on the imaging system. We have discovered, from the images obtained by scanning a 3D geometric calibration phantom made of thin wires and beads, that the image slice is fairly thick. A 2 mm bead is visible on the screen after 4 to 5 mm translation of the scanhead in elevation direction. This suggests much lower elevation resolution than the range resolution. Also, as expected, the sampling function in the far view is fuzzier and larger than in the near view. Though this was not quantitative, this gave us a feel of the shape of the sampling functions to expect.

We are still experimenting with the reconstruction algorithm. One method we have tried is a forward mapping algorithm, where the input sample points in the 2D image slices are distributed to the reconstruction buffer grids around the input sample points using a small spatial filter kernel (Figure 5). The kernel is anisotropic, and its size is determined in the input parameter file to the program. Mostly, we have used a filter kernel with octahedral shape in 3D that has the long-axis of around 6 mm and two short-axes of around 3 mm. This shape resembles the estimated shape of the ultrasound scanner system's sampling function. This corresponds to 12.5 x 6.2 x 6.2 voxels in the reconstruction buffer. Since the sampling function is attached to the image slice's coordinate, the kernel rotates with the input slice. For each orientation of the slice, coefficients of the octahedral kernel are computed on-the-fly on the discretized x-y-z grids of the voxels. The 2D weight buffer in the diagram accumulates the weight of the kernel for later normalization. The age buffer records the staleness of the image in the reconstruction buffer to help determine the weight with which to mix the old value in the buffer with the new contribution.

Another reconstruction algorithm is a backward mapping kind, where the algorithm traverses the reconstruction buffer grid, while the nearby sample values from the last two input image slices are linearly interpolated and collected. This is only C^0 continuous, and its C^1 discontinuity tends to show up in the image by a gradient approximation operator in the shading stage. Though it is fast, this algorithm does not produce as good a result as the one above.

We will continue to do more research in this area to find satisfactory reconstruction with small enough computational cost for interactive visualization.

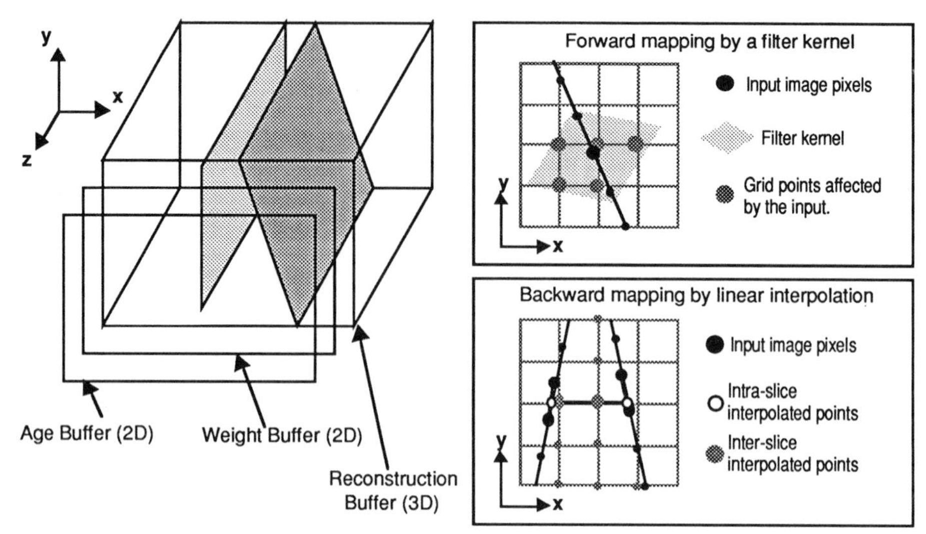

Figure 5. Reconstruction buffer and two reconstruction algorithms implemented; forward mapping using small filter kernel (above) and backward mapping algorithms.

2.2. Incremental Volume Rendering

Once the volume data with regular 3D sample points is reconstructed, it can be rendered using a standard volume rendering algorithm as described in the literature (Upson, 1988) (Sabella, 1988) (Levoy, 1988). As is known, volume rendering can be computationally expensive. We have to make the cost of image generation from reconstructed data low enough to achieve the goal of interactive image generation rate on moderate scale hardware.

The incremental volume rendering algorithm tries to reduce computation by taking advantage of three characteristics of the scanning process: 1) 2DE image slices are acquired incrementally, 2) shading parameters will not change for every few frames, 3) viewpoint will not change every few frames. If we assume these conditions to be true, we can perform incremental rendering. By incrementally shading and ray-sampling per each acquired image slice, the system limits the computation to the 'slab' formed by the last two inserted images, instead of the entire volume of the reconstruction buffer. With fixed viewpoint, incremental ray-sampling is realized by caching the tri-liniear interpolated ray samples from the previous slabs in a 3D array in the 3D screen space called *ray cache*. Ray-cache separates relatively expensive ray-sampling operations from relatively inexpensive ray-compositing operations. Ray-compositing is performed for an entire ray span inside the reconstruction buffer for each frame.

Figure 4 shows the process of incremental volume rendering. After the reconstruction, classification is done by table look up, and then the shading values of color and opacity are computed. The classification and shading take place only inside the slab formed by the current and previous image slices. Currently, we have implemented two shading algorithms: 1) image value shading that directly maps from input scalar value to the color value by table look up, and 2) object-space-gradient Phong shading. The latter performs Phong shading with diffuse and specular components where normal vectors of the field are approximated by finite difference computed in the object space. We included the method 1) for its efficiency, though the method 2) can emulate 1). Shading method 1) combined with additive compositing produce synthetic radiograph image quickly, which is useful to find better viewpoints. The shading results is stored in a shade buffer (3D) that resides in the world coordinate.

The ray-sampling stage performs ray-casting, where a ray is cast from each pixel into the shade buffer, and the shading values are sampled at uniform intervals along the ray. We provide perspective projection which is hoped to give better 3D perception. Perspective projection can cause over and/or under sampling problems in ray-casting. Scheme to correct aliasing due to perspective projection, such as (Novins, 1990), is not employed in our current implementation. A ray sample is the result of tri-linear interpolation of values from eight vertice of cube enclosing the sample point. This stage also works incrementally, and only the same sub-volume processed in the shading stage is sampled. The system clips the ray to the slab (Figure 6a and 6c), to sample only inside the slab. We developed an algorithm that is

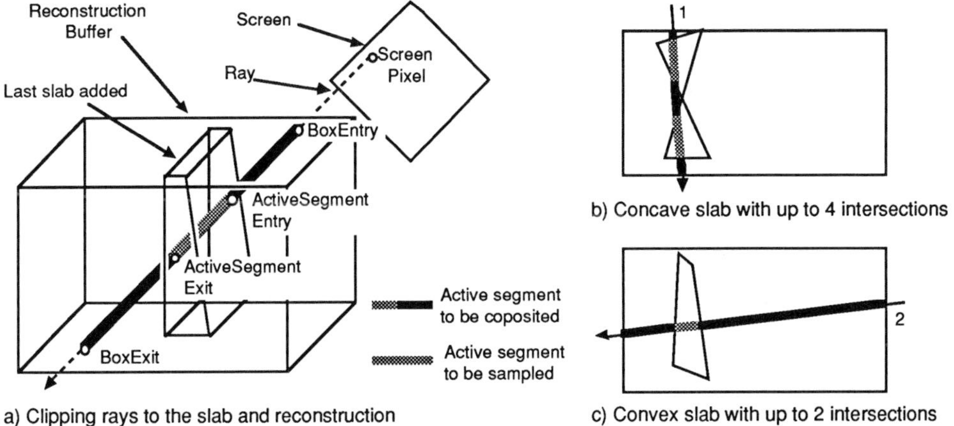

Figure 6a. (Left) A ray from a pixel is clipped to the reconstruction buffer bounding box, and to the last slab for compositing and sampling, respectively.
Figure 6b. (Right Top) Slab can be concave, with up to 4 intersections, or
Figure 6c. (Right Bottom) convex with up to 2 intersections.

essentially the same as the Cyrus-Beck clipping algorithm (Cyrus, 1978) for this clipping. The slab is decomposed into two convex polyhedrons if it is concave (Figure 6b).

A 3D array called *ray-cache* in 3D screen space saves the ray sample values. This is a classic space-time trade off. Each pixel in the frame buffer is associated with a linear array of ray samples along the ray. As the new slab is shaded and sampled, those samples are inserted to the appropriate locations (depth) in the ray-cache, replacing the old value. The shade buffer voxels outside the slab are not sampled and the ray sample values for these remain unchanged. Current choice of compositing include multiplicative compositing as in (Levoy, 1988), additive compositing for X-ray like image, and Maximum Intensity Projection that takes the maximum sample value along the ray as a pixel value.

If multiplicative compositing is used, we must composit an entire span of the ray that lay inside the reconstruction buffer, even if only a portion of it are new samples. (We ignore the adaptive ray termination here for simplicity, though it is implemented.) But since relatively expensive interpolated ray-sampling (and shading) is limited to the inside of the slab, separating ray-sampling from ray-compositing saves time. The span of the ray to be composited is obtained by clipping the ray to the reconstruction buffer bounding box.

Note that the above arguments about incremental volume rendering assumed fixed viewpoint as well as fixed shading parameters. If viewpoint changes, ray-sampling and compositing need be done essentially on the entire volume. If the classification and/or shading parameters changes, classification and shading in addition to ray-sampling and ray-compositing must be done. In volume rendering, change of density classification function to classify desired object (e.g., to render either skin or bone as a surface from X-ray CT images) is as frequent as change of viewpoint. Thus, both of these two operations must be fast. Various forms of coherence, such as image and object coherence as well as temporal coherence can be used to improve performance. But current implementation does not include optimizations for the non-incremental cases.

3. IMPROVING RENDERING PERFORMANCE

Our goal is to make this system as interactive as possible. It should run with the image generation speed (reconstruction and rendering combined) of more than 1 frames/s, at the slowest, on the proof-of-concept system using Pixel-Planes 5.

In the interactive, incremental 3DE system the volume data change every image frame. Many of the conventional optimization schemes that assumed multiple image generation from single data set can not be applied to such a case. For example, skipping empty space by hierarchical space enumeration, using octree, will not be appropriate since octree is usually precomputed. Similar argument goes to the precomputing the gradient vectors or complete shading (color and opacity) values.

Also, use of incremental scheme tipped the scale in relative costs of various stages of volume rendering. A good example is clipping the rays to the volume of interest (i.e., to the reconstruction buffer bounding box). It was a minute part of ray-casting process in the conventional ray-casting based volume rendering. In the incremental algorithm, relative cost of ray-clipping has become a significant part of ray-casting process. This is because a volume to be sampled (i.e., a slab) is much smaller and the clipping need be performed for every input slice. This overhead is expected to worsen if the algorithm is parallelized by subdividing the reconstruction buffer in object space. Shading (Phong shading) and ray-sampling still remains as the other major parts of the rendering process, as well as the reconstruction depending on the reconstruction algorithm used.

3.1. Ray Clipping by Scan Conversion

With the introduction of ray-cache, ray-sampling the unchanged part of the volume outside the slab is obviously wasteful. To sample rays only in the slab formed by latest two slices, each ray from the pixel is clipped to the slab. As mentioned, we used an algorithm similar to Cyrus-Beck (Cyrus, 1978) in the first implementation. As seen in the Table 1, clipping rays to the slab has taken up large portion of rendering time. We needed a faster algorithm.

We have developed a new line-polyhedron intersection algorithm called *D-buffer* algorithm for Distance Buffer. It is not as general as the Cyrus-Beck clipper, but much more efficient if applied to computing intersections of non-trivial number of rays from a screen with polygons. It takes advantage of the efficient polygon scan conversion algorithm to compute intersection distances (see Figure 7a). Following is the sketch of the D-buffer algorithm for ray-clipping.

{ This routine computes the intersection of all the rays from the screen pixels with the slab defined by polygons, using polygon scan-conversion. }
procedure RayClip(Slab)
begin
- If the slab is concave, decompose it into the convex polyhedrons.

for each polyhedrons
- Clip it to the reconstruction buffer bounding box. Clipped faces must be 'closed' by adding new polygons.

for each polygon that forms the polyhedron,
- Transform it into the viewing coordinate, and clip it to the view volume.
- Scan convert the polygon into D-buffer, using PixelUpdate() for each pixel.

end;

{ Polygon scan conversion routine calls PixelUpdate() per scan converted pixel (u, v, n), to update the D-buffer. (u,v) is the location of the pixel on the screen while n is the screen Z value (real) of the scan converted point.}
procedure PixelUpdate(u, v, n)
begin
- Compute the Euclidean distance from projected 2D point (u,v) in 3D screen space, to its back-projected point in 3D screen coordinate (which is the intersection).
- Insert the distance for pixel (u, v) to the corresponding entry of the distance list, which is sorted by the distance.

end.

In the PixelUpdate() routine, if the projection is orthogonal, scan converted screen Z value *is* the Euclidean distance from the pixel to the 3D point projected to that pixel. If the projection is perspective, the distance must be computed by back-projecting the pixel into 3D screen coordinates.

Computed distances from the pixel to the intersections with the polygons are kept in the distance list for each pixel (See Figure 7b). This list is kept in ascending order as intersection distances are inserted. A ray has either two or four intersections with a slab, so our implementation has a 4 position array for the distances. After two or four intersection distances are obtained, we have to determine the interval(s) where the ray is to be sampled. They are the distances in 3D screen space from the screen pixel to the entry and exit intersections of the slab. All the polyhedrons are convex here so that the paring of the intersections to the intervals can be made by the parity rule. This is assured (in non-singular cases) by making the polyhedron closed. To be more general, the 'sidedness' of the polygon (i.e., which side is facing the viewpoint) can be used to pair up the interval. The sidedness is determined by the sign of innerproduct of the polygon's surface normal and the view vector, which is done once per polygon. This is not necessary to clip rays to the slab, but it is useful later when integrating the volume image with the polygonally defined objects using D-buffer.

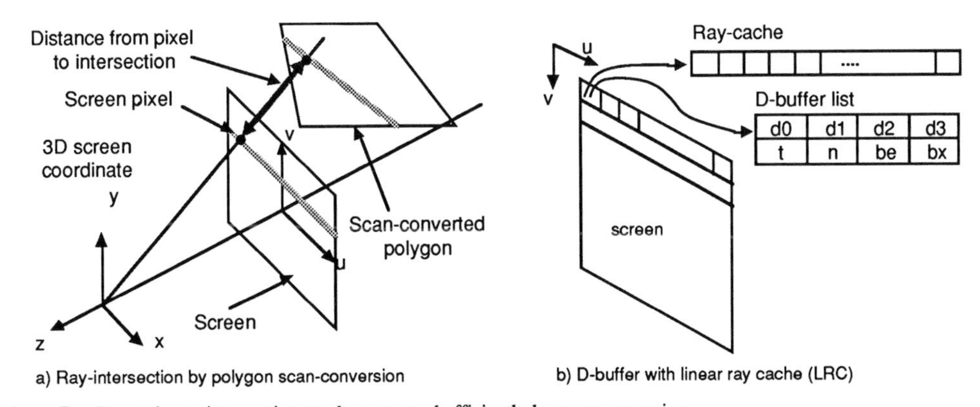

Figure 7a. Ray-polygon intersection can be computed efficiently by scan conversion.
Figure 7b. D-buffer has 1D arrays for ray-cache, as well as distance lists for up to four intersections.

D-buffer structure has a few other fields per pixel, as seen in Figure 7b. One is the ray-cache, a 1D array to store sampled ray values. The interval to be composited is stored in be and bx pair, which are the entry and exit to the reconstruction buffer bounding box. If a ray from a pixel needs compositing, it is marked in field t. Numbers of intersections are stored in n as slab polygons are scan converted.

This D-buffer algorithm exploits the geometric coherence in the polygons that forms the slab being intersected, through the polygon scan conversion algorithm (Foley, 1990). Using polygon scan conversion algorithm limits the computation to only those rays that actually intersect the slab. This is in contrast to the clipping algorithm based on conventional ray-polygon intersection where all the rays must be checked for intersection before they are rejected. Also, due to its incremental nature, for a non-trivial number of intersections, cost per ray is very small.

We have implemented the ray-clipping by D-buffer, as well as the Cyrus-Beck algorithm. Table 1. compares the execution time of various visualization stages per single image generation for three different algorithms: without ray-clipping, ray-clipping by Cyrus-Beck algorithm, and ray-clipping by D-buffer algorithm. In all three cases, reconstruction (backward-mapping algorithm) and shading (Phong shading) are incremental. The backward-mapping reconstruction algorithm used is fast but not of high quality. It is very likely that the computational complexity of the reconstruction will increase in the future as reconstruction quality is improved. The program written in C language is compiled with -O option and executed on the DEC 5810 with 256 MByte of memory which is running Ultrix operating system. As an input, 36 2D image slices of roughly 5 mm intervals are read from a disk file along with their geometries. 34 images of 256 x 256 pixels each (whose 34th image is similar to Figure 3b) are generated. Time for the disk access is excluded from the timing. We measured timings by UNIX system call `times()` for the program to generate 34 images, which is then divided by 34 to obtain average timings per frame.

Table 1 shows great reduction in ray-sampling time by incremental ray-sampling, i.e. by limiting the ray-sampling to inside of the slab by ray-clipping. Table 1 also shows that the D-buffer algorithm virtually removed ray-clipping overhead. Time in the sampling stage is now spent almost entirely in the tri-linear interpolated resampling. Further speed up of ray-sampling will require such optimizations as skipping empty spaces, or adaptively reducing the number of rays cast.

As mentioned above, the relative cost of the reconstruction may increase as the reconstruction

Table 1. Execution time of ray sampling and compositing for the three cases: no-clipping, clipping with Cyrus-Beck clipper, and clipping with D-buffer algorithm. Both reconstruction and shading stages are done incrementally on all three cases. Numbers in parenthesis are the relative time spent in percentile.

Processing Stages	No Clipping	Cyrus-Beck Clipping	D-buffer Clipping
Reconstruction	0.60s (0.3%)	0.60s (3.3%)	0.59s (5.9%)
Shading	2.87s (1.4%)	2.83s (15.4%)	2.81s (28.3%)
Ray-clipping	0.00s (0.0%)	8.40s (45.8%)	0.10s (1.0%)
Ray-sampling	185.56s (91.8%)	4.75s (25.9%)	4.67s (47.1%)
Ray-compositing	13.13s (6.5%)	1.77s (9.6%)	1.75s (17.6%)
Total time	202.18s	18.35s	9.92s

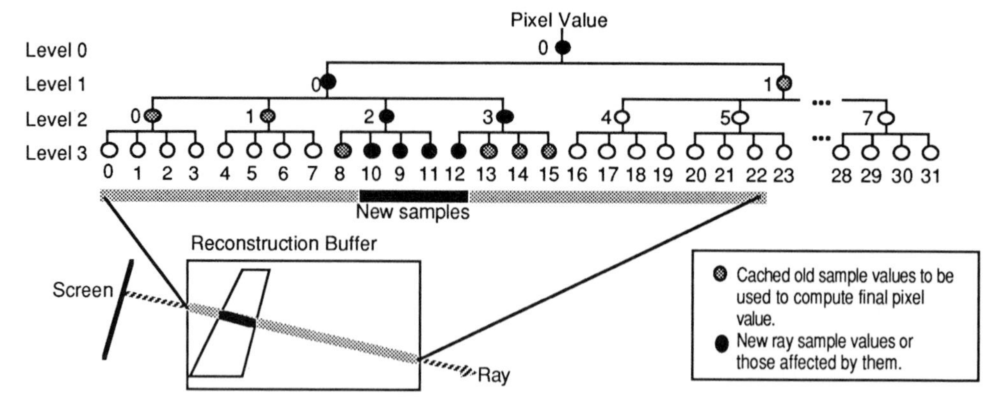

Figure 8. Hierarchical Ray Cache with 32 entries. Changes due to the insertion of a small span of samples are propagated bottom up to the root, which is the pixel value. This takes much smaller number of compositing than with the linear ray-cache.

algorithm becomes sophisticated. The shading stage will be more dominant with the real image data taken from a human target than the doll phantom image scanned in the water tank. Shading computation can be abbreviated if the classification result is near zero, as in the water around the doll. But in the real image, there will be more non-zero values in the data. Time for the compositing stage is small because only the rays with new samples, i.e., those that touch the slab, need be casted, leaving majority of the pixel untouched.

3.2. Hierarchical Ray-Cache

The ray-cache introduced before was a simple 1D array for each pixel, which caches every sample taken along the ray. In this section, we introduce a new ray-cache structure to improve ray-compositing speed, which is currently being implemented along with the polygonal object rendering capability described in the following sections. The new ray-cache is called a *hierarchical ray-cache* (HRC). By saving and reusing the partially composited values stored in tree structured cache, HRC saves significant amount of time compared to a linear ray-cache (LRC) introduced before. This is another case of space-time trade off; a HRC requires even more memory than a LRC, but gains substantially in speed. A HRC is depicted in Figure 8 for 32 entry, 4-ary tree case, though it can have any arity more than 1.

In the HRC, the tree is maintained so that each node has the values of opacity and color that are the result of compositing those values in their child nodes from left to right order of traversal (multiplicative compositing operation does not commute). The leaves of the tree have the opacity and color sample values as they are sampled along the ray. The root of the tree, if the property above is maintained, has the values of opacity and color as the result of correct compositing along entire ray samples; that is, it has the screen pixel value. As a set of new sample values is added to the leaves, updating process takes place from bottom up. For each node, if value of any of its child nodes has changed, a new pair of opacity and color values are computed by compositing. The HRC for a pixel can be embedded in a linear array, and the tree traversal can be done efficiently by index manipulations.

Assuming a binary tree, for a total ray span of 2^n samples, change in one sample requires only n number of compositing to obtain the pixel value. This is the most favorable case for the HRC and is a big improvement over 2^n-1 compositing necessary for the LRC. The worst case possible for the HRC is when the entire span of the ray inside the reconstruction buffer must be sampled anew. In this case, the HRC needs the same 2^n-1 compositing as the LRC; however, the HRC needs more memory write, 2^{n+1}-1, compared to 2^n-1 for the LRC. For most of the cases where the slabs are oblique against the ray, and are thin compared to the reconstruction buffer bounding box dimensions, the HRC will be much more efficient than the LRC.

4. INTEGRATING POLYGONAL OBJECTS

It is obviously advantageous to have geometric objects rendered together with the volume data in the same image. (Levoy, 1990) presents two methods to integrate geometric objects with the volume image in the framework of front-to-back image-order volume renderer. One is to 3D scan convert polygons into

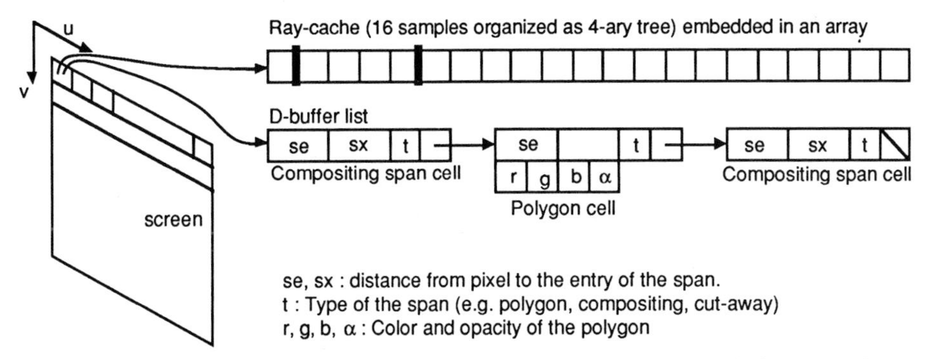

Figure 9. The structure of the D-buffer with hierarchical ray cache. A 16-entry 4-ary tree ray-cache is embedded in a 1D array of size 21.

volume data, with appropriate filtering to band-limit the image of the polygons, and then volume render the combined image as a volume data. The other method, which he called a *hybrid ray-tracer*, runs the ray-casting codes for the volume data and polygons in parallel, combining the result from both for each ray step. The 3D scan conversion method has difficulty in rendering crisp, thin polygons. The hybrid ray-tracer was not fast; the volume must be sampled for its entirety, and lock-step compositing of two ray-casting processes interfered with such optimization as empty space skipping using octree.

In the following, we propose an algorithm that realizes very fast rendering of polygonal and polyhedral objects (including volumes defined by polyhedron), both opaque and transparent, in the volume rendered image with fixed viewpoint. The efficiency (and the limitation of fixed viewpoint) of this algorithm comes from the fact that it works in 3D screen space. It combines the viewing transformed and resampled volume data in the ray-cache with the polygonal and polyhedral objects scan converted into the D-buffer. This algorithm is analogous to Atherton's *object buffer* algorithm (Atherton, 1981). Atherton's algorithm saves a list of all the depth (z) values of scan converted objects, ordered by z and accompanied by the each objects' attributes, in the object buffer. As in Z-buffer, D-buffer algorithm provide visible surface determination. In addition, object buffer provides such effect as cutaway viewing, surface color change, and transparency or translucency of the object, as far as the viewpoint is fixed. (Atherton, 1981) also suggests that the object buffer can have *object buffer matrix*, which is essentially voxel arrays in 3D screen space. This is akin to the Ray-cache in our incremental volume rendering algorithm. His paper suggested application of the algorithm to 3D visualization of CT images and and seismic analysis results. More recently, (Ebert, 1990) added volume rendering capability on top of an A-buffer (Carpenter, 1984) based polygon renderer. His objective was to bring gaseous objects and solid textures to the world of mostly polygonal objects. (Garrity, 1990) combined volume renderer for an irregular geometry volume data with a polygon renderer based on z-buffer.

Figure 9 shows a new frame buffer structure that combines a HRC with a D-buffer. The D-buffer has linked lists of *span-cells*, where each span-cell has its type associated. The type field tells if the span is a sampled volume data to be composited, a volume defined by polygons that is to be rendered with modified color and/or opacity, or a polygon with associated colors and opacity. The HRC's tree structure is embedded in a 1D array as a heap for compactness and fast access. Types of the span-cells are, for example,

1) Compositing span cell : Distances of the volume sample span to be composited,

2) Polygon cell : A polygon distance, color and opacity values.

3) Polyhedra span cell : Distance to the polyhedra span, with its color and opacity modification values.

There is no cell type for cut-away span, since cut-away is done by not compositing a cut-away span, through modification of the sample span cells.

4.1. Integrating Polygons

The incremental volume rendering algorithm keeps ray-samples in the ray-cache to reuse. Thus, for a fixed viewpoint, adding and correctly compositing a polygon to the volume image involves only; 1) scan converting the polygons into the D-buffer, and 2) compositing the cached ray-samples with the scan converted polygons. If the distance value of the scan converted polygon falls inside of a ray span, the ray span is split into two sections. The two sections become two separate compositing span cells in the D-buffer, and a new polygon cell is inserted in between. Notice that there is no ray-sampling involved for image generation (for a fixed viewpoint). Furthermore, with the HRC, compositing a scan converted polygon roughly corresponds to inserting a sample slice of thickness 1 sample and then updating the HRC.

The cost of polygon rendering depends on two major factors; the costs associated with polygon scan conversion (e.g., the cost of shading models used), and the cost of inserting the polygon cell into the D-buffer and updating the HRC. The cost of span-cell list update can be significant if the number of cells is large. The cost of ray-compositing depends both on the number of rays, as well as the cost of compositing a ray. The number of rays is proportional to the sum of the area of polygons projected onto the screen. The cost of compositing a ray using the HRC after insertion of a polygon is proportional to the logarithm of the number of ray-samples.

In implementing the algorithm, there are two sources of aliasing to be taken care of. One is aliasing in 2D screen space that happens at the edge of the polygon. This is known for Z-buffer algorithm, and amenable to such solutions as super sampling and A-buffer (Carpenter, 1984). This problem is not discussed here. The other happens in the depth direction of 3D screen space, and somewhat characteristic to the volume rendering. (In a sense, this depth aliasing corresponds to the Z-buffer resolution problem.) Let's assume a polygon is inserted in between two sample points along the ray. We assume the volume to be a continuous medium of semi-transparent object. Proper compositing of the polygon must consider the opacity contributions of medium in front and back of the polygon in this unit sample interval. Exact determination of these contributions is somewhat costly, thus we have adopted simpler but visually satisfactory solution as proposed in (Levoy, 1990). It treats the polygon near the ray-polygon intersection as a plane perpendicular to the ray. The opacity contribution of the unit sample interval is divided into front and back of the polygon, α_f and α_b respectively, in proportion to their thicknesses. α_f and α_b of the ray u are computed by;

$$\alpha_f(u) = 1-(1-\alpha_v(u))^{t_f^{(u)}/t_v^{(u)}}$$

and

$$\alpha_b(u) = 1-(1-\alpha_v(u))^{t_b^{(u)}/t_v^{(u)}}$$

where t_f and t_b are the thickness of the intervals in front and back of the polygon, while t_v is the thickness of the unit interval. $\alpha_v(u)$ is the opacity of the unit interval, while c_v is the color of the unit interval. Compositing proceeds from front to back; it composits the contribution from the medium in front using c_v and α_f first, the polygon as an object of 0 thickness but finite opacity α_p and color c_p next, and finally the portion in the back of the polygon using α_b.

4.2. Integrating Polyhedral Volume

We can also render a (solid) polyhedral object by scan converting the boundary polygons of the object into D-buffer, and re-compositing those samples inside the object. The ray-clipping algorithm by D-buffer presented before have done this, if you think of the slab to be sampled as a special polyhedron. The object can be general convex polyhedron. This allows cut-away of the volume, highlighting of a volume, or simple rendering of polyhedral (solid) object, along with the volume image. This algorithm is similar to those that render solid objects from *B-reps* (boundary representations) in solid modeling terminology using Z-buffer like technique.

This can provide very useful tools of interaction. Cutting away a section of the volume image is a very useful tool for interaction with the volume image; for example, to see objects obscured by an opaque surface. It may also be used for simulated surgery. Highlighting a selected volume, by changing the color and/or opacity of the volume can be a useful tool to highlight volume of interest. For example, in radiation treatment planning, traces of treatment beam can be highlighted while the other objects are kept dim.

Integration of a polyhedral object is done in the following manner. For each ray, using D-buffer, the entry and exit distances to a span which lies inside the polyhedron is computed and added to the span-cell list. The span can then be used in various ways. The polyhedrally defined volume is cutaway if the samples in the spans are not composited. This can be done very quickly. If the opacities and the colors of the samples in the span are modified and composited, the volume defined by the polyhedron is rendered as a solid polyhedral object. This can be used to highlight the volume if the object rendered is semi-transparent. The opacity and/or color in the ray-cache leaves need not be literally modified; they just have to be modified as they are composited. Two kinds of aliasing mentioned above for polygonal objects also happens in integrating polyhedral objects with the volume image, and they require proper attention.

We should re-iterate that the cut-away of volume and rendering of polygons into a fixed viewpoint volume image is very fast with accurate volume compositing using D-buffer and HRC. Even though highlighting volume of interest by changing color or opacity is not quite as fast as cutaways, it still avoids costly ray sampling process entirely. Thus it will provide a responsive interaction with the volume image. Although only polygonally and polyhedrally defined objects are mentioned in the explanation above, any surfaces or objects that can be rendered efficiently by Z-buffer algorithm may be integrated into this algorithm. Please also note that this technique is not limited to the incremental volume rendering algorithm. It can be applied to other conventional volume rendering algorithms based on ray-casting.

Major limitation of this technique is the fixed viewpoint. Polygons, cut-away volumes or highlighted volume of interest defined in the image can not be viewed from different viewpoints easily, since it exists only in the screen space. To view it from different viewpoint, or to reconstruct the 'sculpting' done to the image by polygonal objects, geometric information of these polygonal objects must be saved separately. In terms of performance, large number of intersections with polygons per ray will degrade the performance of HRC, in addition to obvious overhead of scan-converting many polygons into the span-cell list.

5. CONCLUSIONS

In this paper, we have reported an overview of an incremental, interactive 3D ultrasound echography system, where 2D image slices are acquired, along with their locations and orientations, and then reconstructed and volume rendered. The system generates a new image as a new 2D slice is acquired, to maximize interactivity. The visualization algorithm works in an incremental manner, limiting the computations to the volume where the new input has arrived.

We have established the basics of an efficient incremental volume rendering algorithm, which takes advantage of the incremental nature of the input. We have designed and implemented a faster ray clipping algorithm using polygon scan conversion to clip rays to the slab where the ray should be sampled. The new ray clipping algorithm showed marked improvement over the method we used before.

We have proposed a new ray-caching mechanism called hierarchical ray-cache to speed up the ray-compositing further. We have also proposed an algorithm to render polygonal objects quickly in screen space composited correctly with the volume image. With this proposed algorithm, various operations on the volume image, such as a cutaway of a polyhedral volume, insertion of polygons, and highlighting polyhedral volumes will be possible at an interactive rate.

Clearly, we need more work to get to our goal of an interactive acquisition 3D visualization system. First, economical reconstruction algorithms with good reconstruction quality have to be developed to reconstruct irregular input slices. Current bottlenecks exist in the shading and ray-sampling stages. An interesting low-level speed-up technique used in a parallel volume renderer for Pixel-Planes 5 (Cullip, 1990) is normal coding (Glassner, 1990). With this technique, the shading buffer stores the encoded normal vector and gradient magnitude instead of color and opacity. The shading computation, including the expensive inner product operations, is done by table look-up by the encoded normal/gradient. Reduction of ray-sampling cost by empty space enumeration may be applicable to the incremental volume rendering.

We are planning to parallelize the algorithm to be run on the graphics oriented heterogeneous multicomputer Pixel-Planes 5. We have experimented with a distributed volume rendering algorithm on a set of workstations, which showed promising results. It was parallelized in image space, and based on demand-paged distributed shared memory model, similar to (Badouel, 1990). For the incremental, interactive 3DE system on Pixel-Planes 5, we are planning to use data parallelism in the object space. The D-Buffer algorithm will help reduce the ray-clipping overhead incurred by subdividing the object space among multiple processors. The idea of hierarchical compositing can also be adapted to object space parallelism. We expect to have a proof-of-concept system with Pixel-Planes 5 running sometime in 1991.

Acknowledgments

The authors would like to thank Vern Katz, M.D. for the use of the ultrasound scanner, and Jeff Butterworth for developing the image acquisition program. We also would like to thank the Wake Radiology Association for donating ROHNAR 5580 to us. Suesan Patenaude and Harrison Dekker contributed in improving the manuscript. This research is supported by NSF grant number CDR-86-22201.

References

Atherton, P.R. (1981). A Method of Interactive Visualization of CAD Surface Models on a Color Video Display. ACM Computer Graphics, **15**(3): 279-287.

Badouel, D., Bouatouch, K., Priol, T., (1990). Ray Tracing on Distributed Memory Parallel Computers : strategies for distributing computation and data (Technical Report No. 508). Institut De Recherche en Informatique et Systèmes Aléatoires, Universitaire de Beaulieu, France.

Billion, A.C. (1990) Personal Communication.

Brinkley, J.F., Moritz, W.E., and Baker, D.W. (1978). Ultrasonic Three-Dimensional Imaging and Volume From a Series of Arbitrary Sector Scans. Ultrasound in Med. & Biol., **4**: 317-327.

Carpenter, L. (1984). The A-buffer, an Antializsed Hidden Surface Method. ACM Computer Graphics, **18**(3): 103-108.

Cullip, T. (1990) Personal Communication.

Cyrus, M., and Beck, J. (1978). Generalized Two- and Three-Dimensional Clipping. Computers and Graphics, **3**(1): 23-28.

Ebert, S.D., and Parent, R.E. (1990). Rendering and Animation of Gaseous Phenomena by Combining Volume and Scanline A-buffer Technique. ACM Computer Graphics, **24**(4): 357-366.

Foley, J.D., van Dam, A., Feiner, S. K., and Hughes, J. F. (1990). Computer Graphics Principle and Practice (2'nd Edition, ed.). Addison-Wesley, .

Franke, R. (1982). Scattered Data Interpolation : Tests of Some Methods. Mathematics of Computation, **38**(157): 181-200.

Fuchs, H., Poulton, J., Eyles, J., Greer, T., Goldfeather, J., Ellsworth, D., Molner, S., and Israel, L. (1989). Pixel Planes 5: A Heterogeneous Multiprocessor Graphics System Using Processor-Enhanced Memories. Computer Graphics, **23**(3): 79-88.

Garrity, M.P. (1990). Raytracing Irregular Volume Data. ACM Computer Graphics, **24**(5): 35-40.

Ghosh, A., Nanda, C.N., and Maurer, G. (1982). Three-Dimensional Reconstruction of Echo-Cardiographics Images Using The Rotation Method. Ultrasound in Med. & Biol., **8**(6): 655-661.

Glassner, A. (Ed.). (1990). Graphics Gems. Academic Press, San Diego.

Hottier, F. (1989) Personal Communication.

Lalouche, R.C., Bickmore, D., Tessler, F., Mankovich, H.K., and Kangaraloo, H. (1989). Three-dimensional reconstruction of ultrasound images. In: SPIE`89, Medical Imaging (pp. 59-66).

Levoy, M. (1988). Display of Surface from Volume Data. IEEE CG&A, **8**(5): 29-37.

Levoy, M. (1989) Display of Surfaces From Volume Data. Ph.D Thesis, University of North Carolina at Chapel Hill, Computer Science Department.

Levoy, M. (1990). A Hybrid Ray Tracer for Rendering Polygon and Volume Data. IEEE CG&A, **10**(2): 33-40.

Lin, W., Pizer, S.M., and Johnson, V.E. (1991). Surface Estimation in Ultrasound Images. In: Information Processing in Medical Imaging 1991 (IPMI'91), (In this volume) . Wye, U.K., Springer-Verlag, Heidelberg.

McCann, H.A., Sharp, J.S., Kinter, T.M., McEwan, C.N., Barillot, C., and Greenleaf, J.F. (1988). Multidimensional Ultrasonic Imaging for Cardiology. Proc.IEEE, **76**(9): 1063-1073.

Mills, P.H., and Fuchs, H. (1990). 3D Ultrasound Display Using Optical Tracking. In: 1'st Conference on Visualization for Biomedical Computing, (pp. 490-497). Atlanta, GA, IEEE Computer Society.

Miyazawa, T. (1991). A high-speed integrated rendering for interpreting multiple variable 3D data. In: 1991 SPIE/SPSE Symposium, Extracting Meanging from Complex Data : Processing, Display, Interaction II, . San Jose, CA.,

Nakamura, S. (1984). Three-Dimensional Digital Display of Ultrasonograms. IEEE CG&A, **4**(5): 36-45.

Neeman, H. (1990). A Decomposition Algorithm for Visualizing Irregular Grids. ACM Computer Graphics, **24**(5): 49-56.

Nikravesh, P.E., Skorton, D.J., Chandran, K.B., Attarwala, Y.M., Pandian, N., and Kerber, P.E. (1984). Nikravesh, P.E., Skorton, D.J., Chandran, K.B., Attarwala, Y.M., Pandian, N., and Kerber, P.E. Untrasonic Imaging, **6**: 48-59.

Novins, K.L., François, X. S. and Greenberg, D. P. (1990). An Efficient Method for Volume Rendering using Perspective Projection. ACM Computer Graphics, **24**(5): 95-102.

Ohbuchi, R., and Fuchs, H. (1990). Incremental 3D Ultrasound Imaging from a 2D Scanner. In: First Conference on Visualization in Biomedical Computing, (pp. 360-367). Atlanta, GA, IEEE.

Raichelen, J.S., Trivedi, S.S., Herman, G.T., Sutton, M.G., and Reichek, N. (1986). Dynamic Three Dimensional Reconstruction of the Left Ventricle From Two-Dimensional Echocardiograms. Journal. Amer. Coll. of Cardiology, **8**(2): 364-370.

Sabella, P. (1988). A Rendering Algorithm for Visualizing 3D Scalar Fields. Computer Graphics, **22**(4): 51-58.

Shattuck, D.P., Weishenker, M.D., Smith, S.W., and von Ramm, O.T. (1984). Explososcan: A Parallel Processing Technique for High Speed Ultrasound Imaging with Linear Phased Arrays. JASA, **75**(4): 1273-1282.

Shirley, P., and Tuchman, A. (1990). A polygonal approximation to direct scalar volume rendering. ACM Computer Graphics, **24**(5): 63-70.

Stickels, K.R., and Wann, L.S. (1984). An Analysis of Three-Dimensional Reconstructive Echocardiography. Ultrasound in Med. & Biol., **10**(5): 575-580.

Upson, C., and Keeler, M. (1988). VBUFFER: Visible Volume Rendering. ACM Computer Graphics, **22**(4): 59-64.

Wilhelms, J., Challinger, J., and Vaziri, A. (1990). Direct Volume Rendering of Curvilinear Volumes. ACM Computer Graphics, **24**(5): 41-47.

COMPRESSION OF SEQUENCES OF 3D–VOLUME SURFACE PROJECTIONS

B Gudmundsson and M Randén

Image Processing Laboratory
Department of Electrical Engineering
Linköping University
S–581 83 Linköping, Sweden

Abstract

The large data sets constituting 3D–volumes of reasonable resolution inflict high costs in terms of storage space and computation time. In this paper we describe an algorithm for data compression of sequences of 3D–volume projections where the projection angle increment is small and the surfaces of the objects in the volume are rendered using depth–only shading. The algorithm is based on a ray–casting method for incremental generation of such sequences. Experiments on a specific CT–volume yielded a compression factor of 32 for an angle increment of 1 degree and grid–point accuracy in object surface localization.

Keywords

Data compression; visualization, ray–casting.

1. INTRODUCTION

Visualization of 3D–volumes has attracted considerable interest in the last few years. Although special purpose display devices for true 3D display have begun to emerge, a conventional 2D display device is still the only available option in most cases. Thus, the basic problem to be solved is how to make projections from 3D to 2D in such a way that the 3D–structures in the volume are faithfully revealed to the viewer. A number of methods have been proposed (Herman and Lui, 1979), (Lenz et al, 1986), (Farell and Zapulla, 1989), (Levoy, 1990), (Tiede et al, 1990). Most of them are based on ray–casting, i.e. imaginary rays emanating from grid–points in the projection plane pierce the volume and generate sample points in the projection plane. The viewers perception of the third dimension can be enhanced by means of object surface shading, stereo–viewing and animation. The most prevalent form of animation technique in cases where the object in the volume can be segmented from the background, is rotation of the objects around some axis.

The illusion of a rotating volume is achieved by computing a sequence of projections from different angles and then displaying the sequence, possibly under interactive control. Smooth rotation requires a sequence where two successive projections are only a few degrees apart, and thus a large number of projections is needed. For volumes of reasonable resolution the task of generating such sequences tends to become computationally expensive, unless special–purpose hardware is used. However, the fact that the projections are only a few degrees apart means that there is a high degree of coherence between successive projections in the sense that if an object surface point is visible in one projection, then it is highly probable that it will also be visible in the next. This observation is the basis for an algorithm that computes successive projections in the sequence incrementally (Gudmundsson and Randén, 1990) thereby avoiding a costly full-fledged ray-casting operation for each projection.

In this paper a new method for data compression of projection sequences is presented. It is based on the algorithm for incremental sequence generation which will be briefly described in the next section.

2. INCREMENTAL GENERATION OF SEQUENCES

We assume orthogonal parallel projection and a geometry as shown in Figure 1.

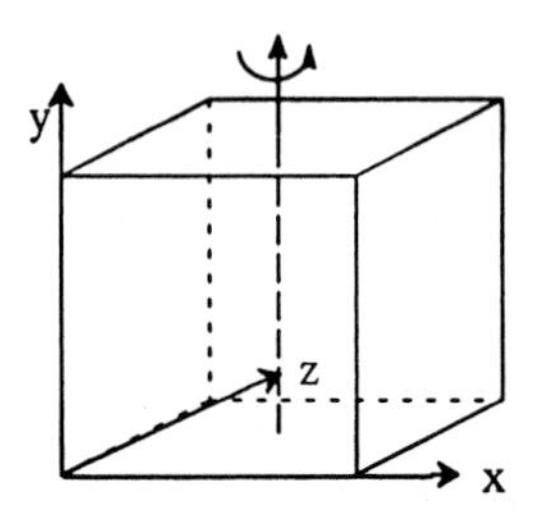

Figure 1

The xy-plane is the image plane (projection plane) and the rotation axis is parallel to the y-axis. The 2-dimensional slices, obtained for example from a CT-scanner, are stacked in the y-direction to form a volume. Parallel projection means that we transform in x and z only and thus we can limit our discussion to a plane of constant y, i.e. a slice. Also, we assume that the volume contains opaque objects whose surfaces we will render.

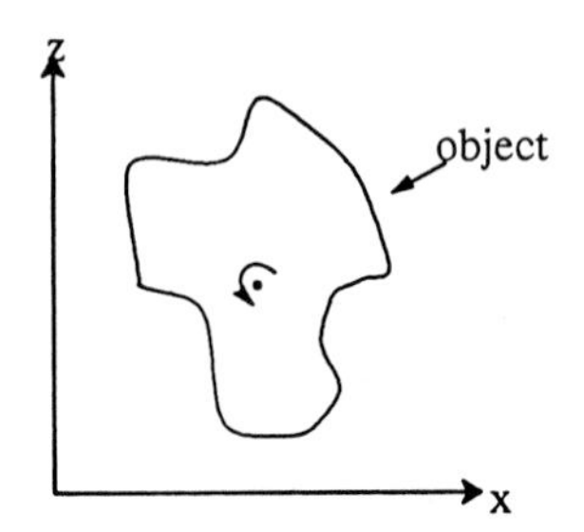

Figure 2

When the volume is rotated a certain angle some points on the surface will become visible while others will become hidden. The incremental algorithm applies a two-dimensional geometric transformation (rotation in x and z) to the currently visible surface points, then it removes those that have become hidden and finally casts a few new rays to detect points that have gone out of hiding, i.e. have become visible. When the rotation angle is small, the vast majority of surface points will remain visible from one projection to the next. More specifically, the algorithm proceeds in the following steps:

1) Detect the visible surface points by means of full ray–casting (one ray per image plane pixel). Save the coordinates of the surface points in a list. This is an initialization step that yields the first projection in the sequence.
2) Transform the points in the list by the rotation matrix

$$R_\phi = \begin{bmatrix} \cos\phi & 0 & -\sin\phi & 0 \\ 0 & 1 & 0 & 0 \\ \sin\phi & 0 & \cos\phi & 0 \\ T_x & 0 & T_z & 1 \end{bmatrix}$$

3) Remove from the list points that have become hidden.
4) Cast new rays to detect points that have become visible and insert them in the list.
5) Interpolate in the list to create an image for display.
6) Goto 2).

Steps 2) – 5) are integrated so that they are all executed in essentially one left–to–right scan of the current list. Note that there is one coordinate list per plane of constant y (slice). In the lists, coordinates are maintained with high precision (in our experiments we have used 32–bit floating point numbers).

The volume is accessed in (u, w)–space (object–space). See Figure 3. A counterclockwise rotation of totally α degrees corresponds to a rotation of the object coordinate system into a system (x, z) (current image space). To obtain the next projection in the sequence the points in the current list are rotated an angle increment ϕ into a new current image space (x', z').

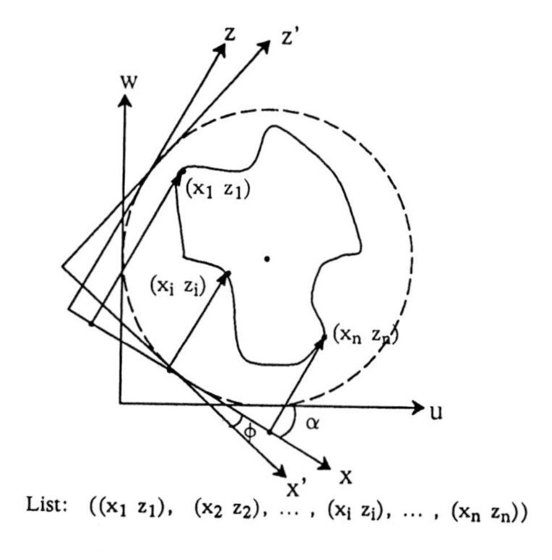

Figure 3

Steps 2)–5) in the algorithm are summarized in the following pseudo-code:

```
display_buffer [] := Background
for each point (xi, zi) in the list do
    (xi', zi') := (xi, zi)·Rφ   {2D geometric transformation}
    if (xi' < xi-1') then
        Remove (x1', z1') from the list.   {It's become hidden}
    else
        x := round(xi')      {nearest neighbor interpolation}*)
        display_buffer [x] := shading (zi')


              {Check if there's visible surface to the left}
        x := x - 1
        while (display_buffer [x] = Background) do
            z := raycast(x)  {returns Background or distance z}
            if (z = Background) then
                exit while loop  {no new visible surface points}
            end
            Insert (x, z) in the list.
            display_ buffer [x] := shading (z)
            x := x-1
        end
    end
end
```

*) To simplify the pseudo-code, we have not shown that if several x_i's round to the same integer x then the (x_i, z_i) whose x_i is closest to x is selected.

Hidden points to be removed from the list are detected by comparing the x-coordinates of two neighboring points after transformation. Since the coordinates in the list are sorted in x, it is obvious that this works if the volume contains one connected object with no holes. If this is not the case, some points that remain visible after the transformation may be thrown away, but this is compensated for by the rule governing the firing of new rays. Points that go out of hiding will appear to the left of visible points, and therefore new rays are cast from positions corresponding to "empty" display buffer positions to the immediate left of inserted points. This also takes care of the missing data problem when points are sliding apart so that gaps develop in the display buffer. Some artifacts in the form of missing data appear for objects with holes when the angle increment is relatively large (4-8 degrees), but this can be taken care of if the algorithm is modified so that the forced exit from the while-loop does not occur until a certain number (>1) of rays have returned *Background*. The shading procedure computes an intensity value that is a decreasing function of z, i.e. depth-only shading (Herman and Liu, 1979). Figure 4 shows four projections generated by the incremental algorithm with different angle increments (ϕ). The surface points that were hit by new rays in the process of computing this projection from the previous one, are marked with white pixels.

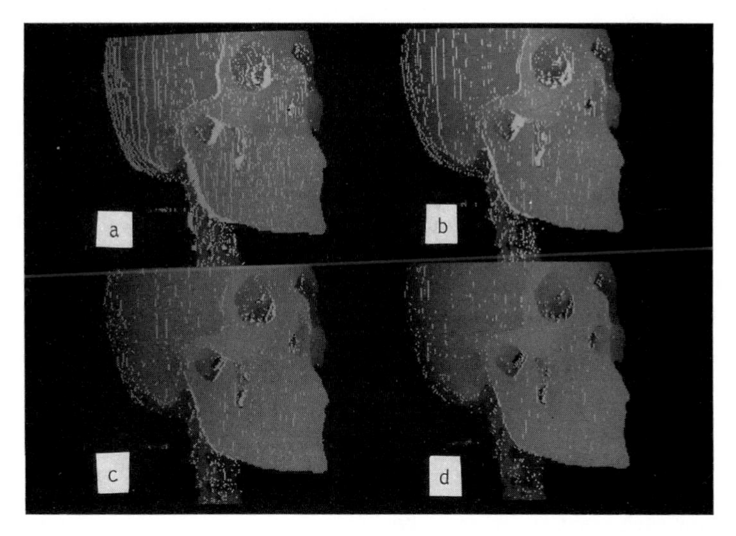

Figure 4 a) ϕ = 8 degrees, b) ϕ = 4 degrees, c) ϕ = 2 degrees, d) ϕ = 1 degree

Experiments have shown that the incremental algorithm yields a speed-up factor of about 10–15 for small angle increments (Table 2 in this paper), (Gudmundsson and Randén, 1990). The algorithm can be extended to include more sophisticated shading methods where the list is supplemented with estimated surface normals.

3. DATA COMPRESSION

It is obvious that an incrementally generated image sequence lends itself to data compression in a natural way. We will now describe a method that allows us to make a reversible compression – as opposed to irreversible compression using techniques such as predictive coding, transform coding, etc – of a projection sequence. The basic idea is that we generate the sequence by means of the incremental algorithm and in so doing save only the new surface points detected when going from one projection to the next.

The incremental generation of coordinate list number k – in which we then interpolate to create projection number k in the sequence – can be described as

$$L_k := RH_\phi(L_{k-1}) \cup N_k \quad (1)$$

where L_k is the coordinate list, RH_ϕ is an operator that rotates the coordinates ϕ degrees and also removes those points that become hidden (see Section 2). N_k is the set of surface point coordinates that were detected by new rays (L_1=N_1). Applying Eq. (1) recursively we get

$$L_k := RH_\phi(RH_\phi(\ldots.RH_\phi(RH_\phi(N_1) \cup N_2)\ldots. \cup N_{k-2}) \cup N_{k-1}) \cup N_k \quad (2)$$

Thus the compressed dataset representing a sequence of n projections is

$$\bigcup_{j=1}^{n} N_j$$

from which we can reconstruct any projection in the sequence by applying Eq. (2). Since $RH_\alpha\, RH_\beta = RH_{\alpha+\beta}$ we can rewrite Eq. (2)

$$L_k := \bigcup_{j=1}^{k-1} \left(RH_{(k-j)\phi}\ (N_j)\right) \cup N_k \tag{3}$$

If we wish to reconstruct each successive projection in a sequence of length Le we apply Eq. (1) sequentially with k=1, 2, ..., Le. On the other hand, if we wish to reconstruct a specific projection in the sequence we should apply Eq. (3). By using Eq. (3) instead of Eq. (2) we save computation time since each surface point is rotated to its final position in only one transformation. However, Eq. (3) implies that the hidden point removal test is applied only locally to each set N_j. To obtain global hidden point removal the transformed N_j's are merged into the display buffer (nearest neighbor interpolation) by means of the z-buffer algorithm (Foley et al, 1990).

The reconstruction process to create projection number k in the sequence using Eq. (3) is explained by the following pseudo-code:

```
display_buffer [] := z_max        {Background}
for j := 1 to k do
    for each point (x_i, z_i) in N_j do
        (x_i', z_i') := (x_i, z_i) · R_(k-j)φ   {2D geometric transformation}
        if (x_i' > x'_i-1) then
            x := round (x_i')*)
            if (display_buffer [x] > z_i') then
                display_buffer [x] := z_i'
            end
        end
    end
end
        *) See note in the pseudo-code in Section 2.
```

The z-values in the display buffer are then used to create a depth-shaded image.

4. RESULTS

The compression method has been tested on a volume obtained from a CT-scanner. The volume contains a human skull from a cadaver which is embedded in a plastic material that has been moulded into the shape of a human head. The volume consists of 99 slices, each with a resolution of 256x256 pixels. Object surfaces were detected by intensity thresholding and the threshold was set so that the skull was made visible (Figure 4).

Sixteen projection sequences with angle increments between 1 and 12 degrees were compressed and reconstructed. The length of a sequence was set so that it approximately covered one full turn around the volume. A sequence with angle increment ϕ has length Le where

$$Le \cdot \phi \leq 360 < (Le + 1) \cdot \phi$$

A compressed sequence consists of lists of surface points in the form of coordinate pairs (x and z, see Figure 3). Table 1 shows the number of such points for some of the compressed sequences that were generated.

Table 1

ϕ (*degrees*)	P (*number of points*)
1 (*Le* = 360)	$134.9 \cdot 10^3$
2 (*Le* = 180)	$125.1 \cdot 10^3$
4 (*Le* = 90)	$113.1 \cdot 10^3$
8 (*Le* = 45)	$99.7 \cdot 10^3$
12 (*Le* = 30)	$90.6 \cdot 10^3$

The amount of storage needed for a compressed sequence is

$$2 \, x \, b \, x \, P \text{ bits} \tag{4}$$

where b is the number of bits used to represent a coordinate value. If the ray-casting procedure uses a nearest-neighbor rule to locate a surface voxel (i.e. grid-point accuracy) then a coordinate value in the 256 x 256 grid of voxels in a slice can be represented by 1 byte (8 bits). Grid-point accuracy is often sufficient for segmentation of bone from soft tissue in CT-volumes (Magnuson et al, 1988). However, if surfaces are localized with sub-voxel accuracy, e.g. by means of linear interpolation, then more bits are necessary to represent the coordinates.

We have compared the storage requirements of the compressed sequences with those of the uncompressed sequences and the original volume. See Figure 5.

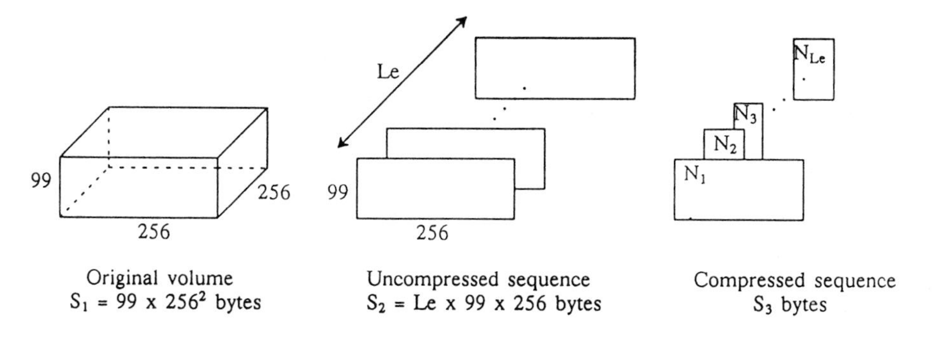

Figure 5

Figure 6 shows the amount of data needed to represent a sequence as a function of Le (and thus of ϕ). The three curves pertain to the original volume (S_1), uncompressed sequences (S_2) and compressed sequences (S_3).

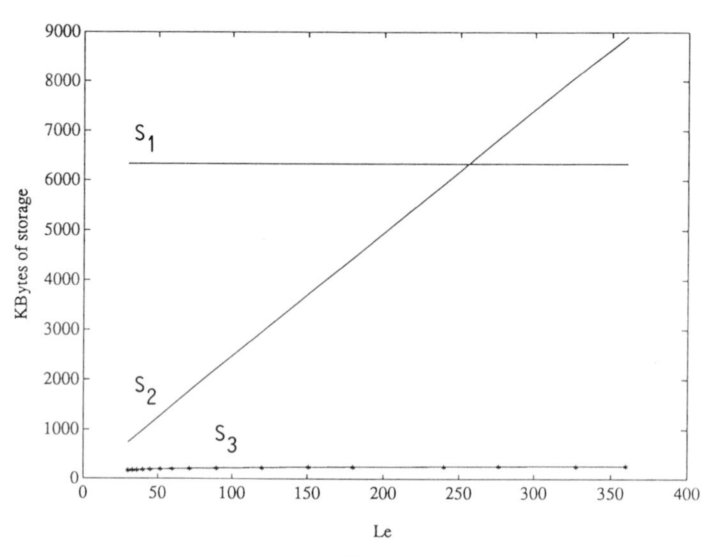

Figure 6

For the compressed sequences, the coordinates were represented by 8-bit bytes, i.e. b=8 in Eq. (4).

As can be seen from Table 1 and S_3 in Figure 6, decreasing the angle increment to obtain denser sequences by going from, say, 2 degrees (Le=180) to 1 degree (Le=360), has a marginal effect in terms of storage space.

In Figure 7 the compression factor S_2/S_3 versus angle increment ϕ is shown.

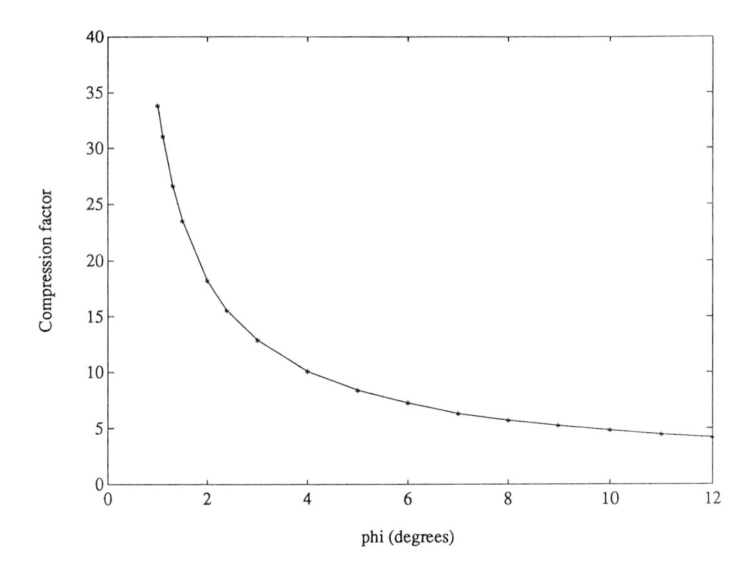

Figure 7

From a strict storage economy view disregarding computational expense, a sequence that is longer than 256 should be represented by the original volume (see Figure 6). A length of 256 corresponds to an angle increment of about 1.4 degrees for one full turn around the volume. However, compression factors for increments smaller than 1.4 degrees have been included in the diagram since one could very well think of generating sequences that cover only a certain sector of a full turn, e.g. 0–180 degrees. The compression factors depicted in Figure 7 can be regarded as averages over a full turn around the volume.

The experiments were carried out on a SUN Sparc1+ computer with enough main memory to hold the entire volume during generation of the sequences. The code was written in C and no in-depth effort was made to optimize it for execution speed. Timing measurements to assess relative speeds were taken. The results are shown in Table 2. The times are given in seconds.

Table 2

ϕ (*degrees*)	T_1	T_2	T_3	T_1/T_2	T_2/T_3
1 (Le = 360)	1756.6	121.2	55.4	14.5	2.2
2 (Le = 180)	878.3	68.2	26.8	12.9	2.5
4 (Le = 90)	437.9	40.8	12.9	10.7	3.2
8 (Le = 45)	219.2	25.7	6.1	8.5	4.2

T_1: Non-incremental generation of sequence (full ray-casting).

T_2: Incremental generation of sequence (compressed and uncompressed sequences are generated simultaneously).

T_3: Reconstruction of sequence from the compressed representation using Eq. (1) for k=1, 2, ..., Le.

Figure 8 clarifies the significance of the timing measurements in Table 2.

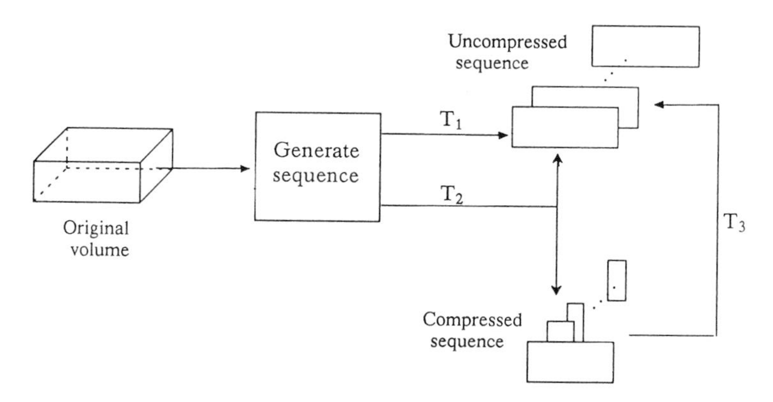

Figure 8

The speed ratios T_1/T_2 were reported in (Gudmundsson and Randén, 1990). Note that the reconstruction process generates one projection in about 0.15 seconds.

5. CONCLUDING REMARKS

We have described an algorithm for reversible data compression of 3D-volume projection sequences with small projection angle increments. Such sequences are required in order to achieve the effect of smooth rotation of the volume. Experiments have indicated that the proposed algorithm performs well in terms of compression factors and speed.

In our experiments we have utilized a projection geometry as shown in Figures 1 and 3. However, with trivial changes in the algorithm any geometry where the axis of rotation is parallel to one of the coordinate axes of the projection plane can be used. This condition implies that surface points will follow horizontal or vertical trajectories in the projection plane during rotation. A somewhat more elaborate modification of the algorithm would allow us to choose a projection geometry where the axis of rotation is parallel to the projection plane.

The object surfaces in the volume have been rendered with depth-only shading. More sophisticated shading methods, e.g. Phong-shading (Foley et al, 1990), require that we compute surface normals. Surface normals can be estimated either from the 2D-gradient in depth-only projections or from the volume 3D-gradient (Magnusson et al, 1988) (Tiede et al, 1990). In the former case the compression factors would not be affected but the time to reconstruct a projection would increase. However, use of the 2D-gradient gives rise to artifacts. If we use 3D-gradients, each point in the lists would have to be supplemented with surface normal components which would decrease the compression factors.

Acknowledgement

This work was supported by the Swedish National Board for Technical Development.

References

Farell E J, Zapulla R A(1989). Three-Dimensional Data Visualization and Biomedical Applications. CRC Critical Reviews in Biomedical Engineering, Vol. 16: 323-363.

Foley J D, van Dam A, Feiner S K, Hughes J F (1990). Computer Graphics: Principles and Practice (2nd Ed.). Addison-Wesley.

Gudmundsson B and Randén M (1990). Incremental Generation of Projections of CT-Volumes. Proc. First Conference on Visualization in Biomedical Computing, Atlanta, Georgia, USA.

Herman, Lui (1979). Three-Dimensional Display of Human Organs from Computed Tomograms. Computer Graphics and Image Processing, Vol. 9:, 1-21.

Levoy M (1990). Efficient Ray Tracing of Volume Data. ACM Transactions on Graphics, Vol. 9: 245-261.

Lenz R, Gudmundsson B, Lindskog B, Danielsson P E (1986). Display of density volumes. IEEE Computer Graphics and Applications. Vol. 6(7): 20-29.

Magnusson M, Lenz R, Danielsson P E (1988). Evaluation of methods for shaded surface display of CT-volumes. Proc. Ninth International Conference on Pattern Recognition, Rome, Italy.

Tiede U, Hoehne K H, Bomans M, Pommert A, Riemer M, Wiebecke G (1990). Investigation of Medical 3D-Rendering Algorithms. IEEE Computer Graphics and Applications. Vol. 10(2): 41-53.

INDEX TO KEYWORDS

Lecture Notes in Computer Science

For information about Vols. 1–420
please contact your bookseller or Springer-Verlag

Vol. 421: T. Onodera, S. Kawai, A Formal Model of Visualization in Computer Graphics Systems. X, 100 pages. 1990.

Vol. 422: B. Nebel, Reasoning and Revision in Hybrid Representation Systems. XII, 270 pages. 1990 (Subseries LNAI).

Vol. 423: L.E. Deimel (Ed.), Software Engineering Education. Proceedings, 1990. VI, 164 pages. 1990.

Vol. 424: G. Rozenberg (Ed.), Advances in Petri Nets 1989. VI, 524 pages. 1990.

Vol. 425: C.H. Bergman, R.D. Maddux, D.L. Pigozzi (Eds.), Algebraic Logic and Universal Algebra in Computer Science. Proceedings, 1988. XI, 292 pages. 1990.

Vol. 426: N. Houbak, SIL – a Simulation Language. VII, 192 pages. 1990.

Vol. 427: O. Faugeras (Ed.), Computer Vision – ECCV 90. Proceedings, 1990. XII, 619 pages. 1990.

Vol. 428: D. Bjørner, C.A.R. Hoare, H. Langmaack (Eds.), VDM '90. VDM and Z – Formal Methods in Software Development. Proceedings, 1990. XVII, 580 pages. 1990.

Vol. 429: A. Miola (Ed.), Design and Implementation of Symbolic Computation Systems. Proceedings, 1990. XII, 284 pages. 1990.

Vol. 430: J.W. de Bakker, W.-P. de Roever, G. Rozenberg (Eds.), Stepwise Refinement of Distributed Systems. Models, Formalisms, Correctness. Proceedings, 1989. X, 808 pages. 1990.

Vol. 431: A. Arnold (Ed.), CAAP '90. Proceedings, 1990. VI, 285 pages. 1990.

Vol. 432: N. Jones (Ed.), ESOP '90. Proceedings, 1990. IX, 436 pages. 1990.

Vol. 433: W. Schröder-Preikschat, W. Zimmer (Eds.), Progress in Distributed Operating Systems and Distributed Systems Management. Proceedings, 1989. V, 206 pages. 1990.

Vol. 434: J.-J. Quisquater, J. Vandewalle (Eds.), Advances in Cryptology – EUROCRYPT '89. Proceedings, 1989. X, 710 pages. 1990.

Vol. 435: G. Brassard (Ed.), Advances in Cryptology – CRYPTO '89. Proceedings, 1989. XIII, 634 pages. 1990.

Vol. 436: B. Steinholtz, A. Sølvberg, L. Bergman (Eds.), Advanced Information Systems Engineering. Proceedings, 1990. X, 392 pages. 1990.

Vol. 437: D. Kumar (Ed.), Current Trends in SNePS – Semantic Network Processing System. Proceedings, 1989. VII, 162 pages. 1990. (Subseries LNAI).

Vol. 438: D.H. Norrie, H.W. Six (Eds.), Computer Assisted Learning – ICCAL '90. Proceedings, 1990. VII, 467 pages. 1990.

Vol. 439: P. Gorny, M. Tauber (Eds.), Visualization in Human-Computer Interaction. Proceedings, 1988. VI, 274 pages. 1990.

Vol. 440: E. Börger, H. Kleine Büning, M.M. Richter (Eds.), CSL '89. Proceedings, 1989. VI, 437 pages. 1990.

Vol. 441: T. Ito, R.H. Halstead, Jr. (Eds.), Parallel Lisp: Languages and Systems. Proceedings, 1989. XII, 364 pages. 1990.

Vol. 442: M. Main, A. Melton, M. Mislove, D. Schmidt (Eds.), Mathematical Foundations of Programming Semantics. Proceedings, 1989. VI, 439 pages. 1990.

Vol. 443: M.S. Paterson (Ed.), Automata, Languages and Programming. Proceedings, 1990. IX, 781 pages. 1990.

Vol. 444: S. Ramani, R. Chandrasekar, K.S.R. Anjaneyulu (Eds.), Knowledge Based Computer Systems. Proceedings, 1989. X, 546 pages. 1990. (Subseries LNAI).

Vol. 445: A.J.M. van Gasteren, On the Shape of Mathematical Arguments. VIII, 181 pages. 1990.

Vol. 446: L. Plümer, Termination Proofs for Logic Programs. VIII ,142 pages. 1990. (Subseries LNAI).

Vol. 447: J.R. Gilbert, R. Karlsson (Eds.), SWAT '90. 2nd Scandinavian Workshop on Algorithm Theory. Proceedings, 1990. VI, 417 pages. 1990.

Vol. 448: B. Simons, A. Spector (Eds.), Fault Tolerant Distributed Computing. VI, 298 pages. 1990.

Vol. 449: M.E. Stickel (Ed.), 10th International Conference on Automated Deduction. Proceedings, 1990. XVI, 688 pages. 1990. (Subseries LNAI).

Vol. 450: T. Asano, T. Ibaraki, H. Imai, T. Nishizeki (Eds.), Algorithms. Proceedings, 1990. VIII, 479 pages. 1990.

Vol. 451: V. Marík, O. Stepánková, Z. Zdráhal (Eds.), Artificial Intelligence in Higher Education. Proceedings, 1989. IX, 247 pages. 1990. (Subseries LNAI).

Vol. 452: B. Rovan (Ed.), Mathematical Foundations of Computer Science 1990. Proceedings, 1990. VIII, 544 pages. 1990.

Vol. 453: J. Seberry, J. Pieprzyk (Eds.), Advances in Cryptology – AUSCRYPT '90 Proceedings, 1990. IX. 462 pages. 1990.

Vol. 454: V. Diekert, Combinatorics on Traces. XII, 165 pages. 1990.

Vol. 455: C.A. Floudas, P.M. Pardalos, A Collection of Test Problems for Constrained Global Optimization Algorithms. XIV, 180 pages. 1990.

Vol. 456: P. Deransart, J. Maluszyn´ski (Eds.), Programming Language Implementation and Logic Programming. Proceedings, 1990. VIII, 401 pages. 1990.

Vol. 457: H. Burkhart (Ed.), CONPAR '90 – VAPP IV. Proceedings, 1990. XIV, 900 pages. 1990.

Vol. 458: J.C.M. Baeten, J.W. Klop (Eds.), CONCUR '90. Proceedings, 1990. VII, 537 pages. 1990.

Vol. 459: R. Studer (Ed.), Natural Language and Logic. Proceedings, 1989. VII, 252 pages. 1990. (Subseries LNAI).

Vol. 460: J. Uhl, H.A. Schmid, A Systematic Catalogue of Reusable Abstract Data Types. XII, 344 pages. 1990.

Vol. 461: P. Deransart, M. Jourdan (Eds.), Attribute Grammars and their Applications. Proceedings, 1990. VIII, 358 pages. 1990.

Vol. 462: G. Gottlob, W. Nejdl (Eds.), Expert Systems in Engineering. Proceedings, 1990. IX, 260 pages. 1990. (Subseries LNAI).

Vol. 463: H. Kirchner, W. Wechler (Eds.), Algebraic and Logic Programming. Proceedings, 1990. VII, 386 pages. 1990.

Vol. 464: J. Dassow, J. Kelemen (Eds.), Aspects and Prospects of Theoretical Computer Science. Proceedings, 1990. VI, 298 pages. 1990.

Vol. 465: A. Fuhrmann, M. Morreau (Eds.), The Logic of Theory Change. Proceedings, 1989. X, 334 pages. 1991. (Subseries LNAI).

Vol. 466: A. Blaser (Ed.), Database Systems of the 90s. Proceedings, 1990. VIII, 334 pages. 1990.

Vol. 467: F. Long (Ed.), Software Engineering Environments. Proceedings, 1969. VI, 313 pages. 1990.

Vol. 468: S.G. Akl, F. Fiala, W.W. Koczkodaj (Eds.), Advances in Computing and Information – ICCI '90. Proceedings, 1990. VII, 529 pages. 1990.

Vol. 469: I. Guessarian (Ed.), Semantics of Systeme of Concurrent Processes. Proceedings, 1990. V, 456 pages. 1990.

Vol. 470: S. Abiteboul, P.C. Kanellakis (Eds.), ICDT '90. Proceedings, 1990. VII, 528 pages. 1990.

Vol. 471: B.C. Ooi, Efficient Query Processing in Geographic Information Systems. VIII, 208 pages. 1990.

Vol. 472: K.V. Nori, C.E. Veni Madhavan (Eds.), Foundations of Software Technology and Theoretical Computer Science. Proceedings, 1990. X, 420 pages. 1990.

Vol. 473: I.B. Damgård (Ed.), Advances in Cryptology – EUROCRYPT '90. Proceedings, 1990. VIII, 500 pages. 1991.

Vol. 474: D. Karagiannis (Ed.), Information Syetems and Artificial Intelligence: Integration Aspects. Proceedings, 1990. X, 293 pages. 1991. (Subseries LNAI).

Vol. 475: P. Schroeder-Heister (Ed.), Extensions of Logic Programming. Proceedings, 1989. VIII, 364 pages. 1991. (Subseries LNAI).

Vol. 476: M. Filgueiras, L. Damas, N. Moreira, A.P. Tomás (Eds.), Natural Language Processing. Proceedings, 1990. VII, 253 pages. 1991. (Subseries LNAI).

Vol. 477: D. Hammer (Ed.), Compiler Compilers. Proceedings, 1990. VI, 227 pages. 1991.

Vol. 478: J. van Eijck (Ed.), Logics in AI. Proceedings, 1990. IX, 562 pages. 1991. (Subseries in LNAI).

Vol. 480: C. Choffrut, M. Jantzen (Eds.), STACS 91. Proceedings, 1991. X, 549 pages. 1991.

Vol. 481: E. Lang, K.-U. Carstensen, G. Simmons, Modelling Spatial Knowledge on a Linguistic Basis. IX, 138 pages. 1991. (Subseries LNAI).

Vol. 482: Y. Kodratoff (Ed.), Machine Learning – EWSL-91. Proceedings, 1991. XI, 537 pages. 1991. (Subseries LNAI).

Vol. 483: G. Rozenberg (Ed.), Advances In Petri Nets 1990. VI, 515 pages. 1991.

Vol. 484: R. H. Möhring (Ed.), Graph-Theoretic Concepts In Computer Science. Proceedings, 1990. IX, 360 pages. 1991.

Vol. 485: K. Furukawa, H. Tanaka, T. Fullsaki (Eds.), Logic Programming '89. Proceedings, 1989. IX, 183 pages. 1991. (Subseries LNAI).

Vol. 486: J. van Leeuwen, N. Santoro (Eds.), Distributed Algorithms. Proceedings, 1990. VI, 433 pages. 1991.

Vol. 487: A. Bode (Ed.), Distributed Memory Computing. Proceedings, 1991. XI, 506 pages. 1991.

Vol. 488: R. V. Book (Ed.), Rewriting Techniques and Applications. Proceedings, 1991. VII, 458 pages. 1991.

Vol. 489: J. W. de Bakker, W. P. de Roever, G. Rozenberg (Eds.), Foundations of Object-Oriented Languages. Proceedings, 1990. VIII, 442 pages. 1991.

Vol. 490: J. A. Bergstra, L. M. G. Feljs (Eds.), Algebraic Methods II: Theory, Tools and Applicatlons. VI, 434 pages. 1991.

Vol. 491: A. Yonezawa, T. Ito (Eds.), Concurrency: Theory, Language, and Architecture. Proceedings, 1989. VIII, 339 pages. 1991.

Vol. 492: D. Sriram, R. Logcher, S. Fukuda (Eds.), Computer-Aided Cooperative Product Development. Proceedings, 1989 VII, 630 pages. 1991.

Vol. 493: S. Abramsky, T. S. E. Maibaum (Eds.), TAPSOFT '91. Volume 1. Proceedings, 1991. VIII, 455 pages. 1991.

Vol. 494: S. Abramsky, T. S. E. Maibaum (Eds.), TAPSOFT '91. Volume 2. Proceedings, 1991. VIII, 482 pages. 1991.

Vol. 495: 9. Thalheim, J. Demetrovics, H.-D. Gerhardt (Eds.), MFDBS '91. Proceedings, 1991. VI, 395 pages. 1991.

Vol. 496: H.-P. Schwefel, R. Männer (Eds.), Parallel Problem Solving from Nature. Proceedings, 1991. XI, 485 pages. 1991.

Vol. 497: F. Dehne, F. Fiala. W.W. Koczkodaj (Eds.), Advances in Computing and Intormation - ICCI '91 Proceedings, 1991. VIII, 745 pages. 1991.

Vol. 498: R. Andersen, J. A. Bubenko jr., A. Sølvberg (Eds.), Advanced Information Systems Engineering. Proceedings, 1991. VI, 579 pages. 1991.

Vol. 499: D. Christodoulakis (Ed.), Ada: The Choice for '92. Proceedings, 1991. VI, 411 pages. 1991.

Vol. 500: M. Held, On the Computational Geometry of Pocket Machining. XII, 179 pages. 1991.

Vol. 501: M. Bidoit, H.-J. Kreowski, P. Lescanne, F. Orejas, D. Sannella (Eds.), Algebraic System Specification and Development. VIII, 98 pages. 1991.

Vol. 502: J. Bārzdiņš, D. Bjørner (Eds.), Baltic Computer Science. X, 619 pages. 1991.

Vol. 503: P. America (Ed.), Parallel Database Systems. Proceedings, 1990. VIII, 433 pages. 1991.

Vol. 504: J. W. Schmidt, A. A. Stogny (Eds.), Next Generation Information System Technology. Proceedings, 1990. IX, 450 pages. 1991.

Vol. 505: E. H. L. Aarts, J. van Leeuwen, M. Rem (Eds.), PARLE '91. Parallel Architectures and Languages Europe, Volume I. Proceedings, 1991. XV, 423 pages. 1991.

Vol. 506: E. H. L. Aarts, J. van Leeuwen, M. Rem (Eds.), PARLE '91. Parallel Architectures and Languages Europe, Volume II. Proceedings, 1991. XV, 489 pages. 1991.

Vol. 507: N. A. Sherwani, E. de Doncker, J. A. Kapenga (Eds.), Computing in the 90's. Proceedings, 1989. XIII, 441 pages. 1991.

Vol. 508: S. Sakata (Ed.), Applied Algebra, Algebraic Algorithms and Error-Correcting Codes. Proceedings, 1990. IX, 390 pages. 1991.

Vol. 509: A. Endres, H. Weber (Eds.), Software Development Environments and CASE Technology. Proceedings, 1991. VIII, 286 pages. 1991.

Vol. 510: J. Leach Albert, B. Monien, M. Rodríguez (Eds.), Automata, Languages and Programming. Proceedings, 1991. XII, 763 pages. 1991.

Vol. 511: A. C. F. Colchester, D.J. Hawkes (Eds.), Information Processing in Medical Imaging. Proceedings, 1991. XI, 512 pages. 1991.